Embedded Controllers:

80186, 80188, and 80386EX

BARRY B. BREY
DeVry Institute of Technology

Pearson
Education

Prentice Hall

Upper Saddle River, New Jersey   *Columbus, Ohio*

Library of Congress Cataloging-in-Publication Data

Brey, Barry B.
    Embedded controllers : 80186, 80188, and 80386EX / Barry B. Brey.
       p.   cm.
    Includes index.
    ISBN 0-13-400136-2
    1. Embedded computer systems.   2. Electronic controllers.
   3. Intel 80xxx series microprocessors.   I. Title.
TK7895.E42B74   1998
004.165--dc21
                                                   97-18174
                                                     CIP

Cover art: Photodisc
Editor: Charles E. Stewart, Jr.
Production Coordination: Tim Flem, Custom Editorial Productions, Inc.
Design Coordinator: Julia Zonneveld Van Hook
Cover Designer: Raymond Hummons
Production Manager: Patricia A. Tonneman
Marketing Manager: Debbie Yarnell

This book was set in Times Roman by Custom Editorial Productions, Inc., and was printed and bound by Courier/Kendallville, Inc. The cover was printed by Phoenix Color Corp.

© 1998 by Prentice-Hall, Inc.
Simon & Schuster/A Viacom Company
Upper Saddle River, New Jersey 07458

Printed in the United States of America

10 9 8 7 6 5 4 3 2 1

ISBN   0-13-400136-2

Prentice-Hall International (UK) Limited, *London*
Prentice-Hall of Australia Pty. Limited, *Sydney*
Prentice-Hall Canada Inc., *Toronto*
Prentice-Hall Hispanoamericana, S. A., *Mexico*
Prentice-Hall of India Private Limited, *New Delhi*
Prentice-Hall of Japan, Inc., *Tokyo*
Simon & Schuster Asia Pte. Ltd., *Singapore*
Editora Prentice-Hall do Brasil, Ltda., *Rio de Janeiro*

This book is dedicated to my loving wife, Sheila.

PREFACE

The Intel embedded microprocessors are widely used in many areas of electronics, communications, and control systems. This is a textbook for the student who needs a thorough knowledge of programming and interfacing of the Intel family of embedded microprocessors; it also serves as a practical reference manual for assembly language programmers.

ORGANIZATION AND COVERAGE

Each chapter of this book begins with a set of objectives that briefly define the contents of the chapter. The body of each chapter includes many programming examples that illustrate the main topics of the chapter. At the end of each chapter is a summary that doubles as a review and study guide. Finally, there are questions and problems to promote practice with the concepts presented in the chapter.

The many example programs provide practice in programming the Intel family of embedded microprocessors. The Microsoft macro assembler program, MASM, is used to develop each program and give you experience using this industry-standard assembly language program. Operation of this programming environment includes the linker, library, macros, DOS function, and BIOS functions. The 80386EX may be used in an environment that includes DOS, BIOS, and Windows on read-only memory.

Also provided is a thorough description of all embedded family members, memory systems, and various I/O systems that include: disk memory, ADC and DAC, 16550 UART, PIAs, timers, keyboard/display controllers, arithmetic coprocessors, and video display systems. Also discussed is the personal computer system, which is used as a launching platform for many embedded designs. Through these systems, you will learn a practical approach to embedded microprocessor interfacing.

APPROACH

Because the Intel family of embedded microprocessors is quite diverse and will eventually include a Pentium-like member, this text initially concentrates on real-mode programming, which is compatible with all versions of the Intel family of microprocessors. Instructions for each family member (80186, 80188, and the 80386EX as well as the 80486 through the Pentium Pro) are compared and contrasted with the 80186/80188 embedded microprocessors. This entire series of microprocessors and embedded controllers is very similar, which allows you to learn more advanced versions once you understand the basic 80186/80188.

In addition to fully explaining the programming and operation of the microprocessor, this text also explains the programming and operation of the numeric coprocessor. The numeric coprocessor provides access to floating-point calculations that are important in applications such as control systems, video graphics, and computer-aided design (CAD). The numeric coprocessor allows a program to access complex arithmetic operations that are otherwise difficult to achieve with the integer instruction set of the embedded microprocessor.

Also described are the pin-outs and functions of the 80186, 80188, and 80386EX embedded microprocessors. Interfacing is first developed using the 80188/80186 with some of the more common peripheral components. After learning the basics, a more advanced emphasis is placed on the 80386EX in the final chapter of the book.

Through this approach, you will gain a practical background in the operation of the microprocessor and programming with the advanced family members, along with interfacing all embedded family members. After studying this textbook, you should be able to:

1. Develop software to control an application interface to the 80186, 80188, and 80386EX embedded microprocessors. These same techniques also apply to the 80486, Pentium, and Pentium Pro microprocessors. This software also includes DOS-based applications so the personal computer system, often a launching pad for development, can be understood.
2. Program using DOS function calls to control the keyboard, video display system, and disk memory in assembly language.
3. Use the BIOS functions to control the keyboard, display, and various other components in the computer system.
4. Develop software that uses macro sequences, procedures, conditional assembly, and flow control assembler directives.
5. Develop software that uses interrupt hooks, and hot keys to gain access to terminate-and-stay resident software.
6. Program the numeric coprocessor to solve complex arithmetic applications.
7. Explain the differences between the embedded family members and highlight the features of each member.
8. Describe and use real- (80186/80188) and protected-mode operation of the 80386EX embedded microprocessor.
9. Interface memory and I/O systems to the embedded microprocessor.
10. Provide a detailed and comprehensive comparison of all family members, their software, and hardware interfaces.
11. Explain the operation of disk and video systems.

CONTENT OVERVIEW

Chapter 1 introduces the Intel family of microprocessors, with an emphasis on the microprocessor-based computer system. This first chapter serves to introduce the microprocessor, its history, its operation, and the methods used to store data in a microprocessor-based system. We also provide number systems for those who are unaware of basic number system theory. Chapter 2 explores the programming model of the microprocessor, embedded microprocessor, and the system architecture. Both the real and protected modes are explained in this second introductory chapter.

Once an understanding of the basic machine is grasped, Chapters 3–6 explain how each instruction functions with the Intel family of microprocessors and embedded microprocessors. As instructions are explained, simple applications are presented to illustrate the operation of the instructions and develop basic programming concepts.

After the basis for programming is developed, Chapters 7 and 8 provide examples that use the assembler program. These applications include programming using DOS and BIOS function calls and the mouse function calls. Disk files are explained as well as keyboard and video operation on a personal computer system. This chapter provides the tools used to develop virtually any program on a personal computer system or for an embedded application. Also introduced is the concept of interrupt hooks and hot keys.

Chapter 9 introduces the 80186 and 80188 family as a foundation for learning basic memory and I/O interfacing that follow in later chapters. This chapter shows the buffered system as well as the system timing.

Chapter 10 provides complete detail on memory interface using both integrated decoders and programmable logic devices. Parity is illustrated as well as dynamic memory systems. The 8- and 16-bit memory systems are provided so that any embedded microprocessor can be interfaced to memory.

Chapter 11 provides a detailed look at basic I/O interfacing that includes PIAs, timers, keyboard/display interfaces, 16550 UART, and ADC/DAC. It also describes the interface of both DC and stepper motors.

Once these basic I/O components and their interface to the microprocessor is understood, Chapters 12 and 13 provide detail on advanced I/O techniques that include interrupts and direct memory access (DMA). Applications include a printer interface, real-time clock, disk memory, and video systems.

Chapter 14 details the operation and programming of arithmetic coprocessors. Today, few applications function efficiently without the power of the arithmetic coprocessor. Remember that all Intel microprocessors since the 80486 contain a coprocessor.

Chapter 15 develops example projects that illustrate the use of the 80186 and 80188 embedded controllers. Each example project is provided with complete hardware and software detail so systems can be developed using the embedded microprocessor.

Chapter 16 provides detail on the advanced 80386EX embedded microprocessor. In this chapter we explore the differences between the 80186/80188 and the 80386EX and compare any enhancements and additional features.

The following appendices are included to enhance the application of the text:

A. Complete listing of the DOS INT 21H function calls. This appendix also details the use of the assembler program and many of the BIOS function calls including BIOS function call INT 10H.
B. Complete listing of all 80186, 80188, and 80386EX instructions including many example instructions and machine coding in hexadecimal as well as clock timing information.
C. Compact list of all the instructions that change the flag bits.

STAY IN TOUCH

You can stay in touch over the Internet. My Internet site contains information about all of my textbooks, and many important links that are specific to the personal computer, microprocessors, embedded microprocessors, hardware, and software. Also available is a lesson that details many of the aspects of the personal computer. The URL is **http://users1.ee.net/brey/**

LAB SUPPLEMENT

A text supplement is available in the form of a lab manual, which shows how to program the embedded microprocessor as well as the personal computer system. It also shows how to use the assembler and CodeView programs that are included with the Microsoft macro assembler program.

The manual, available from Prentice-Hall, is entitled *Laboratory Manual to Accompany the Embedded Controllers: 80186, 80188, and 80386EX* (ISBN 0-02-314254-5) and is written by Barry B. Brey.

CONTENTS

CHAPTER 1

Introduction to the Microprocessor, Embedded Controller, and Personal Computer

INTRODUCTION

This chapter provides an overview of the microprocessor, embedded controller, and personal computer. Included are a discussion of the history of computers as well as the function of the microprocessor in the microprocessor-based computer system (the personal computer). Also introduced are the terms and jargon of the computer field so that you will understand and be able to use computerese when discussing microprocessors and computers.

The block diagram, and a description of the function of each block, detail the operation of a computer system. This chapter also discusses the memory and input/output system of the personal computer.

Finally, we look at the way data is stored in the computer's memory so that you can use each data type in software development. Numeric data is stored as integers, floating-point, and binary-coded decimal (BCD); while ASCII (American Standard Code for Information Interchange) code represents alphanumeric data. We also discuss converting between decimal and any other number base such as binary, octal, and hexadecimal.

CHAPTER OBJECTIVES

Upon completion of this chapter, you will be able to:

1. Use appropriate computer terminology including: bit, byte, data, real memory system, expanded memory system (EMS), extended memory system (XMS), DOS, BIOS, I/O, and so forth.
2. Briefly describe the history of the computer and list applications performed by it.
3. Discuss the pioneering software developers and the programs that they created.
4. Draw the block diagram of a computer system and explain the purpose of each block.
5. Describe the function of the microprocessor and its basic operation.
6. Define the contents of the memory system in the personal computer.
7. Convert between decimal, binary, and hexadecimal data.
8. Differentiate and represent alphanumeric information as integers, floating-point, BCD, and ASCII data.

1–1   A HISTORICAL BACKGROUND

This first section outlines the historical events leading to the development of the microprocessor and, specifically, the extremely powerful and current 80386, 80486, Pentium, and Pentium Pro microprocessors. Although a study of history is not essential to understanding the microprocessor, it is interesting to get a historical retrospective of the fast-paced evolution of the computer.

The Mechanical Age

The idea of a computing system is not new—it was around long before modern electrical and electronic devices. Calculating with a machine dates to before 500 BC when the Babylonians invented the **abacus,** the first mechanical calculator, which used strings of beads to perform calculations. It was first used by the ancient Babylonian priesthood to keep track of their vast storehouses of grain. The abacus, which was used extensively and is still in use today, was not improved until 1642, when mathematician *Blaise Pascal* invented a *calculator* constructed of gears and wheels. Each gear contained ten teeth that, when moved one complete revolution, advanced a second gear one place (see Figure 1–1). This is the same principle employed in the automobile's odometer mechanism and is the basis for all mechanical calculators. Incidentally, the PASCAL programming language is named in honor of Blaise Pascal for his pioneering work in mathematics and the mechanical calculator.

The first practical geared, mechanical machines used to automatically compute information date to the early 1800s. Realize that this is before humans invented the lightbulb or before much was known about electricity. In this dawn of the computer age, humans dreamed of mechanical machines that could compute numerical facts with a program—not merely calculating facts with a calculator.

It was discovered in 1937, through plans and journals, that one early pioneer of mechanical computing machinery was *Charles Babbage,* aided by *Augusta Ada Byron*, the Countess of Lovelace. Babbage was commissioned in 1823 by the Royal Astronomical Society of Great Britain to produce a programmable calculating machine. This machine was to generate navigational tables for the British Royal Navy. He accepted the challenge and began to create what he called his **Analytical Engine.** This engine was a mechanical computer that stored 1000 twenty-digit decimal numbers and a variable program that could modify the function of the machine to perform various calculating tasks. Input to his engine was through punched cards, much as computers in the 1950s and 1960s used punched cards. It is assumed that he obtained the idea of using punched cards from *Joseph Jacquard,* a Frenchman who used punched cards in 1801 as input to a weaving machine that is today called **Jacquard's loom.** Jacquard's loom used punched cards to select intricate weaving patterns in the cloth that it produced. The punched cards *programmed* the loom.

After many years of work, Babbage's dream began to fade when he realized that the machinists of his day were unable to create the mechanical parts needed to complete his work. The

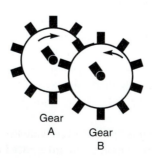

FIGURE 1–1 The gears used by Blaise Pascal for his mechanical calculator.

Gear
A
          Gear
          B

Analytical Engine required more than 50,000 machined parts, which could not be made with enough precision to allow his engine to function reliably.

The Electrical Age

The 1800s saw the advent of the electric motor (conceived by *Michael Faraday*) and with it came a multitude of motor-driven adding machines all based on the mechanical calculator developed by Blaise Pascal. These electrically driven mechanical calculators were common pieces of office equipment until well into the early 1970s, when the small hand-held electronic calculator, first introduced by Bomar as the **Bomar Brain,** appeared. *Monroe* was also a leading pioneer of electronic calculators, but their machines were desktop, four-function models the size of cash registers.

In 1889 *Herman Hollerith* developed the punched card for storing data. He, too, apparently borrowed the idea of a punched card, as did Babbage, from Jacquard. He also developed a mechanical machine—driven by one of the new electric motors—that counted, sorted, and collated information stored on punched cards. The idea of calculating by machinery intrigued the United States government so much that Hollerith was commissioned to use his punched-card system to store and tabulate information for the 1890 Census.

In 1896, Hollerith formed the *Tabulating Machine Company,* which developed a line of machines that used punched cards for tabulation. After a number of mergers, the Tabulating Machine Company became the *International Business Machines Corporation,* now referred to more commonly as IBM. We often refer to the punched cards used in computer systems as **Hollerith cards** in honor of Herman Hollerith. The 12-bit code used on a punched card is called the **Hollerith code.**

Mechanical machines driven by electric motors continued to dominate the information processing world until the advent of the first *electronic calculating machine* in 1942 by a German inventor named *Konrad Zuse.* His calculating computer, the **Z3,** was used in aircraft and missile design during World War II for the German war effort. Note that Zuse began his initial design in 1939 with the Z1. Had Zuse been given adequate funding by the German government, he most likely would have developed a much more powerful computer system. Zuse is finally receiving some belated honor for his pioneering work in the area of digital electronics that began in the 1930s and for his Z3 computer system.

It has been discovered recently (through the declassification of British military documents) that the first truly **electronic computer** was placed into operation in 1943 to break secret German military codes. This first electronic computer system, which used vacuum tubes, was invented by *Alan Turing.* Turing called his machine **Colossus,** most likely because of its size. A problem with Colossus was that its design allowed it to break German military codes generated by the mechanical **Enigma machine,** but it could not solve other problems. Colossus was not programmable; it was a **fixed-program computer system,** which today is called a **special-purpose computer.**

The first general-purpose, programmable, **electronic computer system** was developed in 1946 at the University of Pennsylvania. This first of the modern computers was called the **ENIAC (Electronics Numerical Integrator and Calculator).** The ENIAC was a huge machine containing over 17,000 vacuum tubes and at least 500 miles of wires. This massive machine weighed over 30 tons, yet only performed about 100,000 operations per second. The ENIAC thrust the world into the age of electronic computers. The ENIAC was programmed by rewiring its circuits. This programming required many days to accomplish and many workers to change electrical connections on plug-boards that looked like early telephone switchboards. Another problem with the ENIAC was the short life of the vacuum tube components, which required frequent maintenance.

Breakthroughs that followed were the development of the **transistor** in 1948 at *Bell Labs,* followed by the invention of the **integrated circuit** in 1958 by *Jack Kilby* of Texas Instruments.

The integrated circuit led to the development of digital integrated circuits (RTL or resistor-to-transistor logic) in the 1960s and the first **microprocessor** in 1971 at Intel Corporation. At this time Intel and one of its engineers, *Marcian T. Hoff* (Ted Hoff), developed the 4004 micro-processor—the device that started the microprocessor revolution that continues today at an ever-accelerating pace.

Programming Advancements

Now that programmable machines had been developed, programs and programming languages began to appear. As mentioned, the first truly programmable electronic computer system was pro-grammed by rewiring its circuits. This proved too cumbersome for practical application, so early in the evolution of computer systems, computer languages began to appear to control the com-puter. The first such language was **machine language,** which was constructed of ones and zeros using binary codes that were stored in the memory system in groups called **programs.** This was more efficient than rewiring a machine, but it was still extremely time consuming to develop a program because of the large number of codes required for a program. *John von Neumann,* a mathematician, was the first to develop a system that accepted instructions and stored them in memory. Computers are often called **von Neumann machines** in honor of John von Neumann.

Later, once computer systems such as the UNIVAC I and II became available in the early 1950s, **assembly language** was used to simplify the chore of entering binary code into a com-puter as its instructions. The **assembler,** a program that accepted assembly language, allowed a programmer to use **mnemonic codes,** such as ADD for addition, in place of a cryptic binary number, such as 01000111. Even though assembly language was an aid to programming, it wasn't until 1957 when *Grace Hopper* developed the first **high-level programming language** called **FLOW-MATIC** that computers became easier to program. Later in the same year IBM developed **FORTRAN** (FORMula TRANslator) for its computer systems. The FORTRAN lan-guage allowed programmers to develop programs that used formulas to solve mathematical problems. FORTRAN is still in use by some scientists for computer programming. Another sim-ilar language introduced about a year after FORTRAN was **ALGOL** (ALGOrithmic Language).

The first truly successful, widespread programming language for business applications was **COBOL** (COmputer Business Oriented Language). Although COBOL usage has dimin-ished in recent years, it is still a major player in many large business systems. Another fairly pop-ular business language is **RPG** (Report Program Generator) that allows programming by specifying the form of the input, output, and calculations.

Since these early days of programming, more languages have appeared. Some common ones are BASIC, C/C++, PASCAL, and ADA. The **BASIC** and **PASCAL** languages were both designed as teaching languages, but have escaped from the classroom and are used in many com-puter systems. BASIC is probably the easiest of all to learn. Some estimates indicate that BASIC is used in the personal computer for 80 percent of the programs written by users. Recently, a new version of BASIC, called **VISUAL BASIC,** has made programming in the Windows environ-ment easier. VISUAL BASIC may eventually supplant C/C++ and PASCAL.

In the scientific community **C/C++** and **PASCAL** appear as control programs. Both lan-guages, and especially C/C++, allow the programmer almost complete control over the program-ming environment and computer system. In many cases, C/C++ is replacing some of the low-level, machine-control software normally reserved for assembly language. Even so, as-sembly language still plays an important role in programming. Many video games written for the personal computer are written almost exclusively in assembly language. Assembly language is also interspersed with C/C++ and PASCAL to perform machine-control functions efficiently.

The **ADA** language is used heavily by the Department of Defense. ADA was named after Augusta Ada Byron, Countess of Lovelace. The Countess worked with Charles Babbage in the early 1800s in the development of his Analytical Engine.

The Microprocessor Age

The world's first **microprocessor,** the Intel 4004, is a 4-bit microprocessor—a programmable controller on a chip—that is meager by today's standards. It addressed a mere 4,096 four-bit wide memory locations. (A **bit** is a binary digit with a value of one or zero. A 4-bit wide memory location is often called a **nibble** or nybble.) The 4004 instruction set contained only 45 instructions. It was fabricated with the (then) current state-of-the-art P-channel MOSFET technology, which allowed it to execute instructions at the slow rate of 50 KIPs **(kilo-instructions per second).** This was slow compared to the 100,000 instructions executed per second by the 30-ton ENIAC computer in 1946. The main difference is that the 4004 weighed less than an ounce.

At first, applications abounded for this device. The 4-bit microprocessor debuted in early video game systems and small microprocessor-based control systems. One such early video game, a shuffleboard game, was produced by Balley. The main problems with this early microprocessor were its speed, word width, and memory size. The evolution of the 4-bit microprocessor ended when Intel released the 4040, an updated version of the earlier 4004. The 4040 operated at a higher speed, although it lacked improvements in word width and memory size. Other companies, in particular Texas Instruments, also produced 4-bit microprocessors (TMS-1000). The 4-bit microprocessor still survives in low-end applications such as microwave ovens and small control systems, and is available from some microprocessor manufacturers. Most calculators are also still based on 4-bit microprocessors that process 4-bit BCD **(binary-coded decimal)** codes.

Later, in 1971, realizing that the microprocessor was a commercially viable product, Intel Corporation released the 8008, an extended 8-bit version of the 4004 microprocessor. The 8008 addressed an expanded memory size (16K bytes) and contained additional instructions (a total of 48) that provided an opportunity for its application in more advanced systems. (A **byte** is generally an 8-bit wide binary number and a **computer K** is 1,024. Often, memory size is specified in **K bytes.**)

As engineers developed more demanding uses for the 8008 microprocessor, they discovered that its somewhat small memory size, slow speed, and limited instruction set restricted its usefulness. Intel recognized these limitations and, in 1973, introduced the 8080 microprocessor—the first of the modern 8-bit microprocessors. About six months after Intel released the 8080 microprocessor, Motorola Corporation introduced its MC6800 microprocessor. The floodgates opened and the 8080 and, to a lesser degree, the MC6800 ushered in the age of the microprocessor. Soon other companies began to introduce their own versions of the 8-bit microprocessor. Table 1–1 lists several of these early microprocessors and their manufacturers. Of these early microprocessor producers, only Intel and Motorola continue successfully with newer and improved versions of the microprocessor. Zilog still manufactures microprocessors, but has remained in the background deciding to concentrate, fairly successfully, on microcontrollers and embedded controllers instead of general-purpose microprocessors. Rockwell has all but abandoned microprocessor development in favor of modem circuitry.

TABLE 1–1 Early 8-bit microprocessors

Manufacturer	Part Number
Fairchild	F-8
Intel	8080
MOS Technology	6502
Motorola	MC6800
National Semiconductor	IMP-8
Rockwell International	PPS-8
Zilog	Z-8

What Was Special About the 8080?   Not only could the 8080 address more memory and execute additional instructions, but it executed them 10 times faster than the 8008. An addition instruction that took 20 μs (50,000 instructions per second) on an 8008-based system required only 2.0 μs (500,000 instructions per second) on an 8080-based system. Also, the 8080 was compatible with TTL (**transistor-transistor logic**) and the 8008 was not directly compatible. This made interfacing much easier and less expensive. The 8080 also addresses four times more memory (64K bytes) than the 8008 (16K bytes). These improvements are responsible for ushering in the era of the 8080, and the continuing saga of the microprocessor. Incidentally, the first personal computer, the **MITS Altair 8800,** was released in 1974. (Note that 8800 was probably chosen to avoid copyright violations with Intel.) The BASIC language interpreter, written for the Altair 8800 computer, was developed by *William Gates,* founder of Microsoft Corporation. The assembler program for the Altair 8800 was written by Digital Research Corporation, which now produces DR-DOS for the personal computer.

The 8085 Microprocessor.   In 1977, Intel Corporation introduced an updated version of the 8080—the 8085. This was to be the last 8-bit general-purpose microprocessor developed by Intel. Although only slightly more advanced than an 8080, the 8085 executes software at an even higher speed. An addition instruction that took 2.0 μs (500,000 instructions per second) on the 8080 requires only 1.3 μs (769,230 instructions per second) on the 8085. The main advantages of the 8085 are its internal clock generator, internal system controller, and higher clock frequency. This higher level of component integration reduced its cost and increased the usefulness of the 8085 microprocessor. Intel has managed to sell over 100 million copies of the 8085 microprocessor, its most successful 8-bit microprocessor. Because the 8085 is also manufactured **(second-sourced)** by many other companies, there are over 200 million of these microprocessors in existence. Applications that contain the 8085 are still being used and designed and will likely continue to be popular well into the future. Another company that sold 500 million 8-bit microprocessors is Zilog Corporation, which produced the Z-80 microprocessor. The Z-80 is machine-language code compatible with the 8085, which means that there are over 700 million microprocessors that execute 8085/Z-80 compatible code!

The Modern Microprocessor

In 1978 Intel released the 8086 microprocessor and a year or so later, the 8088. Both devices are 16-bit microprocessors, which execute instructions in as little as 400 ns (2.5 MIPs or 2.5 **millions of instructions per second**). This represented a major improvement over the execution speed of the 8085. Furthermore, the 8086 and 8088 address 1M byte of memory, 16 times more memory than the 8085. (A **1M-byte memory** contains 1,024K byte-sized memory locations or 1,048,576 bytes.) This higher execution speed and larger memory size allowed the 8086 and 8088 to replace smaller minicomputers in many applications. One other feature found in the 8086/8088 is a small 4- or 6-byte instruction **cache** or **queue** that prefetches a few instructions before they are executed. The queue speeds the operation of many sequences of instructions and is the basis for the much larger instruction caches found in modern microprocessors.

The increase in memory size and additional instructions of the 8086 and 8088 have led to many sophisticated applications for microprocessors. Improvements to the instruction set included a multiply and divide instruction, which is missing on earlier microprocessors. The number of instructions increased from 45 on the 4004, to 246 on the 8085, to well over 20,000 variations on the 8086 and 8088 microprocessors. Note that these microprocessors are called CISC (**complex instruction set computer**) because of the number and complexity of instructions. The additional instructions ease the task of developing efficient and sophisticated applications even though their number is at first overwhelming and time-consuming to learn. The 16-bit microprocessor evolved mainly because of the need for larger memory systems. It provides more internal register storage space than the 8-bit microprocessor. The additional registers allow software to be written more efficiently.

The popularity of the Intel family was ensured in 1981 when IBM Corporation decided to use the 8088 microprocessor in its **personal computer.** Applications such as spreadsheets, word processors, spelling checkers, and computer-based thesauruses are memory-intensive. To execute efficiently, they require more than the 64K bytes of memory found in 8-bit microprocessors. The 16-bit 8086 and 8088 provide 1M byte of memory for these applications. Soon, even 1M byte of memory proved limiting for large spreadsheets and other applications. This led to Intel's introduction in 1983 of an updated 8086, the 80286 microprocessor.

The 80286 Microprocessor.   The 80286 (also a 16-bit architecture microprocessor) is almost identical to the 8086 and 8088 except it addresses a 16M-byte memory system instead of 1M byte. The instruction set of the 80286 is almost identical to the 8086 and 8088 except for a few additional instructions that manage the extra 15M bytes of memory. The clock speed of the 80286 is increased so it executes some instructions in as little as 250 ns (4.0 MIPs) with the original release of the 8.0 MHz version. Some changes also occurred in the internal execution of the instructions that led to an eightfold increase in speed for many instructions compared to 8086/8088 instructions.

The 32-Bit Microprocessor.   Applications began to demand faster microprocessor speeds, more memory, and wider data paths. This led to the arrival of Intel's 80386 in 1986. The 80386 represented a major overhaul of the 16-bit 8086/80286 microprocessor's architecture. The 80386 is Intel's first practical 32-bit microprocessor that contains a 32-bit data bus and a 32-bit memory address. (Note that Intel produced an earlier, unsuccessful 32-bit microprocessor, the iapx-432.) Through these 32-bit buses, the 80386 addresses up to 4G bytes of memory. (**One G** of memory contains 1,024M or 1,073,741,824 locations.) A **4G-byte memory** stores an astounding 1,000,000 typewritten, double-spaced pages of data. The 80386 is also available in a few modified versions such as the 80386SX, which addresses 16M bytes of memory through a 16-bit data and 24-bit address bus, and the 80386SL/80386SLC, which addresses 32M bytes of memory through a 16-bit data and 25-bit address bus. An 80386SLC version contains an internal cache memory that allows it to process data at even higher rates. Recently Intel released the **80386EX microprocessor.** The 80386EX is called **an embedded PC** containing all the components of the AT class personal computer on a single integrated circuit. The 80386EX also contains 24 pin connections used for input/output data, a 26-bit address bus, a 16-bit data bus, a DRAM refresh controller, and programmable chip selection logic.

Applications that require higher microprocessor speeds and larger memory systems include software systems that use a GUI (**graphical user interface**). We often call a GUI a WYSIWYG (**what you see is what you get**) display. Modern graphical displays often contain 256,000 or more picture elements (**pixels** or **pels**). The least-sophisticated VGA (**variable graphics array**) video display has a resolution of 640 pixels per scanning line with 480 scanning lines. In order to display one screen of information, each picture element must be changed. This requires a high-speed microprocessor. Many new software packages use this type of video interface. These GUI-based packages require high microprocessor speeds and often accelerated video adapters for quick and efficient manipulation of video text and graphical data. The most striking system, which requires high-speed computing for its graphical display interface, is Microsoft Corporation's Windows.[1]

The 32-bit microprocessor is needed because of the size of its data bus, which transfers real (single-precision floating-point) numbers that require 32-bit wide memory. In order to efficiently process 32-bit real numbers, the microprocessor must efficiently pass them between itself and memory. If they pass through an 8-bit data bus, it takes four read or write cycles, but when passed through a 32-bit data bus only one read or write cycle is required. This significantly

---

[1]Windows is a registered trademark of Microsoft Corporation.

increases the speed of any program that manipulates real numbers. Most high-level languages, spreadsheets, and database management systems use real numbers for data storage. Real numbers are also used in graphical design packages that use vectors to plot images on the video screen. These include CAD (**computer aided drafting/design**) systems such as AUTOCAD, ORCAD, and so forth.

Besides providing higher clocking speeds, the 80386 includes a memory management unit that allows memory resources to be allocated and managed by the operating system. Earlier microprocessors left memory management completely to the software. The 80386 includes hardware circuitry for memory management and memory assignment, which improves its efficiency and reduces software overhead.

The instruction set of the 80386 microprocessor is upward compatible with the earlier 8086, 8088, and 80286 microprocessors. Additional instructions reference the 32-bit registers and manage the memory system. Memory management instructions and techniques used by the 80286 are compatible with the 80386 microprocessor. These features allow older, 16-bit software to operate on the 80386 microprocessor.

The 80486 Microprocessor.   In 1989, Intel released the **80486 microprocessor,** which incorporates an 80386-like microprocessor, an 80387-like numeric coprocessor, and an 8K-byte cache memory system into one integrated package. Although the 80486 is not radically different from the 80386, it does include one substantial change: The internal structure of the 80486 is modified from the 80386 so that about half of its instructions execute in 1 clock instead of 2 clocks. Because the 80486 is available in a 50 MHz version, about half the instructions execute in 25 ns (50 MIPs). The average speed improvement for a typical mix of instructions is about 50 percent over the 80386 operated at the same clock speed. Newer versions of the 80486 execute instructions at even higher speeds with a 66 MHz double-clocked version.

The **double-clocked** 66 MHz version executes instructions at the rate of 66 MHz with memory transfers executed at the rate of 33 MHz. A **triple-clocked** version from Intel, for some reason called the 80486DX4, improves the internal execution speed to 100 MHz with memory transfers at 33 MHz. Note that the 80486DX4 performs integer operations at about the same speed as the 60 MHz Pentium. It also contains an expanded 16K-byte cache in place of the standard 8K-byte cache found on earlier 80486 microprocessors. The future promises to bring microprocessors that internally execute instructions at rates of 250 MHz or higher.

Other versions of the 80486 are called **Overdrive**[2] processors. The Overdrive processor is actually a double-clocked, or recently a triple-clocked, version of the 80486DX that replaces an 80486SX or slower speed 80486DX. When the Overdrive processor is plugged into its socket, it disables or replaces the 80486SX or 80486DX and functions as a doubled-clocked version of the microprocessor. For example, if an 80486SX operating at 25 MHz is replaced with an Overdrive microprocessor, it functions as a 80486DX2 50-MHz microprocessor using a memory transfer rate of 25 MHz.

Table 1–2 lists microprocessors produced by Intel, Motorola, and IBM with information about their word and memory sizes. Other companies produce microprocessors, but none have attained the success of Intel and to a lesser degree Motorola and IBM.

The Pentium Microprocessor.   The **Pentium,** introduced in 1993, is similar to the 80386 and 80486. It was originally labeled the **P5** or **80586,** but Intel decided not to use a number because apparently it is impossible to copyright a number. The two introductory versions of the Pentium operate with a clocking frequency of 60 MHz and 66 MHz and a speed of 110 MIPs, with a higher frequency 100 MHz **one and one-half clocked** version, operating at 150 MIPs. Also **the double-clocked** Pentium operating at 120 MHz will be available soon. Another difference is that

---

[2]Overdrive is a trademark of Intel Corporation.

TABLE 1–2   Modern Microprocessors

Manufacturer	Part	Data Bus Width	Memory Size
Intel	8048	8	2K internal
	8051	8	8K internal
	8085A	8	64K
	8086	16	1M
	8088	8	1M
	8096	16	8K internal
	80186	16	1M
	80188	8	1M
	80286	16	16M
	80386EX	16	64M
	80386DX	32	4G
	80386SL	16	32M
	80386SX	16	16M
	80486DX/DX2	32	4G + 8K cache
	80486DX4	32	4G + 16K cache
	80486SX	32	4G + 8K cache
	Pentium	64	4G + 16K cache
	Pentium Pro	64	4G/64G + 16K L1 + 256K (512K) L2
IBM	80386SLC	16	32M + 1K cache
	Blue Lightning	32	4G + 8K cache
Motorola	6800	8	64K
	6805	8	2K internal
	6809	8	64K
	68000	16	16M
	68008Q	8	1M
	68008D	8	4M
	68010	16	16M
	68020	32	4G
	68030	32	4G + 256 cache
	68040	32	4G + 8K cache
	68060	32	4G + 16K cache
	PowerPC	64	4G + 32K cache

the cache size is increased to 16K bytes from the 8K cache found in the basic version of the 80486. The Pentium contains an 8K-byte instruction cache and an 8K-byte data cache. This allows a program that transfers a large amount of memory data to still benefit from a cache. The memory system contains up to 4G bytes with the data bus width increased from 32 bits, found in the 80386 and 80486, to a full 64 bits. The data bus transfer speed is either 60 or 66 MHz, depending on the version of the Pentium. This wider data bus accommodates double-precision floating-point numbers used for high-speed vector-generated graphical displays. It also transfers data between the memory system and microprocessor at a higher rate. This should allow **virtual reality** and software to operate at more realistic rates on current and future Pentium-based platforms. The widened data bus and higher execution speed of the Pentium should also allow full-frame video displays that operate at scan rates of 30 Hz or higher—comparable to commercial television.

Probably the most ingenious feature of the Pentium is its **dual integer processors.** The Pentium simultaneously executes two instructions, not dependent on each other, because it contains two independent internal integer processors called **superscaler technology.** This allows

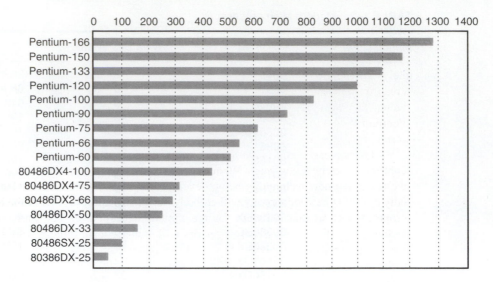

FIGURE 1–2 The Intel iCOMP rating index.

the Pentium to often execute two instructions per clocking period. Another feature that enhances performance is a **jump prediction technology** that speeds the execution of programs that include loops. As with the 80486, the Pentium also employs an internal floating-point coprocessor to handle floating-point data at about a five times speed improvement. These features portend continued success for the Intel family of microprocessors. They also may allow the Pentium to replace some of the RISC (**reduced instruction set computer**) machines that currently execute one instruction per clock, although some newer RISC processors execute more than one instruction per clock through the introduction of superscaler technology. Motorola, Apple, and IBM have recently produced the PowerPC, a RISC microprocessor that has two integer units and one floating-point unit. The PowerPC certainly boosts the performance of the Apple Macintosh,[3] but at present is slow at emulating the Intel family of microprocessors. Tests indicate that the current emulation software executes DOS and Windows applications slower than the 80486SX 25 MHz microprocessor. Therefore, the Intel family should survive for many years in personal computer systems. Note that there are currently 4 million Apple Macintosh systems and over 160 million personal computers based on Intel microprocessors.

Recently, the Pentium Overdrive (P24T) has been made available to upgrade the 80486-based personal computer to the Pentium. The Pentium Overdrive is currently available as a 63 MHz or 83 MHz upgrade. The 80486DX2 50 MHz system is upgraded with the 63 MHz Overdrive and the 80486DX2 66 MHz system with the 83 MHz Overdrive. The 83 MHz Overdrive causes the 80486 system to function at a speed somewhere between the 66 MHz and 75 MHz Pentium.

To compare the speeds of various microprocessors, Intel devised the **iCOMP rating index,** a composite of SPEC92, ZD Bench, and Power Meter. Figure 1–2 compares the relative speed of the 80386DX 25 MHz version at the low end to the Pentium 100 MHz version at the high end of the spectrum.

The Pentium Pro Processor.   In late 1995 Intel released the **Pentium Pro processor,** formerly called the **P6 microprocessor,** which contains over 21 million transistors and 2 integer units as well as a floating-point unit to increase the performance of most software. The basic clock fre-

---

[3]Macintosh is a trademark of Apple Computers Incorporated.

quency is 150 MHz at its introduction. In addition to the internal 16K level one (L1) cache, the P6 also contains a 256K level two (L2) cache. Newer versions run at 200 MHz and contain a 512K L2 cache.

The Pentium Pro executes up to three instructions simultaneously. This, along with its increased speed and optimization for 32-bit instructions, will push the software into the 32-bit arena. The latest 32-bit operation system is Windows 95 or Windows NT. To fully utilize the Pentium Pro, Windows NT is the suggested operating system because the Pentium Pro is optimized to execute the 32-bit instructions of Windows NT, a 32-bit environment.

No one can really make accurate predictions, but the success of the Intel family should continue for quite a few years. What may occur is a change to RISC technology, but more likely a change to a new technology being developed jointly by Intel and Hewlett-Packard. Undoubtedly even this new technology will embody the CISC instruction set of the 80X86 family of microprocessors so that software for the system will survive. The basic premise behind this technology is that many microprocessors will communicate directly with each other, allowing parallel processing without any change to the instruction set or program. Currently, superscaler technology uses many microprocessors, but they all share the same register set. This new technology uses many microprocessors that each contain their own register sets that are linked with the other microprocessors' registers—thus offering the speed of parallel processing without writing any special program.

The Embedded Controller

Now that we have some understanding of the developments that led to the microprocessor, we will discuss the embedded controller. The embedded controller, like the microprocessor, was invented by Intel Corporation. Embedded controllers, sometimes called **embedded microprocessors,** are applied in many control systems because of their size and similarity to the personal computer's 8086/8088 and 80386 microprocessors.

Often an integrated circuit that contains a microprocessor, I/O, read-only memory, and read/write memory is called an **embedded controller.** Likewise, an integrated circuit that contains a microprocessor, I/O, but no memory is called an embedded controller or sometimes an embedded microprocessor. Whatever we label these devices, they are more than just a microprocessor; they are integrated circuits that contain most of the parts found in a complete computer system.

The 8048 Embedded Controller. The first embedded controller, sometimes referred to as a **microcomputer,** was the infamous Intel 8048 family of embedded controllers. The 8048, first introduced in 1978, is a highly integrated version of the 8080 microprocessor. This early embedded controller contained I/O pins, a timer, read-only memory for program storage, and a small scratch-pad RAM for transient data storage. It was a self-contained computer system. Many variations of the 8048 were introduced, with various-sized scratch-pad memories and ROM sizes. The 8048 was used as an embedded controller in dot-matrix printers common in the 1980s. Today the 8048 has been replaced with the 8051 or 80251 for low-end consumer applications. Some systems also use the 8096 embedded controller, which is a modified 16-bit version of the 8051.

The 8-bit embedded controller, produced by companies such as Intel, Motorola, and Zilog, dominate many small systems such as printers, disk controllers, and office copiers. Sophisticated applications are beginning to appear that require more power than available with the 8-bit embedded controller. These newer applications require 16- or 32-bit embedded controllers and much higher clock speeds than found in the earlier 8-bit embedded controllers. These new high-end embedded controllers are also available from Intel and Motorola, but a large percentage are Intel-based applications using the 80186/80188 or 80386 embedded controller. Table 1–3 lists the embedded controllers available from Intel.

TABLE 1–3 Intel embedded controllers and embedded microprocessors

Part	Data Bus Width	Internal Memory Size
8051	8	2K ROM, 256 RAM
8096	16	8K ROM, 232 RAM
80196	16	232 RAM
87196	16	8K ROM, 232 RAM
80186*	16	None
80188*	8	None
80376	16	None
80386CX/80386EX	16	None

*Note: The 80186/80188 family also includes the following members: 80C186XL/ 80C188XL, 80C186EA/80C188EA, 80C186EB/80C188EB, and 80C186EC/80C188EC.

The 80186/80188 Embedded Controllers.   The 80186 and 80188 embedded controllers have been available since 1982. These embedded components are 16-bit microprocessors, with the 80188 containing an 8-bit memory data pathway instead of 16 bits. Both the 80186 and 80188 contain the same register and instruction set as the 8086 and 8088 microprocessors with a few additional instructions that eventually appeared in the 80286 microprocessor. This means that software can be tested and emulated on the personal computer before using it with the 80186 or 80188 embedded controller. This low-cost emulation has made these embedded controllers very popular. Also available are newer, updated CMOS versions labeled 80C186XL/80C188XL, 80C186EA/80C188EA, 80C186EB/80C188EB, and 80C186EC/80C188EC.

In addition to containing the 8086 or 8088 microprocessor, the 80186 and 80188 also contain embedded timers, a clock generator, a programmable interrupt controller, a programmable DMA controller, and a programmable chip selection unit. The 80188EB version also contains a pair of UARTS for serial data communications and parallel port data. The CMOS versions of the 80186 and 80188 also contain a power-saving feature as well as a programmable DRAM refresh circuit. We call a machine that contains a power-saving feature a **green machine.** These controllers and their internal devices are required for many systems, but are ancillary, external components in microprocessor-based personal computer systems.

The 80386CX/80386EX Embedded Controllers.   The most recent additions to the Intel embedded controller family are the 80386CX and 80386EX. These embedded controllers are similar to the 80386SX microprocessor because they address memory through a 16-bit data bus and contain a 32-bit internal architecture. They differ by addressing a 64M memory system in place of the 16M memory system addressed by the 80386SX microprocessor. In addition to addressing extra memory, the 80386EX contains a programmable chip selection unit, a pair of programmable interrupt controllers, a programmable DMA controller, programmable DRAM refresh circuitry, timers, serial interface, and 24 bits of programmable I/O ports.

1–2   THE MICROPROCESSOR-BASED PERSONAL COMPUTER SYSTEM

The computer system has undergone many changes in recent history. Machines that once filled large areas have been reduced to small desktop computer systems because of the microprocessor. Even though these desktop computers are compact, they possess computing power that was only dreamed of a few years ago. Million-dollar mainframe computer systems, developed in the early 1980s, are not as powerful as the 80386, 80486, or Pentium microprocessor-based computers of today. In fact, many smaller companies are replacing their mainframe

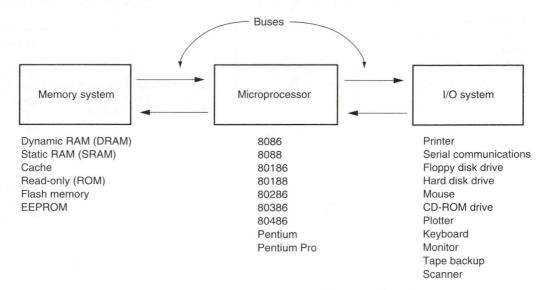

FIGURE 1–3 The block diagram of a microprocessor-based computer system.

computers with microprocessor-based systems. Companies such as DEC (Digital Equipment Corporation) have stopped producing mainframe computer systems in order to concentrate their resources on microprocessor-based computer systems. In this section we discuss the structure of the microprocessor-based personal computer system. This structure includes information about the memory and operating system used in many microprocessor-based computer systems. This section is included here because the personal computer is often used as a development platform for embedded controllers.

Figure 1–3 is a block diagram of the personal computer. This diagram also applies to any computer system from the early mainframe computers to the latest microprocessor-based systems. The block diagram is composed of three blocks that are interconnected by buses. (A **bus** is set of common connections that carry the same type of information. For example, the address bus, which contains 20 to 32 connections, is a bus that conveys the memory address to the memory.) These blocks and their function in a personal computer are discussed in this section.

The Memory and I/O System

The memory structure of all Intel 8086–80486 and Pentium-based personal computer systems is similar. This includes the first personal computers based on the 8088 introduced in 1981 by IBM to the most powerful, high-speed versions of today based on the Pentium microprocessor. Figure 1–4 illustrates the memory map of a personal computer system. This map applies to any IBM personal computer or any of the many IBM-compatible clones that are in existence.

The memory system is divided into three main parts: TPA **(transient program area),** system area, and XMS **(extended memory system).** The type of microprocessor in your computer determines whether an extended memory system exists. If the computer is based on an older 8086 or 8088 (a PC[4] or XT[5]), the TPA and system areas exist, but there is no extended memory area. The PC and XT contain 640K bytes of TPA and 384K bytes of system memory for a total memory size of 1M byte. We often call the first 1M byte of memory the **real memory** because each Intel microprocessor is designed to function in this area in its real mode of operation.

---

[4]PC (personal computer) is a trademark of IBM Corporation.

[5]XT (extended technology) is a trademark of IBM Corporation.

FIGURE 1–4 The memory map of a personal computer system.

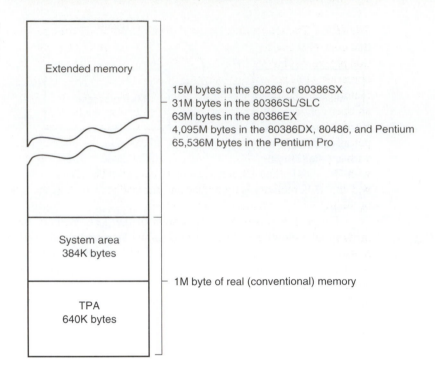

15M bytes in the 80286 or 80386SX
31M bytes in the 80386SL/SLC
63M bytes in the 80386EX
4,095M bytes in the 80386DX, 80486, and Pentium
65,536M bytes in the Pentium Pro

1M byte of real (conventional) memory

Computer systems based on the 80286 through the Pentium not only contain the TPA (640K bytes) and system area (384K bytes), they may also contain extended memory. These machines are often called AT[6] class machines. The PS/1 and PS/2, produced by IBM, are other versions of the same basic memory design. Sometimes these machines are also referred to as ISA **(industry standard architecture)** or EISA **(extended ISA)** machines. The PS/2 is referred to as a micro-channel[7] architecture system or ISA system depending on the model number.

Recently a new bus, the PCI **(peripheral control interconnect)** bus, is beginning to appear in Pentium-based systems. Extended memory contains up to 15M bytes in the 80286- and 80386SX-based computer and up to 4,095M bytes in the 80386DX-, 80486-, and Pentium-based computer in addition to the first 1M byte of real memory. The ISA machine contains an 8-bit peripheral bus used to interface 8-bit devices to the computer in the 8088/8088-based PC or XT computer system. The AT class machine, also an ISA machine, uses a 16-bit peripheral bus for interface and may contain an 80286 or above microprocessor. The EISA bus is a 32-bit peripheral interface bus found in 80386DX through the Pentium-based systems. Note that each of these buses is compatible with the earlier versions. That is, an 8-bit interface card functions in the 16-bit ISA or 32-bit EISA bus standards; likewise, a 16-bit interface card functions in the 16-bit ISA or 32-bit EISA standards.

Another recent bus type is called the VESA[8] **local bus** or **VL bus.** The local bus interfaces disk and video to the microprocessor at the local bus level. This allows 32-bit interfaces to function at the same clocking speed as the microprocessor. A recent modification to the VESA local bus supports the 64-bit data bus of the Pentium microprocessor and competes directly with the PCI bus. The ISA and EISA standards function at only 8 MHz, which reduces the performance of disk and video interfaces using these standards. The PCI bus is either a 32- or 64-bit bus that is specifically designed to function with the Pentium microprocessor at 33 MHz.

---

[6]AT (advanced technology) is a trademark of IBM Corporation.

[7]Micro-channel is a trademark of IBM Corporation.

[8]VESA is the Video Electronics Standards Association.

The TPA. The **transient program area** (TPA) holds the operating system and other programs that control the computer system. The TPA also stores any currently active or inactive application program. The length of the TPA is 640K bytes. As mentioned, this area of memory holds the operating system, which requires a portion of the TPA. In practice, the amount of memory remaining for application software is about 628K bytes if MS-DOS[9] version 5.0 or 6.X is used as an operating system. Earlier versions of MS-DOS required more of the TPA and often left only 530K bytes or less for application programs. Another operating system found in personal computers is PC-DOS.[10] Both PC-DOS and MS-DOS are compatible, so both function in the same with application programs. Windows and OS/2[11] are other operating systems that are compatible with DOS and allow DOS programs to execute. The DOS (disk operating system) controls the way that disk memory is organized and controlled as well as the function and control of the some of the I/O devices connected to the system.

Figure 1–5 shows the organization of the TPA memory map in a computer system. The memory map shows how the many areas of the TPA are used for system programs, data, and drivers. It also shows a large area of memory available for application programs. To the left of each area is a hexadecimal number that represents the memory address that begins and ends each data area. **Hexadecimal memory addresses** or memory locations are used to number each byte of the memory system. (A **hexadecimal number** is a number represented in radix 16 or base 16 with each digit representing a value from 0–9 and A–F. We often end a hexadecimal number with an **H** to indicate that it is a hexadecimal value. For example, 1234H is 1234 hexadecimal.)

The **interrupt vectors** access various features of the DOS, **BIOS (basic I/O system)**, and applications. The BIOS is a collection of programs stored in a read-only (ROM) or flash memory that operate many of the I/O devices connected to your computer system. Note that a flash memory is an EEPROM **(electrically erasable read-only memory)** that is erased in the system electrically, while the ROM is a device that must be programmed in a special machine called an EPROM programmer for an EPROM **(erasable/programmable read-only memory)** or at the factory when a ROM is fabricated. These programs are stored in the system area defined later in this section.

The BIOS and DOS communications areas contain transient data used by programs to access I/O devices and the internal features of the computer system. (Refer to Appendix A for a complete listing of the BIOS and DOS communications areas.) These are stored in the TPA so they can be changed as the system operates. Note that the TPA contains **read/write memory** (called RAM or random access memory) so it can change as a program executes.

The IO.SYS is a program that loads into the TPA from the disk whenever an MS-DOS or PC-DOS system is started. The IO.SYS contains programs that allow DOS to use the keyboard, video display, printer, and other I/O devices often found in the computer system. The IO.SYS program links DOS to the programs stored on the BIOS ROM.

The MS-DOS/PC-DOS program occupies two areas of memory. One area is 16 bytes in length and is located at the top of the TPA and the other is much larger, located near the bottom of the TPA. The MS-DOS program controls the operation of the computer system. The size of the MS-DOS area depends on the version of MS-DOS installed in the computer memory and how it is installed. If DOS is installed in high memory with the **HIMEM.SYS driver,** most of the TPA is free to hold application programs. (High memory is described later in this book and only applies to 80286 or newer microprocessors.)

The size of the driver area and number of drivers change from one computer to another. **Drivers** are programs that control installable I/O devices such as a mouse, disk cache, hand scanner, and CD-ROM memory **(Compact Disk Read-Only Memory). Installable drivers** are

---

[9]MS-DOS (Microsoft Disk Operating System) is a trademark of Microsoft Corporation.

[10]PC-DOS (Personal Computer Disk Operating System) is a trademark of IBM Corporation.

[11]OS/2 (Operating System/version 2) is a trademark of IBM Corporation.

FIGURE 1–5 The memory map of the TPA in a personal computer. (This map will vary between systems.)

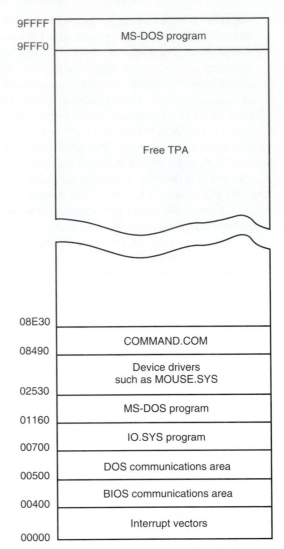

programs that control or drive devices or programs that are added to the computer system. Drivers are normally files that have an extension of .SYS such as MOUSE.SYS, or in DOS version 3.2 and later, .EXE such as **EMM386.EXE.** Because few computer systems are identical, the driver area varies in size and contains different numbers and types of drivers.

The COMMAND.COM program **(command processor)** controls the operation of the computer from the keyboard. The COMMAND.COM program processes the DOS commands as they are typed on the keyboard. For example, if DIR is typed, the COMMAND.COM program displays a directory of the disk files in the current disk directory. If the COMMAND.COM program is erased, the computer cannot be used from the keyboard. Never erase COMMAND. COM, IO.SYS, or MSDOS.SYS to make room for other software or your computer will not function. If erased, these programs can be reloaded to the disk with the SYS.COM program located in the DOS directory.

The free TPA area holds application programs as they are executed. These application programs include word processors, spreadsheet programs, CAD programs, and so forth. The TPA also holds TSR **(terminate and stay resident)** programs that remain in memory in an inactive state until activated by a hot-key sequence or other event such as an interrupt. A calculator program is an

FIGURE 1–6 The system area of a typical personal computer.

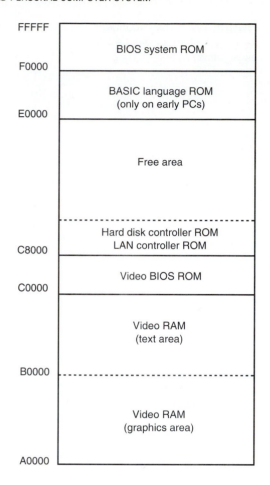

```
FFFFF ┌─────────────────────────┐
      │                         │
      │    BIOS system ROM      │
      │                         │
F0000 ├─────────────────────────┤
      │                         │
      │   BASIC language ROM    │
      │   (only on early PCs)   │
E0000 ├─────────────────────────┤
      │                         │
      │                         │
      │       Free area         │
      │                         │
      │- - - - - - - - - - - - -│
      │  Hard disk controller ROM│
      │    LAN controller ROM   │
C8000 ├─────────────────────────┤
      │     Video BIOS ROM      │
C0000 ├─────────────────────────┤
      │                         │
      │      Video RAM          │
      │      (text area)        │
      │                         │
B0000 │- - - - - - - - - - - - -│
      │                         │
      │      Video RAM          │
      │    (graphics area)      │
      │                         │
A0000 └─────────────────────────┘
```

example of a TSR that activates whenever an ALT-C key (**hot key**) is typed. A hot key is a combination of keys on the keyboard that activate a TSR program. A **TSR program** is also called a pop-up program because, when activated, it appears to pop up inside another program.

The System Area. The system area, although smaller than the TPA, is just as important. The system area contains programs on a read-only memory (ROM) or flash memory and also areas of read/write (RAM) memory for data storage. Figure 1–6 shows the memory map of the system area of a typical computer system. As with the map of the TPA, this map also includes the hexadecimal memory addresses of the various areas.

The first area of the system space contains video display RAM and video control programs on ROM or flash memory. This area generally starts at location A0000H and extends to location C7FFFH. The size and amount of memory used depends on the type of video display adapter attached to the system. Display adapters that are often attached to a computer include the earlier **CGA (color graphics adapter)** and EGA **(enhanced graphics adapter)** or one of the many newer forms of VGA **(variable graphics array).** Generally, the video RAM located at A0000H–AFFFFH stores graphical or bit-mapped data and the memory at B0000H–BFFFFH stores text data. The video BIOS, located on a ROM or flash memory, is found at locations C0000H–C7FFFH and contains programs that control the video display.

If a hard disk memory is attached to the computer, the interface card might contain a ROM. The ROM, often found with older MFM or RLL hard disk drives, holds low-level format software, at location C8000H. The size, location, and presence of the ROM depends on the type of hard disk adapter attached to the computer.

FIGURE 1–7 The expanded
memory system showing a
page frame.

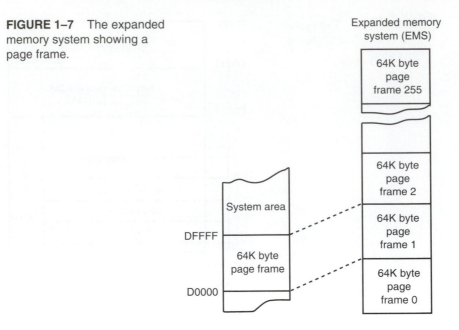

The area at locations C8000H–DFFFFH is often open or free. This area is used for the **expanded memory system** (EMS) in a PC or XT or the upper memory system in an AT system. Its use depends on the system and its configuration. The expanded memory system allows a 64K byte **page frame** of memory to be used by application programs. This 64K byte page frame (usually location D0000H–DFFFFH) is expandable by switching in pages of memory from the EMS into this range of memory addresses.

The information is addressed in the page frame as 16K byte-sized pages of data (each frame holds 4 pages) that are swapped with pages from the EMS. Figure 1–7 shows the expanded memory system. Most application programs that state they are LIM 4.0 driver-compatible can use expanded memory. The LIM 4.0 memory management driver is the result of Lotus, Intel, and Microsoft standardizing access to expanded memory systems. Expanded memory is slow because the change to a new 16K byte memory page requires action by the driver. Expanded memory was designed to expand the memory system of the early 8086/8088-based computer systems. In most cases, except for some DOS-based games that use the sound card, expanded memory should be avoided in 80386 through the Pentium-based systems.

Memory locations E0000H–EFFFFH contain the cassette BASIC language on ROM found in early IBM personal computer systems. This area is often open or free in newer computer systems. In newer systems we often backfill this area with extra RAM, called upper memory. Finally, the system BIOS ROM is located in the top 64K bytes of the system area (F0000H–FFFFFH). This ROM controls the operation of the basic I/O devices connected to the computer system. It does not control the operation of the video system, which has its own BIOS ROM at location C0000H. The first part of the BIOS (F0000H–F7FFFH) contains programs that set up the computer, and the second part contains procedures that control the basic I/O system. Once the system is set up, upper memory blocks at locations F0000H–F7FFFH are available, if EMM386.EXE is installed. Also available for upper memory blocks are locations B0000H–B7FFFH provided black-and-white video is not needed in the CGA mode.

I/O Space. The I/O (*input/output*) space in a computer system extends from I/O port 0000H to port FFFFH. (An **I/O port address** is similar to a memory address except that instead of addressing memory, it addresses an I/O device.) The **I/O devices** allow the microprocessor to communicate between itself and the outside world. The I/O space allows the computer to access up to

FIGURE 1–8 The I/O map of a personal computer showing some of the many areas of I/O devices.

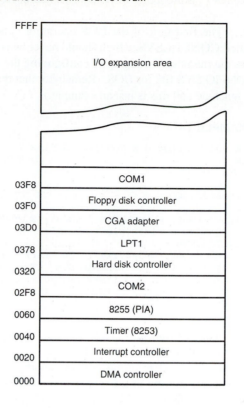

64K different 8-bit I/O devices. A great number of these locations are available for expansion in most computer systems. Figure 1–8 shows the I/O map found in many personal computer systems.

The I/O area contains two major sections: The area below I/O location 0500H is considered reserved for system devices, with many depicted in Figure 1–8. The remaining area is available I/O space for expansion that extends from I/O port 0500H through FFFFH.

Various I/O devices that control the operation of the system are usually not directly addressed. Instead, the BIOS ROM addresses these basic devices, which can vary slightly in location and function from one computer to the next. Access to most I/O devices should always be made through DOS or BIOS function calls to maintain compatibility from one computer system to another.

The DOS Operating System

The operating system is the program that operates the computer. This book assumes that the operating system is either MS-DOS or PC-DOS, which are by far the most common operating systems found in over 160 million personal computers (85 percent according to a recent *PC Magazine*[12] article). According to the same article, the Windows operating system is available to 65 million personal computers. The operating system is stored on a disk in one of the floppy disk drives or on the hard disk drive that is either resident to the computer or to a **local area network** (LAN). Some dedicated systems store the DOS on a ROM, for example the Tandy Corporation personal computer. Each time the computer is powered up or reset, the operating system is read from the disk or LAN. We call this operation **booting the system.** Once DOS is installed in the memory by the boot, it controls the operation of the computer system, its I/O devices, and application programs. In addition to the DOS operating system, other operating systems are sometimes used to control or operate the computer.

---

[12]*PC Magazine* is a Ziff-Davis publication.

The first task of the DOS operating system, after loading into memory, is to use a file called **CONFIG.SYS,** which should never be erased. This file specifies various drivers that load into the memory, setting up or **configuring** the machine for operation. Example 1–1 is a typical CONFIG.SYS file for DOS. (Remember that the statements in this file often vary from machine to machine and this is just an example.)

EXAMPLE 1–1

```
REM DOS VERSION 6.X CONFIG.SYS FILE
REM
FILES=30
BUFFERS=30
STACKS=9,256
FCBS=4
SHELL=C:\DOS\COMMAND.COM C:\DOS\ /E:2048 /P
DEVICE=C:\DOS\HIMEM.SYS
DOS=HIGH,UMB
DEVICE=C:\DOS\EMM386.EXE I=C800-EFFF NOEMS
DEVICEHIGH SIZE C:\LASERLIB\SONY_CDU.SYS
DEVICEHIGH SIZE C:\DOS\SETVER.EXE
DEVICEHIGH SIZE C:\MOUSE1\MOUSE.SYS
LASTDRIVE = F
```

The first four statements in this CONFIG.SYS file set up the number of files, buffers, stacks, and file control blocks required to execute various programs. These settings should be adequate for just about any program loaded into memory using DOS. In general, if a program requires more buffers and so forth, the documentation indicates that the CONFIG.SYS file must be changed to reflect an increased need. Many modern programs automatically adjust the CONFIG.SYS file during installation by changing these parameters or by adding additional statements.

The **SHELL command** specifies the command processor used with DOS. In this example the COMMAND.COM file is the command processor (also selected by default) using the E:2048 /P switches. The E:2048 switch sets the **environment size** to 2,048 bytes. The /P switch tells the command processor to make COMMAND.COM *permanent*. If COMMAND.COM is not permanent, it must be loaded into memory from the disk each time that DOS returns to the command prompt. Although this may free a small amount of memory, the constant access to the disk for the COMMAND.COM program increases wear and tear on the disk drive and also lengthens the time required to return to the DOS prompt.

The first DEVICE **(driver)** loaded into memory in this example is a program called HIMEM.SYS **(high memory driver).** A driver is usually a program that controls an I/O device or a program that must remain in the computer system memory. The HIMEM.SYS program allows a 64K minus 16 byte section of extended memory (100000H–10FFEFH), just above the first 1M byte of memory, to be used for programs in the 80286 through the Pentium-based system. (This extra 64K memory space is supported by most of these computer systems, but not all 80286 systems.) This driver allows a DOS-based system access to 1M plus 64K bytes of memory. This extra 64K byte section of **high memory** holds most of the MS-DOS version 5.0 or 6.2 program freeing additional space in the TPA.

The next command (DOS=HIGH,UMB) tells the computer to load DOS into this high part of memory and also allow the use of **upper memory blocks.** To enable the upper memory blocks, available only in an 80386, 80486, or Pentium-based system, the EMM386.EXE **(extended memory manager)** program is loaded into memory. The extended memory manager is a driver that emulates expanded memory in extended memory and also the extended memory system. This program **backfills** free areas of memory within the system area so programs can be loaded into this area and accessed directly by DOS applications. The I=C800-EFFF switch tells EMM386.EXE to use memory area C8000H–EFFFFH for upper memory or upper memory blocks (UMB). Drivers and programs are loaded into upper memory, freeing even more area in the TPA for application programs. Before using the I=C800-EFFF switch, make sure that your computer does not contain any system ROM/RAM in this area of the memory. Note that NOEMS tells EMM386.EXE to exclude expanded memory. Expanded memory can also be installed by replacing NOEMS with a number that indicates how much extended memory to allocate to LIM 4.0 expanded memory system (EMS).

Today most systems should not use expanded memory. If expanded memory is required, the NOEMS is replaced with RAM 1024 to enable 1,024 bytes of expanded memory. The FRAME= D000 statement places the page frame for expanded memory at location D0000H–DFFFFH if expanded memory is enabled.

The **DEVICEHIGH command** loads drivers and programs into the upper memory blocks allocated by the EMM386.EXE driver. In the CONFIG.SYS file illustrated, three drivers are loaded into upper memory blocks beginning at location C8000H. The first is a program that operates a SONY CD-ROM drive, the second loads a program called SETVER, and the third loads the MOUSE driver.

The last statement in our CONFIG.SYS file is the LASTDRIVE statement. This tells DOS which is the last disk drive connected to your computer system. By using the LASTDRIVE statement, more memory can be freed for use in the TPA. Each drive requires a buffer area and if you use the actual last drive with this statement, extra memory is made available. Other drivers may also be loaded into memory using the CONFIG.SYS file such as a PRINT.SYS driver, ANSI.SYS driver, or any other program that functions as a driver. Driver programs normally contain the DOS extension .SYS used to indicate a system file. Be very careful when changing the CONFIG.SYS file because an error locks up the computer system (except for DOS 6.X, which can exit this type of system lockup). Once the computer is locked up by a CONFIG.SYS error, the only way to recover is to boot off a DOS floppy disk that contains the operating system with a functioning CONFIG.SYS file.

Once the operating system completes its configuration as dictated by CONFIG.SYS, the AUTOEXEC.BAT **(automatic execution batch)** file is executed by the computer. If none exists, the computer asks for the time and date. Example 1–2 shows a typical AUTOEXEC.BAT file for DOS 5.0 or 6.X. This is only an example and often variations occur from system to system. The AUTOEXEC.BAT file contains commands that execute when power is first applied to the computer. These are the same commands that could be typed from the keyboard but AUTOEXEC.BAT saves us from doing so each time the computer is powered up.

EXAMPLE 1–2

```
PATH C:\DOS;C:\;C:\MASM\BIN;C:\MASM\BINB\;C:\UTILITY
PATH C:\WS;C:\LASERLIB
SET BLASTER=A220 I7 D1 T3
SET INCLUDE=C:\MASM\INCLUDE\
SET HELPFILES=C:\MASM\HELP\*.HLP
SET INIT=C:\MASM\INIT\
SET ASMEX=C:\MASM\SAMPLES\
SET TMP=C:\MASM\TMP
SET SOUND=C:\SB
LOADHIGH C:\LASERLIB\MSCDEX.EXE /D:SONY_001 /L:F /M:10
LOADHIGH C:\LASERLIB\LLTSR.EXE ALT-Q
LOADHIGH C:\DOS\FASTOPEN C:=256
LOADHIGH C:\DOS\DOSKEY /BUFSIZE=1024
LOADHIGH C:\LASERLIB\PRINTF.COM
DOSKEY GO=DOSSHELL
DOSSHELL
```

The PATH statement specifies the search paths whenever a program name is typed at the command line. The order of the path search is the same as the order of the paths in the path statement. For example, if PROG is typed at the command line, the machine first searches C:\DOS, then the root directory C:\, then C:\MASM\BIN, and so forth until the program named PROG is found. If it isn't found, the command interpreter (COMMAND.COM) informs the user that the program is not found.

The SET statement sets a variable name to a path, which allows names to be associated with paths for batch programs. It's also used to set command strings (environments) for various programs. The first SET command sets the environment for the sound blaster card. The second SET command sets INCLUDE to the path C:\MASM\INCLUDE\. The SET statements are stored in the DOS environment space that was reserved in the CONFIG.SYS file using the SHELL statement. If the environment becomes too large, you must change the SHELL statement to allow more space.

LOADHIGH or LH places programs into upper memory blocks defined by the EMM386.EXE program. LOADHIGH is used at any DOS command prompt for loading a program into the high (**upper**) memory area as long as the computer is an 80386 or higher. The last command in this AUTOEXEC.BAT file is DOSSHELL. The DOSSHELL program is a menu program included with MS-DOS version 5.0 or 6.X. This last command is replaced by WIN, WIN /S **(standard mode),** WIN /R **(real mode in Windows 3.1),** or WIN /3 **(enhanced mode)** to run Microsoft Windows in place of DOSSHELL.

Once the CONFIG.SYS and AUTOEXEC.BAT files are executed, the program name last shown in the AUTOEXEC.BAT file is executed. In this example, the system operates from the DOSSHELL program.

The Microprocessor

At the heart of the microprocessor-based computer system is the microprocessor integrated circuit. The **microprocessor** (or μp) is the controlling element in a computer system and is sometimes referred to as the CPU **(central processing unit).** The microprocessor controls memory and I/O through a series of connections called **buses.** Buses select an I/O device or a memory location; transfer data between an I/O device or memory and the microprocessor; and control the I/O and memory system. Memory and I/O are controlled through instructions that are stored in the memory and executed by the microprocessor.

The microprocessor performs three main tasks for the computer system: (1) data transfer between itself and the memory or I/O systems, (2) simple arithmetic and logic operations, and (3) program flow via simple decisions. Through these simple tasks, the microprocessor performs virtually any series of operations. The power of the microprocessor is its ability to execute millions of instructions per second from a **program** or software **(group of instructions)** stored in the memory system. This **stored program concept** has made the microprocessor and computer system a very powerful device. Recall that Babbage also wanted to use the stored program concept in his Analytical Engine.

Table 1–4 shows the arithmetic and logic operations that the Intel family of microprocessors can execute. These operations are very basic, but through them, complex problems are solved.

TABLE 1–4　Simple arithmetic and logic operations

Operation	Comment
Addition	
Subtraction	
Multiplication	
Division	
AND	Logical multiplication
OR	Logical addition
NOT	Logical inversion
NEG	Arithmetic negation
Shift	
Rotate	

TABLE 1–5 Decisions made by 8086–80486 and Pentium Pro microprocessors

Decision	Comment
Zero	A number is zero or not zero
Sign	A number is positive or negative
Carry	A carry after addition or a borrow after subtraction
Parity	A number has an even or odd number of ones
Overflow	An invalid signed result

Data is operated on from the memory system or internal registers. Data widths are variable and include a **byte** (8 bits), **word** (16 bits), and **doubleword** (32 bits). Note that only the 80386 through the Pentium directly manipulate 8-, 16-, and 32-bit numbers. The earlier 8086–80286 directly manipulate 8- and 16-bit numbers, but not 32-bit numbers. The 80486DX and Pentium also contain a numeric coprocessor that allows them to perform complex arithmetic using floating-point arithmetic. The numeric coprocessor was an additional component in the 8086 through the 80386-based personal computer.

Another feature that makes the microprocessor powerful is its ability to make simple decisions, which are based on numerical facts. For example, a microprocessor can decide if a number is zero, if it is positive, and so forth. These simple decisions allow the microprocessor to modify the program flow so programs appear to think through these simple decisions. Table 1–5 lists the decision-making abilities of the Intel family of microprocessors.

Buses.   A **bus** is a common group of wires that interconnect components in a computer system. The buses that interconnect the sections of a computer system transfer address, data, and control information between the microprocessor and its memory and I/O systems. In the microprocessor-based computer system, three buses exist for this transfer of information: **address, data,** and **control.** Figure 1–9 shows how these buses interconnect various system components such as the microprocessor, read/write memory (RAM), read-only memory (ROM), and a few I/O devices.

The **address bus** requests a memory location from the memory or an I/O location from the I/O devices. When the I/O system is addressed, the address bus contains a 16-bit I/O address from 0000H through FFFFH. The 16-bit I/O address or port number selects one of 64K different

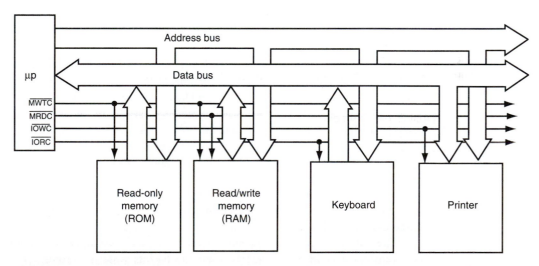

FIGURE 1–9   The block diagram of a microprocessor-based computer system showing the address, data, and control bus structure.

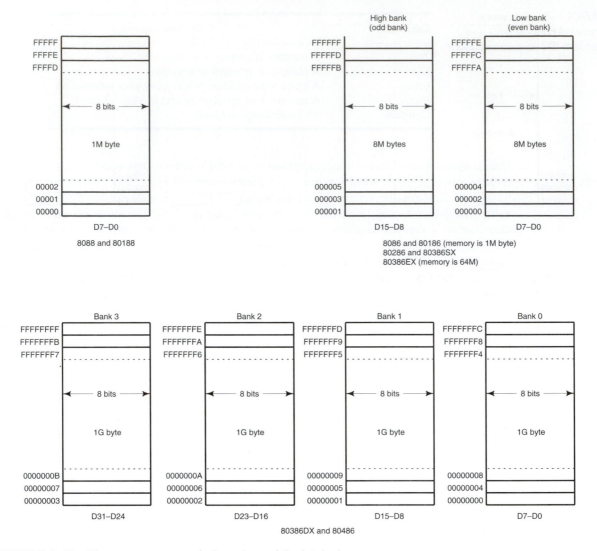

FIGURE 1–10   The memory maps of all versions of the Intel microprocessor.

I/O devices. If memory is addressed, the address bus contains a memory address. The memory address varies in width on the different versions of the microprocessor. The 8086 and 8088 address 1M byte of memory using a 20-bit address that selects locations 00000H–FFFFFH. The 80286 and 80386SX address 16M bytes of memory using a 24-bit address that selects locations 000000H–FFFFFFH. The 80386SL and 80386SLC address 32M bytes of memory using a 25-bit address that selects locations 0000000H–1FFFFFFH. The 80386EX addresses 64M bytes of memory using a 26-bit address that selects locations 0000000H–3FFFFFFH. The 80386DX, 80486SX, and 80486DX address 4G bytes of memory using a 32-bit address that selects locations 00000000H–FFFFFFFFH. The Pentium also addresses 4G bytes of memory, but it uses a 64-bit data bus to access up to 8 bytes of memory at a time. The Pentium Pro processor has a 36-bit address bus, which allows it to address up to 64G bytes of memory.

The **data bus** transfers data between the microprocessor and the memory and I/O systems. Data transfers vary in size from 8 bits wide to 64 bits wide in various versions of the microprocessor. The 8088 and 80188 contain an 8-bit wide data bus used to transfer a byte at a time. The 8086, 80186, 80386SX, and 80386EX transfer two bytes at a time through a 16-bit wide

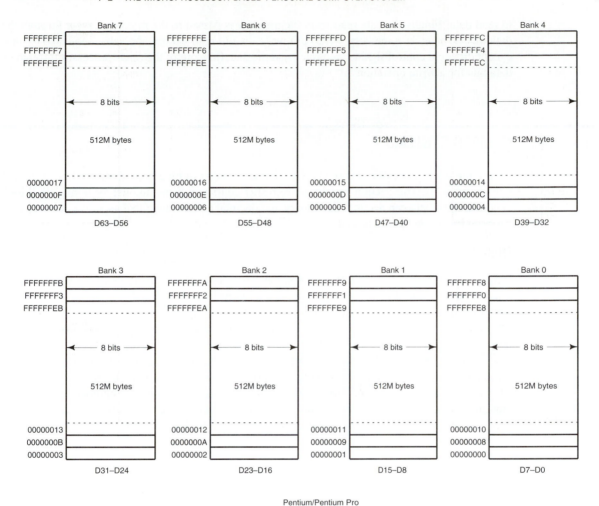

Pentium/Pentium Pro

Note: The memory banks in the Pentium Pro are 8G bytes.

FIGURE 1–10 *(continued)*

data bus; the 80386DX and 80486 transfer four bytes at a time through a 32-bit wide data bus; and the Pentium/Pentium Pro transfers 8 bytes at a time through a 64-bit wide data bus. The advantage of a wider data bus is speed in applications that use wide data and in fetching instructions to the internal cache on some versions of the microprocessor. To fetch a double-precision floating-point number (64-bits wide) in the 8088 or 80188 requires 8 read cycles, where it requires only 1 read cycle in the Pentium. Figure 1–10 shows the memory widths and data bus sizes of the different Intel microprocessor family members. Notice how memory sizes and organizations differ between members. In all cases, the memory locations are numbered by byte.

The **control bus** contains lines that select the memory or I/O and cause them to perform a read or write operation. In most computer systems, there are four control bus connections: $\overline{\text{MRDC}}$ **(memory read control),** $\overline{\text{MWTC}}$ **(memory write control),** $\overline{\text{IORC}}$ **(I/O read control),** and $\overline{\text{IOWC}}$ **(I/O write control).** Note that the overbar indicates that the control signal is active-low; that is, it is active when a logic zero appears on the control line. For example, if $\overline{\text{IOWC}} = 0$, the microprocessor is writing data to an I/O device whose address appears on the address bus. The microprocessor reads the contents of a memory location by sending the memory an address through the address bus. Next it sends the memory read control signal ($\overline{\text{MRDC}}$) to cause memory to read data. Finally the data read from the memory is passed to the microprocessor through the data bus. Whenever a memory write, I/O write, or I/O read occurs, the same sequence ensues, except different control signals are issued and the data flow out of the microprocessor through its data bus for a write operation.

1–3   NUMBER SYSTEMS

To use the microprocessor you need a working knowledge of binary, decimal, and hexadecimal number systems. This section provides a refresher course on these number systems. We'll look at conversions between decimal and binary and decimal and hexadecimal as well as conversions between binary and hexadecimal.

Digits

Before numbers are converted from one number base to another, the digits of a number system must be understood. Early in our education, we learned that a decimal or base 10 number was constructed with 10 digits: 0 through 9. The first digit in any numbering system is always zero. The same rule applies to any other number system. For example, a base 8 *(octal)* number contains 8 digits: 0 through 7; while a base 2 *(binary)* number contains 2 digits: 0 and 1. If the base of a number exceeds 10, the additional digits use the letters of the alphabet, beginning with an A. For example, a base 12 number contains 12 digits: 0 through 9 followed by A for 10 and B for 11. Note that a base 10 number does not contain a 10 digit just as a base 8 number does not contain an 8 digit. The most common number systems in widespread use with computers are *decimal, binary,* occasionally *octal,* and *hexadecimal* (base 16). We will now describe and use each system.

Positional Notation

Once the digits of a number system are understood, larger numbers are constructed using **positional notation.** In grade school, we learned that the position to the left of the units position was the ten's position, the position to the left of the ten's position was the hundred's position, and so forth. What we probably did not learn was that the unit's position has a weight of 10^0 or 1, the ten's position has weight of 10^1 or 10, and the hundred's position has a weight of 10^2 or 100. The powers of the positions are critical in understanding numbers in other numbering systems. The position to the left of the radix **(number base)** point, called a decimal point only in the decimal system, is always the unit's position in any number system. For example, the position to the left of the binary point has a weight of 2^0 or 1, while the position to the left of the octal point has a weight of 8^0 or 1. In any case, any number raised to its zero power is always 1, or in the positional notation system, the unit's position.

The weight of the position immediately to the left of the unit's position is always the number base raised to the first power; in decimal this is 10^1 or 10. In binary it is 2^1 or 2 and in octal 8^1 or 8. Therefore, an 11 decimal has a different value than an 11 binary. The 11 decimal is composed of 1 ten plus 1 unit and has a value of 11 units, while the binary number 11 is composed of 1 two plus 1 unit for a value of 3 units. The number 23 octal has a value of 19 units (2 eight digits plus three units). In any number system, positions to the right of the radix point have **negative powers.** In the decimal system, the first digit to the right of the decimal point has a value of 10^{-1} or 0.1. In the binary system the first digit to the right of the binary point has a value of 2^{-1} or 0.5. In general, the principles that apply to decimal numbers also apply to numbers in any other number system.

Example 1–3 shows the number 110.101 in binary. It also shows the power and weight (value) of each digit position. To convert a binary number to decimal, add the weights of each digit that contains a 1 to form its decimal equivalent. The number 110.101_2 is equivalent to a 6.625 in decimal (4 + 2 + 0.5 + 0.125) because there is one 4 plus one 2 plus one 0.5 plus one 0.125.

EXAMPLE 1–3

```
Power     2²   2¹   2⁰      2⁻¹     2⁻²    2⁻³
Weight    4    2    1       0.5     0.25   .125
Number    1    1    0   .   1       0      1      = 6.625₁₀
```

Suppose that the conversion technique is applied to a base 6 number such as 25.2_6. Example 1–4 shows this number placed under the powers and weights of each position. In this example there are 2 units under the 6^1 which have a value of 12 (2×6) and 5 units under the 6^0 which have a value of 5 (5×1). The whole number portion has a decimal value of 12 + 5 or 17 units. The number to the right of the hex point is 2 units under the 6^{-1} which have a value of .333 ($2 \times .167$). The number 25.2_6 therefore has a decimal value of 17.333.

EXAMPLE 1–4

```
Power     6¹   6⁰   6⁻¹
Weight    6    1    .167
Number    2    5   . 2     =    17.333₁₀
```

Conversion to Decimal

To convert from any number base to decimal, determine the weights of each position of the number and then sum the weights to form the decimal equivalent. Suppose that a 125.7_8 (octal) is converted to decimal. To accomplish this conversion, first write down the weights of each position of the number, as shown in Example 1–5. The value of 125.7_8 is 85.875 decimal, or 1×64 plus 2×8 plus 5×1 plus $7 \times .125$.

EXAMPLE 1–5

```
Power     8²   8¹   8⁰    8⁻¹
Weight    64   8    1     .125
Number    1    2    3   . 7     =    85.875₁₀ or 85.875
```

Here is an algorithm for conversion from any number system to decimal:

1. Find the weight of each position.
2. Multiply the weight by the number located under the weight to find the value of the position.
3. Sum all the weighted values to convert to decimal.

Notice that the weight of the position to the left of the unit's position is 8. This is 8 times 1. Then notice that the weight of the next position is 64, or 8 times 8. If another position existed, it would be 64 times 8, or 512. To find the weight of the next higher-order position multiply the weight of the current position by the number base, which is 8 in this example. To calculate the weights of the position to the right of the radix point, divide by the number base. In the octal system, the position immediately to the right of the octal point is $^1/_8$ or .125. The next position is $.125 \div 8$ or .015625, which can also be written as $^1/_{64}$. The number in Example 1–5 can also be written as the decimal number $85^7/_8$.

Example 1–6 shows the binary number 11011.0111 written with the weights and powers of each position. If these weights are summed, the value of the binary number converted to decimal is 27.4375.

EXAMPLE 1–6

```
Power     2⁴   2³   2²   2¹   2⁰      2⁻¹   2⁻²    2⁻³    2⁻⁴
Weight    16   8    4    2    1       0.5   0.25   .125   .0625
Number    1    1    0    1    1   .   0     1      1      1      =    27.4375₁₀
```

It is interesting to note that 2^{-1} is also $1/2$, 2^{-2} is $1/4$, and so forth. It is also interesting to note that 2^{-4} is $1/16$ or 0.625. The fractional part of this number is $7/16$ or $.4375$ decimal. Notice that 0111 is a 7 in binary code for the numerator and the rightmost one is in the $1/16$ position for the denominator. Other examples are the binary fraction of .101, which is $5/8$, and the binary fraction of .001101, which is $13/64$.

Hexadecimal numbers are often used with computers. A 6A.CH (**H** for hexadecimal) is illustrated with its weights in Example 1–7. The sum of its digits are 106.75 or $106\,3/4$. The whole number part is represented with 6×16 plus 10 (A) $\times$ 1. The fraction part is 12 (C) as a numerator and 16 (16^{-1}) as the denominator or $12/16$, which is $3/4$.

EXAMPLE 1–7

```
Power     16¹   16⁰   16⁻¹
Weight    16    1     .0625
Number    6     A   . C      =    106.75₁₀ or 106.75
```

Conversion from Decimal

Conversions from decimal to other number systems are more difficult than conversions to decimal. To convert the whole number portion of a number to decimal, divide by the radix. To convert the fractional portion, multiply by the radix.

Whole Number Conversion from Decimal.   To convert a decimal whole number to another number system, divide by the radix and save the remainders as significant digits of the result. There is an algorithm for this conversion:

1. Divide the decimal number by the radix (number base).
2. Save the remainder (the first remainder is the least significant digit).
3. Repeat steps 1 and 2 until the quotient is zero.

For example, to convert a 10 decimal to binary, divide it by 2. The result is 5, with a remainder of 0. The first remainder is the unit's position of the result, in this example a 0. Next divide the 5 by 2. The result is 2 with a remainder of 1. The 1 is the value of the two's (2^1) position. Continue the division until the quotient is a zero. Example 1–8 shows this conversion process. The result is written as 1010 binary from the bottom to the top.

EXAMPLE 1–8

```
2) 10   remainder = 0              therefore: 10₁₀ = 1010₂
 2) 5   remainder = 1
  2) 2  remainder = 0
   2) 1 remainder = 1
      0
```

To convert a 10 decimal into base 8, divide by 8 as shown in Example 1–9. A 10 decimal is a 12 octal.

EXAMPLE 1–9

```
8) 10   remainder = 2              therefore: 10₁₀ = 12₈
 8) 1   remainder = 1
     0
```

To convert from decimal to hexadecimal divide by 16. The remainders will range in value from 0 through 15. Any remainder of 10 though 15 is then converted to the letters A through F for the hexadecimal number. Example 1–10 shows the decimal number 109 converted to 6DH.

EXAMPLE 1–10

```
16) 109   remainder = 13 (D)    therefore: 109₁₀ = 6D₁₆
16)   6   remainder = 6
      0
```

Converting from a Decimal Fraction.   To convert from a decimal fraction to another number base multiply by the radix. For example, to convert a decimal fraction into binary, multiply by 2. After the multiplication, the whole number portion of the result is saved as a significant digit of the result and the fractional remainder is again multiplied by the radix. When the fraction remainder is zero, multiplication ends. Note that some numbers are never-ending. That is, a zero is never a remainder.

Here is an algorithm for conversion from a decimal fraction:

1. Multiply the decimal fraction by the radix (number base).
2. Save the whole number portion of the result (even if zero) as a digit. (The first result is written immediately to the right of the radix point.)
3. Repeat steps 1 and 2 using the fractional part of step 2 until the fractional part of step 2 is zero.

Suppose that the number .125 decimal is converted to binary. This is accomplished with multiplication by 2 as illustrated in Example 1–11. Notice that the multiplication continues until the fractional remainder is zero. The whole number portions are written as the binary fraction (0.001) in this example.

EXAMPLE 1–11

```
  .125
x    2
 0.25  digit is 0

  .25
x    2
 0.5   digit is 0

  .5
x  2
 1.0   digit is 1, the result is written as 0.001₂
```

The same technique is used to convert a decimal fraction into any number base. Example 1–12 shows the same decimal fraction of .125 from Example 1–11 converted to octal by multiplying with an 8.

EXAMPLE 1–12

```
  .125
x    8
 1.0 digit is 1, the result is written as 0.1₈
```

Conversion to a hexadecimal fraction appears in Example 1–13. Here the decimal .046875 is converted to hexadecimal by multiplying with a 16. Note that .046875 is a 0.0CH.

EXAMPLE 1–13

```
  .046875
x       16
  0.75 digit is 0
```

```
     .75
x    16
  12.0 digit is 12(C), the result is written as 0.0CH or 0.0C₁₆
```

Binary-Coded Hexadecimal

Binary-coded hexadecimal (BCH) is used to represent hexadecimal data in binary code. A BCH is a hexadecimal number written so each digit is represented by a 4-bit binary number. The values for the BCH digits appear in Table 1–6.

Hexadecimal numbers are represented in BCH code by converting each digit to BCH code with a space between each coded digit. Example 1–14 shows the number $2AC_{16}$ converted to BCH code. Note that each BCH digit is separated by a space.

EXAMPLE 1–14

```
2AC = 0010 1010 1100
```

The purpose of BCH code is to allow a binary version of a hexadecimal number to be written in a form that can easily be converted between BCH and hexadecimal. Example 1–15 shows a BCH-coded number converted back to hexadecimal code.

EXAMPLE 1–15

```
1000 0011 1101 . 1110 = 83D.E
```

Complements

Sometimes data are stored in **complement form** to represent negative numbers. There are two systems that are used to represent negative data: **radix** and **radix –1** complements. The earliest system was the radix –1 complement where each digit of the number is subtracted from the radix to generate the radix –1 complement to represent a negative number.

EXAMPLE 1–16

```
  2 2 2 2   2 2 2 2
- 0 1 0 0   1 1 0 0
  1 0 1 1   0 0 1 1
```

TABLE 1–6 Binary-coded hexadecimal (BCH) code

Hexadecimal Digit	BCH Code
0	0000
1	0001
2	0010
3	0011
4	0100
5	0101
6	0110
7	0111
8	1000
9	1001
A	1010
B	1011
C	1100
D	1101
E	1110
F	1111

Example 1–16 shows how the 8-bit binary number 01001100 is one's (radix −1) complemented to represent it as a negative value. Notice that each digit of the number is subtracted from the radix to generate the radix −1 (one's) complement. In this example, the negative of 01001100 is 10110011. The same technique can be applied to any number system, as illustrated in Example 1–17, where the fifteen's (radix −1) complement of a 5CD hexadecimal is computed by subtracting each digit from a fifteen to generate A32, the negative of 5CD using the radix −1 complement.

EXAMPLE 1–17

```
   15  15  15
 -  5   C   D
    A   3   2
```

Today, the radix −1 complement is not used by itself, but is used as a step for finding the radix complement. The radix complement is the way that negative numbers are represented in modern computer systems, where the radix −1 complement was used in the early days of computer technology. The main problem with the radix −1 complement is that a negative or a positive zero exits, where in the radix complement system, only a positive zero can exist.

To form the radix complement, first find the radix −1 complement and then add a one to the result. Example 1–18 shows how the number 0100 1000 is converted to a negative value by two's (radix) complementing it.

EXAMPLE 1–18

```
   2 2 2 2  2 2 2 2
 - 0 1 0 0  1 0 0 0
   1 0 1 1  0 1 1 1   (one's complement)
 + 0 0 0 0  0 0 0 1
   1 0 1 1  1 0 0 0   (two's complement)
```

To prove that a 0100 1000 is the inverse (negative) of a 1011 0111, add the two numbers together to form an 8-digit result. The ninth digit is dropped and the result is zero because a 0100 1000 is a positive 72, while a 1011 0111 is a negative 72. The same technique applies to any number system. Example 1–19 shows how the inverse of 345 hexadecimal is found by first fifteen's complementing the number and then by adding one to the result to form the sixteen's complement. If the original 3-digit number 345 is added to the inverse of CBB, the result is a 3-digit 000. The fourth bit (carry) is dropped. This proves that 345 is the inverse of CBB. Additional information about one's and two's complements is presented with signed numbers in the next section.

EXAMPLE 1–19

```
   15  15  15
 -  3   4   5
    C   B   A   (fifteen's complement)
 +  0   0   1
    C   B   B   (sixteen's complement)
```

1–4 COMPUTER DATA FORMATS

Successful programming requires a precise understanding of data formats. In this section, we describe many common computer data formats used with the Intel family of microprocessors. Commonly, data appear as ASCII, BCD, signed and unsigned integers, and floating-point numbers (real numbers). Other forms are available, but are not presented here because they are not commonly found.

ASCII Data

ASCII (**American Standard Code for Information Interchange**) data (see Table 1–7) represents alphanumeric characters in the memory of a computer system. The **standard ASCII code** is a 7-bit code with the eighth and most-significant bit used to hold parity in some systems. If ASCII data is used with a printer, the most-significant bit is a 0 for alphanumeric printing, and 1 for graphics printing. In the personal computer, an **extended ASCII code** set is selected by placing a logic 1 in the leftmost bit. Table 1–9 shows the extended ASCII character set using code 80H–FFH. The extended ASCII characters store some foreign letters and punctuation, Greek characters, mathematical characters, box-drawing characters, as well as other special characters. Note that extended characters can vary from one printer to another. This list is designed to be used with the IBM ProPrinter[13] and also matches the special character set found with some word processors.

The ASCII control characters, also listed in Table 1–7, perform control functions in a computer system including: clear screen, backspace, line feed, etc. To enter the control codes through the keyboard, the control key is held down while typing a letter. To obtain the control code 01H, type control A; a 02H is obtained by typing control B, etc. Note that the control codes appear on the screen, from the DOS prompt, as ^A for control A, ^B for control B, and so forth. Also note that the **carriage return** code (CR) is the *enter key* on most modern keyboards. The purpose of CR is to return the cursor or print-head to the left margin. Another code that appears in many programs is the **line feed** code (LF) that moves the cursor down one line. Notice that the first 33 characters of the ASCII code are control characters using two- or three-letter acronyms. Most of these control codes are used with digital communications and also to control the position of the cursor on the video display. See Table 1–8 for a list of these codes and their meaning.

To use Tables 1–7 or 1–9 for converting alphanumeric or control characters into ASCII characters, first locate the alphanumeric code for conversion. Next, find the first digit of the hexadecimal ASCII code. Then find the second digit. For example, the letter A is ASCII code 41H and the letter a is ASCII code 61H.

ASCII data is most often stored in memory using a special directive to the assembler program called **define byte(s)** or DB. (The **assembler** is a program that is used to program a computer in its native binary machine language.) The DB directive, along with several examples of its usage with ASCII-coded character strings, is listed in Example 1–20. Notice how each character string is surrounded by apostrophes ('). Never use the quote ('). Also notice that the assembler lists the ASCII-coded value for each character to the left of the character string. To the far left is the hexadecimal memory location where the character string is first stored in the memory

TABLE 1–7   ASCII code

	X0	X1	X2	X3	X4	X5	X6	X7	X8	X9	XA	XB	XC	XD	XE	XF
First																
0X	NUL	SOH	STX	ETX	EOT	ENQ	ACK	BEL	BS	HT	LF	VT	FF	CR	SO	SI
1X	DLE	DC1	DC2	DC3	DC4	NAK	SYN	ETB	CAN	EM	SUB	ESC	FS	GS	RS	US
2X	SP	!	"	#	$	%	&	'	(	)	*	+	,	-	.	/
3X	0	1	2	3	4	5	6	7	8	9	:	;	<	=	>	?
4X	@	A	B	C	D	E	F	G	H	I	J	K	L	M	N	O
5X	P	Q	R	S	T	U	V	W	X	Y	Z	[	\	]	^	_
6X	`	a	b	c	d	e	f	g	h	i	j	k	l	m	n	o
7X	p	q	r	s	t	u	v	w	x	y	z	{			}	~	:::

---

[13] ProPrinter is a trademark of IBM Corporation.

TABLE 1–8 ASCII control codes

Code	Meaning
nul	null, used to end null strings
soh	start of header
stx	start of text
etx	end of text
eot	end of transmission
enq	enquiry
ack	acknowledge for nak/ack data communications
bel	bell, beeps the speaker
bs	backspaces the cursor
ht	horizontal tab
lf	line feed
vt	vertical tab
ff	form feed
cr	carriage return (enter)
so	shift out
si	shift in
dle	data link escape
dc1	direct control 1
dc2	direct control 2
dc3	direct control 3
dc4	direct control 4
nak	negative acknowledge
syn	synchronous idle
etb	end of transmission block
can	cancel
em	end of medium
sub	substitute
esc	escape
fs	form separator
gs	group separator
rs	record separator
us	unit separator
sp	blank space

TABLE 1–9 Extended ASCII code (as printed by the IBM ProPrinter)

First	X0	X1	X2	X3	X4	X5	X6	X7	X8	X9	XA	XB	XC	XD	XE	XF
0X		☺	☻	♥	♦	♣	♠	●	◘	○	◙	♂	♀	♪	♫	☼
1X	►	◄	↕	‼	¶	§	▬	↨	↑	↓	→	←	∟	↔	▲	▼
8X	Ç	ü	é	â	ä	à	å	ç	ê	ë	è	ï	î	ì	Ä	Å
9X	É	æ	Æ	ô	ö	ò	û	ù	ÿ	Ö	Ü	¢	£	¥	₧	ƒ
AX	á	í	ó	ú	ñ	Ñ	ª	º	¿	⌐	¬	½	¼	¡	«	»
BX	░	▒	▓	│	┤	╡	╢	╖	╕	╣	║	╗	╝	╜	╛	┐
CX	└	┴	┬	├	─	┼	╞	╟	╚	╔	╩	╦	╠	═	╬	╧
DX	╨	╤	╥	╙	╘	╒	╓	╫	╪	┘	┌	█	▄	▌	▐	▀
EX	α	β	Γ	π	Σ	σ	µ	γ	Φ	Θ	Ω	δ	∞	φ	∈	∩
FX	≡	±	≥	≤	⌠	⌡	÷	≈	°	∙	·	√	ⁿ	²	■	

system. For example, the character string WHAT is stored beginning at memory address 001D and the first letter is stored as 57 (W) followed by 68 (H), and so forth.

EXAMPLE 1–20

```
0000    42 61 72 72 79    NAMES   DB    'Barry B. Brey'
        20 42 2E 20 42
        72 65 79
000D    57 68 65 72 65    MESS    DB    'Where can it be?'
        20 63 61 6E 20
        69 74 20 62 65
        3F
001D    57 68 61 74 20    WHAT    DB    'What is on first.'
        69 73 20 6F 6E
        20 66 69 72 73
        74 2E
```

BCD (Binary-Coded Decimal) Data

Binary-coded decimal (BCD) information is stored in either packed or unpacked forms. **Packed BCD data** is stored as 2 digits per byte and **unpacked BCD data** is stored as one digit per byte. The range of a BCD digit extends from 0000_2 to 1001_2, or 0–9 decimal. Unpacked BCD data is often returned from a keypad or keyboard, while packed BCD data is used for some of the instructions included for BCD addition and subtraction in the instruction set of the microprocessor.

Table 1–10 shows some decimal numbers converted to both the packed and unpacked BCD forms. Applications that require BCD data are point-of-sales terminals, and almost any device that performs a minimal amount of simple arithmetic. If a system requires complex arithmetic, BCD data is seldom used because there is no simple and efficient method of performing complex BCD arithmetic.

Example 1–21 shows how to use the assembler to define both packed and unpacked BCD data. In all cases, the convention of storing the least-significant data first is followed. This means that to store an 83 into memory the 3 is stored first, followed by the 8. Also note that with packed BCD data the letter H *(hexadecimal)* follows the number to ensure that the assembler stores the BCD value rather than a decimal value for packed BCD data. Notice how the numbers are stored in memory as unpacked, one digit per byte, or packed as 2 digits per byte.

EXAMPLE 1–21

```
                        ;Unpacked BCD data (least-significant first)
                        ;
0000    03 04 05        NUMB1   DB    3,4,5        ;defines the number 543
0003    07 08           NUMB2   DB    7,8          ;defines the number 87
                        ;
                        ;Packed BCD data (least-significant first)
                        ;
0005    37 34           NUMB3   DB    34H,37H      ;defines the number 3437
0007    03 45           NUMB4   DB    3,45H        ;defines the number 4503
```

TABLE 1–10   Packed and unpacked BCD data

Decimal	Packed		Unpacked		
12	0001 0010		0000 0001	0000 0010	
623	0000 0110	0010 0011	0000 0110	0000 0010	0000 0011
910	0000 1001	0001 0000	0000 1001	0000 0001	0000 0000

Byte-Sized Data

Byte-sized data is stored as unsigned and signed integers. Figure 1–11 illustrates both the unsigned and signed forms of the byte-sized integer. The difference in these forms is the weight of the leftmost bit position. Its value is 128 for the unsigned integer and minus 128 for the signed integer. In the signed integer format, the leftmost bit represents the signed bit of the number as well as a weight of minus 128. For example, an 80H represents a value of 128 as an unsigned number and minus 128 as a signed number. Unsigned integers range in value from 00H–FFH (0–255). Signed integers range in value from −128 to 0 to +127.

Although negative signed numbers are represented in this way, they are stored in the **two's complement form.** The method of evaluating a signed number, using the weights of each bit position, is much easier than the act of two's complementing a number to find its value. This is especially true in the world of calculators designed for programmers.

Whenever a number is two's complemented, its sign changes from negative to positive or positive to negative. For example, the number 00001000_2 is a +8. Its negative value (−8) is found by two's complementing the +8. To form a two's complement, first one's complement the number. To one's complement a number, invert each bit of a number from zero to one, or from one to zero. Once the one's complement is formed, the two's complement is found by adding a one to the one's complement. Example 1–22 shows how numbers are two's complemented using this technique.

EXAMPLE 1–22

```
+8 = 00001000
     11110111 (one's complement)
+           1
-8 = 11111000 (two's complement)
```

Another, and probably simpler, technique for two's complementing a number starts with the rightmost digit. Start writing down the number from right to left. Write the number exactly as it appears until the first one. Write down the first one, and then invert or complement all remaining ones to its left. Example 1–23 shows this technique with the same number used in Example 1–22.

EXAMPLE 1–23

```
+8 = 00001000
         1000 (write number to first 1)
     1111     (invert the remaining bits)
-8 = 11111000
```

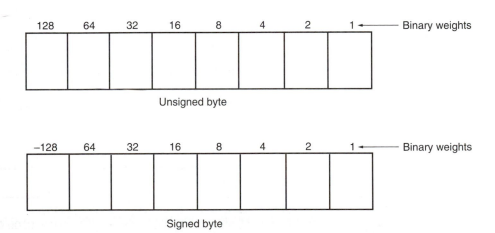

FIGURE 1–11 The unsigned and signed bytes illustrating the weights of each bit position.

To store 8-bit data in memory using the assembler program, use the DB directive as shown in prior examples. Example 1–24 lists many forms of 8-bit numbers stored in memory using the assembler program. Notice in this example that a hexadecimal number is defined with the letter H following the number and that a decimal number is written as is.

EXAMPLE 1–24

```
                        ;Unsigned byte-sized data
                        ;
0000    FE              DATA1   DB    254           ;define 254 decimal
0001    87              DATA2   DB    87H           ;define 87 hexadecimal
0002    47              DATA3   DB    71            ;define 71 decimal
                        ;
                        ;Signed byte-sized data
                        ;
0003    9C              DATA4   DB    -100          ;define a -100 decimal
0004    64              DATA5   DB    +100          ;define a +100 decimal
0005    FF              DATA6   DB    -1            ;define a -1 decimal
0006    38              DATA7   DB    56            ;defines a 56 decimal
```

Word-Sized Data

A **word** (16 bits) is formed with two bytes of data. The least-significant byte is always stored in the lowest-numbered memory location, and the most-significant byte in the highest. This method of storing a number is called the **little endian** format. An alternate method, not used with the Intel family of microprocessors, is called the **big endian** format. With the big endian format, numbers are stored with the lowest location containing the most-significant data. The big endian format is used with the Motorola family of microprocessors. Figure 1–12 (a) shows the weights

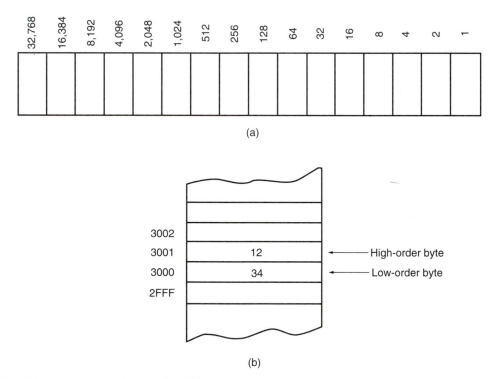

FIGURE 1–12   Word-sized data (a) in any 16-bit register and (b) stored in two consecutive bytes of memory.

of each bit position in a word of data, and Figure 1–12 (b) shows how the number 1234H appears when stored in the memory locations 3000H and 3001H. The only difference between a signed and an unsigned word is the leftmost bit position. In the unsigned form, the leftmost bit is unsigned and in the signed form its weight is a –32,768. As with byte-sized signed data, the signed word is in two's complement form when representing a negative number. Also notice that the low-order byte is stored in the lowest-numbered memory location (3000H) and the high-order byte is stored in the highest-numbered location (3001H).

Example 1–25 shows several signed and unsigned word-sized data stored in memory using the assembler program. Notice that the **define word(s)** directive or DW causes the assembler to store words in the memory instead of bytes as in prior examples. Also notice that the word data is displayed by the assembler in the same form as entered. For example, a 1000H is displayed by the assembler as a 1000. This is for our convenience, because the number is actually stored in the memory as 00 and 10 in two consecutive memory bytes.

EXAMPLE 1–25

```
                    ;Unsigned word-sized data
                    ;
0000   09F0         DATA1   DW   2544        ;define 2544 decimal
0002   87AC         DATA2   DW   87ACH       ;define 87AC hexadecimal
0004   02C6         DATA3   DW   710         ;define 710 decimal
                    ;
                    ;Signed word-sized data
                    ;
0006   CBA8         DATA4   DW   -13400      ;define a -13400 decimal
0008   00C6         DATA5   DW   +198        ;define a +198 decimal
000A   FFFF         DATA6   DW   -1          ;define a -1 decimal
```

Doubleword-Sized Data

Doubleword-sized data requires 4 bytes of memory because it is a 32-bit number. Doubleword data appears as a product after a multiplication and also as a dividend before a division. In the 80386 through the Pentium, memory and registers are also 32 bits in width. Figure 1–13 shows the form used to store doublewords in memory.

When a doubleword is stored in memory, its least-significant byte is stored in the lowest-numbered memory location and its most-significant byte is stored in the highest-numbered memory location using the little endian format. Recall this is also true for word-sized data. For example, a 12345678H that is stored in memory location 00100H–00103H is stored with the 78H in memory location 00100H, the 56H in location 00101H, the 34H in location 00102H, and the 12H in location 00103H.

To define doubleword-sized data, use the assembler directive **define doubleword(s)** of DD. Example 1–26 shows both signed and unsigned numbers stored in memory using the DD directive.

FIGURE 1–13 A doubleword (12345678H) stored in four consecutive bytes of memory.

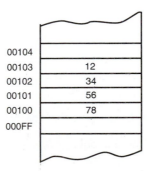

EXAMPLE 1–26

```
                     ;Unsigned doubleword-sized data
                     ;
0000  0003E1C0       DATA1   DD   254400        ;define 254400 decimal
0004  87AC1234       DATA2   DD   87AC1234H     ;define 87AC1234 hexadecimal
0008  00000046       DATA3   DD   70            ;define 70 decimal
                     ;
                     ;Signed doubleword-sized data
                     ;
000C  FFEB8058       DATA4   DD   -1343400      ;define a -1343400 decimal
0010  000000C6       DATA5   DD   +198          ;define a +198 decimal
0014  FFFFFFFF       DATA6   DD   -1            ;define a -1 decimal
```

Integers that are of *any* width may also be stored in memory. The forms listed here are standard forms, but that doesn't mean that a 128-byte wide integer can't be stored in memory. The microprocessor is flexible enough to allow any size data. When nonstandard width numbers are stored in memory, the DB directive is normally used to store them. For example, the 24-bit number 123456H is stored using a DB 56H,34H,12H directive. Note that this conforms to the little endian format.

Real Numbers

Because many high-level languages use the Intel family of microprocessors, real numbers are often encountered. A real number, or as it is often called, a **floating-point number,** contains two parts: a **mantissa (significand or fraction)** and an **exponent.** Figure 1–14 depicts both the 4- and 8-byte forms of real numbers as they are stored in any Intel system. Note that the 4-byte real number is called **single precision** and the 8-byte form is called **double precision.** The form presented here is the same form specified by the IEEE[14] standard, IEEE-754 version 10.0. This standard has been adopted as the standard form of real numbers with virtually all high-level programming languages and many applications packages. The standard also applies to data manipulated by the numeric coprocessor in the personal computer. Figure 1–14 (a) shows the single-precision form that contains a sign-bit, an 8-bit exponent, and a 24-bit fraction (mantissa). Note that because applications often require double-precision floating-point numbers [see Figure 1–14 (b)], the Pentium with its 64-bit data bus performs memory transfers at twice the speed of the 80386/80486 microprocessors.

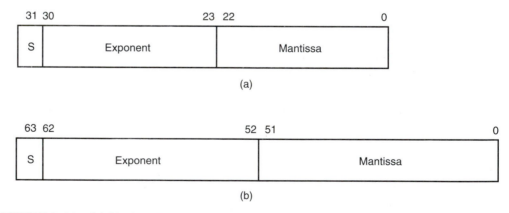

(a)

(b)

FIGURE 1–14   (a) Single-precision floating-point number and (b) double-precision floating-point number.

---

[14]IEEE is the Institute of Electrical and Electronics Engineers.

TABLE 1–11 Examples of single-precision real numbers

Decimal	Binary	Normalized	Sign	Exponent	Mantissa
+12	1100	1.1×2^3	0	1000 0010	1000 0000 0000 0000 0000 0000
−12	1100	1.1×2^3	1	1000 0010	1000 0000 0000 0000 0000 0000
+100	1100100	1.1001×2^6	0	1000 0101	1001 0000 0000 0000 0000 0000
+0.25	0.01	1.0×2^{-2}	0	0111 1101	0000 0000 0000 0000 0000 0000
−1.75	1.11	1.11×2^0	1	0111 1111	1100 0000 0000 0000 0000 0000
0.0	0.0	0.0			0000 0000 0000 0000 0000 0000

In the single-precision real number, simple arithmetic indicates that it should take 33 bits to store all three pieces of data. Not true—the 24-bit mantissa contains an implied *(hidden)* one-bit that allows the mantissa to represent 24 bits while being stored in only 23 bits. The hidden bit is the first bit of the normalized real number. When normalizing a number, it is adjusted so its value is at least 1, but less than 2. For example, if a 12_{10} is converted to binary (1100), it is normalized and the result is a 1.1×2^3. The 1. is not stored in the 23-bit mantissa portion of the number. The 1. is the hidden one-bit. Table 1–11 shows the single-precision form of this number and others.

The **exponent** is stored as a biased exponent. With the single-precision form of the real number, the bias is 127 (7FH) and with the double-precision form it is 1023 (3FFH). The **bias** adds to the exponent before is stored into the exponent portion of the floating-point number. In the previous example, there is an exponent of 2^3, represented as a biased exponent of 127 + 3 or 130 (82H) in the single-precision form or as 1026 (402H) in the double-precision form. There are two exceptions to the rules for floating-point numbers: The number 0.0 is stored as all zeros. The number infinity is stored as all ones in the exponent and all zeros in the mantissa. The sign-bit indicates a ± 0.0 or a ± ∞.

As with other data types, the assembler can be used to define real numbers in both single- and double-precision forms. Because single-precision numbers are 32-bit numbers, use the DD directive or use the **define quadword(s)** or DQ directive to define 64-bit double-precision real numbers. **Optional directives** for real numbers are REAL4, REAL8, and REAL10 for defining single-, double-, and extended-precision real numbers. Example 1–27 shows numbers defined in real number format.

EXAMPLE 1–27

```
                    ;Single-precision real numbers
                    ;
0000   3F9DF3B6     NUMB1   DD      1.234           ;define 1.234
0004   C1BB3333     NUMB2   DD      -23.4           ;define -23.4
0008   43D20000     NUMB3   REAL4   4.2E2           ;define 420
000C   3F9DF3B6     NUMB4   REAL4   1.234           ;define a 4-byte real number

                    ;Double-precision real numbers
                    ;
0010                NUMB5   DQ      123.4           ;define 123.4
     405ED9999999999A
0018                NUMB6   REAL8   -23.4           ;define -23.4
     C1BB333333333333
0028                NUMB7   REAL8   123.4           ;define an 8-byte real number
     405ED9999999999A
                    ;
                    ;Extended-precision real numbers
                    ;
0030                NUMB8   REAL10  123.4           ;define a 10-byte real number
     4005F6CCCCCCCCCCCCCD
```

1–5   SUMMARY

1. The mechanical computer age began with the advent of the abacus in 500 BC. This first mechanical calculator remained unchanged until 1642 when Blaise Pascal improved it. An early first mechanical computer system was the Analytical Engine developed by Charles Babbage in 1823. Unfortunately, this machine never functioned because it was impossible to obtain proper machine parts.

2. The first electronic calculating machine was developed during World War II by Konrad Zuse, an early pioneer of digital electronics. His computer, the Z3, was used in aircraft and missile design for the German war effort.

3. The first electronic computer, which used vacuum tubes, was placed into operation in 1943 to break secret German military codes. This first electronic computer system, the Colossus, was invented by Alan Turing. Its only problem was that the program was fixed and could not be changed.

4. The first general-purpose, programmable electronic computer system was developed in 1946 at the University of Pennsylvania. This first modern computer was called the ENIAC (Electronics Numerical Integrator and Calculator).

5. The first high-level programming language, FLOW-MATIC, was developed for the UNIVAC I computer by Grace Hopper in the early 1950s. This led to FORTRAN and other early programming languages.

6. The world's first microprocessor, the Intel 4004, a 4-bit microprocessor—a programmable controller on a chip—was meager by today's standards. It addressed a mere 4,096 four-bit memory locations. Its instruction set contained only 45 different instructions.

7. Microprocessors that are common today include the 8086/8088 family, which are the first 16-bit microprocessors. Following these early 16-bit machines were the 80286, 80386, 80486, and Pentium microprocessors. With each newer version, improvements followed that increased speed and performance. From all indications, this process of speed and performance improvement will continue.

8. Microprocessor-based personal computers contain memory systems that include three main areas: TPA (transient program area), system area, and extended memory. The TPA holds application programs, the operating system, and drivers. The system area contains memory used for video display cards, disk drives, and the BIOS ROM. The extended memory area is only available to the 80286 through the Pentium microprocessor in an AT-style personal computer system.

9. The 8086, 8088, 80186, and 80188 address 1M byte of memory from location 00000H through location FFFFFH. The 80286 and 80386SX address 16M bytes of memory from location 000000H through FFFFFFH. The 80386SL addresses 32M bytes of memory from location 0000000H through 1FFFFFFH. The 80386EX addresses 64M bytes of memory from location 0000000H through 3FFFFFFH. The 80386DX, 80486, and Pentium address 4G bytes of memory from location 00000000H through FFFFFFFFH. The Pentium Pro addresses 64G of memory through its 36-bit address bus.

10. All versions of the 8086–Pentium Pro microprocessors address 64K bytes of I/O address space. These I/O ports are numbered from 0000H through FFFFH, with I/O ports 0000H–04FFH reserved for use by the personal computer system.

11. The operating system in many personal computers is either MS-DOS (Microsoft disk operating system) or PC-DOS (personal computer disk operating system from IBM). The operating system performs the task of operating or controlling the computer system along with its I/O devices.

12. The microprocessor is the controlling element in a computer system. The microprocessor does data transfers, performs simple arithmetic and logic operations, and makes simple de-

cisions. The microprocessor executes programs stored in the memory system to perform complex operations in short periods of time.

13. All computer systems contain three buses to control memory and I/O. The address bus is used to request a memory location or I/O device. The data bus transfers data between the microprocessor and its memory and I/O spaces. The control bus controls the memory and I/O and requests reading or writing of data. Control is accomplished with $\overline{IORC}$ (I/O read control), $\overline{IOWC}$ (I/O write control), $\overline{MRDC}$ (memory read control), and $\overline{MWTC}$ (memory write control).

14. Numbers are converted from any number base to decimal by noting the weights of each position. The weight of the position to the left of the radix point is always the unit's position in any number system. The position to the left of the unit's position is always the radix times one. Succeeding positions are determined by multiplying by the radix. The weight of the position to the right of the radix point is always determined by dividing by the radix.

15. Conversion from a whole decimal number to any other base is accomplished by dividing by the radix. Conversion from a fractional decimal number is accomplished by multiplying by the radix.

16. Hexadecimal data are represented in hexadecimal form or at times in a code called binary-coded hexadecimal (BCH). A binary-coded hexadecimal number is one that is written with a 4-bit binary number that represents each hexadecimal digit.

17. ASCII code is used to store alphanumeric data. The ASCII code is a 7-bit code and can have an eighth bit used to extend the character set from 128 codes to 256 codes. The carriage return (enter) code returns the print-head or cursor to the left margin. The line feed code moves the cursor or print-head down a line.

18. Binary-coded decimal (BCD) data is sometimes used in a computer system to store decimal data. This data is stored in either packed (2 digits per byte) or unpacked (1 digit per byte) form.

19. Binary data are stored as a byte (8 bits), word (16 bits), or doubleword (32 bits) in a computer system. This data may be unsigned or signed. Signed negative data is always stored in the two's complement form. Data that is wider than 8 bits is always stored using the little endian format.

20. Floating-point data is used in a computer system to store whole, mixed, and fractional numbers. A floating-point number is composed of a sign, mantissa, and an exponent.

21. The assembler directive DB defines bytes, DW defines words, DD defines doublewords, and DQ defines quadwords.

22. Example 1–28 shows the assembly language formats for storing numbers as bytes, words, doublewords, and real numbers. Also shown are ASCII-coded character strings.

EXAMPLE 1–28

```
(Hexadecimal machine                  (Assembly language)
       code)
                            ;ASCII data
                            ;
0000  54 68 69 73 20 69     MES1   DB    'This is a character string in ASCII'
      73 20 61 20 63 68
      61 72 61 63 74 65
      72 20 73 74 72 69
      6E 67 20 69 6E 20
      41 53 43 49 49
0023  53 6F 20 69 73 20     MES2   DB    'So is this'
      74 68 69 73
                            ;BYTE data
                            ;
002D  17                    DATA1  DB    23         ;23 decimal
002E  DE                    DATA2  DB    -34        ;-34 decimal
002F  34                    DATA3  DB    34H        ;34 hexadecimal
```

```
                              ;
                              ;WORD data
                              ;
0030  1000                    DATA4   DW    1000H     ;1000 hexadecimal
0032  FF9C                    DATA5   DW    -100      ;-100 decimal
0034  000C                    DATA6   DW    +12       ;+12 decimal
                              ;
                              ;DOUBLEWORD data
                              ;
0036  00001000                DATA7   DD    1000H     ;1000 hexadecimal
003A  FFFFFED4                DATA8   DD    -300      ;-300 decimal
003E  00012345                DATA9   DD    12345H    ;12345 hexadecimal
                              ;
                              ;Real data
                              ;
0042  4015C28F                DATA10  REAL4 2.34      ;2.34 decimal
0046  C00CCCCD                DATA11  REAL4 -2.2      ;-2.2 decimal
004A                          DATA12  REAL8 100.3     ;100.3 decimal
      4059133333333333
```

1–6   QUESTIONS AND PROBLEMS

1. Who developed the Analytical Engine?
2. The 1890 Census used a new device called a punched card. Who developed the punched card?
3. Who was the founder of IBM Corporation?
4. Who developed the first electronic calculator?
5. The first truly electronic computer system was developed for what purpose?
6. The first general-purpose programmable computer was called the _____.
7. The world's first microprocessor was developed in 1971 by _____.
8. Who was the Countess of Lovelace?
9. Who developed the first high-level programming language called FLOW-MATIC?
10. What is a von Neumann machine?
11. Which 8-bit microprocessor ushered in the age of the microprocessor?
12. The 8085 microprocessor, introduced in 1977, has sold _____ copies.
13. Which Intel microprocessor was the first to address 1M byte of memory?
14. The 80386SL addresses _____ bytes of memory.
15. How much memory is available to the 80486 microprocessor?
16. When did Intel introduce the Pentium microprocessor?
17. When was the Pentium Pro processor introduced by Intel?
18. What is the acronym MIPs?
19. What is the acronym CISC?
20. A binary bit stores a _____ or a _____.
21. A computer K is equal to _____ bytes.
22. A computer M is equal to _____ K bytes.
23. A computer G is equal to _____ M bytes.
24. How many typewritten pages of information are stored in a 4G-byte memory system?
25. The 80186 embedded controller addresses _____ bytes of memory.
26. The 80386EX embedded controller addresses _____ bytes of memory.
27. The first 1M byte of memory in a computer system contains a _____ and a _____ area.
28. How much memory is found in the transient program area?
29. How much memory is found in the system area?
30. The 8086 microprocessor addresses _____ bytes of memory.

31. The 80286 microprocessor addresses _____ bytes of memory.
32. Which microprocessors address 4G bytes of memory?
33. Which microprocessor addresses 64G bytes of memory?
34. Memory above the first 1M byte is called _____ memory.
35. What is the BIOS?
36. What is DOS?
37. What is the difference between an XT and an AT computer system?
38. What is the VESA local bus?
39. What is the PCI bus?
40. The ISA bus holds _____-bit interface cards.
41. What is the XMS?
42. What is the EMS?
43. A driver is stored in the _____ area.
44. What is a TSR?
45. How is a TSR often accessed?
46. What is the purpose of the CONFIG.SYS file?
47. What is the purpose of the AUTOEXEC.BAT file?
48. The COMMAND.COM program processes what information?
49. The personal computer system addresses _____ bytes of I/O space.
50. Where is the high memory located in a personal computer?
51. The DEVICE or DEVICEHIGH statement is found in which file?
52. Where are the upper memory blocks used by MS-DOS version 5.0 or 6.2?
53. Where is the video BIOS?
54. Draw the block diagram of a computer system.
55. What is the purpose of the microprocessor in a microprocessor-based computer system?
56. List the three buses found in all computer systems.
57. Which bus transfers the memory address to the I/O device or to the memory device?
58. Which control signal causes the memory to perform a read operation?
59. What is the purpose of the $\overline{IORC}$ signal?
60. If the $\overline{MRDC}$ signal is a logic 0, which operation is performed by the microprocessor?
61. Convert the following binary numbers into decimal:
    (a) 1101.01
    (b) 111001.0011
    (c) 101011.0101
    (d) 111.0001
62. Convert the following octal numbers into decimal:
    (a) 234.5
    (b) 12.3
    (c) 7767.07
    (d) 123.45
    (e) 72.72
63. Convert the following hexadecimal numbers into decimal:
    (a) A3.3
    (b) 129.C
    (c) AC.DC
    (d) FAB.3
    (e) BB8.0D
64. Convert the following decimal integers into binary, octal, and hexadecimal:
    (a) 23
    (b) 107
    (c) 1238
    (d) 92

    (e) 173
65. Convert the following decimal numbers into binary, octal, and hexadecimal:
    (a) .625
    (b) .00390625
    (c) .62890625
    (d) 0.75
    (e) .9375
66. Convert the following hexadecimal numbers into binary-coded hexadecimal (BCH) code:
    (a) 23
    (b) AD4
    (c) 34.AD
    (d) BD32
    (e) 234.3
67. Convert the following BCH numbers into hexadecimal:
    (a) 1100 0010
    (b) 0001 0000 1111 1101
    (c) 1011.1100
    (d) 0001 0000
    (e) 1000 1011 1010
68. Convert the following binary numbers to the one's complement form:
    (a) 1000 1000
    (b) 0101 1010
    (c) 0111 0111
    (d) 1000 0000
69. Convert the following binary numbers to the two's complement form:
    (a) 1000 0001
    (b) 1010 1100
    (c) 1010 1111
    (d) 1000 0000
70. Define byte, word, and doubleword.
71. Convert the following words into ASCII-coded character strings:
    (a) FROG
    (b) Arc
    (c) Water
    (d) Well
72. What is the ASCII code for the enter key and what is its purpose?
73. Use an assembler directive to store the ASCII-character string 'What time is it?' in memory.
74. Convert the following decimal numbers into 8-bit signed binary numbers:
    (a) +32
    (b) –12
    (c) +100
    (d) –92
75. Convert the following decimal numbers into signed binary words:
    (a) +1000
    (b) –120
    (c) +800
    (d) –3212
76. Use an assembler directive to store –34 into memory.
77. Show how the following 16-bit hexadecimal numbers are stored in the memory system:
    (a) 1234H
    (b) A122H
    (c) B100H

78. What is the difference between the big endian and little endian formats for storing numbers that are larger than 8 bits in width?

79. Use an assembler directive to store a 123A hexadecimal into the memory.

80. Convert the following decimal numbers into both packed and unpacked BCD forms:
    (a) 102
    (b) 44
    (c) 301
    (d) 1000

81. Convert the following binary numbers into signed decimal numbers:
    (a) 10000000
    (b) 00110011
    (c) 10010010
    (d) 10001001

82. Convert the following BCD numbers (assume that these are packed numbers) into decimal numbers:
    (a) 10001001
    (b) 00001001
    (c) 00110010
    (d) 00000001

83. Convert the following decimal numbers into single-precision floating-point numbers:
    (a) +1.5
    (b) −10.625
    (c) +100.25
    (d) −1200

84. Convert the following single-precision floating-point numbers into decimal:
    (a) 0 10000000 11000000000000000000000
    (b) 1 01111111 00000000000000000000000
    (c) 0 10000010 10010000000000000000000
    (d) 1 01111110 11000000000000000000000

85. Use the Internet to locate and write a short paper that describes the accomplishments of Konrad Zuse.

86. Use the Internet to locate and write a short paper that describes the accomplishments of Charles Babbage.

87. Use the Internet to locate the IBM Corporation Web site and write a short paper that describes the types of information available there.

88. Use the Internet to locate the Microsoft Corporation Web site and write a short paper that describes the types of information available there.

CHAPTER 2

The Architecture of the Microprocessor and Embedded Controller

INTRODUCTION

This chapter presents the microprocessor and embedded controller as programmable devices by first looking at their internal programming models. Than we'll look at how the memory space is addressed in both real and protected modes. The architecture of the entire family of microprocessors (8088–Pentium Pro) and embedded controllers (80186, 80188, and 80386CX/80386EX) are presented simultaneously, as are the ways that the family members address the memory system.

The addressing modes are described for both the real and protected modes of operation. Real-mode or conventional-memory exists at locations 00000H–FFFFFH (the first 1M byte of the memory system) and is present on all versions of the microprocessor and embedded controller. Protected-mode memory exists at any location in the entire memory system, but is available only to the 80286 through the Pentium Pro microprocessors and not the earlier 8086 through 80188. Protected-mode memory contains 16M bytes for the 80286; 64M bytes for the 80386CX/80386EX; and 4G bytes for the 80386DX, 80486, and Pentium. The Pentium Pro addresses a very large memory space of 64G bytes through its 36-bit address.

CHAPTER OBJECTIVES

Upon completion of this chapter, you will be able to:

1. Describe the function and purpose of each program-visible register in the 8086 through the Pentium Pro microprocessors, and 80186, 80188, and 80386CX/80386EX embedded controllers.
2. Discuss the flag register and the purpose of each flag bit.
3. Describe how memory is accessed using real- or conventional-mode memory addressing techniques, showing the effect of the segment and offset addresses.
4. Illustrate how memory is accessed using protected- or extended-mode memory addressing techniques, showing how the descriptor describes the memory segment.
5. Describe the program-invisible registers found within the 80286, 80386, 80486, Pentium, and Pentium Pro microprocessors and 80386CX/80386EX embedded controllers.
6. Explain the operation of the memory paging mechanism and show how the linear memory address is converted to a physical memory address.

2–1   THE INTERNAL ARCHITECTURE

Before you can write a program or investigate an instruction, you must understand the internal configuration of the microprocessor. This section details the program-visible internal architecture of the 80186, 80188, and 80386CX/80386EX embedded controllers. We also discuss the function and purpose of each internal register. The 8086 through the Pentium Pro microprocessors share the same programming model as the embedded controllers, but not the same I/O control structure (called a **peripheral control block** or PCB), which is discussed separately.

The Programming Model

The programming models of the 8086 through the Pentium Pro microprocessors and 80186, 80188, and 80386CX/80386EX embedded controllers are considered **program visible** because their registers are used during programming and are specified by the instructions. Other registers, detailed later in this chapter, are considered **program invisible** because they are not normally used during applications programming, but may be used during system programming. Registers, in the **peripheral control block,** are unique to the 80188, 80186, and 80386CX/80386EX embedded controllers and are also introduced here but detailed in later chapters that apply to each control function. Only the 80286 through the Pentium Pro microprocessors and the 80386CX/80386EX embedded controllers contain the program-invisible registers used to control and operate the protected memory system. The protected memory system must be used to access data above the first 1M byte of the real memory system.

Figure 2–1 illustrates the **programming model** of the 8086 through the Pentium microprocessors and 80186, 80188, and 80386CX/880386EX embedded controllers. The earlier 8086–80286 contain **16-bit internal architectures,** a subset of the registers shown in Figure 2–1. The 80386, 80486, and Pentium microprocessors and 80386CX/80386EX embedded controllers contain full **32-bit internal architectures.** The shaded areas in this diagram represent enhanced registers found in the 32-bit architecture machines. The architectures of the earlier 8086–80286 are fully upward compatible to the 80386, 80486, and Pentium microprocessors and 80386CX/80386EX embedded controllers.

The programming model contains 8-, 16-, and 32-bit registers. The **8-bit registers** are AH, AL, BH, BL, CH, CL, DH, and DL and are referred to when an instruction is formed using any of these two-letter designations. For example, the ADD AL,AH instruction adds contents of AH to the contents of AL. The **16-bit registers** are AX, BX, CX, DX, SP, BP, DI, SI, IP, FLAGS, CS, DS, ES, SS, FS, and GS. Likewise, these registers are also referenced with the two-letter designations. For example, an ADD DX,CX instruction adds the contents of CX to the contents of DX. The **extended 32-bit registers** are labeled: EAX, EBX, ECX, EDX, ESP, EBP, EDI, ESI, EIP, and EFLAGS. The extended registers and 16-bit registers FS and GS are only available in the 80386, through the Pentium microprocessors and the 80386CX/80386EX embedded controllers. The new segment registers are referenced using the register designations FS or GS and the extended registers are selected by using any of the three-letter designations. For example, an ADD ECX,EBX instruction adds the contents of EBX to the contents of ECX.

Some registers are general- or multi-purpose registers, while some have special purposes. The *multi-purpose* registers include EAX, EBX, ECX, EDX, EBP, EDI, and ESI. These registers hold various data sizes (bytes, words, or doublewords) and are used for almost any purpose dictated by a program.

FIGURE 2–1   The programming model of the Intel 8086 through the Pentium Pro microprocessors.

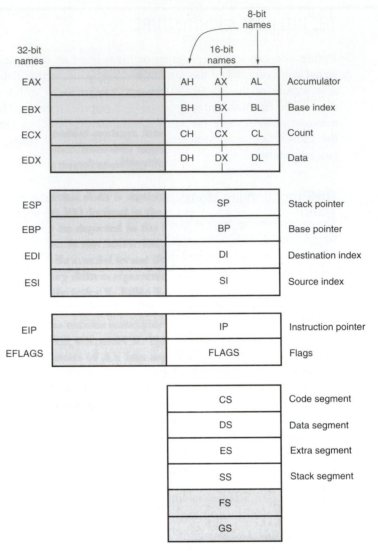

Notes:
1. The shaded areas registers exist ony on the 80386 through the Pentium Pro.

2. The FS and GS registers have no special names.

Multi-Purpose Registers

**EAX
(accumulator)**

Used as a 32-bit register (EAX), as a 16-bit register (AX), or as two 8-bit registers (AH and AL). If an 8- or a 16-bit register is addressed, only that portion of the 32-bit register changes without affecting the remaining bits. The accumulator is used for instructions such as multiplication, division, and some of the adjustment instructions. For these instructions, the accumulator has a special purpose, but is generally considered a multi-purpose register. In the 80386 and above, the EAX register can also address the memory system.

EBX **(base index)**	As with EAX, EBX is also addressable as EBX, BX, BH, or BL. The BX register is sometimes used to address memory in all versions of the microprocessor or embedded controller and in the 80386 and above; EBX can also be used to address memory data.
ECX **(count)**	A general-purpose register that also holds the count for various instructions. In the 80386 and above, the ECX register can also address memory data. Instructions that use a count for a special purpose are the repeated string instructions (REP), shift, rotate, and LOOP/LOOPD instructions. The shift and rotate instructions use CL as the count, the repeated string instructions use CX, and LOOP/LOOPD instructions use either CX or ECX as the count.
EDX **(data)**	A general-purpose register that holds a part of the result from a multiplication or part of the dividend before a division. In the 80386 and above, this register can also address memory data.
EBP **(base pointer)**	Can point to a memory location in all versions of the microprocessor for memory-data transfers. This register is addressed as either BP or EBP.
EDI **(destination index)**	Often addresses string destination data for the string instructions. It also functions as either a 32-bit (EDI) or 16-bit (DI) general-purpose register.
ESI **(source index)**	Used as either ESI or SI. The source index register often addresses source string data for the string instructions. Like EDI, ESI also functions as a general-purpose register. As a 16-bit register it is addressed as SI and as a 32-bit register as ESI.

Special-Purpose Registers.   The special-purpose registers include: EIP, ESP, EFLAGS, and the segment registers CS, DS, ES, SS, FS, and GS.

EIP **(instruction pointer)**	Addresses the next instruction in a section of memory defined as a code segment. This register is IP (16 bits) when the microprocessor operates in the real mode and EIP (32 bits) when the 80386 and above operate in the protected mode. Note that the 8086 through the 80286 do contain EIP, and only the 80286 and above operate in the protected mode. The instruction pointer, which points to the next instruction in a program, is used by the microprocessor to find the next sequential instruction in a program located within the code segment. All steps in a program are ordered and stored in ascending memory locations. The instruction pointer can be modified with a jump or a call instruction to execute instructions located in other sections of the memory.
ESP **(stack pointer)**	Addresses an area of memory called the stack. The stack memory stores data through this pointer and is discussed later in the chapter.
EFLAGS	Indicates the condition of the microprocessor as well as control its operation. Figure 2–2 shows the flag registers of all versions of the microprocessor and embedded controller. Note that the flags are upward compatible from the 8086/8088 through the Pentium microprocessors. The 8086–80286 contain a FLAG register (16 bits), and the 80386 and above contain an EFLAG register (32-bit extended flag register).

FIGURE 2–2 The EFLAG and FLAG register contents for the entire 80X86 and Pentium Pro microprocessor family.

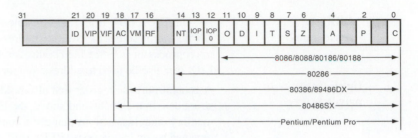

The rightmost five flag bits and the overflow flag change after many arithmetic and logic instructions execute. These flags do not change for data transfer instructions unless the destination is the flag register. Other flags control features found in the microprocessor. As instructions are introduced in subsequent chapters, additional detail on the flag bits is provided. The following is a list of each flag bit with a brief description of its function.

C (carry)	Holds the carry after addition or the borrow after subtraction. The carry flag also indicates error conditions in some programs and procedures. This is especially true of the DOS function calls detailed in later chapters and Appendix A.
P (parity)	Parity is a count of ones in a number expressed as even or odd: a logic 0 for odd parity and a logic 1 for even parity. For example, if a number contains three binary one bits, it has odd parity. If a number contains zero one bits it is considered to have even parity.
A (auxiliary carry)	Holds the carry (half-carry) after addition or the borrow after subtraction between bit positions 3 and 4 of the result. This highly specialized flag bit is tested by the DAA and DAS instructions to adjust the value of AL after a BCD addition or subtraction. Otherwise, the A flag bit is not used by the microprocessor or any other instructions.
Z (zero)	Shows that the result of an arithmetic or logic operation is zero. If Z = 1, the result is zero, and if Z = 0, the result is not zero.
S (sign)	Holds the arithmetic sign of the result after an arithmetic or logic instruction executes. If S = 1, the sign is set or negative, and if S = 0, the sign is cleared or positive.
T (trap)	Enables trapping through an on-chip debugging feature. More detail of this debugging feature is provided later with a discussion of the debug registers and debugging features provided by the microprocessor. (A program is **debugged** to find an error or bug.)
I (interrupt)	Controls the operation of the INTR (interrupt request) input pin. If I = 1, the INTR pin is enabled, and if I = 0, the INTR pin is disabled. The state of the I flag bit is controlled by the STI (set I flag) and CLI (clear I flag) instructions.
D (direction)	Selects either the increment or decrement mode for the DI and/or SI registers during string instructions. If D = 1, the registers are automatically decremented and if D = 0, the registers are automatically incremented. The D flag is set with the STD **(set direction)** and cleared with the CLD **(clear direction)** instructions.
O (overflow)	Occurs when signed numbers are added or subtracted. An overflow indicates that the result has exceeded the capacity of the machine.

For example, if a 7FH (+127) is added, using an 8-bit addition, to a 01H (+1), the result is 80H (−128). This result represents an overflow condition indicated by the overflow flag for signed addition. For unsigned operations, the overflow flag is ignored.

IOPL (input/output privilege level) Used in protected mode operation to select the privilege level for I/O devices. If the current privilege level is higher or more trusted than the IOPL, then I/O executes without hindrance. If the IOPL is lower than the current privilege level, an interrupt occurs causing execution to suspend. Note that an IOPL of 00 is the highest or most trusted, and an IOPL of 11 is the lowest or least trusted.

NT (nested task) Indicates that the current task is nested within another task in protected mode operation. This flag is set when the task is nested by software.

RF (resume) Used with debugging to control the resumption of execution after the next instruction.

VM (virtual mode) Selects virtual mode operation in a protected-mode system. A virtual mode system allows multiple DOS memory partitions that are 1M byte in length to coexist in the memory system. This, in essence, allows the system program to execute multiple DOS programs.

AC (alignment check) Activates if a word or doubleword is addressed on a non-word or non-doubleword boundary. Only the 80486SX microprocessor contains the alignment check bit that is primarily used by its companion numeric coprocessor, the 80487SX, for synchronization.

VIF (virtual interrupt flag) A virtual mode copy of the interrupt flag bit available to the Pentium/Pentium Pro microprocessor.

VIP (virtual interrupt pending) Provides information about a virtual mode interrupt for the Pentium/Pentium Pro microprocessor. This is used in multitasking environments to provide the operating system with virtual interrupt flags and interrupt pending information.

ID (identification) Indicates that the Pentium/Pentium Pro microprocessor supports the CPUID instruction. The CPUID instruction provides the system with information about the Pentium/Pentium Pro microprocessor such as its version number and manufacturer.

Segment Registers.   Additional registers, called **segment registers** generate memory addresses when combined with other registers in the microprocessor. There are either four or six segment registers depending on the version of the microprocessor. A segment register functions differently in the real-mode than in the protected-mode operation of the microprocessor. Detail on their function in real and protected modes is provided later in this chapter. The following is a list of each segment register along with its function in the system.

CS (code) A section of memory that holds the code (programs and procedures) used by the microprocessor. The code segment register defines the starting address of the section of memory that is holding code. In real-mode operation it defines the start of a 64K-byte section of memory, and in protected mode it selects a descriptor that describes the starting address and length of a section of memory holding code. The code segment is limited to 64K bytes in length in the 8088–80286 and 4G bytes in the 80386 and above when operated in the protected mode.

DS (data)   A section of memory that contains most data used by a program. Data is accessed in the data segment by an offset address or the contents of other registers that hold the offset address. As with the code segment and other segments, the length is limited to 64K bytes in the 8086–80286 and 4G bytes in the 80386 through the Pentium Pro.

ES (extra)   An additional data segment that is used by some of the string instructions to hold destination data.

SS (stack)   Defines the area of memory used for the stack. The location of the current entry point in the stack segment is determined by the stack pointer register. The BP register also addresses data within the stack segment.

FS and **GS**   Supplemental segment registers that have no special names. Available in the 80386–Pentium Pro microprocessors allowing programs to access two additional memory segments.

Embedded Controller Registers

In addition to the standard programming model described, the 80186, 80188, and 80386CX/80386EX contain a special set of registers called the **peripheral control block** (PCB). The PCB is addressed as either I/O or memory. The PCB is a special block of registers that control the features added to these embedded microprocessors.

Special 80186 and 80188 Registers.   The 80186 and 80188 are available in four different versions: XL, EA, EB, and EC. Each version of the 80186 and 80188 contain a PCB that controls its interrupt controller, timers, chip selection logic, DMA controller, refresh controller, and power-saving feature. In addition to these basic features, some versions also contain serial ports, parallel ports, and so forth. The contents of the PCB for all versions is illustrated in Figure 2–3. Notice that the address is an offset address beginning with offset 00H and ending with offset FFH.

When power is applied to the 80186 or 80188, the peripheral control block is initialized at I/O locations FF00H through FFFFH. This can be changed by reprogramming word-sized I/O location FFFEH, the **relocation register,** with a new I/O port or memory address to move the PCB to another section of I/O or memory. (Programming the microprocessor through the PCB is relegated to later chapters that detail each hardware feature.)

The following is a list of each area found in the PCB with a brief description of its function in the system (refer to the block diagram of the 80186 embedded controller in Figure 2–4 while reading these descriptions). Not all items in the list appear, because the 80C186XL version does not contain all components.

Relocation register (offset FEH and FFH)   Allows the PCB to be moved to any contiguous 256-byte area of I/O space or memory.

Power-saving feature   In the 80C186 and 80C188 (versions XL, EA, and EC) a power-saving feature is added to reduce the internal clock frequency to the microprocessor and all other internal components during a power-saving mode. This results in a reduction of power consumption.

Refresh control   In the 80C186 and 80C188 generates refresh bus cycles for DRAM connected to the controller. These registers program the refresh partition addresses and the refresh controller.

DMA controller   Controlled to enable DMA (direct memory access) operations and program the DMA transfer address into the DMA controller. C0H through CBH program DMA channel 0, and D0H through CBH program DMA channel 1.

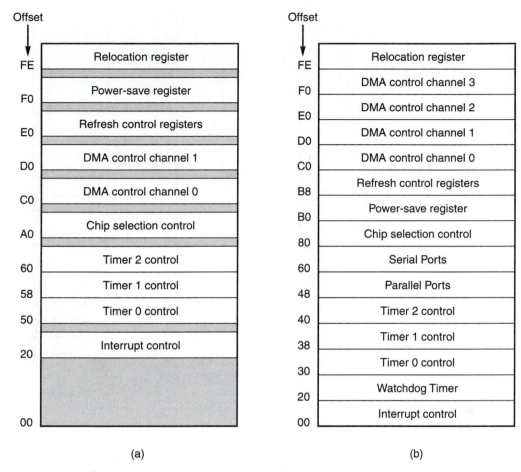

FIGURE 2–3 The peripheral control block for the 80186 and 80188 embedded controllers: (a) 80C186XL/80C188XL or 80C186EA/80C188EA and (b) 80C186EB/80C188EB or 80C186EC/ 80C188EC.

Chip selection registers	Program the location of the upper memory device, lower memory device, four middle memory devices, and the location of up to seven I/O devices. 80C186EB and 80C188EB have enhanced chip selection and use offset 80H–A7H for controlling the memory and I/O devices.
Timer registers	Control the operation of three programmable timers. A **timer** is a programmable divider that can generate wave-forms or cause interrupts.
Interrupt control registers	Control the function and operation of the programmable interrupt controller.
Serial communications control registers	Control the two serial ports available on the EB version.
Watchdog control register	Controls the operation of the watchdog timer found in the EC version.
I/O port control register	Determines the function of the parallel port pins available on the EB and EC versions.

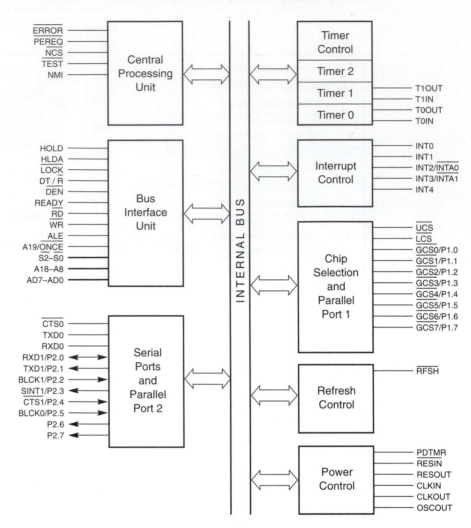

FIGURE 2–4 The internal structure of the 80C188EB embedded controller.

Other versions of the 80186/80188 embedded controllers contain additional features that provide serial and parallel communications as well as enhanced interrupt control. Table 2–1 illustrates each version of the 80186/80188 and the features supported.

Special 80386CX/80386EX Registers. As with the 80186 and 80188, the 80386CX/80386EX contain a peripheral control block for the operation of their internal components as well as other features. Figure 2–5 shows the block diagram of the 80386EX embedded controller. The most significant difference between 80186/80188 and 80386EX is that 80386EX is designed to function as a personal computer. We will learn later that the interrupt, DMA, and timer sections are exactly the same as those inside the personal computer system. Because of this, software from a personal computer easily migrates to the 80386EX embedded system.

The peripheral control block within the 80386EX is larger than that found in the 80186/80188. The PCB in the 80386EX is 1K byte in length as illustrated in Figure 2–6. The peripheral control block can appear as a DOS-compliant block causing all of the internal components to function at the same I/O addresses as found in the personal computer or in three other non-DOS-compliant modes. Figure 2–6 illustrates the peripheral control block for the DOS-compliant mode only; the other modes are described in subsequent chapters that deal with the 80386EX.

TABLE 2–1   The four versions of 80186/80188 embedded controllers

Feature	80C186XL 80C188XL	80C186EA 80C188EA	80C186EB 80C188EB	80C186EC 80C188EC
80286-like instruction set	✔	4	✔	✔
Power-save (green mode)	✔	✔	none	✔
Power down mode		✔	✔	✔
80C187 interface	✔	✔	4	✔
ONCE mode	✔	✔	✔	✔
Interrupt controller	✔	✔	✔	✔
				8259-like
Timer unit	✔	✔	✔	✔
Chip selection unit	✔	✔	✔	✔
			enhanced	enhanced
DMA controller	✔	✔		✔
	2-channel	2-channel	none	4-channel
Serial communications unit			✔	✔
Refresh controller	✔	✔	✔	✔
			enhanced	enhanced
Watchdog timer	none	none	none	✔
I/O ports			✔	✔
	none	none	16 bits	22-bits

FIGURE 2–5   The internal structure of the 80386EX embedded controller.

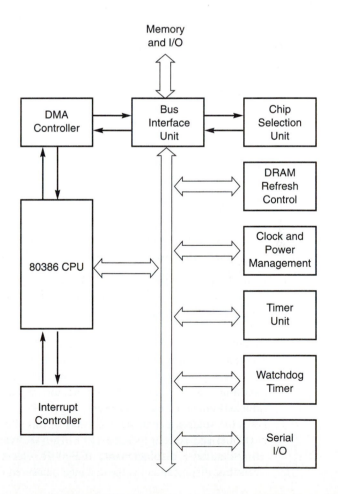

FIGURE 2–6   The peripheral control block for the 80386EX when operated in DOS-compliant mode.

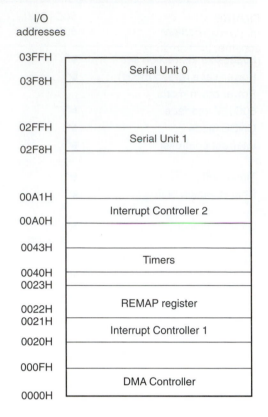

I/O addresses

03FFH	
	Serial Unit 0
03F8H	
02FFH	
	Serial Unit 1
02F8H	
00A1H	
	Interrupt Controller 2
00A0H	
0043H	
	Timers
0040H	
0023H	
	REMAP register
0022H	
0021H	
	Interrupt Controller 1
0020H	
000FH	
	DMA Controller
0000H	

2–2   REAL-MODE MEMORY ADDRESSING

The 80286 through the Pentium Pro operate in either the real or protected mode. The 8086, 8088, 80186, and 80188 operate only in the real mode. This section details the operation of the microprocessor and embedded controller in the real mode. **Real-mode** operation allows the microprocessor or embedded controller to address only the first 1M byte of memory space even if it is the Pentium Pro microprocessor. The first 1M byte of memory is called either the **real** or **conventional** memory system. Both MS-DOS and PC-DOS assume that the microprocessor is operated in the real mode at all times. Real-mode operation allows application software written for the 8086–80188, which contain only 1M byte of memory, to function in the 80286 through the Pentium. The **upward compatibility** of software is partially responsible for the continuing success of the Intel family of microprocessors. In all cases, each of these microprocessors begins operation in the real mode by default whenever power is applied or the microprocessor is reset.

Segments and Offsets

The combination of a segment address and an offset address access a memory location in the real mode. All **real-mode memory addresses** consist of a segment address plus an offset address. The **segment address,** located within one of the segment registers, defines the beginning or starting address of any *64K-byte* memory segment. **The offset address selects** a location within the 64K-byte memory segment. Figure 2–7 shows how the segment plus offset addressing scheme selects a memory location. In this illustration, a memory segment begins at location 10000H and ends at location 1FFFFH–64K bytes in length. It also shows how an offset, sometimes called a **displacement,** of F000H selects location 1F000H in the memory system. The offset or displacement is the distance measured from the start of the segment.

FIGURE 2–7 The real-mode memory-addressing scheme, using a segment address plus an offset.

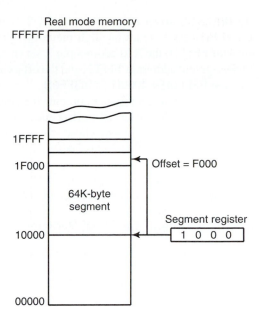

The segment register in Figure 2–7 contains a 1000H, yet it addresses a starting segment at location 10000H. In the real mode, each segment register is internally appended with a **0H** at its rightmost end. This forms a 20-bit memory address allowing it to access the start of a segment at any location within the first 1M byte of memory. For example, if a segment register contains a 1200H, it addresses a 64K-byte memory segment beginning at location 12000H. Likewise, if a segment register contains a 1201H, it addresses a memory segment beginning at location 12010H. Because of the internally appended 0H, real-mode segments can begin only at a 16-byte boundary in the memory system. The 16-byte boundary is called a **paragraph.** They can also overlap completely by using the same segment address in more than one segment register.

Because a real-mode segment of memory is 64K in length, once the beginning address is known, the ending address is found by adding FFFFH. For example, if a segment register contains 3000H, the first address of the segment is 30000H and the last address is 30000H + FFFFH or 3FFFFH. Table 2–2 shows several examples of segment register contents and the starting and ending addresses of the memory segments selected by each segment address.

The offset address adds to the start of the segment to select a memory location from the memory segment. For example, if the segment address is 1000H and the offset address is 2000H, the microprocessor selects memory location 12000H. The segment and offset address is sometimes written as 1000:2000 for a segment address of 1000H with an offset of 2000H. This is especially true if the **DEBUG program** is used to debug software.

In the 80286 (with special external circuitry), 80386, 80486, and Pentium, an extra 64K minus 16 bytes of memory is addressable when the segment address is FFFFH and the

TABLE 2–2 Examples of segment addresses

Segment Register	Starting Address	Ending Address
2000	20000	2FFFF
2001	20010	3000F
2100	21000	30FFF
AB00	AB000	BAFFF
1234	12340	2233F

HIMEM.SYS driver is installed in a DOS-controlled system. This area of memory (FFFF0H–10FFEFH) is referred to as **high memory.** When an address is generated using a segment address of FFFFH, the A20 address pin is set (if supported) when an offset is added. For example, if the segment address is FFFFH and the offset address is 4000H, the machine addresses memory location FFFF0H + 4000H or 103FF0H.

Some addressing modes combine the contents of more than one register to form an offset address. When this occurs, the offset address may exceed FFFFH in value. For example, the location accessed in a segment whose segment address is 3000H and whose offset address is specified as the sum of F000H plus 3000H will select memory location 32000H. When the F000H and 3000H are added together, they form a 16-bit (modulo-16) sum of 2000H used as the offset address. Note that the carry of 1 (F000H + 3000H = 12000H) from the addition is dropped to form the modulo-16 offset address of 2000H.

Default Segment and Offset Registers

The microprocessor has a set of rules that apply to segments whenever memory is addressed. These *rules,* which apply in either the real or protected mode, define the segment register and offset register combinations used by certain addressing modes. For example, the **code segment register** is always used with the **instruction pointer** to address the next instruction in a program. This combination is either CS:IP or CS:EIP, depending upon the microprocessor or embedded controller and their mode of operation. The code segment register defines the start of the code segment and the instruction pointer locates the next instruction within the code segment. This combination (CS:IP or CS:EIP) locates the *next instruction* executed by the microprocessor or embedded controller. For example, if CS = 1400H and IP/EIP = 1200H, the next instruction is fetched from memory location 14000H + 1200H or 15200H.

Another default combination is the stack. **Stack data** is referenced through the stack segment at the memory location addressed by either the stack pointer (SP/ESP) or the base pointer (BP/EBP). These combinations are referred to as SS:SP (SS:ESP) or SS:BP (SS:EBP). For example, if SS = 2000H and BP = 3000H, the selected stack segment memory location is 23000H.

In real mode, only the *rightmost 16 bits* of the extended register address a location within the memory segment. In the 80386–Pentium, never place a number larger than FFFFH into an offset register if the microprocessor or embedded controller is operated in the real mode. A larger number causes the system to halt and indicate an addressing error. Real-mode memory segments have an offset address limit of FFFFH.

Table 2–3 shows the defaults for addressing memory using the 8086 through the Pentium with 16-bit registers. Table 2–4 shows the defaults assumed in the 80386 through the Pentium when using 32-bit registers. Notice that the 80386, 80486, and Pentium have a far greater selection of segment/offset address combinations than do the 8086–80286.

The 8086–80286 allow four memory segments, and the 80386 and above allow six. Figure 2–8 shows a system that contains four memory segments. A memory segment can touch or even overlap if 64K bytes of memory are not required for a segment. Think of segments as *windows* that can be moved over the top of any 64K-byte area of memory to access data or code. A program can have more than four or six segments, but can only access four or six segments at a time.

TABLE 2–3   8086–80486 and Pentium default 16-bit segment and offset address combinations

Segment	Offset	Special Purpose
CS	IP	Instruction address
SS	SP or BP	Stack address
DS	BX, DI, SI, or a 16-bit number offset	Data address
ES	DI for string instructions	String destination address

TABLE 2–4 80386, 80486, and Pentium default 32-bit segment and offset address combinations

Segment	Offset	Special Purpose
CS	EIP	Instruction address
SS	ESP or EBP	Stack address
DS	EAX, EBX, ECX, EDX, EDI, ESI, an 8-bit offset, or a 32-bit offset	Data address
ES	EDI for string instructions	String destination
FS	no default	General address
GS	no default	General address

Suppose an application program requires 1000H bytes of memory for its code, 190H bytes of memory for its data, and 200H bytes of memory for its stack. We will see that this application does not require an extra segment. When this program is placed in the memory system by DOS, it is loaded in the TPA at the first available area of memory above the drivers and

FIGURE 2–8 A memory system showing the placement of four memory segments.

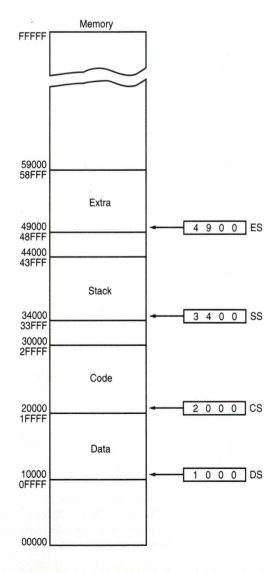

FIGURE 2–9 An application program containing a code, data, and stack segment loaded into a DOS memory system.

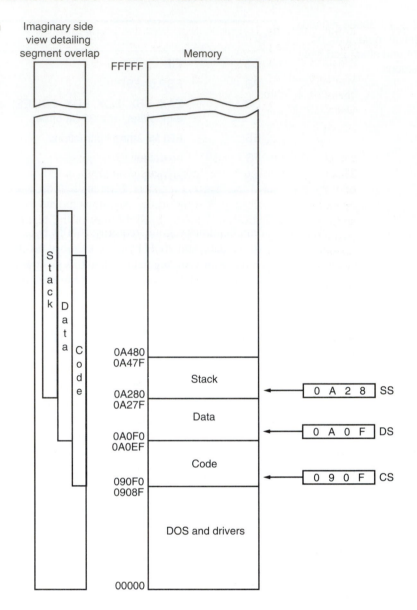

other TPA programs. This area is located by a **free-pointer** that is maintained by DOS. Program loading is handled automatically by the program loader located within DOS. Figure 2–9 shows how this application is stored in the memory system. The segments show an overlap because the amount of data in them does not require 64K bytes of memory. The side view of the segments clearly shows the overlap and how segments can be slid to any area of memory by changing the segment starting address. Fortunately for us, DOS calculates and assigns segment starting addresses. (This is explained in a later chapter that details the operation of the assembler, BIOS, and DOS for an assembly language program.)

Segment and Offset Addressing Allow Relocation

The segment and offset addressing scheme seems unduly complicated. It is complicated, but it also affords an advantage to the system. This complicated scheme of segment plus offset addressing allows programs to be *relocated* in the memory system. It also allows programs, written to function in the real mode, to operate in a protected-mode system. A **relocatable program** is one that can be placed into any area of memory and executed without change. **Relocatable data** is data that can be placed in any area of memory and used without any change to the program. The segment and offset addressing scheme allows both programs and data to be relocated without changing anything in the program or data. This is ideal for a general-purpose computer system where not all machines contain the same memory areas. Because the personal computer memory structure is different from machine to machine, it requires relocatable software and data.

Because memory is addressed within a segment by an offset address, the memory segment can be moved to any place in the memory system without changing any of the offset addresses. This is accomplished by moving the entire program, as a block, to a new area and then changing only the contents of the segment registers. If an instruction is 4 bytes above the start of the segment, its offset address is 4. If the entire program is moved to a new area of memory, this offset address of 4 still points to 4 bytes above the start of the segment. The only things that must be changed are the contents of the segment register to address the program in the new area of memory. This simple form of relocation has made the personal computer based upon Intel microprocessors very powerful and common. Without this feature, a program would have to be extensively rewritten or altered before it is moved.

2–3   PROTECTED-MODE MEMORY ADDRESSING

Protected-mode memory addressing (80286 through the Pentium Pro only) allows access to data and programs located above the first 1M byte of memory as well as within the first 1M byte of memory. Addressing this extended section of the memory system, called **extended memory,** requires a change to the segment plus offset addressing scheme used with real-mode memory addressing. When data and programs are addressed in extended memory, the offset address is still used to access information located within the memory segment. The difference is that the **segment address,** as discussed with real-mode memory addressing, is no longer present in the protected mode. Also, the size of the **offset address** can be 32 bits in the 80386 through the Pentium Pro. In place of the segment address, the segment register contains a **selector** that selects a **descriptor** from a descriptor table. The descriptor describes the memory segment's location, length, and access rights. Because the segment register and offset address still access memory, protected mode instructions look exactly like real-mode instructions. In fact, most programs written to function in the real mode will function in the protected mode without change. The difference between modes is in the way that the segment register is interpreted by the microprocessor or embedded controller to access the memory segment.

Selectors and Descriptors

The *selector,* located in the segment register, selects one of 8,192 descriptors from one of two tables of descriptors. The *descriptor* describes the location, length, and access rights of the segment of memory. Indirectly, the segment register still selects a memory segment, but not directly as in the real mode. For example, in the real mode, if CS = 0008H, the code segment begins at location 00080H. In the protected mode, this segment register contents can address any memory location for a code segment of any length, as explained shortly.

There are two descriptor tables used with the segment registers: one contains global descriptors and the other contains local descriptors. The **global descriptors** contain segment definitions that apply to all programs, while the **local descriptors** are usually unique to an application. Each **descriptor table** contains up to 8,192 descriptors so a total of 16,384 descriptors are available to an application at any time. Because the descriptor describes a memory segment, up to

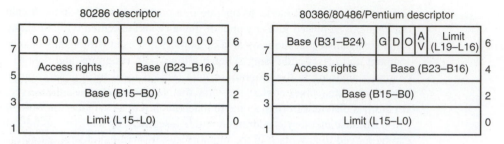

FIGURE 2–10   The descriptor formats for the 80286 and 80386/80486/Pentium microprocessors.

16,384 memory segments can be described for each application. This allows a program to access an astounding *64T* bytes of data because each descriptor can define a segment of *4G* bytes in length. This allows access to 4G times 16K, or 64T bytes of memory.

Figure 2–10 shows the format of a descriptor for the 80286 through the Pentium Pro. Note that each descriptor is 8 bytes in length so the global and local descriptor tables are each a maximum of 64K bytes in length. Descriptors for the 80286 and the 80386 through the Pentium Pro differ slightly, but the 80286 descriptor is upward compatible.

The base address portion of the descriptor indicates the starting location of the memory segment. For the 80286 microprocessor, the **base address** is a 24-bit address so segments begin at any location in its 16M bytes of memory. The 80386 and above use a 32-bit base address that allows segments to begin at any location in its 4G bytes of memory. (The structure is the same in the 80386CX/80386EX embedded controllers, except the size of the memory is 64M bytes.) Notice how the 80286 descriptor's base address is upward compatible to the 80386 through the Pentium Pro descriptor because its most-significant 8 bits are 00H.

The **segment limit** contains the last offset address found in a segment. For example, if a segment begins at memory location F00000H and ends at location F000FFH, the base address is F00000H and the limit is FFH. For the 80286 microprocessor, the base address is F00000H and the limit is 00FFH. For the 80386 and above, the base address is 00F00000H and the limit is 000FFH. Notice the *limit* for the 80286 is a 16-bit limit and 20 bits for the 80386 through the Pentium. The 80286 accesses memory segments that are between 1 and 64K bytes in length. The 80386 through the Pentium access memory segments that are between 1 and 1M byte, or 4K and 4G bytes in length. With the 80386CX/80386EX embedded controllers, the length of the segment is either between 1 and 1M byte, or 4K and 64M bytes.

There is another feature found in the 80386 through the Pentium descriptor that is not found in the 80286 descriptor: the G bit or **granularity bit.** If G = 0, the limit specifies a segment limit of from 1 to 1M byte in length. If G = 1, the value of the limit is multiplied by 4K bytes (appended with 000H). If G = 1, the limit is any multiple of 4K bytes. This allows a segment length of 4K to 4G bytes in steps of 4K bytes. The reason that the segment length is 64K bytes in the 80286 is that the offset address is always 16 bits because of its 16-bit internal architecture. The 80386 and above use a **32-bit architecture,** which allows an offset address of 32 bits. This **32-bit offset address** allows segment lengths of 4G bytes and the **16-bit offset address** allows segment lengths of 64K bytes. In the 80386CX/80386EX embedded controllers, the address bits A26–A31 are not present and are zero in a descriptor.

The **AV bit** in the 80386–Pentium Pro descriptor is used by some operating systems to indicate that the segment is available (AV = 1) or not available (AV = 0). The **D bit** indicates how the 80386–Pentium instructions access register and memory data in the protected or real mode. If D = 0, the 80386–Pentium Pro assumes that the instructions are 16-bit instructions compatible with the 8086–80286. This means that the instructions use 16-bit offset addresses and 16-bit registers by

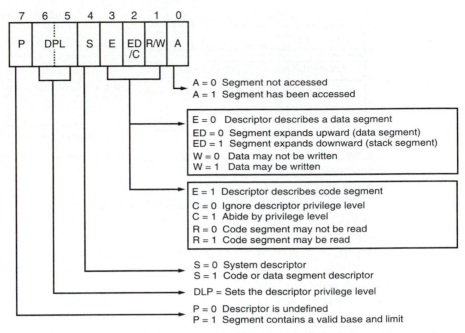

Note: Some of the letters used to describe the bits in the access rights bytes vary in Intel documentation.

FIGURE 2–11   The access rights byte for the 80486, Pentium, and Pentium Pro.

default. This mode is often called the **16-bit instruction mode.** If D = 1, the 80386–Pentium Pro assumes the instructions are 32-bit instructions. The **32-bit instruction mode** assumes all offset addresses are 32 bits as well as all registers by default. Note that default for register size and offset address size can be overridden in both the 16- and 32-bit instruction modes. Therefore, a 32-bit register can be addressed in the 16-bit instruction mode if it is prefixed with an override. MS-DOS and PC-DOS require that the instructions are always used in the 16-bit instruction mode. The Windows 3.X program also requires that the 16-bit instruction mode is selected. 32-bit instruction mode is only accessible in protected-mode systems such as Windows NT, Windows 95, or OS/2. (More detail on these modes and their application to the instruction set appears in chapters 3, 4, and 5.)

The **access rights byte,** shown in Figure 2–11, controls access to the memory segment. This byte describes how the segment functions in the system. Notice how the access rights byte exercises complete control over the segment. If the segment is a data segment, the direction of growth is specified. If the segment grows beyond its limit, the microprocessor's program is interrupted, indicating a general protection fault. These faults occasionally appear in poorly written applications running under Windows. A data segment can be specified as read/write or write-protected. The code segment is controlled in a similar fashion and can have reading inhibited to protect software.

Descriptors are chosen from the descriptor table by the segment register. Figure 2–12 shows how the segment register functions in the protected-mode system. The segment register contains a 13-bit selector field, a table selector bit, and a requested privilege level field. The 13-bit selector chooses one of the 8,192 descriptors from the descriptor table. The *TI bit* selects either the global descriptor table (TI = 0) or the local descriptor table (TI = 1). The **requested privilege level** (RPL) requests the access privilege level of a memory segment. The highest privilege level is 00 and the lowest is 11. If the requested privilege level matches or is higher in priority than the privilege level set by the access rights byte, access is granted. For example, if the requested privilege level is 10 and the access rights byte sets the segment privilege level at 11,

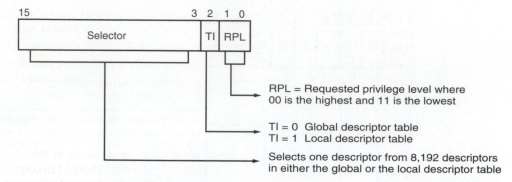

FIGURE 2–12   The contents of a segment register during protected-mode operation of the 80286, 80386, 80486, Pentium, and Pentium Pro microprocessors.

access is granted because 10 is higher in priority than privilege level 11. Privilege levels are used in multi-user environments. If the privilege level is violated, the system normally indicates a privilege violation.

Figure 2–13 shows how the segment register, containing a selector, chooses a descriptor from the global descriptor table. The entry in the global descriptor table selects a segment in the memory system. In this illustration DS contains 0008H, which accesses the descriptor number

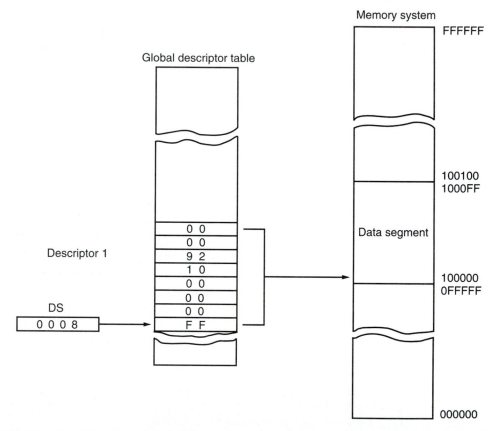

FIGURE 2–13   Using the DS register to select a descriptor from the global descriptor table. In this example, the DS register accesses memory locations 100000H–1000FFH as a data segment.

1 from the global descriptor table using a requested privilege level of 00. Descriptor number 1 contains a descriptor that defines the base address as 00100000H with a segment limit of 000FFH. This means that a value of 0008H loaded into DS causes the microprocessor to use memory locations 00100000H–001000FFH for the data segment with this example descriptor table. Note that descriptor zero is called the null descriptor, which must contain all zeros, and may not be used for accessing memory.

Program-Invisible Registers

The global and local descriptor tables are found in the memory system. In order to access and specify the address of these tables, the 80286 through the Pentium Pro contain **program-invisible registers.** The program-invisible registers are not directly addressed by normal software so they are called invisible. Some of these registers are accessed by the system software. Figure 2–14 illustrates the program-invisible registers as they appear in the 80286 through the Pentium Pro. These registers control how memory is accessed by the microprocessor or embedded controller when operated in the protected mode.

Each of the segment registers contains a program-invisible portion used in the protected mode. This is a **cache** memory that is loaded with the base address, limit, and access rights each time the number is changed in the segment register. (This is not the same cache used to cache instructions and data in the microprocessor-based system.) When a new segment number is placed in a segment register, the microprocessor accesses a descriptor table and loads the descriptor into the program-invisible cache portion of the segment register. It is held there and used to access the memory segment until the segment number is changed again. This allows the microprocessor to repeatedly access a memory segment without referring back to the descriptor table for each access.

The GDTR (**global descriptor table register**) and IDTR (**interrupt descriptor table register**) contain the base address of the descriptor table and its limit. The limit of each descriptor

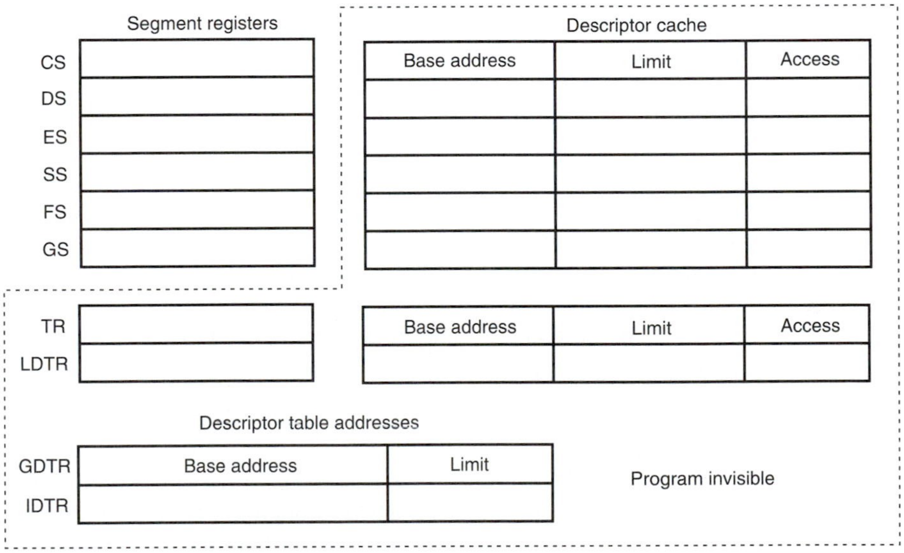

Notes:
1. The 80286 does not contain FS and GS nor the program-invisible portions of these registers.
2. The 80286 contains a base address that is 24 bits and a limit that is 16 bits.
3. The 80386/80486/Pentium contain a base address that is 32 bits and a limit that is 20 bits.
4. The access rights are 8 bits in the 80286 and 12 bits in the 80386/80486/Pentium.

FIGURE 2–14 The program-invisible register within the 80286, 80386, 80486, Pentium, and Pentium Pro microprocessors.

table is 16 bits in the 80286 through the Pentium Pro because the maximum table length is 64K bytes. When protected-mode operation is desired, the address of the global descriptor table and its limit are loaded into the GDTR. Before using protected mode, the interrupt descriptor table and the IDTR must also be initialized. (More detail is provided on protected-mode operation later.)

The location of the local descriptor table is selected from the global descriptor table. One of the global descriptors is set up to addresses in the local descriptor table. To access the local descriptor table, the LDTR **(local descriptor table register)** is loaded with a selector just as a segment register is loaded with a selector. This selector accesses the global descriptor table and loads the base address, limit, and access rights of the local descriptor table into the cache portion of the LDTR.

The **task register** (TR) holds a selector that accesses a descriptor that defines a task. A **task** is usually a procedure or application program. The descriptor for the program is stored in the global descriptor table so access can be controlled through the privilege levels. The task register allows a context or task switch in about 17 μs. Task switching allows the microprocessor or embedded controller to switch between tasks fairly quickly. The task switch allows multi-tasking systems to switch from one task to another in a simple and orderly fashion.

2–4   MEMORY PAGING

The **memory paging mechanism** located within the 80386 through the Pentium Pro allows any physical memory location to be assigned to any linear address. The **linear address** is defined as the address generated by a program. The **physical address** is the actual hardware memory address generated at the microprocessor's address bus connections. With the memory paging unit, the linear address is *invisibly* translated into a physical address. This allows an application written to function at a specific address to be relocated through the paging mechanism. It also allows a program to allocated memory in areas where no memory exists. An example is the **upper memory blocks** provided by EMM386.EXE.

The EMM386.EXE program reassigns extended memory, in 4K blocks, to the system memory between the video BIOS and the system BIOS ROMS to provide upper memory blocks. Without the paging mechanism, this would be impossible.

Because page translations access page translation tables in the memory system, additional time is required for the translation. Realizing that this impedes the microprocessor, Intel has included a page **translation look-aside buffer** (TLB) that caches the most recent translation. The TLB holds 22 entries in the 80486, and there are two in the Pentium and Pentium Pro that each hold 22 entries. The two TLBs in the Pentium and Pentium Pro are for the instruction cache and the data cache. If the page translation exists in the TLB, no access to the page translation tables is required, which speeds software execution.

Paging Registers

The **paging unit** is controlled by the contents of the control registers. Refer to Figure 2–15 for the contents of **control registers** CR0 through CR4. These registers are available only to the 80386 through the Pentium Pro, and only the Pentium and Pentium Pro contain CR4.

The registers that are important to the paging unit are CR0 and CR3. The leftmost bit (PG) position of CR0 selects paging when placed at a logic 1 level. If the PG bit is cleared (0), the linear address generated by the program becomes the physical address used to access memory. If the PG bit is set (1), the linear address is converted to a physical address through the paging mechanism. The paging mechanism functions in both real and protected modes.

The contents of CR3 contain the **page directory base address** and the PCD and PWT bits. The PCD **(processor cache disable)** and PWT **(processor cache write-through)** bits control

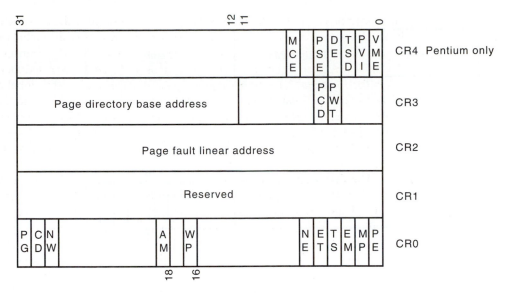

FIGURE 2–15 The control register structure of the microprocessor.

the operation of the PCD and PWT pins on the 80486, Pentium, and Pentium Pro. If PCD is set (1), the PCD pin becomes a logic 1 during bus cycles that are not paged. This allows the external hardware to control the cache memory. The PWT bit also appears on the PWT pin during bus cycles that are not paged to control the write-through cache in the system. The **page directory base address** locates the page directory for the page translation unit. Note that this address locates the page directory at any 4K boundary in the memory system because it is appended internally with a 000H. The **page directory** contains 1,024 directory entries of four bytes each. Each **page directory entry** addresses a page table that contains 1,024 entries.

The linear address, as it is generated by the software, is broken into three sections that dictate the page directory entry, page table entry, and page offset address. Figure 2–16 shows the linear address and its makeup for paging. Notice how the leftmost 10 bits address an entry in the page directory. The first entry in the page directory is accessed for linear addresses 00000000H–003FFFFFH or 4M bytes. This selects a page table that is indexed by the next 10 bits of the linear address (bit positions 12–21). This means that addresses 00000000H–00000FFFH select page directory entry 0 and page table entry 0. Notice that this is a 4K-byte address range. The offset part of the linear address (bit positions 0–11) select a byte in the 4K-byte memory page. If the page table entry 0 contains address 00100000H, then the physical address is 00100000H–00100FFFH for linear address 00000000H–00000FFFH. This means that when the program accesses a location between 00000000H and 00000FFFH, the microprocessor or embedded controller physically addresses location 00100000H–00100FFFH. In a similar fashion, any linear address can be re-paged to any physical address.

The Page Directory and Page Table

Figure 2–17 shows the page directory, a few page tables, and some memory pages. There is only ever one page directory in the system. The page directory contains 1,024 doubleword addresses that locate up to 1,024 page tables. The page directory and each page table are 4K bytes in length. If the entire 4G bytes of memory is paged, the system must allocate 4K bytes of memory for the page directory and 4K times 1,024 or 4M bytes for the 1,024 page tables. In the 80386CX/80386EX, the entire memory is paged with a page directory of 16 entries or 64 bytes and 16 page directories. Paging the entire 80386CX/80386EX requires 64 plus 64K bytes of memory to store the page directory and page tables.

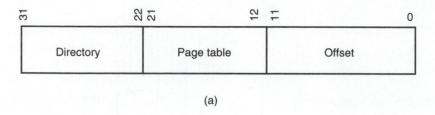

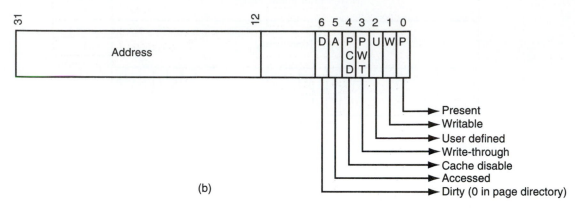

FIGURE 2–16   The format for the linear address (a) and a page directory or page table entry (b).

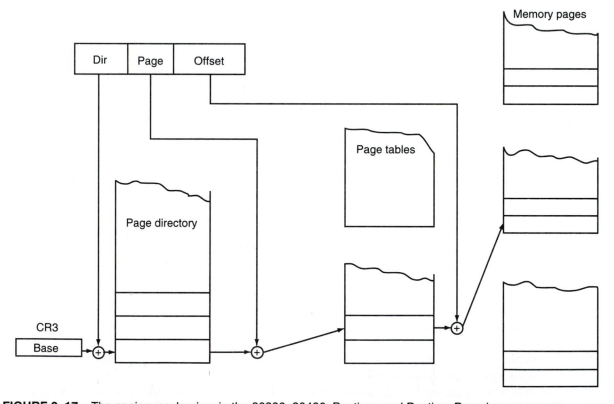

FIGURE 2–17   The paging mechanism in the 80386, 80486, Pentium, and Pentium Pro microprocessors.

FIGURE 2–18 The page directory, page table 0, and two memory pages. Notice how the address of page 000C8000–000C9000 has been moved to 00110000–00110FFF.

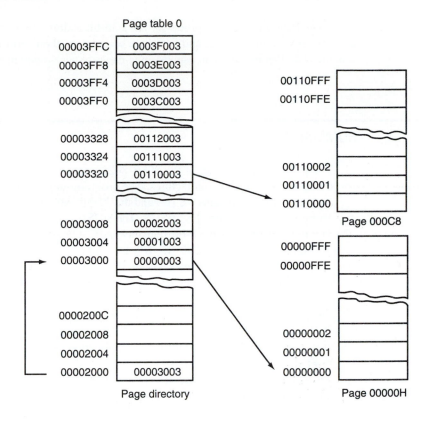

In a personal computer system, DOS uses page tables to redefine the area of memory between locations C8000H and EFFFFH as upper memory blocks. It does this by re-paging extended memory to backfill this part of the conventional memory system to allow DOS access to additional memory. Suppose that the EM386.EXE program allows access to 16M bytes of extended and conventional memory through paging and locations C8000H through EFFFFH must be re-paged to locations 110000 through 138000H with all other areas of memory paged to their normal locations. Such a scheme is depicted in Figure 2–18.

Here the page directory contains four entries. Recall that each entry in the page directory corresponds to 4M bytes of physical memory. The system also contains four page tables with 1,024 entries each. Recall that each entry in the page table re-pages 4K bytes of physical memory. This scheme requires a total of 16K of memory for the four page tables and 16 bytes of memory for the page directory.

As with DOS, the Windows program also re-pages the memory system. At present Windows 3.1 supports paging for only 16M bytes of memory because of the amount of memory required to store the page tables. In the Pentium and Pentium Pro microprocessors, pages can be either 4K or 4M bytes in length. Although no software currently supports the 4M-byte pages, no doubt in the future as the Pentium and Pentium Pro pervade the personal computer operating system, they will begin to support 4M-byte memory pages. (Remember that as of late 1995 the Pentium accounted for 40 percent of all personal computer sales.)

The 4M-byte pages are accessed by using only the page directory to address 4M-byte pages. This means that the linear address is divided into two parts. The leftmost 10 bits address an entry in the page directory. The entry in the page directory in turn addresses a 4M-byte memory page. The rightmost 22 bits of the linear address then address a location in the 4M-byte page.

The Pentium Pro microprocessor has a 36-bit address. In order to access information at addresses between the 4G-byte and 64G-byte boundaries, the Pentium Pro uses yet another paging scheme. The Pentium Pro can use 2M-byte pages to access any memory location.

2–5   SUMMARY

1. The programming model of the 8086–80286 contains 8- and 16-bit registers because these machines have 16-bit architectures. The programming model of the 80386, 80486, Pentium, and Pentium Pro contains 8-, 16-, and the 32-bit extended registers as well as two additional 16-bit segment registers: FS and GS. The 80386 and above are 32-bit architecture machines.

2. The register set contains 8-bit registers (AH, AL, BH, BL, CH, CL, DH, and DL); 16-bit registers (AX, BX, CX, DX, SP, BP, DI, and SI); segment registers (CS, DS, ES, SS, FS, and GS); and 32-bit extended registers (EAX, EBX, ECX, EDX, ESP, EBP, EDI, and ESI). In addition, the microprocessor contains an instruction pointer (IP/EIP) and flag register (FLAGS or EFLAGS).

3. The peripheral control block (PCB) allows the programming of devices located within the 80186 and 80188. The address of the PCB is by default at I/O ports FF00H through FFFFH, unless it is changed by modifying the relocation register at I/O port address FFFEH and FFFFH.

4. All real-mode memory addresses are a combination of a segment address plus an offset address. The starting location of a segment is defined by the 16-bit number in the segment register that is appended with a hexadecimal zero to its rightmost end. The offset address is a 16-bit number added to the 20-bit segment address to form the real-mode memory address.

5. All instructions (code) are accessed by the combination of CS (segment address) plus IP or EIP (offset address).

6. Data is normally referenced through a combination of the DS (data segment) and either an offset address or the contents of a register that contains the offset address. The microprocessors use BX, DI, and SI as default offset registers for data, if 16-bit registers are selected. The 80386 through the Pentium Pro can also use the 32-bit registers EAX, EBX, ECX, EDX, EDI, and ESI as default offset register for data.

7. Protected-mode operation allows memory above the first 1M byte to be accessed by the 80286 through the Pentium Pro. This extended memory system (XMS) is accessed via a segment address plus an offset address just as in the real mode. The difference is that the segment address is not held in the segment register. In the protected mode, the segment starting address is stored in a descriptor that is selected by the segment register.

8. A protected-mode descriptor contains a base address, limit, and access rights byte. The base address locates the starting address of the memory segment. The limit defines the last location of the segment. The access rights byte defines how the memory segment is accessed via a program. The 80286 allows a memory segment to start at any of its 16M bytes of memory using a 24-bit base address. The 80386/80486/Pentium/Pentium Pro allow a memory segment to begin at any of their 4G bytes of memory using a 32-bit base address. The limit is a 16-bit number in the 80286 and a 20-bit number in the 80386–Pentium Pro. This allows an 80286 memory segment limit of 64K bytes and an 80386–Pentium Pro memory segment limit of either 1M byte (G = 0) or 4G bytes (G = 1).

9. The segment register contains three fields of information in the protected mode. The left-most 13 bits of the segment register address one of 8,192 descriptors from a descriptor table. The TI bit accesses either the global descriptor table (TI = 0) or the local descriptor table (TI = 1). The rightmost 2 bits of the segment register select the requested priority level for the memory segment access.

10. The program-invisible registers are used by the 80286–Pentium Pro to access the descriptor tables. Each segment register contains a cache portion that is used in protected mode to hold the base address, limit, and access rights acquired from a descriptor. The cache allows the microprocessor to access the memory segment without referring again to the descriptor table until the segment register's contents are changed.

11. A memory page is 4K bytes in length. The linear address, generated by a program, can be mapped to any physical address through the paging mechanism found within the 80386 through the Pentium Pro.

12. Memory paging is accomplished through control registers CR0 and CR3. The PG bit of CR0 enables paging, and the contents of CR3 addresses the page directory. The page directory contains up to 1,024 page table addresses used to access paging tables. The page table contains 1,024 entries that locate the physical address of a 4K-byte memory page.

2–6   QUESTIONS AND PROBLEMS

1. What are program-visible registers?
2. The 80186 addresses registers that are 8- and _____ - bits wide.
3. The extended registers are addressable by which microprocessors?
4. How is the extended BX register addressed?
5. Which register holds a count for some instructions?
6. What is the purpose of the IP/EIP register?
7. The carry flag bit is set by which arithmetic operations?
8. Will an overflow occur if a signed FFH is added to a signed 01H?
9. A number that contains 3 one bits is said to have _____ parity.
10. Which flag bit controls the INTR pin on the microprocessor?
11. Which microprocessors contain an FS segment register?
12. What is the purpose of a segment register in the real-mode operation of the microprocessor?
13. What is the purpose of the peripheral control block in the 80186 and 80188 embedded controllers?
14. Where is the PCB located, when power is first applied to the 80186 and 80188?
15. In the real mode, show the starting and ending addresses of each segment located by the following segment register values:
    (a) 1000H
    (b) 1234H
    (c) 2300H
    (d) E000H
    (e) AB00H
16. Find the memory address of the next instruction executed by the microprocessor, when operated in the real mode, for the following CS:IP combinations:
    (a) CS = 1000H and IP = 2000H
    (b) CS = 2000H and IP = 1000H
    (c) CS = 2300H and IP = 1A00H
    (d) CS = 1A00H and IP = B000H
    (e) CS = 3456H and IP = ABCDH
17. Real-mode memory addresses allow access to memory below which address?
18. Which register or registers are used as an offset address for string instruction destinations in the 80486 microprocessor?
19. Which 32-bit register or registers are used as an offset address for data segment data in the 80386 microprocessor?

20. The stack memory is addressed by a combination of the _____ segment plus _____ offset.

21. If the base pointer (BP) addresses memory, the _____ segment contains the data.

22. Determine the memory location addressed by the following 80186 register combinations:
    (a) DS = 1000H and DI = 2000H
    (b) DS = 2000H and SI = 1002H
    (c) SS = 2300H and BP = 3200H
    (d) DS = A000H and BX = 1000H
    (e) SS = 2900H and SP = 3A00H

23. Determine the memory location addressed by the following real-mode 80386 register combinations:
    (a) DS = 2000H and EAX = 00003000H
    (b) DS = 1A00H and ECX = 00002000H
    (c) DS = C000H and ESI = 0000A000H
    (d) SS = 8000H and ESP = 00009000H
    (e) DS = 1239H and EDX = 0000A900H

24. Protected-mode memory addressing allows access to which area of the memory in the 80286 microprocessor?

25. Protected-mode memory addressing allows access to which area of the memory in the Pentium microprocessor?

26. What is the purpose of the segment register in protected-mode memory addressing?

27. How many descriptors are accessible in the global descriptor table in the protected mode?

28. For an 80286 descriptor that contains a base address of A00000H and a limit of 1000H, what starting and ending locations are addressed by this descriptor?

29. For an 80486 descriptor that contains a base address of 01000000H, a limit of 0FFFFH, and G = 0, what starting and ending locations are addressed by this descriptor?

30. For a Pentium descriptor that contains a base address of 00280000H, a limit of 00010H, and G = 1, what starting and ending locations are addressed by this descriptor?

31. How many local descriptors are possible in the global descriptor table?

32. If the DS register contains 0020H in a protected-mode system, which global descriptor table entry is accessed?

33. If DS = 0103H, in a protected-mode system, the requested privilege level is _____.

34. If DS = 0105H, in a protected-mode system, which entry, table, and requested privilege level are selected?

35. What is the maximum length of the global descriptor table in the Pentium microprocessor?

36. Code a descriptor that describes a memory segment that begins at location 210000H and ends at location 21001FH. This memory segment is a code segment that can be read. The descriptor is for an 80286 microprocessor.

37. Code a descriptor that describes a memory segment that begins at location 03000000H and ends at location 05FFFFFFH. This memory segment is a data segment that grows upward in the memory system and can be written. The descriptor is for an 80386 microprocessor.

38. Which register locates the global descriptor table?

39. How is the local descriptor table addressed in the memory system?

40. Describe what happens when a new number is loaded into a segment register when the microprocessor is operated in the protected mode.

41. What are the program-invisible registers?

42. What is the purpose of the GDTR?

43. How many bytes are found in a memory page?

44. What register is used to enable the paging mechanism in the 80386, 80486, and Pentium?

45. How many 32-bit addresses are stored in the page directory?

46. Each entry in the page directory translates how much linear memory into physical memory?

47. If the microprocessor sends linear address 00200000H to the paging mechanism, which paging directory entry is accessed, and which page table entry is accessed?

48. What value is placed in the page table to redirect linear address 20000000H to 30000000H?

49. How much memory is required to completely re-page the 80386CX/80386EX memory system?

50. Which microprocessor can address memory in 4M-byte pages?

51. Use the Internet to locate the Texas Instruments Web page and write a report that details the types of memory devices available there.

52. Use the Internet to locate the Intel Web page and list the types of embedded microprocessors/controllers available there.

53. Use the Internet to locate the AMD Web page and list the types of embedded microprocessors/controllers available there.

54. Use the Internet to find a Web site that lists the facts about the Intel microprocessor and write a paper that contrasts at least three family members.

CHAPTER 3

Addressing Modes

INTRODUCTION

Efficient software development for the microprocessor and embedded controller requires a complete familiarity with the addressing modes employed by each instruction. In this chapter, the MOV (move data) instruction is used to describe the data-addressing modes. The MOV instruction transfers bytes or words of data between registers or between registers and memory in the 8086 through the 80286 and bytes, words, or doublewords in the 80386 and above. In describing the program memory-addressing modes, we see how the call and jump instructions modify the flow of the program.

The data-addressing modes include: register, immediate, direct, register indirect, base-plus-index, register relative, and base relative-plus-index in the 8086 through the 80286 microprocessor. The 80386 and above also include a scaled-index mode of addressing memory data. The program memory-addressing modes include: program relative, direct, and indirect. The operation of the stack memory is explained so that you will understand PUSH and POP instructions.

CHAPTER OBJECTIVES

Upon completion of this chapter, you will be able to:

1. Explain the operation of each data-addressing mode.
2. Use the data-addressing modes to write assembly language statements.
3. Explain the operation of each program memory-addressing mode.
4. Use the program memory-addressing modes to write assembly and machine language statements.
5. Select the appropriate addressing mode to accomplish a given task.
6. Explain the difference between addressing memory data using real-mode and protected-mode operations.
7. Describe the sequence of events that places data onto the stack and removes data from the stack.
8. Explain how a data structure is placed in memory and used with software.

3–1 DATA-ADDRESSING MODES

Because MOV is a common, flexible instruction, it will be our basis for understanding data-addressing modes. Figure 3–1 illustrates the MOV instruction and defines the direction of data flow. The source is to the right and the destination is to the left, next to the opcode MOV. (An **opcode** or operation code tells the microprocessor which operation to perform.) This direction of flow, which is applied to all instructions, is at first awkward. We naturally assume things move from left to right, but here they move from right to left. Notice that a comma always separates the destination from the source in an instruction. Also note that memory-to-memory transfers are not allowed by any instruction except for the MOVS instruction.

In Figure 3–1, the MOV AX,BX instruction transfers the word contents of the source register (BX) into the destination register (AX). The source never changes, but the destination almost always changes.[1] It is essential to remember that a MOV instruction always *copies* the source data into the destination. The MOV never actually picks up the data and *moves* it. Also note that the flag register remains unaffected by most data transfer instructions.

Figure 3–2 shows all possible variations of the data-addressing modes using the MOV instruction. This illustration helps to show how each data-addressing mode is formulated with the MOV instruction and also serves as a reference. Note that these are the same data-addressing modes found in all versions of the Intel microprocessor except for the scaled-index addressing mode, which is found only in the 80386 through the Pentium Pro. The following is a list of data-addressing modes:

Register Addressing	Transfers a copy of a byte or word from the source register or memory location to the destination register or memory location. (For example, the MOV CX,DX instruction copies the word-sized contents of register DX into register CX.) In the 80386 and above, a doubleword can be transferred from the source register or memory location to the destination register or memory location. For example, the MOV ECX,EDX instruction copies the doubleword-sized contents of register EDX into register ECX.
Immediate Addressing	Transfers the source of immediate byte or word of data into the destination register or memory location. For example, the MOV AL,22H instruction copies a byte-sized 22H into register AL. In the 80386 and above, a doubleword of immediate data can be transferred into a register or memory location. For example, the MOV EBX,12345678H instruction copies a doubleword-sized 12345678H into the 32-bit wide EBX register.
Direct Addressing	Moves a byte or word between a memory location and a register. The instruction set does not support a memory-to-memory transfer, except

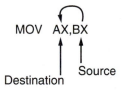

FIGURE 3–1 The MOV instruction showing the source, destination, and direction of data flow.

MOV AX,BX

Destination Source

---

[1]The exceptions are the CMP and TEST instructions that never change the destination. These instructions are described in later chapters.

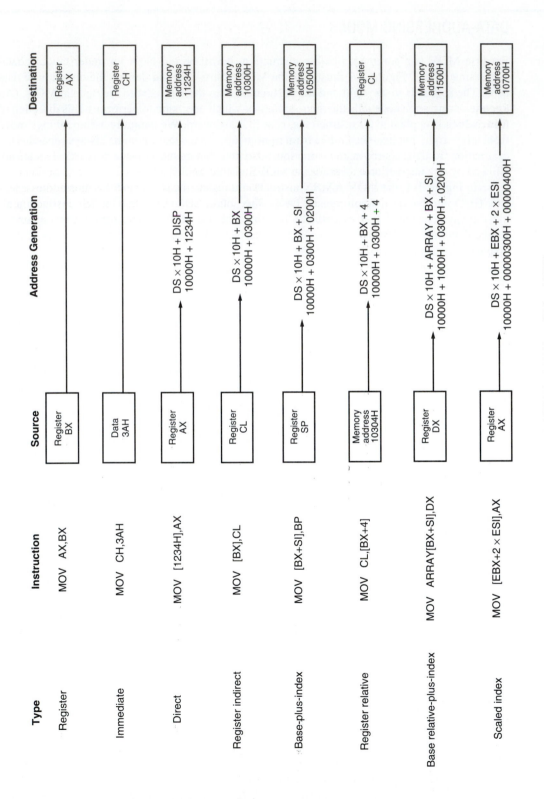

FIGURE 3–2 8086–Pentium data-addressing modes.

76

for the MOVS instruction. For example, the MOV CX,LIST instruction copies the word-sized contents of memory location LIST into register CX. In the 80386 and above, a doubleword-sized memory location can also be addressed. For example, the MOV ESI,LIST instruction copies a 32-bit number, stored in 4 consecutive bytes of memory, from location LIST into register ESI.

Register-Indirect Addressing   Transfers a byte or word between a register and a memory location addressed by an index or base register. The index and base registers are: BP, BX, DI, and SI. For example, the MOV AX,[BX] instruction copies the word-sized data from the data segment offset address indexed by BX into register AX. In the 80386 and above, a byte, word, or double-word is transferred between a register and a memory location addressed by any register: EAX, EBX, ECX, EDX, EBP, EDI, or ESI. For example; the MOV AL,[ECX] instruction loads AL from the data segment offset address selected by the contents of ECX.

Base-Plus-Index Addressing   Transfers a byte or word between a register and the memory location addressed by a base register (BP or BX) plus an index register (DI or SI). For example, the MOV [BX+DI],CL instruction copies the byte-sized contents of register CL into the data segment memory location addressed by BX plus DI. In the 80386 and above, any of the following registers may be combined to generate the memory address: EAX, EBX, ECX, EDX, EBP, EDI, or ESI. For example, the MOV [EAX+EBX],CL instruction copies the byte-sized contents of register CL into the data segment memory location addressed by EAX plus EBX.

Register-Relative Addressing   Moves a byte or word between a register and the memory location addressed by an index or base register plus a displacement. For example, the MOV AX,[BX+4] instruction loads AX from the data segment address formed by BX plus 4. The MOV AX,ARRAY[BX] instruction loads AX from the data segment memory location in ARRAY plus the contents of BX. The 80386 and above use any register to address memory. For example, the MOV AX,[ECX+4] instruction loads AX the data segment address formed by ECX plus 4. The MOV AX,ARRAY[EBX] instruction loads AX from the data segment memory location ARRAY plus the contents of EBX.

Base Relative-Plus-Index Addressing   Transfers a byte or word between a register and the memory location addressed by a base and an index register plus a displacement. For example, the MOV AX,ARRAY[BX+DI] or MOV AX,[BX+DI+4] instructions both load AX from a data segment memory location. The first instruction uses an address formed by adding ARRAY, BX, and DI and the second by adding BX, DI, and 4. For 80386 and above MOV EAX,ARRAY[EBX+ECX] loads EAX from the data segment memory location accessed by the sum of ARRAY, EBX, and ECX.

Scaled-Index Addressing   Available only in the 80386 through the Pentium Pro microprocessor. The second register of a pair of registers is modified by the scale factor of 2X, 4X, or 8X to generate the operand memory address. For example, a MOV EDX,[EAX+4*EBX] instruction loads EDX from the data segment memory location addressed by EAX plus 4 times EBX. Scaling allows access to word (2X), doubleword (4X), or quadword (8X) memory array data. Note that a scaling factor of 1X also exists, but it is normally implied and does not appear in the instruction. The MOV AL,[EBX+ECX] instruction is an example where the scaling factor is a one. Alternately, the instruction can be rewritten as MOV AL,[EBX+1*ECX]. Another example is a MOV AL,[2*EBX] instruction, which uses only one scaled register to address memory.

Register Addressing

Register addressing is the most basic form of data addressing, once the register names are learned. The microprocessor contains the following 8-bit registers used with register addressing: AH, AL, BH, BL, CH, CL, DH, and DL. Also present are the following 16-bit registers: AX, BX, CX, DX, SP, BP, SI, and DI. In the 80386 and above, the extended 32-bit registers are: EAX, EBX, ECX, EDX, ESP, EBP, EDI, and ESI. Some MOV instructions and the PUSH and POP instructions address the 16-bit segment registers (CS, ES, DS, SS, FS, and GS) with register addressing. It is important that instructions to use registers are the same size. Never mix an 8-bit register with a 16-bit register, an 8-bit register with a 32-bit register, or a 16-bit register with a 32-bit register because this is not allowed by the microprocessor and results in an error when assembled. This is true even when a MOV AX,AL or a MOV EAX,AL instruction may seem to make sense. Of course the MOV AX,AL or MOV EAX,AL instruction is not allowed because these registers are of different sizes. Note that a few instructions, such as SHL DX,CL, take exception to this rule, as discussed in later chapters. It is also important to note that none of the MOV instructions affect the flag bits.

Table 3–1 shows many versions of register move instructions. It is impossible to show all variations because there are so many possible combinations. For example, just the 8-bit subset of the MOV instruction has 64 variations. About the only type of register MOV instruction not allowed is a segment-to-segment register MOV instruction. Also note that the code segment register may not be changed by a MOV instruction because the address of the next instruction is found in both IP/EIP and CS. If only CS is changed, the address of the next instruction is unpredictable. Therefore, changing the CS register with a MOV is not allowed.

Figure 3–3 shows the operation of a MOV BX,CX instruction. Note that the source register's contents do not change, but the destination register's contents do change. This instruction moves a 1234H from register CX into register BX. This erases the old contents (76AFH) of register BX, but the contents of CX remain unchanged. The contents of the destination register or destination

TABLE 3–1 Examples of the register-addressed instructions

Assembly Language	Operation
MOV AL,BL	Copies BL into AL
MOV CH,CL	Copies CL into CH
MOV AX,CX	Copies CX into AX
MOV SP,BP	Copies BP into SP
MOV DS,AX	Copies AX into DS
MOV SI,DI	Copies DI into SI
MOV BX,ES	Copies ES into BX
MOV ECX,EBX	Copies EBX into ECX
MOV ESP,EDX	Copies EDX into ESP
MOV ES,DS	Not allowed (segment-to-segment)
MOV BL,DX	Not allowed (mixed sizes)
MOV CS,AX	Not allowed (the code segment register may be used only as the destination)

FIGURE 3–3 The effect of executing the MOV BX,CX instruction at the point just before the BX register changes. Note that only the rightmost 16 bits of register EBX change.

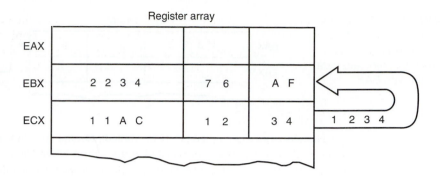

memory location change for all instructions except the CMP and TEST instructions. The MOV BX,CX instruction does not affect the leftmost 16 bits of register EBX.

Example 3–1 shows a short sequence of assembled instructions that copy various data between 8-, 16-, and 32-bit registers. As we mentioned, the act of moving data from one register to another only changes the destination register, never the source. The last instruction (MOV CS,AX) in this example assembles without error, but will cause problems if executed. If only the contents of CS change without changing IP, the next step in the program is unknown and therefore causes the program to go awry.

EXAMPLE 3–1

```
0000 8B C3        MOV   AX,BX      ;copy contents of BX into AX
0002 8A CE        MOV   CL,DH      ;copy the contents of DH into CL
0004 8A CD        MOV   CL,CH      ;copy the contents of CH into CL
0006 66| 8B C3    MOV   EAX,EBX    ;copy the contents of EBX into EAX
0009 66| 8B D8    MOV   EBX,EAX    ;copy EAX into EBX, ECX, and EDX
000C 66| 8B C8    MOV   ECX,EAX
000F 66| 8B D0    MOV   EDX,EAX
0012 8C C8        MOV   AX,CS      ;copy CS into DS
0014 8E D8        MOV   DS,AX
0016 8E C8        MOV   CS,AX      ;assembles, but will cause problems
```

Immediate Addressing

Another data-addressing mode is **immediate addressing.** The term *immediate* implies that the data immediately follow the hexadecimal opcode in the memory. Also note that immediate data is constant data, while the data transferred from a register is variable data. Immediate addressing operates upon a byte or word of data. In the 80386 through the Pentium Pro immediate addressing also operates on doubleword data. The MOV immediate instruction transfers a copy of the immediate data into a register or a memory location. Figure 3–4 shows the operation of a MOV EAX,13456H instruction. This instruction copies the 13456H from the instruction, located in the memory immediately following the hexadecimal opcode, into register EAX. As with the MOV instruction illustrated in Figure 3–3, the source data overwrites the destination data.

In symbolic assembly language the symbol # precedes immediate data in some assemblers[2]. The MOV AX,#3456H instruction is an example. Most assemblers *do not* use the # symbol but represent immediate data as in the MOV AX,3456H instruction. In this book the # symbol is not used for immediate data. The most common assemblers, Intel ASM, Microsoft MASM[3], and Borland TASM[4] do not use the # symbol for immediate data, but an older assembler used with the HP64000 logic development system does, as may others.

---

[2]This is true for the assembler provided with the HP64100 logic development system manufactured by Hewlett-Packard, Inc.

[3]MASM (macro assembler) is a trademark of Microsoft Corporation.

[4]TASM (turbo assembler) is a trademark of Borland Corporation.

FIGURE 3–4 The operation of the MOV EAX,13456H instruction. This instruction copies the immediate data (13456H) into EAX.

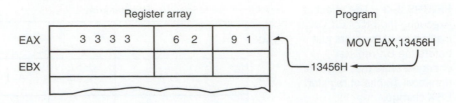

The symbolic assembler portrays immediate data in many ways. The *letter H* appends hexadecimal data. If hexadecimal data begins with a letter, the assembler requires that it start with a 0. For example, to represent a $F2_{16}$, a 0F2H is used in assembly language. In some assemblers, but not MASM or TASM or this book, hexadecimal data is represented with an 'h as in MOV AX,#'h1234. **Decimal data** is represented as is and requires no special codes or adjustments. An example is the 100 decimal in the MOV AL,100 instruction. An **ASCII-coded character** or characters may be depicted in the immediate form if the ASCII data are enclosed in apostrophes. An example is the MOV BH,'A' instruction, which moves an ASCII-coded A (41H) into register BH. Be careful to use the apostrophe (') for ASCII data and not the single quotation mark ('). **Binary data** is represented if the binary number is followed by the letter B, or, in some assemblers, the letter Y. Table 3–2 shows many variations of MOV instructions that apply immediate data.

Example 3–2 shows various immediate instructions in a short program that places a 0000H into the 16-bit registers AX, BX, and CX. This is followed by an instruction that uses register addressing to copy the contents of AX into registers SI, DI, and BP. This is a complete program that uses **programming models** for assembly and execution. The **.MODEL TINY statement** directs the assembler to assemble the program into a single segment. The **.CODE statement** or directive indicates the start of the code segment; the **.STARTUP statement** indicates the starting instruction in the program; the **.EXIT statement** causes the program to exit to DOS. The END statement indicates the end of the program file. This program can be assembled with MASM and executed with CodeView (CV). To store the program into the system, use either the DOS EDIT program or Programmer's WorkBench (PWB). Note that a TINY program assembles as a command (.COM) program.

EXAMPLE 3–2

```
                          .MODEL TINY      ;choose single segment model
0000                      .CODE            ;indicate start of code segment

                          .STARTUP         ;indicate start of program

0100  B8 0000             MOV   AX,0       ;place 0000H into AX
0103  BB 0000             MOV   BX,0000H   ;place 0000H into BX
0106  B9 0000             MOV   CX,0       ;place 0000H into CX

0109  8B F0               MOV   SI,AX      ;copy AX into SI
010B  8B F8               MOV   DI,AX      ;copy AX into DI
010D  8B E8               MOV   BP,AX      ;copy AX into BP

                          .EXIT            ;exit to DOS
                          END              ;end of file
```

Each **statement** in a program consists of four parts or fields as illustrated in Example 3–3. The leftmost field is called the **label** and is used to store a symbolic name for the memory location that it represents. All labels begin with a letter or one of the following special characters: @, $, _, or ?. A label may be any length from 1 to 35 characters. The label appears in a program to identify the name of a memory location for storing data and for other purposes that we explain as

TABLE 3–2 Examples of immediate addressing using the MOV instruction

Assembly Language	Operation
MOV BL,44	Moves a 44 decimal (2CH) into BL
MOV AX,44H	Moves a 44 hexadecimal into AX
MOV SI,0	Moves a 0000H into SI
MOV CH,100	Moves a 100 (64H) into CH
MOV AL,'A'	Moves an ASCII A (41H) into AL
MOV AX,'AB'	Moves an ASCII BA* (4241H) into AX
MOV CL,11001110B	Moves a binary 11001110 into CL
MOV EBX,12340000H	Moves a 12340000H into EBX
MOV ESI,12	Moves a 12 decimal into ESI
MOV EAX,100Y	Moves a 100 binary into EAX

Note: This is not an error; the ASCII characters are stored as a BA; be careful when using a word-sized pair of ASCII characters.

they appear. The next field is called the **opcode field** and is designed to hold the instruction or opcode. The MOV instruction is an example of an opcode. To the right of the opcode field is the **operand** field that contains information used by the opcode. For example, the MOV AL,BL instruction has the opcode MOV and operands AL and BL. Note that some instructions contain between zero and three operands. The final field, the **comment** field, contains a comment about an instruction or a group of instructions. A comment always begins with a semicolon (;).

EXAMPLE 3–3

```
LABEL       OPCODE   OPERAND    COMMENT

DATA1       DB       23H        ;define DATA1 as a byte of 23H
DATA2       DW       1000H      ;define DATA2 as a word of 1000H

START:      MOV      AL,BL      ;copy BL into AL
            MOV      BH,AL      ;copy AL into BH
            MOV      CX,200     ;copy 200 decimal into CX
```

When the program is assembled and the list (.LST) file is viewed, it appears as the program listed in Example 3–2. The hexadecimal number at the far left is the **offset address** of the instruction or data. This number is generated by the assembler. The number or numbers to the right of the offset address are the **machine-coded instructions** or **data** that is also generated by the assembler. For example, if the instruction MOV AX,0 appears in a file and it is assembled, it appears in memory location 0100 in Example 3–2. Its hexadecimal machine language form is B8 0000. When the program was written, only the MOV AX,0 was typed into the editor, the assembler generated the machine code and address and stored the program in a file ending with the extension .LST. Note that all programs shown in this text are in the form generated by the assembler.

Direct-Data Addressing

Most instructions can use the **direct-data addressing** mode. In fact, direct-data addressing is applied to many instructions in a typical program. There are two basic forms of direct-data addressing: (1) direct addressing that applies to a MOV between a memory location and AL, AX, or EAX; and (2) displacement addressing for almost any instruction in the instruction set. In either case, the address is formed by adding the displacement to the default data segment address or an alternate segment address.

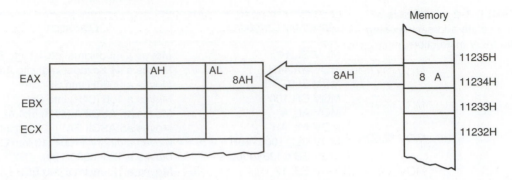

FIGURE 3–5   The operation of the MOV AL,[1234H] instruction when DS = 1000H.

Direct Addressing.   **Direct addressing** with a MOV instruction transfers data between a memory location, located within the data segment, and the AL (8-bit), AX (16-bit), or EAX (32-bit) register. This instruction is usually a 3-byte long instruction. (In the 80386 and above, a register size prefix may appear before the instruction causing it to exceed 3 bytes in length.)

The MOV AL,DATA instruction, as represented by most assemblers, loads AL from data segment memory location DATA (1234H). Memory location DATA is a symbolic memory location, while the 1234H is the actual hexadecimal location. With many assemblers this instruction is represented as a MOV AL,[1234H].[5] The [1234H] is an absolute memory location that is not allowed by all assembler programs. Note that this may need to be formed as MOV AL,DS:[1234H] with some assemblers to show that the address is in the data segment. Figure 3–5 shows how this instruction transfers a copy of the byte-sized contents of memory location 11234H into AL. The effective address is formed by adding 1234H (the offset address) to 10000H (the data segment address) in a system operating in the real mode.

Table 3–3 lists the three forms of direct addressed instructions. Because these instructions often appear in programs, Intel decided to make them special 3-byte long instructions to reduce the length of programs. All other instructions that move data from a memory location

TABLE 3–3   Direct addressed instructions using EAX, AX and AL

Assembly Language	Operation
MOV AL,NUMBER	Copies the byte contents of data segment memory address NUMBER into AL
MOV AX,COW	Copies the word contents of data segment memory address COW into AX
MOV EAX,WATER	*Copies the doubleword contents of data segment memory WATER into EAX
MOV NEWS,AL	Copies AL into data segment memory address NEWS
MOV THERE,AX	Copies AX into data segment memory address THERE
MOV HOME,EAX	*Copies EAX into data segment memory address HOME

*Note: The 80386/80486/Pentium/Pentium Pro microprocessors will at times use more than 3 bytes of memory for the 32-bit move between EAX and memory.

---

[5]This form may be used with MASM, but most often appears when a program is entered or listed by DEBUG, a debugging tool provided with the disk operating system.

TABLE 3–4 Examples of direct-data addressing using a displacement

Assembly Language	Operation
MOV CH,DOG	Loads CH with the contents of data segment memory location DOG
MOV CH,[1000H]	*Loads CH with the contents of data segment memory location 1000H
MOV ES,DATA6	Loads ES with the word-sized contents of data segment memory location DATA6
MOV DATAS,BP	BP is copied into data segment memory location DATAS
MOV NUMBER,SP	SP is copied into data segment memory location NUMBER
MOV DATA1,EAX	EAX is copied into data segment memory location DATA1
MOV EDI,SUM1	Loads EDI with the contents of data segment memory location SUM1

*Note: This form of addressing is seldom used with most assemblers because an actual numeric offset address is rarely accessed.

to a register, called *displacement-addressed instructions,* require 4 or more bytes of memory for storage in a program.

Displacement Addressing. **Displacement addressing** is almost identical to direct addressing except the instruction is 4 bytes wide instead of 3. In the 80386 through the Pentium Pro, this instruction can be up to 7 bytes wide if a 32-bit register and a 32-bit displacement are specified. This type of direct-data addressing is much more flexible because most instructions use it.

If you compare the operation of the MOV CL,[1234H] instruction to that of the MOV AL,[1234H] instruction of Figure 3–5, you'll see that both perform basically the same operation except for the destination register (CL versus AL). Another difference becomes apparent only upon examination of the assembled versions of these two instructions. The MOV AL,[1234H] instruction is 3 bytes in length and the MOV CL,[1234H] instruction is 4 bytes, as illustrated in Example 3–4. This example shows how the assembler converts these two instructions into hexadecimal machine language.

EXAMPLE 3–4

```
0000 A0 1234 R          MOV  AL,[1234H]
0003 8A 0E 1234 R       MOV  CL,[1234H]
```

Table 3–4 lists some MOV instructions using the displacement form of direct addressing. Not all variations are listed because there are many MOV instructions of this type. Please notice that the segment registers can be stored or loaded from memory.

Example 3–5 is a short program using models that address information in the data segment. The data segment begins with a **.DATA statement** to inform the assembler where the data segment begins. The model size is adjusted from TINY, as in Example 3–3, to SMALL so that a data segment can be included. The **SMALL model** allows one data segment and one code segment. The SMALL model is often used whenever memory data is required for a program. A SMALL model program assembles as an execute (.EXE) program. Notice how this example allocates memory locations in the data segment using the DB and DW directives. Here the .STARTUP statement not only indicates the start of the code, but it also loads the data segment register with the address of the data segment. If this program is assembled and executed with CodeView, the instructions can be viewed as they execute and change registers and memory locations.

EXAMPLE 3–5

```
                        .MODEL SMALL        ;select SMALL model
0000                    .DATA               ;indicate start of DATA segment

0000 10        DATA1    DB   10H            ;place 10H in DATA1
0001 00        DATA2    DB   0              ;place 0 in DATA2
0002 0000      DATA3    DW   0              ;place 0 in DATA3
0004 AAAA      DATA4    DW   0AAAAH         ;place AAAAH in DATA4

0000                    .CODE               ;indicate start of CODE segment
                        .STARTUP            ;indicate start of program

0017 A0 0000 R          MOV  AL,DATA1       ;copy DATA1 to AL
001A 8A 26 0001 R       MOV  AH,DATA2       ;copy DATA2 to AH
001E A3 0002 R          MOV  DATA3,AX       ;save AX at DATA3
0021 8B 1E 0004 R       MOV  BX,DATA4       ;load BX with DATA4

                        .EXIT               ;exit to DOS
                        END                 ;end file
```

Register-Indirect Addressing

Register-indirect addressing allows data to be addressed at any memory location through the offset address held in any of the following registers: BP, BX, DI, and SI. For example, if register BX contains a 1000H and the MOV AX,[BX] instruction executes, the word contents of data segment memory offset address 1000H is copied into register AX. If the microprocessor is operated in the real mode and DS = 0100H, this instruction addresses a word stored at memory bytes 2000H and 2001H and transfers them into register AX (see Figure 3–6). Note that the contents of 2000H are moved into AL and the contents of 2001H are moved into AH. The [ ] symbols denote indirect addressing in assembly language. In addition to using the BP, BX, DI, and SI registers to indirectly address memory, the 80386 and above allow register-indirect addressing with any extended register except ESP. Some typical instructions using indirect addressing appear in Table 3–5.

The **data segment** is used by default with register-indirect addressing or any other addressing mode that uses BX, DI, or SI to address memory. If register BP addresses memory, the

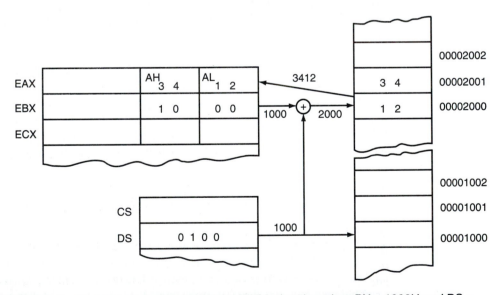

FIGURE 3–6   The operation of the MOV AX,[BX] instruction when BX = 1000H and DS = 0100H. Note that this instruction is shown after the contents of memory are transferred to AX.

TABLE 3–5 Examples of instructions using register-indirect addressing

Assembly Language	*Operation
MOV CX,[BX]	CX is loaded with a word from the data segment memory location addressed by BX
MOV [BP],DL	A byte is copied from register DL into the stack segment memory location addressed by BP
MOV [DI],BH	A byte is copied from register BH into the data segment memory location addressed by DI
MOV [DI],[BX]	Memory-to-memory transfers are not allowed except with the string instructions
MOV DI,[DI]	This instruction is not allowed because the register used to indirectly address memory may not be changed by the instruction
MOV AL,[EDX]	AL is loaded from the data segment memory location addressed by EDX
MOV ECX,[EBX]	ECX is loaded with a doubleword from the data segment memory location addressed by EBX

*Note: Data addressed by BP or EBP is by default located in the stack segment, while all other indirect addressing modes use the data segment by default.

stack segment is used by default. These are considered the default settings for these four index and base registers. For the 80386 and above, EBP addresses memory, by default, in the stack segment while EAX, EBX, ECX, EDX, EDI, and ESI address memory in the data segment by default. When using a 32-bit register to address memory in the real mode, the contents of the 32-bit register must never exceed 0000FFFFH. In the protected mode, any value can be used in a 32-bit register used to indirectly address memory as long as it does not access a location outside of the segment as dictated by the access rights byte. For example, the 80386–Pentium Pro instruction MOV EAX,[EBX] loads EAX with the doubleword-sized number stored at the data segment offset address indexed by EBX.

In some cases, indirect addressing requires that the size of the data are specified with the special assembler directive BYTE PTR, WORD PTR, or DWORD PTR. These directives indicate the size of the memory data addressed by the memory pointer (PTR). For example, the MOV AL,[DI] instruction is clearly a byte-sized move instruction, but the MOV [DI],10H instruction is ambiguous. Does the MOV [DI],10H instruction address a byte-, word-, or doubleword-sized memory location? The assembler can't determine the size of the 10H. The instruction MOV BYTE PTR [DI],10H clearly designates the location addressed by DI as a byte-sized memory location. Likewise, the MOV DWORD PTR [DI],10H clearly identifies the memory location as doubleword-sized. The BYTE PTR, WORD PTR, and DWORD PTR directives are only used with instructions that address a memory location through a pointer or index register with immediate data and for a few other instructions that are described in subsequent chapters.

Indirect addressing often allows a program to refer to tabular data located in the memory system. For example, suppose that you must create a table of information with 50 samples taken from memory location 0000:046C. Location 0000:046C contains a counter that is maintained by the personal computer's real-time clock. Figure 3–7 shows the table and the BX register used to address each location in the table sequentially. To accomplish this task, load the starting location of the table into the BX register with a MOV immediate instruction, After initializing the starting address of the table, use register-indirect addressing to store the 50 samples sequentially.

FIGURE 3–7 An array
(TABLE) containing 50 bytes
that are indirectly addressed
through register BX.

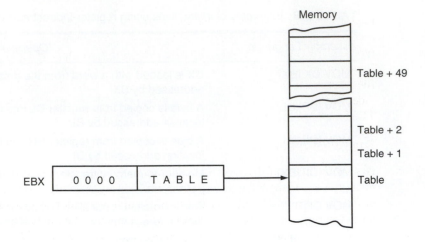

EXAMPLE 3–6

```
                                .MODEL  SMALL           ;select SMALL model
0000                            .DATA                   ;start DATA segment

0000 0032 [       DATAS   DW    50 DUP (?)              ;setup array of 50 bytes
          0000
              ]

0000                            .CODE                   ;start CODE segment
                                .STARTUP                ;indicate start of program

0017 B8 0000                    MOV    AX,0
001A 8E C0                      MOV    ES,AX            ;address segment 0000 with ES

001C BB 0000 R                  MOV    BX,OFFSET DATAS  ;address DATAS array
001F B9 0032                    MOV    CX,50            ;load counter with 50
0022                AGAIN:
0022 26: A1 046C                MOV    AX,ES:[046CH]    ;get clock value
0026 89 07                      MOV    [BX],AX          ;save clock value in DATAS
0028 43                         INC    BX               ;increment BX to next element
0029 E2 F7                      LOOP   AGAIN            ;repeat 50 times

                                .EXIT                   ;exit to DOS
                                END                     ;end file
```

The sequence shown in Example 3–6 loads register BX with the starting address of the table and initializes the count, located in register CX, to 50. The OFFSET directive tells the assembler to load BX with the offset address of memory location TABLE and not the contents of TABLE. For example, the MOV BX,DATAS instruction copies the contents of memory location DATAS into BX while the MOV BX,OFFSET DATAS instruction copies the offset address of DATAS into BX. When the OFFSET directive is used with the MOV instruction, the assembler calculates the offset address and then uses a move immediate instruction to load the address into the specified 16-bit register.

Once the counter and pointer are initialized, a repeat-until CX = 0 loop executes. Here data is read from extra segment memory location 46CH with the MOV AX,ES:[046CH] instruction and stored in memory that is indirectly addressed by the offset address located in register BX. Next, BX is incremented (one is added to BX) to the next table location and finally the LOOP instruction repeats the LOOP 50 times. The LOOP instruction decrements (subtracts one from) the counter (CX) and if CX is not zero, LOOP causes a jump to memory location AGAIN. If CX becomes zero, no jump occurs and this sequence of instructions ends. This example copies the most recent 50 values from the clock into the memory array DATAS. This program will often show the same data in each location because the contents of the clock are changed only 18.2 times per second. Use the CodeView program to view the program and its execution. To use CodeView, type CV FILE.EXE or CV FILE.COM or access it as DEBUG from the Programmer's Work-Bench program under the RUN menu. Note that CodeView only functions with .EXE or .COM files. Some useful CodeView switches are /50 for a fifty-line display and /S for high-resolution video displays in an application. To debug the file TEST.COM with 50 lines, type CV /50 TEST.COM at the DOS prompt.

Base-Plus-Index Addressing

Base-plus-index addressing is similar to indirect addressing because it indirectly addresses memory data. In the 8086 through the 80286, this type of addressing uses **one base register** (BP or BX) and **one index register** (DI or SI) to indirectly address memory. Often the base register holds the beginning location of a memory array, while the index register holds the relative position of an element in the array. Remember that whenever BP addresses memory data, both the stack segment register and BP generate the effective address.

In the 80386 and above, this type of addressing allows the combination of any two 32-bit extended registers except ESP. For example, the MOV DL,[EAX+EBX] instruction is an example using EAX (as the base) plus EBX (as the index). If the EBP register is used, the data is located in the stack segment instead of the data segment.

Locating Data with Base-Plus-Index Addressing. Figure 3–8 shows how data is addressed by the MOV DX,[BX+DI] instruction when the microprocessor operates in the real mode. In this example, BX = 1000H, DI = 0010H, and DS = 0100H, which translate into memory address 02010H. This instruction transfers a copy of the word from location 02010H into the DX register. Table 3–6 lists some instructions used for base-plus-index addressing. Note that the Intel assembler requires that this addressing mode appear as [BX][DI] instead of [BX+DI]. The MOV DX,[BX+DI] instruction is MOV DX,[BX][DI] for a program written for the Intel ASM assembler. This book uses the first form in all example programs, but the second form can be used in many assemblers, including MASM from Microsoft.

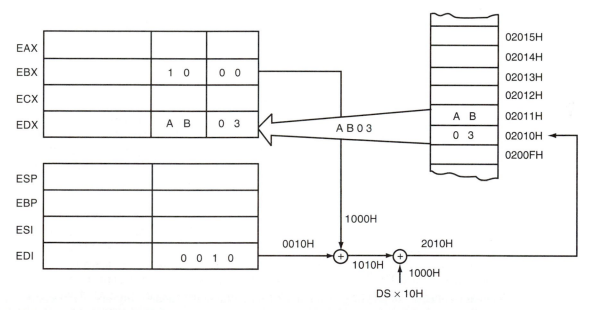

FIGURE 3–8 An example showing how the base-plus-index addressing mode functions for the MOV DX,[BX+DI] instruction. Notice that memory address 02010H is accessed because DS = 0100H, BX = 1000H, and DI = 0010H.

TABLE 3–6 Examples of base-plus-index addressing

Assembly Language	Operation
MOV CX,[BX+DI]	Loads CX with the word contents of the data segment memory location addressed by the sum of BX and DI
MOV CH,[BP+SI]	Loads CH with the byte contents of the stack segment memory location addressed by the sum of BP and SI
MOV [BX+SI],SP	Stores the word contents of SP into the data segment memory location addressed by the sum of BX and SI
MOV [BP+DI],CX	Stores the word contents of CX into the stack segment memory location addressed by the sum of BP and DI
MOV CL,[EDX+EDI]	Loads CL with the byte contents of the data segment memory location addressed by the sum of EDX and EDI
MOV [EAX+EBX],ECX	Stores the doubleword contents of ECX into the data segment memory location addressed by the sum of EAX and EBX

Locating Array Data Using Base-Plus-Index Addressing.   A major use of the base-plus-index addressing mode is to address elements in a memory array. Suppose that the elements in an array located in the data segment at memory location ARRAY must be accessed. To accomplish this, load the BX register (base) with the beginning address of the array and the DI register (index) with the element number to be accessed. Figure 3–9 shows the use of BX and DI to access an element in an array of data.

A short program listed in Example 3–7 moves array element 10H into array element 20H. Notice that the array element number, loaded into the DI register, addresses the array element. Also notice how the contents of the ARRAY have been initialized so element 10H contains a 29H.

EXAMPLE 3–7

```
                        .MODEL  SMALL           ;select SMALL model
0000                    .DATA                   ;start of DATA segment

0000 0010 [     ARRAY   DB    16 DUP (?)        ;setup ARRAY
        00
          ]
0010 29                 DB    29H               ;sample data at element 10H
0011 001E [             DB    30 DUP (?)
        00
          ]
0000                    .CODE                   ;start of CODE segment
                        .STARTUP                ;indicate start of program

0017 BB 0000 R          MOV   BX,OFFSET ARRAY   ;address ARRAY
001A BF 0010            MOV   DI,10H            ;address element 10H
001D 8A 01              MOV   AL,[BX+DI]        ;get element 10H
001F BF 0020            MOV   DI,20H            ;address element 20H
0022 88 01              MOV   [BX+DI],AL        ;save in element 20H

                        .EXIT                   ;exit to DOS
                        END                     ;indicate end of file
```

Register-Relative Addressing

Register-relative addressing is similar to base-plus-index addressing and displacement addressing discussed earlier. In register-relative addressing, the data in a segment of memory is addressed by adding the displacement to the contents of a base or an index register (BP, BX, DI,

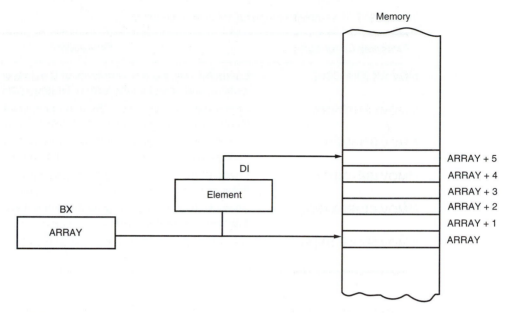

FIGURE 3–9 An example of the base-plus-index addressing mode. Here an element (DI) of an ARRAY (BX) is addressed.

or SI). Figure 3–10 shows the operation of the MOV AX,[BX+1000H] instruction. (This instruction can also be represented as MOV AX,[BX][1000H] for the Intel assembler.) In this example, BX = 0100H and DS = 0200H, so the address generated is the sum of DS × 10H, BX, and the displacement of 1000H or 03100H. Remember that BX, DI, or SI address the data segment and BP addresses the stack segment. In the 80386 and above, the displacement can be a 32-bit number and the register can be any 32-bit register except the ESP register. Remember that the size of a real-mode segment is 64K bytes in length. Table 3–7 lists a few instructions that use register-relative addressing.

The **displacement** can be a number added to the register within the [ ], as in the MOV AL,[DI+2] instruction, or it can be a displacement subtracted from the register, as in MOV AL,[SI–1]. A displacement also can be an offset address appended to the front of the [ ] as in MOV AL,DATA[DI]. Both forms of displacements also can appear simultaneously as in the MOV AL,DATA[DI+3] instruction. In all cases, both forms of the displacement add to the base

FIGURE 3–10 The operation of the MOV AX,[BX+1000H] instruction, when BX = 0100H and DS = 0200H.

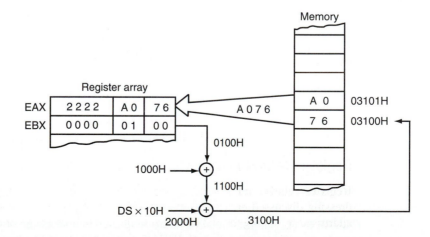

TABLE 3–7 Examples of register-relative addressing

Assembly Language	Operation
MOV AX,[DI+100H]	Loads AX with the word contents of the data segment memory location addressed by the sum of DI and 100H
MOV ARRAY[SI],BL	Stores the byte contents of BL into the data segment memory location addressed by the sum of ARRAY and SI
MOV LIST[SI+2],CL	Stores the byte contents of CL into the data segment memory location addressed by the sum of LIST, SI, and 2
MOV DI,SETX[BX]	Loads DI from the data segment memory location addressed by the sum of SETX and BX
MOV DI,[EAX+100H]	Loads DI from the data segment memory location addressed by the sum of EAX and 100H
MOV ARRAY[EBX],AL	Stores the contents of AL into the data segment memory location addressed by the sum of ARRAY and EBX

or base and index register within the [ ]. In the 8086–80286 microprocessor the value of the displacement is limited to a 16-bit signed number with a value ranging between +32,767 (7FFFH) and –32,768 (8000H) and in the 80386 and above, a 32-bit displacement is allowed ranging in value from +2,147,483,647 (7FFFFFFFH) and –2,147,483,648 (80000000H).

Addressing Array Data with Register Relative.   It is possible to address array data with register-relative addressing much as you do with base-plus-index addressing. In Figure 3–11, register-relative addressing is illustrated with the same example as for base-plus-index addressing. This shows how the displacement ARRAY adds to index register DI to generate a reference to an array element.

Example 3–8 shows how this new addressing mode can transfer the contents of array element 10H into array element 20H. Notice the similarity between this example and Example 3–7.

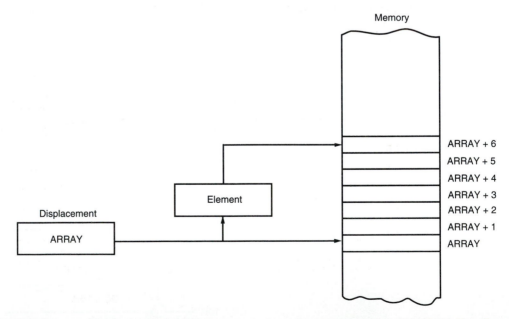

FIGURE 3–11 Register-relative addressing used to address an element of ARRAY. The displacement addresses the start of ARRAY, and DI accesses an element.

The main difference is that in Example 3–8 register BX is not used to address memory area ARRAY; instead ARRAY is used as a displacement to accomplish the same task.

EXAMPLE 3–8

```
                          .MODEL SMALL        ;select SMALL model
0000                      .DATA               ;start of DATA segment

0000 0010 [     ARRAY  DB    16 DUP (?)       ;setup ARRAY
          00
              ]
0010 29                DB    29H              ;sample data at element 10H
0011 001E [            DB    30 DUP (?)
          00
              ]
0000                      .CODE               ;start of CODE segment
                          .STARTUP            ;indicate start of program

0017 BF 0010           MOV   DI,10H           ;address element 10H
001A 8A 85 0000 R      MOV   AL,ARRAY[DI]     ;get element 10H
001E BF 0020           MOV   DI,20H           ;address element 20H
0021 88 85 0000 R      MOV   ARRAY[DI],AL     ;save in element 20H

                          .EXIT               ;exit to DOS
                          END                 ;indicate end of file
```

Base Relative-Plus-Index Addressing

The **base relative-plus-index addressing** mode is similar to the base-plus-index addressing mode, but adds a displacement besides using a base register and an index register to form the memory address. This type of addressing mode often addresses a two-dimensional array of memory data.

Addressing Data with Base Relative-Plus-Index. Base relative-plus-index addressing is the least-used addressing mode. Figure 3–12 shows how data is referenced if the instruction executed by the microprocessor is a MOV AX,[BX+SI+100H]. The displacement of 100H adds to BX and SI to form the offset address within the data segment. Registers BX = 0020H, SI = 0010H, and DS = 1000H so the effective address for this instruction is 10130H, the sum of these registers plus a displacement of 100H. This addressing mode is too complex for frequent use in a program. Some typical instructions using base relative-plus-index addressing appear in Table 3–8.

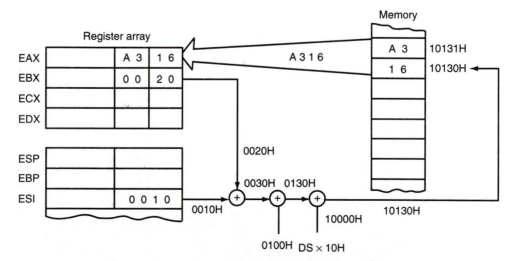

FIGURE 3–12 An example of base relative-plus-index addressing using a MOV AX,[BX+SI+100H] instruction. *Note:* DS = 1000H.

TABLE 3–8 Examples of base relative-plus-index instructions

Assembly Language	Operation
MOV DH,[BX+DI+20H]	Loads DH from the data segment memory location addressed by the sum of BX, DI, and 20H
MOV AX,FILE[BX+DI]	Loads AX from the data segment memory location addressed by the sum of FILE, BX, and DI
MOV LIST[BP+DI],CL	Stores CL at the stack segment memory location addressed by the sum of LIST, BP, and DI
MOV LIST[BP+SI+4],DH	Stores DH at the stack segment memory location addressed by the sum of LIST, BP, SI, and 4
MOV AL,FILE[EBX+ECX+2]	Loads AL from the data segment memory location addressed by the sum of FILE, EBX, ECX, and 2

Notice that with the 80386 and above, the effective address is generated by the sum of two 32-bit registers plus a 32-bit displacement.

Addressing Arrays with Base Relative-Plus-Index. Suppose that a file of many records exists in memory and that each record contains many elements. This displacement addresses the file; the base register addresses a record; the index register addresses an element of a record. Figure 3–13 illustrates this very complex form of addressing.

Example 3–9 is a program that copies element 0 of record A into element 2 of record C using the base relative-plus-index mode of addressing. This example FILE contains four records and each record contains ten elements. Notice how the THIS BYTE statement is used to define the label FILE and RECA at the same memory location.

EXAMPLE 3–9

```
                             .MODEL SMALL        ;indicate SMALL model
0000                         .DATA               ;start of DATA segment

0000 = 0000      FILE    EQU   THIS BYTE         ;assign FILE to this byte

0000 000A [      RECA    DB    10 DUP (?)        ;reserve 10 bytes for RECA
         00
         ]
000A 000A [      RECB    DB    10 DUP (?)        ;reserve 10 bytes for RECB
         00
         ]
0014 000A [      RECC    DB    10 DUP (?)        ;reserve 10 bytes for RECC
         00
         ]
001E 000A [      RECD    DB    10 DUP (?)        ;reserve 10 bytes for RECD
         00
         ]

0000                         .CODE               ;start of CODE segment
                             .STARTUP            ;indicate start of program

0017 BB 0000 R           MOV   BX,OFFSET RECA    ;address RECA
001A BF 0000             MOV   DI,0              ;address element 0
001D 8A 81 0000 R        MOV   AL,FILE[BX+DI]    ;get data
0021 BB 0014 R           MOV   BX,OFFSET RECC    ;address RECC
0024 BF 0002             MOV   DI,2              ;address element 2
0027 88 81 0000 R        MOV   FILE[BX+DI],AL    ;save data

                             .EXIT               ;exit to DOS
                             END                 ;indicate end of file
```

FIGURE 3–13 Base relative-plus-index addressing used to access a FILE that contains multiple records (REC).

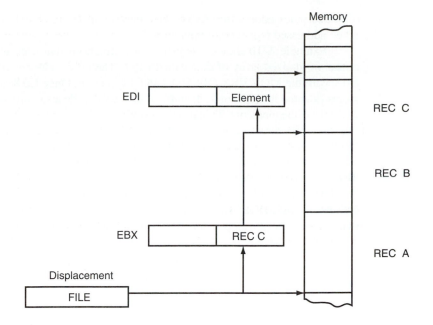

Scaled-Index Addressing

Scaled-index addressing is the last type of data-addressing mode discussed. This data-addressing mode is unique to the 80386 through the Pentium Pro microprocessors. Scaled-index addressing uses two 32-bit registers (a base register and an index register) to access the memory. The second register (index) is multiplied by a **scaling factor.** The scaling factor can be 1X, 2X, 4X, or 8X. A scaling factor of 1X is implied and need not be included in the assembly language instruction (MOV AL,[EBX+ECX]). A scaling factor of 2X is used to address word-sized memory arrays; a scaling factor of 4X is used with doubleword-sized memory arrays; a scaling factor of 8X is used with quadword-sized memory arrays.

For example, the instruction MOV AX,[EDI+2*ECX] uses a scaling factor of 2X, which multiplies the contents of ECX by 2 before adding it to the EDI register to form the memory address. If ECX contains a 00000000H, word-sized memory element 0 is addressed. If ECX contains a 00000001H, word-sized memory element 1 is accessed, and so forth. This scales the index (ECX) by a factor of 2 for a word-sized memory array. Refer to Table 3–9 for some examples

TABLE 3–9 Examples of instructions that use scaled-index addressing

Assembly Language	Operation
MOV EAX,[EBX+4*ECX]	Loads EAX from the data segment memory location addressed by the sum of EBX and four times ECX
MOV [EAX+2*EBX],CX	Stores CX at the data segment memory location addressed by EAX plus 2 times EBX
MOV AX,[EBP+2*EDI+100H]	Loads AX from the stack segment memory location addressed by EBP plus 2 times EDI plus 100H
MOV LIST[EAX+2*EBX+10H],DX	Stores DX in the data segment memory locations addressed by LIST plus 2 times EBX plus 10H
MOV EAX,ARRAY[4*ECX]	Loads EAX with the contents of the doubleword data segment memory location addressed by ARRAY plus 4 times ECX

of scaled-index addressing. As you can imagine, there are an extremely large number of scaled-index addressed register combinations.

Example 3–10 shows a sequence of instructions that uses scaled-index addressing to access a word-sized array of data called LIST. The offset address of LIST is loaded into register EBX with the MOV EBX,OFFSET LIST instruction. Once EBX addresses array LIST, the elements (located in ECX) of 2, 4, and 7 of this word-wide array are added using a scaling factor of 2 to access the elements. This program stores the 2 at element 2 into elements 4 and 7. Also notice the .386 directive to select the 80386 microprocessor. This directive must follow the .MODEL statement for the assembler to process 80386 instructions for DOS. If the 80486 is in use, the **.486 directive** appears after the .MODEL statement and if the Pentium is in use, the **.586 directive** appears after .MODEL. If the microprocessor selection directive appears before the .MODEL statement, the microprocessor executes instructions in the 32-bit mode, which is not compatible with DOS. The Pentium Pro has yet to be defined for the assembler. It is suggested that software be developed for the 80386, which essentially provides just about all of the instructions available to the 80486 through the Pentium Pro. In this book, because many of the programs are written for the 80186/80188, the .186 directive appears in many subsequent examples.

EXAMPLE 3–10

```
                                .MODEL SMALL            ;select SMALL model
                                .386                    ;use the 80386
0000                            .DATA                   ;start of the DATA segment

0000 0000 0001 0002   LIST  DW  0,1,2,3,4               ;define array list
     0003 0004
000A 0005 0006 0007         DW  5,6,7,8,9
     0008 0009

0000                            .CODE                   ;start of the CODE segment
                                .STARTUP                ;indicate start of program
0010 66| BB 00000000 R          MOV EBX,OFFSET LIST     ;address array LIST

0016 66| B9 00000002            MOV ECX,2               ;get element 2
001C 67& 8B 04 4B               MOV AX,[EBX+2*ECX]

0020 66| B9 00000004            MOV ECX,4               ;store in element 4
0026 67& 89 04 4B               MOV [EBX+2*ECX],AX

002A 66| B9 00000007            MOV ECX,7               ;store in element 7
0030 67& 89 04 4B               MOV [EBX+2*ECX],AX
                                .EXIT                   ;exit to DOS
                                END                     ;end of file
```

Data Structures

A **data structure** is used to specify how information is stored in a memory array and can be quite useful with applications that use arrays. It is best to think of a data structure as a template for data. The start of a structure is identified with the STRUC assembly language directive and the end with the ENDS statement. A typical data structure is defined and used three times in Example 3–11. Notice that the name of the structure appears with the STRUC statement and also with the ENDS statement.

EXAMPLE 3–11

```
                    ;Define INFO data structure
                    ;
0057                INFO    STRUC

0000 0020 [         NAMES   DB    32 DUP (?)       ;32 bytes for name
```

```
              00
                  ]
0020 0020 [          STREET DB    32 DUP (?)          ;32 bytes for street
              00
                  ]
0040 0010 [          CITY   DB    16 DUP (?)          ;16 bytes for city
              00
                  ]
0050 0002 [          STATE  DB    2  DUP (?)          ;2 bytes for state
              00
                  ]
0052 0005 [          ZIP    DB    5  DUP (?)          ;5 bytes for zip code
              00
                  ]
                     INFO   ENDS

0000 42 6F 62 20 53 6D NAME1 INFO <'Bob Smith','123 Main Street','Wanda','OH','44444'>
     69 74 68
     0017 [
              00
                  ]
     31 32 33 20 4D
     61 69 6E 20 53 74
     72 65 65 74
     0011 [
              00
                  ]
     57 61 6E 64 61
     000B [
              00
                  ]
     4F 48 34 34 34
     34 34

0057 53 74 65 76 65 20 NAME2 INFO <'Steve Doe','222 Mouse Lane','Miller','PA','18100'>
     44 6F 65
     0017 [
              00
                  ]
     32 32 32 20 4D
     6F 75 73 65 20 4C
     61 6E 65
     0012 [
              00
                  ]
     4D 69 6C 6C 65
     72
     000A [
              00
                  ]
     50 41 31 38 31
     30 30

00AE 42 65 6E 20 44 6F NAME3 INFO <'Ben Dover','303 Main Street','Orender','CA','90000'>
     76 65 72
     0017 [
              00
                  ]
     33 30 33 20 4D
     61 69 6E 20 53 74
     72 65 65 74
     0011 [
              00
                  ]
     4F 72 65 6E 64
     65 72
```

```
0009 [
        00
        ]
43 41 39 30 30
30 30
```

The data structure in Example 3–11 defines five fields of information. The first is 32 bytes in length and holds a name; the second is 32 bytes in length and holds a street address; the third is 16 bytes in length for the city; the fourth is 2 bytes for the state; and the fifth is 5 bytes for the ZIP code. Once the structure is defined (INFO), it can be filled as illustrated with names and addresses. Three examples of use for INFO are illustrated. Carefully note that literals are surrounded with apostrophes and the entire field is surrounded with < > symbols when the data structure is used to define data.

When data is addressed in a structure, use the structure name and the field name to select a field from the structure. For example, to address the STREET in NAME2 use the operand NAME2.STREET, where the name of the structure is first followed by a period and then the name of the field. Likewise, use NAME3.CITY to refer to the city in structure NAME3.

EXAMPLE 3–12

```
                        ;Clear names in array NAME1

0000 B9 0020                MOV    CX,32
0003 B0 00                  MOV    AL,0
0005 BE 0000 R              MOV    SI,OFFSET NAME1.NAMES
0008 F3/ AA                 REP    STOSB

                        ;Clear street in array NAME2

000A B9 0020                MOV    CX,32
000D B0 00                  MOV    AL,0
0010 BE 0077 R              MOV    SI,OFFSET NAME2.STREET
0013 F3/ AA                 REP    STOSB

                        ;Clear zip code in array NAME3

0015 B9 0005                MOV    CX,5
0018 B0 00                  MOV    AL,0
001A BE 0100 R              MOV    SI,OFFSET NAME3.ZIP
001D F3/ AA                 REP    STOSB
```

A short sequence of instructions appears in Example 3–12 that clears the name field in structure NAME1, the address field of structure NAME2, and the ZIP code field of structure NAME3. The function and operation of the instructions in this program are defined in later chapters. You may want to refer back to this example, once you learn these instructions.

3–2   PROGRAM MEMORY-ADDRESSING MODES

Program memory-addressing modes, used with the JMP and CALL instructions, consist of three distinct forms: direct, relative, and indirect. This section introduces these three addressing forms, using the JMP instruction to illustrate their operation.

Direct Program Memory-Addressing

Direct program memory-addressing is what many early microprocessors used for all jumps and calls. Direct program memory-addressing is also used in high-level languages such as the

FIGURE 3–14 The 5-byte machine language version of a JMP [10000H] instruction.

Opcode	Offset (low)	Offset (high)	Segment (low)	Segment (high)
E A	0 0	0 0	0 0	1 0

BASIC language GOTO and GOSUB instructions. The microprocessor uses this form of addressing, but doesn't use it nearly as often as relative and indirect program memory-addressing.

The instructions for direct program memory-addressing store the address with the opcode. For example, if a program jumps to memory location 10000H for the next instruction, the address (10000H) is stored following the opcode in the memory. Figure 3–14 shows the direct intersegment JMP instruction and the 4 bytes required to store the address 10000H. This JMP instruction loads CS with 1000H and IP with 0000H to jump to memory location 10000H for the next instruction. (An **intersegment jump** is a jump to any memory location within the entire memory system.) The **direct jump** is often called a *far jump* because it can jump to any memory location for the next instruction. In the real mode, a far jump accesses any location within the first 1M byte of memory by changing both CS and IP. In protected-mode operation, the far jump accesses a new code segment descriptor from the descriptor table allowing it to jump to any memory location in the entire 4G-byte address range in an 80386 through Pentium Pro microprocessor.

The only other instruction that uses direct program addressing is the intersegment or far CALL instruction. Usually, the name of a memory address, called a **label,** refers to the location that is called or jumped to instead of the actual numeric address. When using a label with the CALL or JMP instruction, most assemblers select the best form of program addressing.

Relative Program Memory-Addressing

Relative program memory-addressing is not available in all early microprocessors, but it is available to this family of microprocessors. The term *relative* means "relative to the instruction pointer (IP)." For example, if a JMP instruction skips the next 2 bytes of memory, the address in relation to the instruction pointer is a 2 that adds to the instruction pointer. This develops the address of the next program instruction. An example of the relative JMP instruction is shown in Figure 3–15. Notice that the JMP instruction is a 1-byte instruction with a 1- or 2-byte displacement that adds to the instruction pointer. A 1-byte displacement is used in *short* jumps and a 2-byte displacement is used with *near* jumps and calls. Both types are considered intrasegment jumps. (An **intrasegment jump** is a jump anywhere within the current code segment.) In the 80386 and above, the displacement can also be a 32-bit value allowing them to use relative addressing to any location within its up to 4G-byte code segment.

Relative JMP and CALL instructions contain either an 8-bit or a 16-bit signed displacement that allows a forward memory reference or a reverse memory reference. (The 80386 and above can have an 8-bit or 32-bit displacement.) All assemblers automatically calculate the distance for the displacement and select the proper 1-, 2-, or 4-byte form. If the distance is too far for a 2-byte displacement in an 8086 through 80286 microprocessor, some assemblers use the direct jump. An 8-bit displacement *(short)* has a jump range of between +127 and –128 byte from the next instruction, while a 16-bit displacement *(near)* has a range of ±32K bytes. In the 80386 and above, a 32-bit displacement allows a range of ±2G bytes.

FIGURE 3–15 A JMP [2] instruction. This instruction skips over the 2 bytes of memory that follow the JMP instruction.

10000	EB
10001	02
10002	—
10003	—
10004	←

JMP [2]

TABLE 3–10 Examples of program indirect addressing

Assembly Language	Operation
JMP AX	Jump to the location addressed by AX in the current code segment
JMP CX	Jump to the location addressed by CX in the current code segment
JMP NEAR PTR [BX]	Jump to the current code segment location using the address stored at the data segment addressed by BX
JMP NEAR PTR [DI+2]	Jump to the current code segment location using the address stored at the data segment location addressed by the sum of DI and 2
JMP TABLE [BX]	Jump to the current code segment location addressed by the contents of data segment TABLE plus BX
JMP ECX	Jump to the current code segment location addressed by ECX

Indirect Program Memory-Addressing

The microprocessor allows several forms of *indirect program memory-addressing* for the JMP and CALL instruction. Table 3–10 lists some acceptable program indirect jump instructions, which can use any 16-bit register (AX, BX, CX, DX, SP, BP, DI, or SI); any relative register ([BP], [BX], [DI], or [SI]); and any relative register with a displacement. In the 80386 and above, an extended register can also be used to hold the address or indirect address of a relative JMP or CALL. For example, a JMP EAX jumps to the location address by register EAX.

If a 16-bit register holds the address of a JMP instruction, the jump is near. For example, if the BX register contains a 1000H and a JMP BX instruction executes, the microprocessor jumps to offset address 1000H in the current code segment.

If a relative register holds the address, the jump is also considered an indirect jump. For example, a JMP [BX] refers to memory location within the data segment at the offset address contained in BX. At this offset address is a 16-bit number that is used as the offset address in the intrasegment jump. This type of jump may be called an *indirect-indirect* or double-indirect jump.

Figure 3–16 shows a jump table that is stored beginning at memory location TABLE. This jump table is referenced by the short program in Example 3–13. In this example, the BX register is loaded with a 4, so when it combines in the JMP TABLE[BX] instruction with TABLE, the effective address is the contents of the second entry in the jump table.

EXAMPLE 3–13

```
                    ;Using indirect addressing for a jump
                    ;
0000 BB 0004            MOV  BX,4          ;address LOC2
0003 FF A7 23A1 R       JMP  TABLE[BX]     ;jump to LOC2
```

FIGURE 3–16 A jump table that stores addresses of various programs. The exact address chosen from the TABLE is determined by an index stored with the jump instruction.

```
TABLE DW  LOC0
      DW  LOC1
      DW  LOC2
      DW  LOC3
```

3–3   STACK MEMORY-ADDRESSING

The stack plays an important role in all microprocessors. It holds data temporarily and stores return addresses for procedures. The **stack memory** is a LIFO **(last-in, first-out)** memory, which describes the way that data is stored and removed from the stack. Data is placed onto the stack with a **PUSH** instruction and removed with a **POP** instruction. The CALL instruction also uses the stack to hold the return address for procedures and a RET (return) instruction to remove the return address from the stack.

The stack memory is maintained by two registers: the **stack pointer** (SP or ESP) and the **stack segment** register (SS). Whenever a word of data is pushed onto the stack [see Figure 3–17 (a)], the high-order 8 bits are placed in the location addressed by SP – 1. The low-order 8 bits are placed in the location addressed by SP – 2. The SP is then decremented by 2 so the next word of data is stored in the next available stack memory location. The SP/ESP register always points to an area of memory located within the stack segment. The SP/ESP register adds to SS × 10H to form the stack memory address in the real mode. In protected-mode operation, the SS register holds a selector that accesses a descriptor for the base address of the stack segment.

Whenever data is popped from the stack [see Figure 3–17 (b)], the low-order 8 bits are removed from the location addressed by SP. The high-order 8 bits are removed from the location

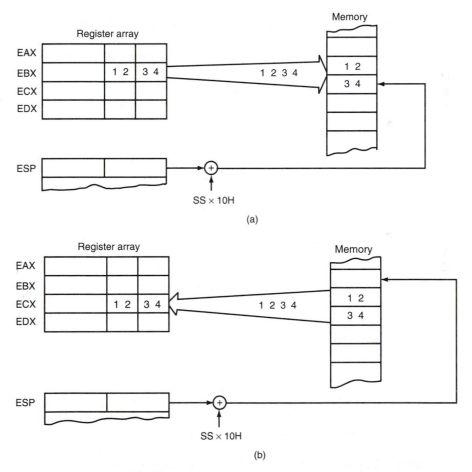

FIGURE 3–17 The PUSH and POP instructions. (a) PUSH BX places the contents of BX onto the stack. (b) POP CX removes data from the stack and places it into CX. Both instructions are shown after execution.

TABLE 3–11 Examples of PUSH and POP instructions

Assembly Language	Operation
POPF	Removes a word from the stack and places it into the flags
POPFD	Removes a doubleword from the stack and places it into the EFLAG register
PUSHF	Pushes the contents of the flag register onto the stack
PUSHFD	Pushes the contents of the EFLAG register onto the stack
PUSH AX	Stores a copy of AX on the stack
POP BX	Removes a word from the stack and places it into BX
PUSH DS	Stores a copy of DS on the stack
PUSH 1234H	Stores a 1234H on the stack
POP CS	Illegal instruction
PUSH [BX]	Stores a copy of the word contents of the data segment memory location addressed by BX on the stack
PUSHA	Stores a copy of registers AX, CX, DX, BX, SP, BP, SI, and SI on the stack
PUSHAD	Stores a copy of registers EAX, ECX, EDX, EBX, ESP, EBP, EDI, and ESI on the stack
POPA	Removes data from the stack and places it into SI, DI, BP, SP, BX, DX, CX, and AX
POPAD	Removes data from the stack and places it into ESI, EDI, EBP, ESP, EBX, EDX, ECX, and EAX
POP EAX	Removes data from the stack and places it into EAX
PUSH EDI	Stores a copy of EDI on the stack

addressed by SP + 1. The SP register is then incremented by 2. Table 3–11 lists some of the PUSH and POP instructions available to the microprocessor. Notice that PUSH and POP always store or retrieve words of data—never bytes—in the 8086 through the 80286 microprocessors. The 80386 and above allow words or doublewords to be transferred to and from the stack. Data may be pushed onto the stack from any 16-bit register or segment register, and in the 80386 and above, any 32-bit extended register. Data may be popped off the stack into any 16-bit register or any segment register except CS. The reason that data may not be popped from the stack into CS is that this only changes part of the address of the next instruction.

The PUSHA and POPA instructions either push or pop all of the registers, except the segment registers, on the stack. These instructions are not available on the early 8086/8088 microprocessors. The push immediate instruction is also new to the 80286 through the Pentium Pro microprocessors. The examples in Table 3–11 show the order of the registers transferred by the PUSHA and POPA instructions. The 80386 and above also allow extended registers to be pushed or popped.

Example 3–14 lists a short program that pushes the contents of AX, BX, and CX onto the stack. The first POP retrieves the value that was pushed onto the stack from CX and places it into AX. The second POP places the original value of BX into CX. The last POP places the original value of AX into BX.

EXAMPLE 3–14

```
                          .MODEL TINY      ;select TINY model
0000                      .CODE            ;start CODE segment
                          .STARTUP         ;start of program
```

```
0100 B8 1000              MOV   AX,1000H          ;load test data
0103 BB 2000              MOV   BX,2000H
0106 B9 3000              MOV   CX,3000H

0109 50                   PUSH AX                 ;1000H to stack
010A 53                   PUSH BX                 ;2000H to stack
010B 51                   PUSH CX                 ;3000H to stack

010C 58                   POP   AX                ;3000H to AX
010D 59                   POP   CX                ;2000H to CX
010E 5B                   POP   BX                ;1000H to BX
                          .EXIT                   ;exit to DOS
                          END                     ;end of file
```

3–4   SUMMARY

1. The data-addressing modes include: register, immediate, direct, register indirect, base-plus index, register relative, and base relative-plus-index addressing. In the 80386 through the Pentium Pro microprocessors an additional addressing mode, called scaled-index addressing, exists.

2. The program memory-addressing modes include: direct, relative, and indirect addressing.

3. Table 3–12 lists all real mode data-addressing modes available to the 8086 through the 80286 microprocessors. Notice that the 80386 and above also use these modes plus the many defined throughout this chapter. In the protected mode, the function of the segment register is to address a descriptor that contains the base address of the memory segment.

4. The 80386, 80486, Pentium, and Pentium Pro microprocessors have additional addressing modes that allow the extended registers EAX, EBX, ECX, EDX, EBP, EDI, and ESI to address memory. These addressing modes are too numerous to list in tabular form, but in general any of these registers function in the same way as those listed in Table 3–12. For example, the MOV AL,TABLE[EBX+2*ECX+10H] is a valid addressing mode for the 80386–Pentium Pro microprocessors.

5. The MOV instruction copies the contents of the source operand into the destination operand. The source never changes for any instruction.

6. Register addressing specifies any 8-bit register (AH, AL, BH, BL, CH, CL, DH, or DL) or any 16-bit register (AX, BX, CX, DX, SP, BP, SI, or DI). The segment registers (CS, DS, ES, or SS) are also addressable for moving data between a segment register and a 16-bit register/memory location or for PUSH and POP. In the 80386 through the Pentium Pro microprocessors, the extended registers also are used for register addressing and consist of: EAX, EBX, ECX, EDX, ESP, EBP, EDI, and ESI. Also available to the 80386 and above are the FS and GS segment registers.

7. The MOV immediate instruction transfers the byte or word, immediately following the opcode, into a register or a memory location. Immediate addressing manipulates constant data in a program. In the 80386 and above, a doubleword immediate data may also be loaded into a 32-bit register or memory location.

8. The .MODEL statement is used with assembly language to identify the start of a file and the type of memory model used with the file. If the size is TINY, the program exists in one segment, the code segment, and is assembled as a command (.COM) program. If the SMALL model is used, the program uses a code and data segment and assembles as an execute (.EXE) program. Other model sizes and their attributes are listed in Appendix A.

9. Direct addressing occurs in two forms in the microprocessor: (1) direct addressing and (2) displacement addressing. Both forms of addressing are identical except direct addressing is used to transfer data between EAX, AX, or AL and memory while displacement addressing is used with any register-memory transfer. Direct addressing requires 3 bytes of memory,

TABLE 3–12 Real mode
data-addressing modes

Assembly Language	Address Generation
MOV AL,BL	8-bit register addressing
MOV DI,BP	16-bit register addressing
MOV BX,DS	16-bit segment register addressing
MOV AL,LIST	(DS x 10H) + LIST
MOV CH,DATA1	(DS x 10H) + DATA1
MOV ES,DATA2	(DS x 10H) + DATA2
MOV AL,12	Immediate data of 12 decimal
MOV AL,[BP]	(SS x 10H) + BP
MOV AL,[BX]	(DS x 10H) + BX
MOV AL,[DI]	(DS x 10H) + DI
MOV AL,[SI]	(DS x 10H) + SI
MOV AL,[BP+2]	(SS x 10H) + BP + 2
MOV AL,[BX+4]	(DS x 10H) + BX + 4
MOV AL,[DI+1000H]	(DS x 10H) + DI + 1000H
MOV AL,[SI+300H]	(DS x 10H) + SI + 300H
MOV AL,LIST[BP]	(SS x 10H) + BP + LIST
MOV AL,LIST[BX]	(DS x 10H) + BX + LIST
MOV AL,LIST[DI]	(DS x 10H) + DI + LIST
MOV AL,LIST[SI]	(DS x 10H) + SI + LIST
MOV AL,LIST[BP+2]	(SS x 10H) + BP + 2
MOV AL,LIST[BX+6]	(DS x 10H) + BX + 6
MOV AL,LIST[DI+100H]	(DS x 10H) + DI + 100H
MOV AL,LIST[SI+20H]	(DS x 10H) + SI + 20H
MOV AL,[BP+DI]	(SS x 10H) + BP + DI
MOV AL,[BP+SI]	(SS x 10H) + BP + SI
MOV AL,[BX+DI]	(DS x 10H) + BX + DI
MOV AL,[BX+SI]	(DS x 10H) + BX + SI
MOV AL,[BP+DI+2]	(SS x 10H) + BP + DI + 2
MOV AL,[BP+SI−4]	(SS x 10H) + BP + SI − 4
MOV AL,[BX+DI+30H]	(DS x 10H) + BX + DI + 30H
MOV AL,[BX+SI+10H]	(DS x 10H) + BX + SI + 10H
MOV AL,LIST[BP+DI]	(SS x 10H) + BP + DI + LIST
MOV AL,LIST[BP+SI]	(SS x 10H) + BP + SI + LIST
MOV AL,LIST[BX+DI]	(DS x 10H) + BX + DI + LIST
MOV AL,LIST[BX+SI]	(DS x 10H) + BX + SI + LIST
MOV AL,LIST[BP+DI+2]	(SS x 10H) + BP + DI + LIST + 2
MOV AL,LIST[BP+SI−7]	(SS x 10H) + BP + SI + LIST − 7
MOV AL,LIST[BX+DI−10H]	(DS x 10H) + BX + DI + LIST − 10H
MOV AL,LIST[BX+SI+1AFH]	(DS x 10H) + BX + SI + LIST + 1AFH

while displacement addressing requires 4 bytes. Some of these instructions in the 80386 and above may require additional bytes in the form of prefixes for register and operand sizes.

10. Register-indirect addressing allows data to be addressed at the memory location pointed to by either a base (BP and BX) or index register (DI and SI). In the 80386 and above, extended registers (EAX, EBX, ECX, EDX, EBP, EDI, and ESI) are used to address memory data.

11. Base-plus-index addressing often addresses data in an array. The memory address for this mode is formed by adding a base register, index register, and the contents of a segment register times 10H. In the 80386 and above, the base and index registers may be any 32-bit register except EIP and ESP.

12. Register-relative addressing uses either a base or index register plus a displacement to access memory data.

13. Base relative-plus-index addressing is useful for addressing a two-dimensional memory array. The address is formed by adding a base register, an index register, displacement, and the contents of a segment register times 10H.

14. Scaled-index addressing is unique to the 80386 through the Pentium Pro. The second of two registers (index) is scaled by a factor of 2X, 4X, or 8X to access words, doublewords, or quadwords in memory arrays. MOV AX,[EBX+2*ECX] and MOV [4*ECX],EDX are examples of scaled-index instructions.

15. Data structures are templates for storing arrays of data and are addressed by array name and field. For example, array NUMBER and field TEN of array NUMBER is addressed as NUMBER.TEN.

16. Direct program memory-addressing is allowed with the JMP and CALL instructions to any location in the memory system. With this addressing mode, the offset address and segment address are stored with the instruction.

17. Relative program addressing allows a JMP or CALL instruction to branch forward or backward in the current code segment by ±32K bytes. In the 80386 and above, the 32-bit displacement allows a branch to any location in the current code segment using a displacement value of ±2G bytes.

18. Indirect program addressing allows the JMP or CALL instructions to address another portion of the program or subroutine indirectly through a register or memory location.

19. The PUSH and POP instructions transfer a word between the stack and a register or memory location. A PUSH immediate instruction is available to place immediate data on the stack. The PUSHA and POPA instructions transfer AX, CX, DX, BX, BP, SP, SI, and DI between the stack and these registers. In the 80386 and above, the extended register and extended flags can also be transferred between registers and the stack. A PUSHFD stores the EFLAGS, while a PUSHF stores the FLAGS.

20. Example 3–15 shows many of the addressing modes presented in this chapter. This example program fills the ARRAY1 from locations 0000:0000 through 0000:0009. It then fills ARRAY2 with 0 through 9. Finally, it exchanges the contents of ARRAY1 element 2 with ARRAY2 element 3.

EXAMPLE 3–15

```
                                .MODEL SMALL          ;select SMALL model
0000                            .DATA                 ;start of DATA segment

0000 000A [          ARRAY1 DB    10 DUP (?)          ;reserve 10 bytes (ARRAY1)
          00
                   ]
000A 000A [          ARRAY2 DB    10 DUP (?)          ;reserve 10 bytes (ARRAY2)
          00
                   ]
0000                            .CODE                 ;start of CODE segment
                                .STARTUP              ;indicate start of program

0017 B8 0000                    MOV   AX,0            ;segment 0000 using ES
001A 8E C0                      MOV   ES,AX

001C BF 0000                    MOV   DI,0            ;address element 0
001F B9 000A                    MOV   CX,10           ;count of 10
0022                 LAB1:
0022 26: 8A 05                  MOV   AL,ES:[DI]      ;copy 0000:0000-0000:0009
0025 88 85 0000 R               MOV   ARRAY1[DI],AL   ;into ARRAY1
0029 47                         INC   DI
002A E2 F6                      LOOP LAB1

002C BF 0000                    MOV   DI,0            ;address element 0
002F B9 000A                    MOV   CX,10           ;count of 10
0032 B0 00                      MOV   AL,0            ;initial value
```

```
0034                    LAB2:
0034 88 85 000A R               MOV   ARRAY2[DI],AL      ;fill ARRAY2
0038 FE C0                      INC   AL
003A 47                        INC   DI
003B E2 F7                     LOOP  LAB2

003D BF 0003                   MOV   DI,3                ;exchange array data
0040 8A 85 0000 R               MOV   AL,ARRAY1[DI]
0044 8A A5 000B R               MOV   AH,ARRAY2[DI+1]
0048 88 A5 0000 R               MOV   ARRAY1[DI],AH
004C 88 85 000B R               MOV   ARRAY2[DI+1],AL

                               .EXIT                     ;exit to DOS
                               END                       ;end of file
```

3–5   QUESTIONS AND PROBLEMS

1. What do the following MOV instructions accomplish?
   (a) MOV AX,BX
   (b) MOV BX,AX
   (c) MOV BL,CH
   (d) MOV ESP,EBP
   (e) MOV AX,CS
2. List the 8-bit registers that are used for register addressing.
3. List the 16-bit registers that are used for register addressing.
4. List the 32-bit registers that are used for register addressing in the 80386 through the Pentium Pro microprocessors.
5. List the 16-bit segment registers used with register addressing by MOV, PUSH, and POP.
6. What is wrong with a MOV BL,CX instruction?
7. What is wrong with a MOV DS,SS instruction?
8. Select an instruction for each of the following tasks:
   (a) copy EBX into EDX
   (b) copy BL into CL
   (c) copy SI into BX
   (d) copy DS into AX
   (e) copy AL into AH
9. Select an instruction for each of the following tasks:
   (a) move a 12H into AL
   (b) move a 123AH into AX
   (c) move a 0CDH into CL
   (d) move a 1000H into SI
   (e) move a 1200A2H into EBX
10. What special symbol is sometimes used to denote immediate data?
11. What is the purpose of the .MODEL TINY statement?
12. What assembly language directive indicates the start of the CODE segment?
13. What is a label?
14. The MOV instruction is placed in which field of a statement?
15. A label may begin with which characters?
16. What is the purpose of the .EXIT directive?
17. The .MODEL TINY statement causes a program to assemble an execute program. (True or False)
18. The .STARTUP directive accomplishes which tasks in the small memory model?

19. What is a displacement? How does it determine the memory address in a MOV [2000H],AL instruction?
20. What do the symbols [ ] indicate?
21. Given that DS = 0200H, BX = 0300H, and DI = 400H, determine the memory address accessed by each of the following instructions assuming real-mode operation:
    (a) MOV AL,[1234H]
    (b) MOV EAX,[BX]
    (c) MOV [DI],AL
22. What is wrong with a MOV [BX],[DI] instruction?
23. Choose an instruction that requires BYTE PTR.
24. Choose an instruction that requires WORD PTR.
25. Choose an instruction that requires DWORD PTR.
26. Explain the difference between the MOV BX,DATA and MOV BX,OFFSET DATA instructions.
27. Given that DS = 1000H, SS = 2000H, BP = 1000H, and DI = 0100H, determine the memory address accessed by each of the following assuming real-mode operation:
    (a) MOV AL,[BP+DI]
    (b) MOV CX,[DI]
    (c) MOV EDX,[BP]
28. What, if anything, is wrong with a MOV AL,[BX][SI] instruction?
29. Given that DS = 1200H, BX = 0100H, and SI = 0250H, determine the address accessed by each of the following instructions assuming real-mode operation:
    (a) MOV [100H],DL
    (b) MOV [SI+100H],EAX
    (c) MOV DL,[BX+100H]
30. Given that DS = 1100H, BX = 0200H, LIST = 0250H, and SI = 0500H, determine the address accessed by each of the following instructions assuming real-mode operation:
    (a) MOV LIST[SI],EDX
    (b) MOV CL,LIST[BX+SI]
    (c) MOV CH,[BX+SI]
31. Given that DS = 1300H, SS = 1400H, BP = 1500H, and SI = 0100H, determine the address accessed by each of the following instructions assuming real-mode operation:
    (a) MOV EAX,[BP+200H]
    (b) MOV AL,[BP+SI-200H]
    (c) MOV AL,[SI-0100H]
32. Which base register addresses data in the stack segment?
33. Given that EAX = 00001000H, EBX = 00002000H, and DS = 0010H, determine the addresses accessed by each of the following instructions assuming real-mode operation:
    (a) MOV ECX,[EAX+EBX]
    (b) MOV [EAX+2*EBX],CL
    (c) MOV DH,[EBX+4*EAX+1000H]
34. Develop a data structure that has five fields of one word each named F1, F2, F3, F4, and F5 with a structure name of FIELDS.
35. Show how field F3 of the data structure constructed in question 34 is addressed in a program.
36. List all three program memory-addressing modes.
37. How many bytes of memory store a far direct jump instruction? What is stored in each of the bytes?
38. What is the difference between an intersegment and intrasegment jump?
39. If a near jump uses a signed 16-bit displacement, how can it jump to any memory location within the current code segment?
40. The 80386 and above use a _____-bit displacement to jump to any location within the 4G-byte code segment.

41. What is a far jump?

42. If a JMP instruction is stored at memory location 100H within the current code segment, it cannot be a _____ jump if it is jumping to memory location 200H within the current code segment.

43. Show which JMP instruction (short, near, or far) assembles if the JMP THERE instruction is stored at memory address 10000H and the address of THERE is:
    (a) 10020H
    (b) 11000H
    (c) 0FFFEH
    (d) 30000H

44. Write a JMP instruction that jumps to the address pointed to by the BX register.

45. Select a JMP instruction that jumps to the location stored in memory at a location table. Assume that it is a near JMP.

46. How many bytes are stored on the stack by PUSH instructions?

47. Explain how the PUSH [DI] instruction functions.

48. Which registers are placed in what order on the stack by the PUSHA instruction?

49. What does the PUSHAD instruction accomplish?

50. Which instruction places the EFLAGS on the stack in the Pentium microprocessor?

51. Use the Internet to write a report detailing the addressing modes available in the Intel 80196 embedded controller.

52. Use the Internet to write a report detailing the addressing modes available in the Intel 8051 embedded controller.

53. Use the Internet to write a report detailing the addressing modes available in the Motorola 6811 embedded controller.

CHAPTER 4

Data Movement Instructions

INTRODUCTION

This chapter concentrates on the data movement instructions: MOV, MOVSX, MOVZX, PUSH, POP, BSWAP, XCHG, XLAT, IN, OUT, LEA, LDS, LES, LFS, LGS, LSS, LAHF, and SAHF, and the string instructions: MOVS, LODS, STOS, INS, and OUTS. The data movement instructions are presented first because they are more commonly used and easier to understand.

The microprocessor requires an assembler program that generates machine language instructions, which are too complex to efficiently generate by hand. This chapter describes assembly language syntax and some of its directives. We assume that you are developing software on an IBM personal computer or clone. We recommend that you use the Microsoft macro assembler (MASM) as your development tool, but the Intel Assembler (ASM), Borland Turbo assembler (TASM), or similar software functions equally well. This book presents information that functions with the Microsoft MASM assembler, but most programs assemble without modification with other assemblers. Appendix A explains the Microsoft assembler, the linker program, and Programmer's Workbench.

CHAPTER OBJECTIVES

Upon completion of this chapter, you will be able to:

1. Explain the operation of each data movement instruction with applicable addressing modes.
2. Explain the purposes of the assembly language pseudo-operations and keywords such as: ALIGN, ASSUME, DB, DD, DW, END, ENDS, ENDP, EQU, .MODEL, OFFSET, ORG, PROC, PTR, SEGMENT, USE16, USE32, and USES.
3. Given a specific data movement task, select the appropriate assembly language instruction to accomplish it.
4. Given a hexadecimal machine language instruction, determine the symbolic opcode, source, destination, and addressing mode.
5. Use the assembler to set up a data segment, stack segment, and code segment.
6. Show how to set up a procedure using PROC and ENDP.
7. Explain the difference between memory models and full segment definitions for the MASM assembler.

4–1   MOV REVISITED

The **MOV instruction,** introduced in Chapter 3, explains the diversity of 8086–80486 and Pentium–Pentium Pro addressing modes. In this chapter, the MOV instruction introduces the machine language instructions available with various addressing modes and instructions. **Machine code** is introduced because it may occasionally be necessary to interpret machine language programs generated by an assembler. Interpretation of the machine's native language **(machine language)** allows debugging or modification at the machine language level. Occasionally machine language patches are made using the DEBUG program available with DOS, which requires some knowledge of machine language. Conversion between machine and assembly language instructions are illustrated using Appendix B.

Machine Language

Machine language is the native binary code that the microprocessor understands and uses as its instructions to control its operation. Machine language instructions, for the 8086 through the Pentium Pro, vary in length from 1 to 13 bytes. There are over 100,000 variations of machine language instructions, which means there is no complete list of these variations. Because of this, some binary bits in a machine language instruction are given, and the remainder are determined for each variation of the instruction.

Instructions for the 8086 through the 80286 are 16-bit mode instructions that take the form found in Figure 4–1 (a). The **16-bit mode instructions** are compatible with the 80386 and above if they are programmed to operate in the 16-bit instruction mode, but may be prefixed as in Figure 4–1 (b). The 80386 through the Pentium Pro assume that all instructions are 16-bit mode instructions when the machine is operated in the real mode. In the protected mode, the upper byte of the descriptor contains the D-bit that selects either the 16- or 32-bit instruction mode. At present only Windows NT and OS/2 operate in 32-bit instruction mode. The **32-bit mode instructions** are shown in Figure 4–1 (b). These instructions occur in 16-bit instruction mode, by the use of prefixes, which are explained later.

The first two bytes of the 32-bit instruction mode format are called **override prefixes** because they are not always present. The first prefix modifies the size of the operand address used by the instruction and the second modifies the register size. If the 80386 through the Pentium Pro operate as a 16-bit instruction mode machine (real or protected mode) and a 32-bit register is used, the **register-size prefix** (66H) is appended to the front of the instruction. If operated in the 32-bit instruction mode (protected mode only) and a 32-bit register is used, the register-size

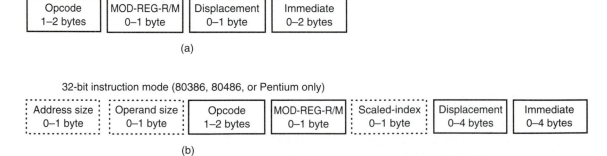

FIGURE 4–1   The formats of the 8086–Pentium instructions. (a) The 16-bit form and (b) the 32-bit form.

FIGURE 4–2 Byte 1 of many machine language instructions, showing the position of the D and W bits.

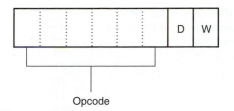

Opcode

prefix is absent. If a 16-bit register appears in an instruction in the 32-bit instruction mode, the register-size prefix is present to select a 16-bit register. The **address size-prefix** (67H) is used in a similar fashion, as explained later. The prefixes toggle the size of the register and operand address from 16-bit to 32-bit or 32-bit to 16-bit for the prefixed instruction. Note that the 16-bit instruction mode uses 8- and 16-bit registers and addressing modes, while the 32-bit instruction mode uses 8- and 32-bit registers and addressing modes by default. The prefixes override these defaults so a 32-bit register can be used in the 16-bit mode or a 16-bit register can be used in the 32-bit mode. The mode of operation (16 or 32 bits) should be selected to conform with the application at hand. If 8- and 32-bit data pervade the application, then the 32-bit mode should be selected; likewise, if 8- and 16-bit data pervade, then the 16-bit mode should be selected. Normally, mode selection is a function of the operating system.

The Opcode.   The **opcode** selects the operation (addition, subtraction, move, etc.) performed by the microprocessor. The opcode is either one or two bytes in length for most machine language instructions. Figure 4–2 illustrates the general form of the first opcode byte of many, but not all, machine language instructions. Here the first 6 bits of the first byte are the binary opcode. The remaining 2 bits indicate the **direction** (D), not to be confused with the instruction mode bit (16/32) or direction flag bit (used with string instructions), of the data flow and whether the data is a byte or a **word** (W). In the 80386 and above, words and doublewords are both specified when W = 1. The instruction mode and register-size prefix (66H) determine whether W represents a word or a doubleword.

   If the direction bit (D) = 1, data flows to the **register** (REG) field from the **R/M field** located in the second byte of an instruction. If the D-bit = 0 in the opcode, data flows to the R/M field from the REG field. If the W-bit = 1, the data size is a word or doubleword, and if the W-bit = 0, the data size is a byte. The W-bit appears in most instructions while the D-bit mainly appears with the MOV and some other instructions. Refer to Figure 4–3 for the binary bit pattern of the second opcode byte (reg-mod-r/m) of many instructions. Figure 4–3 shows the location of the MOD **(mode)**, REG (register), and R/M (register/memory) fields.

MOD Field.   The **MOD field** specifies the addressing mode (MOD) for the selected instruction. This field selects the type of addressing and whether a displacement is present with the selected type. Table 4–1 lists the operand forms available to the MOD field for 16-bit instruction mode unless the operand address-size override prefix (67H) appears. If the MOD field contains a 11 it selects the register-addressing mode. Register addressing uses the R/M field to specify a register instead of a memory location. If the MOD field contains a 00, 01, or 10, the R/M field selects one of the data memory-addressing modes. When MOD selects a data memory-addressing mode, it indicates whether the addressing mode contains no displacement (00), an 8-bit sign-extended

FIGURE 4–3 Byte 2 of many machine language instructions, showing the position of the MOD, REG, and R/M fields.

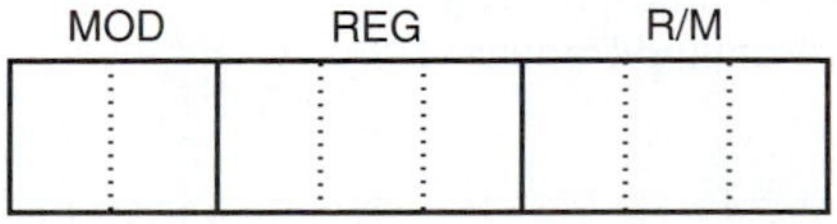

MOD	REG	R/M

TABLE 4–1   MOD field specifications for the 16-bit instruction mode

MOD	Function
00	No displacement
01	8-bit sign-extended displacement
10	16-bit displacement
11	R/M is a register

displacement (01), or a 16-bit displacement (10). The MOV AL,[DI] instruction is an example showing no displacement; a MOV AL,[DI + 2] instruction uses an 8-bit displacement (+ 2); a MOV AL,[DI + 1000H] instruction uses a 16-bit displacement (+ 1000H).

All 8-bit displacements are **sign-extended** into 16-bit displacements when the microprocessor executes the instruction. If the 8-bit displacement is 00H–7FH (positive), it is sign-extended to 0000H–007FH before adding to the offset address. If the 8-bit displacement is 80H–FFH (negative), it is sign-extended to FF80H–FFFFH. To sign-extend a number, its sign-bit is copied to the next higher-order byte, which generates either a 00H or an FFH in the higher-order byte. Note that some assembler programs do not use the 8-bit displacements.

In the 80386 through the Pentium Pro microprocessor, the MOD field may be the same as Table 4–1 or, if the instruction mode is 32-bits, it is as appears in Table 4–2. The MOD field is interpreted as selected by the address-size override prefix or the operating mode of the microprocessor. This change in the interpretation of the MOD field and instruction supports many of the numerous additional addressing modes allowed in the 80386 through the Pentium Pro. The main difference is when the MOD field is a 10; this causes the 16-bit displacement to become a 32-bit displacement to allow any protected-mode memory locations (4G bytes) to be accessed. The 80386 and above only allow an 8- or 32-bit displacement when operated in the 32-bit instruction mode unless the address-size override prefix appears. If an 8-bit displacement is selected, it is sign-extended into a 32-bit displacement by the microprocessor.

Register Assignments.   Table 4–3 lists the **register assignments** for the REG and R/M fields (MOD = 11). This table contains three lists of register assignments: one is used when the W-bit = 0 (bytes), and the other two are used when the W-bit = 1 (words or doublewords). Note that doubleword registers are only available to the 80386 through the Pentium Pro.

Suppose that a two-byte instruction, 8BECH, appears in a machine language program. Because neither a 67H (operand address-size override prefix) nor a 66H (register-size override prefix) appears as the first byte, the first byte is the opcode. Assuming that the microprocessor is operated in the 16-bit instruction mode, this instruction is converted to binary and placed in the instruction format of bytes 1 and 2 as illustrated in Figure 4–4. The opcode is 100010. If you refer to Appendix B, which lists the machine language instructions, you will find that this is the opcode for a MOV instruction. Also notice that both the D and W bits are a logic 1, which means that a word moves into the destination register specified in the REG field. The REG field contains a 101, indicating register BP, so the MOV instruction moves data into register BP. Because the MOD field contains a 11, the R/M field also indicates a register. Here, R/M = 100 (SP);

TABLE 4–2   MOD field specifications for the 32-bit instruction mode (80386/ 80486/Pentium/Pentium Pro)

MOD	Function
00	No displacement
01	8-bit sign-extended displacement
10	32-bit displacement
11	R/M is a register

TABLE 4–3 REG and R/M (when MOD = 11) assignments

Code	W = 0 (Byte)	W = 1 (Word)	W = 1 (Doubleword)
000	AL	AX	EAX
001	CL	CX	ECX
010	DL	DX	EDX
011	BL	BX	EBX
100	AH	SP	ESP
101	CH	BP	EBP
110	DH	SI	ESI
111	BH	DI	EDI

therefore, this instruction moves data from SP into BP and is written in symbolic form as a MOV BP,SP instruction.

Suppose that a 668BE8H instruction appears in an 80386 or above operated in the 16-bit instruction mode. The first byte (66H) is the register-size override prefix that selects 32-bit register operands for the 16-bit instruction mode. The remainder of the instruction indicates that the opcode is a MOV with a source operand of EAX and a destination operand of EBP. This instruction is a MOV EBP,EAX. The same instruction becomes a MOV BP,AX instruction in the 80386 and above if it is operated in the 32-bit instruction mode because the register-size override prefix selects a 16-bit register. Luckily, the assembler program keeps track of the register- and address-size prefixes and the mode of operation. Recall that if the .386 switch is placed before the .MODEL statement, the 32-bit mode is selected, and if it is placed after the .MODEL statement, the 16-bit mode is selected.

R/M Memory Addressing. If the MOD field contains a 00, 01, or 10, the R/M field takes on a new meaning. Table 4–4 lists the memory-addressing modes for the R/M field when MOD is a 00, 01, or 10 for the 16-bit instruction mode.

All of the 16-bit addressing modes presented in Chapter 3 appear in Table 4–4. The displacement, discussed in Chapter 3, is defined by the MOD field. If MOD = 00 and R/M = 101, the addressing mode is [DI]. If MOD = 01 or 10, the addressing mode is [DI + 33H] or LIST [DI + 22H] for the 16-bit instruction mode. This example uses LIST, 33H, and 22H as arbitrary values for the displacement.

Figure 4–5 illustrates the machine language version of the 16-bit instruction MOV DL,[DI] or instruction (8A15H). This instruction is 2 bytes long and has an opcode 100010, D = 1 (to REG from R/M), W = 0 (byte), MOD = 00 (no displacement), REG = 010 (DL), and R/M = 101 ([DI]). If the instruction changes to MOV DL,[DI + 1], the MOD field changes to 01 for an 8-bit displacement, but the first two bytes of the instruction remain the same. The instruction

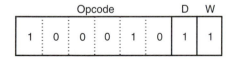

Opcode = MOV
D = Transfer to register (REG)
W = Word
MOD = R/M is a register
REG = BP
R/M = SP

FIGURE 4–4 The 8BEC instruction placed into byte 1 and 2 formats from Figures 4–2 and 4–3. This instruction is a MOV BP,SP.

TABLE 4–4 The 16-bit R/M memory-addressing modes

Code	Addressing Mode
000	DS:[BX+SI]
001	DS:[BX+DI]
010	SS:[BP+SI]
011	SS:[BP+DI]
100	DS:[SI]
101	DS:[DI]
110	SS:[BP]*
111	DS:[BX]

Note: See text under Special Addressing Mode.

now becomes 8A5501H instead of 8A15H. Notice that the 8-bit displacement appends to the first two bytes of the instruction to form a 3-byte instruction. If the instruction is changed again to a MOV DL,[DI + 1000H], the machine language form becomes an 8A750010H. Here the 16-bit displacement of 1000H (coded as 0010H) appends the opcode.

Special Addressing Mode.   There is a **special addressing mode** (not listed in Tables 4–2, 4–3, or 4–4) that occurs whenever memory data is referenced by only the displacement mode of addressing for 16-bit instructions. Examples are the MOV [1000H],DL and MOV NUMB,DL instructions. The first instruction moves the contents of register DL into data segment memory location 1000H. The second instruction moves register DL into symbolic data segment memory location NUMB.

Whenever an instruction has only a displacement, the MOD field is always a 00 and the R/M field is always a 110. This combination normally shows that the instruction contains no displacement and uses addressing mode [BP]. You cannot actually use addressing mode [BP] without a displacement in machine language. The assembler takes care of this by using an 8-bit displacement (MOD = 01) of 00H whenever the [BP] addressing mode appears in an instruction. This means that the [BP] addressing mode assembles as a [BP + 0] even though a [BP] is used in the instruction. The same special addressing mode is also available to the 32-bit mode. Figure 4–6 shows the binary bit pattern required to encode the MOV [1000H],DL instruction in machine language. If you were translating this symbolic instruction into machine language and did not know about the special addressing mode, you would incorrectly translate to a MOV [BP],DL instruction. Figure 4–7 shows the actual form of the MOV [BP],DL instruction. Notice that this is a 3-byte instruction with a displacement of 00H.

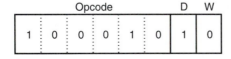
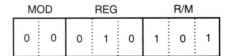

Opcode						D	W
1	0	0	0	1	0	1	0

MOD		REG			R/M		
0	0	0	1	0	1	0	1

Opcode = MOV
D = Transfer to register (REG)
W= Byte
MOD = No displacement
REG = DL
R/M = DS:[DI]

FIGURE 4–5 A MOV DL,[DI] instruction converted to its machine-language form.

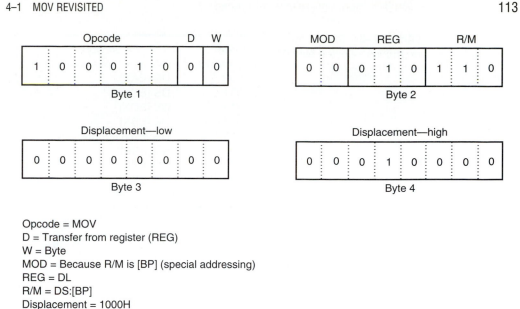

Opcode = MOV
D = Transfer from register (REG)
W = Byte
MOD = Because R/M is [BP] (special addressing)
REG = DL
R/M = DS:[BP]
Displacement = 1000H

FIGURE 4–6 The MOV [1000H],DL instruction uses the special addressing mode.

32-Bit Addressing Modes. The **32-bit addressing modes** found in the 80386 and above are obtained by running these machines either in the 32-bit instruction mode or the 16-bit instruction mode by using the address-size prefix 67H. Table 4–5 shows the coding for R/M used to specify the 32-bit addressing modes. Notice that when R/M = 100 an additional byte, called a scaled-index byte, appears in the instruction. The scaled-index byte indicates the additional forms of scaled-index addressing that do not appear in Table 4–5. The scaled-index byte is used mainly when two registers are added to specify the memory address in an instruction. Because the

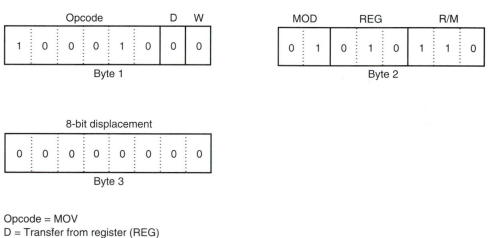

Opcode = MOV
D = Transfer from register (REG)
W = Byte
MOD = Because R/M is [BP] (special addressing)
REG = DL
R/M = DS:[BP]
Displacement = 00H

FIGURE 4–7 The MOV [BP],DL instruction converted to binary machine language.

TABLE 4–5 32-bit addressing modes selected by R/M

Code	Function
000	DS:[EAX]
001	DS:[ECX]
010	DS:[EDX]
011	DS:[EBX]
100	Uses scaled-index byte
101	SS:[EBP]*
110	DS:[ESI]
111	DS:[EDI]

Note: If the MOD bits are 00, this addressing mode uses a 32-bit displacement without register EBP. This is similar to the special addressing mode for the16-bit instruction mode.

scaled-index byte is added to the instruction, there are 7 bits in the opcode to define and 8 bits in the scaled-index byte. This means that a scaled-index instruction has 2^{15} (32K) possible combinations. There are over 32,000 variations of the MOV instruction in the 80386 through the Pentium Pro microprocessor.

Figure 4–8 shows the format of the scaled-index byte as selected by a value of 100 in the R/M field of an instruction when the 80386 and above use a 32-bit address. The leftmost 2 bits select a scaling factor (multiplier) of 1×, 2×, 4×, or 8×. Note that a scaling factor of 1× is implicit, if none is used, in an instruction that contains two 32-bit indirect address registers. Both the index and base fields contain register numbers as indicated in Table 4–3 for 32-bit registers.

The instruction MOV EAX,[EBX+4*ECX] is encoded as 67668B048BH. Notice that both the address-size (67H) and register-size (66H) override prefixes appear for this instruction. This means that the instruction is 67668B048BH when operated with the 80386 and above in the 16-bit instruction mode. If the microprocessor is operated in the 32-bit instruction mode, both prefixes disappear, so the instruction becomes a 8B048BH instruction. The use of the prefixes depends on the mode of operation of the microprocessor. Scaled-index addressing can also use a single register multiplied by a scaling factor. An example is the MOV AL,[2*ECX] instruction. The contents of the data segment location addressed by two times ECX is copied into AL.

An Immediate Instruction.   Suppose the MOV WORD PTR [BX + 1000H],1234H instruction is chosen as an example of a 16-bit instruction using **immediate addressing.** This instruction moves a 1234H into the word-sized memory location addressed by the sum of 1000H, BX, and DS × 10H. This 6-byte instruction uses two bytes for the opcode, W, MOD, and R/M fields. Two of the six bytes are the data of 1234H and two of the six bytes are the displacement of 1000H. Figure 4–9 shows the binary bit pattern for each byte of this instruction.

FIGURE 4–8 The scaled-index byte of some 32-bit addressing modes.

ss
00 = × 1
01 = × 2
10 = × 4
11 = × 8

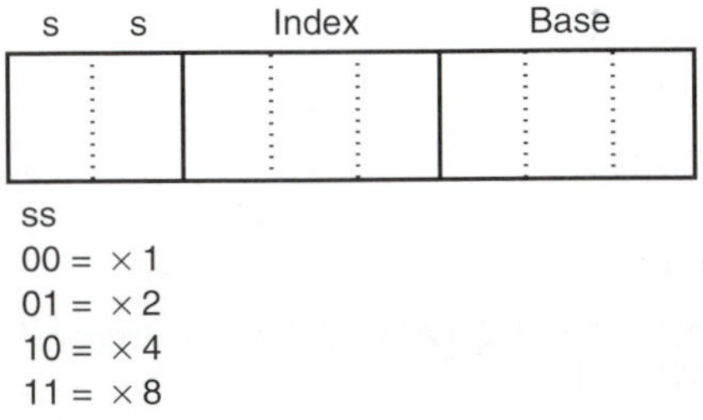

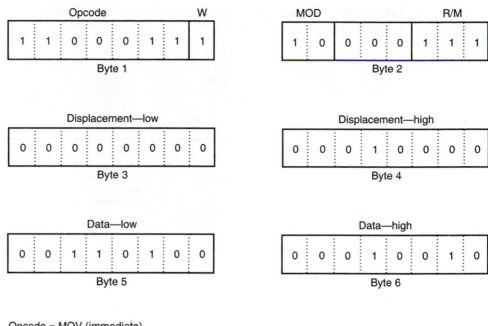

Opcode = MOV (immediate)
W = Word
MOD = 16-bit displacement
REG = 000 (not used in immediate addressing)
R/M = DS:[BX]
Displacement = 1000H
Data = 1234H

FIGURE 4–9 A MOV WORD PTR [BX+1000H],1234H instruction converted to binary machine language.

This instruction, in symbolic form, includes WORD PTR. The WORD PTR directive indicates to the assembler that the instruction uses a word-sized memory pointer. If the instruction moves a byte of immediate data, then BYTE PTR replaces WORD PTR in the instruction. Likewise, if the instruction uses a doubleword of immediate data, the DWORD PTR directive replaces BYTE PTR. Most instructions that refer to memory through a pointer do not need the BYTE PTR, WORD PTR, or DWORD PTR directives. These are only necessary when it is not clear if the operation is a byte or a word. The MOV [BX],AL instruction is clearly a byte move, while the MOV [BX],1 instruction is not exact and could therefore be a byte-, word-, or double-word-sized move. Here the instruction must coded as MOV BYTE PTR [BX],1; MOV WORD PTR [BX],1; or MOV DWORD PTR [BX],1. If not, the assembler flags it as an error because it cannot determine the intent of this instruction.

Segment MOV Instructions. If the contents of a segment register are moved by the MOV, PUSH, or POP instructions, a special set of register bits (REG field) selects the segment register (see Table 4–6).

Figure 4–10 shows a MOV BX,CS instruction converted to binary. The opcode for this type of MOV instruction is different from the prior MOV instructions. Segment registers can be moved between any 16-bit register or 16-bit memory location. For example, the MOV [DI],DS instruction stores the contents of DS into the memory location addressed by DI in the data segment. An immediate segment register MOV is not available in the instruction set. To load a segment register with immediate data, first load another register with the data and then move it to a segment register.

TABLE 4–6 Segment register selection bits

Code	Segment Register
000	ES
001	CS*
010	SS
011	DS
100	FS
101	GS

Note: MOV CS,R/M(16) and POP CS are not allowed by the microprocessor. The FS and GS segments are only available to the 80386/ 80486/Pentium/Pentium Pro microprocessors.

Opcode							
1	0	0	0	1	1	0	0

MOD		REG			R/M		
1	1	0	0	1	0	1	1

Opcode = MOV
MOD = R/M is a register
REG = CS
R/M = BX

FIGURE 4–10   A MOV BX,CS instruction converted to binary machine language.

Although this has not been a complete coverage of machine language coding, it should give you a good start in machine language programming. Remember that a program written in symbolic assembly language (assembly language) is rarely assembled by hand into binary machine language. Instead, an assembler converts symbolic assembly language into machine language. With the microprocessor and its over 100,000 instruction variations, let us hope that an assembler is available for the conversion, because the process is very time consuming, although not impossible.

4–2   PUSH/POP

PUSH and POP are important instructions that store and retrieve data from the LIFO (**last-in, first-out**) stack memory. The microprocessor has six forms of PUSH/POP instructions: register, memory, immediate, segment register, flags, and all registers. The PUSH and POP immediate and the PUSHA and POPA (all registers) forms are not available in the earlier 8086/8088 microprocessors, but are available to the 80286 through the Pentium Pro.

Register addressing allows the contents of any 16-bit register to be transferred to or from the stack. In the 80386 and above, the 32-bit extended registers and flags (EFLAGS) can also be pushed or popped from the stack. Memory-addressing PUSH and POP instructions store the contents of a 16-bit memory location (or 32 bits in the 80386 and above) on the stack or stack data into a memory location. Immediate addressing allows immediate data to be pushed onto the stack, but not popped off the stack. Segment register addressing allows the contents of any segment register to be pushed onto the stack or removed from the stack (CS may be pushed, but data from the stack may never be popped into CS). The flags may be pushed or popped from that stack, and the contents of all the registers may be pushed or popped.

PUSH

The 8086–80286 **PUSH** instruction always transfers two bytes of data to the stack, and the 80386 and above transfer two or four bytes, depending on the register or size of the memory location. The source of the data may be any internal 16-bit/32-bit register, immediate data, any segment register, or any two bytes of memory data. There is also a PUSHA instruction that copies the contents of the internal register set, except the segment registers, to the stack. The PUSHA (push all) instruction copies the registers to the stack in the following order: AX, CX, DX, BX, SP, BP, SI, and DI. The value of SP pushed to the stack is whatever it was before the PUSHA instruction executes. The PUSHF **(push flags)** instruction copies the contents of the flag register to the stack. The PUSHAD and POPAD instructions push and pop the contents of the 32-bit register set found in the 80386 through the Pentium Pro.

Whenever data is pushed onto the stack, the first (most-significant) data byte moves into the stack segment memory location addressed by SP − 1. The second (least-significant) data byte moves into the stack segment memory location addressed by SP − 2. After the data is stored by a PUSH, the contents of the SP register decrement by 2. The same is true for a doubleword push except four bytes are moved to the stack memory (most-significant byte first), then the stack pointer decrements by 4. Figure 4–11 shows the operation of the PUSH AX instruction. This instruction copies the contents of AX onto the stack where address SS:[SP − 1] = AH, SS:[SP − 2] = AL, and afterwards SP = SP − 2.

The **PUSHA instruction** pushes all the internal 16-bit registers onto the stack as illustrated in Figure 4–12. This instruction requires 16 bytes of stack memory space to store all eight

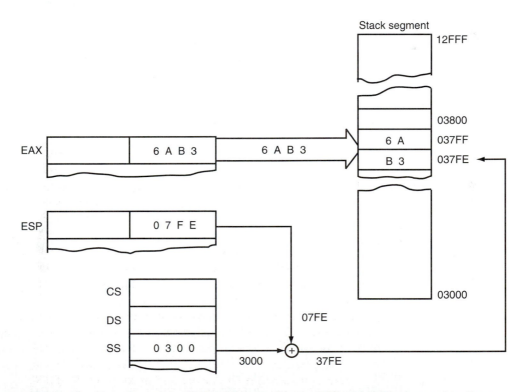

FIGURE 4–11 The effect of the PUSH AX instruction on ESP and stack memory locations 37FFH and 37 FEH. This instruction is shown at the point after execution.

FIGURE 4–12   The operation of the PUSHA instruction, showing the location and order of stack data.

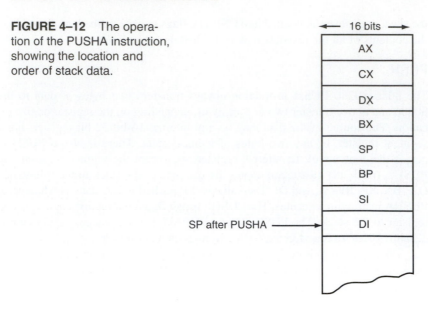

16-bit registers. After all registers are pushed, the contents of the SP register are decremented by 16. The PUSHA instruction is very useful when the entire register set (**microprocessor environment**) of the 80286 and above must be saved during a task. The PUSHAD instruction places the 32-bit register on the stack in the 80386 through the Pentium Pro. PUSHAD requires 32 bytes of stack storage space.

The **PUSH immediate** data instruction has two different opcodes, but in both cases a 16-bit immediate number moves onto the stack, or if PUSHD is used, a 32-bit immediate datum is pushed. If the value of the immediate data is 00H–FFH, the opcode is a 6AH and if the data is 0100H–FFFFH, the opcode is 68H. The PUSH 8 instruction, which pushes a 0008H onto the stack, assembles as a 6A08H and the PUSH 1000H instruction assembles as a 680010H. Another example of PUSH immediate is the PUSH 'A' instruction, which pushes a 0041H onto the stack. Here the 41H is the ASCII code for the letter A.

Table 4–7 lists the PUSH instructions that include PUSHA and PUSHF. Notice how the instruction set is used to specify different data sizes with the assembler.

TABLE 4–7   The PUSH instructions

Assembly Language	Example	Note
PUSH reg16	PUSH BX	16-bit register
PUSH reg32	PUSH EBX	32-bit register
PUSH mem16	PUSH WORD PTR [BX]	16-bit data
PUSH mem32	PUSH DWORD PTR [BX]	32-bit data
PUSH seg	PUSH DS	Segment register
PUSH imm8	PUSH 12H	8-bit immediate data
PUSH imm16	PUSH 1000H	16-bit immediate data
PUSHD imm32	PUSHD 12345678H	32-bit immediate data
PUSHA	PUSHA	Save 16-bit registers
PUSHAD	PUSHAD	Save 32-bit registers
PUSHF	PUSHF	Save FLAGS
PUSHFD	PUSHFD	Save EFLAGS

Note: the 80386/80486/Pentium/Pentium Pro are required to operate with 32-bit addresses, registers, and immediate data.

POP

The **POP instruction** performs the inverse operation of a PUSH instruction. POP removes data from the stack and places it into the target 16-bit register, segment register, or 16-bit memory location. In the 80386 and above, a POP can also remove 32-bit data from the stack and use a 32-bit address. The POP instruction is not available as an immediate POP. The POPF (pop flags) instruction removes a 16-bit number from the stack and places it into the flag register, and the POPFD removes a 32-bit number from the stack and places it into the extended flag register. The POPA (pop all) instruction removes 16 bytes of data from the stack and places it into the following registers in the order shown: DI, SI, BP, SP, BX, DX, CX, and AX. This is the reverse order from the way they are placed on the stack by the PUSHA instruction, causing the same data to return to the same registers. In the 80386 and above, a POPAD instruction reloads the 32-bit registers from the stack.

Suppose that a POP BX instruction executes. The first byte of data removed from the stack (the memory location addressed by SP in the stack segment) moves into register BL. The second byte is removed from stack segment memory location SP + 1, and placed into register BH. After both bytes are removed from the stack, the SP register increments by 2. Figure 4–13 shows how the POP BX instruction removes data from the stack and places it into register BX.

The opcodes used for the POP instruction, and all its variations, appear in Table 4–8. Note that a POP CS instruction is not a valid instruction in the instruction set. If a POP CS instruction executes, only a portion of the address (CS) of the next instruction changes. This makes the POP CS instruction unpredictable and therefore not allowed.

Initializing the Stack

When the stack area is initialized, you must load both the **stack segment** (SS) register and the **stack pointer** (SP) register. It is normal to designate an area of memory as the stack segment by loading SS with the bottom location of the stack segment.

For example, if the stack segment is to reside in memory locations 10000H–1FFFFH, load SS with a 1000H. (Recall that the rightmost end of the stack segment register is appended with a

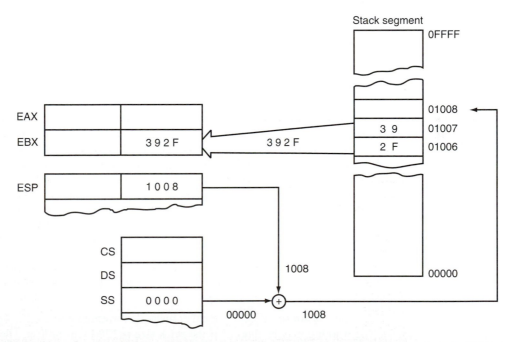

FIGURE 4–13 The POP BX instruction, showing how data is removed from the stack. This instruction is shown after execution.

TABLE 4–8   The POP instructions

Assembly Language	Example	Note
POP reg16	POP DI	16-bit register
POP reg32	POP ESI	32-bit register
POP mem16	POP DELTA	16-bit memory
POP mem32	POP DWORD PTR [EBX]	32-bit memory
POP seg	POP GS	Segment register
POPA	POPA	16-bit registers
POPAD	POPAD	32-bit registers
POPF	POPF	FLAGS
POPFD	POPFD	EFLAGS

Note: The 80386/80486/Pentium/Pentium Pro are required to operate with 32-bit address-es and registers.

0H for real-mode addressing.) To start the stack at the **top** of this 64K-byte stack segment, the stack pointer (SP) is loaded with a 0000H. Likewise, to address the top of the stack at location 10FFFH, use a value of 1000H in SP. Figure 4–14 shows how this value causes data to be pushed onto the top of the stack segment with a PUSH CX instruction. Remember that all segments are cyclic in nature—that is, the top location of a segment is contiguous with the bottom location of the segment.

In assembly language, a stack segment is set up as illustrated in Example 4–1. The first statement identifies the start of the stack segment and the last statement identifies the end of the stack segment. The assembler and linker program place the correct stack segment address in SS and the length of the segment (top of the stack) into SP. There is no need to load these registers in your program unless for some reason you wish to change the initial values.

EXAMPLE 4–1

```
0000                    STACK_SEG     SEGMENT STACK

0000  0100[                           DW    100H DUP (?)
        ????
              ]
0200                    STACK_SEG     ENDS
```

An alternative method for defining the stack segment is used with one of the memory models for the MASM assembler only (refer to Appendix A). Other assemblers do not use models, or if they do, they are not exactly the same as with MASM. In Example 4–2 the .STACK statement, followed by the number of bytes allocated to the stack, defines the stack area. The function is identical to Example 4–1. The .STACK statement also initializes both SS and SP. (These examples use memory models designed to be used with the Microsoft Macro Assembler program, MASM.)

EXAMPLE 4–2

```
                .MODEL SMALL
                .STACK 200H         ;set stack size
```

If the stack is not specified using either method, a warning will appear when the program is linked. The warning may be ignored if the stack size is 128 bytes or less. The system automati-cally assigns (through DOS) at least 128 bytes of memory to the stack. This memory section is lo-cated in the **program segment prefix** (PSP), which is appended to the beginning of each program

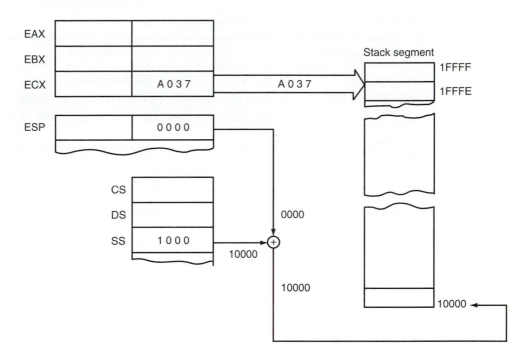

FIGURE 4–14   The PUSH CX instruction showing the cyclic nature of the stack segment. This instruction is shown just before execution, to illustrate that the stack bottom is contiguous to the top.

file. If you use more memory for the stack, you will erase information in the PSP that is critical to the operation of your program and the computer. This error often causes the computer program to crash. If the TINY memory model is used, the stack is automatically located at the very end of the segment, which allows for a larger stack area.

4–3   LOAD-EFFECTIVE ADDRESS

There are several load-effective address instructions in the microprocessor instruction set. The LEA instruction loads any 16-bit register with the address determined by the addressing mode selected for the instruction. The LDS and LES variations load any 16-bit register with the offset address retrieved from a memory location and then load either DS or ES with a segment address retrieved from memory. In the 80386 and above, LFS, LGS, and LSS are added to the instruction set and a 32-bit register can be selected to receive a 32-bit offset from memory. Table 4–9 lists the load-effective address instructions.

LEA

The **LEA instruction** loads a 16- or 32-bit register with the offset address of the data specified by the operand. As the first example in Table 4–9 shows, the operand address NUMB is loaded into register AX, not the contents of address NUMB. Note that the source data must never be a register for the LEA instruction.

By comparing LEA with MOV, it is observed that: LEA BX,[DI] loads the offset address specified by [DI] (contents of DI) into the BX register; MOV BX,[DI] loads the data stored at the memory location addressed by [DI] into register BX.

TABLE 4–9   The load-effective address instructions

Assembly Language	Operation
LEA AX,NUM	AX is loaded with the offset address of NUMB
LEA EAX,NUMB	EAX is loaded with the offset address of NUMB
LDS DI,LIST	DS and DI are loaded with the segment and 16-bit offset address stored at LIST
LDS EDI,LIST	DS and EDI are loaded with the segment and 32-bit offset address stored at LIST
LES BX,CAT	ES and BX are loaded with the segment and 16-bit offset address stored at CAT
LFS DI,DATA1	FS and DI are loaded with the segment and 16-bit offset address stored at DATA1
LGS SI,DATA5	GS and SI are loaded with the segment and 16-bit offset address stored at DATA5
LSS SP,MEM	SS and SP are loaded with the segment and 16-bit offset address stored at MEM

Earlier we saw several examples using the OFFSET directive. The **OFFSET directive** performs the same function as an LEA instruction if the operand is a displacement. For example, the MOV BX,OFFSET LIST performs the same function as LEA BX,LIST. Both instructions load the offset address of memory location LIST into the BX register. Example 4–3 is a short program that loads SI with the address of DATA1 and DI with the address of DATA2. It then exchanges the contents of these memory locations. Note that the LEA and MOV with OFFSET instruction are both the same length (3 bytes).

EXAMPLE 4–3

```
                            .MODEL SMALL            ;select SMALL model
0000                        .DATA                   ;start of DATA segment
0000 2000        DATA1      DW    2000H             ;define DATA1
0002 3000        DATA2      DW    3000H             ;define DATA2
0000                        .CODE                   ;start of CODE segment
                            .STARTUP                ;indicate start of program
0017 BE 0000 R              LEA   SI,DATA1          ;address DATA1 with SI
001A BF 0002 R              MOV   DI,OFFSET DATA2   ;address DATA2 with DI

001D 8B 1C                  MOV   BX,[SI]           ;exchange DATA1 with DATA2
001F 8B 0D                  MOV   CX,[DI]
0021 89 0C                  MOV   [SI],CX
0023 89 1D                  MOV   [DI],BX
                            .EXIT                   ;exit to DOS
                            END                     ;indicate end of file
```

But why is the LEA instruction available if the OFFSET directive accomplishes the same task? First, OFFSET only functions with simple operands such as LIST. It may not be used for an operand such as [DI], LIST [SI], etc. The OFFSET directive is **more efficient** than the LEA instruction for simple operands. It takes the microprocessor longer to execute the LEA BX,LIST instruction than the MOV BX,OFFSET LIST. The 80486 microprocessor, for example, requires two clocks to execute the LEA BX,LIST instruction and only one clock to execute MOV BX,OFFSET LIST. The MOV BX,OFFSET LIST instruction executes faster because the assembler calculates the offset address of LIST; with the LEA instruction, the microprocessor does the calculation as it executes the instruction. The MOV BX,OFFSET LIST instruction is actually assembled as a move immediate instruction and is more efficient.

Suppose that the microprocessor executes an LEA BX,[DI] instruction and DI contains a 1000H. Because DI contains the offset address, the microprocessor transfers a copy of DI into BX. A MOV BX,DI instruction performs this task in less time and is often preferred to the LEA BX,[DI] instruction.

Another example is LEA SI,[BX + DI]. This instruction adds BX to DI and stores the sum in the SI register. The sum generated by this instruction is a modulo-64K sum. If BX = 1000H and DI = 2000H, the offset address moved into SI is 3000H. If BX = 1000H and DI = FF00H, the offset address is 0F00H instead of 10F00H. Notice that the second result is a modulo-64K sum of 0F00H. (A modulo-64K sum drops any carry out of the 16-bit result.)

LDS, LES, LFS, LGS, and LSS

The LDS, LES, LFS, LGS, and LSS instructions load any 16-bit or 32-bit register with an offset address and the DS, ES, FS, GS, or SS segment register with a segment address. These instructions use any of the memory-addressing modes to access a 32- or 48-bit section of memory that contains both the segment and offset addresses. The 32-bit section of memory contains a 16-bit offset and segment address, while the 48-bit section contains a 32-bit offset and a segment address. The offset data is stored in the location addressed by the source, which is followed by the segment data. These instructions may not use the register-addressing mode (MOD = 11). The LFS, LGS, and LSS instructions are available only on 80386 and above, as are the 32-bit registers.

Figure 4–15 illustrates an example LDS BX,[DI] instruction. This instruction transfers the 32-bit number, addressed by DI in the data segment, into the BX and DS registers. The LDS, LES, LFS, LGS, and LSS instructions obtain a new far address from memory. The offset address appears first, followed by the segment address. This format is used for storing all 32-bit memory addresses.

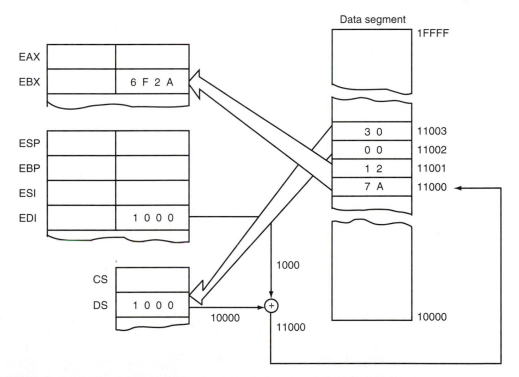

FIGURE 4–15 The LDS BX,[DI] instruction loads register BX from addresses 11000H and 11001H and register DS from locations 11002H and 11003H. This instruction is shown at the point just before DS changes to 3000H and BX changes to 127AH.

A far address can be stored in memory by the assembler. For example, the ADDR DD FAR PTR FROG instruction stores the offset and segment addresses (far address) of FROG in 32 bits of memory at location ADDR. The DD directive tells the assembler to store a doubleword (32-bit number) in memory address ADDR.

In the 80386 and above, an LDS EBX,[DI] instruction loads EBX from the four-byte section of memory addressed by DI in the data segment. Following this four-byte offset is a word that is loaded to the DS register. Notice that instead of addressing a 32-bit section of memory, the 80386 and above address a 48-bit section of the memory whenever a 32-bit offset address is loaded to a 32-bit register. The first four bytes contain the offset value loaded to the 32-bit register and the last two bytes contain the segment address.

The most useful of the load instructions is the LSS instruction. Example 4–4 is a short program that creates a new stack area after saving the address of the old stack area. After executing some dummy instructions, the old stack area is reactivated by loading both SS and SP with the LSS instruction. The CLI **(disable interrupt)** and STI **(enable interrupt)** instructions must be included to disable interrupts, a topic discussed near the end of this chapter. Because the LSS instruction functions in the 80386 or above, the .386 statement appears after the .MODEL statement to select the 80386 microprocessor. The WORD PTR directive is used to override the doubleword (DD) definition for the old stack address memory location. If an 80386 or newer microprocessor is in use, it is suggested that the .386 switch be used to develop software for the 80386 microprocessor. This is true even if the microprocessor is a Pentium Pro. The reason is that the 80486–Pentium Pro offer only a few additional instructions to the 80386 instruction set, which are seldom used in software development. If you need to use any of the CMPXCHG, CMPXCHG8 (new to the Pentium), XADD, or BSWAP instructions, then select the .486 switch for the 80486 microprocessor or the .586 switch for the Pentium.

EXAMPLE 4–4

```
                                .MODEL SMALL            ;select SMALL model
                                .386                    ;select 80386 microprocessor
0000                            .DATA                   ;start of DATA segment
0000 00000000   SADDR    DD    ?                        ;old stack address
0004 1000 [     SAREA    DW    1000H DUP (?)            ;new stack area
        0000
      ]
2004 = 2004     STOP     EQU   THIS WORD                ;define new stack
0000                            .CODE                   ;start of CODE segment
                                .STARTUP                ;indicate start of program

0010 FA                         CLI                     ;disable interrupt

0011 8B C4                      MOV   AX,SP             ;save old SP
0013 A3 0000 R                  MOV   WORD PTR SADDR,AX
0016 8C D0                      MOV   AX,SS             ;save old SS
0018 A3 0002 R                  MOV   WORD PTR SADDR+2,AX

001B 8C D8                      MOV   AX,DS             ;load new SS
001D 8E D0                      MOV   SS,AX
001F B8 2004 R                  MOV   AX,OFFSET STOP    ;load new SP
0022 8B E0                      MOV   SP,AX

0024 FB                         STI                     ;enable interrupt

0025 8B C0                      MOV   AX,AX             ;do dummy instructions
0027 8B C0                      MOV   AX,AX             ;anything can appear here

0029 0F B2 26 0000 R            LSS   SP,SADDR          ;load old SS and SP

                                .EXIT                   ;exit to DOS
                                END                     ;end of file
```

4–4   STRING DATA TRANSFERS

The five string data transfer instructions (LODS, STOS, MOVS, INS, and OUTS) allow data transfers that are either a single byte, word, or doubleword; or if repeated, a block of bytes, words, or doublewords. Before we discuss string instructions, we must understand the operation of the D flag-bit (direction), DI, and SI as they apply to string instructions.

The Direction Flag

The **direction flag** (D) selects **auto-increment** (D = 0) or **auto-decrement** (D = 1) operations for the DI and SI registers during string operations. The direction flag is used only with the string instructions. The **CLD** instruction clears the D flag (D = 0) and the **STD** instruction sets it (D = 1). Therefore, the CLD instruction selects the auto-increment mode (D = 0) and STD selects the auto-decrement mode (D = 1).

     Whenever a string instruction transfers a byte, the contents of DI and/or SI increment or decrement by 1. If a word is transferred, the contents of DI and/or SI increment or decrement by 2. Doubleword transfers cause DI and/or SI to increment or decrement by 4. Only the actual registers used by the string instruction increment or decrement. For example, the STOSB instruction uses the DI register to address a memory location. When STOSB executes, only DI increments or decrements without affecting SI. The same is true of the LODSB instruction, which uses the SI register to address memory data. LODSB only increments/decrements SI without affecting DI.

DI and SI

During the execution of a string instruction, memory accesses occur through the DI and/or SI registers. The **DI offset address** accesses data in the extra segment for all string instructions that use it. The **SI offset address** accesses data, by default, in the data segment. The segment assignment of SI may be changed with a segment override prefix as described later in this chapter. The DI segment assignment is always in the extra segment when a string instruction executes. This assignment cannot be changed. One pointer addresses data in the extra segment and the other in the data segment so that the MOVS instruction can move 64K bytes of data from one segment of memory to another.

LODS

The **LODS instruction** loads AL, AX, or EAX with data stored at the data segment offset address indexed by the SI register. (Only the 80386, 80486, and Pentium can use EAX.) After loading AL with a byte, AX with a word, or EAX with a doubleword, the contents of SI increment, if D = 0, or decrement, if D = 1. A one is added to or subtracted from SI for a byte-sized LODS; a 2 is added or subtracted for a word-sized LODS; and a 4 is added or subtracted for a doubleword-sized LODS.

     Table 4–10 lists the permissible forms of the LODS instruction. The LODSB (loads a byte) instruction causes a byte to be loaded into AL; the LODSW (loads a word) instruction causes a word to be loaded into AX; the LODSD (loads a doubleword) instruction causes a doubleword to be loaded into EAX. Although rare, as an alternative to LODSB, LODSW, and LODSD the LODS instruction may be followed by a byte-, word-, or doubleword-sized operand to select a byte, word, or doubleword transfer. Operands are often defined as bytes with DB, as words with DW, and as doublewords with DD. The DB pseudo-operation defines byte(s); the DW pseudo-operation defines word(s); the DD pseudo-operation defines doubleword(s).

     Figure 4–16 shows the effect of executing the LODSW instruction if the D flag = 0, SI = 1000H, and DS = 1000H. Here a 16-bit number, stored at memory locations 11000H and

TABLE 4–10   Forms of the LODS instruction

Assembly Language	Operation
LODSB	AL = DS:[SI]; SI = SI ± 1
LODSW	AX = DS:[SI]; SI = SI ± 2
LODSD	EAX = DS:[SI]; SI = SI ± 4
LODS LIST	AL = DS:[SI]; SI = SI ± 1 (if LIST is a byte)
LODS DATA1	AX = DS:[SI]; SI = SI ± 2 (if LIST is a word)
LODS ES:DATA4	EAX = ES:[SI]; SI = SI ± 4 (if DATA4 is a doubleword)

Note: Only the 80386/80486/Pentium/Pentium Pro use doublewords. The segment can be overridden with a segment override prefix as in LODS ES:DATA4.

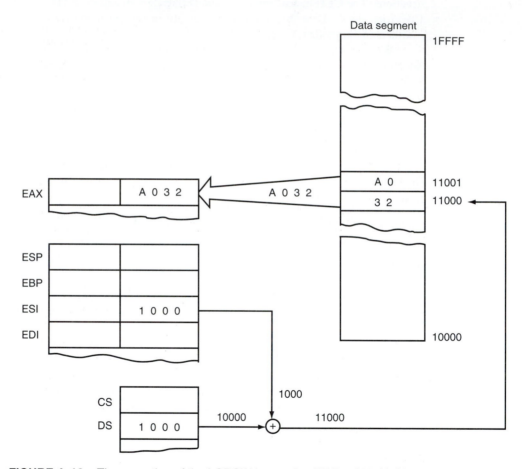

FIGURE 4–16   The operation of the LODSW instruction if DS = 1000H, SI = 1000H, D = 0, 11000H = 32, and 11001H = A0. This instruction is shown after AX is loaded from memory, but before SI increments by 2.

11001H moves into AX. Because D = 0 and this is a word transfer, the contents of SI increment by 2 after AX loads with memory data.

STOS

The **STOS instruction** stores AL, AX, or EAX at the extra segment memory location addressed by the DI register. (Only the 80386–Pentium Pro can use EAX and doublewords.) Table 4–11

TABLE 4–11 Forms of the STOS instruction

Assembly Language	Operation
STOSB	ES:[DI] = AL; DI = DI ± 1
STOSW	ES:[DI] = AX; DI = DI ± 2
STOSD	ES:[DI] = EAX; DI = DI ± 4
STOS LIST	ES:[DI] = AL; DI = DI ± 1 (if LIST is a byte)
STOS DATA1	ES:[DI] = AX; DI = DI ± 2 (if DATA1 is a word)
STOS DATA4	ES:[DI] = EAX; DI = DI ± 4 (if DATA4 is a doubleword)

Note: Doublewords are used only by the 80386, 80486, Pentium, or Pentium Pro microprocessors.

lists all forms of the STOS instruction. As with LODS, a STOS instruction may be appended with a B, W, or D for a byte, word, or doubleword transfers. The STOSB (stores a byte) instruction stores the byte in AL at the extra segment memory location addressed by DI. The STOSW (stores a word) instruction stores AX in the extra segment memory location addressed by DI. A doubleword is stored in the extra segment location addressed by DI with the STOSD (stores a doubleword) instruction. After the byte (AL), word (AX), or doubleword (EAX) is stored, the contents of DI increments or decrements.

STOS with a REP.   The **repeat prefix** (REP) is added to any string data transfer instruction except the LODS instruction. It doesn't make any sense to perform a repeated LODS operation. The REP prefix causes CX to decrement by 1 each time the string instruction executes. After CX decrements, the string instruction repeats. If CX reaches a value of 0, the instruction terminates and the program continues with the next sequential instruction. Thus, if CX is loaded with a 100 and a REP STOSB instruction executes, the microprocessor automatically repeats the STOSB instruction 100 times. Since the DI register is automatically incremented or decremented after each datum is stored, this instruction stores the contents of AL in a block of memory instead of a single byte of memory.

Suppose that the STOSW instruction is used to clear the video text display (see Example 4–5). This is accomplished by addressing video text memory that begins at memory location B800:0000. Each character position on the 25-line by 80-character per line display is comprised of two bytes. The first byte contains the ASCII-coded character and the second contains the color and attributes of the character. In this example AL is the ASCII code for space (20H) and AH is the color white text on a black background (07H). Notice how this program uses a count of 25*80 and the REP STOSW instruction to clear the screen with ASCII spaces.

The operands in a program can be modified by using arithmetic or logic operators such as multiplication (*). Other operators appear in Table 4–12.

EXAMPLE 4–5

```
                              .MODEL TINY            ;select TINY model
0000                          .CODE                  ;start of CODE segment
                              .STARTUP               ;indicate start of program
0100 FC                       CLD                    ;select increment mode
0101 B8 B800                  MOV   AX,0B800H         ;address segment B800
0104 8E C0                    MOV   ES,AX

0106 BF 0000                  MOV   DI,0              ;address offset 0000
0109 B9 07D0                  MOV   CX,25*80          ;load count
010C B8 0720                  MOV   AX,0720H          ;load data

010F F3/ AB                   REP   STOSW             ;clear the screen
                              .EXIT                  ;exit to DOS
                              END                    ;end of file
```

TABLE 4–12   Common operators

Operator	Example	Comment
+	MOV AL,6+3	Copies 9 into AL
−	MOV AL,8–2	Copies 6 into AL
*	MOV CX,4*3	Copies 12 into CX
/	MOV AX,12/5	Copies 2 into AX (remainder is dropped)
MOD	MOV AX,12 MOD 7	Copies 5 into AX (quotient is dropped)
AND	MOV AX,12 AND 4	Copies 4 into AX (1100 AND 0100 = 0100)
OR	MOV AX,12 OR 1	Copies 13 into AX (1100 OR 0001 = 1101
NOT	MOV AL,NOT 1	Copies 254 into AL (NOT 0000 0001 = 1111 1110 or 254)

The REP prefix precedes the STOSW instruction in both assembly language and hexadecimal machine language. In machine language, the F3H is the REP prefix and ABH is the STOSW opcode.

If the value loaded to AX is changed to 0731H, the video display fills with white ones on a black background. If AX is changed to 0132H, the video display fills with blue twos on a black background. By changing the value loaded to AX, the display can be filled with any character and any color combination. We'll see more information in a later chapter on accessing the video display.

MOVS

One of the more useful string data transfer instructions is **MOVS** because it transfers data from one memory location to another. This is the only **memory-to-memory transfer** allowed in the 8086–Pentium Pro microprocessors. The MOVS instruction transfers a byte, word, or doubleword from the data segment location addressed by SI, to the extra segment location addressed by DI. As with the other string instructions, the pointers then increment or decrement as dictated by the direction flag. Table 4–13 lists all the permissible forms of the MOVS instruction. Note that only the source operand (SI), located in the data segment, may be overridden so another segment can be used. The destination operand (DI) must always be located in the extra segment.

Suppose that the video display needs to be scrolled up one line. Because we now know the location of the video display, a repeated MOVSW instruction can be used to scroll the video dis-

TABLE 4–13   Forms of the MOVS instruction

Assembly Language	Operation
MOVSB	ES:[DI] = DS:[SI]; DI = DI ± 1; SI = SI ± 1 (byte is transferred)
MOVSW	ES:[DI] = DS:[SI]; DI = DI ± 2; SI = SI ± 2 (word is transferred)
MOVSD	ES:[DI] = DS:[SI]; DI = DI ± 4; SI = SI ± 4 (doubleword is transferred)
MOVS BYTE1,BYTE2	ES:[DI] = DS:[SI]; DI = DI ± 1; SI = SI ± 1 (if BYTE1 and BYTE2 are bytes)
MOVS WORD1,WORD2	ES:[DI] = DS:[SI]; DI = DI ± 2; SI = SI ± 2 (if WORD1 and WORD2 are words)
MOVS DWORD1,DWORD2	ES:[DI] = DS:[SI]; DI = DI ± 4; SI = SI ± 4 (if DWORD1 and DWORD2 are doublewords)

play up a line. Example 4–6 is a short program that addresses the video text display beginning at location B800:0000 with the DS:SI register combination and location B800:00A0 with the ES:DI register combination. Next, the REP MOVSW instruction is executed 24*80 times to scroll the display up a line. This is followed by a sequence that addresses the last line of the display so it can be cleared. The last line is cleared in this example by storing spaces on a black background. The last line could be cleared by changing only the ASCII code to a space without modifying the attribute by reading the code and attribute into a register. Once in a register, the code is modified, and both the code and attribute are stored in memory.

EXAMPLE 4–6

```
                            .MODEL TINY       ;select TINY model
0000                        .CODE             ;start of CODE segment
                            .STARTUP          ;indicate start of program
0100 FC                     CLD               ;select increment
0101 B8 B800                MOV   AX,0B800H    ;load ES and DS with B800
0104 8E C0                  MOV   ES,AX
0106 8E D8                  MOV   DS,AX

0108 BE 00A0                MOV   SI,160       ;address line 1
010B BF 0000                MOV   DI,0         ;address line 0
010E B9 0780                MOV   CX,24*80     ;load count
0111 F3/ A5                 REP   MOVSW        ;scroll screen

0113 BF 0F00                MOV   DI,24*80*2   ;clear bottom line
0116 B9 0050                MOV   CX,80
0119 B8 0720                MOV   AX,0720H
011C F3/ AB                 REP   STOSW
                            .EXIT             ;exit to DOS
                            END               ;end of file
```

INS

The INS (**input string**) instruction (not available on the 8086/8088 microprocessors) transfers a byte, word, or doubleword of data from an I/O device into the extra segment memory location addressed by the DI register. The I/O address is contained in the DX register. This instruction is useful for inputting a block of data from an external I/O device directly into memory. One application transfers data from a disk drive to memory. Disks drives are often considered and interfaced as I/O devices in a computer system.

As with the prior string instructions, there are two basic forms of the INS. The INSB instruction inputs data from an 8-bit I/O device and stores it in the byte-sized memory location indexed by SI. The INSW instruction inputs 16-bit I/O data and stores it in a word-sized memory location. The INSD instruction inputs a doubleword. These instructions can be repeated using the REP prefix. This allows an entire block of input data to be stored in the memory from an I/O device. Table 4–14 lists the various forms of the INS instruction.

TABLE 4–14   Forms of the INS instruction

Assembly Language	Operation
INSB	ES:[DI] = [DX]; DI = DI ± 1 (byte transferred)
INSW	ES:[DI] = [DX]; DI = DI ± 2 (word transferred)
INSD	ES:[DI] = [DX]; DI = DI ± 4 (doubleword transferred)
INS LIST	ES:[DI] = [DX]; DI = DI ± 1 (if LIST is a byte)
INS DATA1	ES:[DI] = [DX]; DI = DI ± 2 (if DATA1 is a word)
INS DATA4	ES:[DI] = [DX]; DI = DI ± 4 (if DATA4 is a doubleword)

Note: [DX] indicates that DX contains the I/O device address. These instructions are not available on the 8086/8088 microprocessors, and only the 80386/80486/Pentium/Pentium Pro use doublewords.

TABLE 4–15   Forms of the OUTS instruction

Assembly Language	Operation
OUTSB	[DX] = DS:[SI]; SI = SI ± 1 (byte transferred)
OUTSW	[DX] = DS:[SI]; SI = SI ± 2 (word transferred)
OUTSD	[DX] = DS:[SI]; SI = SI ± 4 (doubleword transferred)
OUTS LIST	[DX] = DS:[SI]; SI = SI ± 1 (if LIST is a byte)
OUTS DATA1	[DX] = DS:[SI]; SI = SI ± 2 (if DATA1 is a word)
OUTS DATA4	[DX] = DS:[SI]; SI = SI ± 4 (if DATA4 is a doubleword)

Note: [DX] indicates that DX contains the I/O device address. These instructions are not available on the 8086/8088 microprocessors, and only the 80386/80486/Pentium/Pentium Pro use doublewords.

Example 4–7 is a sequence of instructions that input 50 bytes of data from an I/O device whose address is 03ACH and stores the data in extra segment memory array LISTS. This software assumes that data is available from the I/O device at all times. Otherwise, the software must check to see if the I/O device is ready to transfer data precluding the use of a REP prefix.

EXAMPLE 4–7

```
                    ;Using the REP INSB to input data to a memory array
                    ;
0000 BF 0000 R          MOV     DI,OFFSET LISTS  ;address array
0003 BA 03AC            MOV     DX,3ACH          ;address I/O
0006 FC                 CLD                      ;auto-increment
0007 B9 0032            MOV     CX,50            ;load count
000A F3/6C              REP     INSB             ;input data
```

OUTS

The OUTS (**output string**) instruction (not available on the 8086/8088 microprocessors) transfers a byte, word, or doubleword of data from the data segment memory location address by SI to an I/O device. The I/O device is addressed by the DX register as it is with the INS instruction. Table 4–15 shows the variations available for the OUTS instruction.

Example 4–8 is a short sequence of instructions that transfer data from a data segment memory array (ARRAY) to an I/O device at I/O address 3ACH. This software assumes that the I/O device is always ready for data.

EXAMPLE 4–8

```
                    ;Using the REP OUTS to output data from a memory array
                    ;
0000 BE 0064 R          MOV     SI,OFFSET ARRAY  ;address array
0003 BA 03AC            MOV     DX,3ACH          ;address I/O
0006 FC                 CLD                      ;auto-increment
0007 B9 0064            MOV     CX,100           ;load count
000A F3/6E              REP     OUTSB
```

4–5   MISCELLANEOUS DATA TRANSFER INSTRUCTIONS

Don't be fooled by the term "miscellaneous"; these instructions are important because they are used in programs. The data transfer instructions discussed in this section are: XCHG, LAHF, SAHF, XLAT, IN, OUT, BSWAP, MOVSX, and MOVZX.

XCHG

The **exchange** (XCHG) instruction exchanges the contents of a register with the contents of any other register or memory location. The XCHG instruction cannot exchange segment registers or memory-to-memory data. Exchanges are byte-, word-, or doubleword-sized (80386 and above) and use any addressing mode discussed in Chapter 3 except immediate addressing. Table 4–16 shows the forms available for the XCHG instruction.

The XCHG instruction, using the 16-bit AX register with another 16-bit register, is the most efficient exchange. This instruction occupies one byte of memory. Other XCHG instructions require two or more bytes of memory depending on the addressing mode selected.

When using a memory addressing mode and the assembler, it doesn't matter which operand addresses memory. The XCHG AL,[DI] instruction is identical to the XCHG [DI],AL instruction as far as the assembler is concerned.

If the 80386 through the Pentium Pro microprocessors are available, the XCHG instruction can exchange doubleword data. For example, the XCHG EAX,EBX instruction exchanges the contents of the EAX register with the EBX register.

LAHF and SAHF

The LAHF and SAHF instructions are seldom used because they were designed as bridge instructions. These instructions allowed 8085 (an early 8-bit microprocessor) software to be translated into 8086 software by a translation program. Because any software that required translation was probably completed many years ago, these instructions have little application today. The LAHF instruction transfers the rightmost 8 bits of the flag register into the AH register. The SAHF instruction transfers the AH register into the rightmost 8 bits of the flag register.

At times the SAHF instruction may find some application with the numeric coprocessor. The numeric coprocessor contains a status register that is copied into the AX register with the FSTSW instruction. The SAHF instruction is used to copy from AH into the flag register; the flags are then tested for some of the conditions of the numeric coprocessor. This is detailed in the chapter that explains the operation and programming of the numeric coprocessor.

XLAT

The XLAT (**translate**) instruction converts the contents of the AL register into a number stored in a memory table. This instruction performs the direct table lookup technique often used to convert one code to another. An XLAT instruction first adds the contents of AL to BX to form a memory address within the data segment. It then copies the contents of this address into AL. This is the only instruction that adds an 8-bit number to a 16-bit number.

Suppose that a seven-segment LED display lookup table is stored in memory at address TABLE. The XLAT instruction then translates the BCD number in AL to a seven-segment code in AL. Exam-ple 4–9 is a short program that converts from a BCD code to a seven-segment code. Figure 4–17 shows the operation of this example program if TABLE = 1000H, DS = 1000H, and the initial value of AL = 05H (a 5 BCD). After the translation, AL = 6DH.

TABLE 4–16 Forms of the XCHG instruction

Assembly Language	Operation
XCHG reg,reg	Exchanges byte, word, or doubleword register contents
XCHG reg,mem	Exchanges byte, word, or doubleword memory data with register data

Note: Only the 80386/80486/Pentium/Pentium Pro use doubleword data.

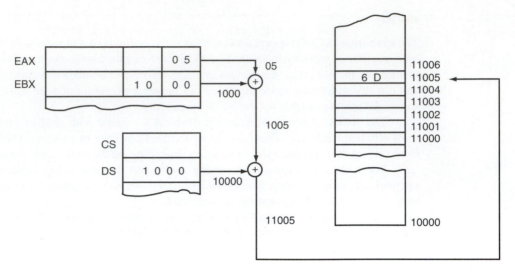

FIGURE 4–17   The operation of the XLAT instruction at the point just before 6DH is loaded into AL.

EXAMPLE 4–9

```
                          ;Using an XLAT to convert from BCD to 7-segment code
                          ;
                                .MODEL SMALL            ;select SMALL model
0000                            .DATA                   ;start of DATA segment
0000 3F 06 5B 4F    TABLE   DB     3FH,6,5BH,4FH     ;7-segment lookup table
0004 66 6D 7D 27            DB     66H,6DH,7DH,27H
0008 7F 6F                  DB     7FH,6FH
000A 00             CODE7   DB     ?                 ;reserve place for result
0000                            .CODE                   ;start of CODE segment
                                .STARTUP                ;indicate start of program
0017 B0 04                  MOV  AL,4                 ;load test data
0019 BB 0000 R              MOV  BX,OFFSET TABLE      ;address lookup table
001C D7                     XLAT                      ;convert AL to 7-segment code
001D A2 000A R              MOV  CODE7,AL             ;save 7-segment code
                                .EXIT                   ;exit to DOS
                                END                     ;end of file
```

IN and OUT

Table 4–17 lists the forms of the IN and OUT instructions that perform I/O operations. Notice that the contents of only AL, AX, or EAX are transferred between the I/O device and the microprocessor. An **IN instruction** transfers data from an external I/O device to AL, AX, or EAX, and an **OUT instruction** transfers data from AL, AX, or EAX to an external I/O device. (Only the 80386 and above contain EAX.)

Two forms of I/O device (port) addresses exist for IN and OUT: fixed-port and variable-port. **Fixed-port addressing** allows data transfer between AL, AX, or EAX using an 8-bit I/O port address. It is called fixed-port addressing because the port number follows the instruction's opcode. Often instructions are stored in a ROM. A fixed port instruction stored in a ROM has its port number permanently fixed because of the nature of read-only memory. If the fixed-port address is stored in a RAM, it is possible to modify it, but such a modification does not conform to good programming practices.

The port address appears on the address bus during an I/O operation. For the 8-bit fixed-port I/O instructions, the 8-bit port address is zero-extended into a 16-bit address. For example, if the IN AL,6AH instruction executes, data from I/O address 6AH is input to AL. The address appears as a 16-bit 006AH on pins A0–A15 of the address bus. Address bus bits A16–A19 (8086/

TABLE 4–17 IN and OUT instructions

Assembly Language	Operation
IN AL,p8	8-bit data is input to AL from port p8
IN AX,p8	16-bit data is input to AX from port p8
IN EAX,p8	32-bit data is input to EAX from port p8
IN AL,DX	8-bit data is input to AL from port DX
IN AX,DX	16-bit data is input to AX from port DX
IN EAX,DX	32-bit data is input to EAX from port DX
OUT p8,AL	8-bit data is sent to port p8 from AL
OUT p8,AX	16-bit data is sent to port p8 from AX
OUT p8,EAX	32-bit data is sent to port p8 from EAX
OUT DX,AL	8-bit data is sent to port DX from AL
OUT DX,AX	16-bit data is sent to port DX from AX
OUT DX,EAX	32-bit data is sent to port DX from EAX

Note: p8 = an 8-bit I/O port number and DX = the 16-bit port address held in DX.

8088/80186/80188), A16–A23 (80286/80386SX), A16–A24 (80386SL/80386SLC/80386EX), A16–A32 (80386/80486/Pentium), and A16–A35 (Pentium Pro) are undefined for an IN or OUT instruction. Intel reserves the last 16 I/O ports for use with some of its peripheral components.

Variable-port addressing allows data transfers between AL, AX, or EAX and a 16-bit port address. It is called variable-port addressing because the I/O port number is stored in register DX, which can be changed (varied) during the execution of a program. The 16-bit I/O port address appears on the address bus pin connections A0–A15. The IBM PC uses a 16-bit port address to access its I/O space. The I/O space for a PC is located at I/O port 0000H–03FFH. Some plug-in adapter cards may use I/O addresses above 03FFH.

Figure 4–18 illustrates the execution of the OUT 19H,AX instruction, which transfers the contents of AX to I/O port 19H. Notice that the I/O port number appears as a 0019H on the 16-bit address bus and that the data from AX appears on the data bus of the microprocessor. The system control signal, $\overline{\text{IOWC}}$ (**I/O write control**), is a logic zero to enable the I/O device.

Example 4–10 is a short program that clicks the speaker in the personal computer. The speaker is controlled by accessing I/O port 61H. If the rightmost two bits of this port are set (11) and then cleared (00), a click is heard on the speaker. This program uses a logical OR instruction

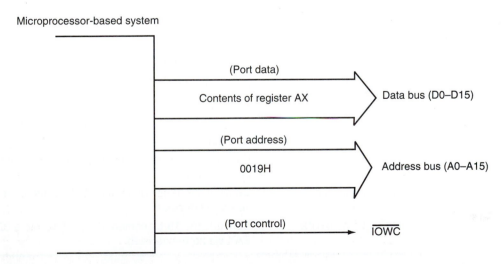

FIGURE 4–18 The signals found in the microprocessor-based system for an OUT 19H,AX instruction.

to set these two bits and a logical AND instruction to clear them. (These logic operation instructions are described in the next chapter.) The MOV CX,1000H instruction followed by the LOOP L1 instruction is used as a time delay. If the count is increased, the click will become longer and if shortened, the click will become shorter.

EXAMPLE 4–10

```
                        .MODEL TINY         ;select TINY model
0000                    .CODE               ;start of code segment
                        .STARTUP            ;indicate start of program
0100 E4 61              IN   AL,61H         ;read port 61H
0102 0C 03              OR   AL,3           ;set rightmost two bits
0104 E6 61              OUT  61H,AL         ;speaker is on

0106 B9 1000            MOV  CX,1000H       ;delay count
0109           L1:
0109 E2 FE              LOOP L1             ;time delay

010B E4 61              IN   AL,61H         ;read port 61H
010D 24 FC              AND  AL,0FCH        ;clear rightmost two bits
010F E6 61              OUT  61H,AL         ;speaker is off
                        .EXIT               ;exit to DOS
                        END                 ;end of file
```

MOVSX and MOVZX

The MOVSX (**move and sign-extend**) and MOVZX (**move and zero-extend**) instructions are found in the 80386, 80486, Pentium, and Pentium Pro instruction sets. These instructions move data and at the same time either sign- or zero-extend it. Table 4–18 gives several examples of these instructions.

When a number is zero-extended, the most-significant part fills with zeros. For example, if an 8-bit 34H is zero-extended into a 16-bit number, it becomes 0034H. Zero-extension is often used to convert unsigned 8- or 16-bit numbers into unsigned 16- or 32-bit numbers using the MOVZX instruction.

TABLE 4–18 The MOVSX and MOVZX instructions

Assembly Language	Operation
MOVSX CX,BL	Converts the 8-bit contents of BL into a 16-bit number in CX via sign-extension
MOVSX BX,DATA	Converts the 8-bit contents of DATA into a 16-bit number in BX via sign extension
MOVSX EAX,[EDI]	Converts the 16-bit contents of the data segment memory location address by EDI into a 32-bit number in EAX via sign-extension
MOVZX DX,AL	Converts the 8-bit contents of AL into a 16-bit number in DX via zero-extension
MOVZX EBP,DI	Converts the 16-bit contents of DI into a 32-bit number in EBP via zero-extension
MOVZX DX,DATA1	Converts the 8-bit contents of DATA1 into a 16-bit number in DX via zero-extension
MOVZX EAX,DATA2	Converts the 16-bit contents of DATA2 into a 32-bit number in EAX via zero-extension

A number is sign-extended when its sign-bit is copied into the most-significant part. For example, if an 8-bit 84H is sign-extended into a 16-bit number, it becomes FF84H. The sign-bit of an 84H is a one, which is copied into the most-significant part of the sign-extended result. Sign-extension is most often used to convert 8- and 16-bit signed numbers into 16- and 32-bit signed numbers using the MOVSX instruction.

BSWAP

The **byte swap instruction** (BSWAP) is available only in the 80486–Pentium Pro microprocessors. This instruction takes the contents of any 32-bit register and swaps the first byte with the fourth, and the second with the third. For example, the BSWAP EAX instruction with EAX = 00112233H swaps bytes in EAX so it results in EAX = 33221100H. Notice that the order of all four bytes is reversed by this instruction. This instruction is used to convert data from big endian form to little endian form or vise versa.

4–6   SEGMENT OVERRIDE PREFIX

The **segment override prefix,** which may be added to almost any instruction in any memory-addressing mode, allows the programmer to deviate from the default segment. The segment override prefix is an additional byte that appends the front of an instruction to select an alternate segment register. About the only instructions that cannot be prefixed are the jump and call instructions that must use the code segment register for address generation. The segment override is also used to select the FS and GS segments in the 80386 through the Pentium Pro microprocessors.

For example, the MOV AX,[DI] instruction accesses data within the data segment by default. If required by a program, this can be changed by prefixing the instruction. Suppose that the data is in the extra segment instead of the data segment. This instruction addresses the extra segment if changed to MOV AX,ES:[DI].

Table 4–19 shows some altered instructions that address different memory segments than normal. Each time an instruction is prefixed with a segment override prefix, the instruction becomes one byte longer. Although this is not a serious change to the length of the instruction, it does add to its execution time. Usually, it is customary to limit the use of the segment override prefix and remain in the default segments to write shorter and more efficient software.

TABLE 4–19 Instructions that include segment override prefixes

Assembly Language	Segment Accessed	Default Segment
MOV AX,DS:[BP]	Data	Stack
MOV AX,ES:[BP]	Extra	Stack
MOV AX,SS:[DI]	Stack	Data
MOV AX,CS:[SI]	Code	Data
MOV AX,ES:LIST	Extra	Data
LODS ES:DATA1	Extra	Data
MOV EAX,FS:DATA2	FS	Data
MOV BL,GS:[ECX]	GS	Data

Note: Only the 80386, 80486, Pentium, and Pentium Pro allow the use of the FS and GS segments.

4–7   ASSEMBLER DETAIL

The assembler[1] for the microprocessor can be used in two ways: (1) with models (used in most examples in this book) that are unique to a particular assembler and (2) with full segment definitions that allow complete control over the assembly process and are universal to all assemblers. This section presents both methods and explains how to organize a program's memory space using the assembler. It also explains the purpose and use of some of the more important directives used with this assembler. Appendix A provides additional detail about the assembler.

Assembler Function

The assembler is a program that converts **source code** (programs in symbolic form) into **object code** (hexadecimal machine language). Modern assembler programs are multi-pass programs that pass through the source file many times as they assemble the object file.

During the first pass, the assembler program places the labels used in the program into a table along with the address of the label. The label table is very important, because it allows the assembler program to use forward references. Early assembler programs were one-pass systems that allowed only reverse references to labels.

During the second pass, the assembler looks up the opcodes for each instruction as well as any references to a label from the label table and generates the hexadecimal object program. The second pass is also where the assembler converts decimal and ASCII data into its hexadecimal representation. The object file, using the extension .OBJ, is in a form that cannot be executed. Instead, the form of the object program is such that it can be converted to an executable file by the linker program. This allows many programming modules (in object form) to be connected and linked with the linker program.

Directives

Before we discuss the format of an assembly language program, we must learn some details about the directives **(pseudo-operations)** that control the assembly process. Some common assembly language directives are shown in Table 4–20. Directives indicate how an operand or section of a program is to be processed by the assembler. Some directives generate and store information in the memory, while others do not. The DB **(define byte)** directive stores bytes of data in the memory, while the BYTE PTR directive never stores data. The **BYTE PTR directive** indicates the size of the data referenced by a pointer or index register.

By default the assembler accepts only 8086/8088 instructions unless a program is preceded by the .386 or .386P directive or one of the other microprocessor selection switches. The .386 directive tells the assembler to use the 80386 instruction set in the real mode, while the .386P directive tells the assembler to use the 80386 protected-mode instruction set.

Storing Data in a Memory Segment.   The DB **(define byte),** DW **(define word),** and DD **(define doubleword)** directives, first presented in Chapter 1, are most often used with the microprocessor to define and store memory data. If a numeric coprocessor is present in the system, the DQ **(define quadword)** and DT **(define ten bytes)** directives are also common. These directives use a label to identify a memory location and the directive (DB, DW, etc.) to indicate the size of the location.

Example 4–11 is a memory segment that contains various forms of data definition directives. It also shows the full segment definition with the first SEGMENT statement to indicate the

---

[1]The assembler used in this text is the Microsoft macro assembler (MASM).

TABLE 4–20 Common assembler directives

Directive	Function
.186	Selects the 80186/80188 instruction set
.286	Selects the 80286 instruction set
.286P	Selects the protected-mode 80286 instruction set
.386	Selects the 80386 instruction set
.386P	Selects the protected-mode 80386 instruction set
.486	Selects the 80486 instruction set
.486P	Selects the protected-mode 80486 instruction set
.586	Selects the Pentium instruction set*
.586P	Selects the protected-mode Pentium instruction set*
.187	Selects the 80187 numeric coprocessor
.287	Selects the 80287 numeric coprocessor
.387	Selects the 80387 numeric coprocessor
.CODE	Used with programming models to indicate the start of the code segment
.DATA	Used with programming models to indicate the start of the data segment
.EXIT	Used with programming models to exit to DOS
.MODEL	Selects the programming model
.STARTUP	Used with programming models to indicate the start of the program
ALIGN	Used with full segment definitions to align data
ASSUME	Indicates the names of each segment to the assembler—it does not load the segment registers
BYTE	Indicates a byte-sized operand as in BYTE PTR or THIS BYTE
DB	Defines byte(s) (8 bits)
DD	Defines doubleword(s) (32 bits)
DQ	Defines quadword(s) (64 bits)
DT	Defines ten byte(s) (80 bits)
DUP	Generates duplicates of characters or numbers
DW	Defines word(s) (16 bits)
DWORD	Indicates a doubleword-sized operand as in DWORD PTR
END	Indicates the end of the program file
ENDM	Indicates the end of a macro sequence
ENDP	Indicates the end of a procedure
ENDS	Indicates the end of a segment or structure
EQU	Equates data to a label
FAR	Specifies a far address as in FAR PTR
MACRO	Defines the name, parameters, and start of a macro sequence
NEAR	Specifies a near address as in NEAR PTR
OFFSET	Specifies the offset address
ORG	Sets the origin offset address within a segment
PROC	Defines the beginning of a procedure
PTR	Indicates a memory pointer
SEGMENT	Defines the start of a memory segment
STACK	Indicates that a segment is the stack segment
STRUC	Defines the start of a data structure
THIS	Used with EQU to set a label to a byte, word, or doubleword
USES	An MASM version 6.X directive that saves registers used within a procedure
USE16	Directs the assembler to use the 16-bit instruction mode
USE32	Directs the assembler to use the 32-bit instruction mode
WORD	Acts as a word operand as in WORD PTR

*Note: For version 6.11 of MASM most of these directives function with most versions of the assembler.

start of the segment and its symbolic name. Alternatively, as in past examples in this and prior chapters, the SMALL model can be used with the .DATA statement. The last statement in this example contains the ENDS directive that indicates the end of the segment. The name of the segment (LIST_SEG) can be anything that the programmer desires to call it. This allows a program to contain as many segments as required.

EXAMPLE 4–11

```
                        ;Using the DB, DW, and DD directives
                        ;
0000                    LIST_SEG   SEGMENT

0000 01 02 03   DATA1   DB    1,2,3                ;define bytes
0003 45                 DB    45H                  ;hexadecimal
0004 41                 DB    'A'                  ;ASCII
0005 F0                 DB    11110000B            ;binary
0006 000C 000D  DATA2   DW    12,13                ;define words
000A 0200               DW    LIST1                ;symbolic
000C 2345               DW    2345H                ;hexadecimal
000E 00000300   DATA3   DD    300H                 ;hexadecimal
0012 4007DF3B           DD    2.123                ;real
0016 544269E1           DD    3.34E+12             ;real
001A 00         LISTA   DB    ?                    ;reserve 1 byte
001B 000A[      LISTB   DB    10 DUP (?)           ;reserve 10 bytes
        ??
        ]
0025 00                 ALIGN 2                    ;set word boundary

0026 0100[      LISTC   DW    100H DUP (0)         ;word array
     0000
        ]
0226 0016[      LIST_9  DD    22 DUP (?)           ;doubleword array
     ????????
        ]
027E 0064[      SIXES   DB    100 DUP (6)          ;byte array
        06
        ]
02E2            LIST_SEG ENDS
```

Example 4–11 shows various forms of data storage for bytes at DATA1. More than one byte can be defined on a line in binary, hexadecimal, decimal, or ASCII code. The DATA2 label shows how to store various forms of word data. Doublewords are stored at DATA3 and include floating-point, single-precision real numbers.

Memory can be reserved for use in the future by using a ? as an operand for a DB, DW, or DD directive. When a ? is used in place of a numeric or ASCII value, the assembler sets aside a location and does not initialize it to any specific value. (Actually the assembler stores a zero into locations specified with a ?.) The DUP **(duplicate)** directive creates an array as shown in several ways in Example 4–11. A 10 DUP (?) reserves 10 locations of memory, but stores no specific value in any of the 10 locations. If a number appears within the ( ) part of the DUP statement, the assembler initializes the reserved section of memory with the data indicated. For example, the DATA1 DB 10 DUP (2) instruction reserves 10 bytes of memory for array DATA1 and initializes each location with a 02H.

The **ALIGN directive,** used in Example 4–11, makes sure that the memory arrays are stored on word boundaries. An ALIGN 2 places data on word boundaries and an ALIGN 4 places them on doubleword boundaries. In the Pentium, quadword data for double-precision floating-point numbers should use ALIGN 8. It is important that word-sized data be placed at word boundaries and doubleword-sized data at doubleword boundaries. If not, the microprocessor spends additional time accessing these data types. A word stored at an odd-numbered memory location takes twice as long to access as a word stored at an even-numbered memory location. Note that the ALIGN directive cannot be used with memory models, because the size of the model determines the data alignment. In a memory model, place the largest data sizes first and the data will be aligned on the proper boundaries.

EQU and THIS.   The **equate directive** (EQU) equates a numeric, ASCII, or label to another label. Equate statements make a program clearer and simplify debugging. Example 4–12 contains several equate statements and a few instructions that show how they function in a program.

EXAMPLE 4–12

```
                    ;Using equate directive
                    ;
= 000A              TEN     EQU  10
= 0009              NINE    EQU  9

0000 B0 0A                  MOV  AL,TEN
0002 04 09                  ADD  AL,NINE
```

The **THIS directive** always appears as THIS BYTE, THIS WORD, or THIS DWORD. In certain cases, data must be referred to as both a byte and a word. The assembler can assign only a byte or a word address to a label. To assign a byte label to a word, use the software shown in Example 4–13.

EXAMPLE 4–13

```
                    ;Using the THIS and ORG directives
                    ;
0000                DATA_SEG SEGMENT

0100                         ORG  100H

= 0100              DATA1    EQU  THIS BYTE
0100 0000           DATA2    DW   ?

0102                DATA_SEG ENDS

0000                CODE_SEG SEGMENT 'CODE'

                             ASSUME  CS:CODE_SEG,DS:DATA_SEG

0000 8A 1E 0100 R            MOV  BL,DATA1
0004 A1 0100 R               MOV  AX,DATA2
0007 8A 3E 0101 R            MOV  BH,DATA1+1

000B                CODE_SEG ENDS
```

Example 4–13 also illustrates how the ORG (**origin**) statement changes the starting offset address of the data in the data segment to location 100H. At times, the origin of the data or code must be assigned to an absolute offset address with the ORG statement. The **ASSUME statement** tells the assembler which names have been chosen for the code, data, extra, and stack segments. Without the ASSUME statement, the assembler assumes nothing and automatically uses a segment override prefix on all instructions that address memory data. The ASSUME statement is only used with full-segment definitions as described later in this section.

PROC and ENDP.   The PROC and ENDP directives indicate the start and end of a procedure (subroutine). These directives force structure because the procedure is clearly defined. If structure is to be violated, use the CALLF, CALLN, RETF, and RETN instructions. Both the PROC and ENDP directives require a label to indicate the name of the procedure. The PROC directive, which indicates the start of a procedure, must be followed with a NEAR or FAR. A NEAR procedure is one that resides in the same code segment as the program. A FAR procedure may reside at any location in the memory system. Often the call NEAR procedure is considered local, and the call FAR procedure global. The term global denotes a procedure that can be used by any program, while local defines a procedure that is used only by the current program. Any labels that are defined within the procedure block are also defined as either local (NEAR) or global (FAR).

Example 4–14 shows a procedure that adds BX, CX, and DX and stores the sum in register AX. Although this procedure is short, and may not be particularly useful, it does illustrate how to use the PROC and ENDP directives to delineate the procedure. Information about the operation of the procedure should appear as a grouping of comments that tell you which registers were changed by the procedure and the result of the procedure.

EXAMPLE 4–14

```
                        ;A procedure that adds BX, CX, and DX with the sum
                        ;stored in AX.
                        ;
0000                    ADDEM    PROC FAR              ;start procedure

0000 03 D9                       ADD    BX,CX
0002 03 DA                       ADD    BX,DX
0004 8B C3                       MOV    AX,BX
0006 CB                          RET

0007                    ADDEM    ENDP                  ;end procedure
```

If version 6.X of the Microsoft MASM assembler program is available, the PROC directive specifies and automatically saves any registers used within the procedure. The USES statement indicates which registers are used by the procedure so the assembler can automatically save them before your procedure begins and restore them before the procedure ends with the RET instruction. For example, the ADDS PROC USES AX BX CX statement automatically pushes AX, BX, and CX on the stack before the procedure begins and pops them from the stack before the RET instruction executes at the end of the procedure. Example 4–15 is a procedure (written using MASM 6.X) that shows the USES statement. Note that the registers in the list are separated by spaces, not commas, and the PUSH and POP instructions are displayed in the procedure listing because it was assembled with the .LISTALL directive. The instructions prefaced with an asterisk are inserted by the assembler. The USES statement does not appear elsewhere in this book so compatibility with MASM version 5.10 is maintained. If version 6.X is available, you should include the USES statement is with most procedures.

EXAMPLE 4–15

```
                        ;A procedure that includes the USES directive to save
                        ;BX, CX, and DX on the stack and restore them before
                        ;the RET instruction.
                        ;
0000                    ADDS     PROC NEAR  USES BX CX DX

0000 53          *               push bx
0001 51          *               push cx
0002 52          *               push dx
0003 03 D8                       ADD    BX,AX
0005 03 CB                       ADD    CX,BX
0007 03 D1                       ADD    DX,CX
0009 8B C2                       MOV    AX,DX
                                 RET
000B 5A          *               pop    dx
000C 59          *               pop    cx
000D 5B          *               pop    bx
000E C3          *               ret    00000h

000F                    ADDS     ENDP
```

Memory Organization

The assembler uses two basic formats for developing software: One method uses models and the other uses full segment definitions. Memory models, as presented in this section and also briefly in Chapter 3, are unique to the MASM assembler program. The TASM assembler also uses memory models, but they differ somewhat from the MASM models. The full-segment definitions are common to most assemblers, including the Intel assembler, and are often used for software development. The models are easier to use for simple tasks. The full-segment definitions offer better control over the assembly language task and are recommended for complex programs. The model was used in previous chapters because it is easier to understand for the beginning programmer. Models are also used with assembly language procedures in high-level languages such as C/C++. We fully develop and use the memory model definitions for our programming examples, but realize that full-segment definitions offer some advantages over memory models, as discussed later in this section.

Models.   There are many models available to the MASM assembler, from tiny to huge. Appendix A contains a table that lists all the models available for use with the assembler. To designate a model, use the **.MODEL statement** followed by the size of the memory system. The tiny model requires that all software and data fit into one 64K-byte memory segment and is useful for many small programs. The small model requires that only one data segment is used with one code segment for a total of 128K bytes of memory. Other models are available up to the huge model.

Example 4–16 illustrates how the .MODEL statement defines the parameters of a short program that copies the contents of a 100-byte block of memory (LISTA) into a second 100-byte block of memory (LISTB). It also shows how to define the stack, data, and code segments. The .EXIT 0 directive returns to DOS with an error code of 0 (no error). If no parameter is added to .EXIT, it still returns to DOS, but the error code is not defined. Also note that special directives such as @DATA (see Appendix A) are used to identify various segments. If the .STARTUP directive is used (MASM version 6.X), the MOV AX,@DATA followed by MOV DS,AX statements can be eliminated. Models are important with both Microsoft C/C++ and Borland C/C++ development systems if assembly language is included with C/C++ programs. Both development systems use in-line assembly programming for adding assembly language instructions and require an understanding of programming models. Refer to the respective C/C++ language reference for each system to determine the model protocols.

EXAMPLE 4–16

```
                        .MODEL SMALL
                        .STACK 100H            ;define stack
                        .DATA                  ;define data segment

0000 0064[      LISTA   DB    100 DUP (?)
      ??
         ]
0064 0064[      LISTB   DB    100 DUP (?)
      ??
          ]

                        .CODE                  ;define code segment

0000 B8 ---- R  HERE:   MOV   AX,@DATA         ;load ES, DS
0003 8E C0              MOV   ES,AX
0005 8E D8              MOV   DS,AX

0007 FC                 CLD                    ;move data
0008 BE 0000 R          MOV   SI,OFFSET LISTA
000B BF 0064 R          MOV   DI,OFFSET LISTB
```

```
000E B9 0064                MOV  CX,100
0011 F3/A4                  REP  MOVSB

0013                        .EXIT  0                ;exit to DOS
                            END  HERE
```

Full Segment Definitions.   Example 4–17 illustrates the program in Example 4–16 using full-segment definitions. **Full-segment definitions** are also used with the Borland and Microsoft C/C++ environments for procedures developed in assembly language. The program in Example 4–17 appears longer than the one in Example 4–16, but it is more structured than the model method of setting up a program. The first segment defined is the STACK_SEG that is clearly delineated with the SEGMENT and ENDS directives. Within these directives a DW 100 DUP (?) sets aside 100H words for the stack segment. Because the word STACK appears next to SEGMENT, the assembler and linker automatically load both the stack segment register (SS) and stack pointer (SP).

   Next the data is defined in the DATA_SEG. Here two arrays of data appear as LISTA and LISTB. Each array contains 100 bytes of space for the program. The names of the segments in this program can be changed to any name. Always include the group name 'DATA' so the Microsoft program CodeView can be used effectively to symbolically debug this software. CodeView is part of the MASM package used to debug software. To access CodeView, type CV followed by the file name at the DOS command line; if operating from Programmer's Workbench, select Debug under the Run menu. If the group name is not placed in a program, CodeView can still be used to debug a program, but the program will not be debugged in symbolic form. Other group names such as 'STACK', 'CODE', and so forth are listed in Appendix A. You must at least place the word 'CODE' next to the code segment SEGMENT statement if you want to view the program symbolically in CodeView.

EXAMPLE 4–17

```
0000                        STACK_SEG   SEGMENT   STACK

0000 0100[                              DW    100H DUP (?)
        ????
              ]

0200                        STACK_SEG   ENDS

0000                        DATA_SEG    SEGMENT   'DATA'

0000 0064[                  LISTA       DB    100 DUP (?)
        ??
              ]
0064 0064[                  LISTB       DB    100 DUP (?)
        ??
              ]

00C8                        DATA_SEG    ENDS

0000                        CODE_SEG    SEGMENT   'CODE'

                                        ASSUME CS:CODE_SEG,DS:DATA_SEG
                                        ASSUME SS:STACK_SEG

0000                        MAIN        PROC FAR

0000 B8 ---- R                          MOV  AX,DATA_SEG      ;load DS and ES
0003 8E C0                              MOV  ES,AX
0005 8E D8                              MOV  DS,AX

0007 FC                                 CLD                   ;move data
0008 BE 0000 R                          MOV  SI,OFFSET LISTA
000B BF 0064 R                          MOV  DI,OFFSET LISTB
```

```
000E B9 0064                    MOV   CX,100
0011 F3/A4                      REP   MOVSB

0013 B4 4C                      MOV   AH,4CH          ;exit to DOS
0015 CD 21                      INT   21H

0017                    MAIN    ENDP

0017                    CODE_SEG ENDS

                                END   MAIN
```

The CODE_SEG is organized as a far procedure because most software is procedure oriented. Before the program begins, the code segment contains the ASSUME statement. The ASSUME statement tells the assembler and linker the name used for the code segment (CS) is CODE_SEG. It also tells the assembler and linker that the data segment is DATA_SEG and the stack segment is STACK_SEG. Notice that the group name 'CODE' is used for the code segment for use by CodeView. Other group names appear in Appendix A with the models.

After the program loads both the extra segment register and data segment register with the location of the data segment, it transfers 100 bytes from LISTA to LIST B. Following this is a sequence of two instructions that return control back to DOS (the disk operating system). Note that the program loader does not automatically initialize DS and ES. These registers must be loaded with the desired segment addresses in the program.

The last statement in the program is END MAIN. The END statement indicates the end of the program and the location of the first instruction executed. Here we want the machine to execute the main procedure, so a label follows the END directive.

In the 80386 through the Pentium Pro microprocessors, an additional directive is found attached to the code segment. The USE16 or USE32 directive tells the assembler to use either the 16- or 32-bit instruction modes for the microprocessor. Software developed for the DOS environment must use the USE16 directive for the 80386 through the Pentium to function correctly, because MASM assumes that all segments are 32 bits and all instruction modes are 32 bits by default. In fact, any program designed to execute in the real mode must include the **USE16 directive** to deviate from the default 8086/8088. Example 4–18 shows how the same software in Example 4–17 is formed for the 80386 microprocessor. If models are in use, the .386 following the model statement selects 16-bit instruction mode; if the .386 precedes the model statement, the 32-bit instruction mode is selected.

EXAMPLE 4–18

```
                                .386                    ;select the 80386
0000                    STACK_SEG SEGMENT STACK

0000 0100[              DW    100H DUP (?)
        ????
            ]

0200                    STACK_SEG ENDS

0000                    DATA_SEG  SEGMENT  'DATA'

0000 0064[     LISTA    DB    100 DUP (?)
        ??
          ]
0064 0064[     LISTB    DB    100 DUP (?)
        ??
          ]

00C8                    DATA_SEG  ENDS

0000                    CODE_SEG  SEGMENT  USE16 'CODE'
```

```
                              ASSUME CS:CODE_SEG,DS:DATA_SEG
                              ASSUME SS:STACK_SEG

0000              MAIN        PROC FAR
0000 B8 ---- R                MOV  AX,DATA_SEG      ;load DS and ES
0003 8E C0                    MOV  ES,AX
0005 8E D8                    MOV  DS,AX
0007 FC                       CLD                   ;move data
0008 BE 0000 R                MOV  SI,OFFSET LISTA
000B BF 0064 R                MOV  DI,OFFSET LISTB
000E B9 0064                  MOV  CX,100
0011 F3/A4                    REP  MOVSB
0013 B4 4C                    MOV  AH,4CH           ;exit to DOS
0015 CD 21                    INT  21H

0017              MAIN        ENDP

0017              CODE_SEG    ENDS
                              END  MAIN
```

A Sample Program

Example 4–19 is a sample program using full-segment definitions that reads a character from the keyboard and displays it on the CRT screen. Although this program is trivial, it does illustrate a complete workable program that functions on any personal computer using DOS from the earliest 8088-based system to the latest Pentium-based system. This program also illustrates the use of a few DOS function calls. Appendix A lists the DOS function calls with their parameters. The BIOS function calls allow the use of the keyboard, printer, disk drives, and everything else that is available in your computer system.

EXAMPLE 4–19

```
                  ;An example program that reads a key and displays it.
                  ;Note that an @ key ends the program.
                  ;
0000              CODE_SEG    SEGMENT   'CODE'

                              ASSUME    CS:CODE_SEG

0000              MAIN        PROC FAR

0000 B4 06                    MOV  AH,6             ;read key
0002 B2 FF                    MOV  DL,0FFH
0004 CD 21                    INT  21H
0006 74 F8                    JE   MAIN             ;if no key

0008 3C 40                    CMP  AL,'@'           ;test for @
000A 74 08                    JE   MAIN1           ;if @

000C B4 06                    MOV  AH,6             ;display key
000E 8A D0                    MOV  DL,AL
0010 CD 21                    INT  21H
0012 EB EC                    JMP  MAIN             ;repeat
0014              MAIN1:
0014 B4 4C                    MOV  AH,4CH           ;exit to DOS
0016 CD 21                    INT  21H

0018              MAIN        ENDP

0018              CODE_SEG    ENDS

                              END  MAIN
```

Example 4–19 uses only a code segment because there is no data. A stack segment should appear, but has been left out because DOS automatically allocates a 128-byte stack for all programs. The only time that the stack is used in this example is for the INT 21H instructions that call a procedure in DOS. When this program is linked, the linker signals that no stack segment is present. This warning may be ignored in this example because the stack is less than 128 bytes.

Notice that the entire program is placed into a far procedure called MAIN. It is good programming practice to write all software in procedural form. This allows the program to be used as a procedure if necessary at some future time. It is also fairly important to document register use and any parameters required for the program in the program header. The program header is a section of comments that appear at the start of the program.

The program uses DOS functions 06H and 4CH. The function number is placed in AH before the INT 21H instructions execute. The 06H function reads the keyboard if DL = 0FFH or displays the ASCII contents of DL if it is not 0FFH. Upon close examination, the first section of the program moves a 06H into AH and a 0FFH into DL so a key is read from the keyboard. The INT 21H tests the keyboard; if no key is typed, it returns equal. The JE instruction tests the equal condition and jumps to MAIN if no key is typed.

When a key is typed, the program continues to the next step. This step compares the contents of AL with an @ symbol. Upon return from the INT 21H, the ASCII character of the typed key is found in AL. In this program if an @ symbol is typed, the program ends. If the @ symbol is not typed, the program continues by displaying the character typed on the keyboard with the next INT 21H instruction.

The second INT 21H instruction moves the ASCII character into DL so it can be displayed on the CRT screen. After displaying the character a JMP executes. This causes the program to continue at MAIN where it repeats reading a key.

If the @ symbol is typed, the program continues at MAIN1 where it executes the DOS function code number 4CH. This causes the program to return to the DOS prompt (A:\) so the computer can be used for other tasks.

More information about the assembler and its application appears in Appendix A and in the next several chapters. Appendix A provides a complete overview of the assembler, linker, and DOS functions. It also provides a list of the BIOS (basic I/O system) functions. The following chapters describe how to use the assembler for certain tasks at different levels.

4-8 SUMMARY

1. Data movement instructions transfer data between registers, a register and memory, a register and the stack, memory and the stack, the accumulator and I/O, and the flags and the stack. Memory-to-memory transfers are only allowed with the MOVS instruction.

2. Data movement instructions include: MOV, PUSH, POP, XCHG, XLAT, IN, OUT, LEA, LDS, LES, LSS, LGS, LFS, LAHF, and SAHF, and the string instructions: LODS, STOS, MOVS, INS, and OUTS.

3. The first byte of an instruction contains the opcode, which specifies the operation performed by the microprocessor. The opcode may be preceded by one or more override prefixes in some forms of instructions.

4. The D-bit, located in many instructions, selects the direction of data flow. If D = 0, the data flows from the REG field to the R/M field of the instruction. If D = 1, the data flows from the R/M field to the REG field.

5. The W-bit, found in most instructions, selects the size of the data transfer. If W = 0, the data is byte-sized and if W = 1, the data is word-sized. In the 80386 and above, W = 1 specifies either a word or doubleword register.

6. MOD selects the addressing mode of operation for a machine language instruction's R/M field. If MOD = 00, there is no displacement; if a 01, an 8-bit sign-extended displacement

appears; if a 10, a 16-bit displacement occurs; and if an 11, a register is used instead of a memory location. In the 80386 and above, the MOD bits also specify a 32-bit displacement.

7. A 3-bit binary register code specifies the REG and R/M fields when the MOD = 11. The 8-bit registers are: AH, AL, BH, BL, CH, CL, DH, and DL, and the 16-bit registers are: AX, BX, CX, DX, SP, BP, DI, and SI. The 32-bit registers are EAX, EBX, ECX, EDX, ESP, EBP, EDI, and ESI.

8. When the R/M field depicts a memory mode, a 3-bit code selects one of the following modes: [BX+DI], [BX+SI], [BP+DI], [BP+SI], [BX], [BP], [DI], or [SI] for 16-bit instructions. In the 80386 and above, the R/M field specifies EAX, EBX, ECX, EDX, EBP, EDI, and ESI or one of the scaled index modes of addressing memory data. If the scaled index mode is selected (R/M = 100), an additional byte (scaled index byte) is added to the instruction to specify the base register, index register, and the scaling factor.

9. All memory addressing modes, by default, address data in the data segment unless BP or EBP addresses memory. The BP or EBP register addresses data in the stack segment.

10. The segment registers are addressed only by the MOV, PUSH, and POP instructions. The MOV instruction may transfer a segment register to a 16-bit register or vice versa. A MOV CS,reg or POP CS instruction is not allowed because it changes only part of the address. In the 80386 through the Pentium Pro, there are two additional segment registers, FS and GS.

11. Data is transferred between a register or a memory location and the stack by the PUSH and POP instructions. Variations of these instructions allow immediate data to be pushed onto the stack, the flags to be transferred between the stack, and all the 16-bit registers can be transferred between the stack and the registers. When data is transferred to the stack, two bytes (8086–80286) always move with the least-significant byte placed at the SP location −1 byte and the most-significant byte placed at the SP location −2 bytes. After placing the data on the stack, SP decrements by 2. In the 80386–Pentium Pro, four bytes of data from a memory location or register may also be transferred to the stack.

12. Opcodes that transfer data between the stack and the flags are PUSHF and POPF. Opcodes that transfer all the 16-bit registers between the stack and the registers are PUSHA and POPA. In the 80386 and above, a PUSHFD and POPFD transfer the contents of the EFLAGS between the microprocessor and the stack.

13. LEA, LDS, and LES instructions load a register or registers with an effective address. The LEA instruction loads any 16-bit register with an effective address, while LDS and LES load any 16-bit register and either DS or ES with the effective address. In the 80386 and above, additional instructions include: LFS, LGS, and LSS that load a 16-bit register and FS, GS, or SS.

14. String data transfer instructions use DI and/or SI to address memory. The DI offset address is located in the extra segment and the SI offset address is located in the data segment.

15. The direction flag (D) chooses the auto-increment or auto-decrement mode of operation for DI and SI for string instructions. To clear D to 0, use the CLD instruction to select the auto-increment mode; to set D to 1, use the STD instruction to select the auto-decrement mode. DI and/or SI increment/decrement by 1 for a byte operation, by 2 for a word operation, and by 4 for a doubleword operation.

16. LODS loads AL, AX, or EAX with data from the memory location addressed by SI; STOS stores AL, AX, or EAX in the memory location addressed by DI; MOVS transfers a byte or a word from the memory location addressed by SI into the location addressed by DI.

17. INS inputs data from an I/O device addressed by DX and stores it in the memory location addressed by DI. OUTS outputs the contents of the memory location addressed by SI and sends it to the I/O device addressed by DX.

18. The REP prefix may be attached to any string instruction to repeat it. The REP prefix repeats the string instruction the number of times found in register CX.

19. Arithmetic and logic operators can be used in assembly language. An example is MOV AX,34*3, which loads AX with 102.

20. Translate (XLAT) converts the data in AL into a number stored at the memory location address by BX plus AL.

21. IN and OUT transfer data between AL, AX, or EAX and an external I/O device. The address of the I/O device is stored either with the instruction (fixed port) or in register DX (variable port).

22. The segment override prefix selects a different segment register for a memory location than the default segment. For example, the MOV AX,[BX] instruction uses the data segment, but the MOV AX,ES:[BX] instruction uses the extra segment because of the ES: prefix. The segment override prefix is the only way that the FS and GS segments are addressed in the 80386 through the Pentium Pro.

23. The MOVZX (move and zero-extend) and MOVSX (move and sign-extend) instructions, found in the 80386 and above, increase the size of a byte to a word and a word to a doubleword. The zero-extend version increases the size of the number by inserting leading zeros. The sign-extend version increases the size of the number by copying the sign-bit into the most-significant bits of the number.

24. Assembler directives DB, (define byte), DW (define word), DD (define doubleword), and DUP (duplicate) store data in the memory system.

25. The EQU (equate) directive allows data or labels to be equated to labels.

26. The SEGMENT directive identifies the start of a memory segment and ENDS identifies the end of a segment when full-segment definitions are in use.

27. The ASSUME directive tells the assembler which segment names you have assigned to CS, DS, ES, and SS when full-segment definitions are in effect. In the 80386 and above, ASSUME also indicates the segment name for FS and GS.

28. The PROC and ENDP directives indicate the start and end of a procedure. The USES directive (MASM version 6.X) automatically saves and restores any number of register on the stack if it appears with the PROC directive.

29. The assembler assumes that software is being developed for the 8086/8088 microprocessors unless the .186, .286, .386, .486, or .586 directive is used to select one of these other microprocessors. This directive follows the .MODEL statement to use the 16-bit instruction mode and precedes it for the 32-bit instruction mode.

30. Memory models can be used to shorten the program slightly, but they can cause problems for very large programs. Also be aware that memory models are not compatible with all assembler programs.

4–9 QUESTIONS AND PROBLEMS

1. The first byte of an instruction is the _____ unless it contains one of the override prefixes.

2. Describe the purpose of the D- and W-bits found in some machine language instructions.

3. The MOD field in a machine language instruction specifies what information?

4. If the register field (REG) of an instruction contains a 010 and W = 0, which register is selected, assuming that the instruction is a 16-bit mode instruction?

5. How are the 32-bit registers selected for the 80486 microprocessor?

6. What memory-addressing mode is specified by R/M = 001 with MOD = 00 for a 16-bit instruction?

7. Identify the default segment register assigned to:
   (a) SP
   (b) EBX
   (c) DI
   (d) EBP
   (e) SI
8. Convert an 8B07H from machine language to assembly language.
9. Convert an 8B1E004CH from machine language to assembly language.
10. If a MOV SI,[BX + 2] instruction appears in a program, what is its machine language equivalent?
11. If a MOV ESI,[EAX] instruction appears in a program for the Pentium microprocessor operated in the 16-bit instruction mode, what is its machine language equivalent?
12. What is wrong with a MOV CS,AX instruction?
13. Form a short sequence of instructions that load the data segment register with a 1000H.
14. The PUSH and POP instructions always transfer a _____-bit number between the stack and a register or memory location in the 8086–80286 microprocessors.
15. Which segment register may not be popped from the stack?
16. Which registers move onto the stack with the PUSHA instruction?
17. Which registers move onto the stack with the PUSHAD instruction?
18. Describe the operation of each of the following instructions:
   (a) PUSH AX
   (b) POP ESI
   (c) PUSH [BX]
   (d) PUSHFD
   (e) POP DS
   (f) PUSHD 4
19. Explain what happens when the PUSH BX instruction executes. Be sure to show where BH and BL are stored. (Assume that SP = 0100H and SS = 0200H.)
20. Repeat Question 19 for the PUSH EAX instruction.
21. The 16-bit POP instruction (except for POPA) increments SP by _____.
22. What values appear in SP and SS if the stack is addressed at memory location 02200H?
23. Compare the operation of a MOV DI,NUMB instruction with an LEA DI,NUMB instruction.
24. What is the difference between an LEA SI,NUMB instruction and a MOV SI,OFFSET NUMB instruction?
25. Which is more efficient, a MOV with an OFFSET or an LEA instruction?
26. Describe how the LDS BX,NUMB instruction operates.
27. What is the difference between the LDS and LSS instructions?
28. Develop a sequence of instructions that move the contents of data segment memory locations NUMB and NUMB + 1 into BX, DX, and SI.
29. What is the purpose of the direction flag?
30. Which instructions set and clear the direction flag?
31. The string instructions use DI and SI to address memory data in which memory segments?
32. Explain the operation of the LODSB instruction.
33. Explain the operation of the STOSW instruction.
34. Explain the operation of the OUTSB instruction.
35. What does the REP prefix accomplish and what type of instruction is it used with?
36. Develop a sequence of instructions that copies 12 bytes of data from an area of memory addressed by SOURCE into an area of memory addressed by DEST.
37. Where is the I/O address (port number) stored for an INSB instruction?
38. Select an assembly language instruction that exchanges the contents of the EBX and ESI registers.

39. Would the LAHF and SAHF instructions normally appear in software?

40. Explain how the XLAT instruction transforms the contents of the AL register.

41. Write a short program that uses the XLAT instruction to convert the BCD numbers 0–9 into ASCII-coded numbers 30H–39H. Store the ASCII-coded data in a TABLE located within the data segment.

42. Explain what the IN AL,12H instruction accomplishes.

43. Explain how the OUT DX,AX instruction operates.

44. What is a segment override prefix?

45. Select an instruction that moves a byte of data from the memory location addressed by the BX register, in the extra segment, into the AH register.

46. Develop a sequence of instructions that exchanges the contents of AX with BX, ECX with EDX, and SI with DI.

47. What is an assembly language directive?

48. Describe the purpose of the following assembly language directives: DB, DW, and DD.

49. Select an assembly language directive that reserves 30 bytes of memory for array LIST1.

50. Describe the purpose of the EQU directive.

51. What is the purpose of the .186 directive?

52. What is the purpose of the .MODEL directive?

53. If the start of a segment is identified with .DATA, what type of memory organization is in effect?

54. If the SEGMENT directive identifies the start of a segment, what type of memory organization is in effect?

55. What does the INT 21H accomplish if AH contains a 4CH?

56. What directives indicate the start and end of a procedure?

57. Explain the purpose of the USES statement as it applies to a procedure with version 6.X of MASM.

58. How is the 80486 microprocessor instructed to use the 16-bit instruction mode?

59. Develop a near procedure that stores AL in four consecutive memory locations, within the data segment, as addressed by the DI register.

60. Develop a far procedure that copies contents of the word-sized memory location CS:DATA1 into AX, BX, CX, DX, and SI.

61. Use the Internet to write a report that details the data transfer instructions found in the 80186 embedded controller.

62. Use the Internet to write a report that compares the data transfer operations found in the 80186 with those found in the 8051.

63. Use the Internet to write a report that details the various embedded controllers manufactured by a company called Zilog.

64. Use the Internet to access the Borland Web site and access information about the TASM program.

CHAPTER 5

Arithmetic and Logic Instructions

INTRODUCTION

In this chapter, we examine the arithmetic and logic instructions. Arithmetic instructions include: addition, subtraction, multiplication, division, comparison, negation, increment, and decrement. Logic instructions include: AND, OR, exclusive-OR, NOT, shift, rotate, and the logical compare (TEST). Also presented are the 80386 through the Pentium Pro instructions: XADD, SHRD, SHLD, bit tests, and bit scans.

    We also introduce string comparison instructions, which are used for scanning tabular data and for comparing sections of memory data. Both tasks perform efficiently with the string scan (SCAS) and string compare (CMPS) instructions.

    If you are familiar with an 8-bit microprocessor, you will recognize that the 8086 through the Pentium Pro instruction set is superior to most 8-bit microprocessors because most of the instructions have two operands instead of one. Even if this is your first microprocessor, you will quickly learn that this microprocessor possesses a powerful set of arithmetic and logic instructions that are easy to use.

CHAPTER OBJECTIVES

Upon completion of this chapter, you will be able to:

1. Use the arithmetic and logic instructions to accomplish simple binary, BCD, and ASCII arithmetic.
2. Use AND, OR, and exclusive-OR to accomplish binary bit manipulation.
3. Use the shift and rotate instructions.
4. Explain the operation of the 80386 through the Pentium Pro exchange and add, compare and exchange, double-precision shift, bit test, and bit scan instructions.
5. Check the contents of a table for a match with the string instructions.

5–1   ADDITION, SUBTRACTION, AND COMPARISON

The arithmetic instructions found in any microprocessor include addition, subtraction, and comparison. In this section we look at instructions and use them to manipulate register and memory data.

Addition

Addition appears in many forms in the microprocessor. This section details the use of the ADD instruction for 8-, 16-, and 32-bit binary addition. Another form of addition, called add-with-carry is introduced with the ADC instruction. Finally, we look at the increment instruction (INC), a special type of addition that adds a one to a number. In Section 5–3 other forms of addition are examined, such as BCD and ASCII. Also described is the XADD instruction found in the 80486 and Pentium Pro microprocessors.

Table 5–1 illustrates the addressing modes available to the ADD instruction. (These addressing modes include almost all those mentioned in Chapter 3.) However, since there are over 32,000 variations of the ADD instruction in the instruction set, it is impossible to list them all in this table. The only types of addition not allowed are **memory-to-memory** and **segment register.** The segment registers can only be moved, pushed, or popped. Note that as with all other instructions, the 32-bit registers are only available with the 80386 through the Pentium Pro.

Register Addition. Example 5–1 is a simple procedure that uses **register addition** to add the contents of several registers. In this example, the contents of AX, BX, CX, and DX are added to form a 16-bit result stored in the AX register. Here a procedure is used, because assembly language is procedure-oriented, as are most languages.

TABLE 5–1 Addition instructions

Assembly Language	Operation
ADD AL,BL	AL = AL + BL
ADD CX,DI	CX = CX + DI
ADD EBP,EAX	EBP = EBP + EAX
ADD CL,44H	CL = CL + 44H
ADD BX,35AFH	BX = BX + 35AFH
ADD EDX,12345H	EDX = EDX + 00012345H
ADD [BX],AL	AL adds to the contents of the data segment offset location addressed by BX and the result is stored in the same memory location
ADD CL,[BP]	The contents of the stack segment location addressed by BP adds to CL and the result is stored in CL
ADD AL,[EBX]	The contents of the data segment offset location addressed by EBX adds to AL and the result is stored in AL
ADD BX,[SI+2]	The word-sized contents of the data segment location addressed by SI plus 2 adds to BX and the result is stored in BX
ADD CL,TEMP	The contents of data segment location TEMP add to CL with the result stored in CL
ADD BX,TEMP[DI]	The word-sized contents of the data segment location addressed by TEMP plus DI adds to BX and the result is stored in BX
ADD [BX+DI],DL	DL adds to the contents of the data segment memory location addressed by the sum of BX and DI and the result is stored into the same memory location
ADD BYTE PTR [DI],3	A 3 is added to the byte-sized contents of the data segment memory location addressed by DI
ADD BX,[EAX+2*ECX]	The contents of the data segment word addressed by the sum of 2 times ECX plus EAX adds to BX

EXAMPLE 5–1

```
                                ;A procedure that sums AX, BX, CD, and DX
                                ;result is returned in AX.
                                ;
0000                            ADDS    PROC NEAR

0000 03 C3                              ADD   AX,BX
0002 03 C1                              ADD   AX,CX
0004 03 C2                              ADD   AX,DX
0006 C3                                 RET

0007                            ADDS    ENDP
```

Whenever arithmetic and logic instructions execute, the contents of the **flag register** change. Note that the contents of the interrupt, trap, and other flags do not change due to arithmetic and logic instructions. Only the flags located in the rightmost eight bits of the flag register and the overflow flag change. These rightmost flags denote the result of an arithmetic or a logic operation. Any ADD instruction modifies the contents of the sign, zero, carry, auxiliary carry, parity, and overflow flags. The flag bits never change for most of the data transfer instructions presented in Chapter 4.

Immediate Addition.   **Immediate addition** is employed whenever constant or known data are added. An 8-bit immediate addition is shown in Example 5–2. In this example, load DL is first loaded with a 12H by using an immediate move instruction. Next a 33H is added to the 12H in DL using an immediate addition instruction. After the addition, the sum (45H) moves into register DL and the flags change as follows:

Z = 0 (result not zero)

C = 0 (no carry)

A = 0 (no half-carry)

S = 0 (result positive)

P = 0 (odd parity)

O = 0 (no overflow)

EXAMPLE 5–2

```
0006 B2 12                      MOV   DL,12H
0008 80 C2 33                   ADD   DL,33H
```

Memory-to-Register Addition.   Suppose an application requires that memory data add to the AL register. Example 5–3 adds 2 consecutive bytes of data, stored at the data segment offset locations NUMB and NUMB + 1, to the AL register.

EXAMPLE 5–3

```
                                ;A procedure that sums data in locations NUMB and NUMB+1.
                                ;The result is returned in AX.
                                ;
0000                            SUMS    PROC NEAR

0000 BF 0000 R                          MOV   DI,OFFSET NUMB    ;address NUMB
0003 B0 00                              MOV   AL,0              ;clear sum
0005 02 05                              ADD   AL,[DI]           ;add NUMB
0007 02 45 01                           ADD   AL,[DI+1]         ;add NUMB+1
000A C3                                 RET

000B                            SUMS    ENDP
```

The procedure in Example 5–3 first loads the contents of the destination index register (DI) with offset address NUMB. The DI register addresses data in the data segment beginning at memory location NUMB. In most cases loading the address inside a procedure is poor programming practice. It is usually better to load the address outside the procedure and then CALL the procedure with the address in place. Next, the ADD AL,[DI] instruction adds the contents of memory location NUMB to AL. Note that AL is initialized to zero because DI addresses memory location NUMB, and the instruction adds its contents to AL. Finally, the ADD AL,[DI + 1] instruction adds the contents of memory location NUMB plus one byte to the AL register. After both ADD instructions execute, the result appears in the AL register as the sum of the contents of NUMB plus the contents of NUMB + 1.

Array Addition. **Memory arrays** are sequential lists of data. Suppose that an array of data (ARRAY) contains 10 bytes numbered from element 0 through element 9. Example 5–4 is a procedure that adds the contents of array elements 3, 5, and 7. (The procedure and the array elements it adds demonstrate the use of some of the addressing modes for the microprocessor.)

EXAMPLE 5–4

```
                          ;A procedure that sums ARRAY elements 3, 5, and 7.
                          ;The result is returned in AL.
                          ;
                          ;Note this procedure destroys the contents of SI.
                          ;
0000                      SUM       PROC NEAR

0000 B0 00                          MOV   AL,0            ;clear sum
0002 BE 0003                        MOV   SI,3            ;address element 3
0005 02 84 0002 R                   ADD   AL,ARRAY[SI]    ;add element 3
0009 02 84 0004 R                   ADD   AL,ARRAY[SI+2]  ;add element 5
000D 02 84 0006 R                   ADD   AL,ARRAY[SI+4]  ;add element 7
0011 C3                             RET

0012                      SUM       ENDP
```

Example 5–4 first clears AL to zero so it can be used to accumulate the sum. Next, register SI is loaded with a 3 to initially address array element 3. The ADD AL,ARRAY[SI] instruction adds the contents of array element 3 to the sum in AL. The instructions that follow add array elements 5 and 7 to the sum in AL, using a 3 in SI plus a displacement of 2 to address element 5, and a displacement of 4 to address element 7.

Suppose that an array of data contains 16-bit numbers used to form a 16-bit sum in register AX. Example 5–5 is a procedure written for the 80386 and above showing the scaled-index form of addressing to add elements 3, 5, and 7 to an area of memory called ARRAY. In this example, EBX is loaded with the address ARRAY, and ECX holds the array element number. Notice how the scaling factor is used to multiply the contents of the ECX register by 2 to address words of data. Recall that words are two bytes in length.

EXAMPLE 5–5

```
                          ;procedure that sums ARRAY elements 3, 5, and 7
                          ;result is returned in AX
                          ;
                          ;Note that the contents of EBX and ECX are destroyed.
                          ;
0000                      SUM       PROC NEAR

0000 66| BB 00000000 R              MOV   EBX,OFFSET ARRAY ;address ARRAY
0006 66| B9 00000003                MOV   ECX,3            ;address element 3
000C 67& 8B 04 4B                   MOV   AX,[EBX+2*ECX]   ;get element 3
0010 66| B9 00000005                MOV   ECX,5            ;address element 5
0016 67& 03 04 4B                   ADD   AX,[EBX+2*ECX]   ;add element 5
```

```
001A 66| B9 00000007        MOV   ECX,7              ;address element 7
0020 67& 03 04 4B           ADD   AX,[EBX+2*ECX]     ;add element 7
0024 C3                     RET

0025                 SUM     ENDP
```

Increment Addition.   **Increment addition** (INC) adds 1 to a register or a memory location. The INC instruction can add 1 to any register or memory location except a segment register. Table 5–2 lists some possible forms of the increment instruction available to the 8086–80486 and Pentium Pro microprocessors. As with other instructions presented thus far, it is impossible to show all variations of the INC instruction because of the large number available.

With indirect memory increments, the size of the data must be described using the BYTE PTR, WORD PTR, or DWORD PTR directives because the assembler program cannot determine if, for example, the INC [DI] instruction is a byte-, word-, or doubleword-sized increment. The INC BYTE PTR [DI] instruction clearly indicates byte-sized memory data; the INC WORD PTR [DI] instruction unquestionably indicates a word-sized memory data; the INC DWORD PTR [DI] instruction indicates doubleword-sized data.

Example 5–6 shows how the procedure of Example 5–3 is modified to use the increment instruction for addressing NUMB and NUMB + 1. Here, an INC DI instruction changes the contents of register DI from offset address NUMB to offset address NUMB + 1. Both procedures of Examples 5–3 and 5–6 add the contents of NUMB and NUMB + 1. The difference between these examples is the way that the data's address is formed through the contents of the DI register using the increment instruction.

EXAMPLE 5–6

```
                         ;procedure that sums NUMB and NUMB+1
                         ;result is returned in AL
                         ;
                         ;Note that the contents of DI are destroyed.
                         ;
0000                 SUMS    PROC  NEAR

0000 BF 0000 R               MOV   DI,OFFSET NUMB   ;address NUMB
0003 B0 00                   MOV   AL,0             ;clear sum
0005 02 05                   ADD   AL,[DI]          ;add NUMB
0007 47                      INC   DI               ;address NUMB+1
0008 02 05                   ADD   AL,[DI]          ;add NUMB+1
000A C3                      RET

000B                 SUMS    ENDP
```

TABLE 5–2   Increment instructions

Assembly Language	Operation
INC BL	BL = BL + 1
INC SP	SP = SP + 1
INC EAX	EAX = EAX + 1
INC BYTE PTR [BX]	Byte contents of the data segment memory location addressed by BX increment
INC WORD PTR [SI]	Word contents of the data segment memory location addressed by SI increment
INC DWORD PTR [ECX]	Doubleword contents of the data segment memory location addressed by ECX increment
INC DATA1	Contents of data segment location DATA1 increment

TABLE 5–4 Subtraction instructions

Assembly Language	Operation
SUB CL,BL	CL = CL – BL
SUB AX,SP	AX = AX – SP
SUB ECX,EBP	ECX = ECX – EBP
SUB DH,6FH	DH = DH – 6FH
SUB AX,0CCCCH	AX = AX – 0CCCCH
SUB EAX,23456H	EAX = EAX – 00023456H
SUB [DI],CH	CH is subtracted from the byte contents of the data segment memory location addressed by DI; the difference is stored in the same memory location
SUB CH,[BP]	Byte contents of the stack segment memory location addressed by BP is subtracted from CH
SUB AH,TEMP	Byte contents of data segment memory location TEMP are subtracted from AH
SUB SI,TEMP[BX]	Word contents of the data segment memory location addressed by TEMP plus BX are subtracted from SI
SUB ECX,DATA1	Doubleword contents of the data segment location DATA1 are subtracted from ECX

Note: Only the 80386/80486/Pentium/Pentium Pro use 32-bit registers and addressing modes.

Register Subtraction.   Example 5–9 shows a sequence of instructions that perform **register subtraction.** This example subtracts the 16-bit contents of registers CX and DX from the contents of register BX. After each subtraction, the microprocessor modifies the contents of the flag register. The flags change for most arithmetic and logic operations.

EXAMPLE 5–9

```
0000 2B D9              SUB    BX,CX
0002 2B DA              SUB    BX,DX
```

Immediate Subtraction.   As with addition, the microprocessor also allows immediate operands for the subtraction of constant data. Example 5–10 is a short sequence of instructions that subtract a 44H from a 22H. Here, we first load the 22H into CH using an immediate move instruction. Next, the SUB instruction, using immediate data 44H, subtracts a 44H from the 22H. After the subtraction, the difference (DEH) moves into the CH register. The flags change as follows for this subtraction:

Z = 0 (result not zero)

C = 1 (borrow)

A = 1 (half-borrow)

S = 1 (result negative)

P = 1 (even parity)

O = 0 (no overflow)

EXAMPLE 5–10

```
0000 B5 22              MOV    CH,22H
0002 80 ED 44           SUB    CH,44H
```

Both **carry flags** (C and A) hold borrows after a subtraction instead of carries, as after an addition. Notice in this example there is no overflow. We subtracted a 44H (+68) from a 22H (+34) resulting in a DEH (−34). Because the correct 8-bit signed result is a −34, there is no overflow in this example. An 8-bit overflow only occurs if the signed result is greater than +127 or less than −128.

Decrement Subtraction.   **Decrement subtraction** (DEC) subtracts a 1 from a register or the contents of a memory location. Table 5–5 lists some decrement instructions that illustrate register and memory decrements.

The decrement indirect memory data instructions require BYTE PTR, WORD PTR, or DWORD PTR because the assembler cannot distinguish a byte from a word when an index register addresses memory. For example, DEC [SI] is vague because the assembler cannot determine if the location addressed by SI is a byte or a word. Using DEC BYTE PTR [SI], DEC WORD PTR [DI], or DEC DWORD PTR [SI] reveals the size of the data.

Subtract with Borrow.   A **subtraction-with-borrow** (SBB) instruction functions as a regular subtraction, except the carry flag (C), which holds the borrow, also subtracts from the difference. The most common use for this instruction is for subtractions that are wider than 16 bits in the 8086–80286 microprocessors or wider than 32 bits in the 80386 through the Pentium Pro. Wide subtractions require that borrows propagate through the subtraction, just as wide additions propagate the carry.

Table 5–6 lists many SBB instructions with comments that define their operation. Like the SUB instruction, SBB affects the flags. Notice that the subtract from memory immediate instruction in this table requires a BYTE PTR, WORD PTR, or DWORD PTR directive.

When the 32-bit number held in BX and AX is subtracted from the 32-bit number held in SI and DI, the carry flag propagates the borrow between the two 16-bit subtractions required to perform this operation in the microprocessor. Figure 5–2 shows how the borrow propagates through the carry flag (C) for this task. Example 5–11 shows how this subtraction is performed by a program. With wide subtraction, the least-significant 16- or 32-bit data is subtracted with the SUB instruction. All subsequent and more-significant data are subtracted using the SBB instruction. Example 5–11 uses the SUB instruction to subtract DI from AX then SBB to subtract-with-borrow SI from BX.

EXAMPLE 5–11

```
0004 2B C7              SUB   AX,DI
0006 1B DE              SBB   BX,SI
```

TABLE 5–5   Decrement instructions

Assembly Language	Operation
DEC BH	BH = BH − 1
DEC DI	DI = DI − 1
DEC BYTE PTR [DI]	Byte contents of the data segment memory location addressed by DI decrement
DEC WORD PTR [BP]	Word contents of the stack segment memory location addressed by BP decrement
DEC DWORD PTR [EBX]	Doubleword contents of the data segment memory location addressed by EBX decrement
DEC NUMB	Decrements the contents of data segment memory location NUMB. The way NUMB is defined determines its size.

Note: Only the 80386/80486/Pentium/Pentium Pro use 32-bit registers and addressing modes.

TABLE 5–6 Subtract-with-borrow instructions

Assembly Language	Operation
SBB AH,AL	AH = AH – AL – carry
SBB AX,BX	AX = AX – BX – carry
SBB EAX,EBX	EAX = EAX – EBX – carry
SBB CL,3	CL = CL – 3 – carry
SBB BYTE PTR [DI],3	A 3 and carry are subtracted from the byte contents of the data segment memory location addressed by DI
SBB [DI],AL	AL and carry are subtracted from the byte contents of the data segment memory location addressed by DI
SBB DI,[BP+2]	The word contents of the stack segment memory location addressed by BP plus 2 and carry subtracts from DI
SBB AL [EBX+ECX]	The byte contents of the data segment memory location addressed by the sum of EBX and ECX are subtracted with carry from AL

Note: Only the 80386/80486/Pentium/Pentium Pro use 32-bit registers and addressing modes.

FIGURE 5–2 Subtraction with borrow showing how the carry flag (C) propagates the borrow.

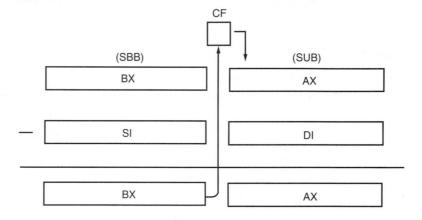

Comparison

The **comparison** (CMP) instruction is a subtraction that changes only the flag bits. A comparison is useful for checking the entire contents of a register or a memory location against another value. A CMP is normally followed by a conditional jump instruction, which tests the condition of the flag bits.

Table 5–7 lists a variety of comparison instructions that use the same addressing modes as the addition and subtraction instructions already presented. Similarly, the only disallowed forms of compare are memory-to-memory and segment register compares.

Example 5–12 shows a comparison followed by a conditional jump instruction. In this example, the contents of AL are compared with a 10H. Conditional jump instructions that often follow the compare are JA (jump above) or JB (jump below). If the JA follows the compare, the jump occurs if the value in AL is above 10H. If the JB follows the compare, the jump occurs if the value in AL is below 10H. In Example 5–12, the JAE instruction follows the compare. This instruction causes the program to continue at memory location SUBER if the value in AL is 10H or above. There is also a JBE (jump below or equal) instruction that could follow the compare to jump if the outcome is below or equal to 10H. Chapter 6 provides more detail on the compare and conditional jump instructions.

EXAMPLE 5-12

```
0000 3C 10              CMP   AL,10H              ;compare with 10H
0002 73 1C              JAE   SUBER               ;if 10H or above
```

Compare and Exchange (80486–Pentium Pro).   The **compare and exchange** (CMPXCHG) instruction found only in the 80486–Pentium Pro instruction sets compares the destination operand with the accumulator. If they are equal, the source operand is copied into the destination. If they are not equal, the destination operand is copied into the accumulator. This instruction functions with 8-, 16-, or 32-bit data.

The CMPXCHG CX,DX instruction is an example of the compare and exchange instruction. This instruction first compares the contents of CX with AX. If CX equals AX, DX is copied into AX. If CX is not equal to AX, CX is copied into AX. This instruction also compares AL with 8-bit data and EAX with 32-bit data if the operands are either 8- or 32-bit.

In the Pentium and Pentium Pro only, a CMPXCHG8B instruction is available that compares two quadwords. In fact this is the only new data manipulation instruction provided to the Pentium. The Pentium Pro contains a CMOV instruction that is presented in the next chapter because of its conditional nature. The compare and exchange 8-byte instruction compares the 64-bit value located in EDX:EAX with a 64-bit number located in memory. An example is CMPXCHG8B TEMP. If TEMP equals EDX:EAX, TEMP is replaced with the value found in ECX:EBX. If TEMP does not equal EDX:EAX, then the number found in TEMP is loaded into EDX:EAX. The zero flag bit indicates that the values are equal after the comparison.

TABLE 5–7   Comparison instructions

Assembly Language	Operation
CMP CL,BL	Subtracts BL from CL; neither BL nor CL change, but the flags change
CMP AX,SP	Subtracts SP from AX; neither AX nor SP change, but the flags change
CMP EBP,ESI	Subtracts ESI from EBP; neither ESI nor EBP change, but the flags change
CMP AX,3	Subtracts 3 from AX; AX does not change, but the flags change
CMP [DI],CH	Subtracts CH from the byte contents of the data segment memory location addressed by DI; neither CH nor memory change, but the flags change
CMP CL,[BP]	Subtracts the byte contents of the stack segment memory location addressed by BP from CL; neither CL nor memory change, but the flags change
CMP AH,TEMP	Subtracts the byte contents of data segment location TEMP from AH, neither AH nor memory change, but the flags change
CMP DI,TEMP[BX]	Subtracts the word contents of the data segment memory location addressed by the sum of TEMP and BX from DI; neither DI nor memory change, but the flags change
CMP AL,[EDI+ESI]	Subtracts the byte contents of the data segment memory location addressed by the sum of EDI and ESI from AL; neither AL nor memory change, but the flags change

Note: Only the 80386/80486/Pentium/Pentium Pro use 32-bit registers and addressing modes.

5–2   MULTIPLICATION AND DIVISION

Only modern microprocessors contain multiplication and division instructions. Earlier 8-bit microprocessors could not multiply or divide without a program that multiplied or divided using a series of shifts and additions or subtractions. Because microprocessor manufacturers were aware of this inadequacy, they incorporated multiplication and division instructions into the instruction sets of the newer microprocessors. In fact, the Pentium and Pentium Pro both contain special circuitry that performs a multiplication in as little as one clocking period, while it takes over 40 clocking periods to perform the same multiplication in earlier Intel microprocessors.

Multiplication

Multiplication is performed on bytes, words, or doublewords and can be signed integer (IMUL or unsigned (MUL). Note that only the 80386 through the Pentium Pro multiply 32-bit doublewords. The product after a multiplication is always a double-width product. If two 8-bit numbers are multiplied, they generate a 16-bit product; if two 16-bit numbers are multiplied, they generate a 32-bit product; if two 32-bit numbers are multiplied, they generate a 64-bit product.

Some **flag bits** (O and C) change when the multiply instruction executes and produce predictable outcomes. The other flags also change, but their results are unpredictable and therefore are unused. In an 8-bit multiplication, if the most-significant 8 bits of the result are 0, both C and O flag bits equal 0. These flag bits show that the result is 8 bits wide (C = 0) or 16 bits wide (C = 1). In a 16-bit multiplication, if the most-significant 16 bits of the product are 0, both C and O clear to 0. in a 32-bit multiplication, both C and O indicate that the most-significant 32 bits of the product are zero.

8-Bit Multiplication.   With **8-bit multiplication,** whether signed or unsigned, the multiplicand is always in the AL register. The multiplier can be any 8-bit register or memory location. Immediate multiplication is not allowed unless the special signed immediate multiplication instruction, discussed later in this section, appears in a program. The multiplication instruction contains one operand because it always multiplies the operand times the contents of register AL. An example is the MUL BL instruction, which multiplies the unsigned contents of AL by the unsigned contents of BL. After the multiplication, the unsigned product is placed in AX, a double-width product. Table 5–8 lists some 8-bit multiplication instructions.

Suppose that BL and CL each contain two 8-bit unsigned numbers and these numbers must be multiplied to form a 16-bit product stored in DX. This procedure cannot be accomplished by

TABLE 5–8   8-bit multiplication instructions

Assembly Language	Operation
MUL CL	AL is multiplied by CL; this unsigned multiplication returns the product in AX
IMUL DH	AL is multiplied by DH; this signed multiplication returns the product in AX
IMUL BYTE PTR [BX]	AL is multiplied by the byte contents of the data segment memory location addressed by BX; this signed multiplication returns the product in AX
MUL TEMP	AL is multiplied by the contents of data segment memory location TEMP; if TEMP is defined as an 8-bit number, the unsigned product returns in AX

a single instruction because we can only multiply a number times the AL register for an 8-bit multiplication. Example 5–13 is a short program that generates: DX = BL × CL. This example loads register BL and CL with example data 5 and 10. The product, a 50, moves into DX from AX after the multiplication by using the MOV DX,AX instruction.

EXAMPLE 5–13

```
0000 B3 05            MOV   BL,5            ;load data
0002 B1 0A            MOV   CL,10
0004 8A C1            MOV   AL,CL           ;position data
0006 F6 E3            MUL   BL              ;multiply
0008 8B D0            MOV   DX,AX           ;position product
```

For signed multiplication, the product is in true binary form, if positive; in two's complement form, if negative. These are the same forms used to store all positive and negative signed numbers used by the microprocessor. If the program in Example 5–13 multiplies two signed numbers, only the MUL instruction is changed to IMUL.

16-Bit Multiplication.   **Word multiplication** is very similar to byte multiplication. The difference is that AX contains the multiplicand instead of AL, and the product appears in DX – AX instead of in AX. The DX register always contains the most-significant 16 bits of the product and AX the least-significant 16 bits. As with 8-bit multiplication, the choice of the multiplier is up to you. Table 5–9 shows several different 16-bit multiplication instructions.

A Special Immediate 16-Bit Multiplication.   The 8086/8088 microprocessors could not perform immediate multiplication, but the 80186 through the Pentium can by using a special version of the multiply instruction. **Immediate multiplication** must be signed multiplication; the instruction format is different because it contains three operands. The first operand is the 16-bit destination register; the second operand is a register or memory location that contains the 16-bit multiplicand; the third operand is either an 8-bit or 16-bit immediate data used as the multiplier.

The IMUL CX,DX,12H instruction multiplies 12H times DX and leaves a 16-bit signed product in CX. If the immediate data is 8 bits, it sign-extends into a 16-bit number before the multiplication occurs. Another example is IMUL BX,NUMBER,1000H, which multiplies NUMBER times 1000H and leaves the product in BX. Both the destination and multiplicand must be 16-bit numbers. Although this is immediate multiplication, the restrictions placed upon it limit its utility, especially the fact that it is a signed multiplication and the product is 16 bits wide.

32-Bit Multiplication.   In the 80386 and above, **32-bit multiplication** is allowed because these microprocessors contain 32-bit registers. As with 8- and 16-bit multiplication, 32-bit multiplication can be signed or unsigned by using the IMUL and MUL instructions. With 32-bit multiplication, the contents of EAX are multiplied by the operand specified with the instruction. The product (64 bits wide) is found in EDX–EAX, where EAX contains the least-significant 32 bits of the product. Table 5–10 lists some of the 32-bit multiplication instructions found in the 80386 and above instruction set.

TABLE 5–9   16-bit multiplication instructions

Assembly Language	Operation
MUL CX	AX is multiplied by CX; the unsigned product is found in DX – AX
IMUL DI	AX is multiplied by DI; the signed product is found in DX – AX
MUL WORD PTR [SI]	AX is multiplied by the word contents of the data segment memory location addressed by SI; the unsigned product is found in DX – AX

TABLE 5–10   32-bit multiplication instructions

Assembly Language	Operation
MUL ECX	EAX is multiplied by ECX; the unsigned product is found in EDX – EAX
IMUL EDI	EAX is multiplied by EDI; the signed product is found in EDX – EAX
MUL DWORD PTR [ECX]	EAX is multiplied by the doubleword contents of the data segment memory location addressed by ECX; the unsigned product is found in EDX – EAX

Division

As with multiplication, **division** occurs on 8- or 16-bit numbers and 32-bit numbers in the 80386 through the Pentium Pro. These numbers are signed (IDIV) or unsigned (DIV) integers. The **dividend** is always a double-width dividend that is divided by the operand. This means that an 8-bit division divides a 16-bit number by an 8-bit number; a 16-bit division divides a 32-bit number by a 16-bit number; a 32-bit division divides a 64-bit number by a 32-bit number. There is no immediate division instruction available to any microprocessor.

None of the flag bits change predictably for a division. A division can result in two different types of errors: One of these is an attempt to divide by zero and the other is a divide overflow. A **divide overflow** occurs when a small number divides into a large number. For example, suppose that AX = 3,000 and that it is divided by 2. Because the quotient for an 8-bit division appears in AL, the result of 1,500 causes a divide overflow because the 1,500 does not fit into AL. In both cases the microprocessor generates an interrupt if a divide error occurs. In most cases, a divided error interrupt displays an error message on the video screen. The divide-error-interrupt and all other interrupts for the microprocessor are explained in a later chapter.

8-Bit Division.   An **8-bit division** uses the AX register to store the dividend that is divided by the contents of any 8-bit register or memory location. The **quotient** moves into AL after the division with AH containing a whole number **remainder.** For a signed division, the quotient is positive or negative and the remainder always assumes the sign of the dividend and is always an integer. For example, if AX = 0010H (+16) and BL = FDH (–3) and the IDIV BL instruction executes, AX = 01FBH. This represents a quotient of –5 (AL) with a remainder of 1 (AH). On the other hand, if a –16 is divided by a +3, the result will be a quotient of –5 (AL) with a remainder of –1 (AH). Table 5–11 lists some of the 8-bit division instructions.

With 8-bit division, the numbers are usually 8 bits wide. This means that one of them, the dividend, must be converted to a 16-bit wide number in AX. This is accomplished differently for

TABLE 5–11   8-bit division instructions

Assembly Language	Operation
DIV CL	AX is divided by CL; the unsigned quotient is in AL and the remainder is in AH
IDIV BL	AX is divided by BL; the signed quotient is in AL and the remainder is in AH
DIV BYTE PTR [BP]	AX is divided by the byte contents of the stack segment memory location addressed by BP; the unsigned quotient is in AL and the remainder is in AH

signed and unsigned numbers. For the unsigned number, the most-significant 8 bits must be cleared to zero (**zero-extended**). The MOVZX instruction described in Chapter 4 can be used to zero-extend a number in the 80386 through the Pentium Pro microprocessors. For the signed number, the least-significant 8 bits are sign-extended into the most-significant 8 bits. In the microprocessor a special instruction exists that sign-extends AL into AH, or converts an 8-bit signed number in AL into a 16-bit signed number in AX. The CBW (**convert byte to word**) instruction performs this conversion. In the 80386 through the Pentium Pro, a MOVSX instruction (see Chapter 4) can sign-extend a number.

Example 5–14 divides the unsigned byte contents of memory location NUMB by the unsigned contents of memory location NUMB1. Here the quotient is stored in location ANSQ and the remainder in location ANSR. Notice how the contents of location NUMB are retrieved from memory and then zero-extended to form a 16-bit unsigned number for the dividend.

EXAMPLE 5–14

```
0000 A0 0000 R            MOV   AL,NUMB         ;get NUMB
0003 B4 00                MOV   AH,0            ;zero-extend
0005 F6 36 0002 R         DIV   NUMB1           ;divide by NUMB1
0009 A2 0003 R            MOV   ANSQ,AL         ;save quotient
000C 88 26 0004 R         MOV   ANSR,AH         ;save remainder
```

Example 5–15 is basically the same program except that the numbers are signed numbers. This means that instead of zero-extending AL into AH, it is sign-extended with the CBW instruction.

EXAMPLE 5–15

```
0000 A0 0000 R            MOV   AL,NUMB         ;get NUMB
0003 98                   CBW                   ;sign-extend
0004 F6 3E 0002 R         IDIV  NUMB1           ;divide by NUMB1
0008 A2 0003 R            MOV   ANSQ,AL         ;save quotient
000B 88 26 0004 R         MOV   ANSR,AH         ;save remainder
```

16-Bit Division.   **Sixteen-bit division** is similar to 8-bit division except that instead of dividing into AX, the 16-bit number is divided into DX – AX, a 32-bit dividend. The quotient appears in AX and the remainder in DX after a 16-bit division. Table 5–12 lists some of the 16-bit division instructions.

As with 8-bit division, numbers must often be converted to the proper form for the dividend. If a 16-bit unsigned number is placed in AX, then DX must be cleared to 0. In the 80386 and above, the number is zero-extended using the MOVZX instruction, If AX is a 16-bit signed number, the CWD (**convert word to doubleword**) instruction sign-extends it into a

TABLE 5–12   16-bit division instructions

Assembly Language	Operation
DIV CX	DX – AX is divided by CX, the unsigned quotient is in AX and the remainder is in DX
IDIV SI	DX – AX is divided by SI; the signed quotient is in AX and the remainder is in DX
DIV NUMB	DX – AX is divided by the word contents of data segment memory location NUMB; the unsigned quotient is in AX and remainder is in DX

signed 32-bit number. If the 80386 and above are available, the MOVSX instruction can also be used to sign-extend a number.

Example 5–16 shows the division of two 16-bit signed numbers. Here a –100 in AX is divided by a +9 in CX. The CWD instruction converts the –100 in AX to a –100 in DX – AX before the division. After the division, the results appear in DX – AX as a quotient of –11 in AX and a remainder of –1 in DX.

EXAMPLE 5–16

```
0000 B8 FF9C          MOV  AX,-100        ;load -100
0003 B9 0009          MOV  CX,9           ;load +9
0006 99               CWD                 ;sign-extend
0007 F7 F9            IDIV CX
```

32-Bit Division.   The 80386 through the Pentium Pro perform **32-bit division** on signed or unsigned numbers. The 64-bit contents of EDX–EAX are divided by the operand specified by the instruction leaving a 32-bit quotient in EAX and a 32-bit remainder in EDX. Other than the size of the registers, this instruction functions in the same manner as the 8- and 16-bit divisions. Table 5–13 shows some 32-bit division instructions. The **convert doubleword-to-quadword** instruction (CDQ) is used before a signed division to convert the 32-bit contents of EAX into a 64-bit signed number in EDX – EAX.

The Remainder.   What is done with the remainder after a division? There are a few possible choices. The **remainder** could be used to round the result or just dropped to truncate the result. If the division is unsigned, rounding requires that the remainder be compared with half the divisor to decide whether to round-up the quotient. The remainder could also be converted to a fractional remainder.

Example 5–17 is a sequence of instructions that divide AX by BL and round the result. This program doubles the remainder before comparing it with BL to decide whether to round the quotient. Here, an INC instruction rounds the contents of AL after the compare.

EXAMPLE 5–17

```
0000 F6 F3            DIV  BL             ;divide
0002 02 E4            ADD  AH,AH          ;double remainder
0004 3A E3            CMP  AH,BL          ;test for rounding
0006 72 02            JB   NEXT
0008 FE C0            INC  AL             ;round
000A        NEXT:
```

TABLE 5–13   32-bit division instructions

Assembly Language	Operation
DIV ECX	EDX – EAX is divided by ECX; the unsigned quotient is in EAX and the remainder is in EDX
DIV DATA2	EDX – EAX is divided by the doubleword contents of the data segment memory location DATA2; the unsigned quotient is in EAX and the remainder is in EDX
IDIV DWORD PTR [EDI]	EDX – EAX is divided by the doubleword contents of the data segment memory location addressed by EDI; the signed quotient is in EAX and the remainder is in EDX

Suppose that a fractional remainder is required instead of an integer remainder. A fractional remainder is obtained by saving the quotient. Next the AL register is cleared to zero. The number remaining in AX is now divided by the original operand to generate a fractional remainder.

Example 5–18 shows how a 13 is divided by a 2. The 8-bit quotient is saved in memory location ANSQ and then AL is cleared. Next the contents of AX are again divided by 2 to generate a fractional remainder. After the division, the AL register equals an 80H. This is a 10000000_2. If the binary point (radix) is placed before the leftmost bit of AL, the fractional remainder in AL is 0.10000000_2 or 0.5 decimal. The remainder is saved in memory location ANSR in this example.

EXAMPLE 5–18

```
0000 B8 000D            MOV   AX,13            ;load 13
0003 B3 02              MOV   BL,2             ;load 2
0005 F6 F3              DIV   BL               ;13/2
0007 A2 0003 R          MOV   ANSQ,AL          ;save quotient
000A B0 00              MOV   AL,0             ;clear AL
000C F6 F3              DIV   BL               ;generate remainder
000E A2 0004 R          MOV   ANSR,AL          ;save remainder
```

5–3 BCD AND ASCII ARITHMETIC

The microprocessor allows arithmetic manipulation of both **binary-coded decimal** (BCD) and ASCII data. This is accomplished by instructions that adjust the numbers for BCD and ASCII arithmetic.

The BCD operations occur in systems such as point-of-sales terminals (cash registers) and others that seldom require arithmetic. The ASCII operations are performed on ASCII data used by many programs. BCD and ASCII arithmetic are rarely used today.

BCD Arithmetic

Two arithmetic techniques operate with BCD data: addition and subtraction. The instruction set provides two instructions that correct the result of a BCD addition and a BCD subtraction. The DAA **(decimal adjust after addition)** instruction follows BCD addition and DAS **(decimal adjust after subtraction)** follows BCD subtraction. Both instructions correct the result of the addition or subtraction so it is a BCD number.

For BCD data, the numbers always appear in the packed BCD form and are stored as two BCD digits per byte. The adjust instructions function only with the AL register after BCD addition and subtraction.

DAA Instruction. The **DAA instruction** follows the ADD or ADC instruction to adjust the result into a BCD result. Suppose that DX and BX each contain four-digit packed BCD numbers. Example 5–19 is a short program that adds the BCD numbers in DX and BX and stores the result in CX.

EXAMPLE 5–19

```
0000 BA 1234            MOV   DX,1234H          ;load 1,234
0003 BB 3099            MOV   BX,3099H          ;load 3,099
0006 8A C3              MOV   AL,BL             ;sum BL with DL
0008 02 C2              ADD   AL,DL
000A 27                 DAA                     ;adjust
000B 8A C8              MOV   CL,AL             ;answer to CL
000D 8A C7              MOV   AL,BH             ;sum BH, DH, and carry
000F 12 C6              ADC   AL,DH
0011 27                 DAA                     ;adjust
0012 8A E8              MOV   CH,AL             ;answer to CH
```

Because the DAA instruction functions only with the AL register, this addition must occur 8 bits at a time. After adding the BL and DL registers, the result is adjusted with a DAA instruction before being stored in CL. Next, add BH and DH registers with carry and the result again is adjusted with DAA before being stored in CH. In this example a 1,234 adds to a 3,099 to generate a sum of 4,333 that moves into CX after the addition. Note that 1234 BCD is the same as 1234H.

DAS Instruction. The **DAS instruction** functions as does the DAA instruction except it follows a subtraction instead of an addition. Example 5–20 is basically the same as Example 5–19 except that it subtracts instead of adds DX and BX. The main difference in these programs is that the DAA instructions change to DAS, and the ADD and ADC instructions change to SUB and SBB instructions.

EXAMPLE 5–20

```
0000 BA 1234          MOV   DX,1234H        ;load 1,234
0003 BB 3099          MOV   BX,3099H        ;load 3,099
0006 8A C3            MOV   AL,BL           ;subtract DL from BL
0008 2A C2            SUB   AL,DL
000A 2F               DAS                   ;adjust
000B 8A C8            MOV   CL,AL           ;answer to CL
000D 8A C7            MOV   AL,BH           ;subtract DH
000F 1A C6            SBB   AL,DH
0011 2F               DAS                   ;adjust
0012 8A E8            MOV   CH,AL           ;answer to CH
```

ASCII Arithmetic

The ASCII arithmetic instructions function with ASCII-coded numbers. These numbers range in value from 30H to 39H for the numbers 0–9. There are four instructions used with ASCII arithmetic operations: AAA **(ASCII adjust after addition),** AAD **(ASCII adjust before division),** AAM **(ASCII adjust after multiplication),** and AAS **(ASCII adjust after subtraction).** These instructions use register AX as the source and destination.

AAA Instruction. The addition of two 1-digit ASCII-coded numbers will not result in any useful data. For example, if a 31H and 39H are added, the result is a 6AH. This ASCII addition (1 + 9) should produce a 2-digit ASCII result equivalent to a 10 decimal, which is a 31H and a 30H in ASCII code. If the **AAA instruction** is executed after this addition, the AX register will contain a 0100H. Although this is not ASCII code, it can be converted to ASCII code by adding 3030H, which generates 3130H. The AAA instruction clears AH if the result is less than 10 and adds a 1 to AH if the result is greater than 10.

Example 5–21 shows how ASCII addition functions in the microprocessor. Please note that AH is cleared before the addition by using the MOV AX,31H instruction. The operand of 0031H places a 00H in AH and a 31H into AL.

EXAMPLE 5–21

```
0000 B8 0031          MOV   AX,31H          ;load ASCII 1
0003 04 39            ADD   AL,39H          ;add ASCII 9
0005 37               AAA                   ;adjust
0006 05 3030          ADD   AX,3030H        ;answer to ASCII
```

AAD Instruction. Unlike all the other adjust instructions, the **AAD instruction** appears *before* a division. The AAD instruction requires that the AX register contain a 2-digit unpacked BCD number (not ASCII) before executing. After adjusting the AX register with AAD, it is divided by an unpacked BCD number to generate a single-digit result in AL with any remainder in AH.

Example 5–22 illustrates how a 72 in unpacked BCD is divided by 9 to produce a quotient of 8. The 0702H loaded into the AX register is adjusted by the AAD instruction to 0048H. Notice that this converts a 2-digit unpacked BCD number into a binary number so it can be divided with the binary division instruction (DIV). The AAD instruction converts the unpacked BCD numbers between 00 and 99 into binary.

EXAMPLE 5–22

```
0000 B3 09                MOV   BL,9         ;load divisor
0002 B8 0702              MOV   AX,0702H     ;load dividend
0005 D5 0A                AAD                ;adjust
0007 F6 F3                DIV   BL
```

AAM Instruction.   The **AAM instruction** follows the multiplication instruction after multiplying two 1-digit unpacked BCD numbers. Example 5–23 is a short program that multiplies a 5 times a 5. The result after the multiplication is 0019H in the AX register. After adjusting the result with the AAM instruction, AX contains a 0205H. This is an unpacked BCD result of 25. If 3030H adds to 0205H, this becomes an ASCII result of 3235H.

EXAMPLE 5–23

```
0000 B0 05                MOV   AL,5         ;load multiplicand
0002 B1 05                MOV   CL,5         ;load multiplier
0004 F6 E1                MUL   CL
0006 D4 0A                AAM                ;adjust
```

The AAM instruction accomplishes this conversion by dividing AX by 10. The remainder is found in AL and the quotient is in AH. The second byte of the instruction contains a 0AH; if the 0AH is changed to another value, AAM divides by the new value. For example, if the second byte is changed to a 0BH, the AAM instruction divides by an 11.

One side-benefit of the AAM instruction is that AAM converts from binary to unpacked BCD. If a binary number between 0000H and 0063H appears in the AX register, the AAM instruction converts it to BCD. For example, if AX contains a 0060H before AAM, it will contain a 0906H after AAM executes. This is the unpacked BCD equivalent of 96 decimal. If 3030H is added to 0906H, the result changes to ASCII code.

Example 5–24 shows how the 16-bit binary contents of AX are converted to a 4-digit ASCII character string by using division and the AAM instruction. Note that this works for numbers between 0 and 9,999. First DX is cleared and then DX – AX is divided by 100. For example, if AX = 245, after the division, AX = 2 and DX = 45. These separate halves are converted to BCD using AAM and then a 3030H is added to convert to ASCII code.

EXAMPLE 5–24

```
0000 33 D2                XOR   DX,DX        ;clear DX register
0002 B9 0064              MOV   CX,100       ;divide DX–AX by 100
0005 F7 F1                DIV   CX           ;AX = quotient and
                                             ;DX = remainder
0007 D4 0A                AAM                ;convert quotient to BCD
0009 05 3030              ADD   AX,3030H     ;convert to ASCII
000C 92                   XCHG  AX,DX        ;repeat for remainder
000D D4 0A                AAM
000F 05 3030              ADD   AX,3030H
```

Example 5–25 uses the DOS 21H function AH = 02H to display a sample number in decimal on the video display using the AAM instruction. Notice how AAM is used to convert AL into BCD. Next ADD AX,3030H converts the BCD code in AX into ASCII for display with DOS INT 21H. Once the data is converted to ASCII code it is displayed by loading DL with the most-significant digit from AH. Next the least-significant digit is displayed from AL. Notice that the DOS INT 21H function calls change AL.

EXAMPLE 5–25

```
                              ;A program that displays the number loaded into AL,
                              ;with the first instruction (48H) as a decimal number.
                              ;
                              .MODEL TINY        ;select TINY model
0000                          .CODE              ;start of CODE segment
                              .STARTUP           ;indicate start of program
0100 B0 48                    MOV   AL,48H        ;load AL with test data
0102 B4 00                    MOV   AH,0          ;clear AH
0104 D4 0A                    AAM                 ;convert to BCD
0106 05 3030                  ADD   AX,3030H      ;convert to ASCII
0109 8A D4                    MOV   DL,AH          ;display digit
010B B4 02                    MOV   AH,2
010D 50                       PUSH AX             ;save digit
010E CD 21                    INT   21H
0110 58                       POP   AX            ;restore AL
0111 8A D0                    MOV   DL,AL          ;display digit
0113 CD 21                    INT   21H
                              .EXIT               ;exit to DOS
                              END                 ;end of file
```

AAS Instruction. Like other ASCII adjust instructions, **AAS** adjusts the AX register after an ASCII subtraction. For example, suppose that a 35H subtracts from a 39H. The result will be a 04H, which requires no correction. Here AAS will modify neither AH or AL. On the other hand, if 38H subtracts from 37H, then AL will equal 09H and the number in AH will decrement by 1. This decrement allows multiple-digit ASCII numbers to be subtracted from each other.

5–4   BASIC LOGIC INSTRUCTIONS

The basic logic instructions include: AND, OR, Exclusive-OR, and NOT. Another logic instruction is TEST, which is a special form of the AND instruction. We also look at the NEG instruction, which is similar to the NOT instruction.

Logic operations provide binary bit control in low-level software. The logic instructions allow bits to be set, cleared, or complemented. Low-level software appears in machine language or assembly language form and often controls the I/O devices in a system. All logic instructions affect the flag bits. Logic operations always clear the carry and overflow flags, while the other flags change to reflect the condition of the result.

When binary data is manipulated in a register or memory location, the rightmost bit position is always numbered bit 0. Bit position numbers increase from bit 0 toward the left, to bit 7 for a byte, and to bit 15 for a word. A doubleword (32 bits) uses bit position 31 as its leftmost bit.

AND

The **AND operation** performs logical multiplication as illustrated by the truth table in Figure 5–3. Here two bits, A and B, are ANDed to produce the result X. As indicated by the truth table, X is a logic 1 only when both A and B are logic 1's. For all other input combinations of A and B, X is a logic 0. It is important to remember that 0 AND anything is always 0 and 1 AND 1 is always a 1.

The AND instruction can replace discrete AND gates if the speed required is not too great, although this is normally reserved for embedded control applications. (Intel has released the 80386EX embedded controller, which embodies the basic structure of the personal computer system.) With the 8086 microprocessor, the AND instruction often executes in about a microsecond. With newer versions the execution speed is greatly increased. If the circuit that the AND instruction replaces operates at a much slower speed than the microprocessor, the

FIGURE 5–3   (a) The truth table for the AND operation and (b) the logic symbol of an AND gate.

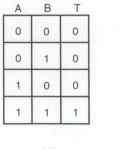

A	B	T
0	0	0
0	1	0
1	0	0
1	1	1

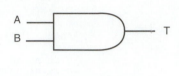

(a)   (b)

AND instruction is a logical replacement. This replacement can save a considerable amount of money. A single AND gate integrated circuit (7408) costs approximately 40¢, while it costs less than 1/100¢ to store the AND instruction in a read-only memory. Note that such logic circuit replacement appears only in control systems based on microprocessors and does not generally find application in the personal computer.

The AND operation also clears bits of a binary number. The task of clearing a bit in a binary number is called *masking*. Figure 5–4 illustrates the process of masking. Notice that the left-most four bits clear to 0, because 0 AND anything is 0. The bit positions of AND with 1's do not change. This occurs because if a 1 ANDs with a 1, a 1 results; if a 1 ANDs with a 0, a 0 results.

The AND instruction uses any addressing mode except memory-to-memory and segment register addressing. Refer to Table 5–14 for a list of some AND instructions and comments about their operation.

An ASCII-coded number can be converted to BCD by using the AND instruction to mask off the leftmost four binary bit positions. This converts the ASCII 30H to 39H to 0–9. Example 5–26

FIGURE 5–4   The operation of the AND function showing how bits of a number are cleared to zero.

```
    x x x x  x x x x   Unknown number
•   0 0 0 0  1 1 1 1   Mask
    ─────────────────
    0 0 0 0  x x x x   Result
```

TABLE 5–14   AND instructions

Assembly Language	Operation
AND AL,BL	AL = AL AND BL
AND CX,DX	CX = CX AND DX
AND ECX,EDI	ECX = ECX AND EDI
AND CL,33H	CL = CL AND 33H
AND DI,4FFFH	DI = DI AND 4FFFH
AND ESI,34H	ESI = ESI AND 00000034H
AND AX,[DI]	AX is ANDed with the word contents of the data segment memory location addressed by DI; the result is in AX
AND ARRAY[SI],AL	The byte contents of the data segment memory location addressed by ARRAY plus SI is ANDed with AL; the result is in memory
AND [EAX],CL	The byte contents of CL is ANDed with the byte contents of the data segment memory location addressed by EAX; the result is in memory

FIGURE 5–5 (a) The truth table for the OR operation and (b) the logic symbol of an OR gate.

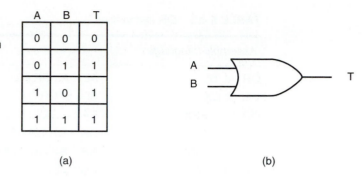

A	B	T
0	0	0
0	1	1
1	0	1
1	1	1

(a)

(b)

is a short program that converts the ASCII contents of BX into BCD. The AND instruction in this example converts two digits from ASCII to BCD simultaneously.

EXAMPLE 5–26

```
0000 BB 3135          MOV  BX,3135H        ;load ASCII
0003 81 E3 0F0F       AND  BX,0F0FH        ;mask BX
```

OR

The **OR operation** performs logical addition and is often called the inclusive-OR function. The OR function generates a logic 1 output if any inputs are 1. A 0 appears at the output only when all inputs are 0. The truth table for the OR function appears in Figure 5–5. Here the inputs A and B OR together to produce the X output. It is important to remember that 1 ORed with anything yields a 1.

In embedded controller applications, the OR instruction can also replace discrete OR gates. This results in a considerable savings, because a quad, 2-input OR gate (7432) costs about 40¢, while the OR instruction costs less than 1/100¢ to store in read-only memory.

Figure 5–6 shows how the OR gate sets (1) any bit of a binary number. Here an unknown number (XXXX XXXX) ORs with a 0000 1111 to produce a result of XXXX 1111. The rightmost four bits set, while the leftmost four bits remain unchanged. The OR operation sets any bit and the AND operation clears any bit.

The OR instruction uses any of the addressing modes allowed to any other instruction except segment register addressing. Table 5–15 lists several OR instructions with comments about their operation.

Suppose two BCD numbers are multiplied and adjusted with the AAM instruction. The result appears in AX as a 2-digit unpacked BCD number. Example 5–27 illustrates this multiplication and shows how to change the result into a 2-digit ASCII-coded number using the OR instruction. Here, OR AX,3030H converts the 0305H found in AX to 3335H. The OR operation can be replaced with an ADD AX,3030H to obtain the same results.

EXAMPLE 5–27

```
0000 B0 05            MOV  AL,5            ;load data
0002 B3 07            MOV  BL,7
0004 F6 E3            MUL  BL
0006 D4 0A            AAM                  ;adjust
0008 0D 3030          OR   AX,3030H        ;to ASCII
```

FIGURE 5–6 The operation of the OR function showing how bits of a number are set to one.

```
    x x x x  x x x x   Unknown number
+   0 0 0 0  1 1 1 1   Mask
    ─────────────────
    x x x x  1 1 1 1   Result
```

TABLE 5–15   OR instructions

Assembly Language	Operation
OR AL,BL	AL = AL OR BL
OR SI,DX	SI = SI OR DX
OR EAX,EBX	EAX = EAX OR EBX
OR DH,0A3H	DH = DH OR 0A3H
OR SP,990DH	SP = SP OR 990DH
OR EBP,10	EBP = EBP OR 0000000AH
OR DX,[BX]	DX is ORed with the word contents of the data segment memory location addressed by BX; the result is in DX
OR DATES[DI+2],AL	The data segment memory location addressed by DATES plus DI plus 2 is ORed with AL

Exclusive-OR

The **exclusive-OR** (XOR) instruction differs from inclusive-OR (OR). The difference is that a 1,1 condition of the OR function produces a 1, while the 1,1 condition of the XOR produces a 0. The exclusive-OR operation excludes this condition, while the inclusive-OR includes it.

Figure 5–7 shows the truth table of the exclusive-OR function. (Compare this with Figure 5–5 to appreciate the difference between these two OR functions.) If the inputs of the exclusive-OR function are both 0 or both 1, the output is 0. If the inputs are different, the output is a 1. Because of this, the exclusive-OR is sometimes called a **comparator.**

The XOR instruction uses any addressing mode except segment register addressing. Table 5–16 lists various forms of the exclusive-OR instruction with comments about their operation.

As with the AND and OR functions, exclusive-OR can replace discrete logic circuitry in embedded applications. The 7486 quad, 2-input exclusive-OR gate is replaced by one XOR instruction. The 7486 costs about 40¢, while the instruction costs less than 1/100¢ to store in the memory. Replacing just one 7486 saves a considerable amount of money, especially if many systems are built.

The exclusive-OR instruction is useful if some bits of a register or memory location must be inverted. This instruction allows part of a number to be inverted or complemented. Figure 5–8 shows how just part of an unknown quantity can be inverted by XOR. Notice that when a 1 exclusive-ORs with X, the result is X̄. If a 0 exclusive-ORs with X, the result is X.

Suppose that the leftmost ten bits of the BX register must be inverted without changing the rightmost six bits. The XOR BX,0FFC0H instruction accomplishes this task. The AND instruction clears (0) bits; the OR instruction sets (1) bits; now the exclusive-OR instruction inverts bits.

FIGURE 5–7   (a) The truth table for the exclusive-OR operation and (b) the symbol of an exclusive-OR gate.

A	B	T
0	0	0
0	1	1
1	0	1
1	1	0

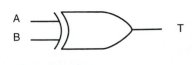

(a)   (b)

TABLE 5–16 Exclusive-OR instructions

Assembly Language	Operation
XOR CH,DL	CH = CH XOR DL
XOR SI,BX	SI = SI XOR BX
XOR EBX,EDI	EBX = EBX XOR EDI
XOR CH,0EEH	CH = CH XOR 0EEH
XOR DI,00DDH	DI = DI XOR 00DDH
XOR ESI,100	ESI = ESI XOR 00000064H
XOR DX,[SI]	DX is XORed with the word contents of data segment memory addressed by SI; the result is in DX
XOR TEMP[DI],AL	The byte contents of the data segment memory location addressed by the sum of TEMP and DI are XORed with AL

FIGURE 5–8 The operation of the exclusive-OR function showing how bits of a number are inverted.

```
x x x x  x x x x   Unknown number
⊕ 0 0 0 0  1 1 1 1   Mask
─────────────────
x x x x  x̄ x̄ x̄ x̄   Result
```

These three instructions allow a program to gain complete control over any bit stored in any register or memory location. This is ideal for control system applications where equipment must be turned on (1), turned off (0), and toggled from on to off.

A fairly common use for the exclusive-OR instruction is to clear a register to zero. For example, the XOR CH,CH instruction clears register CH to 00H and requires two bytes of memory to store the instruction. Likewise, the MOV CH,00H instruction also clears CH to 00H, but requires three bytes of memory. Because of this savings, the XOR instruction is used to clear a register in place of a move immediate.

Example 5–28 is a short sequence of instructions that clear bits 0 and 1 of CX, set bits 9 and 10 of CX, and invert bit 12 of CX. The OR instruction is used to set bits; the AND instruction is used to clear bits; the XOR instruction inverts bits.

EXAMPLE 5–28

```
0000 81 C9 0600        OR   CX,0600H     ;set bits 9 and 10
0004 83 E1 FC          AND  CX,0FFFCH    ;clear bits 0 and 1
0007 81 F1 1000        XOR  CX,1000H     ;invert bit 12
```

Test and Bit Test Instructions

The **TEST instruction** performs the AND operation. The difference is that the AND instruction changes the destination operand, while the TEST instruction does not. A TEST only affects the condition of the flag register, which indicates the result of the test. The TEST instruction uses the same addressing modes as the AND instruction. Table 5–17 lists some forms of the TEST instruction with comments about their operation.

The TEST instruction functions in the same manner as a CMP instruction. The *difference* is that the TEST instruction normally tests a single bit (or occasionally multiple bits), while the CMP instruction tests the entire byte, word, or doubleword. The zero flag (Z) is a logic 1 (indicating a zero result) if the bit under test is a zero, and Z = 0 (indicating a non-zero result) if the bit under test is not zero.

Usually the TEST instruction is followed by either the JZ (jump zero) or JNZ (jump not zero) instruction. The destination operand is normally tested against immediate data. The

TABLE 5–17 TEST instructions

Assembly Language	Operation
TEST DL,DH	DL is ANDed with DH; neither DL nor DH change; only the flags change
TEST CX,BX	CX is ANDed with BX; neither CX nor BX change; only the flags change
TEST EDX,ECX	EDX is ANDed with ECX; neither EDX nor ECX change; only the flags change
TEST AH,4	AH is ANDed with 4; AH does not change; only the flags change
TEST DL,TEMPS	DL is ANDed with the byte contents of data segment memory location TEMPS

values of immediate data are 1 to test the rightmost bit position, 2 to test the next bit, 4 for the next, etc.

Example 5–29 is a short program that tests the rightmost and leftmost bit positions of the AL register. Here, 1 selects the rightmost bit and 128 selects the leftmost bit. The JNZ instruction follows each test to jump to different memory locations depending on the outcome of the tests. The JNZ instruction jumps to the operand address (RIGHT or LEFT in the example) if the bit under test is not zero.

EXAMPLE 5–29

```
0000 A8 01            TEST    AL,1           ;test right bit
0002 75 1C            JNZ     RIGHT          ;if set
0004 A8 80            TEST    AL,128         ;test left bit
0006 75 38            JNZ     LEFT           ;if set
```

Bit Test Instructions

The 80386–Pentium Pro contain additional test instructions that test bit positions. Table 5–18 lists the four **bit test instructions** available to these microprocessors.

All four forms of the bit test instruction test the bit position in the destination operand selected by the source operand. For example, the BT AX,4 instruction tests bit position 4 in AX. The result of the test is located in the carry flag bit. If bit position 4 is a 1, carry is set; if bit position 4 is a zero, carry is cleared.

The remaining three bit test instructions also place the bit under test into the carry flag, and afterwards, change the bit under test. The BTC AX,4 instruction complements bit position 4 after

TABLE 5–18 Bit test instructions

Assembly Language	Operation
BT	Test the bit in the destination operand specified by the source operand
BTC	Test and complement the bit in the destination operand specified by the source operand
BTR	Test and reset the bit in the destination operand specified by the source operand
BTS	Test and set the bit in the destination operand specified by the source operand

TABLE 5–19 NOT and NEG instructions

Assembly Language	Operation
NOT CH	CH is one's complemented
NEG CH	CH is two's complemented
NEG AX	AX is two's complemented
NOT EBX	EBX is one's complemented
NEG ECX	ECX is two's complemented
NOT TEMP	The contents of data segment memory location TEMP is one's complemented; the size of TEMP is determined by how TEMP is defined
NOT BYTE PTR [BX]	The byte contents of the data segment memory location addressed by BX is one's complemented

testing it; the BTR AX,4 instruction clears it (0) after the test; the BTS AX,4 instruction sets it (1) after the test.

Example 5–30 repeats the sequence of instructions listed in Example 5–28. Here the BTR instruction clears bits in CX; BTS sets bits in CX; BTC inverts bits in CX.

EXAMPLE 5–30

```
0000 0F BA E9 09        BTS   CX,9        ;set bit 9
0004 0F BA E9 0A        BTS   CX,10       ;set bit 10
0008 0F BA F1 00        BTR   CX,0        ;clear bit 0
000C 0F BA F1 01        BTR   CX,1        ;clear bit 1
0010 0F BA F9 0C        BTC   CX,12       ;invert bit 12
```

NOT and NEG

Logical inversion or the **one's complement** (NOT) and arithmetic sign inversion or the **two's complement** (NEG) are two of a few instructions that contain only one operand. Table 5–19 lists some variations of the NOT and NEG instructions. As with most other instructions, NOT and NEG can use any addressing mode except segment register addressing.

The NOT instruction inverts all bits of a byte, word, or doubleword. The NEG instruction two's complements a number, which means that the arithmetic sign of a signed number changes from positive to negative or negative to positive. The NOT function is considered logical and the NEG function is considered an arithmetic operation.

5–5 SHIFT AND ROTATE

Shift and rotate instructions manipulate binary numbers at the binary bit level, as do the AND, OR, exclusive-OR, and NOT instructions. Shift and rotate find their most common application in low-level software used to control I/O devices. The microprocessor contains a complete set of shift and rotate instructions used to shift or rotate any memory data or register.

Shift

Shift instructions position or move numbers to the left or right within a register or memory location. They also perform simple arithmetic such as multiplication by powers of 2^{+n} (**left shift**) and division by powers of 2^{-n} (**right shift**). The microprocessor's instruction set contains four

FIGURE 5–9 The shift instructions showing the operation and direction of the shift.

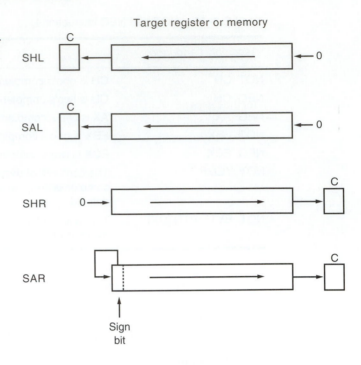

different shift instructions: two are logical shifts and two are arithmetic shifts. All four shift operations appear in Figure 5–9.

Notice in Figure 5–9 that there are two right shifts and two left shifts. The logical shifts move a 0 into the rightmost bit position for a logical left shift and a 0 into the leftmost bit position for a logical right shift. There are also two arithmetic shifts. The arithmetic and logical left shifts are identical. The arithmetic and logical right shifts are different because the arithmetic right shift copies the sign bit through the number while the logical right shift copies a 0 through the number.

Logical shift operations function with unsigned numbers and arithmetic shifts function with signed numbers. Logical shifts multiply or divide unsigned data and **arithmetic shifts** multiply or divide signed data. A shift left always multiplies by 2 for each bit position shifted and a shift right always divides by 2 for each bit position shifted. Shifting a number 2 places multiplies or divides by 4.

Table 5–20 lists some addressing modes allowed for the various shift instructions. There are two different forms of shifts that allow any register (except the segment register) or memory

TABLE 5–20 Shift instructions

Assembly Language	Operation
SHL AX,1	Logically shift AX left 1 place
SHR BX,12	Logically shift BX right 12 places
SHR ECX,10	Logically shift ECX right 10 places
SAL DATA1,CL	Arithmetically shift DATA1, in the data segment, left the number of places specified by the contents of CL
SAR SI,2	Arithmetically shift SI right 2 places
SAR EDX,14	Arithmetically shift EDX right 14 places

Note: The 8086/8088 microprocessors allow an immediate shift count of 1 only.

location to be shifted. One mode uses an immediate shift count, and the other uses register CL to hold the shift count. Note that CL must hold the shift count. When CL is the shift count, it does not change when the shift instruction executes. The shift count is a modulo-32 count. This means that a shift count of 33 will shift the data one place (33 / 32 = remainder of 1).

Example 5–31 shows how to shift the DX register left 14 places in two ways: The first method uses an immediate shift count of 14. The second method loads a 14 into CL and then uses CL as the shift count. Both instructions shift the contents of the DX register logically to the left 14 binary bit positions or places.

EXAMPLE 5–31

```
0000 C1 E2 0E              SHL   DX,14

              or

0003 B1 0E                 MOV   CL,14
0005 D3 E2                 SHL   DX,CL
```

Suppose that the contents of AX must be multiplied by 10, as in Example 5–32. This can be done in two ways: by the MUL instruction or by shifts and additions. A number is doubled when it shifts left one binary place. When a number is doubled, then added to the number times 8, the result is 10 times the number. The number 10 decimal is 1010 in binary. A logic 1 appears in both the 2's and 8's positions. If 2 times the number is added to 8 times the number, the result is 10 times the number. Using this technique, a program can be written to multiply by any constant. This technique often executes faster than the multiply instruction in earlier versions of the Intel microprocessor.

EXAMPLE 5–32

```
                        ;Multiply AX by 10 (1010).
                        ;
0000 D1 E0                 SHL   AX,1         ;AX times 2
0002 8B D8                 MOV   BX,AX
0004 C1 E0 02              SHL   AX,2         ;AX times 8
0007 03 C3                 ADD   AX,BX        ;10 times AX
                        ;
                        ;Multiply AX by 18 (10010).
                        ;
0009 D1 E0                 SHL   AX,1         ;AX times 2
000B 8B D8                 MOV   BX,AX
000D C1 E0 03              SHL   AX,3         ;AX times 16
0010 03 C3                 ADD   AX,BX        ;18 times AX
                        ;
                        ;Multiply AX by 24 (11000).
                        ;
0012 C1 E0 03              SHL   AX,3         ;AX times 8
0015 8B D8                 MOV   BX,AX
0017 D1 E0                 SHL   AX,1         ;AX times 16
0019 03 C3                 ADD   AX,BX        ;24 times AX
```

Double-Precision Shifts (80386/80486/Pentium/Pentium Pro). The 80386 and above contain two **double-precision shifts,** SHLD (**shift left**) and SHRD (**shift right**). Each instruction contains three operands instead of the two found with the other shift instructions. Both instructions function with two 16- or 32-bit registers or with one 16- or 32-bit memory location and a register.

The SHRD AX,BX,12 instruction is an example of the double-precision shift right instruction. This instruction logically shifts AX right by 12-bit positions. The rightmost 12 bits of BX shift into the leftmost 12 bits of AX. The contents of BX remain unchanged by this instruction. The shift count can be an immediate count, as in this example, or can be found in register CL as with other shift instructions.

FIGURE 5–10   The rotate instructions showing the direction and operation of each rotate.

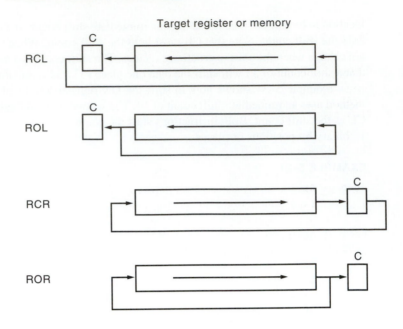

The SHLD EBX,ECX,16 instruction shifts EBX left. The leftmost 16 bits of ECX fill the rightmost 16 bits of EBX after the shift. As before, the contents of ECX, the second operand, remain unchanged. This instruction as well as SHRD affect the flag bits.

Rotate

Rotate instructions position binary data by rotating the information in a register or memory location either from one end to another or through the carry flag. They are often used to shift or position numbers that are wider than 16 bits in the 8086–80286 microprocessors or wider than 32 bits in the 80386 through the Pentium. The four rotate instructions are shown in Figure 5–10.

Numbers rotate through a register or a memory location and the C flag (carry) or through a register or memory location only. With either type of rotate instruction, the programmer can select either a left or a right rotate. Addressing modes used with rotate are the same as those used with shifts. A rotate count can be immediate or located in register CL. Table 5–21 lists some of the possible rotate instructions. If CL is used for a rotate count, it does not change. As with shifts, the count in CL is a modulo-32 count.

Rotates are often used to shift wide numbers to the left or right. The program in Example 5–33 shifts the 48-bit number in registers DX, BX, and AX left one binary place. Notice that the least-significant 16 bits (AX) are shifted left first. This moves the leftmost bit of AX into the carry flag

TABLE 5–21   Rotate instructions

Assembly Language	Operation
ROL SI,4	SI rotates left 4 places
RCL BL,6	BL rotates left through carry 6 places
ROL ECX,18	ECX rotates left 18 places
ROR AH,CL	AH rotates right the number of places specified by CL
ROR WORD PTR [BP],2	The word contents of the stack segment memory location addressed by BP rotate right 2 places

Note: The 8086/8088 microprocessors can use only an immediate rotate count of 1.

bit. Next the rotate BX instruction rotates carry into BX and its leftmost bit moves into carry. The last instruction rotates carry into DX and the shift is complete.

EXAMPLE 5–33

```
0000 D1 E0          SHL  AX,1
0002 D1 D3          RCL  BX,1
0004 D1 D2          RCL  DX,1
```

Bit Scan Instructions

Although the **bit scan instructions** don't shift or rotate numbers, they do scan through a number searching for a 1-bit. Because this is accomplished within the microprocessor by shifting the number, bit scan instructions are discussed here.

The bit scan instructions **BSF** (bit scan forward) and **BSR** (bit scan reverse) are only available in the 80386–Pentium Pro. Both forms scan through a number searching for the first 1-bit encountered. The BSF instruction scans the number from the rightmost bit towards the left and BSR scans the number from the leftmost bit towards the right. If a 1-bit is encountered by either bit scan instruction, the zero flag is set and the bit position of the 1-bit is placed into the destination operand. If no 1-bit is encountered (i.e., the number contains all zeros), the zero flag is cleared. This means that the result is not-zero and no 1-bit is encountered.

For example, if EAX = 60000000H and the BSF EBX,EAX instruction executes, the number is scanned from the rightmost bit toward the left. The first 1-bit encountered is at bit position 29, which is placed into EBX and the zero flag bit is set. If the same value for EAX is used for the BSR instruction, the EBX register is loaded with a 30 and the zero flag bit is set.

5–6   STRING COMPARISONS

As discussed in Chapter 4, the string instructions are very powerful because they allow you to manipulate large blocks of data with relative ease. Block data manipulation occurs with the string instructions: MOVS, LODS, STOS, INS, and OUTS.

In this section, we discuss additional string instructions that allow a section of memory to be tested against a constant or against another section of memory. To accomplish these tasks we use the SCAS (string scan) or CMPS (string compare) instructions.

SCAS

The **string scan** instruction (SCAS) compares the AL register with a byte block of memory, the AX register with a word block of memory, or the EAX register (80386–Pentium Pro only) with a doubleword block of memory. The SCAS instruction subtracts memory from AL, AX, or EAX without affecting either the register or the memory location. The opcode used for byte comparison is SCASB; the opcode used for the word comparison is SCASW; the opcode used for a doubleword comparison is SCASD. In all cases the contents of the extra segment memory location addressed by DI are compared with AL, AX, or EAX. Recall that this default segment (ES) cannot be changed with a segment override prefix.

Like the other string instructions, SCAS instructions use the direction flag (D) to select either an auto-increment or auto-decrement operation for DI. They also repeat if prefixed by a conditional repeat prefix.

Suppose a section of memory is 100 bytes in length and begins at location BLOCK. This section of memory must be tested to see if any location contains a 00H. The program in Example 5–34 searches this part of memory for a 00H using the SCASB instruction. In this example, the SCASB instruction is prefixed with an REPNE (**repeat while not equal**). The REPNE prefix causes the SCASB instruction to repeat until either the CX register reaches 0, or until an equal condition exists as the outcome of the SCASB instruction's comparison. Another conditional repeat prefix is REPE (**repeat while equal**). With either repeat prefix, the contents of CX decrements without affecting the flag bits. The SCASB instruction and the comparison it makes change the flags.

EXAMPLE 5–34

```
0000 BF 0011 R          MOV    DI,OFFSET BLOCK  ;address data
0003 FC                 CLD                     ;auto-increment
0004 B9 0064            MOV    CX,100           ;load counter
0007 32 C0              XOR    AL,AL            ;clear AL
0009 F2/AE              REPNE SCASB             ;search
```

Suppose you must develop a program that skips ASCII-coded spaces in a memory array (see Example 5–35). This procedure assumes that the DI register already addresses the ASCII-coded character string, and that the length of the string is 256 bytes or less. Because this program is to skip spaces (20H), the REPE (repeat while equal) prefix is used with a SCASB instruction. The SCASB instruction repeats the comparison, searching for a 20H, as long as an equal condition exists.

EXAMPLE 5–35

```
0000                    SKIP   PROC FAR

0000 FC                        CLD              ;auto-increment
0001 B9 0100                   MOV  CX,256      ;counter
0004 B0 20                     MOV  AL,20H      ;get space
0006 F3/AE                     REPE  SCASB      ;search
0008 CB                        RET
0009                    SKIP   ENDP
```

CMPS

The **compare strings** instruction (CMPS) always compares two sections of memory data as bytes (CMPSB), words (CMPSW), or doublewords (CMPSD). (Only the 80386–Pentium Pro can use doublewords.) The contents of the data segment memory location addressed by SI is compared with the contents of the extra segment memory location addressed by DI. The CMPS instruction increments or decrements both SI and DI. The CMPS instruction is normally used with either the REPE or REPNE prefix. Alternatives to these prefixes are REPZ (**repeat while zero**) and REPNZ (**repeat while not zero**), but usually the REPE or REPNE are used in programming.

Example 5–36 compares two sections of memory searching for a match. The CMPSB instruction is prefixed with a REPE. This causes the search to continue as long as an equal condition exists. When the CX register becomes 0 or an unequal condition exists, the CMPSB instruction stops execution. After the CMPSB instruction ends, the CX register is zero or the flags indicate an equal condition when the two strings match. If CX is not zero or the flags indicate a not-equal condition, the strings do not match.

EXAMPLE 5–36

```
0000                    MATCH  PROC FAR

0000 BE 0075 R                 MOV  SI,OFFSET LINE   ;address LINE
0003 BF 007F R                 MOV  DI,OFFSET TABLE  ;address TABLE
0006 FC                        CLD                   ;auto-increment
```

```
0007 B9 000A              MOV   CX,10              ;counter
000A F3/A6                REPE  CMPSB              ;search
000C CB                   RET

000D            MATCH   ENDP
```

5–7 SUMMARY

1. Addition (ADD) can be 8-, 16-, or 32-bit. The ADD instruction allows any addressing mode except segment register addressing. Most flags (C, A, S, Z, P, and O) change when the ADD instruction executes. A different type of addition, add-with-carry (ADC), adds two operands and the contents of the carry flag bit (C). The 80486–Pentium Pro microprocessors have an additional instruction (XADD) that combines an addition with an exchange.

2. The increment instruction (INC) adds 1 to the byte, word, or doubleword contents of a register or memory location. The INC instruction affects the same flag bits as ADD except the carry flag. The BYTE PTR, WORD PTR, and DWORD PTR directives appear with the INC instruction when the contents of a memory location are addressed by a pointer.

3. Subtraction (SUB) is a byte, word, or doubleword and is performed on a register or a memory location. The only form of addressing not allowed by the SUB instruction is segment register addressing. The subtract instruction affects the same flags as ADD and subtracts carry if the SBB form is used.

4. The decrement (DEC) instruction subtracts 1 from the contents of a register or a memory location. The only addressing modes not allowed with DEC are immediate or segment register addressing. The DEC instruction does not affect the carry flag and is often used with BYTE PTR, WORD PTR, or DWORD PTR.

5. The compare (CMP) instruction is a special form of subtraction that does not store the difference; instead the flags change to reflect the difference. Compare is used to compare an entire byte or word located in any register (except segment) or memory location. In the 80486–Pentium Pro, the CMPXCHG is a combination of compare and exchange instructions. In the Pentium and Pentium Pro microprocessors, the CMPXCHG8B instruction compares and exchanges quadword data.

6. Multiplication is byte, word, or doubleword and can be signed (IMUL) or unsigned (MUL). The 8-bit multiplication always multiplies register AL by an operand with the product found in AX. The 16-bit multiplication always multiplies register AX by an operand with the product found in DX–AX. The 32-bit multiply always multiplies register EAX by an operand with the product found in EDX–EAX. A special IMUL immediate instruction exists on the 80186–Pentium Pro that contains three operands. For example, the IMUL BX,CX,3 instruction multiplies CX by 3 and leaves the product in BX.

7. Division is byte, word, or doubleword and can be signed (IDIV) or unsigned (DIV). For an 8-bit division, the AX register divides by the operand, after which the quotient appears in AL and remainder in AH. In the 16-bit division, the DX–AX register divides by the operand, after which the AX register contains the quotient and DX the remainder. In the 32-bit division, the EDX–EAX register is divided by the operand, after which the EAX register contains the quotient and the EDX register contains the remainder. Remember that the remainder after a signed division always assumes the sign of the dividend.

8. BCD data add or subtract in packed form by adjusting the result of the addition with DAA or the subtraction with DAS. ASCII data are added, subtracted, multiplied, or divided when the operations are adjusted with AAA, AAS, AAM, or AAD.

9. The AAM instruction has an interesting added feature that allows it to convert a binary number into unpacked BCD. This instruction converts a binary number between 00H–63H

into unpacked BCD in AX. The AAM instruction divides AX by 10, and leaves the remainder in AL and the quotient in AH.

10. The AND, OR, and exclusive-OR instructions perform logic functions on a byte, word, or doubleword stored in a register or memory location. All flags change with these instructions with carry (C) and overflow (O) cleared.

11. The TEST instruction performs the AND operation, but the logical product is lost. This instruction changes the flag bits to indicate the outcome of the test.

12. The NOT and NEG instructions perform logical inversion and arithmetic inversion. The NOT instruction one's complements an operand and the NEG instruction two's complements an operand.

13. There are eight different shift and rotate instructions. Each of these instructions shifts or rotates a byte, word, or doubleword register or memory data. These instructions have two operands: the first is the location of the data shifted or rotated, and the second is an immediate shift or rotate count of CL. If the second operand is CL, the CL register holds the shift or rotate count. In the 80386–Pentium Pro, two additional double-precision shift (SHRD and SHLD) instructions exist.

14. The scan string (SCAS) instruction compares AL, AX, or EAX with the contents of the extra segment memory location addressed by DI.

15. The string compare (CMPS) instruction compares the byte, word, or doubleword contents of two sections of memory. One section is addressed by DI, in the extra segment; the other by SI, in the data segment.

16. The SCAS and CMPS instructions repeat with the REPE or REPNE prefixes. The REPE prefix repeats the string instruction while an equal condition exists and the REPNE repeats the string instruction while a not equal condition exists.

17. Example 5–37 uses some of the instructions in this chapter to search the video display (beginning at address B800:000) to find if it contains the word BUG. If the word BUG is found, the program displays a Y. If BUG is not found, it displays an N. Notice how the CMPSB instruction is used to search for BUG.

EXAMPLE 5–37

```
                        ;A program that tests the video display for the word BUG.
                        ;If BUG appears anywhere on the display, the program
                        ;displays Y.
                        ;If BUG does not appear, the program displays N.
                        ;
                                .MODEL SMALL            ;select SMALL model
0000                            .DATA                   ;start of DATA segment
0000 42 55 47           DATA1   DB    'BUG'             ;define BUG
0000                            .CODE                   ;start of CODE segment
                                .STARTUP                ;indicate start of program
0017 B8 B800                    MOV   AX,0B800H         ;address segment B800
001A 8E C0                      MOV   ES,AX
001C B9 07D0                    MOV   CX,25*80          ;set count
001F FC                         CLD                     ;select increment
0020 BF 0000                    MOV   DI,0              ;address display
0023                   L1:
0023 BE 0000 R                  MOV   SI,OFFSET DATA1   ;address BUG
0026 57                         PUSH  DI                ;save display address
0027 A6                         CMPSB                   ;test for B
0028 75 0A                      JNE   L2                ;if display is not B
002A 47                         INC   DI                ;address next display
002B A6                         CMPSB                   ;test for U
002C 75 06                      JNE   L2                ;if display is not U
002E 47                         INC   DI                ;address next display
002F A6                         CMPSB                   ;test for G
0030 B2 59                      MOV   DL,'Y'            ;load Y for possible BUG
0032 74 09                      JE    L3                ;if BUG is found
0034                   L2:
```

```
0034 5F              POP   DI          ;restore display address
0035 83 C7 02        ADD   DI,2        ;point to next position
0038 E2 E9           LOOP  L1          ;repeat for entire screen
003A 57              PUSH  DI          ;save display address
003B B2 4E           MOV   DL,'N'      ;indicate N if not BUG
003D          L3:
003D 5F              POP   DI          ;clear stack
003E B4 02           MOV   AH,2        ;display DL function
0040 CD 21           INT   21H         ;display ASCII from DL
                     .EXIT             ;exit to DOS
                     END               ;end of file
```

5–8 QUESTIONS AND PROBLEMS

1. Select an ADD instruction that will:
   (a) add BX to AX
   (b) add 12H to AL
   (c) add EDI and EBP
   (d) add 22H to CX
   (e) add the data addressed by SI to AL
   (f) add CX to the data stored at memory location FROG
2. What is wrong with the ADD ECX,AX instruction?
3. Is it possible to add CX to DS with the ADD instruction?
4. If AX = 1001H and DX = 20FFH, give the sum and the contents of each flag register bit (C, A, S, Z, and O) after the ADD AX,DX instruction executes.
5. Develop a short sequence of instructions that adds AL, BL, CL, DL, and AH. Save the sum in the DH register.
6. Develop a short sequence of instructions that adds AX, BX, CX, DX, and SP. Save the sum in the DI register.
7. Develop a short sequence of instructions that adds ECX, EDX, and ESI. Save the sum in the EDI register.
8. Select an instruction that adds BX to DX and that also adds the contents of the carry flag (C) to the result.
9. Choose an instruction that adds a 1 to the contents of the SP register.
10. What is wrong with the INC [BX] instruction?
11. Select a SUB instruction that will:
    (a) subtract BX from CX
    (b) subtract 0EEH from DH
    (c) subtract DI from SI
    (d) subtract 3322H from EBP
    (e) subtract the data address by SI from CH
    (f) subtract the data stored 10 words after the location addressed by SI from DX
    (g) subtract AL from memory location FROG
12. If DL = 0F3H and BH = 72H, give the difference after BH subtracts from DL and show the contents of the flag register bits.
13. Write a short sequence of instructions that subtracts the numbers in DI, SI, and BP from the AX register. Store the difference in register BX.
14. Choose an instruction that subtracts 1 from register EBX.
15. Explain what the SBB [DI – 4],DX instruction accomplishes.
16. Explain the difference between the SUB and CMP instructions.
17. When two 8-bit numbers are multiplied, where is the product found?

18. When two 16-bit numbers are multiplied, which two registers hold the product? Show which register contains the most- and least-significant portions of the product.

19. When two numbers multiply, what happens to the O and C flag bits?

20. Where is the product stored for the MUL EDI instruction?

21. What is the difference between the IMUL and MUL instructions?

22. Write a sequence of instructions that cubes the 8-bit number found in DL. Load DL with a 5 initially and make sure your result is a 16-bit number.

23. Describe the operation of the IMUL BX,DX,100H instruction.

24. When 8-bit numbers are divided, in which register is the dividend found?

25. When 16-bit numbers are divided, in which register is the quotient found?

26. What type of errors are detected during a division?

27. Explain the difference between the IDIV and DIV instructions.

28. Where is the remainder found after an 8-bit division?

29. Write a short sequence of instructions that divides the number in BL by the number in CL and then multiplies the result by 2. The final answer must be a 16-bit number stored in the DX register.

30. Which instructions are used with BCD arithmetic operations?

31. Which instructions are used with ASCII arithmetic operations?

32. Explain how the AAM instruction converts from binary to BCD.

33. Develop a sequence of instructions that converts the unsigned number in AX (values of 0–65535) into a 5-digit BCD number stored in memory beginning at the location addressed by the BX register in the data segment. (Note that the most-significant character is stored first and no attempt is made to blank leading zeros.)

34. Develop a sequence of instructions that adds the 8-digit BCD number in AX and BX to the 8-digit BCD number in CX and DX. (AX and CX are the most-significant registers. The result must be found in CX and DX after the addition.)

35. Select an AND instruction that will:
    (a) AND BX with DX and save the result in BX
    (b) AND 0EAH with DH
    (c) AND DI with BP and save the result in DI
    (d) AND 1122H with EAX
    (e) AND the data addressed by BP with CX and save the result in memory
    (f) AND the data stored in four words before the location addressed by SI with DX and save the result in DX
    (g) AND AL with memory location WHAT and save the result at location WHAT

36. Develop a short sequence of instructions that clears (0) the three leftmost bits of DH without changing the remainder DH and stores the result in BH.

37. Select an OR instruction that will:
    (a) OR BL with AH and save the result in AH
    (b) OR 88H with ECX
    (c) OR DX with SI and save the result in SI
    (d) OR 1122H with BP
    (e) OR the data addressed by BX with CX and save the result in memory
    (f) OR the data stored 40 bytes after the location addressed by BP with AL and save the result in AL
    (g) OR AH with memory location WHEN and save the result in WHEN

38. Develop a short sequence of instructions that sets (1) the rightmost five bits of DI without changing the remaining bits of DI. Save the result in SI.

39. Select the XOR instruction that will:
    (a) XOR BH with AH and save the result in AH
    (b) XOR 99H with CL

    (c) XOR DX with DI and save the result in DX

    (d) XOR 1A23H with ESP

    (e) XOR the data addressed by EBX with DX and save the result in memory

    (f) XOR the data stored 30 words after the location addressed by BP with DI and save the result in DI

    (g) XOR DI with memory location WELL and save the result in DI

40. Develop a sequence of instructions that sets (1) the rightmost four bits of AX, clears (0) the leftmost three bits of AX, and inverts bits 7, 8, and 9 of AX.

41. Describe the difference between the AND and TEST instructions.

42. Select an instruction that tests bit-position 2 of register CH.

43. What is the difference between the NOT and NEG instructions?

44. Select the correct instruction to perform each of the following tasks:

    (a) shift DI right three places with zeros moved into the leftmost bit

    (b) move all bits in AL left one place, making sure that a 0 moves into the rightmost bit position

    (c) rotate all the bits of AL left three places

    (d) rotate carry right one place through EDX

    (e) move the DH register right one place making sure that the sign of the result is the same as the sign of the original number

45. What does the SCASB instruction accomplish?

46. For string instructions, DI always addresses data in the _____ segment.

47. What is the purpose of the D flag bit?

48. Explain what the REPE prefix does when coupled with the SCASB instruction.

49. What condition or conditions will terminate the repeated string instruction REPNE SCASB?

50. Describe what the CMPSB instruction accomplishes.

51. Develop a sequence of instructions that scans through a 300H-byte section of memory called LIST located in the data segment searching for a 66H.

52. What happens if AH = 02H and DL = 43H when the INT 21H instruction is executed?

53. Use the Internet to research three different embedded microprocessors and describe the differences between the arithmetic instructions provided by each.

54. Use the Internet to write a paper that compares the multiplication speeds of at least three different microprocessors.

55. Use the Internet to write a report that describes the arithmetic instruction set of any RISC microprocessor such as the PowerPC.

CHAPTER 6

Program Control Instructions

INTRODUCTION

The program control instructions direct the flow of a program and allow it to change. A change in flow often occurs after a decision, made with the CMP or TEST instruction, is followed by a conditional jump instruction. This chapter explains the program control instructions including the jumps, calls, returns, interrupts, and machine control instructions.

Also presented are some of the new features provided by the latest version of the Microsoft assembler (MASM) for conditional assembly of programs. Although these new tools are not essential, they make programming in assembly language much easier.

CHAPTER OBJECTIVES

Upon completion of this chapter, you will be able to:

1. Use both conditional and unconditional jump instructions to control the flow of a program.
2. Use the call and return instructions to include procedures in the program structure.
3. Explain the operation of the interrupts and interrupt control instructions.
4. Use machine control instructions to modify the flag bits.
5. Use ENTER and LEAVE to enter and leave programming structures.
6. Explain the operation of the Pentium Pro conditional move instructions.
7. Use .IF, .ENDIF, .ELSE, .WHILE, and .REPEAT when developing software.

6–1 THE JUMP GROUP

The main program control instruction, **jump** (JMP), allows you to skip sections of a program and branch to any part of the memory for the next instruction. A conditional jump instruction lets you make decisions based on numerical tests. The results of these numerical tests are held in the flag bits, which are then tested by conditional jump instructions. We also look at another instruction similar to the conditional jump, the conditional set.

In this section, all jump instructions are illustrated in sample programs. We also revisit the LOOP and conditional LOOP instructions, first presented in Chapter 3, because they are also forms of the jump instruction.

Unconditional Jump (JMP)

Three types of **unconditional jump** instructions (refer to Figure 6–1) are available to the microprocessor: short jump, near jump, and far jump. The **short jump** is a 2-byte instruction that allows jumps or branches to memory locations within +127 and –128 bytes from the address following the jump. The 3-byte **near jump** allows a branch or jump within ±32K bytes (or anywhere in the current code segment) from the instruction in the current code segment. Remember, the segments are cyclic in nature, which means that one location above offset address FFFFH is offset address 0000H. For this reason if you jump 2 bytes ahead in memory and the instruction pointer addresses offset address FFFFH, the flow continues at offset address 0001H. Thus a displacement of ±32K bytes allows a jump to any location within the current code segment. Finally, the 5-byte **far jump** allows a jump to any memory location within the entire real memory system. The short and near jumps are often called **intrasegment** jumps and the far jumps are often called **intersegment** jumps.

In the 80386 through the Pentium microprocessors, the near jump is within ±2G if the machine is operated in the protected mode with a code segment of 4G bytes in length and ±32K bytes if operated in the real mode. In the protected mode, the 80386 and above use a 32-bit displacement that is not shown in Figure 6–1. The 80386 through the Pentium Pro far jump instruction allows a branch to any location within the 4G-byte address range of these microprocessors.

Short Jump.   Short jumps are also called **relative jumps** because they can be moved, along with their related software, to any location in current code segment without a change. This is because the jump address is not stored with the opcode. Instead of a jump address, a distance or displacement follows the opcode. The short jump displacement is a distance represented by a 1-byte signed number whose value ranges between +127 and –128. The short jump instruction is shown in Figure 6–2. When the microprocessor executes a short jump, the displacement is sign-extended and added to the instruction pointer (IP/EIP) to generate the jump address within the current code segment. The short jump instruction branches to this new address for the next instruction in the program.

Example 6–1 shows how short jump instructions pass control from one part of the program to another. It also illustrates the use of a label with the jump instruction. Notice how one jump (JMP SHORT NEXT) uses the **SHORT directive** to force a short jump, while the other does not. Most assembler programs choose the best form of the jump instruction so the second jump instruction (JMP START) also assembles as a short jump. If the address of the next instruction (0009H) is added to the sign-extended displacement (0017H) of the first jump, the address of NEXT is at location 0017H + 0009H or 0020H.

FIGURE 6–1   The three main forms of the JMP instruction. Note that Disp is either an 8- or a 16-bit signed displacement or distance.

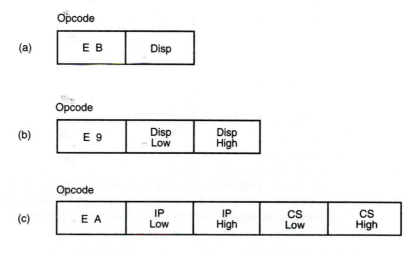

FIGURE 6–2   A short jump to four memory locations beyond the address of the next instruction.

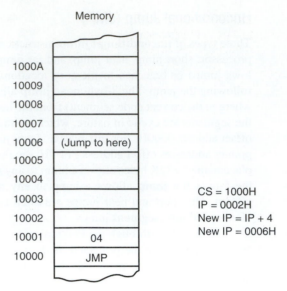

Memory

1000A	
10009	
10008	
10007	
10006	(Jump to here)
10005	
10004	
10003	
10002	
10001	04
10000	JMP

CS = 1000H
IP = 0002H
New IP = IP + 4
New IP = 0006H

EXAMPLE 6–1

```
0000 33 DB                   XOR   BX,BX

0002 B8 0001      START:     MOV   AX,1
0005 03 C3                   ADD   AX,BX
0007 EB 17                   JMP   SHORT NEXT

0020 8B D8        NEXT:      MOV   BX,AX
0022 EB DE                   JMP   START
```

Whenever a jump instruction references an address, a **label** normally identifies the address. The JMP NEXT is an example that jumps to label NEXT for the next instruction. It is very rare to use an actual hexadecimal address with any jump instruction, but the assembler supports addressing in relation to the instruction pointer by using the **$** + a displacement. For example, a JMP $ + 2 jumps over the next two memory locations following the JMP instruction. The label NEXT must be followed by a colon (NEXT:) to allow an instruction to reference it for a jump. If a colon does not follow a label, you cannot jump to it. Note that the only time a colon is used after a label is when the label is used with a jump or call instruction.

Near Jump.   The near jump is similar to the short jump except the distance is farther. A **near jump** passes control to an instruction in the current code segment located within ±32K bytes from the near jump instruction or ±2G in the 80386 and above operated in protected mode. The near jump is a 3-byte instruction that contains an opcode followed by a signed 16-bit displacement. In the 80386 through the Pentium Pro, the displacement is 32 bits and the near jump is 5 bytes in length. The signed displacement adds to the instruction pointer (IP or EIP) to generate the jump address. Because the signed displacement is in the range of ±32K, a near jump can jump to any memory location within the current real-mode code segment. The protected-mode code segment in the 80386 and above can be 4G bytes in length so the 32-bit displacement allows a near jump to any location within ±2G bytes. Figure 6–3 illustrates the operation of the real-mode near jump instruction.

The near jump is also relocatable, as was the short jump, because it is also a relative jump. If the code segment moves to a new location in the memory, the distance between the jump instruction and the operand address remains the same. This allows a code segment to be relocated

FIGURE 6–3 A near jump that adds the displacement (0002H) to the contents of IP.

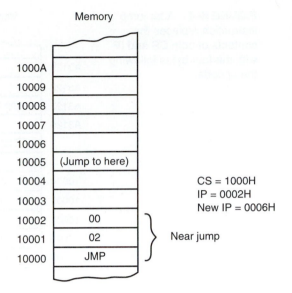

Memory

1000A	
10009	
10008	
10007	
10006	
10005	(Jump to here)
10004	
10003	
10002	00
10001	02
10000	JMP

CS = 1000H
IP = 0002H
New IP = 0006H

Near jump

by simply moving it. This feature, along with the relocatable data segments, makes the Intel family of microprocessors ideal for use in a general-purpose computer system. Software can be written and loaded anywhere in the memory; it functions without modification because of the relative jumps and relocatable data segments.

Example 6–2 is the same basic program that appeared in Example 6–1, except the jump distance is greater. The first jump (JMP NEXT) passes control to the instruction at offset memory location 0200H within the code segment. Notice that the instruction assembles as an E9 0200 R. The letter R denotes a relocatable jump address of 0200H. The relocatable address of 0200H is for the assembler program's internal use only. The actual machine language instruction assembles as an E9 F6 01, which does not appear in the assembler listing. The actual displacement is a 01F6H for this jump instruction. The assembler lists the jump address as 0200 R so the address is easier to interpret as software is developed. If the linked execution file (.EXE) is displayed in hexadecimal, the jump instruction appears as an E9 F6 01.

EXAMPLE 6–2

```
0000 33 DB                   XOR   BX,BX

0002 B8 0001      START:     MOV   AX,1
0005 03 C3                   ADD   AX,BX
0007 E9 0200 R               JMP   NEXT

0200 8B D8        NEXT:      MOV   BX,AX
0202 E9 0002 R               JMP   START
```

Far Jump. The **far jump** instruction (see Figure 6–4) obtains a new segment and offset address to accomplish the jump. Bytes 2 and 3, of this 5-byte instruction, contain the new offset address and bytes 4 and 5 contain the new segment address. If the microprocessor (80286 through the Pentium Pro) is operated in the protected mode, the segment address accesses a descriptor that contains the base address of the far jump segment. The offset address, which is either 16 or 32 bits, contains the offset location within the new code segment.

Example 6–3 is a short program that uses a far jump instruction. The far jump instruction sometimes appears with the FAR PTR directive as illustrated. Another way to obtain a far jump is to define a label as a far label. A **label** is far only if it is external to the current code segment or

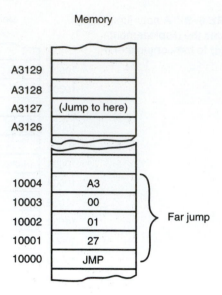

FIGURE 6–4 A far jump instruction replaces the contents of both CS and IP with the four bytes following the opcode.

procedure. The JMP UP instruction in Example 6–3 references a far label. The label UP is defined as a far label by the EXTRNUP:FAR directive. External labels appear in programs that contain more than one program file. Another way of defining a label as global is to use a **double colon** (LABEL::) following the label in place of the single colon. This is required inside procedure blocks that are defined as near if the label is accessed from outside the procedure block.

EXAMPLE 6–3

```
                                 EXTRN  UP:FAR

0000 33 DB                       XOR    BX,BX

0002 B8 0001         START:      MOV    AX,1
0005 03 C3                       ADD    AX,BX
0007 E9 0200 R                   JMP    NEXT

0200 8B D8           NEXT:       MOV    BX,AX
0202 EA 0002 ---- R              JMP    FAR PTR START

0207 EA 0000 ---- E              JMP    UP
```

When the program files are joined, the linker inserts the address for the UP label into the JMP UP instruction. It also inserts the segment address in the JMP START instruction. The segment address in JMP FAR PTR START is listed as – – – – R for relocatable and the segment address in JMP UP is listed as – – – – E for external. In both cases the – – – – is filled in by the linker when it links or joins the program files.

Jumps with Register Operands. The jump instruction can also use a 16- or 32-bit register as an operand. This automatically sets up the instruction as an indirect jump. The address of the jump is in the register specified by the jump instruction. Unlike the displacement associated with the near jump, the contents of the register are transferred directly into the instruction pointer. It does not add to the instruction pointer as with the short and near jumps. The JMP AX instruction, for example, copies the contents of the AX register into the IP when the jump occurs. This allows a jump to any location within the current code segment. In the 80386 and above, a JMP EAX instruction also jumps to any location within the current code segment. The difference is that in protected mode the code segment can be 4G bytes in length so a 32-bit offset address is needed.

Example 6–4 shows how the JMP AX instruction accesses a jump table in the code segment. This program reads a key from the keyboard and then modifies the ASCII code to a 00H in AL for a '1', a 01H for a '2', and a 02H for a '3'. If a '1', '2', or '3' is typed, AH is cleared to 00H. Because the jump table contains 16-bit offset addresses, the contents of AX are doubled to 0, 2, or 4 so a 16-bit entry in the table can be accessed. Next, the offset address of the start of the jump table is loaded to SI and AX is added to form the reference to the jump address. The MOV AX,[SI] instruction then fetches an address from the jump table so the JMP AX instruction jumps to the addresses (ONE, TWO, or THREE) stored in the jump table.

EXAMPLE 6–4

```
                        ;A program that reads only the number 1, 2, or 3
                        ;from the keyboard. If a 1, 2, or 3 is typed, the
                        ;program displays a 1, 2, or 3.
                        ;
                                .MODEL SMALL            ;select SMALL model
0000                            .DATA                   ;start of DATA
0000 0030 R     TABLE   DW      ONE                     ;define lookup table
0002 0034 R             DW      TWO
0004 0038 R             DW      THREE
0000                            .CODE                   ;start of CODE
                                .STARTUP                ;start of program
0017            TOP:
0017 B4 01              MOV     AH,1                    ;read key into AL
0019 CD 21              INT     21H

001B 2C 31              SUB     AL,31H                  ;convert from ASCII
001D 72 F8              JB      TOP                     ;if below '1'
001F 3C 02              CMP     AL,2
0021 77 F4              JA      TOP                     ;if above '3'

0023 B4 00              MOV     AH,0                    ;double value
0025 03 C0              ADD     AX,AX
0027 BE 0000 R          MOV     SI,OFFSET TABLE         ;address lookup table
002A 03 F0              ADD     SI,AX                   ;add to table
002C 8B 04              MOV     AX,[SI]                 ;get address
002E FF E0              JMP     AX                      ;jump to address
0030            ONE:
0030 B2 31              MOV     DL,'1'                  ;load '1' for display
0032 EB 06              JMP     BOT                     ;go display '1'
0034            TWO:
0034 B2 32              MOV     DL,'2'                  ;load '2' for display
0036 EB 02              JMP     BOT                     ;go display '2'
0038            THREE:
0038 B2 33              MOV     DL,'3'                  ;load '3' for display
003A            BOT:
003A B4 02              MOV     AH,2                    ;display ASCII
003C CD 21              INT     21H
                                .EXIT                   ;exit to DOS
                                END                     ;end of file
```

Indirect Jumps Using an Index. The jump instruction may also use the [ ] form of addressing to directly access the jump table. The jump table can contain offset addresses for near indirect jumps or segment and offset addresses for far indirect jumps. (This type of jump is also known as a **double-indirect** jump if the register jump is called an indirect jump.) The assembler assumes that the jump is near unless the FAR PTR directive indicates a far jump instruction. Example 6–5 repeats Example 6–4 by using the JMP TABLE [SI] instead of JMP AX. This reduces the length of the program.

EXAMPLE 6–5

```
                                .MODEL SMALL            ;select SMALL model
0000                            .DATA                   ;start of DATA
0000 002D R        TABLE    DW    ONE                   ;lookup table
0002 0031 R                 DW    TWO
0004 0035 R                 DW    THREE
0000                            .CODE                   ;start of CODE
                                .STARTUP                ;start of program
0017               TOP:
0017 B4 01                  MOV    AH,1                 ;read key to AL
0019 CD 21                  INT    21H

001B 2C 31                  SUB    AL,31H               ;test for below '1'
001D 72 F8                  JB     TOP                  ;if below '1'
001F 3C 02                  CMP    AL,2
0021 77 F4                  JA     TOP                  ;if above '3'
0023 B4 00                  MOV    AH,0                 ;calculate address
0025 03 C0                  ADD    AX,AX
0027 03 F0                  ADD    SI,AX
0029 FF A4 0000 R           JMP    TABLE [SI]           ;jump to address
002D               ONE:
002D B2 31                  MOV    DL,'1'               ;load DL with '1'
002F EB 06                  JMP    BOT
0031               TWO:
0031 B2 32                  MOV    DL,'2'               ;load DL with '2'
0033 EB 02                  JMP    BOT
0035               THREE:
0035 B2 33                  MOV    DL,'3'               ;load DL with '3'
0037               BOT:
0037 B4 02                  MOV    AH,2                 ;display ASCII
0039 CD 21                  INT    21H
                                .EXIT                   ;exit to DOS
                                END                     ;end of file
```

The mechanism used to access the jump table is identical to a normal memory reference. The JMP TABLE [SI] instruction points to a jump address stored at a code segment offset location addressed by SI. It jumps to the address stored in the memory at this location. Both the register and indirect indexed jump instructions usually address a 16-bit offset. This means that both types of jumps are near jumps. If a JMP FAR PTR [SI] or JMP TABLE [SI] with TABLE data defined with the DD directive appear in a program, the microprocessor assumes that the jump table contains doubleword, 32-bit addresses (IP and CS).

Conditional Jumps and Conditional Sets

Conditional jump instructions are always short jumps in the 8086 through the 80286 microprocessors. This limits the range of the jump to within +127 bytes and –128 bytes from the location following the conditional jump. In the 80386 and above, conditional jumps are either short or near jumps, allowing these microprocessors to use a conditional jump to any location within the current code segment. Table 6–1 lists all the conditional jump instructions with their test conditions. The Microsoft MASM version 6.X assembler automatically adjusts conditional jumps if the distance is too great.

The conditional jump instructions test the following flag bits: sign (S), zero (Z), carry (C), parity (P), and overflow (O). If the condition under test is true, a branch to the label associated with the jump instruction occurs. If the condition is false, the next sequential step in the program executes. For example, a JC will jump if the carry bit is set.

The operation of most conditional jump instructions is straightforward because they often test just one flag bit, although some test more than one flag. Relative magnitude comparisons require more complicated conditional jump instructions that test more than one flag bit.

TABLE 6–1 Conditional jump instructions

Assembly Language	Condition Tested	Operation
JA	C = 0 and Z = 0	Jump if above
JAE	C = 0	Jump if above or equal
JB	C = 1	Jump if below
JBE	C = 1 or Z = 1	Jump if below or equal
JC	C = 1	Jump if carry set
JE or JZ	Z = 1	Jump if equal or jump if zero
JG	Z = 0 and S = O	Jump if greater than
JGE	S = O	Jump if greater than or equal
JL	S ≠ O	Jump if less than
JLE	Z = 1 or S ≠ O	Jump if less than or equal
JNC	C = 0	Jump if carry cleared
JNE or JNZ	Z = 0	Jump if not equal or jump if not zero
JNO	O = 0	Jump if no overflow
JNS	S = 0	Jump if no sign
JNP or JPO	P = 0	Jump if no parity or jump if parity odd
JO	O = 1	Jump if overflow set
JP or JPE	P = 1	Jump if parity or jump if parity even
JS	S = 1	Jump if sign is set
JCXZ	CX = 0	Jump if CX = 0
*JECXZ	ECX = 0	Jump if ECX = 0

*Note: 80386/80486/Pentium/Pentium only.

Because both signed and unsigned numbers are used in programming, and because the order of these numbers is different, there are two sets of conditional jump instructions for magnitude comparisons. Figure 6–5 shows the order of both signed and unsigned 8-bit numbers. The 16- and 32-bit numbers follow the same order as the 8-bit numbers except they are larger. Notice that an FFH (255) is above the 00H in the set of unsigned numbers, but an FFH (–1) is less than 00H for signed numbers. Therefore, an unsigned FFH is above 00H, but a signed FFH is less than 00H.

When **signed numbers** are compared, use the JG, JL, JGE, JLE, JE, and JNE instructions. The terms greater than and less than refer to signed numbers. When **unsigned numbers**

FIGURE 6–5 Signed and unsigned numbers follow different orders.

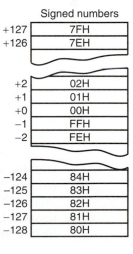

are compared, use the JA, JB, JAE, JBE, JE, and JNE instructions. The terms above and below refer to unsigned numbers.

The remaining conditional jumps test individual flag bits such as overflow and parity. Notice that JE has an alternative opcode JZ. All instructions have alternates, but many aren't used in programming because they don't usually fit the condition under test. (The alternates appear in Appendix B with the instruction set listing.) For example, the JA instruction (jump if above) has the alternate JNBE (jump if not below or equal). A JA functions exactly as a JNBE, but JNBE is awkward in many cases when compared to JA.

The conditional jump instructions all test flag bits except for JCXZ (jump if CX = 0) and JECXZ (jump if ECX = 0). Instead of testing flag bits, JCXZ directly tests the contents of the CX register without affecting the flag bits, and JECXZ tests the contents of the ECX register. For the JCXZ instruction, if CX = 0, a jump occurs; if CX ≠ 0, no jump occurs. Likewise for the JECXZ instruction, if ECX = 0, a jump occurs; if CX ≠ 0, no jump occurs.

Example 6–6 is a program that uses JCXZ. Here the SCASB instruction searches a table for a 0AH. Following the search, a JCXZ instruction tests CX to see if the count has reached zero. If the count is zero, the 0AH is not found in the table. The carry flag is used in this example to pass the not found condition back to the calling program. Another method used to test if the data is found is the JNE instruction. If JNE replaces JCXZ, it performs the same function. After the SCASB instruction executes, the flags indicate a not equal condition if the data was not found in the table.

EXAMPLE 6–6

```
                        ;A procedure that searches a table of 100 bytes for
                        ;0AH. The address of the TABLE is transferred to
                        ;the procedure through the SI register.
                        ;
0017                    SCAN    PROC NEAR

0017 B9 0064                    MOV   CX,100         ;load count of 100
001A B0 0A                      MOV   AL,0AH         ;load AL with 0AH
001C FC                         CLD                  ;select increment
001D F2/ AE                     REPNE SCASB          ;test 100 bytes
001F F9                         STC                  ;set carry
0020 E3 01                      JCXZ NOT_FOUND       ;if not found
0022 F8                         CLC                  ;clear carry if found

0023                    NOT_FOUND:

0023 C3                         RET                  ;return

0024                    SCAN    ENDP
```

The Conditional Set Instructions.   In addition to the conditional jump instructions, the 80386 through the Pentium Pro contain **conditional set** instructions. The conditions tested by conditional jumps are put to work with the conditional set instructions. The conditional set instructions set a byte to either a 01H or clear a byte to 00H depending on the outcome of the condition under test. Table 6–2 lists the available forms of the conditional set instructions.

These instructions are useful where a condition must be tested at a point much later in the program. For example, a byte can be set to indicate that the carry is cleared some point in the program by using the SETNC MEM instruction. This instruction places a 01H into memory location MEM if carry is cleared and a 00H into MEM if carry is set. The contents of MEM can be tested at a later point in the program to determine if carry is cleared at the point where the SETNC MEM instruction executed.

TABLE 6–2 The conditional set instructions

Assembly Language	Condition Tested	Operation
SETA	C = 0 and Z = 0	Set byte if above
SETAE	C = 0	Set byte if above or equal
SETB	C = 1	Set byte if below
SETBE	C = 1 or Z = 1	Set byte if below or equal
SETC	C = 1	Set byte if carry is set
SETE or SETZ	Z = 1	Set byte if equal or if zero
SETG	Z = 0 and S = O	Set byte if greater than
SETGE	S = O	Set byte if greater than or equal
SETL	S ≠ O	Set byte if less than
SETLE	Z = 1 or S ≠ O	Set byte if less than or equal
SETNC	C = 0	Set byte if carry is cleared
SETNE or SETNZ	Z = 0	Set byte if not equal or not zero
SETNO	C = 0	Set byte if no overflow
SETNS	S = 0	Set byte if no sign
SETNP	P = 0	Set byte if no parity
SETO	O = 1	Set byte if overflow
SETP	P = 1	Set byte if parity set
SETS	S = 1	Set byte if sign set

LOOP

The LOOP instruction is a combination of a decrement CX and the JNZ conditional jump. In the 8086 through the 80286, LOOP decrements CX and if CX ≠ 0, it jumps to the address indicated by the label. If CX becomes a 0, the next sequential instruction executes. In the 80386 and above, LOOP decrements either CX or ECX depending on the instruction mode. If the 80386 through the Pentium Pro operate in the 16-bit instruction mode, LOOP uses CX; if operated in the 32-bit instruction mode, LOOP uses ECX. This default is changed by the LOOPW (using CX) and LOOPD (using ECX) instructions in the 80386 through the Pentium Pro.

Example 6–7 shows how data in one block of memory (BLOCK1) adds to data in a second block of memory (BLOCK2) using LOOP to control how many numbers add. The LODSW and STOSW instructions access the data in BLOCK1 and BLOCK2. The ADD AX,ES:[DI] instruction accesses the data in BLOCK2 located in the extra segment. The only reason that BLOCK2 is in the extra segment is that DI addresses extra segment data for the STOSW instruction. The .STARTUP directive loads only DS with the address of the data segment. In Example 6–7, the extra segment also addresses data in the data segment so the contents of DS are copied to ES through the accumulator. Unfortunately, there is no direct move from the segment register-to-segment register instruction.

EXAMPLE 6–7

```
                    ;A program that sums the contents of BLOCK1 and BLOCK2.
                    ;and stores the results on top of the data in BLOCK2.
                    ;
                            .MODEL SMALL            ;select SMALL model
0000                        .DATA                   ;start of DATA
0000 0064 [     BLOCK1  DW    100 DUP (?)           ;reserve 100 bytes
         0000
              ]
00C8 0064 [     BLOCK2  DW    100 DUP (?)           ;reserve 100 bytes
         0000
```

```
                          ]
0000                              .CODE                  ;start of CODE
                                  .STARTUP               ;start of program
0017 8C D8                        MOV  AX,DS             ;overlap DS and ES
0019 8E C0                        MOV  ES,AX

001B FC                           CLD                    ;select increment
001C B9 0064                      MOV  CX,100            ;load count of 100
001F BE 0000 R                    MOV  SI,OFFSET BLOCK1  ;address BLOCK1
0022 BF 00C8 R                    MOV  DI,OFFSET BLOCK2  ;address BLOCK2

0025                 L1:
0025 AD                           LODSW                  ;load AX with BLOCK1
0026 26: 03 05                    ADD  AX,ES:[DI]        ;add BLOCK2
0029 AB                           STOSW                  ;store sum in BLOCK2
002A E2 F9                        LOOP L1                ;repeat 100 times
                                  .EXIT                  ;exit to DOS
                                  END                    ;end of file
```

Conditional LOOPs.   As with REP, the LOOP instruction also has conditional forms: LOOPE and LOOPNE. The LOOPE (loop while equal) instruction jumps if CX ≠ 0 while an equal condition exists. It will exit the loop if the condition is not equal or if the CX register decrements to 0. The LOOPNE (loop while not equal) instruction jumps if CX ≠ 0 while a not equal condition exists. It will exit the loop if the condition is equal or if the CX register decrements to 0. In the 80386–Pentium Pro, the conditional LOOP instruction can use either CX or ECX as the counter. The LOOPEW, LOOPED, LOOPNEW, or LOOPNED override the instruction mode if needed.

As with the conditional repeat instructions, alternates exist for LOOPE and LOOPNE. The LOOPE instruction is the same as LOOPZ and the LOOPNE is the same as LOOPNZ. In most programs only the LOOPE and LOOPNE apply.

6–2   PROCEDURES

The procedure or subroutine is an important part of any computer system's architecture. A **procedure** is a group of instructions that usually performs one task. A procedure is a reusable section of the software that is stored in memory once, but used as often as necessary. This saves memory space and makes it easier to develop software. The only disadvantage of a procedure is that it takes the computer a small amount of time to link to the procedure and return from it. The CALL instruction links to the procedure and the RET (return) instruction returns from the procedure.

The stack stores the return address whenever a procedure is called during the execution of a program. The CALL instruction pushes the address of the instruction following the CALL (return address) on the stack. The RET instruction removes an address from the stack so the program returns to the instruction following the CALL.

With the assembler, there are specific rules for the storage of procedures. A **procedure** begins with the PROC directive and ends with the ENDP directive. Each directive appears with the name of the procedure. This programming structure makes it easy to locate the procedure in a program listing. The PROC directive is followed by the type of procedure: NEAR or FAR. Example 6–8 shows how the assembler uses the definition of both a near (intrasegment) and far (intersegment) procedure. In MASM version 6.X, the NEAR or FAR type can be followed by the USES statement, which allows any number of registers to be automatically pushed to the stack and popped from the stack within the procedure. The USES statement is also illustrated in Example 6–8.

EXAMPLE 6-8

```
0000                    SUMS    PROC NEAR

0000 03 C3                      ADD   AX,BX
0002 03 C1                      ADD   AX,CX
0004 03 C2                      ADD   AX,DX
0006 C3                         RET

0007                    SUMS    ENDP

0007                    SUMS1   PROC FAR

0007 03 C3                      ADD   AX,BX
0009 03 C1                      ADD   AX,CX
000B 03 C2                      ADD   AX,DX
000D CB                         RET

000E                    SUMS1   ENDP

000E                    SUMS2   PROC NEAR    USES BX CX DX

0011 03 C3                      ADD   AX,BX
0013 03 C1                      ADD   AX,CX
0015 03 C2                      MOV   AX,DX
                                RET

001B                    SUMS2   ENDP
```

When these two procedures are compared, the only difference is the opcode of the return instruction. The near return instruction uses opcode C3H and the far return uses opcode CBH. A near return removes a 16-bit number from the stack and places it into the instruction pointer to return from the procedure in the current code segment. A far return removes 32 bits from the stack and places it into both IP and CS to return from the procedure to any memory location.

Procedures that are to be used by all software **(global)** should be written as far procedures. Procedures that are used by a given task **(local)** are normally defined as near procedures.

CALL

The CALL instruction transfers the flow of the program to the procedure. The CALL instruction differs from the jump instruction because a CALL saves a return address on the stack. The return address returns control to the instruction that immediately follows the CALL in a program when a RET instruction executes.

Near CALL. The near CALL instruction is 3 bytes long, with the first byte containing the opcode and the second and third bytes containing the displacement or distance of ±32K in the 8086 through the 80286. This is identical to the form of the near jump instruction. The 80386 and above use a 32-bit displacement when operated in the protected mode to allow a distance of ±2G bytes. When the near CALL executes, it first pushes the offset address of the next instruction on the stack. The offset address of the next instruction appears in the instruction pointer (IP or EIP). After saving this return address, it then adds the displacement from bytes 2 and 3 to the IP to transfer control to the procedure. There is no short CALL instruction. A variation on the opcode exists as CALLN, but you should avoid this and use the PROC statement to define the CALL as near.

Why save the IP or EIP on the stack? The instruction pointer always points to the next instruction in the program. For the CALL instruction, the contents of IP/EIP are pushed onto the stack so program control passes to the instruction following the CALL after a procedure ends. Figure 6-6 shows the return address (IP) stored on the stack, and the call to the procedure.

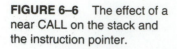

FIGURE 6–6   The effect of a near CALL on the stack and the instruction pointer.

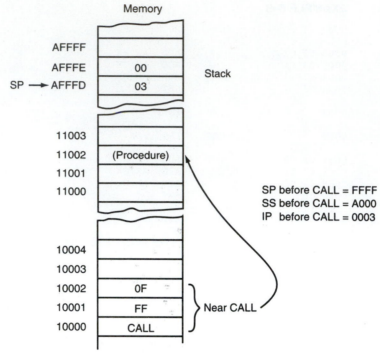

Far CALL.   The **far** CALL instruction is like a far jump because it can CALL a procedure stored in any memory location in the system. The far CALL is a 5-byte instruction that contains an opcode followed by the next value for the IP and CS registers. Bytes 2 and 3 contain the new contents of the IP, and bytes 4 and 5 contain the new contents for CS.

The far CALL instruction places the contents of both IP and CS on the stack before jumping to the address indicated by bytes 2–5 of the instruction. This allows the far CALL to call a procedure located anywhere in the memory and return from that procedure.

Figure 6–7 shows how the far CALL instruction calls a far procedure. Here the contents of IP and CS are pushed onto the stack. Next the program branches to the procedure. A variant of the far CALL exists as CALLF, but you should avoid this and define the type of CALL instruction with the PROC statement.

CALLs with Register Operands.   Like jumps, CALLs also may contain a register operand. An example is the CALL BX instruction, which pushes the contents of IP onto the stack. It then jumps to the offset address, located in register BX, in the current code segment. This type of CALL always uses a 16-bit offset address stored in any 16-bit register except the segment registers.

Example 6–9 illustrates the use of the CALL register instruction to call a procedure that begins at offset address DISP. (This call could also directly call the procedure by using the CALL DISP instruction.) The OFFSET address DISP is placed into the BX register and then the CALL BX instruction calls the procedure beginning at address DISP. This program displays an "OK" on the monitor screen.

EXAMPLE 6–9

```
                    ;a program that displays OK on the monitor screen
                    ;using procedure DISP
                    ;
                            .MODEL TINY        ;select TINY model
0000                        .CODE              ;start of CODE
```

```
                                .STARTUP                    ;start of program

0100 BB 0110 R                  MOV   BX,OFFSET DISP        ;address DISP with BX
0103 B2 4F                      MOV   DL,'O'                ;display 'O'
0105 FF D3                      CALL  BX
0107 B2 4B                      MOV   DL,'K'                ;display 'K'
0109 FF D3                      CALL  BX

                                .EXIT                       ;exit to DOS
                        ;
                        ;a procedure that displays the ASCII contents of DL
                        ;on the monitor screen
                        ;
0110                    DISP    PROC NEAR

0110 B4 02                      MOV   AH,2                  ;select function 02H
0112 CD 21                      INT   21H                   ;execute DOS function
0114 C3                         RET                         ;return

0115                    DISP    ENDP

                                END                         ;end of file
```

CALLs with Indirect Memory Addresses. A CALL, with an indirect memory address, is particularly useful whenever different subroutines need to be chosen in a program. This selection process is often keyed with a number that addresses a CALL address in a lookup table.

Example 6–10 shows three separate subroutines referenced by the numbers 1, 2, and 3 as read from the keyboard on a personal computer. The calling sequence adjusts the value of AL and extends it to a 16-bit number before adding it to the location of the lookup table. This references one of the three subroutines using the CALL TABLE [BX] instruction. When this program

FIGURE 6–7 The effect of a far CALL instruction.

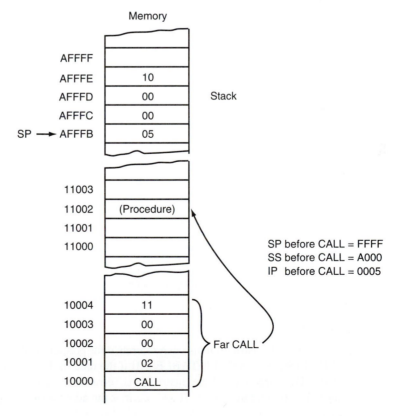

executes, the letter A is displayed when a 1 is typed; the letter B is displayed when a 2 is typed; the letter C is displayed when a 3 is typed.

EXAMPLE 6–10

```
                        ;A program that uses a CALL lookup table to access one of
                        ;three different procedures: ONE, TWO, or THREE.
                        ;
                                .MODEL SMALL            ;select SMALL model
0000                            .DATA                   ;start of DATA
0000 0000 R     TABLE   DW      ONE                     ;define lookup table
0002 0007 R             DW      TWO
0004 000E R             DW      THREE
0000                            .CODE                   ;start of CODE

0000            ONE     PROC NEAR

0000 B4 02              MOV     AH,2                    ;display letter A
0002 B2 41              MOV     DL,'A'
0004 CD 21              INT     21H
0006 C3                 RET

0007            ONE     ENDP

0007            TWO     PROC NEAR

0007 B4 02              MOV     AH,2                    ;display letter B
0009 B2 42              MOV     DL,'B'
000B CD 21              INT     21H
000D C3                 RET

000E            TWO     ENDP

000E            THREE   PROC NEAR

000E B4 02              MOV     AH,2                    ;display letter C
0010 B2 43              MOV     DL,'C'
0012 CD 21              INT     21H
0014 C3                 RET

0015            THREE   ENDP


                                .STARTUP                ;start of program
002C            TOP:
002C B4 01              MOV     AH,1                    ;read key into AL
002E CD 21              INT     21H

0030 2C 31              SUB     AL,31H                  ;convert from ASCII
0032 72 F8              JB      TOP                     ;if below 0
0034 3C 02              CMP     AL,2                    ;if above 2
0036 77 F4              JA      TOP

0038 B4 00              MOV     AH,0                    ;form lookup address
003A 8B D8              MOV     BX,AX
003C 03 DB              ADD     BX,BX
003E FF 97 0000 R       CALL    TABLE [BX]              ;call procedure

                                .EXIT                   ;exit to DOS
                                END                     ;end of file
```

The CALL instruction also can reference far pointers if the instruction appears as a CALL FAR PTR [SI] or as a CALL TABLE [SI] if the data in the table is defined as doubleword data with the DD directive. These instructions retrieve a 32-bit address from the data segment memory location addressed by SI and use it as the address of a far procedure.

RET

The **return instruction** (RET) removes either a 16-bit number (near return) from the stack and places it into IP or a 32-bit number (far return) and places it into IP and CS. The near and far return instructions are both defined in the procedure's PROC directive. This automatically selects the proper return instruction. In the 80386 through the Pentium Pro operated in the protected mode, the far return removes 6 bytes from the stack; the first four contain the new value for EIP and the last two contain the new value for CS. In the 80386 and above, a protected mode near return removes 4 bytes from the stack and places them into EIP.

When IP/EIP or IP/EIP and CS are changed, the address of the next instruction is at a new memory location. This new location is the address of the instruction that immediately follows the most recent CALL to a procedure. Figure 6–8 shows how the CALL instruction links to a procedure and how the RET instruction returns in the 8086–80286.

There is another form of the return instruction that adds a number to the contents of the stack pointer (SP) after the return address is removed from the stack. A return that uses an immediate operand is ideal for a system that uses the C or PASCAL calling convention. (This is true even though the C and PASCAL calling conventions require the caller to remove stack data for many functions.) These conventions push parameters on the stack before calling a procedure. If the parameters are to be discarded upon return, the return instruction contains a number that represents the number of bytes pushed to the stack as parameters.

Example 6–11 shows how this type of return erases the data placed on the stack by a few pushes. The RET 4 adds a 4 to SP after removing the return address from the stack. Since the PUSH AX and PUSH BX together place 4 bytes of data on the stack, this return effectively deletes AX and BX from the stack. This type of return rarely appears in assembly language programs, but is used in higher-level programs to clear stack data after a procedure. Notice how the parameters are addressed on the stack by using the BP register, which by default addresses the stack segment. Parameter stacking is common in procedures written for C or PASCAL using the C or PASCAL calling conventions. (More detail on C and PASCAL interfacing is provided in another chapter.)

FIGURE 6–8 The effect of a near return instruction on the stack and instruction pointer.

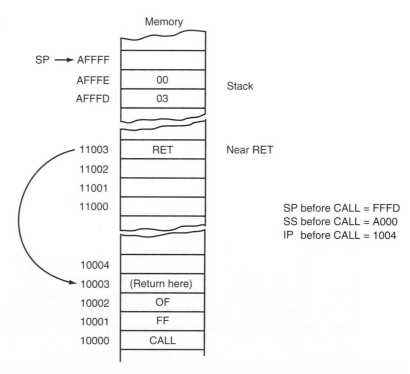

EXAMPLE 6–11

```
0000 B8 001E              MOV   AX,30
0003 BB 0028              MOV   BX,40
0006 50                   PUSH  AX                   ;stack parameter 1
0007 53                   PUSH  BX                   ;stack parameter 2

0008 E8 0066              CALL  ADDM                 ;add stack parameters
                            .     .
                            .     .                  ;program continues
                            .     .
0071              ADDM     PROC  NEAR

0071 55                   PUSH  BP                   ;save BP
0072 8B EC                MOV   BP,SP                ;address stack [BP]
0074 8B 46 04             MOV   AX,[BP+4]            ;get parameter 1
0077 03 46 06             ADD   AX,[BP+6]            ;add parameter 2
007A 5D                   POP   BP                   ;restore BP

007B C2 0004              RET   4                    ;return

007E              ADDM     ENDP
```

As with the CALLN and CALLF instructions, there are also variants of the return instruction: RETN and RETF. You should avoid these and use the PROC statement to define the type of call and return.

6–3

INTRODUCTION TO INTERRUPTS

An **interrupt** is either a hardware-generated CALL (externally derived from a hardware signal) or a software-generated CALL (internally derived from the execution of an instruction or by some other internal event). At times an internal interrupt is called an *exception*. Either type interrupts the program by calling an **interrupt service procedure** or **interrupt handler**.

This section explains software interrupts, which are special types of CALL instructions. We look at the three types of software interrupt instructions (INT, INTO, and INT 3), provide a map of the interrupt vectors, and explain the purpose of the special interrupt return instruction (IRET).

Interrupt Vectors

An **interrupt vector** is a 4-byte number stored in the first 1,024 bytes of the memory (000H–3FFH) when the microprocessor operates in the real mode. In the protected mode, the vector table is replaced by an interrupt descriptor table that uses 8-byte descriptors to describe each of the interrupts. There are 256 different interrupt vectors. Each vector contains the address of an interrupt service procedure. Table 6–3 lists the interrupt vectors with a brief description and the memory location of each vector for the real mode. Each vector contains a value for IP and CS that forms the address of the interrupt service procedure. The first 2 bytes contain the IP and the last 2 bytes contain the CS.

Intel reserves the first 32 interrupt vectors for present and future microprocessor products. The remaining interrupt vectors (32–255) are available to the programmer. Some of the reserved vectors are for errors that occur during the execution of software such as the divide error interrupt. Some vectors are reserved for the coprocessor. Still others occur for normal events in the system. In a personal computer, the reserved vectors are used for system functions, as we'll see

TABLE 6–3 Interrupt vectors

Number	Address	Microprocessor	Function
0	0H–3H	All	Divide error
1	4H–7H	All	Single-step
2	8H–BH	All	NMI (hardware)
3	CH–FH	All	Breakpoint
4	10H–13H	All	Interrupt on overflow
5	14H–17H	80186–Pentium Pro	BOUND interrupt
6	18H–1BH	80186–Pentium Pro	Invalid opcode
7	1CH–1FH	80286–Pentium Pro	Coprocessor emulation
8	20H–23H	80386–Pentium Pro	Double fault
9	24H–2BH	80386	Coprocessor segment overrun
A	28H–2BH	80386–Pentium Pro	Invalid task state segment
B	2CH–2FH	80386–Pentium Pro	Segment not present
C	30H–33H	80386–Pentium Pro	Stack fault
D	34H–37H	80386–Pentium Pro	General protection fault
E	38H–3BH	80386–Pentium Pro	Page fault
F	3CH–3FH	All	Reserved*
10	40H–43H	80286–Pentium Pro	Floating-point error
11	44H–47H	80486SX	Alignment error
12	48H–4BH	Pentium–Pentium Pro	Machine check exception
19–31	50H–7FH	All	Reserved*
32–255	80H–3FFH	All	User interrupts

*Note: Some of these interrupts will appear on newer versions of the 8086–Pentium Pro when they become available. Also, some interrupts may differ on the 80186/80188/80386EX embedded controllers.

later in this section. Vectors 1–6, 7, 9, 16, and 17 function in the real mode and protected mode; the remaining vectors function only in the protected mode.

Interrupt Instructions

The microprocessor has three different interrupt instructions available to the programmer: INT, INTO, and INT 3. In the real mode, each of these instructions fetches a vector from the vector table and then calls the procedure stored at the location addressed by the vector. In the protected mode, each of these instructions fetches an interrupt descriptor from the interrupt descriptor table. The descriptor specifies the address of the interrupt service procedure. The interrupt call is similar to a far CALL instruction because it places the return address (IP/EIP and CS) on the stack.

INTs. There are 256 different software interrupt (INT) instructions available to the programmer. Each INT instruction has a numeric operand whose range is 0 to 255 (00H–FFH). For example, the INT 100 uses interrupt vector 100, which appears at memory address 190H–193H. The address of the interrupt vector is determined by multiplying the interrupt type number times four. For example, the INT 10H instruction calls the interrupt service procedure whose address is stored beginning at memory location 40H (10H × 4) in the real mode. In protected mode, the interrupt descriptor is located by multiplying the type number by 8 instead of 4 because each descriptor is 8 bytes long.

Each INT instruction is 2 bytes in length. The first byte contains the opcode and the second byte contains the vector type number. The only exception to this is INT 3, a 1-byte special software interrupt used for breakpoints.

Whenever a software interrupt instruction executes, it

1. pushes the flags onto the stack
2. clears the T and I flag bits
3. pushes CS onto the stack
4. fetches the new value for CS from the vector
5. pushes IP/EIP onto the stack
6. fetches the new value for IP/EIP from the vector
7. jumps to the new location addressed by CS and IP/EIP

The INT instruction performs as a far CALL except that it not only pushes CS and IP onto the stack, it also pushes the flags onto the stack. The INT instruction performs the operation of a PUSHF, followed by a far CALL instruction.

Notice that when the INT instruction executes, it clears the interrupt flag (I), which controls the external hardware interrupt input pin INTR (interrupt request). When I = 0, the microprocessor disables the INTR pin and when I = 1, the microprocessor enables the INTR pin.

Software interrupts are most commonly used to call system procedures because the address of the system function need not be known. The system procedures are common to all system and application software. The interrupts often control printers, video displays, and disk drives. Besides relieving the program from remembering the address of the system call, the INT instruction replaces a far CALL that would otherwise be used to call a system function. The INT instruction is 2 bytes in length while the far CALL is 5 bytes. Each time that the INT instruction replaces a far CALL it saves 3 bytes of memory in a program. This can amount to a sizable savings if the INT instruction appears often in a program, as it does for system calls.

IRET/IRETD. The **interrupt return** (IRET) instruction is used only with software or hardware interrupt service procedures. Unlike a simple return instruction (RET), the IRET instruction will (1) pop stack data back into the IP, (2) pop stack data back into CS, and (3) pop stack data back into the flag register. The IRET instruction accomplishes the same tasks as the POPF, followed by a far RET instruction.

Whenever an IRET instruction executes, it restores the contents of I and T from the stack. This is important because it preserves the state of these flag bits. If interrupts were enabled before an interrupt service procedure, they are automatically re-enabled by the IRET instruction because it restores the flag register.

In the 80386 through the Pentium Pro, the IRETD instruction is used to return from an interrupt service procedure that is called in the protected mode. It differs from the IRET because it pops a 32-bit instruction pointer (EIP) from the stack. The IRET is used in the real mode and the IRETD is used in the protected mode.

INT 3. An INT 3 instruction is a special software interrupt designed to be used as a breakpoint. The difference between it and the other software interrupts is that INT 3 is a 1-byte instruction while the others are 2-byte instructions.

It is common to insert an INT 3 instruction in software to interrupt or break the flow of the software. This function is called a *breakpoint*. A breakpoint occurs for any software interrupt, but because INT 3 is 1-byte long, it is easier to use for this function. Breakpoints help to debug faulty software.

INTO. **Interrupt on overflow** (INTO) is a conditional software interrupt that tests the overflow flag (O). If O = 0, the INTO instruction performs no operation, but if O = 1 and an INTO instruction executes, an interrupt occurs via vector type number 4.

The INTO instruction appears in software that adds or subtracts signed binary numbers. With these operations, it is possible to have an overflow. Either the JO or INTO instruction detects the overflow condition.

An Interrupt Service Procedure

Suppose that, in a particular system, a procedure is required to add the contents of DI, SI, BP, and BX and save the sum in AX. Because this is a common task in this system, it may occasionally be worthwhile to develop the task as a software interrupt. Realize that interrupts are usually reserved for system events and this is merely an example showing how an interrupt service procedure appears. Example 6–12 shows this software interrupt. The main difference between this procedure and a normal far procedure is that it ends with the IRET instruction instead of the RET instruction and the contents of the flag register are saved on the stack during its execution.

EXAMPLE 6–12

```
0000                    INTS        PROC FAR

0000 03 C3                          ADD   AX,BX
0002 03 C5                          ADD   AX,BP
0004 03 C7                          ADD   AX,DI
0006 03 C6                          ADD   AX,SI
0008 CF                             IRET

0009                    INTS        ENDP
```

Interrupt Control

Although this section does not explain hardware interrupts, we introduce two instructions are that control the INTR pin. The **set interrupt flag** instruction (STI) places a 1 into the I flag bit, which enables the INTR pin. The **clear interrupt flag** instruction (CLI) places a 0 into the I flag bit, which disables the INTR pin. The STI instruction enables INTR and the CLI instruction disables INTR. In a software interrupt service procedure, hardware interrupts are enabled as one of the firsts steps. This is accomplished by the STI instruction. Interrupts are enabled early in an interrupt service procedure because just about all of the I/O devices in the personal computer are interrupt-processed. If the interrupts are disabled too long, severe system problems result.

Interrupts in the Personal Computer

The interrupts found in the personal computer differ somewhat from the ones presented in Table 6–3. The reason that they differ is that the original personal computers are 8086/8088-based systems. Therefore, they contained only Intel-specified interrupts 0–4. This design is carried forward so newer systems are compatible with the early personal computers.

Because the personal computer is operated in the real mode, the interrupt vector table is located at addresses 00000H–003FFH. The assignments used by the personal computer system are listed in Table 6–4. Notice that these differ somewhat from the assignments in Table 6–3. Some of the interrupts shown in this table are used in example programs in later chapters. An example is the clock tick, which is extremely useful for timing events because it occurs a constant 18.2 times a second in all personal computers.

Interrupts 00H–1FH and 70H–77H are present in the computer no matter which operating system is installed. If DOS is installed, interrupts 20H–2FH are also present. The BIOS uses interrupts 11H through 1FH; the video BIOS uses INT 10H; the hardware in the system uses interrupts 00H–0FH and 70H–77H.

TABLE 6–4 The hexadecimal interrupt assignments for the personal computer

Number	Purpose
0	Divide error
1	Single-step (debug)
2	NMI interrupt pin
3	Breakpoint
4	Arithmetic overflow (INTO instruction)
5	Print screen key (BOUND)
6	Illegal instruction
7	Coprocessor not present interrupt
8	Timer tick (18.2 Hz)
9	Keyboard interrupt
A	Hardware interrupt 2 (cascade in AT)
B–F	Hardware interrupts 3–7
10	Video BIOS
11	Equipment environment
12	Conventional memory size
13	Direct disk services
14	Serial COM port services
15	Miscellaneous services
16	Keyboard services
17	Parallel port (LPT) services
18	ROM BASIC
19	Reboot
1A	Clock services
1B	Control-break handler
1C	User timer services
1D	Pointer for video parameter table
1E	Pointer for disk parameter table
1F	Pointer to graphics character pattern table
20	Terminate program
21	DOS services
22	Program termination handler
23	Control C handler
24	Critical error handler
25	Read disk
26	Write disk
27	Terminate and stay resident
28	DOS idle
2F	Multiplex handler
31	DPMI
33	Mouse driver
67	VCPI
70–77	Hardware interrupts 8–15

6–4   MACHINE CONTROL AND MISCELLANEOUS INSTRUCTIONS

The last category of real-mode instructions are the machine control and miscellaneous group. These instructions provide control of the carry bit, sample the $\overline{BUSY}/\overline{TEST}$ pin, and perform various other functions. Because many of these instructions are used in hardware control, they need only be explained briefly at this point.

Controlling the Carry Flag Bit

The **carry flag** (C) propagates the carry or borrow in multiple-word/doubleword addition and subtraction. It also indicates errors in procedures. There are three instructions that control the contents of the carry flag: STC **(set carry),** CLC **(clear carry),** and CMC **(complement carry).**

Because the carry flag is seldom used, except with multiple-word addition and subtraction, it is available for other uses. The most common task for the carry flag is to indicate error upon return from a procedure. Suppose that a procedure reads data from a disk memory file. This operation can be successful or an error can occur, such as file-not-found. Upon return from this procedure, if C = 1, an error has occurred; if C = 0, no error occurred. Most of the DOS and BIOS procedures use the carry flag to indicate error conditions.

WAIT

The WAIT instruction monitors the hardware $\overline{\text{BUSY}}$ pin on the 80286 and 80386 and the $\overline{\text{TEST}}$ pin on the 8086/8088. The name of this pin was changed in the 80286 microprocessor from $\overline{\text{TEST}}$ to $\overline{\text{BUSY}}$. If the WAIT instruction executes while the $\overline{\text{BUSY}}$ pin = 1, nothing happens and the next instruction executes. If the $\overline{\text{BUSY}}$ pin = 0 when the WAIT instruction executes, the microprocessor waits for the $\overline{\text{BUSY}}$ pin to return to a logic 1. This pin indicates a busy condition when at a logic 0 level.

The $\overline{\text{BUSY/TEST}}$ pin of the microprocessor is usually connected to the $\overline{\text{BUSY}}$ pin of the 8087 through the 80387 numeric coprocessors. This connection allows the microprocessor to wait until the coprocessor finishes a task. Because the coprocessor is inside an 80486–Pentium Pro, the $\overline{\text{BUSY}}$ pin is not present in these microprocessors.

HLT

The **halt** instruction (HLT) stops the execution of software. There are three ways to exit a halt: by an interrupt, by a hardware reset, or during a DMA operation. This instruction normally appears in a program to wait for an interrupt. It often synchronizes external hardware interrupts with the software system.

NOP

When the microprocessor encounters a **no operation** instruction (NOP), it takes a short time to execute. In early years, before software development tools were available, a NOP, which performs absolutely no operation, was often used to pad software with space for future machine language instructions. If you are developing machine language programs, which is extremely rare, it is recommended that you place 10 or so NOPs in your program at 50-byte intervals. This is done in case you need to add instructions at some future point. An NOP may also find application in time delays to waste short periods of time. Realize that an NOP used for timing is not very accurate because of the cache and pipelines in modern microprocessors.

Lock Prefix

The LOCK prefix appends an instruction and causes the LOCK pin to become a logic 0. The LOCK pin often disables external bus masters or other system components. The LOCK prefix causes the lock pin to activate for the duration of a locked instruction. If more than one sequential instruction is locked, the LOCK pin remains a logic 0 for the duration of the sequence of locked instructions. The LOCK:MOV AL,[SI] instruction is an example of a locked instruction.

ESC

The **escape** (ESC) instruction passes information to the 8087–80387 numeric coprocessors. Whenever an ESC instruction executes, the microprocessor provides the memory address, if required, but otherwise performs an NOP. The 8087–80387 use 6 bits of the ESC instruction to obtain their opcodes and begin executing a coprocessor instruction.

The ESC opcode never appears in a program as ESC. In its place are a set of coprocessor instructions (FLD, FST, FMUL, etc.) that assemble as ESC instructions for the coprocessor. This instruction is considered obsolete except as a command to the coprocessor. More detail is provided in the chapter that details the 8087–80387, 80487, Pentium, and Pentium Pro coprocessors.

Bound

The BOUND instruction, first made available in the 80186 microprocessor, is a compare instruction that may cause an interrupt (vector type number 5). This instruction compares the contents of any 16- or 32-bit register against the contents of two words or doublewords of memory: an upper and a lower boundary. If the value in the register compared with memory is not within the upper and lower boundaries, a type 5 interrupt ensues. If it is within the boundary, the next instruction in the program executes.

For example, if the BOUND SI,DATA instruction executes, word-sized location DATA contains the lower boundary, and word-sized location DATA + 2 bytes contains the upper boundary. If the number contained in SI is less than memory location DATA or greater than memory location DATA + 2 bytes, a type 5 interrupt occurs. When this interrupt occurs, the return address points to the BOUND instruction, not to the instruction following BOUND. This differs from a normal interrupt where the return address points to the next instruction in the program.

Enter and Leave

The ENTER and LEAVE instructions, first made available to the 80186 microprocessor, are used with stack frames. A **stack frame** is a mechanism used to pass parameters to a procedure through the stack memory. The stack frame also holds local memory variables for the procedure. Stack frames provide dynamic areas of memory for procedures in multi-user environments.

The ENTER instruction creates a stack frame by pushing BP onto the stack and then loading BP with the uppermost address of the stack frame. This allows stack frame variables to be accessed through the BP register. The ENTER instruction contains two operands: the first operand species the number of bytes to reserve for variables on the stack frame, and the second specifies the level of the procedure.

Suppose that an ENTER 8,0 instruction executes. This instruction reserves 8 bytes of memory for the stack frame and the zero specifies level 0. Figure 6–9 shows the stack frame set up by this instruction. Note that this instruction stores BP onto the top of the stack. It then subtracts 8 from the stack pointer, leaving 8 bytes of memory space for temporary data storage. The uppermost location of this 8-byte temporary storage area is addressed by BP. The LEAVE instruction reverses this process by reloading both SP and BP with their prior values.

Example 6–13 shows how the ENTER instruction creates a stack frame so two 16-bit parameters are passed to a system level procedure. Notice how the ENTER and LEAVE instructions appear in this program, and how the parameters pass through the stack frame to and from the procedure. This procedure uses two parameters that pass to it and returns two results through the stack frame.

EXAMPLE 6–13

```
;A sequence used to call system software that
;uses parameters stored in a stack frame.
;
```

```
0000 C8 0004 00                    ENTER 4,0                    ;create 4-byte frame

0004 A1 00C8 R                     MOV   AX,DATA1
0007 89 46 FC                      MOV   [BP-4],AX              ;save para 1
000A A1 00CA R                     MOV   AX,DATA2
000D 89 46 FE                      MOV   [BP-2],AX              ;save para 2

0010 E8 0100 R                     CALL SYS                     ;call subroutine

0013 8B 46 FC                      MOV   AX,[BP-4]              ;get result 1
0016 A3 00C8 R                     MOV   DATA1,AX               ;save result 1
0019 8B 46 FE                      MOV   AX,[BP-2]              ;get result 2
001C A3 00CA R                     MOV   DATA2,AX               ;save result 2

001F C9                            LEAVE
                                      .      .
                                      .      .
                                   (other software continues here)
                                      .      .
                                      .      .
                                   ;system subroutine that uses the stack frame
                                   ;
0100                      SYS      PROC NEAR

0100 60                            PUSHA

0101 8B 46 FC                      MOV  AX,[BP-4]               ;get para 1
0104 8B 5E FE                      MOV  BX,[BP-2]               ;get para 2
                                      .      .

                                   (software that uses the parameters)
                                      .      .
                                      .      .
0130 89 46 FC                      MOV  [BP-4],AX               ;save result 1
0133 89 5E FE                      MOV  [BP-2],BX               ;save result 2

0136 61                            POPA
0137 C3                            RET

0138                      SYS      ENDP
```

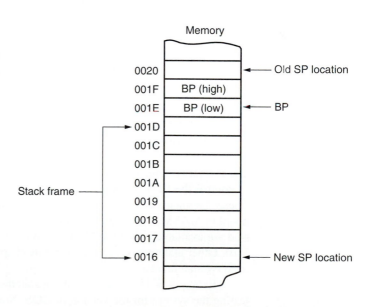

FIGURE 6–9 The stack frame created by the ENTER 8.0 instruction. Notice that BP is stored beginning at the top of the stack frame. This is followed by an 8-byte area called a stack frame.

TABLE 6–5   The conditional move instructions

Assembly Language	Condition Tested	Operation
CMOVB	C = 1	Move if below
CMOVAE	C = 0	Move if above or equal
CMOVBE	Z = 1 or C = 1	Move if below or equal
CMOVA	Z = 0 and C = 0	Move if above
CMOVE or CMOVZ	Z = 1	Move if equal or set if zero
CMOVNE or CMOVNZ	Z = 0	Move if not equal or set if not zero
CMOVL	S <> 0	Move if less than
CMOVLE	Z = 1 or S <> 0	Move if less than or equal
CMOVG	Z = 0 and S = 0	Move if greater than
CMOVGE	S = 0	Move if greater than or equal
CMOVS	S = 1	Move if sign (negative)
CMOVNS	S = 0	Move if no sign (positive)
CMOVC	C = 1	Move if carry
CMOVNC	C = 0	Move if no carry
CMOVO	O = 1	Move if overflow
CMOVNO	O = 0	Move if no overflow
CMOVP or CMOVPE	P = 1	Move if parity or set if parity even
CMOVNP or CMOVPO	P = 0	Move if no parity or set if parity odd

CMOV (Pentium Pro Only)

The CMOV (conditional move) class of instruction is new to the Pentium Pro processor instruction set. Actually there are many variations of the CMOV instruction. Table 6–5 lists these variations of CMOV. These instructions move the data only if the condition is true. For example, the CMOVZ instruction only moves data if the result from some prior instruction is a zero. The destination may be only a 16- or 32-bit register, but the source can be a 16- or 32-bit register or memory location.

Because this is a new instruction, you cannot use it with the assembler until a .686 switch is provided. In the interim, the instruction can be coded in hexadecimal form using the DB directive. The opcode for the CMOV instruction is a 0F4XH, where X is the condition code 0000–1111 (refer to Appendix B for the codes). This is followed by a mod-reg-r/m byte. Example 6–14 shows how the CMOVB instruction is coded into hexadecimal using the DB directive.

EXAMPLE 6–14

```
0000 0F 42 C3        DB    0FH,42H,0C3H       ;same as a CMOVB AX,BX
```

6–5   CONTROLLING THE FLOW OF AN ASSEMBLY LANGUAGE PROGRAM

The assembly language statements .IF, .ELSE, .ELSEIF, and .ENDIF are more useful directives to control the flow of the program than trying to find the correct conditional jump statement. These statements always begin with a period, which indicates a special assembly language command to MASM. Note that the control flow assembly language statements beginning with a period are available only to MASM version 6.X and not earlier versions of the assembler such as 5.10. Other statements developed in this chapter include the .REPEAT-.UNTIL and .WHILE-.ENDW statements.

Example 6–15 shows how these statements are used to control the flow of a program by testing the system for the version of DOS. Notice that in this example DOS INT 21H, function number 30H is used to read the DOS version. The version is tested to determine if it is above or below version 3.3. If below version 3.3, the program terminates, using DOS INT 21H function number 4CH.

EXAMPLE 6–15 (A)

```
                        ;source program sequence
                        ;
                            MOV   AH,30H
                            INT   21H                  ;get DOS version

                        .IF    AL<3 && AH<30

                            MOV   AH,4CH      ;terminate program
                            INT   21H

                        .ENDIF
```

EXAMPLE 6–15 (B)

```
                        ;assembled listing file of Example 6-15 (A)
                        ;
0000  B4 30                 MOV   AH,30H
0002  CD 21                 INT   21H                  ;get DOS version

                        .IF    AL<3 && AH<30
0004  3C 03        *        cmp   al,003h
0006  73 09        *        jae   @C0001
0008  80 FC 1E     *        cmp   ah,01Eh
000B  73 04        *        jae   @C0001

000D  B4 4C                 MOV   AH,4CH      ;terminate program
000F  CD 21                 INT   21H

                        .ENDIF
0011            *  @C0001:
```

Example 6–15 (A) shows the source program sequence as it was typed and Example 6–15 (B) shows the fully expanded assembled output generated by the assembler program. Notice that assembler-generated and -inserted statements begin with an asterisk (*) in the listing. The .IF AL<3 && AH<30 statement tests for DOS version 3.30. If the major version number (AL) is less than 3 AND the minor version number (AH) is less than 30, the MOV AH,0 and INT 21H instructions execute. Notice how the && symbol represents the word AND in the IF statement. Refer to Table 6–6 for a complete list of relational operators used with the .IF statement. Many of these conditions (such as &&) are also used by high-level languages such as C/C++.

TABLE 6–6 Relational operators used with the .IF statement

Operator	Function
==	Equal or the same as
!=	Not equal
>	Greater than
>=	Greater than or equal
<	Less than
<=	Less than or equal
&	Bit test
!	Logical inversion
&&	Logical AND
||	Logical OR

Example 6–16 shows another example of the conditional .IF directive that converts all ASCII-coded letters to uppercase. First the keyboard is read without echo using DOS INT 21H function 06H, and then the .IF statement converts the character into uppercase if needed. In this example the logical AND function (&&) is used to determine if the character is lowercase. This program reads a key from the keyboard and converts it to uppercase before displaying it. Notice how the program terminates when the control C key (ASCII = 03H) is typed. The .LISTALL directive causes all assembler-generated statements to be listed, including the label @Startup generated by the .STARTUP directive. The .EXIT directive also is expanded by .LISTALL to show the use of the DOS INT 21H function 4CH, which returns control to DOS.

EXAMPLE 6–16

```
                        ;A program that reads the keyboard and converts all
                        ;lowercase data to uppercase before displaying it.
                        ;
                        ;This program uses a control C for termination.
                        ;
                                .MODEL TINY             ;select TINY model
                                .LISTALL                ;list all statements
0000                            .CODE                   ;start CODE segment
                                .STARTUP                ;start program
0100            *       @Startup:
0100                    MAIN1:
0100 B4 06                      MOV   AH,6              ;read key without echo
0102 B2 FF                      MOV   DL,0FFH
0104 CD 21                      INT   21H
0106 74 F8                      JE    MAIN1             ;if no key typed
0108 3C 03                      CMP   AL,3              ;test for control C key
010A 74 10                      JE    MAIN2             ;if control C key

                                .IF   AL>='a' && AL<='z'

010C 3C 61           *          cmp   al, 'a'
010E 72 06           *          jb    @C0001
0110 3C 7A           *          cmp   al, 'z'
0112 77 02           *          ja    @C0001

0114 2C 20                      SUB   AL,20H

                                .ENDIF
0116            *       @C0001:

0116 8A D0                      MOV   DL,AL             ;echo character to display
0118 CD 21                      INT   21H
011A EB E4                      JMP   MAIN1             ;repeat
011C                    MAIN2:
                                .EXIT                   ;exit to DOS on control C

011C B4 4C           *          mov   ah, 04Ch
011E CD 21           *          int   021h

                                END                     ;end of file
```

In this program a lowercase letter is converted to uppercase by the use of the .IF AL >= 'a' && AL <= 'z' statement. If AL contains a value that is greater than or equal to a lowercase a, and less than or equal to a lowercase z (a value of a–z), the statement between the .IF and .ENDIF executes. This statement (SUB AL,20H) subtracts 20H from the lowercase letter to change it to an uppercase letter. Notice how the assembler program implements the .IF statement (see lines that begin with *). The label @C0001 is an assembler-generated label used by the conditional jump statements placed in the program by the .IF statement.

Example 6–17 shows another example using the conditional .IF statement. This program reads a key from the keyboard and then converts it to hexadecimal code. (This program is not listed in expanded form.)

In Example 6–17, the .IF AL >='a' && AL<= 'f' statement causes the next instruction (SUB AL,57H) to execute if AL contains letters a–f converting them to hexadecimal. If it is not between letters a and f, the next .ELSEIF statement tests it for the letters A–F. If it is letters A–F, a 37H is subtracted from AL. If neither of these is true, a 30H is subtracted from AL before AL is stored at data segment memory location TEMP.

EXAMPLE 6–17

```
                        ;A program that reads a key and stores its hexadecimal
                        ;value in memory location TEMP.
                        ;
                                .MODEL SMALL            ;select SMALL model
0000                            .DATA                   ;start DATA segment
0000 00                 TEMP    DB    ?                 ;define TEMP
0000                            .CODE                   ;start CODE segment
                                .STARTUP                ;start program
0017 B4 01                      MOV   AH,1              ;read key
0019 CD 21                      INT   21H

                                .IF   AL>='a' && AL<='f'  ;if lowercase
0023 2C 57                      SUB   AL,57H

                                .ELSEIF AL>='A' && AL<='F'  ;if uppercase
002F 2C 37                      SUB   AL,37H

                                .ELSE                   ;otherwise
0033 2C 30                      SUB   AL,30H

                                .ENDIF

0035 A2 0000 R                  MOV   TEMP,AL
                                .EXIT                   ;exit to DOS
                                END                     ;end of file
```

Do-While Loops

As with most higher-level languages, the assembler also provides the DO-WHILE loop construct, available to MASM version 6.X. The .WHILE statement is used with a condition to begin the loop and the .ENDW statement to end the loop.

Example 6–18 shows how the .WHILE statement is used to read data from the keyboard and store it into an array called BUF until the enter key (0DH) is typed. This program assumes that BUF is stored in the extra segment because the STOSB instruction is used to store the keyboard data in memory. The .WHILE loop portion of the program is shown in expanded form so that the statements inserted by the assembler (beginning with a *) can be studied. After the enter key (0DH) is typed, the string is appended with a $ so it can be displayed with DOS INT 21H function number 9.

EXAMPLE 6–18

```
                        ;A program that reads a character string from the
                        ;keyboard and, after enter is typed, displays it again.
                        ;
                                .MODEL SMALL            ;select small model
0000                            .DATA                   ;indicate DATA segment
0000 0D 0A              MES     DB    13,10             ;return & line feed
0002 0100 [             BUF     DB    256 DUP (?)       ;character string buffer
       00
     ]
0000                            .CODE                   ;start of CODE segment
                                .STARTUP                ;start of program
0017 8C D8                      MOV   AX,DS             ;make ES overlap DS
0019 8E C0                      MOV   ES,AX

001B FC                         CLD                     ;select increment
```

```
001C BF 0002 R                  MOV DI,OFFSET BUF        ;address buffer

                                .WHILE AL != 0DH         ;loop while AL not enter

001F EB 05      *               jmp @C0001
0021           * @C0002:

0021 B4 01                          MOV AH,1             ;read key with echo
0023 CD 21                          INT 21H
0025 AA                             STOSB                ;store key code

                                .ENDW                    ;end while loop

0026           * @C0001:
0026 3C 0D      *               cmp  al,00Dh
0028 75 F7      *               jne  @C0002

002A C6 45 FF 24                MOV BYTE PTR [DI-1],'$'  ;make it $ string
002E BA 0000 R                  MOV DX,OFFSET MES        ;address MES
0031 B4 09                      MOV AH,9                 ;display MES
0033 CD 21                      INT  21H
                                .EXIT                    ;exit to DOS
                                END
```

The program in Example 6–18 functions perfectly as long as we arrive at the .WHILE statement with AL containing some value other than 0DH. This can be corrected by adding a MOV AL,0DH instruction before the .WHILE statement. A better way of handling this problem is illustrated in Example 6–19. In this example, the .BREAK statement is used to break out of the .WHILE loop. A .WHILE 1 creates an infinite loop and the .BREAK statement tests for a value of 0DH (enter) in AL. If AL = 0DH, the program breaks out of the infinite loop, correcting the problem exhibited in Example 6–18. Notice that the .BREAK statement causes the break to occur at the point where it appears in the program. This is important, because it allows the point of the break to be selected by the programmer.

Not illustrated in Example 6–19 is the .CONTINUE statement, which can be used to allow the DO-WHILE loop to continue if a certain condition is met. For example, a .CONTINUE .IF AL == 15 allows the loop to continue if AL equals 15. The .BREAK and .CONTINUE commands function in the same manner in a C-language program.

EXAMPLE 6–19

```
                                .MODEL SMALL
0000                            .DATA
0000 0D 0A       MES       DB    13,10                   ;define string
0002 0100 [      BUF       DB    256 DUP (?)             ;memory for string
        00
          ]
0000                            .CODE
                                .STARTUP
0017 8C D8                      MOV  AX,DS               ;make ES overlap DS
0019 8E C0                      MOV  ES,AX

001B FC                         CLD                      ;select increment
001C BF 0002 R                  MOV  DI,OFFSET BUF       ;address BUF

                                .WHILE 1                 ;create an infinite loop
001F           * @C0001:

001F B4 01                          MOV  AH,1            ;read key
0021 CD 21                          INT  21H
0023 AA                             STOSB                ;store key code in BUF

                                .BREAK .IF AL == 0DH     ;breaks loop for a 0DH
0024 3C 0D      *               cmp  al,00Dh
```

```
0026 74 02        *          je    @C0002

                             .ENDW
0028 EB F5        *          jmp   @C0001
002A             * @C0002:

002A C6 45 FF 24             MOV   BYTE PTR [DI-1],'$'  ;make it a $ string
002E BA 0000 R              MOV   DX,OFFSET MES        ;display string
0031 B4 09                  MOV   AH,9
0033 CD 21                  INT   21H
                             .EXIT
                             END
```

Example 6-20 is a practical example using the DO-WHILE construct to display the contents of EAX in decimal on the video display. The EAX register is initialized with a number (123455) to test this program. Two infinite loops are used to convert EAX to decimal. The first divides EAX by 10 until the quotient is zero. After each division, the remainder is saved on the stack as a significant digit in the result. Also located within the first infinite loop is a comma counter stored in CL. Each time that the quotient is not zero, the comma counter increments. If the comma counter reaches 3, a comma is pushed onto the stack for later display, and the comma counter is reset to zero. The final infinite loop displays the result. After each POP DX instruction, the break statement checks DX to find if it contains a 10. The 10 was pushed on the stack to indicate the end of the number. If it does contain a 10, the loop breaks; if it doesn't, a decimal digit or a comma is displayed. This procedure can be added to any program where a decimal number of up to four billion must be displayed with commas at the correct places.

EXAMPLE 6–20

```
                   ;A program that displays the contents of EAX in decimal.
                   ;This program inserts commas between thousands,
                   ;millions, and billions.
                   ;
                             .MODEL TINY
                             .386                    ;select 80386
0000                         .CODE
                             .STARTUP

0100 66| B8 0001E23F   MOV   EAX,123455              ;load test data
0106 E8 0004           CALL  DISPE                   ;display EAX in decimal
                             .EXIT
                   ;
                   ;The DISPE procedure displays EAX in decimal format.
                   ;
010D               DISPE   PROC NEAR

010D 66| BB 0000000A        MOV   EBX,10             ;load 10 for decimal
0113 53                     PUSH  BX                 ;save end of number
0114 B1 00                  MOV   CL,0               ;load comma counter

                             .WHILE 1                ;first infinite loop

0116 66| BA 00000000          MOV   EDX,0            ;clear EDX
011C 66| F7 F3                DIV   EBX              ;divide EDX:EAX by 10
011F 80 C2 30                 ADD   DL,30H           ;convert to ASCII
0122 52                       PUSH  DX               ;save remainder

                             .BREAK .IF EAX == 0     ;break if quotient zero

0128 FE C1                    INC   CL               ;increment comma counter

                             .IF CL == 3             ;if comma count is 3
012F 6A 2C                      PUSH ','             ;save comma
0131 B1 00                      MOV  CL,0            ;clear comma counter
```

```
                                   .ENDIF

                                   .ENDW                       ;end first loop

                                   .WHILE 1                    ;second infinite loop
0135 5A                                POP  DX                 ;get remainder
                                   .BREAK .IF DL == 10         ;break if remainder is 10

013B B4 02                             MOV  AH,2               ;display decimal digit
013D CD 21                             INT  21H

                                   .ENDW

0141 C3                            RET

0142               DISPE           ENDP
                                   END
```

Repeat-Until Loops

Also available to the assembler is the REPEAT-UNTIL construct. A series of instructions is repeated until come condition occurs. The .REPEAT statement defines the start of the loop and the end is defined with the .UNTIL statement that contains a condition. Note that .REPEAT and .UNTIL are available to version 6.X of MASM.

Example 6–18 can be reworked once again using the REPEAT-UNTIL construct, which appears to be the best solution. Refer to Example 6–21 for a program that reads keys from the keyboard and stores keyboard data into extra segment array BUF until the enter key is pressed. This program also fills the buffer with keyboard data until the enter key (0DH) is typed. Once enter is typed, the program displays the character string using DOS INT 21H function number 9 after appending the buffer data with the required dollar sign. Notice how the .UNTIL AL == 0DH statement generates code (statement beginning with *) to test for the enter key.

EXAMPLE 6–21

```
                                   .MODEL SMALL
0000                               .DATA
0000 0D 0A       MES           DB    13,10           ;define MES
0002 0100 [      BUF           DB    256 DUP (?)      ;reserve memory for BUF
      00
        ]
0000                               .CODE
                                   .STARTUP
0017 8C D8                         MOV  AX,DS          ;overlap DS and ES
0019 8E C0                         MOV  ES,AX
001B FC                            CLD                 ;select increment
001C BF 0002 R                     MOV  DI,OFFSET BUF  ;address BUF

                                   .REPEAT
001F             * @C0001:

001F B4 01                             MOV  AH,1       ;read key with echo
0021 CD 21                             INT  21H
0023 AA                                STOSB           ;save key code in BUF

                                   .UNTIL  AL == 0DH
0024 3C 0D       *                 cmp  al, 00Dh
0026 75 F7       *                 jne  @C0001

0028 C6 45 FF 24                   MOV  BYTE PTR [DI-1],'$'  ;make $ string
002C B4 09                         MOV  AH,9                 ;display MES and BUF
002E BA 0000 R                     MOV  DX,OFFSET MES
0031 CD 21                         INT  21H
                                   .EXIT
                                   END
```

There is also an .UNTILCXZ instruction available that uses the LOOP instruction to check for the until condition. The .UNTILCXZ can have a condition or may just use the CX register as a counter to repeat a loop a fixed number of times. Example 6–22 shows a sequence of instructions that use the .UNTILCXZ instruction to add the contents of byte-sized array ONE to byte-sized array TWO. The sums are stored in array THREE. Each array contains 100 bytes of data, so the loop is repeated 100 times. This example assumes that array THREE is in the extra segment and that arrays ONE and TWO are in the data segment.

EXAMPLE 6–22

```
012C B9 0064            MOV   CX,100            ;set count
012F BF 00C8 R          MOV   DI,OFFSET THREE   ;address arrays
0132 BE 0000 R          MOV   SI,OFFSET ONE
0135 BB 0064 R          MOV   BX,OFFSET TWO

                        .REPEAT

0138          *  @C0001:

0138 AC                 LODSB
0139 02 07              ADD   AL,[BX]
013B AA                 STOSB
013C 43                 INC   BX

                        .UNTILCXZ

013D E2 F9    *         loop  @C0001
```

6–6 SUMMARY

1. There are three types of unconditional jump instructions: short, near, and far. The short jump allows a branch to within +127 and –128 bytes. The near jump (using a displacement of ±32K) allows a jump to anywhere in the current code segment (intrasegment). The far jump allows a jump to any location in the memory system (intersegment). The near jump in an 80386 through a Pentium Pro is within ±2G bytes because these microprocessors can use a 32-bit signed displacement.

2. Whenever a label appears with a JMP instruction or conditional jump, the label, located in the label field, must be followed by a colon (LABEL:). The JMP DOGGY instruction jumps to memory location DOGGY:.

3. The displacement that follows a short or near jump is the distance from the next instruction to the jump location.

4. Indirect jumps are available in two forms: (1) jump to the location stored in a register and (2) jump to the location stored in a memory word (near indirect) or doubleword (far indirect).

5. Conditional jumps are all short jumps that test one or more of the flag bits: C, Z, O, P, or S. If the condition is true, a jump occurs; if the condition is false, the next sequential instruction executes. Note that the 80386 and above also allow a 16-bit signed displacement for the conditional jump instructions.

6. A special conditional jump instruction (LOOP) decrements CX and jumps to the label when CX is not 0. Other forms of loops include: LOOPE, LOOPNE, LOOPZ, and LOOPNZ. The LOOPE instruction jumps if CX is not 0, and if an equal condition exists. In the 80386 through the Pentium Pro, the LOOPD, LOOPED, and LOOPNED instructions also use the ECX register as a counter.

7. In the 80386 through the Pentium Pro, a group of conditions exist that set a byte either to 01H or clear it to 00H. If the condition under test is true, the operand byte is set to a 01H; if the condition under test is false, the operand byte is cleared to 00H.

8. Procedures are groups of instructions that perform one task and are used from any point in a program. The CALL instruction links to a procedure and the RET instruction returns from a procedure. In assembly language, the PROC directive defines the name and type of procedure. The ENDP directive declares the end of the procedure.

9. The CALL instruction is a combination of a PUSH and a JMP instruction. When CALL executes, it pushes the return address on the stack and then jumps to the procedure. A near CALL places the contents of IP on the stack and a far CALL places both IP and CS on the stack.

10. The RET instruction returns from a procedure by removing the return address from the stack and placing it into IP (near return) or IP and CS (far return).

11. Interrupts are either software instructions similar to CALL or hardware signals used to call procedures. This process interrupts the current program and calls a procedure. After the procedure, a special IRET instruction returns control to the interrupted software.

12. Real-mode interrupt vectors are four bytes in length and contain the address (IP and CS) of the interrupt service procedure. The microprocessor contains 256 interrupt vectors in the first 1K byte of memory. The first 32 are defined by Intel; the remaining 224 are user interrupts. In protected-mode operation, the interrupt vector is 8 bytes in length; the interrupt vector table may be relocated to any section of the memory system.

13. Whenever an interrupt is accepted by the microprocessor, the flags, IP, and CS are pushed on the stack. Besides pushing the flags, the T and I flag bits are cleared to disable both the trace function and the INTR pin. The final event that occurs for the interrupt is that the interrupt vector is fetched from the vector table and a jump to the interrupt service procedure occurs.

14. Software interrupt instructions (INT) often replace system calls. Software interrupts save 3 bytes of memory each time they replace CALL instructions.

15. A special return instruction (IRET) must be used to return from an interrupt service procedure. The IRET instruction not only removes IP and CS from the stack, it also removes the flags from the stack.

16. Interrupt on an overflow (INTO) is a conditional interrupt that calls an interrupt service procedure if the overflow flag (O) = 1.

17. The interrupt enable flag (I) controls the INTR pin connection on the microprocessor. If the STI instruction executes, it sets I to enable the INTR pin. If the CLI instruction executes, it clears I to disable the INTR pin.

18. The carry flag bit (C) is clear, set, and complemented by the CLC, STC, and CMC instructions.

19. The WAIT instruction tests the condition of the $\overline{\text{BUSY}}$ or $\overline{\text{TEST}}$ pin on the microprocessor. If $\overline{\text{BUSY}}$ or $\overline{\text{TEST}}$ = 1, WAIT does not wait, but if $\overline{\text{BUSY}}$ or $\overline{\text{TEST}}$ = 0, WAIT continues testing the $\overline{\text{BUSY}}$ or $\overline{\text{TEST}}$ pin until it becomes a logic 1. Note that the 8086/8088 contain the $\overline{\text{TEST}}$ pin, while the 80286–80386 contain the $\overline{\text{BUSY}}$ pin. The 80486–Pentium Pro do not contain a $\overline{\text{BUSY}}$ or $\overline{\text{TEST}}$ pin.

20. The LOCK prefix causes the LOCK pin to become a logic 0 for the duration of the locked instruction. The ESC instruction passes instructions to the numeric coprocessor.

21. The BOUND instruction compares the contents of any 16-bit register against the contents of two words of memory: an upper and a lower boundary. If the value in the register compared with memory is not within the upper and lower boundaries, a type 5 interrupt ensues.

22. The ENTER and LEAVE instructions are used with stack frames. A stack frame is a mechanism used to pass parameters to a procedure through the stack memory. The stack frame also holds local memory variables for the procedure. The ENTER instructions create the stack frame, and the LEAVE instruction removes the stack frame from the stack. The BP register addresses stack frame data.

23. The CMOV instruction, new to the Pentium Pro, is a conditional move that uses the same condition as the conditional jump and conditional set instructions. If the condition is true, the source operand is copied into the destination. If the condition is false, no move occurs.

24. The .IF and .ENDIF statements are useful in assembly language for making decisions. These instructions cause the assembler to generate conditional jump statements that modify the flow of the program.

25. The .WHILE and .ENDW statements allow an assembly language program to use the DO-WHILE construction; and the .REPEAT and .UNTIL statements allow an assembly language program to use the REPEAT-UNTIL construct.

26. Example 6–23 is a program that uses some of the instructions presented in this chapter as well as those presented in prior chapters. This example contains a procedure that displays a character string on the monitor. As a test of the program, a few sample lines are displayed. The character string is called a null string because it ends with a null (00H).

EXAMPLE 6–23

```
                            ;A program that displays a string of characters
                            ;using the procedure STRING.
                            ;
                                    .MODEL SMALL            ;select SMALL model
0000                                .DATA                   ;start of DATA
0000 0D 0A 0A 00    MES1    DB      13,10,10,0
0004 54 68 69 73 20 MES2    DB      'This is a sample line.',0
     69 73 20 61 20
     73 61 6D 70 6C
     65 20 6C 69 6E
     65 2E 00
0000                                .CODE                   ;start of CODE
                            ;
                            ;A procedure that displays the character string
                            ;address by SI in the data segment.  The
                            ;character string must end with a null.
                            ;
                            ;This procedure changes AX, DX, and SI.
                            ;
0000                        STRING PROC NEAR

0000 AC                             LODSB                   ;get character
0001 3C 00                          CMP  AL,0               ;test for null
0003 74 08                          JE   STRING1            ;if null
0005 8A D0                          MOV  DL,AL              ;move code to DL
0007 B4 02                          MOV  AH,2               ;select function 02H
0009 CD 21                          INT  21H                ;access DOS
000B EB F3                          JMP  STRING             ;repeat until null
000D                        STRING1:
000D C3                             RET                     ;return

000E                        STRING ENDP
                                    .STARTUP                ;start of program
0025 FC                             CLD                     ;select increment
0026 BE 0000 R                      MOV  SI,OFFSET MES1     ;address MES1
0029 E8 FFD4                        CALL STRING             ;display MES1
002C BE 0004 R                      MOV  SI,OFFSET MES2     ;address MES2
002F E8 FFCE                        CALL STRING             ;display MES2
                                    .EXIT                   ;exit to DOS
                                    END                     ;end of file
```

6–7 QUESTIONS AND PROBLEMS

1. What is a short JMP?
2. Which type of JMP is used when jumping to any location within the current code segment?
3. Which JMP instruction allows the program to continue execution at any memory location in the system?
4. Which JMP instruction is 5 bytes long?

5. What is the range of a near jump in the 80386–Pentium Pro microprocessors?
6. Which type of JMP instruction (short, near, or far) assembles for the following:
   (a) if the distance is 0210H bytes
   (b) if the distance is 0020H bytes
   (c) if the distance is 10000H bytes
7. What can be said about a label that is followed by a colon?
8. What can be said about a label that is followed by a double colon?
9. The near jump modifies the program address by changing which register or registers?
10. The far jump modifies the program address by changing which register or registers?
11. Explain what the JMP AX instruction accomplishes. Also identify it as a near or a far jump instruction.
12. Contrast the operation of a JMP DI with a JMP [DI].
13. Contrast the operation of a JMP [DI] with a JMP FAR PTR [DI].
14. List the five flag bits tested by the conditional jump instructions.
15. Describe how the JA instruction operates.
16. When will the JO instruction jump?
17. Which conditional jump instructions follow the comparison of signed numbers?
18. Which conditional jump instructions follow the comparison of unsigned numbers?
19. Which conditional jump instructions test both the Z and C flag bits?
20. When does the JCXZ instruction jump?
21. Which SET instruction is used to set AL if the flag bits indicate a zero condition?
22. The 8086 LOOP instruction decrements register _____ and tests it for a 0 to decide if a jump occurs.
23. The Pentium LOOPD instruction decrements register _____ and tests it for a 0 to decide if a jump occurs.
24. Explain how the LOOPE instruction operates.
25. Develop a short sequence of instructions that stores a 00H into 150H bytes of memory beginning at extra segment memory location DATA. You must use the LOOP instruction to help perform this task.
26. Develop a sequence of instructions that searches through a block of 100H bytes of memory. This program must count all the unsigned numbers that are above 42H and all that are below 42H. Byte-sized data segment memory location UP must contain the count of numbers above 42H and data segment location DOWN must contain the count of numbers below 42H.
27. What is a procedure?
28. Explain how the near and far CALL instructions function.
29. How does the near RET instruction function?
30. The last executable instruction in a procedure must be a _____.
31. Which directive identifies the start of a procedure?
32. How is a procedure identified as near or far?
33. Explain what the RET 6 instruction accomplishes.
34. Write a near procedure that cubes the contents of the CX register. This procedure may not affect any register except CX.
35. Write a procedure that multiplies DI by SI and then divides the result by 100H. Make sure that the result is left in AX upon returning from the procedure. This procedure may not change any register except AX.
36. Write a procedure that sums EAX, EBX, ECX, and EDX. If a carry occurs, place a logic 1 in EDI. If no carry occurs, place a 0 in EDI. The sum should be found in EAX after the execution of your procedure.
37. What is an interrupt?
38. Which software instructions call an interrupt service procedure?
39. How many different interrupt types are available in the microprocessor?

40. What is the purpose of interrupt vector type number 0?
41. Illustrate the contents of an interrupt vector and explain the purpose of each part.
42. How does the IRET instruction differ from the RET instruction?
43. What is the IRETD instruction?
44. The INTO instruction interrupts the program for only what condition?
45. The interrupt vector for an INT 40H instruction is stored at which memory locations?
46. Which instructions control the function of the INTR pin?
47. Which personal computer interrupt services the parallel LPT port?
48. Which personal computer interrupt services the keyboard?
49. Which instruction tests the $\overline{\text{TEST}}$ pin?
50. When will the BOUND instruction interrupt a program?
51. An ENTER 16,0 instruction creates a stack frame that contains _____ bytes.
52. Which register moves to the stack when an ENTER instruction executes?
53. Which instruction passes opcodes to the numeric coprocessor?
54. The CMOV instruction is available on which microprocessor?
55. Explain how the CMOVNE EAX,EBX instruction functions.
56. Show which assembly language instructions are generated by the following sequence:

```
.IF AL==3
        ADD AL,2
.ENDIF
```

57. What happens if the .WHILE 1 instruction is placed in a program?
58. Develop a short sequence of instructions that uses the REPEAT-UNTIL construct to copy the contents of byte-sized memory BLOCKA into byte-sized memory BLOCKB until a 00H is moved.
59. What is the purpose of the .BREAK directive?
60. Using the DO-WHILE construct develop a sequence of instructions that add the byte-sized contents of BLOCK A to BLOCKB while the sum is not a 12H.
61. What is a null string?
62. Explain how the STRING procedure operates in Example 6–23.
63. Rewrite Example 6–23 so it displays your name.
64. Use the Internet to write a report that compares the jump instruction of the 80186 with the branch instructions of the 8051.
65. Use the Internet to locate data on the 6811 and describe its program control instructions.
66. Use the Internet to locate data on the 80251 and compare the procedure mechanism with that of the 80186.

CHAPTER 7

Programming with DOS and BIOS Functions

INTRODUCTION

This chapter develops programs and programming techniques using the MASM macro assembler, the DOS function calls, and the BIOS function calls. We'll use some of the DOS and BIOS function calls here, but they are all explained in complete detail in Appendix A. Please review the function calls as required as you read this chapter. The MASM assembler has already been explained and demonstrated in prior chapters, yet there are still many more features to learn at this point.

Many embedded applications use BIOS functions to perform basic I/O operations; with the advent of the 80386EX embedded PC, the DOS functions are also becoming commonplace in embedded systems.

Some programming techniques explained in this chapter include: macro sequences, keyboard and display manipulation, program modules, library files, data conversion, and other important programming techniques. This chapter is meant as an introduction to programming. It discusses valuable programming techniques you can use to efficiently develop programs for the personal computer using either PC-DOS or MS-DOS as a springboard.

CHAPTER OBJECTIVES

Upon completion of this chapter, you will be able to:

1. Use the MASM assembler and linker program to create programs that contain more than one module.
2. Explain the use of EXTRN and PUBLIC as they apply to modular programming.
3. Set up a library file that contains commonly used subroutines.
4. Write and use MACRO and ENDM to develop macro sequences used with linear programming.
5. Develop programs using DOS function calls.
6. Differentiate a DOS function call from a BIOS function call.
7. Display and read numeric data in decimal and hexadecimal formats.
8. Use a lookup table to convert between data formats.

7–1   MODULAR PROGRAMMING

Most programs are too large to be developed by one person; instead they are often developed by teams of programmers. The linker program is provided with MS-DOS and PC-DOS so programming modules can be linked together into a complete program. This section describes the linker, the linking task, library files, EXTRN, and PUBLIC as they apply to program modules and modular programming.

The Assembler and Linker

The **assembler** program converts a symbolic **source** module or file into a hexadecimal **object** file. We have seen many examples of symbolic source files in prior chapters. Example 7–1 shows the assembler dialog that appears as a source module named FILE.ASM is assembled. Whenever you create a source file, it usually contains the extension ASM to designate the file as an assembly language source file. Note that the extension is not typed into the assembler prompt when assembling a file. Source files are created using the editor that comes with the assembler or by almost any other editor or word processor that is capable of generating an ASCII file. If you use MASM version 6.X, use the Programmer's Workbench editor to create the source file.

EXAMPLE 7–1

```
A:\>MASM
Microsoft (R) Macro Assembler Version 5.10
Copyright (C) Microsoft Corp 1981, 1989. All rights reserved.

Source filename [.ASM]: FILE
Object filename [FILE.OBJ]: FILE
Source listing [NUL.LST]: FILE
Cross reference [NUL.CRF]: FILE
```

If version 5.10 of MASM is in use, the assembler program (MASM) asks for the source filename, the object filename, the list filename, and a cross-reference filename. In most cases, the name for each of these will be the same as the source file. The object file (.OBJ) is not executable, but is designed as an input file to the linker. The source listing file (.LST) contains the assembled version of the source file and its hexadecimal machine language equivalent. The cross-reference file (.CRF) lists all labels and pertinent information required for cross-referencing.

If version 6.X of MASM is in use, the project is built using the build option after the file is saved as FILE.ASM and the option menu selects the type of project to build. In many cases, for simple assembly language programs the standard DOS .EXE file option is selected.

The **linker** program reads the object files created by the assembler program and links them together into a single execution file. With MASM version 6.X, the linker is included and executed when a program is built using the build option. Either the linker program (MASM version 5.10) or build generates the execution file with the filename extension EXE. Execution files are executed by typing the filename at the DOS prompt (A:\>). An example execution file is FROG.EXE that is executed by typing FROG at the DOS command prompt.

If a file is short enough, less than 64K bytes in length, it can be converted from an execution file to a command file (.COM). The command file is slightly different from an execution file in that the program must be originated at location 100H before it can execute. In many cases, software assembled for the 80186/80188 will be in the command file format. If MASM version 5.10 is in use, the program EXE2BIN is used for converting an execution file into a command file. The main advantage of a command file is that it loads off the disk into the computer much more quickly than an execution file. It also requires slightly less disk storage space than the execution file. If MASM version 6.X is in use, the EXE2BIN program is not needed. The Programmer's Workbench editor program for version 6.X creates the command file directly if the command file (.COM) option is selected. Workbench also builds the file by assembling it and linking it.

Example 7–2 shows the protocol involved with the linker program to link the files FROG, WHAT, and DONUT. The linker also links the library files (LIBS) so the procedures located within LIBS can be used with the linked execution file. To invoke the linker, type LINK at the DOS command prompt as illustrated in Example 7–2. Before files can be linked, they must first be assembled and they must be error-free. Note that version 6.X of MASM allows a project with multiple files to be built using the build option.

In Example 7–2, after typing LINK, the linker program asks for the object modules, which are created by the assembler. This example has three object modules: FROG, WHAT, and DONUT. If more than one object file exists, the main program file (FROG in this example) is typed first followed by any other supporting modules. (A plus sign is used to separate module names.)

EXAMPLE 7–2

```
A:\>LINK

Microsoft (R) Overlay Linker Version 3.64
Copyright (C) Microsoft Corp 1983-1988. All rights reserved.

Object Modules [.OBJ]: FROG+WHAT+DONUT
Run File [FROG.EXE]: FROG
List File [NUL.MAP]: FROG
Libraries [.LIB]: LIBS
```

After the program module names are typed, the linker suggests that the execution (**run-time**) filename is FROG.EXE. This may be selected by typing the same name or if desired by typing the enter key. It may also be changed to any other name at this point.

The list file is where a map of the program segments appear as created by the linking. If enter is typed, no list file is created; if a name is typed, the list file appears on the disk.

Library files are entered in the last line. Example 7–2 uses the library filename LIBS. This library contains procedures used by the other program modules.

If the Programmer's Workbench is not used to assemble and link software with version 6 X of the MASM macro assembler, the sequence listed in Example 7–3 is used to both assemble and link a program file. The example shows how the file TEST.ASM is assembled and linked using the ML program provided with version 6.X of MASM. The /Fl switch (must be uppercase F, lowercase l) generates the list file TEST.LST in this example. If a .COM file must be generated at the command line, use the /AT switch.

EXAMPLE 7–3

```
C:\>ML /FlTEST.LST TEST.ASM

Microsoft (R) Macro Assembler Version 6.11
Copyright (C) Microsoft Corp 1981-1993. All rights reserved.

    Assembling: TEST.ASM

Microsoft (R) Segmented-Executable Linker Version 5.13
Copyright (C) Microsoft Corp 1984-1993. All rights reserved.

Object Modules [.OBJ]: TEST.obj/t
Run File [TEST.com]: "TEST.com"
List File [NUL.MAP]: NUL
Libraries [.LIB]:
Definitions File [NUL.DEF]: ;
```

PUBLIC and EXTRN

The PUBLIC and EXTRN directives are very important to modular programming. The PUBLIC directive declares that labels of code, data, or entire segments are available to other program modules. The EXTRN (external) directive (EXTERN is used as an alternate in version 6.X of the assembler) declares that labels are external to a module. Without these statements, modules could not be linked together to create a program using modular programming techniques.

The PUBLIC directive is normally placed in the opcode field of an assembly language statement to define a label as public so it can be used by other modules. This label can be a jump address or a data address, or an entire segment can be made public. Example 7–4 shows how the PUBLIC statement used to define some labels public to other modules is a program that uses full-segment definitions. When segments are made public, they are combined with other public segments that contain data with the same segment name. Note that with models, the segments are always public and always combined.

EXAMPLE 7–4

```
                DAT1    SEGMENT  PUBLIC          ;declare segment public

                        PUBLIC   DATA1           ;declare DATA1 public
                        PUBLIC   DATA2           ;declare DATA2 public

0000 0064[      DATA1   DB    100 DUP (?)        ;global
        ??
            ]
0064 0064[      DATA2   DB    100 DUP (?)        ;global
        ??
            ]

00C8            DAT1    ENDS

0000            CODES   SEGMENT 'CODE'

                        ASSUME   CS:CODES,DS:DAT1

                        PUBLIC   READ            ;declare READ public

0000            READ    PROC FAR

0000 B4 06              MOV   AH,6               ;read keyboard
0002 B2 FF              MOV   DL,0FFH            ;no echo
0004 CD 21              INT   21H
0006 74 F8              JE    READ
0008 CB                 RET

0009            READ    ENDP

0009            CODES   ENDS

                        END
```

Example 7–4 shows how the procedure READ is made public so it is available to other program modules. The READ procedure reads a key without echo back to the video display. It also shows how DATA1 and DATA2 are made public and available to other modules.

The EXTRN statement appears in both data and code segments to define labels as external to the segment. If data is defined as external, its size must be represented as BYTE, WORD, or DWORD. If a jump or call address is external, it must be represented as NEAR or FAR. Example 7–5 shows how the external statement is used to indicate that several labels are external to the program listed using full-segment definitions. Notice in this example that any external address or data is defined with the letter E in the hexadecimal assembled listing. In this example, the READ procedure from Example 7–4 is called to read 10 keys from the keyboard and store them in an array called DATA1.

EXAMPLE 7–5

```
0000                    DAT1    SEGMENT   PUBLIC        ;declare segment public

                                EXTRN   DATA1:BYTE
                                EXTRN   DATA2:BYTE
                                EXTRN   DATA3:WORD
                                EXTRN   DATA4:DWORD

0000                    DAT1    ENDS

0000                    CODES   SEGMENT  'CODE'

                                ASSUME   CS:CODES,ES:DAT1

                                EXTRN   READ:FAR

0000                    MAIN    PROC FAR

0000 B8 ---- R                  MOV   AX,DAT1
0003 8E C0                      MOV   ES,AX

0005 BF 0000 E                  MOV   DI,OFFSET DATA1
0008 B9 000A                    MOV   CX,10
000B                    MAIN1:
000B 9A 0000 ---- E              CALL   READ
0010 AA                         STOSB
0011 E2 F8                      LOOP   MAIN1
0013 CB                         RET

0014                    MAIN    ENDP

0014                    CODES   ENDS

                                END   MAIN
```

PUBLIC and EXTRN are also used with models. Example 7–6 shows how to declare variables public and how to use external variables with models. (Note that this is a repeat of Example 7–5.)

EXAMPLE 7–6

```
                                .MODEL SMALL          ;select SMALL model
0000                            .DATA                 ;start of DATA segment
                                EXTRN DATA1:BYTE      ;declare DATA1 external
                                EXTRN DATA2:BYTE      ;declare DATA2 external
                                EXTRN DATA3:WORD      ;declare DATA3 external
                                EXTRN DATA4:DWORD     ;declare DATA4 external
0000                            .CODE                 ;start of CODE segment
                                EXTRN READ:FAR        ;declare READ far
                                .STARTUP              ;indicate start of program
0017 8C D8                      MOV   AX,DS           ;load ES with DS
0019 8E C0                      MOV   ES,AX

001B BF 0000 E                  MOV   DI,OFFSET DATA1 ;address DATA1
001E B9 000A                    MOV   CX,10           ;load count with 10
0021                    MAIN1:
0021 9A ---- 0000 E              CALL READ            ;read a key without echo
0026 AA                         STOSB                 ;store key code in DATA1
0027 E2 F8                      LOOP MAIN1            ;repeat 10 times
                                .EXIT                 ;exit to DOS
                                END                   ;end of file
```

Libraries

Library files are collections of procedures that can be used by many different programs. These procedures are assembled and compiled into a library file by the LIB program that accompanies the MASM assembler program. Libraries allow common procedures to be collected into one place so they can be used by many different applications. The library file (FILENAME.LIB) is invoked when a program is linked with the linker program.

Why bother with library files? A library file is a good place to store a collection of related procedures. When the library file is linked with a program, only the procedures required by the program are removed from the library file and added to the program. If assembly language programming is to be accomplished efficiently, a good set of library files is essential.

Creating a Library File. A library file is created with the LIB command typed at the DOS prompt. A library file is a collection of assembled .OBJ files that each perform one procedure. Example 7–7 shows two separate files that contain the READ_KEY and ECHO procedures. When these files are edited, the one containing READ_KEY is named READ_KEY.ASM and the one containing ECHO is named ECHO.ASM. Both files are used to structure a library file. The name of the procedure must be declared PUBLIC in a library file and does not necessarily need to match the filename, although it does in this example. The READ_KEY procedure uses DOS INT 21H function number 06H to read a key from the keyboard without echo. The ECHO procedure also uses DOS INT 21H function number 06H, but to display the contents of the AL register.

EXAMPLE 7–7

```
                        ;The first library module is called READ_KEY.
                        ;This procedure reads a key from the keyboard
                        ;and returns with the ASCII character in AL.
                        ;
0000                    LIB         SEGMENT 'CODE'          ;start of LIB segment
                                    ASSUME  CS:LIB          ;indicate that CS is LIB

                                    PUBLIC READ_KEY         ;define READ_KEY public

0000                    READ_KEY    PROC FAR                ;start of procedure

0000 52                             PUSH DX                 ;save DX

0001                    READ_KEY1:

0001 B4 06                          MOV  AH,6               ;select DOS function 06H
0003 B2 FF                          MOV  DL,0FFH            ;read key without echo
0005 CD 21                          INT  21H                ;access DOS
0007 74 F8                          JE   READ_KEY1          ;if no key is typed
0009 5A                             POP  DX                 ;restore DX
000A CB                             RET                     ;return from procedure

000B                    READ_KEY    ENDP                    ;end of procedure

000B                    LIB         ENDS                    ;end of LIB segment
                                    END


                        ;
                        ;This second library module is called ECHO.
                        ;This procedure displays the ASCII character
                        ;in AL on the CRT screen.
                        ;
0000                    LIB         SEGMENT 'CODE'          ;start of LIB segment
                                    ASSUME  CS:LIB          ;indicate that CS is LIB

                                    PUBLIC  ECHO            ;indicate ECHO is public

0000                    ECHO        PROC FAR                ;start of procedure
```

```
0000 52                      PUSH DX                  ;save DX
0001 B4 06                   MOV  AH,6                 ;select function 06H
0003 8A D0                   MOV  DL,AL                ;display AL
0005 CD 21                   INT  21H                  ;access DOS
0007 5A                      POP  DX                   ;restore DX
0008 CB                      RET                       ;return from procedure

0009             ECHO        ENDP                      ;end of procedure

0009             LIB         ENDS                      ;end of LIB segment
                             END                       ;end of file
```

After each file is assembled, the LIB program is used to combine them into a library file. The LIB program prompts for information as illustrated in Example 7–8 where these files are combined to form the library IO.

EXAMPLE 7–8

```
A:\>LIB

Microsoft (R) Library Manager Version 3.10
Copyright (C) Microsoft Corp 1983-1988. All rights reserved.

Library name:IO
Library file does not exist. Create? Y
Operations:READ_KEY+ECHO
List file:IO
```

The LIB program begins with the copyright message from Microsoft, followed by the prompt Library name:. The library name chosen is IO for the IO.LIB file. Because this is a new file, the library program asks if we wish to create the library file. The Operations: prompt is where the library module names are typed. In this case a library is created using two procedure files (READ_KEY and ECHO). The list file shows the contents of the library and is illustrated for this library in Example 7–9. The list file shows the sizes and names of the files used to create the library and the public label (procedure name) that is used in the library file.

EXAMPLE 7–9

```
ECHO..............ECHO                   READ_KEY..........READ_KEY

READ_KEY          Offset: 00000010H  Code and data size: BH
  READ_KEY

ECHO              Offset: 00000070H  Code and data size: 9H
    ECHO
```

If you must add additional library modules at a later time, type the name of the library file after invoking LIB. At the Operations: prompt, type the new module name preceded with a plus sign to add a new procedure. If you must delete a library module, use a minus sign before the operation filename.

Once the library file is linked to your program file, only the library procedures actually used by your program are placed in the execution file. Don't forget to use the label EXTRN when specifying library calls from your program module.

Macros

A **macro** is a group of instructions that perform one task just as a procedure performs one task. The difference is that a procedure is accessed via a CALL instruction, while a macro is inserted in the program by using the name of the macro, which represents a sequence of instructions. A macro is similar to creating a new instruction or directive for the assembler. Macro sequences execute faster than procedures because there is no CALL and RET instruction to execute. The instructions of the macro sequence are placed in the program at the point where they are invoked. When software is developed using macro sequences, it flows from top-down.

Macro sequences are ideal for systems that contain cache memory or are required to execute software with maximum speed and efficiency. Programs that use macro sequences in place of procedures are often called *linear programs* because they flow from the top to the bottom without any calls. If a cache memory is used in the computer system, it is often desirable to develop programs that use macro sequences because the entire program can often be executed from the cache. This significantly increases its execution speed. If a procedure is called, the cache must often be reloaded, requiring additional time. If a macro is invoked, no additional time is needed because no jump to the procedure occurs. Cache memory looks ahead, in a sense, and often loads instructions from the memory system before they are executed by the microprocessor. It does this by loading the cache with the next sequential section of four 32-bit doublewords. In a macro sequence, the next four 32-bit doublewords will often contain the entire macro if it is short. If a procedure is called, it is usually not in the next four 32-bit doublewords of memory, which requires an access to another section of the memory for caching. This requires additional time on the part of the memory system, and thus is less efficient.

The MACRO and ENDM directives are used to delineate a macro sequence. The first statement of a macro is the MACRO statement that contains the name of the macro and any parameters associated with it. An example is MOVE MACRO A,B that defines the macro as MOVE. This new macro name is used like an opcode or directive that contains two parameters: A and B. These parameters are often called *replaceable* parameters because as the macro is expanded for storage in the program, they are replaced. Macros can contain up to 32 replaceable parameters. The last statement of a macro is the ENDM instruction on a line by itself without a label.

Example 7–10 shows how a macro is created and used in a program. This is not a complete program and will not execute as it is printed; it is included to show how macro sequences are created and generate code when invoked. This macro moves the word-sized contents of memory location B into word-sized memory location A. After the macro is defined with the first 6 statements, it is used twice. The macro is expanded in this example to show how it assembles and generates the move. Notice how parameters A and B are replaced by VAR1 and VAR2 the first time the macro is used, and with VAR3 and VAR4 the second time. Any hexadecimal machine language statement followed by a 1 is a macro expansion statement. If a macro uses a second macro (nesting), the statement is followed by a 2, and so forth. The expansion statements were not typed in the source program. The expansion statements are generated each time a macro sequence is invoked. Notice that the comment in the macro is preceded with a ;; instead of ; as is customary. To cause the macro expansion statements to list, place the .LISTALL statement at the start of the program.

EXAMPLE 7–10

```
                        MOVE      MACRO A,B              ;;moves word from B to A

                                  PUSH  AX
                                  MOV   AX,B
                                  MOV   A,AX
                                  POP   AX

                                  ENDM

                                  MOVE  VAR1,VAR2         ;use macro MOVE

0000 50                 1         PUSH  AX
0001 A1 0002 R          1         MOV   AX,VAR2
```

```
0004 A3 0000 R    1          MOV    VAR1,AX
0007 58           1          POP    AX

                             MOVE   VAR3,VAR4        ;use macro MOVE

0008 50           1          PUSH   AX
0009 A1 0006 R    1          MOV    AX,VAR4
000C A3 0004 R    1          MOV    VAR3,AX
000F 58           1          POP    AX
```

Local Variable in a Macro.   Sometimes macros must contain local variables. A local variable is one that appears in the macro, but is not available outside the macro. To define a local variable, use the LOCAL directive. Example 7–11 shows how a local variable, used as a jump address, appears in a macro definition. If this jump address is not defined as local, the assembler will flag it with a duplicate label error on the second and subsequent attempts to use the macro in a program.

Example 7–11 reads two characters from the keyboard and stores them in the byte-sized memory locations indicated as a parameter with the macro. Notice how the local label READ1 is treated in the expanded macros. Local variables always appear as ??nnnn, where nnnn is a number identifying it as a unique label.

The LOCAL directive must always immediately follow the MACRO directive without any intervening blank lines or comments. If a comment or blank line appears between the MACRO and LOCAL statements, the assembler indicates an error and will not accept the variable as local.

EXAMPLE 7–11

```
                      ;A program that uses a macro to read 2 keys from the
                      ;keyboard and echo them to the display.  The first key
                      ;code is stored at VAR5 and the second at VAR6.
                      ;
                            .MODEL SMALL          ;select SMALL model
0000                        .DATA                 ;start of DATA segment
0000 00        VAR5    DB    0                     ;define VAR5
0001 00        VAR6    DB    0                     ;define VAR6
0000                        .CODE                 ;start of CODE segment

               READ    MACRO A               ;;read keyboard with echo
                       LOCAL  READ1          ;;define READ1 as local

                       PUSH DX
               READ1:
                       MOV  AH,6             ;;read key
                       MOV  DL,0FFH
                       INT  21H
                       JE   READ1
                       MOV  A,AL             ;;save key code
                       MOV  DL,AL            ;;echo key
                       INT  21H
                       POP  DX

                       ENDM
                       .STARTUP              ;start of program

                       READ VAR5             ;read key to VAR5

0000 52        1            PUSH DX
0001           1 ??0000:
0001 B4 06     1            MOV  AH,6
0003 B2 FF     1            MOV  DL,0FFH
0005 CD 21     1            INT  21H
0007 74 F8     1            JE   ??0000
0009 A2 0000 R 1            MOV  VAR5,AL
000C 8A D0     1            MOV  DL,AL
000E CD 21     1            INT  21H
```

```
0010 5A              1           POP   DX

                                 READ  VAR6                 ;read key to VAR6

0011 52              1           PUSH  DX
0012                 1 ??0001:
0012 B4 06           1           MOV   AH,6
0014 B2 FF           1           MOV   DL,0FFH
0016 CD 21           1           INT   21H
0018 74 F8           1           JE    ??0001
001A A2 0001 R       1           MOV   VAR6,AL
001D 8A D0           1           MOV   DL,AL
001F CD 21           1           INT   21H
0021 5A              1           POP   DX
                                 .EXIT                      ;exit to DOS
                                 END                        ;end of file
```

Placing Macro Definitions in Their Own Module.   Macro definitions are placed in the program file as shown, or are placed in their own macro module. A file that contains only macros that are to be included with other program files is called an INCLUDE file. The INCLUDE directive indicates that a program file will include a module that contains external macro definitions. Although this is not a library file, for all practical purposes it functions as a library of macro sequences. A macro include file is an unassembled ASCII file generated by an editor or word processor. A good collection of macro statements is invaluable when programming in assembly language because it eliminates the task of writing the same sequences of instructions many times.

When macro sequences are placed in a file (often with the extension INC or MAC) they do not contain PUBLIC statements. If a file called MACRO.MAC contains macro sequences, the INCLUDE statement is placed in the program file as INCLUDE C:\ASSM\MACRO.MAC. Notice that the macro file is on drive C, subdirectory ASSM in this example. The INCLUDE statement includes these macros just as if you had typed them into the file. No EXTRN statement is needed to access the macro statements that have been included.

Using Condition Statements in Macros.   There are some useful conditional statements that apply to macro sequences. The most common are the IF and ENDIF statements. The IF statement allows decisions that control which instructions are generated by the macro when it is invoked. The ENDIF statement indicates the end of an IF sequence. Table 7–1 lists the various forms of the IF statement as well as other conditional macros used with the IF statement such as ENDIF.

Suppose that a macro is needed to display an ASCII-coded character or a carriage return/line feed sequence. This can be accomplished by using DOS INT 21H function 02H, which displays

TABLE 7–1   The conditional IF and related statements

Assembly Language	Operation
IF expression	Assembles if expression is true
IFE expression	Assembles if expression is false
IFDEF name	Assembles if name is defined
IFNDEF name	Assembles if name is not defined
IFB argument	Assembles if argument is blank
IFNB	Assembles if argument is not blank
IFIDN arg1,arg2	Assembles if arg1 equals arg2
IFDIF arg1,arg2	Assembles if arg1 does not equal arg2
ENDIF	Closes an IF sequence
ELSE	Allows the opposite choice for IF
ELSEIF	Allows an additional condition

the ASCII contents of DL and one of the IF statements from Table 7–1. Example 7–12 is a macro called DISP that displays the parameter following DISP. If the parameter is blank, the macro displays a carriage return/line feed combination. The IFB statement is used to test for a blank parameter in this example. The ELSE statement is also used, in case the parameter is not blank, and then can be displayed. If the parameter (PARA) is blank, the instruction is required to display a carriage return and line feed assemble. If PARA is not blank, the ASCII-coded value of PARA is displayed. PARA can be a numeric value or any 8-bit register or memory location. Notice that arguments must be surrounded with < >.

EXAMPLE 7–12

```
DISP    MACRO    PARA         ;;display PARA

        MOV  AH,2             ;;select function 2
        IFB  <PARA>           ;;test for blank PARA
            MOV  DL,0DH       ;;display carriage return
            INT  21H
            MOV  DL,0AH       ;;display line feed
        ELSE                  ;;if PARA not blank
            MOV  DL,PARA
        ENDIF
            INT  21H
        ENDM
```

In addition to the IFB and IFNB statements, other IF statements allow comparison of data in parameters. The relational operators used in IF statements are listed in Table 7–2. These operators are used only with the IF, IFE, ENDIF, and ELSEIF statements.

Example 7–13 uses the IF statement to test the value of a parameter (PARA). In this example, if PARA is a zero, a key is read from the keyboard using DOS INT 21H function 1. If PARA is blank, a carriage return/line feed combination is displayed using DOS INT 21H function 2. If PARA is an ASCII-coded character, the character is displayed. This example illustrates the power of conditional assembly within a macro sequence.

EXAMPLE 7–13

```
IO    MACRO  PARA

      IFB     <PARA>        ;;if PARA is blank
         MOV  AH,2
         MOV  DL,0DH
         INT  21H
         MOV  DL,0AH
      ELSEIF  PARA EQ 0     ;;if PARA = 0
         MOV  AH,1
      ELSE                  ;;else display ASCII
         MOV  AH,2
         MOV  DL,PARA
      ENDIF
         INT  21H
         ENDM
```

The Modular Programming Approach

The modular programming approach often involves a team of people, each assigned to a different programming task. The team manager assigns various portions of the program to different team members by developing the system flow-chart or shell and then dividing it into modules.

A team member might be assigned the task of developing a macro definition file. This file might contain macro definitions that handle the I/O operations for the system. Another team

TABLE 7–2 Relational operators used with IF statements

Operator	Meaning
EQ	Equal to
NE	Not equal to
LT	Less than
LE	Less than or equal
GT	Greater than
GE	Greater than or equal

member might be assigned the task of developing the procedures used for the system. In most cases the procedures are organized as a library file that is linked to the program modules. Finally, several program files or modules might be used for the final system, each developed by different team members.

This approach requires good communication among team members and good documentation so that modules interface correctly.

7–2 USING THE KEYBOARD AND VIDEO DISPLAY

Today most programs make use of the keyboard and video display on a personal computer. This section explains how to use the keyboard and video display connected to the personal computer running under either MS-DOS or PC-DOS. We have seen some examples using the keyboard and video display earlier in this chapter and in prior chapters.

Reading the Keyboard with DOS Functions

The keyboard on the personal computer is often read via a DOS function call. A complete listing of the DOS function calls appears in Appendix A. In this section we use INT 21H with various DOS function calls to read the keyboard. Data read from the keyboard is either in ASCII-coded or extended ASCII-coded form. The exact form depends on which keys are typed on the keyboard.

The ASCII-coded data appears as outlined in Tables 1–7 and 1–9 in Section 1–4. Notice that these codes correspond to most of the keys on the keyboard. The extended ASCII codes listed in Table 1–9 apply to the printer or video screen. Also available through the keyboard is a different set of extended ASCII-coded keyboard data. Table 7–3 lists most of the extended ASCII-codes obtained with various keys and key combinations. It also lists the scan codes read directly from the key before conversion to ASCII code. These codes are different from the extended ASCII printer/video display codes in Chapter 1. Notice that most keys on the keyboard have alternate key codes. The function keys have four sets of codes selected by the function keys, the shift function keys, alternate function keys, and the control function keys.

There are three ways to read the keyboard with DOS: The first method reads a key and echoes (displays) the key on the video screen. A second way just tests to see if a key is pressed and if it is pressed, it reads the key; otherwise, it returns without any key. The third way allows an entire character line to be read from the keyboard.

Reading a Key with an Echo. Example 7–14 reads a key from the keyboard and echoes (sends) it back out to the video display. Although this is the easiest way to read a key, it is also the most limited because it always echoes (displays) the character to the screen even if it is an unwanted

TABLE 7–3   The keyboard scanning and extended ASCII codes as returned from the keyboard

Scan Key	Extended Code	ASCII Code with ... Nothing	Shift	Control	Alternate
ESC	01				01
1	02				78
2	03			03	79
3	04				7A
4	05				7B
5	06				7C
6	07				7D
7	08				7E
8	09				7F
9	0A				80
0	0B				81
−	0C				82
+	0D				83
BKSP	0E				0E
TAB	0F	0F	94		A5
Q	10				10
W	11				11
E	12				12
R	13				13
T	14				14
Y	15				15
U	16				16
I	17				17
O	18				18
P	19				19
[	1A				1A
]	1B				1B
Large Enter	1C				1C
Small Enter	1C				A6
Left Control	1D				
Rght Control	1D				
A	1E				1E
S	1F				1F
D	20				20
F	21				21
G	22				22
H	23				23
J	24				24
K	25				25
L	26				26
;	27				27
'	28				28
`	29				29
L Shift	2A				
\	2B				
Z	2C				2C
X	2D				2D
C	2E				2E
V	2F				2F
B	30				30

TABLE 7–3   *(continued)*

Scan Key	Extended Code	ASCII Code with ...			
		Nothing	Shift	Control	Alternate
N	31				31
M	32				32
,	33				33
.	34				34
/	35				35
Gray /	35			95	A4
R Shift	36				
PRTSC	E0 2A EO				
	37				
L ALT	38				
R ALT	39				
Space	39				
Caps	3A				
F1	3B	3B	54	5E	68
F2	3C	3C	55	5F	69
F3	3D	3D	56	60	6A
F4	3E	3E	57	61	6B
F5	3F	3F	58	62	6C
F6	40	40	59	63	6D
F7	41	41	5A	64	6E
F8	42	42	5B	65	6F
F9	43	43	5C	66	70
F10	44	44	5D	67	71
F11	57	85	87	89	8B
F12	58	86	88	8A	8C
NUM	45				
Scroll	46				
Home	E0 47	47	47	77	97
Up	48	48	48	8D	98
Page Up	E0 49	49	49	84	99
Gray −	4A				
Left	4B	4B	4B	73	9B
Center	4C				
Right	4D	4D	4D	74	9D
Gray +	4E				
End	E0 4F	4F	4F	75	9F
Down	E0 50	50	50	91	A0
Page Down	E0 51	51	51	76	A1
Insert	E0 52	52	52	92	A2
Delete	E0 53	53	53	93	A3
Pause	E0 10 45				

character. The DOS function number 01H also responds to the control C key and exits the program to DOS if it is typed.

EXAMPLE 7–14

```
0000                 KEY    PROC    FAR

0000 B4 01                  MOV    AH,1        ;function 01H
0002 CD 21                  INT    21H         ;read key
```

```
0004 0A C0                  OR     AL,AL              ;test for 00H, clear carry
0006 75 03                  JNZ    KEY1
0008 CD 21                  INT    21H                ;get extended
000A F9                     STC                       ;indicate extended
000B            KEY1:
000B CB                     RET

000C            KEY     ENDP
```

To read and echo a character, the AH register is loaded with DOS function number 01H. This is followed by the INT 21H instruction. (All DOS functions use AH to hold the function number before the INT 21H DOS function call.) Upon return from the INT 21H, the AL register contains the ASCII character typed and the screen also shows the typed character. The return from this DOS function call does not occur until a key is typed. If AL = 0 after the return, the INT 21H instruction must be executed again to obtain the extended ASCII-coded character. This procedure in Example 7–14 returns with carry set (1) to indicate an extended-ASCII character and carry cleared (0) to indicate a normal ASCII character.

Reading a Key Without an Echo.   The best single-character key-reading function is function number 06H. This function reads a key without an echo to the screen. It also returns with extended-ASCII characters and does not respond to the control C key. This function uses AH for the function number (06H) and DL = 0FFH to indicate that the function call (INT 21H) will read the keyboard without an echo.

Example 7–15 is a procedure that uses function number 06H to read the keyboard. This performs the same as Example 7–14, except that no character is echoed to the video display.

EXAMPLE 7–15

```
0000           KEYS    PROC   FAR

0000 B4 06                  MOV    AH,6               ;function 06H
0002 B2 FF                  MOV    DL,0FFH
0004 CD 21                  INT    21H                ;read key
0006 74 F8                  JE     KEYS               ;if no key
0008 0A C0                  OR     AL,AL              ;test for 00H, clear carry
000A 75 03                  JNE    KEYS1
000C CD 21                  INT    21H                ;get extended
000E F9                     STC                       ;indicate extended
000F           KEYS1:
000F CB                     RET

0010           KEYS    ENDP
```

If you examine this procedure, there is one other difference from the procedure in Example 7–14. Function call number 06H returns from the INT 21H even if no key is typed, while function call 01H waits for a key to be typed. This is an important difference; this feature allows software to perform other tasks between checking the keyboard for a character.

Read an Entire Line with an Echo.   Sometimes it is advantageous to read an entire line of data with one function call. Function call number 0AH reads an entire line of information—up to 255 characters—from the keyboard. It continues to acquire keyboard data until the enter key (0DH) is typed. This function requires that AH = 0AH and DS:DX addresses the keyboard buffer (a memory area where the ASCII data are stored). The first byte of the buffer area contains the maximum number of keyboard characters read by this function. If the number typed exceeds this maximum number, the DOS function returns just as if the enter key were typed. The second byte of the buffer contains the count of the actual number of characters typed and the remaining locations in the buffer contain the ASCII keyboard data.

Example 7–16 uses function 0AH to read two lines of information into two memory buffers (BUF1 and BUF2). Before the call to the DOS function through procedure LINE, the first byte of the buffer is loaded with a 255 so up to 255 characters can be typed. If you assemble and execute this program, the first line is accepted and so is the second. The only problem is that the second line appears on top of the first line. In the next section, we explain how to output characters to the video display to solve this problem and to display memory data on the video screen.

EXAMPLE 7–16

```
                        ;A program that reads two lines of data from the keyboard
                        ;using DOS INT 21H function number 0AH.
                        ;***uses***
                        ;LINE procedure to read a line.
                        ;
                                .MODEL SMALL            ;select SMALL model
0000                            .DATA                   ;start DATA segment
0000 0101 [             BUF1    DB    257 DUP (?)        ;define BUF1
        00
           ]
0101 0101 [             BUF2    DB    257 DUP (?)        ;define BUF2
        00
           ]
0000                            .CODE                   ;start CODE segment
                                .STARTUP                ;start program
0017 C6 06 0000 R FF            MOV  BUF1,255           ;character count of 255
001C BA 0000 R                  MOV  DX,OFFSET BUF1     ;address BUF1
001F E8 000F                    CALL LINE               ;read a line

0022 C6 06 0101 R FF            MOV  BUF2,255           ;character count of 255
0027 BA 0101 R                  MOV  DX,OFFSET BUF2     ;address BUF2
002A E8 0004                    CALL LINE               ;read a line
                                .EXIT                   ;exit to DOS
                        ;
                        ;The LINE procedure uses DOS INT 21H function 0AH to read
                        ;and echo an entire line from the keyboard.
                        ;***parameters***
                        ;DX must contain the data segment offset address of the
                        ;buffer. The first location in the buffer contains the
                        ;number of characters to be read for the line.
                        ;Upon return the second location in the buffer contains
                        ;the line length.
                        ;
0031                    LINE    PROC NEAR

0031 B4 0A                      MOV  AH,0AH             ;select function 0AH
0033 CD 21                      INT  21H                ;access DOS
0035 C3                         RET                     ;return from procedure

0036                    LINE    ENDP
                                END                     ;end of file
```

Reading a Key with BIOS INT 16H. The BIOS can also be used to read a key from the keyboard through BIOS INT 16H. If AH = 00H (83-key keyboard) or 10H (101-key keyboard), the INT 16H instruction waits for a key to be typed and returns the ASCII character in AL and scan code in AH. (The scan codes are given in Table 7–3.) If the status of the keyboard is checked, function 01H (83 key) or 11H (101 key) is used to test the keyboard. If the zero flag is set (1), no key was typed. If the zero flag is cleared (0), a key was typed that can be read with function 00H (10H). For additional detail on INT 16H, refer to Appendix A. The two procedures in Example 7–17 illustrate the use of INT 16H for reading the keyboard. Note that no echo is provided by INT 16H.

EXAMPLE 7–17

```
0000                    KEYS    PROC NEAR

0000 B4 10                      MOV  AH,10H             ;read 101-key keyboard
0002 CD 16                      INT  16H                ;wait for key, no echo
0004 C3                         RET
```

```
0005                    KEYS    ENDP

0005                    STAT    PROC NEAR

0005 B4 11                      MOV   AH,11H          ;check 101-key keyboard
0007 CD 16                      INT   16H             ;get status
0009 C3                         RET

000A                    STAT    ENDP
```

Writing to the Video Display with DOS Functions

With almost any program, data must be displayed on the video display. Video data is displayed in a number of different ways with DOS function calls. Function 02H or 06H displays one character at a time, and function 09H displays an entire string of characters. Because functions 02H and 06H are identical, function 06H is often used because it is also used to read a key.

Displaying One ASCII Character.   We discuss DOS functions 02H and 06H together because they are identical for displaying ASCII data. Example 7–18 shows how this function displays a carriage return or enter (0DH) and a line feed (0AH). Here a macro, called DISP (display), displays the carriage return and line feed. The combination of a carriage return and a line feed moves the cursor to the next line at the left margin of the video screen. This two-step process corrects the problem that occurred between the lines typed through the keyboard in Example 7–16.

EXAMPLE 7–18

```
                        ;A program that displays a carriage return and a line feed
                        ;using the DISP macro.
                        ;
                                .MODEL TINY             ;select TINY model
                                .CODE                   ;start CODE segment
                        DISP    MACRO A                 ;;display A macro

                                MOV  AH,06H             ;;DOS function 06H
                                MOV  DL,A               ;;place parameter A in DL
                                INT  21H                ;;display parameter A

                                ENDM

                                .STARTUP                ;start program

                                DISP  0DH               ;display carriage return

0100 B4 06           1          MOV  AH,06H
0102 B2 0D           1          MOV  DL,0DH
0104 CD 21           1          INT  21H

                                DISP  0AH               ;display line feed

0106 B4 06           1          MOV  AH,06H
0108 B2 0A           1          MOV  DL,0AH
010A CD 21           1          INT  21H
                                .EXIT                   ;exit to DOS
                                END                     ;end of file
```

Display a Character String.   A **character string** is a series of ASCII-coded characters that end with a $ (24H) when used with DOS function call number 09H. Other endings are possible, such as a null (00H) for a null character string. Example 7–19 displays a message at the current cursor position on the video display. Function call number 09H requires that DS:DX address the character string before executing the INT 21H.

EXAMPLE 7–19

```
                                .MODEL  SMALL            ;select SMALL model
0000                            .DATA                    ;start DATA segment
0000  0D 0A 0A 54     MES       DB    13,10,10,'This is a test line.$'
      68 69 73 20
      69 73 20 61
      20 74 65 73
      74 20 6C 69
      6E 65 2E 24
0000                            .CODE                    ;start CODE segment
                                .STARTUP                 ;start program

0017  B4 09                     MOV   AH,9               ;select function 09H
0019  BA 0000 R                 MOV   DX,OFFSET MES      ;address character string
001C  CD 21                     INT   21H                ;access DOS

                                .EXIT                    ;exit to DOS
                                END                      ;end of file
```

Example 7–19 can be entered into the assembler, linked, and executed to produce "This is a test line." on the video display. As always, enter only the source code and not the hexadecimal addresses or data.

Consolidated Read Key and Echo Macros.   Example 7–20 illustrates two macro sequences that can be added to an include file or to any program. The READ macro reads a keyboard character and returns with either the ASCII or extended ASCII character in AL. If carry is set, AL contains the extended ASCII code; if carry is cleared, it contains the standard ASCII code. The ECHO macro displays ASCII coded characters located next to the word *echo* as a parameter at the current cursor position. (The disk included with this book contains many useful macros in a macro include file that can be used when developing software using MASM.)

EXAMPLE 7–20

```
        ;
        ;A macro that reads a key from the keyboard without echo.
        ;If carry = 0, AL contains the standard ASCII key code.
        ;If carry = 1, AL contains the extended ASCII key code.
        ;
READ        MACRO                   ;;read key macro
            LOCAL READ1,READ2
            PUSH DX                 ;;save DX
            MOV  AH,6               ;;select function 06H
            MOV  DL,0FFH            ;;select read key
READ1:
            INT  21H                ;;access DOS
            JE   READ1              ;;if no key typed
            OR   AL,AL              ;;test for extended key
            JNZ  READ2              ;;if standard ASCII key
            INT  21H                ;;access DOS
            STC                     ;;if extended key code
READ2:
            POP  DX                 ;;restore DX
            ENDM
        ;
        ;A macro that displays an ASCII character from parameter.
        ;CHAR
        ;
ECHO        MACRO  CHAR             ;;display CHAR macro
            PUSH DX                 ;;save DX
            MOV  AH,6               ;;select function 06H
            MOV  DL,CHAR            ;;character to DL
            INT  21H                ;;access DOS
            ENDM
```

TABLE 7–4   BIOS function INT 10H, sub-functions 02H and 03H

AH	Description	Parameters
02H	Sets cursor position	DH = row DL = column BH = page number
03H	Reads cursor position	DH = row DL = column BH = page number

Using the Video BIOS Function Calls

In addition to the DOS function call INT 21H, the video BIOS (basic I/O system) uses function calls to INT 10H. The DOS function calls allow a key to be read and a character to be displayed with ease, but the cursor is difficult to position at the desired screen location. The video BIOS function calls allow more complete control over the video display than do the DOS function calls. The video BIOS function calls also require less time to execute than the DOS function calls.

Cursor Position.   Before any information is placed on the video screen, you should know the position of the cursor. This allows the screen to be cleared and the displayed information to be placed at any location on the video screen. The BIOS INT 10H function number 03H allows the cursor position to be read from the video interface. The BIOS INT 10H function number 02H allows the cursor to be placed at any screen position. Table 7–4 shows the contents of various registers for both video BIOS INT 10H function calls, 02H and 03H.

The page number, in register BH, should be 0 before setting the cursor position. Most software does not access the other pages (1–7) for the video display. If the display adapter is a VGA display, always use page 0. The page number is often ignored after a cursor read. The 0 page is available in the CGA (color graphics adapter), EGA (enhanced graphics adapter), and VGA (variable graphics array) text modes. The other pages are available in some VGA and EGA modes and all CGA modes.

The cursor position assumes that the left-hand page column is column 0, progressing across a line to column 79. The row number corresponds to the character line number on the screen. Row 0 is the uppermost line, while row 24 is the last line on the screen. This assumes that the text mode selected for the video adapter is 80 characters per line by 25 lines (80×25). Other text modes are available such as 40×25 and 96×43.

Example 7–21 shows how the INT 10H BIOS function call is used to clear the video screen. This is just one method of clearing the screen. Notice that the first function call positions the cursor at row 0 and column 0, which is called the home position. Next the DOS function call writes 2000 (80 characters per line $\times$ 25 character lines) blank spaces (20H) on the video display; then the cursor is again moved to the home position.

EXAMPLE 7–21

```
              ;A program that clears the screen and homes the
              ;cursor to the upper-left corner of the screen.
              ;
                      .MODEL TINY          ;select TINY model
0000                  .CODE                ;start CODE segment
              HOME    MACRO                ;;home cursor macro
                      MOV   AH,2           ;;function 02H
                      MOV   BH,0           ;;page 0
                      MOV   DX,0           ;;row 0, line 0
                      INT   10H            ;;home cursor
                      ENDM
```

```
                                    .STARTUP                 ;start program
                                    HOME                     ;home cursor
0100 B4 02          1               MOV   AH,2
0102 B7 00          1               MOV   BH,0
0104 BA 0000        1               MOV   DX,0
0107 CD 10          1               INT   10H
0109 B9 07D0                        MOV   CX,25*80           ;load character count
010C B4 06                          MOV   AH,6               ;select function 06H
010E B2 20                          MOV   DL,' '             ;select a space
0110                     MAIN1:
0110 CD 21                          INT   21H                ;display a space
0112 E2 FC                          LOOP  MAIN1              ;repeat 2000 times
                                    HOME                     ;home cursor
0114 B4 02          1               MOV   AH,2
0116 B7 00          1               MOV   BH,0
0118 BA 0000        1               MOV   DX,0
011B CD 10          1               INT   10H
                                    .EXIT                    ;exit to DOS
                                    END                      ;end of file
```

When Example 7–21 is assembled, linked, and executed, a problem surfaces: This program is far too slow to be useful in most cases. To correct this situation, another video BIOS function call is used. The scroll function (06H) clears the screen at a much higher speed.

Function 06H is used with a 00H in AL to blank the entire screen. This allows Example 7–21 to be rewritten so that the screen clears at a much higher speed. See Example 7–22 for a better clear and home cursor program. Here function call number 08H reads the character attributes for blanking the screen. Next, they are positioned in the correct registers and DX is loaded with the screen size, 4FH (79) and 19H (25). If this program is assembled, linked, executed, and compared with Example 7–21, there is a big difference in the speed at which the screen is cleared. Please refer to Appendix A for other useful video BIOS INT 10H function calls. Also in Appendix A is a complete list of all the INT functions available in most computers.

EXAMPLE 7–22

```
                   ;A program that clears the screen and homes the cursor.
                   ;
                                    .MODEL TINY              ;select TINY model
0000                                .CODE                    ;start code segment
                                    HOME MACRO               ;;home cursor
                                    MOV   AH,2
                                    MOV   BH,0
                                    MOV   DX,0
                                    INT   10H
                                    ENDM

                                    .STARTUP                 ;start program
0100 B7 00                          MOV   BH,0
0102 B4 08                          MOV   AH,8
0104 CD 10                          INT   10H                ;read video attribute

0106 8A DF                          MOV   BL,BH              ;load page number
0108 8A FC                          MOV   BH,AH
010A B9 0000                        MOV   CX,0               ;load attributes
010D BA 194F                        MOV   DX,194FH           ;line 25, column 79
0110 B8 0600                        MOV   AX,600H            ;select scroll function
0113 CD 10                          INT   10H                ;scroll screen
                                    HOME                     ;home cursor
0115 B4 02          1               MOV   AH,2
0117 B7 00          1               MOV   BH,0
0119 BA 0000        1               MOV   DX,0
011C CD 10          1               INT   10H

                                    .EXIT                    ;exit to DOS
                                    END                      ;end program
```

Display Macro

One of the more useful macro sequences is illustrated in Example 7–23. Although it is simple and has been presented before, it saves much typing when creating programming that must display many individual characters. What makes this macro so useful is that a register can be specified as the argument, an ASCII character in quotes, or the numeric value for an ASCII character.

EXAMPLE 7–23

```
                        ;A program that displays AB followed by a carriage return
                        ;and line feed combination using the DISP macro.
                        ;
                                .MODEL TINY             ;select TINY model
                                .CODE                   ;start CODE segment
                        DISP    MACRO VAR               ;;display VAR macro
                                MOV   DL,VAR
                                MOV   AH,6
                                INT   21H
                                ENDM
                                .STARTUP                ;start program
                                DISP  'A'               ;display 'A'
0100 B2 41          1           MOV   DL,'A'
0102 B4 06          1           MOV   AH,6
0104 CD 21          1           INT   21H

0106 B0 42                      MOV   AL,'B'            ;load AL with 'B'
                                DISP  AL                ;display 'B'
0008 8A D0          1           MOV   DL,AL
000A B4 06          1           MOV   AH,6
000C CD 21          1           INT   21H

                                DISP  13                ;display carriage return
000E B2 0D          1           MOV   DL,13
0010 B4 06          1           MOV   AH,6
0012 CD 21          1           INT   21H

                                DISP  10                ;display line feed
0014 B2 0A          1           MOV   DL,10
0016 B4 06          1           MOV   AH,6
0018 CD 21          1           INT   21H

                                .EXIT                   ;exit to DOS
                                END                     ;end of file
```

7–3   DATA CONVERSIONS

In computer systems and embedded applications, data is seldom in the correct form. One main task of the system is to convert data from one form to another. This section describes conversions between binary and ASCII. Binary data is removed from a register or memory and converted to ASCII for the video display. In many cases, ASCII data is converted to binary as it is typed on the keyboard. We also explain conversions between ASCII and hexadecimal data.

Converting from Binary to ASCII

Conversion from binary to ASCII is accomplished in two ways: (1) by the AAM instruction if the number is less than 100 or (2) by a series of decimal divisions (divide by 10). Both techniques are presented in this section.

The AAM instruction converts the value in AX into a two-digit unpacked BCD number in AX. If the number in AX is 0062H (98 decimal) before AAM executes, AX contains a 0908H after AAM executes. This is not ASCII code, but it is converted to ASCII code by adding a 3030H to AX. Example 7–24 is a procedure that processes the binary value in AL (0–99) and displays it on the video screen as decimal. This procedure blanks a leading zero, which occurs for the numbers 0–9, with an ASCII space code. The program displays the hexadecimal contents of AL, in this case a 4AH as a decimal 74 on the video display.

EXAMPLE 7–24

```
                        ;A program that uses the DISP procedure to display 74
                        ;decimals on the video display.
                        ;
                                .MODEL TINY             ;select TINY mode
0000                            .CODE                   ;start code segment
                                .STARTUP                ;start program
0100 B0 4A                      MOV  AL,4AH             ;load test data to AL
0102 E8 0004                    CALL DISP               ;display decimal test data
                                .EXIT                   ;exit to DOS
                        ;
                        ;DISP procedure displays AL (0 to 99) as a decimal number.
                        ;AX is destroyed by this procedure.
                        ;
0109                    DISP    PROC NEAR

0109 52                         PUSH DX                 ;save DX
010A B4 00                      MOV  AH,0               ;clear AH
010C D4 0A                      AAM                     ;convert to BCD
010E 80 C4 20                   ADD  AH,20H
0111 80 FC 20                   CMP  AH,20H             ;test for leading zero
0114 74 03                      JE   DISP1              ;if leading zero
0116 80 C4 10                   ADD  AH,10H             ;convert to ASCII
0119                    DISP1:
0119 8A D4                      MOV  DL,AH              ;display first digit
011B B4 06                      MOV  AH,6
011D 50                         PUSH AX
011E CD 21                      INT  21H
0120 58                         POP  AX
0121 8A D0                      MOV  DL,AL
0123 80 C2 30                   ADD  DL,30H             ;convert second digit
0126 CD 21                      INT  21H                ;display second digit
0128 5A                         POP  DX                 ;restore DX
0129 C3                         RET

012A                    DISP    ENDP
                                END                     ;end of file
```

The reason that AAM converts any number between 0 and 99 to a two-digit unpacked BCD number is because it divides AX by 10. The result is left in AX, so AH contains the quotient and AL the remainder. This same scheme of dividing by 10 can be expanded to convert any whole number from binary to an ASCII-coded character string that can be displayed on the video screen. Here is an algorithm for converting from binary to ASCII.

1. Divide by 10 and save the remainder on the stack as a significant BCD digit.
2. Repeat step 1 until the quotient is a 0.
3. Retrieve each remainder and add a 30H to convert to ASCII before displaying or printing.

Example 7–25 shows how the unsigned 16-bit contents of AX is converted to ASCII and displayed on the video screen as an unsigned integer. Here AX is divided by 10 and the remainder is saved on the stack after each division for later conversion to ASCII. The reason that data is stored on the stack is because the least-significant digit is returned first by the division. The stack is used to reverse the order of the data so it can be displayed correctly from the most-significant digit to the least. After all the digits have been converted by division, the result is displayed on the video screen by removing the remainders from the stack and converting them to ASCII code. This procedure also blanks any leading zeros that occur.

EXAMPLE 7–25

```
                        ;A program that uses DISPX to display AX in decimal.
                        ;
                                .MODEL TINY              ;select TINY model
0000                            .CODE                    ;start CODE segment
                                .STARTUP                 ;start program
0100 B8 04A3                    MOV   AX,4A3H            ;load AX with test data
0103 E8 0004                    CALL  DISPX             ;display AX in decimal
                                .EXIT                    ;exit to DOS
                        ;
                        ;The DISPX procedure displays AX in decimal.
                        ;AX is destroyed.
                        ;
010A                    DISPX     PROC NEAR

010A 52                           PUSH DX                ;save DX, CX, and BX
010B 51                           PUSH CX
010C 53                           PUSH BX
010D B9 0000                      MOV  CX,0              ;clear digit counter
0110 BB 000A                      MOV  BX,10             ;set for decimal
0113                    DISPX1:
0113 BA 0000                      MOV  DX,0              ;clear DX
0116 F7 F3                        DIV  BX                ;divide DX:AX by 10
0118 52                           PUSH DX                ;save remainder
0119 41                           INC  CX                ;count remainder
011A 0B C0                        OR   AX,AX             ;test for quotient of zero
011C 75 F5                        JNZ  DISPX1            ;if quotient is not zero
011E                    DISPX2:
011E 5A                           POP  DX                ;get remainder
011F B4 06                        MOV  AH,6              ;select function 06H
0121 80 C2 30                     ADD  DL,30H            ;convert to ASCII
0124 CD 21                        INT  21H               ;display digit
0126 E2 F6                        LOOP DISPX2            ;repeat for all digits

0128 5B                           POP  BX                ;restore BX, CX, and DX
0129 59                           POP  CX
012A 5A                           POP  DX
012B C3                           RET

012C                    DISPX     ENDP
                                  END                    ;end of file
```

It is interesting to note that the same steps to convert from binary to ASCII-coded decimal can also be used to convert from binary to any other number base. If, for example, step 1 is changed to divide by 7, the converted ASCII data will be in number base 7. Likewise, if step 1 is changed to divide by 3, the output will be in number base 7. To convert to a number base larger than 10, the technique must include changing the remainders that are larger than 9 into the letters A, B, and so forth; that is, add 37H in place of 30H. The technique required to display hexadecimal data appears in the following pages.

Converting from ASCII to Binary

Conversions from ASCII to binary usually start with keyboard data entry. If a single key is typed, the conversion is accomplished by subtracting a 30H from the number. If more than one key is typed, conversion from ASCII to binary still requires that 30H be subtracted, but there is one additional step. After subtracting 30H, the number is added to the result after the prior result is first multiplied by 10. Here is an algorithm to convert ASCII to binary.

1. Begin with a binary result of 0.
2. Subtract 30H from the character typed on the keyboard to convert it to BCD.

3. Multiply the binary result by 10 and add the new BCD digit.
4. Repeat steps 2 and 3 until the character typed is not an ASCII-coded number of 30H–39H.

Example 7–26 implements the ASCII-to-binary conversion algorithm. Here the binary number returns in the AX register as a 16-bit result. If a larger result is required, the procedure must be reworked for 32-bit arithmetic. Each time this procedure is called, it reads a number from the keyboard until any key other than 0 through 9 is typed. It then returns with the binary equivalent in the AX register.

EXAMPLE 7–26

```
                        ;A program that reads one decimal number from the
                        ;keyboard and stores the binary value at memory word TEMP.
                        ;
                                .MODEL SMALL            ;select TINY model
0000                            .DATA                   ;start DATA segment
0000 0000       TEMP    DW   ?                          ;define TEMP
0000                            .CODE                   ;start CODE segment
                                .STARTUP                ;start program
0017 E8 0007                    CALL READN              ;read a number
001A A3 0000 R                  MOV  TEMP,AX            ;save it in TEMP
                                .EXIT                   ;exit to DOS
                        ;
                        ;READN procedure reads a decimal number from the keyboard
                        ;and returns its binary value in AX.
                        ;
0021            READN   PROC NEAR

0021 53                         PUSH BX                 ;save BX and CX
0022 51                         PUSH CX
0023 B9 000A                    MOV  CX,10              ;load 10 for decimal
0026 BB 0000                    MOV  BX,0               ;clear result
0029            READN1:
0029 B4 01                      MOV  AH,1               ;read key with echo
002B CD 21                      INT  21H

002D 3C 30                      CMP  AL,'0'
002F 72 14                      JB   READN2             ;if below '0'
0031 3C 39                      CMP  AL,'9'
0033 77 10                      JA   READN2             ;if above '9'

0035 2C 30                      SUB  AL,'0'             ;convert to ASCII

0037 50                         PUSH AX                 ;save digit
0038 8B C3                      MOV  AX,BX              ;multiply result by 10
003A F7 E1                      MUL  CX
003C 8B D8                      MOV  BX,AX
003E 58                         POP  AX
003F B4 00                      MOV  AH,0
0041 03 D8                      ADD  BX,AX              ;add digit value to result
0043 EB E4                      JMP  READN1             ;repeat
0045            READN2:
0045 8B C3                      MOV  AX,BX              ;get binary result into AX
0047 59                         POP  CX                 ;restore CX and BX
0048 5B                         POP  BX
0049 C3                         RET

004A            READN   ENDP
                        END                             ;end of file
```

As with conversion to ASCII-coded decimal, this procedure also converts from number bases other than base 10. If step 3 is changed to multiply by 8, the result will be octal to binary conversion. The checks for numbers between 0 and 9 must also be changed to test for a number between 0 and 7 for octal.

Displaying and Reading Hexadecimal Data

Hexadecimal data is easier to read and display than decimal data. Hexadecimal data is not used at the applications level, but at the system level. System level data is often hexadecimal and must be displayed either in hexadecimal form or read from the keyboard as hexadecimal data.

Reading Hexadecimal Data.   Hexadecimal data appears as 0 to 9 and A to F. The ASCII codes obtained from the keyboard for hexadecimal data are 30H to 39H for the numbers 0 through 9 and 41H to 46H (A–F) and 61H to 66H (a–f) for the letters. To be useful, a procedure that reads hexadecimal data must be able to accept both lowercase (a–f) and uppercase (A–F) letters.

Example 7–27 shows two procedures: one converts the contents of the data in AL from an ASCII-coded character to a single hexadecimal digit, and the other reads a 4-digit hexadecimal number from the keyboard and returns with it in register AX. The second procedure can be modified to read any size hexadecimal number from the keyboard. Notice how the letters a through f are converted to uppercase (subtract 20H) in the first procedure. The program that uses these procedures stores the result in memory location TEMP.

EXAMPLE 7–27

```
                        ;A program that reads a 4-digit hexadecimal number from
                        ;the keyboard and stores the result in word-sized
                        ;location TEMP.
                        ;
                                .MODEL SMALL            ;select SMALL model
0000                            .DATA                   ;start DATA segment
0000 0000       TEMP    DW    ?                         ;define TEMP
0000                            .CODE                   ;start CODE segment
                                .STARTUP                ;start program
0017 E8 0007                    CALL READH              ;read hexadecimal number
001A A3 0000 R                  MOV  TEMP,AX            ;save it at TEMP
                                .EXIT                   ;exit to DOS
                        ;
                        ;The READH procedure that reads a 4-digit hexadecimal
                        ;number from the keyboard and returns it in AX.
                        ;This procedure next checks for errors and uses CONV.
                        ;
0021            READH   PROC NEAR

0021 51                         PUSH CX                 ;save BX and CX
0022 53                         PUSH BX
0023 B9 0004                    MOV  CX,4               ;load CX and SI with 4
0026 8B F1                      MOV  SI,CX
0028 BB 0000                    MOV  BX,0               ;clear result
002B            READH1:
002B B4 01                      MOV  AH,1               ;read a key with echo
002D CD 21                      INT  21H
002F E8 000A                    CALL CONV               ;convert to binary
0032 D3 E3                      SHL  BX,CL
0034 02 D8                      ADD  BL,AL              ;form result in BX
0036 4E                         DEC  SI
0037 75 F2                      JNZ  READH1             ;repeat 4 times
0039 8B C3                      MOV  AX,BX              ;move result to AX
003B 5B                         POP  BX                 ;restore BX and CX
003C 59                         POP  CX
003D C3                         RET
003E            READH   ENDP
                        ;
                        ;The CONV procedure converts AL into hexadecimal.
                        ;
003E            CONV    PROC NEAR

003E 3C 39                      CMP  AL,'9'
0040 76 08                      JBE  CONV2              ;if 0 through 9
0042 3C 61                      CMP  AL,'a'
0044 72 02                      JB   CONV1              ;if uppercase A through F
```

```
0046 2C 20                  SUB  AL,20H            ;convert to uppercase
0048             CONV1:
0048 2C 07                  SUB  AL,7
004A             CONV2:
004A 2C 30                  SUB  AL,30H
004C C3                     RET
004D             CONV     ENDP
                 END                               ;end of file
```

The CONV procedure first tests AL for a 0 through 9. If the value of AL is 0 through 9, the procedure subtracts 30H and returns. If the value in AL is A through F, the CMP AL, 'a' instruction returns a below result and a 07H followed by a 30H are subtracted from AL to produce 0AH through 0FH. If the letters a through f appear in AL, an additional 20H is subtracted.

Displaying Hexadecimal Data. To display hexadecimal data, a number is separated into 4-bit segments that are converted to hexadecimal digits. Conversion is accomplished by adding a 30H to the numbers 0–9 and a 37H to the letters A–F.

Example 7–28 displays the contents of the AX register on the video display. Here the number is rotated left so the leftmost digit is displayed first. Because AX contains a 4-digit hexadecimal number, the procedure displays 4 hexadecimal digits. This procedure can be modified to display wider hexadecimal numbers. The program that uses this procedure displays a test value loaded into AX before calling DISPH.

EXAMPLE 7–28

```
                 ;A program that displays the hexadecimal value loaded
                 ;into AX. This program uses DISPH to display a 4-digit
                 ;value.
                 ;
                          .MODEL TINY           ;select TINY model
0000                      .CODE                 ;start CODE segment
                          .STARTUP              ;start program
0100 B8 0ABC              MOV  AX,0ABCH         ;load AX with test data
0103 E8 0004              CALL DISPH            ;display AX in hexadecimal
                          .EXIT                 ;exit to DOS
                 ;
                 ;The DISPH procedure displays AX in hexadecimal.
                 ;
010A             DISPH    PROC NEAR

010A 53                   PUSH BX               ;save BX and CX
010B 51                   PUSH CX
010C B1 04                MOV  CL,4             ;load rotate count
010E B5 04                MOV  CH,4             ;load digit count
0110             DISPH1:
0110 D3 C0                ROL  AX,CL            ;position digit
0112 50                   PUSH AX
0113 24 0F                AND  AL,0FH           ;convert it to ASCII
0115 04 30                ADD  AL,30H
0117 3C 39                CMP  AL,'9'
0119 76 02                JBE  DISPH2
011B 04 07                ADD  AL,7
011D             DISPH2:
011D B4 02                MOV  AH,2             ;display hexadecimal digit
011F 8A D0                MOV  DL,AL
0121 CD 21                INT  21H
0123 58                   POP  AX
0124 FE CD                DEC  CH
0126 75 E8                JNZ  DISPH1           ;repeat for 4 digits
0128 59                   POP  CX               ;restore registers
0129 5B                   POP  BX
012A C3                   RET

012B             DISPH    ENDP
                 END                            ;end of file
```

Using Lookup Tables for Data Conversions

Lookup tables are often used to convert from one data form to another. A lookup table is formed in the memory as a list of data that is referenced by a procedure to perform conversions. In the case of many lookup tables, the XLAT instruction is used to look up data, provided that the table contains 8-bit wide data and its length is less than or equal to 256 bytes.

Converting from BCD to Seven-Segment Code.   One simple application that uses a lookup table is BCD to seven-segment code conversion. Example 7–29 illustrates a lookup table that contains the seven-segment codes for the numbers 0–9. These codes are used with the seven-segment display pictured in Figure 7–1. This seven-segment display uses active high (logic 1) inputs to light a segment. The code is arranged so that the *a* segment is in bit position 0 and the *g* segment is in bit position 6. Bit position seven is zero in this example, but can be used for displaying a decimal point.

EXAMPLE 7–29

```
0000                    SEG7    PROC    FAR

0000 53                         PUSH    BX
0001 BB 0008 R                  MOV     BX,OFFSET TABLE
0004 2E: D7                     XLAT    CS:TABLE        ;see text
0006 5B                         POP     BX
0007 CB                         RET

0008 3F                 TABLE   DB      3FH             ;0
0009 06                         DB      6               ;1
000A 5B                         DB      5BH             ;2
000B 4F                         DB      4FH             ;3
000C 66                         DB      66H             ;4
000D 6D                         DB      6DH             ;5
000E 7D                         DB      7DH             ;6
000F 07                         DB      7               ;7
0010 7F                         DB      7FH             ;8
0011 6F                         DB      6FH             ;9

0012                    SEG7    ENDP
```

The procedure (see Example 7–9) that performs the conversion contains only two instructions and assumes that AL contains the BCD digit to be converted to seven-segment code. One of the instructions addresses the lookup table by loading its address into BX; the other performs the conversion and returns the seven-segment code in AL.

Because the lookup table is located in the code segment, and the XLAT instruction accesses the data segment by default, the XLAT instruction includes a segment override. Notice that a dummy operand (TABLE) is added to the XLAT instruction so the (CS:) code segment override prefix can be added to the instruction. Normally XLAT does not contain an operand

FIGURE 7–1   The seven-segment display.

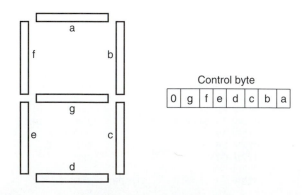

unless its default segment must be overridden. The LODS and MOVS instructions are also overridden in the same manner as XLAT by using a dummy operand.

Using a Lookup Table to Access ASCII Data. Some programming techniques require that numeric codes be converted to ASCII character strings. For example, suppose that you need to display the days of the week for a calendar program. Because the number of ASCII characters in each day is different, some type of lookup table must be used to reference the ASCII-coded days of the week.

Example 7–30 is a table that references ASCII-coded character strings located in the code segment. Each character string contains an ASCII-coded day of the week. The table references each day of the week. The procedure that accesses the day of the week uses the AL register and the numbers 0–6 to refer to Sunday through Saturday. If AL contains a 2 when this procedure is called, the word "Tuesday" is displayed on the video screen.

This procedure first accesses the table by loading the table address into the SI register. Next, the number in AL is converted to a 16-bit number and doubled because each table entry is 2 bytes in length. This index is now added to SI to address the correct entry in the lookup table. The address of the ASCII character string is then loaded into DX by the MOV DX,[SI] instruction.

EXAMPLE 7–30

```
                                ;A program that displays the current day of the
                                ;week by using the system clock/calendar.
                                ;
                                        .MODEL SMALL            ;select SMALL model
0000                                    .DATA                   ;start DATA segment
0000 000E R 0015 R      DTAB    DW      SUN,MON,TUE,WED,THU,FRI,SAT
     001C R 0024 R
     002E R 0037 R
     003E R
000E 53 75 6E 64 61 79  SUN     DB      'Sunday$'
     24
0015 4D 6F 6E 64 61 79  MON     DB      'Monday$'
     24
001C 54 75 65 73 64 61  TUE     DB      'Tuesday$'
     79 24
0024 57 65 64 6E 65 73  WED     DB      'Wednesday$'
     64 61 79 24
002E 54 68 75 72 73 64  THU     DB      'Thursday$'
     61 79 24
0037 46 72 69 64 61 79  FRI     DB      'Friday$'
     24
003E 53 61 74 75 72 64  SAT     DB      'Saturday$'
     61 79 24
0000                                    .CODE                   ;start CODE segment
                                        .STARTUP                ;start program
0017 B4 2A                              MOV  AH,2AH             ;get day of week
0019 CD 21                              INT  21H               ;access DOS
001B E8 0004                            CALL DAYS              ;display day of week
                                        .EXIT                   ;exit to DOS

0022                    DAYS    PROC NEAR

0022 52                                 PUSH DX                ;save DX and SI
0023 56                                 PUSH SI
0024 BE 0000 R                          MOV  SI,OFFSET DTAB    ;address lookup table
0027 B4 00                              MOV  AH,0              ;find day of week
0029 03 C0                              ADD  AX,AX
002B 03 F0                              ADD  SI,AX
002D 8B 14                              MOV  DX,[SI]           ;get address of day
002F B4 09                              MOV  AH,9              ;display day string
0031 CD 21                              INT  21H
0033 5E                                 POP  SI                ;restore registers
0034 5A                                 POP  DX
```

```
0035 C3                           RET

0036                      DAYS     ENDP
                                   END                    ;end of file
```

Before the INT 21H DOS function is called, the DS register is placed on the stack and loaded with the segment address of CS. This allows DOS function number 09H (display a string) to be used to display the day of the week. This procedure converts the numbers 0–6 to the days of the week.

The program that uses the procedure in Example 7–30 accesses the system clock by using a DOS function call to read the day of the week. Refer to Appendix A for the operation of DOS INT 21H function call number 2AH.

An Example Program Using Data Conversions

Let's look at an example that combines some of the data conversion DOS functions we've discussed. Suppose that you must display the time and date on the video screen. Example 7–31 displays the time as 10:34 A.M. and the date as Tuesday, July 4, 1997. The program is long because it calls a procedure that displays the time and a second procedure that displays the date.

EXAMPLE 7–31

```
                         ;A program that displays the time and date in the
                         ;following form:
                         ;10:34 A.M.,  Tuesday July 4, 1997
                         ;
                              .MODEL SMALL            ;select SMALL model
                              .NOLISTMACRO            ;don't expand macros
0000                          .DATA                   ;start CODE segment
0000 0026 R 002F R       DTAB     DW    SUN,MON,TUE,WED,THU,FRI,SAT
     0038 R 0042 R
     004E R 0059 R
     0062 R
000E 006D R 0076 R       MTAB     DW    JAN,FEB,MAR,APR,MAY,JUN
     0080 R 0087 R
     008E R 0093 R
001A 0099 R 009F R                DW    JUL,AUG,SEP,OCT,NOV,DCE
     00A7 R 00B2 R
     00BB R 00C5 R
0026 53 75 6E 64 61 79   SUN      DB    'Sunday, $'
     2C 20 24
002F 4D 6F 6E 64 61 79   MON      DB    'Monday, $'
     2C 20 24
0038 54 75 65 73 64 61   TUE      DB    'Tuesday, $'
     79 2C 20 24
0042 57 65 64 6E 65 73   WED      DB    'Wednesday, $'
     64 61 79 2C 20 24
004E 54 68 75 72 73 64   THU      DB    'Thursday, $'
     61 79 2C 20 24
0059 46 72 69 64 61 79   FRI      DB    'Friday, $'
     2C 20 24
0062 53 61 74 75 72 64   SAT      DB    'Saturday, $'
     61 79 2C 20 24
006D 4A 61 6E 75 61 72   JAN      DB    'January $'
     79 20 24
0076 46 65 62 72 75 61   FEB      DB    'February $'
     72 79 20 24
0080 4D 61 72 63 68 20   MAR      DB    'March $'
     24
0087 41 70 72 69 6C 20   APR      DB    'April $'
     24
008E 4D 61 79 20 24      MAY      DB    'May $'
0093 4A 75 6E 65 20 24   JUN      DB    'June $'
```

```
0099 4A 75 6C 79 20 24  JUL      DB     'July $'
009F 41 75 67 75 73 74  AUG      DB     'August $'
     20 24
00A7 53 65 70 74 65 6D  SEP      DB     'September $'
     62 65 72 20 24
00B2 4F 63 74 6F 62 65  OCT      DB     'October $'
     72 20 24
00BB 4E 6F 76 65 6D 62  NOV      DB     'November $'
     65 72 20 24
00C5 44 65 63 65 6D 62  DCE      DB     'December $'
     65 72 20 24
0000                              .CODE               ;start CODE segment
                        DISP     MACRO  CHAR
                                 PUSH AX              ;;save AX and DX
                                 PUSH DX
                                 MOV  DL,CHAR         ;;display character
                                 MOV  AH,2
                                 INT  21H
                                 POP  DX              ;;restore AX and DX
                                 POP  AX
                                 ENDM
                                 .STARTUP            ;start program
0017 E8 0007                     CALL TIMES           ;display time
001A E8 00A3                     CALL DATES           ;display date
                                 .EXIT                ;exit to DOS

0021                    TIMES    PROC NEAR

0021 B4 2C                       MOV  AH,2CH          ;get time from DOS
0023 CD 21                       INT  21H
0025 B7 41                       MOV  BH,'A'          ;set 'A' for AM
0027 80 FD 0C                    CMP  CH,12
002A 72 05                       JB   TIMES1          ;if below 12:00 noon
002C B7 50                       MOV  BH,'P'          ;set 'P' for PM
002E 80 ED 0C                    SUB  CH,12           ;adjust to 12 hours
0031                    TIMES1:
0031 0A ED                       OR   CH,CH           ;test for 0 hour
0033 75 02                       JNE  TIMES2          ;if not 0 hour
0035 B5 0C                       MOV  CH,12           ;change 0 to 12 hours
0037                    TIMES2:
0037 8A C5                       MOV  AL,CH
0039 B4 00                       MOV  AH,0
003B D4 0A                       AAM                  ;convert hours to BCD
003D 0A E4                       OR   AH,AH
003F 74 0D                       JZ   TIMES3          ;if no tens of hours
0041 80 C4 30                    ADD  AH,'0'          ;convert to ASCII
                                 DISP AH              ;display hours
004E                    TIMES3:
004E 04 30                       ADD  AL,'0'          ;convert to ASCII
                                 DISP AL              ;display units
                                 DISP ':'             ;display colon
0064 8A C1                       MOV  AL,CL           ;get minutes
0066 B4 00                       MOV  AH,0
0068 D4 0A                       AAM                  ;convert to BCD
006A 05 3030                     ADD  AX,3030H        ;convert to ASCII
006D 50                          PUSH AX
                                 DISP AH              ;display minutes
0078 58                          POP  AX
                                 DISP AL              ;display units
                                 DISP ' '             ;display space
                                 DISP BH              ;display 'A' or 'P'
                                 DISP '.'             ;display .
                                 DISP 'M'             ;display M
                                 DISP '.'             ;display .
                                 DISP ' '             ;display space
00BF C3                          RET

00C0                    TIMES    ENDP
```

```
00C0                              DATES    PROC NEAR

00C0 B4 2A                                 MOV  AH,2AH            ;get date from DOS
00C2 CD 21                                 INT  21H
00C4 52                                    PUSH DX
00C5 B4 00                                 MOV  AH,0              ;get day of week
00C7 03 C0                                 ADD  AX,AX
00C9 BE 0000 R                             MOV  SI,OFFSET DTAB    ;address day table
00CC 03 F0                                 ADD  SI,AX
00CE 8B 14                                 MOV  DX,[SI]           ;address day of week
00D0 B4 09                                 MOV  AH,9              ;display day of week
00D2 CD 21                                 INT  21H
00D4 5A                                    POP  DX
00D5 52                                    PUSH DX
00D6 8A C6                                 MOV  AL,DH             ;get month
00D8 FE C8                                 DEC  AL
00DA B4 00                                 MOV  AH,0
00DC 03 C0                                 ADD  AX,AX
00DE BE 000E R                             MOV  SI,OFFSET MTAB    ;address month table
00E1 03 F0                                 ADD  SI,AX
00E3 8B 14                                 MOV  DX,[SI]           ;address month
00E5 B4 09                                 MOV  AH,9              ;display month
00E7 CD 21                                 INT  21H
00E9 5A                                    POP  DX
00EA 8A C2                                 MOV  AL,DL             ;get day of month
00EC B4 00                                 MOV  AH,0
00EE D4 0A                                 AAM                    ;convert to BCD
00F0 0A E4                                 OR   AH,AH
00F2 74 0D                                 JZ   DATES1            ;if tens of day is 0
00F4 80 C4 30                              ADD  AH,30H            ;convert to ASCII
                                           DISP AH               ;display tens of day
0101                    DATES1:
0101 04 30                                 ADD  AL,30H            ;convert to ASCII
                                           DISP AL               ;display of day
                                           DISP ','              ;display comma
                                           DISP ' '              ;display space
0121 81 F9 07D0                            CMP  CX,2000           ;test for year 2000
0125 72 19                                 JB   DATES2            ;if below year 2000
0127 83 E9 64                              SUB  CX,100            ;scale to 1900 - 1999
                                           DISP '2'              ;display 2
                                           DISP '0'              ;display 0
013E EB 14                                 JMP  DATES3
0140                    DATES2:
                                           DISP '1'              ;display 1
                                           DISP '9'              ;display 9
0154                    DATES3:
0154 81 E9 076C                            SUB  CX,1900           ;scale to 00 - 99
0158 8B C1                                 MOV  AX,CX
015A D4 0A                                 AAM                    ;convert to BCD
015C 05 3030                               ADD  AX,3030H          ;convert to ASCII
                                           DISP AH               ;display tens of year
                                           DISP AL               ;display units
0173 C3                                    RET

0174                    DATES    ENDP
                                 END                             ;end of file
```

The time is available from DOS using INT 21H function call number 2CH. This function returns with the hours in CH and minutes in CL. Also available are seconds in DH and hundredths of seconds in DL. The date is available using INT 21H function call number 2AH. This leaves the day of the week in AL, the year in CX, the day of the month in DH, and the month in DL.

The DATES procedure uses two ASCII lookup tables that convert the day and month to ASCII character strings. It also uses the AAM instruction to convert from binary to BCD for the time and date. Data display is handled in two ways: by character string (function 09H) and by single character (function 02H) located in a macro called DISP.

7–4   SUMMARY

1. The assembler program assembles modules that contain PUBLIC variables and segments plus EXTRN (external) variables. The linker program links modules and library files to create a run-time program executed from the DOS command line. The run-time program usually has the extension EXE.

2. The MACRO and ENDM directives create a new opcode to use in programs. These macros are similar to procedures except there is no call or return. In place of them, the assembler inserts the code of the macro sequence into a program each time it is invoked. Macros can include variables that pass information and data to the macro sequence.

3. The DOS INT 21H function call provides a method of using the keyboard and video display. Function number 06H, placed into register AH, provides an interface to the keyboard and display. If DL = 0FFH, this function tests the keyboard for a keystroke. If no keystroke is detected, it returns equal. If a keystroke is detected, the standard ASCII character returns in AL. If an extended ASCII character is typed, it returns with AL = 00H, where the function must be called again to return with the extended ASCII character in AL. To display a character, DL is loaded with the character and AH with 06H before the INT 21H is used in a program.

4. Character strings are displayed using function number 09H. The DS:DX register combination addresses the character string, which must end with a $.

5. The INT 10H instruction accesses BIOS (basic I/O system) procedures that control the video display and keyboard. The BIOS functions are independent of DOS and function with any operating system.

6. Data conversion from binary to BCD is accomplished with the AAM instruction for numbers that are less than 100 or by repeated division by 10 for larger numbers. Once converted to BCD, a 30H is added to convert each digit to ASCII code for the video display.

7. When converting from an ASCII number to BCD, a 30H is subtracted from each digit. To obtain the binary equivalent, multiply by 10.

8. Lookup tables are used for code conversion with the XLAT instruction if the code is an 8-bit code. If the code is wider than 8 bits, then a short procedure that accesses a lookup table provides the conversion. Lookup tables are also used to hold addresses so that different parts of a program or different procedures can be selected.

7–5   QUESTIONS AND PROBLEMS

1. The assembler converts a source file into an _____ file.
2. Which files are generated from the source file TEST.ASM as it is processed by MASM?
3. The linker program links object files and _____ files to create an execution file.
4. What is the difference between an .EXE and a .COM file?
5. What does the PUBLIC directive indicate when placed in a program module?
6. What does the EXTRN directive indicate when placed in a program module?
7. What directives appear with labels defined external?
8. Describe how a library file works when it is linked to other object files by the linker program.
9. What is a macro sequence?
10. What assembler language directives delineate a macro sequence?
11. How are parameters transferred to a macro sequence?
12. Develop a macro called ADD32 that adds the 32-bit contents of DX-CX to the 32-bit contents of BX–AX.

13. Develop a macro called STUB that sign-extends the 8-bit number in AL into a 64-bit sign-extended number in ECX–EBX. (Note that ECX is the most-significant part of the 64-bit result.)

14. How is the LOCAL directive used within a macro sequence?

15. Develop a macro called ADDLIST PARA1,PARA2 that adds the contents of PARA1 to PARA2. Each of these parameters represents an area of memory. The number of bytes added are indicated by register CX before the macro is invoked.

16. Develop a macro that sums a list of byte-sized data invoked by the macro ADDM LIST,LENGTH. The label LIST is the starting address of the data block and length is the number of data added. The result must be a 16-bit sum found in AX at the end of the macro sequence.

17. What is the purpose of the INCLUDE directive?

18. Develop a procedure called RANDOM. This procedure must return an 8-bit random number in register CL at the end of the subroutine. (One way to generate a random number is to increment CL each time the DOS function 06H tests the keyboard and finds no keystroke. In this way a random number is generated.)

19. Modify the procedure of Question 18 so the random number ranges in value from 1 through and including 6.

20. Develop a procedure that displays a character string that ends with a 00H. Your procedure must use the DS:DX register to address the start of the character string.

21. Develop a procedure that reads a key and displays the hexadecimal value of an extended ASCII-coded keyboard character if it is typed. If a normal character is typed, the procedure ignores it.

22. Use BIOS INT 10H to develop a procedure that positions the cursor at line 3, column 6.

23. When a number is converted from binary to BCD, the _____ instruction accomplishes the conversion provided the number is less than 100 decimal.

24. How is a large number (over 100 decimal) converted from binary to BCD?

25. A BCD digit is converted to ASCII code by adding a _____.

26. An ASCII-coded number is converted to BCD by subtracting _____.

27. Develop a procedure that reads an ASCII number from the keyboard and stores it as a BCD number into memory array DATA. The number ends when anything other than a number is typed.

28. Explain how a 3-digit ASCII-coded number is converted to binary.

29. Develop a procedure that converts all lowercase ASCII-coded letters into uppercase ASCII-coded letters. Your procedure may not change any character except a letter a–z.

30. Develop a lookup table that converts hexadecimal data 00H–0FH into the ASCII-coded characters that represent the hexadecimal digits. Make sure to show the lookup table and any software required for the conversion.

31. Develop a program sequence that jumps to memory location ONE if AL = 6; TWO if AL = 7; THREE if AL = 8.

32. Show how to use the XLAT instruction to access a lookup table called LOOK that is located in the stack segment.

33. Write a program that reads any decimal number between 0 and 65,535 and displays the 16-bit binary version on the video display.

34. Write a program that displays the binary powers of two (in decimal) on the video screen for the powers 0 through 7. Your display shows 2^n = value for each power of 2.

35. Develop a program that accepts a four-digit number in hexadecimal code and displays it as a base 7 code.

36. Using the technique learned in Question 18, develop a program that displays random numbers between 1 and 47 (or whatever) that your state's lottery uses for its range of numbers.

37. Develop a program that displays the hexadecimal contents of a block of 256 bytes of memory. Your software must be able to accept the starting address as a hexadecimal number between 00000H and FFF00H.
38. Develop a program that displays the hexadecimal number 123AH as an octal number on the video display. Use the technique learned for displaying decimal data, but divide by 8 instead of 10.
39. Use the Internet to locate software written in assembly language for the 80X88 microprocessor. Visit at least four different sites.
40. Use the Internet to locate sorting techniques and write a report describing one such technique.
41. Use the Internet to obtain information about assembly language programs other than MASM. Write a report that describes at least two such programs and contrasts them with MASM.

CHAPTER 8

Mouse, Disk, and Video Programming

INTRODUCTION

This chapter presents more advanced programming for the personal computer. Although this information may not be required for programming an embedded controller, it may be needed for programming the embedded PC using the 80386EX.

Detailed are the application of software for programming the video display using the text-mode memory system as well as the graphics memory. We will explore modes including video display mode 3, 12H, and 13H. We also discuss techniques that modify the character set displayed by the video BIOS. The disk system is detailed to show how a disk is formatted and how software can be developed to create, access, and modify disk file data. Finally we look at the operation of the mouse for applications that are mouse driven.

CHAPTER OBJECTIVES

Upon completion of this chapter, you will be able to:

1. Detect the mouse and use the mouse in both text and graphics modes.
2. Detail the purpose of a track, sector, cluster, cylinder, FAT, and other terms that apply to disk memory systems.
3. Write software that creates, reads, writes, and modifies files on either the hard disk or floppy disk drive.
4. Explain how to modify the text-mode character display by directly accessing the memory.
5. Develop software that uses the 320×200–256 color graphics display.
6. Develop software that uses the 640×480–16 color graphics display.

8–1 USING THE MOUSE

The **mouse pointing device** is controlled with the INT 33H instruction. Refer to Appendix A for a list of the Microsoft-compatible mouse functions associated with INT 33H. Unlike the DOS INT 21H function number, the mouse function number is selected through the AL register; AH is usually set to 00H before the INT 33H is executed. There are about 50 mouse functions available, of which only the main functions are described here.

Testing for a Mouse

Before the mouse can be used in an application, you must determine if a mouse driver is installed in the system. The presence of the mouse driver is tested by examining interrupt vector 33H. If interrupt vector 33H contains a 0000:0000, the mouse driver is not installed in the system. In some systems, a vector exists even though no mouse driver is present. In this instance, the INT 33H vector address points to an IRET instruction (CFH) somewhere in the system memory. The interrupt vector address is retrieved by using the DOS INT 21H function 35H. The address is then tested for 0000:0000 and if it contains another value, the contents of the address pointed to by interrupt vector 33H are tested for CFH. Example 8–1 is a procedure that tests for the existence of the mouse driver.

Once it is determined that a possible mouse driver exists, the mouse is reset to make certain it is connected to the system and functioning. The mouse reset is accomplished by using mouse function 00H. The return from function 00H is AX = 0000H if no mouse is present. The **CHKM procedure** returns if the mouse exists with carry cleared and if no mouse exists with carry set.

EXAMPLE 8–1

```
                        ;A procedure that tests for the presence of a mouse driver.
                        ;***Output parameters***
                        ;Carry = 1, if no mouse present.
                        ;Carry = 0, if mouse is present.
                        ;
0000            CHKM    PROC NEAR

0000 B8 3533            MOV  AX,3533H          ;get INT 33H vector
0003 CD 21              INT  21H               ;returns vector in ES:BX

0005 8C C0              MOV  AX,ES
0007 0B C3              OR   AX,BX             ;test for 0000:0000
0009 F9                 STC                    ;indicate no mouse
000A 74 13              JZ   CHKM1             ;if no mouse driver

000C 26: 80 3F CF       CMP  BYTE PTR ES:[BX],0CFH
0010 F9                 STC
0011 74 0C              JE   CHKM1             ;if no mouse driver

0013 B8 0000            MOV  AX,0
0016 CD 33              INT  33H               ;reset mouse
0018 83 F8 00           CMP  AX,0
001B F9                 STC
001C 74 01              JZ   CHKM1             ;if no mouse
001E F8                 CLC
001F            CHKM1:
001F C3                 RET

0020            CHKM    ENDP
```

Which Mouse and Driver?

The mouse function interrupt determines the type of mouse connected to the system and the driver version number. Example 8–2 displays the mouse type and driver version number after a test is made to determine if the mouse is present by using the procedure of Example 8–1. Here mouse INT 33H, function 24H locates the mouse driver version number and mouse driver type. The return from function 24H leaves the mouse driver number in BX (BH = major and BL = minor) and the mouse type in CH. If the mouse driver version is 8.00, then BH = 08H and BL = 00H. The mouse types that are returned in register CH are currently: bus = 1, serial = 2, InPort = 3, PS/2 = 4, and Hewlett-Packard = 5. As time passes, this list of mouse types will grow as new mouse types are developed.

EXAMPLE 8–2

```
                              ;A program that displays the mouse driver version
                              ;number and the type of mouse installed.
                              ;
                                      .MODEL SMALL
0000                                  .DATA
0000 0D 0A 4E 6F 20 4D   MES1    DB    13,10,'No MOUSE or MOUSE DRIVER found.$'
     4F 55 53 45 20 6F
     72 20 4D 4F 55 53
     45 20 44 52 49 56
     45 52 20 66 6F 75
     6E 64 2E 24
0022 0D 0A 4D 6F 75 73   MES2    DB    13,10,'Mouse driver version'
     65 20 64 72 69 76
     65 72 20 76 65 72
     73 69 6F 6E 20
0039 20 20 20 20 20 20   M1      DB    '        ',13,10,'$'
     20 0D 0A 24
0043 004D R 0051 R       TYPES   DW    T1,T2,T3,T4,T5
     0058 R 005F R
     0064 R
004D 42 75 73 24         T1      DB    'Bus$'
0051 53 65 72 69 61 6C   T2      DB    'Serial$'
     24
0058 49 6E 50 6F 72 74   T3      DB    'InPort$'
     24
005F 50 53 2F 32 24      T4      DB    'PS/2$'
0064 48 50 24            T5      DB    'HP$'
0067 20 6D 6F 75 73 65   MES3    DB    'mouse installed.',13,10,'$'
     20 69 6E 73 74 61
     6C 6C 65 64 2E 0D
     0A 24
0000                             .CODE
                                 .STARTUP
0017 E8 0041                     CALL CHKM              ;test for mouse
001A 73 05                       JNC  MAIN1             ;if mouse present
001C BA 0000 R                   MOV  DX,OFFSET MES1
001F EB 32                       JMP  MAIN2             ;if no mouse
0021                     MAIN1:
0021 B8 0024                     MOV  AX,24H
0024 CD 33                       INT  33H               ;get driver type
0026 BF 0039 R                   MOV  DI,OFFSET M1
0029 8A C7                       MOV  AL,BH             ;save major version
002B E8 004D                     CALL DISP
002E C6 05 2E                    MOV  BYTE PTR [DI],'.'  ;save period
0031 47                          INC  DI

0032 8A C3                       MOV  AL,BL             ;save minor version
0034 E8 0044                     CALL DISP

0037 BA 0022 R                   MOV  DX,OFFSET MES2    ;display version
003A B4 09                       MOV  AH,9
003C CD 21                       INT  21H

003E BE 0043 R                   MOV  SI,OFFSET TYPES   ;index type
0041 B4 00                       MOV  AH,0
0043 8A C5                       MOV  AL,CH
0045 48                          DEC  AX
0046 03 F0                       ADD  SI,AX
0048 03 F0                       ADD  SI,AX
004A 8B 14                       MOV  DX,[SI]           ;display type
004C B4 09                       MOV  AH,9
004E CD 21                       INT  21H
0050 BA 0067 R                   MOV  DX,OFFSET MES3
0053                     MAIN2:
0053 B4 09                       MOV  AH,9
```

```
0055 CD 21                              INT    21H
                                        .EXIT
                                ;A procedure that tests for the presence of a mouse
                                ;driver.
                                ;***Output parameters***
                                ;Carry = 1, if no mouse present.
                                ;Carry = 0, if mouse is present.
                                ;
005B                            CHKM   PROC NEAR

005B B8 3533                            MOV    AX,3533H            ;get INT 33H vector
005E CD 21                              INT    21H                 ;vector in ES:BX

0060 8C C0                              MOV    AX,ES
0062 0B C3                              OR     AX,BX               ;test for 0000:0000
0064 F9                                 STC
0065 74 13                              JZ     CHKM1               ;if no mouse driver
0067 26: 80 3F CF                       CMP    BYTE PTR ES:[BX],0CFH
006B F9                                 STC
006C 74 0C                              JE     CHKM1               ;if no mouse driver
006E B8 0000                            MOV    AX,0
0071 CD 33                              INT    33H                 ;reset mouse
0073 83 F8 00                           CMP    AX,0
0076 F9                                 STC
0077 74 01                              JZ     CHKM1               ;if no mouse
0079 F8                                 CLC
007A                            CHKM1:
007A C3                                 RET

007B                            CHKM   ENDP
                                ;
                                ;Save the ASCII-coded version number.
                                ;***input parameters***
                                ;AL = version
                                ;DS:DI = address where stored
                                ;***output parameters***
                                ;ASCII version number stored at DS:DI
                                ;
007B                            DISP   PROC NEAR

007B B4 00                              MOV    AH,0
007D D4 0A                              AAM                        ;convert to BCD
007F 05 3030                            ADD    AX,3030H
0082 80 FC 30                           CMP    AH,30H              ;save ASCII version
0085 74 03                              JE     DISP1               ;suppress zero
0087 88 25                              MOV    [DI],AH
0089 47                                 INC    DI
008A                            DISP1:
008A 88 05                              MOV    [DI],AL
008C 47                                 INC    DI
008D C3                                 RET

008E                            DISP   ENDP
                                        END
```

Using the Mouse

The mouse functions in either text or graphics mode. This section illustrates how to enable the mouse with a text-mode program. It also functions in a graphics mode, but instead of displaying a block as a cursor or mouse pointer, an arrow is displayed in the graphics mode. As with the prior examples, the first step is to check for the presence of a mouse driver. Example 8–3 uses the CHKM procedure to test for the presence of the mouse. If no mouse is present, a return from TM_ON occurs with the carry flag set. If the mouse is present, the cursor is displayed and a return occurs with the carry flag cleared.

EXAMPLE 8–3

```
                        ;The TM_ON procedure tests for the presence of a mouse
                        ;and enables mouse pointer.
                        ;uses the CHKM (check for mouse) procedure
                        ;
                        ;***output parameters***
                        ;Carry = 0, if mouse is present pointer enabled.
                        ;Carry = 1, if no mouse present.
                        ;
0000                    TM_ON   PROC NEAR

0000 E8 FFDD                    CALL CHKM               ;test for mouse
0003 72 06                      JC   TM_ON1             ;if no mouse
0005 B8 0001                    MOV  AX,1               ;show mouse pointer
0008 CD 33                      INT  33H
000A F8                         CLC                     ;show mouse present
000B            TM_ON1:
000B C3                         RET

000C                    TM_ON   ENDP
```

The procedure of Example 8–3 enables only the mouse and displays the mouse cursor. To use the mouse, a program like the one in Example 8–4 must be written that tracks the mouse and its position.

EXAMPLE 8–4

```
                        ;A program that displays the mouse pointer and its
                        ;X and Y position.
                        ;
                                .MODEL SMALL
0000                            .DATA
0000 0D 58 20 50 6F 73  MES     DB   13,'X Position= '
     69 74 69 6F 6E 3D
     20
000D 20 20 20 20 20 20  MX      DB   '      '
0013 59 20 50 6F 73 69          DB   'Y Position= '
     74 69 6F 6E 3D 20
001F 20 20 20 20 20 20  MY      DB   '      $'
     24
0026 0000               X       DW   ?                  ;X position
0028 0000               Y       DW   ?                  ;Y position
0000                            .CODE
                                .STARTUP
0017 E8 006D                    CALL TM_ON              ;enable mouse
001A 72 47                      JC   MAIN4              ;if no mouse
001C            MAIN1:
001C B8 0003                    MOV  AX,3               ;get mouse status
001F CD 33                      INT  33H
0021 83 FB 01                   CMP  BX,1
0024 74 38                      JE   MAIN3              ;for left button
0026 3B 0E 0026 R               CMP  CX,X
002A 75 06                      JNE  MAIN2              ;if X changed
002C 3B 16 0028 R               CMP  DX,Y
0030 74 EA                      JE   MAIN1              ;if Y did not change
0032            MAIN2:
0032 89 0E 0026 R               MOV  X,CX               ;save new position
0036 89 16 0028 R               MOV  Y,DX
003A BF 000D R                  MOV  DI,OFFSET MX
003D 8B C1                      MOV  AX,CX
003F E8 0051                    CALL PLACE              ;store ASCII X
0042 BF 001F R                  MOV  DI,OFFSET MY
0045 A1 0028 R                  MOV  AX,Y
0048 E8 0048                    CALL PLACE              ;store ASCII Y

004B B8 0002                    MOV  AX,2
004E CD 33                      INT  33H                ;hide mouse pointer
```

```
0050 B4 09                           MOV   AH,9
0052 BA 0000 R                       MOV   DX,OFFSET MES
0055 CD 21                           INT   21H               ;display position

0057 B8 0001                         MOV   AX,1
005A CD 33                           INT   33H               ;show mouse pointer

005C EB BE                           JMP   MAIN1             ;do again
005E                    MAIN3:
005E B8 0000                         MOV   AX,0              ;reset mouse
0061 CD 33                           INT   33H

0063                    MAIN4:
                                     .EXIT
                        ;
                        ;A procedure that tests for the presence of a mouse
                        ;driver.
                        ;***Output parameters***
                        ;Carry = 1, if no mouse present.
                        ;Carry = 0, if mouse is present.
                        ;
0067                    CHKM    PROC NEAR

0067 B8 3533                         MOV   AX,3533H          ;get INT 33H vector
006A CD 21                           INT   21H               ;vector in ES:BX

006C 8C C0                           MOV   AX,ES
006E 0B C3                           OR    AX,BX             ;test for 0000:0000
0070 F9                              STC
0071 74 13                           JZ    CHKM1             ;if no mouse driver
0073 26: 80 3F CF                    CMP   BYTE PTR ES:[BX],0CFH
0077 F9                              STC
0078 74 0C                           JE    CHKM1             ;if no mouse driver
007A B8 0000                         MOV   AX,0
007D CD 33                           INT   33H               ;reset mouse
007F 83 F8 00                        CMP   AX,0
0082 F9                              STC
0083 74 01                           JZ    CHKM1             ;if no mouse
0085 F8                              CLC
0086                    CHKM1:
0086 C3                              RET

0087                    CHKM    ENDP
                        ;
                        ;The TM_ON procedure tests for the presence of a
                        ;mouse and enables mouse pointer.
                        ;uses the CHKM (check for mouse) procedure
                        ;
                        ;***output parameters***
                        ;Carry = 0, if mouse is present pointer enabled.
                        ;Carry = 1, if no mouse present.
                        ;
0087                    TM_ON   PROC NEAR

0087 E8 FFDD                         CALL CHKM               ;test for mouse
008A 72 06                           JC   TM_ON1
008C B8 0001                         MOV  AX,1               ;show mouse pointer
008F CD 33                           INT  33H
0091 F8                              CLC
0092                    TM_ON1:
0092 C3                              RET

0093                    TM_ON   ENDP
                        ;
                        ;The PLACE procedure converts the contents of AX
                        ;into a decimal ASCII-coded number stored at the
                        ;memory location addressed by DS:DI.
                        ;***input parameters***
                        ;AX = number to be converted to decimal ASCII code
```

```
                              ;DS:DI = address where number is stored
                              ;
0093                          PLACE     PROC NEAR

0093 B9 0000                            MOV  CX,0           ;clear count
0096 BB 000A                            MOV  BX,10          ;set divisor
0099                          PLACE1:
0099 BA 0000                            MOV  DX,0           ;clear DX
009C F7 F3                              DIV  BX             ;divide by 10
009E 52                                 PUSH DX
009F 41                                 INC  CX
00A0 83 F8 00                           CMP  AX,0
00A3 75 F4                              JNE  PLACE1         ;repeat until 0
00A5                          PLACE2:
00A5 BB 0005                            MOV  BX,5
00A8 2B D9                              SUB  BX,CX
00AA                          PLACE3:
00AA 5A                                 POP  DX
00AB 80 C2 30                           ADD  DL,30H         ;convert to ASCII
00AE 88 15                              MOV  [DI],DL        ;store digit
00B0 47                                 INC  DI
00B1 E2 F7                              LOOP PLACE3
00B3 83 FB 00                           CMP  BX,0
00B6 74 08                              JE   PLACE5
00B8 8B CB                              MOV  CX,BX
00BA                          PLACE4:
00BA C6 05 20                           MOV  BYTE PTR [DI],20H
00BD 47                                 INC  DI
00BE E2 FA                              LOOP PLACE4
00C0                          PLACE5:
00C0C3                                  RET

00C1                          PLACE     ENDP
                                        END
```

The program in Example 8–4 displays the mouse cursor by placing a 0001H into AX, followed by the INT 33H instruction. Next the status of the mouse is read with function AX = 0003H. The status function returns with the status of the mouse buttons in BX, the X coordinate of the mouse pointer in CX, and the Y coordinate in DX. (Refer to Appendix A for more complete information on the status for the mouse.) In this example program, if the left mouse button is pressed, the program terminates; otherwise, the coordinates are compared with the prior values saved in X and Y. If a change has occurred in these coordinates, the new coordinates are calculated and displayed. Notice that before the video display is accessed, the mouse pointer is hidden using INT 33H with AX = 0002H. This is *very important*. If you don't hide the mouse pointer, the display will become unstable and the computer may even re-boot. In most cases, a copy of the mouse pointer remains on the screen if data are displayed without turning it off.

8–2   DISKS AND DISK ORGANIZATION

Data is stored on the disk in files that contain either data or programs. Before we examine files, we'll look at the way information is placed onto the surface of the disk. The data on a disk is organized in four main areas: the boot sector, the file allocation table (FAT), the root directory, and the data storage area. The first sector on the disk is called the boot sector. The **boot sector** contains a program called a **bootstrap loader** that reads the disk operating system (usually DOS) from the disk into the memory system when power is applied to the computer. The **FAT** stores codes that indicate the disk clusters or sectors that contain data—which are free and which are bad. All references to disk files are handled through the FAT and a directory entry. Directory names, files, and programs are referenced through the **root directory.** The disk files are all considered sequential access files, meaning they are accessed one byte at a time from the beginning of the file to the end.

Filenames

Files and programs are stored on a disk and referenced by a filename and an extension to the filename. With the DOS operating system, the **filename** can be any size from one to eight characters in length. The filename can contain just about any ASCII character except for spaces or the " \ . / [ ] * , : < > | ; ? = characters. In addition to the filename, the file can have an optional 1- to 3-digit **extension** to the filename. Table 8–1 lists a few filenames with and without extensions. The name of a file and its extension are always separated by a period. If you use Windows 95, the filename can be any length (up to 255 characters) and can even contain spaces. This is an improvement over the eight-character filename limitation of DOS.

Directory and Subdirectory Names.   The DOS file management system arranges the data and programs on a disk into directories and subdirectories. The rules that apply to filenames also apply to directory and subdirectory names. That is, the name can be up to eight characters in length and an extension can appear that is up to three characters in length. Most of the time an extension is not used with a directory name, but it can be used if necessary. The disk is structured so it contains a root directory when first formatted. The root directory for a hard disk used as drive C is C:\. Any other directory is placed in the root directory. For example, C:\DATA is directory DATA in the root directory. Each directory placed in the root directory can have subdirectories. Examples are the subdirectories C:\DATA\AREA1 and C:\DATA\AREA2 where the directory DATA contains two subdirectories: AREA1 and AREA2. Subdirectories can also have additional subdirectories. For example, C:\DATA\AREA2\LIST depicts directory DATA and subdirectory AREA, which contains a subdirectory LIST.

Disk Organization

Figure 8–1 illustrates the organization of data stored in sectors and tracks on the surface of a $5^1/4''$ floppy disk. This arrangement applies to both floppy and hard or fixed disk memory systems, except the $3^1/2''$ floppy disk and hard disks do not have an index hole. The surface of the disk is divided into concentric, non-spiraling rings of data called **tracks.** The **outer track** is always track 0 and the inner track is 39, on a double-density, $5^1/4''$ floppy disk, or 79 on all other floppy disks. The **inner track** on a hard disk is determined by the disk size and could be 1,000 or higher for very large hard disks. Note that the current version of Windows can have a problem creating a permanent virtual swap file on hard disks that contains more than 1,024 tracks.

TABLE 8–1 DOS Filenames and Extensions

Name.Extension
TEST.TXT
READ-ME.DOC
ALWAY12.COM
RUN_IT~.EXE
CHAPTER.02
NOEXTEN1
1.2

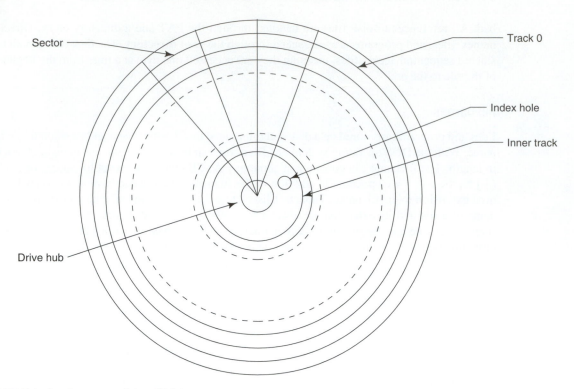

FIGURE 8–1   Structure of the 5^1/$_4$″ floppy disk.

Each track is divided into sections or groups of data areas called **sectors.** A sector almost always contains 512 bytes of data. When a program or a file containing data is stored on the disk, it is stored in sectors that may or may not be contiguous to each other. It is not unusual to find that one program or data file occupies many sectors scattered about the surface of the disk. This, of course, is not desirable, as will be explained later.

Floppy Disks.   Modern floppy disks are available in two different sizes: the 5^1/$_4$″ **mini-floppy** and the 3^1/$_2$″ **micro-floppy.** The 5^1/$_4$″ mini-floppy disk, which is being phased out, is currently available as a **double-sided, double-density** (DSDD) disk that stores 360K bytes and a **high-density** (HD) version that stores 1.2M bytes. Early 5^1/$_4$″ floppy disks were single-sided and stored either 180K (double-density) or 92K (single-density) bytes. Still earlier versions of the floppy were the once standard 8″ size, which stored 256K bytes of data. The 3^1/$_2$″ micro-floppy is available as a double-sided, double-density (DSDD) disk that stores 720K bytes; a high-density (HD) version that stores 1.44 M bytes; an **extra-high-density** (EHD or ED) version that stores 2.88M bytes. At present the 2.88M-byte floppy disks aren't very popular.

In all cases each type of disk memory organizes its surface into tracks that contain sectors. Each sector on a disk usually contains 512 bytes of data. **Clusters** are groups of sectors used to access information on the disk. The number of sectors and tracks varies from one disk type to another. For example, the 5^1/$_4$″ double-density, double-sided disk contains 40 tracks with nine sectors per track ($40 \times 9 \times 512 = 180$K bytes) per side. A pair of tracks (top and bottom) is called a **cylinder.** We often use the term cylinder in place of track and upper or lower head in place of side. Some fixed or hard disk drives have up to 16 **heads** (tracks) per cylinder because they contain more than one disk platter. Table 8–2 lists the common sizes of floppy disks and information about the number of tracks, sectors, clusters, root directory size, and disk capacities. The capacity is unformatted, which means that the actual capacity is less, due to a loss of space to the boot sector, FAT, root directory, and operating system if present.

TABLE 8–2 5^1/$_4$″ and 3^1/$_2$″ floppy disks

Size	Tracks per Side	Sectors per Track	Sectors per Cluster	Bytes in Root	Unformatted Capacity
5^1/$_4$″ DSDD	40	9	2	3,584	368,640 (360K)
5^1/$_4$″ HD	80	15	1	7,168	1,228,800 (1.2M)
3^1/$_2$″ DSDD	80	9	2	3,584	737,280 (720K)
3^1/$_2$″ HD	80	18	1	7,168	1,474,560 (1.44M)
3^1/$_2$″ ED	80	36	1	7,168	2,949,120 (2.88M)

Note: All sectors are 512 bytes in length.

Figure 8–2 shows the organization of major areas on the disk. The length of the FAT (**file allocation table**) is determined by the size of the disk because each disk sector or group of sectors is represented in the FAT by a 16-bit number. Some extremely large fixed disks use a 32-bit number as entries in the FAT. Another variation is a 12-bit FAT used on many floppy disks. The length of the root directory is usually fixed at 3,584 bytes on a 5^1/$_4$″ DSDD floppy disk. The root directory is larger on a 3^1/$_2$″ floppy and hard disk. The boot sector is always a single 512-byte long sector located in the outer track (track 0) at sector 0, the first sector on the disk.

The **boot sector** contains a bootstrap loader program, which is read into RAM when the system is powered-up. After the bootstrap loader is read into RAM, it loads the IO.SYS and MS-DOS.SYS programs into RAM. Next, the bootstrap loader passes control to the MS-DOS control program, which uses the CONFIG.SYS file to configure the system. After the CONFIG.SYS file configures the system, the MSDOS.SYS program loads the DOS command processor COM-MAND.COM into memory. This is where the AUTOEXEC.BAT file is executed and control is passed to the user. User control is often provided at the DOS prompt, from DOSSHELL (available with DOS version 5.0), or from Windows.

The File Allocation Table (FAT). The **FAT** indicates which sectors or clusters of sectors are free, which are corrupted (unusable), and which contain valid data or programs. Data is accessed by cluster number on all hard disk drives. The FAT is referenced each time that DOS writes data to the disk so that it can find a free or empty sector. The FAT is also accessed for a read operation to determine the location of the next cluster in a chain. Each free cluster is indicated by a 0000H in the FAT, and each occupied cluster is indicated by the cluster number of the next cluster in the file chain or a last cluster code of FFFFH. A cluster can be any size from one or two sectors on many floppy disks, to up to 16 sectors per clusters on a hard disk memory. Many hard disk memory systems use four or more sectors per cluster, which means the smallest file allocation unit is 512 × 4 or 2,048 bytes in length if a cluster contains four sectors. Clusters are also referred

FIGURE 8–2 Main data storage areas on a disk.

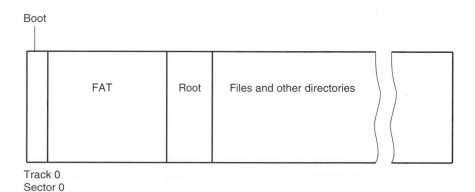

Boot

FAT | Root | Files and other directories

Track 0
Sector 0

to as **file allocation units.** The larger the hard disk drive, the more sectors in a cluster. A 16-bit FAT table accesses 64K, or 17 clusters. A 0000H code indicates a free cluster; FFF7H indicates a bad cluster; FFFFH indicates the end of a file chain; the remaining cluster numbers 0001H–FFF0H are available. If a cluster is four sectors of 512 bytes, a fixed disk containing a 16-bit FAT accesses a maximum of 65,519 clusters. Because a sector is 512 bytes and a cluster is 2,048 bytes or four sectors, this hard disk drive contains a maximum of 134,182,912 bytes of unformatted data. If a cluster contains eight sectors (4,096 bytes), then the maximum capacity of the hard disk drive is 268,365,824 bytes.

With 16 sectors per cluster, the hard disk can be a maximum of 536,731,648 bytes. If a larger hard disk exists, either more sectors per cluster are allocated or the FAT is increased to 24 bits. With DOS, the FAT is currently 16 bits for most hard disks and 12 bits for most floppy disks. If the cluster size is 4K bytes, the smallest file allocated by DOS is 4K bytes even if it only contains 1 byte of data.

Directories and Subdirectories.   Figure 8–3 shows the format of each **directory entry** in the root or any other directory or subdirectory. An example DOS command-line entry is C:\GAME\ TEST.TXT, where the disk drive is C, the directory is GAMES, and the file is TEST.TXT. Each entry contains the name, extension, attribute, time, date, location, and length. The length of the file is stored as a 32-bit number. This means that a file can have a maximum length of 4G bytes. The location is the starting cluster number. The root directory, on a $5\frac{1}{4}''$ DSDD floppy disk, has room for 112 of these 32-byte directory entries. On other floppy disks and on a hard disk drive, the root directory contains 224 or more entries. The root directory is a fixed length area on the disk. When the root directory is full, the disk is full. This is why it is important to use directories and subdirectories, which can contain any number of files. A single directory requires one entry in the root directory. The entry points to another cluster or clusters that contain the directory. A directory can be any length. A subdirectory entry or a file is stored in the directory created and not in the root directory. This allows an almost infinite combination of files stored in any number of directories and subdirectories (see Figure 8–4). The same filename can appear in different directories and subdirectories and is considered different.

The location area in the directory entry holds the first cluster number where the file is stored on the disk. The FAT location that pertains to this cluster address holds either the next cluster

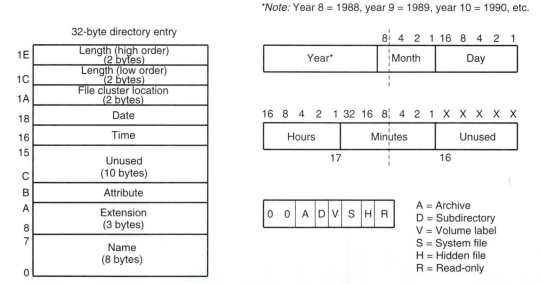

FIGURE 8–3   Format for any directory or subdirectory entry.

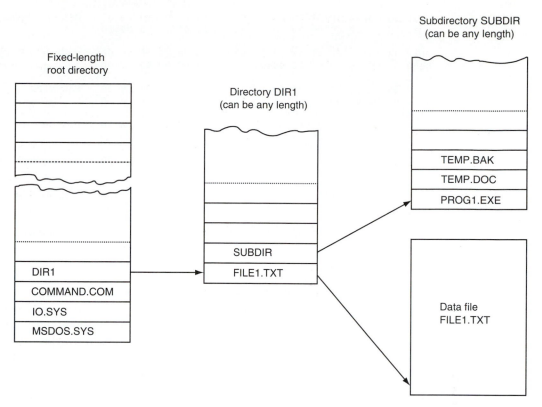

FIGURE 8–4 The root directory, a directory, a subdirectory, and a data file. Notice how the root directory entry addresses the directory, and how the directory addresses the subdirectory and a data file FILE1.TXT.

number where the file continues or an FFFFH to indicate the end of the file. This is required so DOS can store files on the disk in **fragments** (pieces) as mentioned earlier. Fragments cause problems after a disk is in use for a long period of time. As programs and files are stored and erased, they always seem to occupy a larger number of smaller and smaller fragments of disk space. Eventually a file that requires four clusters, will be stored in four clusters located at four different places throughout the surface of the disk. This significantly increases the time required to access and load large files. Fragmentation is corrected by using a defragmentation program such as PCTOOLS[1] to compress or optimize space or the DOS DEFRAG program. It is recommended that the hard disk be defragmented occasionally, especially if load time becomes a noticeable problem. In practice, it is recommended that the hard disk be defragmented no more than once a month because of the time it takes to defragment a disk.

Example 8–5 shows how part of the root directory appears in a hexadecimal dump. Try to identify the date, time, location, and length of each entry. Also identify the attribute for each entry. The example shows hexadecimal data and also ASCII data as is customary in most computer hexadecimal dumps.

EXAMPLE 8–5

```
0000 49 4F 20 20 20 20 20 20 53 59 53 07 00 00 00 00    IO      SYS
0010 00 00 00 00 00 00 00 00 93 11 02 00 39 82 00 00
```

---

[1]PCTOOLS is a registered trademark of Central Point Software.

```
0020  4D 53 44 4F 53 20 20 20 53 59 53 07 00 00 00 00    MSDOS    SYS
0030  00 00 00 00 00 00 C0 44 93 12 13 00 92 00 00 00

0040  43 4F 4D 4D 41 4E 44 20 43 4F 4D 00 00 00 00 00    COMMAND  COM
0050  00 00 00 00 00 00 00 00 93 11 26 00 B5 92 00 00

0060  42 41 52 52 59 20 42 52 45 59 20 28 00 00 00 00    BARRY BREY
0070  00 00 00 00 00 00 E0 AD 6A 13 00 00 00 00 00 00

0080  50 43 54 4F 4F 4C 53 20 20 20 20 10 00 00 00 00    PCTOOLS
0090  00 00 00 00 00 00 80 AE 6A 13 5C 00 00 00 00 00

00A0  44 4F 53 20 20 20 20 20 20 20 20 10 00 00 00 00    DOS
00B0  00 00 00 00 00 00 E0 B0 6A 13 4E 00 00 00 00 00

00C0  52 55 4E 5F 46 57 20 20 42 41 54 00 00 00 00 00    FUN_FW   BAT
00D0  00 00 00 00 00 00 40 BD 6A 13 97 0F 4A 00 00 00

00E0  46 4F 4E 54 57 41 52 45 20 20 20 10 00 00 00 00    FONTWARE
00F0  00 00 00 00 00 00 60 BD 6A 13 6E 00 00 00 00 00
```

Example 8–6 shows how two long filenames appear in the Windows 95 directory. The first file is CHAPTER1.DOC and the second entry is PRENTICE HALL CAREER 1.DOC. Notice that the first entry starts at offset address 0020H, while the long filename is stored beginning at offset address 0000H. The filename stored in the normal directory entry is CHAPE~1.DOC stored in ASCII code for compatibility with the older file system. The long filename is stored in the Unicode, which is a 16-bit code that uses ASCII as it first 256 combinations. The starting sector for a long filename entry contains a 0000H. This is how the long filename is detected by Windows 95 or any software that uses long filenames.

EXAMPLE 8–6

```
0000  41 43 00 68 00 61 00 70-00 74 00 0F 00 42 65 00    AC.h.a.p.t...Be.
0010  72 00 20 00 31 00 2E 00-64 00 00 00 6F 00 63 00    r. .1...d...o.c.

0020  43 48 41 50 54 45 7E 31-44 4F 43 20 00 54 4D 70    CHAPTE~1DOC .TMp
0030  9B 20 9B 20 00 00 91 78-91 20 48 00 00 F6 03 00    . . ...x. H.....

0040  42 20 00 43 00 41 00 52-00 45 00 0F 00 A0 45 00    B .C.A.R.E...E.
0050  52 00 20 00 31 00 2E 00-64 00 00 00 6F 00 63 00    R. .1...d...o.c.

0060  01 50 00 52 00 45 00 4E-00 54 00 0F 00 A0 49 00    .P.R.E.N.T....I.
0070  43 00 45 00 20 00 48 00-41 00 00 00 4C 00 4C 00    C.E. .H.A...L.L.

0080  50 52 45 4E 54 49 7E 31-44 4F 43 20 00 7C CB 6D    PRENTI~1DOC .|.m
0090  9B 20 9B 20 00 00 58 7A-68 20 02 00 00 8C 00 00    . . ..Xzh ......
```

Compact Disk Read-Only Memory (CD-ROM). Another common storage medium is the **compact disk read-only memory** (CD-ROM) or optical disk memory. The CD-ROM is a low cost, read-only, optical disk memory. Access times for a single-speed CD-ROM are typically 300 ms or longer—about the same as a floppy disk. For comparison, a hard disk magnetic memory can have access times as little as 7 ms. Data transfer rates are 150K bytes per second for the single-speed CD-ROM, 300K bytes for the double-speed, 450K bytes for the triple-speed, and 600K bytes for the quad-speed version. Data transfer speeds on a hard disk memory are often well over 1M byte per second and can be much higher—up to 132M bytes per second in a PCI or VESA local bus system. The CD-ROM software is available for most programs and applications. The CD-ROM is available with large volume data storage such as the Bible, encyclopedia, clip art, magazine articles, etc. These applications have wide appeal at the current prices. A CD-ROM stores 660M bytes of data or a combination of data and musical passages. The data stored on the CD-ROM is in the same form as that stored on a floppy or hard disk drive. The CD-ROM contains a FAT, a root directory, and data organized in clusters. The musical passages are always stored using the 150K bytes per second transfer rate; one CD-ROM can hold up to 70 minutes of music. Because this transfer rate is limited, music is often stored in a digital compressed form that must be decoded by a digital sound card. The data is stored in files on a CD-ROM drive in the same manner as on a floppy or hard disk memory. CD-ROM musical passages are stored in the same format as that found on an audio CD. As systems develop and become more visually active, the use of the CD-ROM drive will become a system requirement.

Figure 8–5 illustrates the operation of the CD-ROM drive. A solid-state laser beam is focused on the surface of the CD-ROM through a set of lenses. If the laser light strikes a pit (a depression in the surface of the CD-ROM), little light is reflected through the lenses back to a photo-detector and no voltage is generated. If the laser light strikes a land (an unpitted surface), most of it is reflected through the lenses back to the photo-detector to produce a voltage. The lack of a voltage at the photo-detector represents a logic 0 and the presence of a voltage represents a logic 1. In this way binary digital data is read from the surface of the CD-ROM.

Still another type of optical memory is the WORM (**write-once/read mostly**) drive, which has some commercial application. The problem is that its application is very specialized due to the nature of the WORM. Because data may be written only once, the main application is in the banking industry, insurance industry, and other massive data storage organizations. The WORM is normally used to form an audit trail of transactions that are spooled onto the WORM and retrieved only during an audit. You might call the WORM an archiving device. Another problem with the WORM drive is that it is not compatible with the CD-ROM.

The **read/write** CD-ROM is also available. These presently cost four to six times more than the read-only CD-ROM and are recorded in the same format as the CD-ROM for a computer or in the same format as an audio CD.

The main advantage of the optical disk is its durability. Because a solid-state laser beam is used to read the data from the disk, and the focus point is below a protective plastic coating, the

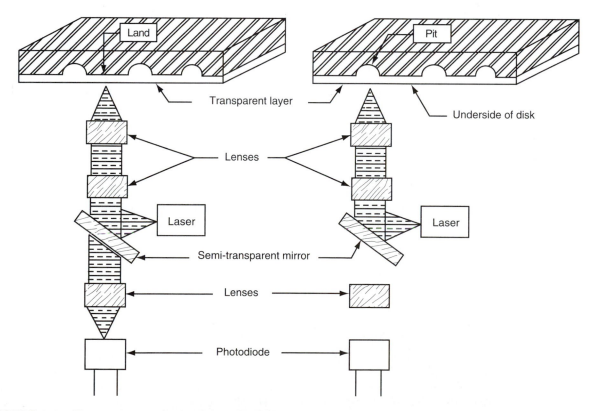

FIGURE 8–5 The mechanism for reading a CD-ROM.

surface of the disk may contain small scratches and dirt particles and still be read correctly. This feature allows less care of the optical disk than a comparable floppy disk. About the only way to destroy data on an optical disk is to break it or deeply scar it.

The DVD (digital video disk), which is available for video, will be adapted to use with the computer. The DVD can store many gigabytes of data, yet it is the same size as the CD-ROM.

8–3   SEQUENTIAL ACCESS DISK FILES

Files are normally accessed through DOS INT 21H function calls. Although files may also be accessed through BIOS INT 13H, it is fairly difficult to accomplish because BIOS INT 13H does not use the DOS directory structure and can be destructive. There are two ways to access a file using an INT 21H. One method uses a file control block and the other uses a file handle. Today, all software file accesses are via a **file handle,** so we use file handles for file access here. Also note that even though this chapter presents access to disk files using the assembler, such access is often accomplished with high-level programming languages. **File control blocks** are a carry-over from an earlier operating system called CP/M[2] **(control program/microprocessor),** which was designed for 8-bit computer systems based on the Z80, 8080, or 8085 microprocessor.

All DOS files are sequential access files. A **sequential access file** is stored and addressed from its beginning to its end. This means that the first byte, and all bytes between it and the last, must be accessed to read the last byte. Fortunately, files are read and written with the DOS INT 21H function calls (refer to Appendix A), which simplifies file access and data manipulation. This section describes how to create, read, write, delete, and rename a sequential access file by using the DOS INT 21H function calls.

File Creation

Before a file is used, it must exist on the disk. A file is created with the INT 21H, function call number 3CH. To **create a file,** the name of the file and its extension are stored in memory as an ASCII-Z string at a location addressed by DS:DX before calling the function. The CX register must also contain the **attribute** of the file created by function 3CH. DOS function 5BH also creates a file and uses the same parameters as function 3CH. The difference is that function 3CH erases the file if it already exists, while 5BH signals an error if the file already exists. Which function is used depends on the program and whether the file that has the same name must be deleted or erased.

A filename is always stored as an ASCII-Z string and may contain the disk drive letter and directory path(s) if needed. Example 8–7 shows several example ASCII-Z string filenames stored in memory for access by file utilities. An **ASCII-Z string** is a character string that ends with a null character (00H).

EXAMPLE 8–7

```
0000 44 4F 47 2E 54 58   FILE1   DB     'DOG.TXT',0        ;file name DOG.TXT
     54 00
0008 43 3A 44 41 54 41   FILE2   DB     'C:DATA.DOC',0     ;file C:DATA.DOC
     2E 44 4F 43 00
0013 43 3A 5C 44 52 45   FILE3   DB     'C:\DREAD\ERROR.FIL',0
     41 44 5C 45 52 52
     4F 52 2E 46 49 4C
     00
```

---

[2]CP/M is a registered trademark of Digital Research Corporation.

Suppose that a 256 memory buffer area is filled with data that must be stored in a new file called DATA.NEW on the default disk drive and current directory path. Before data is written to this new file, it must first be created. Example 8–7 is a short program that creates this new file on the disk.

Whenever a file is created, the CX register must contain the attributes or **characteristics** of the file. In Example 8–8 a file is created without any attributes, which is often the case. Table 8–3 lists and defines the attribute bit positions. A logic 1 in a bit selects the attribute, while a logic 0 does not. For example, to create a hidden archive file, CX is loaded with a 0022H. The hidden and archive bits are set by 0022H or 02H plus 20H.

EXAMPLE 8–8

```
                        ;A program that creates file DATA.NEW.
                        ;DO NOT RUN this program because it does not close the
                        ;file.
                        ;
                                .MODEL SMALL
0000                            .DATA
0000 44 41 54 41  FILEN  DB    'DATA.NEW',0      ;filename
     2E 4E 45 57
     00
0000                            .CODE
                                .STARTUP
0017 B4 3C                      MOV   AH,3CH           ;create file function
0019 B9 0000                    MOV   CX,0             ;normal file attribute
001C BA 0000 R                  MOV   DX,OFFSET FILEN  ;address filename
001F CD 21                      INT   21H              ;access DOS
                                .EXIT
                                END
```

After returning from the INT 21H, the carry flag indicates whether an error occurred (C = 1) during the creation of the file. Some errors that occur during file creation and are obtained if needed by INT 21H function call number **59H,** are: path not found, no file handles available, and media error. If carry is cleared (C = 0), no error occurred during file creation and the AX register contains a file handle. The **file handle** is a number that refers to the file after it is created or opened. The file handle allows a file to be accessed without using the ASCII-Z string name of the file, speeding the operation for subsequent file accesses.

In DOS there are as many file handles available as required, but the number is normally restricted to 50 or less. File handles specify newly created files or opened files. When a file is closed, the file handle is released for the next creation or open. File handles also specify the common I/O devices connected to the personal computer. Table 8–4 lists the first five file handles

TABLE 8–3 File attribute definitions

Bit Position	Value	Attribute	Function
0	01H	Read-only	Read-only file or subdirectory
1	02H	Hidden	Prevents a file or directory name from appearing in a directory list
2	04H	System	Specifies a system file
3	08H	Volume	Names the disk
4	10H	Subdirectory	Specifies a subdirectory name
5	20H	Archive	Indicates that a file has changed since the last backup

TABLE 8–4   The first five file handles

Handle	Function
0000H	Reads the keyboard (CON:)
0001H	Displays on the video display (CON:)
0002H	Error display (CON:)
0003H	Uses the COM1 port (AUX:)
0004H	Uses the printer port (PRN:)

and the devices that are addressed by them. If needed, data can be sent to the video display through file handle 0001H or the keyboard can be read using file handle 0000H. File handles are not normally used for the keyboard and display, but they can be.

Writing to a File

Now that a new file (FILE.NEW) is created in Example 8–8, data can be written to it. Before writing to a file, it must be created or opened. When a file is created or opened, the file handle returns in the AX register. The **file handle** is used to refer to the file whenever data is written.

Function number **40H** writes data to an opened or newly created file. In addition to loading a 40H into AH, BX is loaded with the file handle, CX with the number of bytes to be written, and DS:DX with the address of the memory buffer data area to be written to the disk.

Suppose that 256 bytes are written from data segment memory area BUFFER to a file. This is accomplished using function 40H as illustrated in Example 8–9. If an error occurs during a write operation, the carry flag is set. If no error occurs, the carry flag is cleared and the number of bytes written to the file are returned in the AX register. Errors that occur during write operations usually indicate that the disk is full or that there is some type of media error. Media errors are caused by a bad disk (floppy) or if there is a problem with the disk electronics. They may also occur if the disk drive door is left open or no disk is placed in the drive.

EXAMPLE 8–9

```
                           .        .
                           .        .
                           .        .
0010 8B D8              MOV    BX,AX              ;move handle to BX
0012 B4 40              MOV    AH,40H             ;load write function
0014 B9 0100            MOV    CX,256             ;load count
0017 BA 0009 R          MOV    DX,OFFSET BUFFER   ;address BUFFER
001A CD 21              INT    21H                ;write 256 bytes

001C 72 32              JC     ERROR1             ;on write error
                           .        .
                           .        .
                           .        .
```

Opening, Reading, and Closing a File

To read an existing file, it first must be **opened.** When a file is opened, DOS checks the directory to determine if the file exists and returns the DOS file handle in register AX. The DOS file handle must be used for reading, writing, and closing a file.

Example 8–10 is a program that opens a file, reads 256 bytes from the file into memory area BUF, and then closes the file. When a file is opened (AH = 3DH), the AL register specifies the type of operation allowed for the opened file. If AL = 00H, the file is opened for a read operation; if AL = 01H, the file is opened for a write operation; if AL = 02H, the file is opened for a read or a write operation. In this example it is opened for a read/write operation.

EXAMPLE 8–10

```
                              ;A program that opens the file TEMP.ASM and reads the
                              ;first 256 bytes into an area of memory called BUF.
                              ;
                                      .MODEL SMALL
0000                                  .DATA
0000 54 45 4D 50     FILEN    DB      'TEMP.ASM',0    ;filename
     2E 41 53 4D
     00
0009 0100 [          BUF      DB      256 DUP (?)     ;buffer area
          00
        ]
0000                                  .CODE
                                      .STARTUP
0017 B8 3D02                  MOV     AX,3D02H        ;open file function
001A BA 0000 R                MOV     DX,OFFSET FILEN ;address filename
001D CD 21                    INT     21H             ;access DOS
001F 8B D8                    MOV     BX,AX           ;file handle to BX

0021 B4 3F                    MOV     AH,3FH          ;read file function
0023 B9 0100                  MOV     CX,256          ;read 256 bytes
0026 BA 0009 R                MOV     DX,OFFSET BUF   ;store data at BUF
0029 CD 21                    INT     21H             ;access DOS

002B B4 3E                    MOV     AH,3EH          ;close file function
002D CD 21                    INT     21H             ;access DOS
                                      .EXIT
                                      END
```

Function number **3FH** causes a file to be read. As with the write function, BX contains the file handle, CX contains the number of bytes to be read, and DS:DX contains the location of a memory area where the data are stored. As with all disk functions, the carry flag indicates an error when C = 1. If C = 0, the AX register indicates the number of bytes read from the file. In Example 8–10, the errors are ignored by not testing the carry flag after the INT 21H instructions. This is done to shorten the example.

Closing a file is very important. If a file is left open, some serious problems occur that can actually destroy the disk and all its data. If a file is written and not closed, the FAT can become corrupted, making it difficult or impossible to retrieve data from the disk. Always be certain to close the file after it is read or written. If you suspect that a file has been written without closing, you should execute the DOS utility program CHKDSK **(check disk)** before any subsequent write to the disk. The CHKDSK program tests the disk and looks for lost file chains. If it detects a lost file chain (usually caused by forgetting to close a file), it can correct the problem. Run the CHKDSK /F command at the DOS prompt to fix any lost file chains. You can also use the SCANDISK program with DOS version 6.X. Be careful to *never use* CHKDSK from Windows. This also applies to the MS-DOS icon in Windows.

The File Pointer

Whenever a file is opened, written, or read, the file pointer addresses the current location in the sequential file. When a file is opened, the file pointer always addresses the first byte of the file. If a file is 1,024 bytes in length and a read function (3FH) reads 1,023 bytes, the file pointer addresses the last byte of the file, but not the end of the file.

The **file pointer** is a 32-bit number that addresses any byte in a file. Once a file is opened, the file pointer can be changed with the move file pointer function number 42H to access any location in the file. A file pointer can be moved from the start of the file (AL = 00H), from the current location (AL = 01H), or from the end of the file (AL = 02H). In practice, all three directions are used to access different parts of the file. The distance moved by the file pointer is specified by registers CX and DX. The DX register holds the least-significant part and CX the most-significant part of the distance. Register BX must contain the file handle before using function 42H to move the file pointer.

Suppose that a file exists on the disk and that you must append the file with 256 bytes of new information. When the file is opened, the file pointer addresses the first byte of the file. If you attempt to write new data without moving the file pointer to the end of the file, the new data will overwrite the first 256 bytes of the file. Example 8–11 opens a file, moves the file pointer to the end of the file, writes 256 bytes of data, and then closes the file. This **appends** the file with 256 new bytes of data.

EXAMPLE 8–11

```
                        ;A program that opens FILE.NEW and appends it with 256
                        ;bytes of data from BUF.
                        ;
                                .MODEL SMALL
0000                            .DATA
0000  46 49 4C 45   FILEN   DB     'FILE.NEW',0      ;filename
      2E 4E 45 57
      00
0009  0100 [        BUF     DB    256 DUP (?)        ;buffer
        00
             ]
0000                            .CODE
                                .STARTUP
0017  B8 3D02                MOV   AX,3D02H          ;open FILE.NEW
001A  BA 0000 R             MOV   DX,OFFSET FILEN
001D  CD 21                 INT   21H
001F  8B D8                 MOV   BX,AX

0021  B8 4202                MOV   AX,4202H          ;move file pointer to end
0024  BA 0000               MOV   DX,0
0027  B9 0000               MOV   CX,0
002A  CD 21                 INT   21H

002C  B4 40                 MOV   AH,40H             ;write BUF to end of file
002E  B9 0100               MOV   CX,256
0031  BA 0009 R             MOV   DX,OFFSET BUF
0034  CD 21                 INT   21H

0036  B4 3E                 MOV   AH,3EH             ;close file
0038  CD 21                 INT   21H
                            .EXIT
                            END
```

One of the more difficult file maneuvers is inserting new data into the middle of a file. Figure 8–6 shows how this is accomplished by creating a new file. Notice that the part of the file before the insertion point is copied into the new file. This is followed by the new information before the remainder of the file is appended to the end of the new file. Once the new file is complete, the old file is deleted and the new file is renamed to the old filename.

Example 8–12 shows a program that inserts new data into an old file. The new data comes from the DATA.NEW file. This information, 256 bytes in length, is added between the first 256 bytes of file DATA.OLD and the remainder of the file. Notice how this is accomplished by creating a new file on the disk called DATA.TMP. The first 256 bytes of DATA.OLD are read and stored in DATA.TMP. Afterwards, the 256 bytes from DATA.NEW are read and stored in DATA.TMP. Next, the remainder of DATA.OLD is read and stored in DATA.TMP in 256-byte sections until no more data is available. Finally, the DATA.OLD file is deleted and the DATA.TMP file is renamed to DATA.OLD. This is a fairly long program, but it is clearly listed and commented so you can understand it.

EXAMPLE 8–12

```
;A program that adds the 256-byte contents of the file
;DATA.NEW to DATA.OLD at a point between the first 256
;bytes of DATA.OLD and the remainder of the file.
```

```
                          ;
                                    .MODEL SMALL
0000                                .DATA
0000 0000           HAN1    DW    ?                    ;file handle for DATA.TMP
0002 0000           HAN2    DW    ?                    ;file handle for DATA.OLD
0004 44 41 54 41    FILE1   DB    'DATA.TMP',0
     2E 54 4D 50
     00
000D 44 41 54 4     FILE2   DB    'DATA.OLD',0
     2E 4F 4C 44
     00
0016 44 41 54 41    FILE3   DB    'DATA.NEW',0
     2E 4E 45 57
     00
001F 0100 [         BUF     DB    256 DUP (?)          ;data buffer area
        00
            ]
0000                                .CODE
                                    .STARTUP
0017 B4 3C                  MOV   AH,3CH               ;create DATA.TMP
0019 B9 0000                MOV   CX,0
001C BA 0004 R              MOV   DX,OFFSET FILE1
001F CD 21                  INT   21H
0021 A3 0000 R              MOV   HAN1,AX              ;save handle at HAN1

0024 B8 3D02                MOV   AX,3D02H             ;open DATA.OLD
0027 BA 000D R              MOV   DX,OFFSET FILE2
002A CD 21                  INT   21H
002C 8B D8                  MOV   BX,AX
002E A3 0002 R              MOV   HAN2,AX              ;save handle at HAN2

0031 B4 3F                  MOV   AH,3FH               ;read 256 bytes of DATA.OLD into BUF
0033 B9 0100                MOV   CX,256
0036 BA 001F R              MOV   DX,OFFSET BUF
0039 CD 21                  INT   21H

003B B4 40                  MOV   AH,40H               ;write BUF to DATA.TMP
003D 8B 1E 0000 R           MOV   BX,HAN1              ;get handle
0041 B9 0100                MOV   CX,256
0044 BA 001F R              MOV   DX,OFFSET BUF
0047 CD 21                  INT   21H

0049 B8 3D02                MOV   AX,3D02H             ;open DATA.NEW
004C BA 0016 R              MOV   DX,OFFSET FILE3
004F CD 21                  INT   21H
0051 8B D8                  MOV   BX,AX

0053 B4 3F                  MOV   AH,3FH               ;read 256 bytes to BUF
0055 B9 0100                MOV   CX,256
0058 BA 001F R              MOV   DX,OFFSET BUF
005B CD 21                  INT   21H

005D B4 3E                  MOV   AH,3EH               ;close DATA.NEW
005F CD 21                  INT   21H

0061 B4 40                  MOV   AH,40H               ;write BUF to DATA.TMP
0063 8B 1E 0000 R           MOV   BX,HAN1              ;get handle
0067 B9 0100                MOV   CX,256
006A BA 001F R              MOV   DX,OFFSET BUF
006D CD 21                  INT   21H
006F                MAIN1:
006F B4 3F                  MOV   AH,3FH               ;read 256 bytes to BUF
0071 8B 1E 0002 R           MOV   BX,HAN2
0075 B9 0100                MOV   CX,256
0078 BA 001F R              MOV   DX,OFFSET BUF
007B CD 21                  INT   21H
007D 0B C0                  OR    AX,AX                ;test for zero byte read
```

```
007F 74 10                      JZ     MAIN2              ;if file empty
0081 B4 40                      MOV    AH,40H             ;write BUF to DATA.TMP
0083 8B 1E 0000 R               MOV    BX,HAN1
0087 B9 0100                    MOV    CX,256
008A BA 001F R                  MOV    DX,OFFSET BUF
008D CD 21                      INT    21H
008F EB DE                      JMP    MAIN1
0091                  MAIN2:
0091 B4 3E                      MOV    AH,3EH             ;close DATA.OLD
0093 CD 21                      INT    21H

0095 B4 41                      MOV    AH,41H             ;delete DATA.OLD
0097 BA 000D R                  MOV    DX,OFFSET FILE2
009A CD 21                      INT    21H

009C B4 3E                      MOV    AH,3EH             ;close DATA.TMP
009E 8B 1E 0000 R               MOV    BX,HAN1
00A2 CD 21                      INT    21H

00A4 8C D8                      MOV    AX,DS
00A6 8E C0                      MOV    ES,AX              ;overlap DS and ES

00A8 B4 56                      MOV    AH,56H             ;rename DATA.TMP
00AA BA 0004 R                  MOV    DX,OFFSET FILE1    ;old name
00AD BF 000D R                  MOV    DI,OFFSET FILE2    ;new name
00B0 CD 21                      INT    21H
                                .EXIT
                                END
```

This program uses two new INT 21H function calls: The **delete** (41H) and **rename** (56H) function calls are used to delete the old file before the temporary file is renamed to the old filename. These functions and all other DOS file functions are listed in Appendix A for reference.

Another method exists for creating a temporary file for this program. The DOS function **5AH creates a unique filename** that can be used as the name of a temporary file. Upon entry, CX = attribute code and DS:DX is an ASCII-Z string that contains the path name ending with a backslash (\). Upon exit, DOS returns a unique filename appended to the end of the path. It also creates and opens this file.

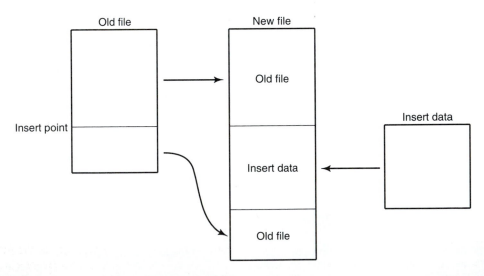

FIGURE 8–6   Inserting new data in an old file.

8–4   RANDOM ACCESS FILES

Random access files are developed through software. A **random access file** is a DOS sequential access file that is indexed by record numbers to represent a random access file. A random access file is addressed by a record number rather than by passing through the file searching for data. The move pointer function is very important when random access files are created and a record is addressed. Random access files are much easier to use for large volumes of data than sequential access files. The only disadvantage to a random access file is that it requires more room on the disk because it often contains many empty records.

Creating a Random Access File

Planning is paramount to creating a random access file system. Suppose that a random access file is required for storing customer information. Each customer record requires 16 bytes for the last name, 16 bytes for the first name, and 1 byte for the middle initial. Each customer record contains two street address lines of 32 bytes each, a city line of 16 bytes, 2 bytes for the state code, and 9 bytes for the zip code. Just the basic customer information requires 105 bytes. Additional information expands the record to 256 bytes in length. Because the business is growing, provisions are made for 5,000 customers. This means that the total required random access file is 1,280,000 bytes in length.

Example 8–13 creates a file called CUST.FIL and inserts 5,000 blank records of 256 bytes each. A blank record contains 00H in each of its bytes. This appears to be a large file, but it fits on a single high-density $5^{1}/_{4}''$ or $3^{1}/_{2}''$ floppy disk drive; in fact, this program assumes that the disk is in drive A. Note that this program takes a considerable amount of time to execute, because it writes data to virtually every byte on the floppy disk. If executed to create a file on a hard disk drive, the program takes much less time.

EXAMPLE 8–13

```
                        ;A program that creates CUST.FIL and then fills 5,000
                        ;records of 256 bytes each with zeros.
                        ;
                                .MODEL SMALL
0000                            .DATA
0000 43 55 53 54   FILE1   DB      'CUST.FIL',0      ;filename
     2E 46 49 4C
     00
0009 0100 [        BUF     DB      256 DUP (0)       ;buffer
          00
             ]
0000                            .CODE
                                .STARTUP
0017 B4 3C                      MOV     AH,3CH            ;create CUST.FIL
0019 B9 0000                    MOV     CX,0
001C BA 0000 R                  MOV     DX,OFFSET FILE1
001F CD 21                      INT     21H
0021 8B D8                      MOV     BX,AX             ;handle to BX

0023 BD 1388                    MOV     BP,5000           ;record counter
0026               MAIN1:
0026 B4 40                      MOV     AH,40H            ;write record
0028 B9 0100                    MOV     CX,256
002B BA 0009 R                  MOV     DX,OFFSET BUF
002E CD 21                      INT     21H
0030 4D                         DEC     BP                ;decrement record count
0031 75 F3                      JNZ     MAIN1             ;for 5000 records
```

```
0033 B4 3E              MOV   AH,3EH          ;close file
0035 CD 21              INT   21H
                        .EXIT
                        END
```

Reading and Writing a Record

Whenever a record is read, the record number is loaded into the BP register before the procedure in Example 8–14 is called. This procedure (READ) assumes that FIL contains the file handle number and that the CUST.FIL remains open at all times. An example might be the database in a videotape rental store. As long as the store is open, the CUST.FIL stays open for access to customer records.

EXAMPLE 8–14

```
                 ;The READ procedure reads one record from the opened
                 ;CUST.FIL.
                 ;Input parameters are:
                 ;FIL (word) = CUST.FIL handle
                 ;BP = record number
                 ;Output parameters are:
                 ;BUFFER (256 bytes) = customer record
                 ;
0000             READ    PROC FAR

0000 8B 1E 0100 R        MOV   BX,FIL          ;get handle
0004 B8 0100             MOV   AX,256          ;multiply by 256
0007 F7 E5               MUL   BP
0009 8B CA               MOV   CX,DX
000B 8B D0               MOV   DX,AX
000D B8 4200             MOV   AX,4200H        ;move pointer
0010 CD 21               INT   21H

0012 B4 3F               MOV   AH,3FH          ;read record
0014 B9 0100             MOV   CX,256
0017 BA 0000 R           MOV   DX,OFFSET BUFFER
001A CD 21               INT   21H
001C CB                  RET

001D             READ    ENDP
```

Notice how the record number is multiplied by 256 to obtain a count for the move pointer function. In each case, the file pointer is moved from the start of the file to the desired record before it is read into memory area BUFFER. Although not shown, writing a record is performed in the same manner as reading. Note that DOS does not read the disk information for the move file pointer. The move file pointer causes the disk read/write head to be positioned over the area that contains the information. This operation is called a *seek*.

Cataloging Random Access Files

It obviously takes considerable time to search each record for a customer name or address. Suppose a second file called CATNAME is created to hold all the customer names in the random access file and the record number where the data for each customer name appears. This **catalog** is a short and quickly searched file that improves search times whenever a customer must be found in the database. The same is true for the address or any other field in the database. This type of file is often called an index file.

The first example in this section used a database with 5,000 entries of 256 bytes each that contained various information. The names were stored in the first 32-byte section of each field.

The database is stored in a huge file that is 5,000 × 256, or 1,280,000 bytes in length. A catalog file called CATNAME requires a second file of 5,000 × 34, or 170,000 bytes of disk space, but requires much less time to load and search than the entire database. The entire CATNAME file could be loaded into memory and be searched in one operation, which takes little time. Each entry contains 32 bytes for the customer name; 2 bytes of each entry are required to hold the record number. Another advantage of the CATNAME catalog is that it can be sorted in alphabetical order without moving any of the records in the database, which saves a tremendous amount of time if an alphabetic list is needed.

Example 8–15 opens, reads, and searches the CATNAME file. The name to be located in CATNAME is transferred to the procedure in the DS:SI register and the name of the CATNAME file is stored at address DS:DX, which is also transferred to the procedure. A buffer area for file storage must be 8,192 bytes, whose address is passed to the procedure in DS:BP. Each entry in CATNAME, and the name to be searched for, are padded with 00H if the name is not 32 bytes in length. The last two bytes of each entry in CATNAME contain the record number that is returned to the calling program in the BP register with carry cleared if the name is found. If the name is not found, a return with carry set occurs.

EXAMPLE 8–15

```
                        ;CALLING PARAMETERS:  DS:BP = buffer address (8192 bytes)
                        ;DS:SI = search name address, and DS:DX = filename
                        ;address
                        ;RETURN PARAMETER: BP = record number
                        ;
                                PUBLIC SEARCH           ;declare search public

0000            SEARCH  PROC FAR

0000 B8 3D02            MOV   AX,3D02H           ;open file
0003 CD 21              INT   21H
0005 72 43              JC    SEARCH5            ;if error
0007 8B D8              MOV   BX,AX              ;get handle
0009 06                 PUSH  ES                 ;make ES = DS
000A 8C D8              MOV   AX,DS
000C 8E C0              MOV   ES,AX
000E            SEARCH1:
000E B4 3F              MOV   AH,3FH
0010 B9 2000            MOV   CX,8192
0013 8B D5              MOV   DX,BP
0015 CD 21              INT   21H                ;read file (8192 bytes)
0017 72 2B              JC    SEARCH4            ;if error
0019 83 F8 00           CMP   AX,0
001C 74 26              JE    SEARCH4            ;if no more data
001E B1 22              MOV   CL,34
0020 F6 F1              DIV   CL                 ;How many records are read?
0022 8A D0              MOV   DL,AL              ;save it as count
0024 8B FD              MOV   DI,BP
0026            SEARCH2:
0026 56                 PUSH  SI
0027 57                 PUSH  DI
0028 B9 0020            MOV   CX,32              ;record length
002B F3/ A6             REPE  CMPSB              ;compare for name
002D E3 0B              JCXZ  SEARCH3            ;if found
002F 5F                 POP   DI
0030 5E                 POP   SI
0031 83 C7 22           ADD   DI,34              ;go to next record
0034 FE CA              DEC   DL                 ;decrement count
0036 75 EE              JNZ   SEARCH2            ;look again
0038 EB D4              JMP   SEARCH1            ;read more of file
003A            SEARCH3:
003A 8B 2D              MOV   BP,[DI]            ;get record number
003C 5F                 POP   DI
```

```
003D 5E                      POP   SI                  ;clear stack
003E B4 3E                   MOV   AH,3EH
0040 CD 21                   INT   21H                 ;close file
0042 07                      POP   ES                  ;restore ES
0043 CB                      RET
0044               SEARCH4:
0044 B4 3E                   MOV   AH,3EH              ;close file
0046 CD 21                   INT   21H
0048 F9                      STC                       ;show error
0049 07                      POP   ES                  ;restore ES
004A               SEARCH5:
004ACB                       RET

004B               SEARCH    ENDP
```

8–5   THE DISK DIRECTORY

So far we have discussed files along with manipulating file data. Another feature of the disk
memory system is the **directory,** which is also controlled using DOS function INT 21H calls.
Functions used for directory control are: **create subdirectory** (39H), **erase subdirectory**
(3AH), **change subdirectory** (3BH), **read current directory** (47H), **select default disk drive**
(0EH), **find first matching file** (4EH), and **find next matching file** (4FH). Although the last
two functions do not change a directory, they are used to search a directory for a filename, which
is often useful.

Using the Directory Commands

To illustrate how some of the directory functions are used, we will develop a program that
changes to the root directory of drive C and then creates the subdirectory name typed at the command line. This illustrates the use of the change directory function and also the create subdirectory function. It also allows the use of the command line to place parameters into a program via
the PSP **(program segment prefix).** This is important because many programs use command-
line parameters.

      Example 8–16 extracts a parameter from the command line. This example procedure uses
two .REPEAT–.UNTIL loops. The .REPEAT, .IF, and .UNTIL commands function only with
MASM 6.X or later. The first skips over any ASCII character that is a space or below, except for
enter. The second stores uppercase characters from the command line in memory addressed by
ES:DI and is terminated by an enter (13), comma (2CH), or any character equal to or below a
space. Fortunately DOS provides the command-line information in the PSP. Refer to Appendix
A, Figure A–5, for the contents of the PSP. The PSP appends the front of an execute file and is
256 bytes in length. When a program is started, both the DS and ES segment registers address the
PSP. The command line is stored as it was typed beginning at offset address 81H, with room for
127 characters. The command line is partially parsed at offsets 5CH and 6CH. The first word on
the command line is placed in memory beginning at 5CH, and the second at 6CH. These areas
were designed for use before DOS used subdirectories and are only 16 bytes in length. Because
there is not enough room to store any possible filename with subdirectories in these areas, they
are seldom used.

EXAMPLE 8–16

```
            ;
            ;The EX procedure copies the first parameter from the
            ;command line into the location addressed by ES:DI.
            ;***Input parameters***
            ;ES:DI = address for storage of command-line parameter
```

```
                                   ;DS = segment of PSP
                                   ;***Output parameters***
                                   ;Carry = 0 for normal return.
                                   ;Carry = 1 if no parameter is found.
                                   ;
0000                   EX          PROC NEAR

0000 BE 0081                       MOV  SI,81H                 ;address command line

                                   .REPEAT                     ;skip spaces and tabs
0003 AC                                LODSB
                                   .UNTIL  AL > ' '  ||  AL == 13

000C 3C 0D                         CMP  AL,13                  ;test for enter
000E F9                            STC                         ;set no parameter error
000F 74 1A                         JE   EX1                    ;if enter
0011 4E                            DEC  SI                     ;adjust pointer

                                   .REPEAT                     ;store uppercase parameter
0012 AC                                LODSB
                                       .IF   AL >= 'a' && AL <= 'z'
001B 2C 20                                 SUB  AL,20H
                                       .ENDIF
001D AA                                STOSB
                                   .UNTIL  AL == 13 || AL <= ' ' || AL == 2CH

0026 4F                            DEC  DI                     ;adjust pointer
0027 B0 00                         MOV  AL,0                   ;make it an ASCII-Z string
0029 AA                            STOSB
002A F8                            CLC                         ;indicate parameter
002B                   EX1:
002B C3                            RET

002C                   EX          ENDP
```

Figure 8–7 shows the organization of the memory and the locations addressed by various segment and pointer registers when DOS executes a program. Notice that both the CS and SS registers address the start of the code segment that is placed in memory by the linker immediately

FIGURE 8–7 The loader program (a part of DOS) places the program and its segment in memory following the program segment prefix.

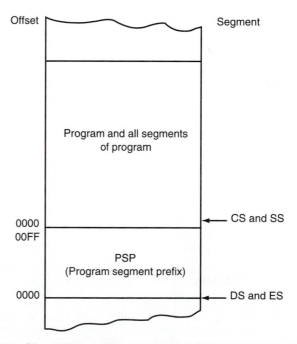

Note: SP = 0000H and IP = the address following the END directive.

following the PSP. This assumes that no stack segment has been designated in the program; otherwise, the stack segment appears between the program and the PSP. Also note that the contents of SP are 0000H and the contents of IP are loaded with the address specified by the END directive in the program.

The **linker program** stores the segments in memory exactly as they appear in the assembler file. In our examples, this is (1) stack, (2) data and extra, and (3) code. This can be changed, but it is usually not necessary. An optional convention that Microsoft uses with many assembly language programs is called the Microsoft DOS segment order. If you need to generate a program that follows the DOS segment order, place the .DOSSEG directive as the first line of the program. This causes segments of the program to be stored in the memory in the following order: (1) code, (2) data and extra segments, and (3) the stack segment.

The **PSP** contains whatever information was entered on the command line beginning at offset address 81H. (Note that 81H almost always contains a 20H or space, but to make sure, start at offset 81H.) For example, if a program named FOO is executed at the DOS prompt as c:\>FOO BETTER, memory location 82H contains the letter B (42H), location 83H contains the E (45H), and so forth. The last entry following the letter R (52H) is a carriage return or 0DH. The case of the command prompt is preserved in the memory beginning at 81H.

Because both the DS and ES registers address the PSP when a program is started, it is fairly simple to extract the command line information beginning at offset address 81H. The procedure in Example 8–16 extracts the command-line parameter and stores it into data segment location addressed by DI. Leading spaces are skipped and the end of the command line is indicated by any space (20H), carriage return (0DH), comma (2CH), or tab (09H). If more than a single parameter exists, the procedure must be modified to extract additional parameters.

To illustrate the procedure in Example 8–16 and the original programming concept described at the beginning of this section, Example 8–17 shows the program that changes to drive C, selects the root directory, and creates a new directory. The program name is NEWSUB and is followed by the name of the directory created in the root. An example is NEWSUB FROG that creates the FROG directory in the root directory.

EXAMPLE 8–17

```
                    ;A program that creates a new directory in the root of
                    ;drive C.  The name of the directory is a parameter.
                    ;
                                .MODEL SMALL
0000                            .DATA
0000 0080 [         FILE1    DB    128 DUP (0)        ;space for parameter
        00
        ]
0080 43 3A 5C 00    ROOTS    DB    'C:\',0            ;root directory
0000                            .CODE
                                .STARTUP
0017 1E                     PUSH DS                   ;exchange ES with DS
0018 06                     PUSH ES
0019 1F                     POP  DS
001A 07                     POP  ES
001B BF 0000 R              MOV  DI,OFFSET FILE1      ;address FILE1
001E E8 001E                CALL EX                   ;get command-line parameter
0021 72 18                  JC   ERR                  ;if no parameter
0023 8C C0                  MOV  AX,ES
0025 8E D8                  MOV  DS,AX                ;make ES and DS overlap

0027 B4 0E                  MOV  AH,0EH               ;log to drive C
0029 B2 02                  MOV  DL,2                 ;select drive C (2)
002B CD 21                  INT  21H

002D B4 3B                  MOV  AH,3BH               ;log to root
002F BA 0080 R              MOV  DX,OFFSET ROOTS
```

```
0032 CD 21                       INT   21H

0034 B4 39                       MOV   AH,39H              ;make new directory
0036 BA 0000 R                   MOV   DX,OFFSET FILE1
0039 CD 21                       INT   21H
003B                   ERR:
                                 .EXIT
                       ;
                       ;The EX procedure copies the first parameter from the
                       ;command line into the location addressed by ES:DI.
                       ;***Input parameters***
                       ;ES:DI = address for storage of command-line parameter
                       ;DS = segment of PSP
                       ;***Output parameters***
                       ;Carry = 0 for normal return.
                       ;Carry = 1 if no parameter is found.
                       ;
003F                   EX    PROC NEAR

003F BE 0081                     MOV  SI,81H               ;address command line

                                 .REPEAT
0042 AC                              LODSB
                                 .UNTIL  AL > ' ' || AL == 13

004B 3C 0D                       CMP   AL,13               ;test for enter
004D F9                          STC
004E 74 1E                       JE    EX1                 ;if enter
0050 4E                          DEC   SI                  ;adjust pointer

                                 .REPEAT
0051 AC                              LODSB
                                     .IF   AL >= 'a' && AL <= 'z'
005A 2C 20                              SUB   AL,20H
                                     .ENDIF
005C AA                              STOSB
                                 .UNTIL  AL == 13 || AL <= ' ' || AL == 2CH

0069 4F                          DEC   DI                  ;adjust pointer
006A B0 00                       MOV   AL,0                ;make it an ASCII-Z string
006C AA                          STOSB
006D F8                          CLC
006E                   EX1:
006E C3                          RET

006F                   EX    ENDP
                                 END
```

8–6   LOW RESOLUTION GRAPHICS DISPLAYS

Because most modern systems contain some form of VGA graphics display, we include a section on this topic as an introduction to graphics display systems. This topic could consume an entire textbook, so only the basics are explained and applied here as a launching point to spark your interest in graphics displays.

The Basic VGA Display System

The basic VGA display system operates in several modes, but we shall concentrate on the most common modes used with DOS: the 16-color, 640 × 480 display and the 256-color, 320 × 200 display. This section develops the 256-color, 320 × 200 **low resolution** video display. Other

FIGURE 8–8   The attribute byte for text mode. Note that the text color is any color from Table 8–5 and the background color is 000–111.

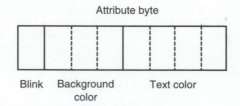

Attribute byte

Blink   Background   Text color
color

resolutions are available, but not on all display adapters. These two basic graphics modes are used for many applications except the ones that require the display of extremely high resolution video images. To display a high resolution video image, use the 256-color, 800 × 600 or 1024 × 768 display modes, which are not available to all VGA displays. The most common resolutions are available on more advanced VGA display adapters: 256-color, 640 × 480; 16-color, 800 × 600; 256-color, 800 × 600; 16-color, 1024 × 768; 256-color, 1024 × 768. In all cases these displays are bit-mapped or graphics displays instead of character-mode displays. This means that the display data is sent to the video adapter as a series of bits, or bit combinations, instead of ASCII characters as with the DOS function calls described in the previous chapter.

Video memory exists at memory locations A0000H–AFFFFH for the VGA graphics modes. (The text modes use video memory at locations B0000H–B7FFFH or B8000H–BFFFFH. If mode 7 is selected, the CGA black and white mode is activated.)

When the 256-color, 320 × 200 mode is selected, a byte in the video memory selects a single color (1 of 256) for a single PEL, or picture element. This type of display is often called a bit-mapped display. This **graphics display mode** (13H) uses 64,000 bytes of memory, slightly less than 64K bytes, which are directly addressed at locations A0000H–AF9FFH. Although this display mode has a lower resolution, it displays 256 colors and is still used by many video games for the personal computer. This area of memory is organized differently than for the 16-color, 640 × 480 VGA display mode that is described in the next section. The first byte of graphics memory (A000:0000) holds the upper-left PEL and the last byte (A000:F9FF) holds the lower-right PEL.

The most common **text mode,** mode 3, uses memory beginning at location B8000H to store ASCII-coded text and attributes for each character. The **first byte** at location B8000H holds the ASCII character displayed in the upper-left corner of the video display. The **second byte** at location B8001H holds the attribute that describes this character. The second character and its attribute are stored at location B8002H and B8003H. This continues until the last character and its attribute are stored at memory locations B8F9EH and B8F9FH for this 25-line by 80-character per line mode. Figure 8–8 shows the contents of the **attribute byte** of the text display mode. Notice that it contains two sections: one for the text color and brightness and the other for the background color and a bit that blinks the line. The 4-bit text color corresponds to the codes listed in Table 8–5. The background color corresponds to the first eight entries in Table 8–5.

TABLE 8–5   Default color codes for all text and the first 16 colors for all graphics modes

Code	Color	Code	Color
0000	Black	1000	Gray
0001	Blue	1001	Bright blue
0010	Green	1010	Bright green
0011	Cyan	1011	Bright cyan
0100	Red	1100	Bright red
0101	Magenta	1101	Bright magenta
0110	Brown	1110	Yellow
0111	White	1111	Bright white

Example 8–18 is a short program that addresses the text memory. This program first changes the attributes to black letters on a red video screen and stores blank spaces (20H) in all character display positions. Then the program moves the cursor (using BIOS INT 10H) to the upper-left corner and displays the message "This is black on red." using DOS INT 21H function number 9. In most systems, type **MODE co80** to return to the normal white on black display.

EXAMPLE 8-18

```
                               ;A program that blanks the test mode screen and
                               ;makes it red.  It then displays the message
                               ;"This is a test line." before returning to
                               ;DOS.
                               ;
                                       .MODEL SMALL
0000                                   .DATA
0000 54 68 69 73 20 69    MES   DB     'This is a test line.$'
     73 20 61 20 74 65
     73 74 20 6C 69 6E
     65 2E 24
0000                                   .CODE
                                       .STARTUP
0017 B8 B800                   MOV     AX,0B800H         ;address text
001A 8E C0                     MOV     ES,AX
001C FC                        CLD                       ;increment mode
001D BF 0000                   MOV     DI,0              ;address text
0020 B4 40                     MOV     AH,40H            ;set attribute
0022 B0 20                     MOV     AL,20H            ;load space
0024 B9 07D0                   MOV     CX,25*80          ;set count
0027 F3/ AB                    REP     STOSW             ;change screen

0029 B4 02                     MOV     AH,2              ;home cursor
002B B7 00                     MOV     BH,0              ;page 0
002D BA 0000                   MOV     DX,0              ;row 0, char 0
0030 CD 10                     INT     10H

0032 BA 0000 R                 MOV     DX,OFFSET MES     ;display MES
0035 B4 09                     MOV     AH,9
0037 CD 21                     INT     21H
                                       .EXIT
                                       END
```

Programming the 256-Color, 320 × 200 Mode

It is fairly easy to write software for this bit-mapped 256-color, 320 × 200 display mode because each byte represents a single PEL (**picture element**) on the video display. Before any information is displayed, the display adapter (assume VGA is present) must be switched to mode 13H, the 256-color, 320 × 200 mode. This is accomplished with video BIOS INT 10H, function 00H. To select a new video mode, place a 00H into AH and the new mode number in AL. This is followed by the video BIOS function call INT 10H to select the new mode. When a new mode is selected, the video display is cleared. The normal DOS video mode is 03H.

Once in the new mode, a program can begin to display graphics information. Example 8–19 switches the display to the 256-color, 320 × 200 VGA mode and then displays four bands of a vertical color bar pattern on the screen to show the 256 colors programmed by default into the VGA adapter. Here five PELS are used for each color across the entire width of the display. The colors displayed on the screen are the default colors that are programmed into the VGA display adapter. The first band contains colors 0 through 63, the second colors 64 to 127, and so forth. After displaying the four bands, the program waits for any key to be typed on the keyboard before returning to DOS with mode 3 selected. The first band contains the default colors listed in Table 8–5. The next 16 colors (16–31) are shades of gray, and the remaining colors are selected by the video display adapter.

The BAND procedure displays a color bar that displays 64 colors across the screen. Each color in the band is displayed with the width of 5 PELs; the height of the band is 40 pels. At the end of the BAND procedure, a return is made allowing 10 PELs between bands.

EXAMPLE 8–19

```
                        ;A program that displays all of 256 colors
                        ;available to the 320 x 200 video mode (13H).
                        ;***uses***
                        ;BAND procedure: displays 64 colors at a time
                        ;
                                .MODEL TINY
0000                            .CODE
                                .STARTUP
0100 B8 0013                    MOV  AX,13H            ;select mode 13H
0103 CD 10                      INT  10H

0105 B8 A000                    MOV  AX,0A000H         ;address segment A000
0108 8E C0                      MOV  ES,AX
010A FC                         CLD                    ;select increment
010B BF 0000                    MOV  DI,0              ;address offset 0000

010E B0 00                      MOV  AL,0              ;load starting color
0110 E8 001C                    CALL BAND              ;display one band

0113 B0 40                      MOV  AL,64             ;load starting color
0115 E8 0017                    CALL BAND              ;display one band

0118 B0 80                      MOV  AL,128            ;load starting color
011A E8 0012                    CALL BAND              ;display one band

011D B0 C0                      MOV  AL,192            ;load starting color
011F E8 000D                    CALL BAND              ;display one band

0122 B4 01                      MOV  AH,1              ;wait for any key
0124 CD 21                      INT  21H

0126 B8 0003                    MOV  AX,3              ;switch to DOS mode
0129 CD 10                      INT  10H
                                .EXIT
                        ;
                        ;The BAND procedure displays a band of 64 colors.
                        ;***input parameters***
                        ;AL = starting color number
                        ;ES = A000H
                        ;DI = starting offset address for display
                        ;
012F                    BAND    PROC NEAR

012F B7 28                      MOV  BH,40             ;load line count
0131                    BAND1:
0131 50                         PUSH AX                ;save starting color
0132 B9 0040                    MOV  CX,64             ;load line count
0135                    BAND2:
0135 B3 05                      MOV  BL,5              ;load display count
0137                    BAND3:
0137 AA                         STOSB                  ;store color
0138 FE CB                      DEC  BL
013A 75 FB                      JNZ  BAND3             ;repeat 5 times
013C FE C0                      INC  AL                ;change to next color
013E E2 F5                      LOOP BAND2             ;repeat for 64 colors
0140 58                         POP  AX                ;restore color
0141 FE CF                      DEC  BH
0143 75 EC                      JNZ  BAND1             ;repeat for 40 lines
0145 81 C7 0C80                 ADD  DI,320*10         ;skip 10 lines
0149 C3                         RET

014A                    BAND    ENDP
                                END
```

What if you want to change the default colors? This is accomplished by reprogramming a series of **palette registers** located on the video display card. There are 256 different locations in the palette that each represent a video display color. Each picture element is composed of the three basic video colors of red, green, and blue. These are the primary colors of light. To display a color, the monitor uses a combination of these three primary colors. The **palette memory** contains three 6-bit numbers (one for each primary color) that change the brightness of each primary color. The program in Example 8–19 uses the default palette, which has colors set so the first 16 are compatible with other display modes; the second 16 colors are shades of gray; the remaining are various colors.

Video BIOS INT 10H, function AH = 10H, AL = 10H, and BX = color number (00H–FFH) allow a palette location (color number) to be selected and changed to CH = green, CL = blue, and DH = red. The color values in CH, CL, and DH are 6-bit numbers ranging from 00H to 3FH. A value of 00H is off and 3FH is maximum brightness. For example, a very bright cyan (blue-green) is CH = 3FH, CL = 3FH, and DH = 00H. Varying these color amplitudes and various combinations of the three colors, varies the brightness and hue of a color number.

To illustrate how the palette is changed, Example 8–20 alters color numbers 80H–BFH to every possible brightness of red. It then displays a vertical color bar pattern, containing 64 bars, to illustrate each intensity of red. This program can be changed to display other colors by modifying the values loaded into the palette registers. Note that the BAND procedure is used again to display the band of colors in this example.

EXAMPLE 8–20

```
                        ;A program that displays all the possible
                        ;brightness levels of the color red for the
                        ;320 x 200, 256-color mode (13H).
                        ;
                                .MODEL TINY
0000                            .CODE
                                .STARTUP
0100 B8 0013                    MOV  AX,13H           ;switch to mode 13H
0103 CD 10                      INT  10H

0105 B8 A000                    MOV  AX,0A000H        ;address segment A000
0108 8E C0                      MOV  ES,AX
010A FC                         CLD                   ;select increment

010B B5 00                      MOV  CH,0             ;green value
010D B1 00                      MOV  CL,0             ;blue value
010F B6 00                      MOV  DH,0             ;red value
0111 BB 0080                    MOV  BX,80H           ;color number 80H
0114 B8 1010                    MOV  AX,1010H         ;change palette color
0117 B2 40                      MOV  DL,64            ;color count
0119                    PROG1:
0119 CD 10                      INT  10H              ;change a color value
011B FE C6                      INC  DH               ;next color of red
011D 43                         INC  BX               ;next color register
011E FE CA                      DEC  DL
0120 75 F7                      JNZ  PROG1            ;repeat for 64 colors

0122 BF 0000                    MOV  DI,0             ;address offset 0000
0125 B0 80                      MOV  AL,80H           ;starting color
0127 E8 000D                    CALL BAND             ;display 64 colors

012A B4 01                      MOV  AH,1             ;wait for any key
012C CD 21                      INT  21H

012E B8 0003                    MOV  AX,3             ;switch to DOS mode
0131 CD 10                      INT  10H
                                .EXIT
                        ;
                        ;The BAND procedure displays a color band of 64
                        ;colors.
```

```
                              ;***input parameters***
                              ;AL = starting color number
                              ;ES = A000H
                              ;DI = starting offset address for display
                              ;
0137                   BAND     PROC NEAR

0137 B7 28                      MOV  BH,40              ;line count of 40
0139                   BAND1:
0139 50                         PUSH AX                 ;save starting color
013A B9 0040                    MOV  CX,64              ;color count of 64
013D                   BAND2:
013D B3 05                      MOV  BL,5               ;load color count
013F                   BAND3:
013F AA                         STOSB                   ;store color
0140 FE CB                      DEC  BL
0142 75 FB                      JNZ  BAND3              ;repeat 5 times
0144 FE C0                      INC  AL                 ;get next color
0146 E2 F5                      LOOP BAND2              ;repeat for 64 colors
0148 58                         POP  AX                 ;restore color
0149 FE CF                      DEC  BH
014B 75 EC                      JNZ  BAND1              ;repeat for 40 lines
014D 81 C7 0C80                 ADD  DI,320*10          ;skip 10 raster lines
0151 C3                         RET

0152                   BAND     ENDP
                                END
```

To utilize a graphics display, many procedures are needed to draw shapes and forms on the screen. We cannot show all variations of these programs, but will show one at this point. Example 8–21 draws a box on the 256-color, 20 × 200 VGA display. The box is any size and can be placed at any location on the screen. Parameters are used to transfer size, color, and location through registers to this procedure. To illustrate the operation of the BOX procedure, the example includes a program that draws one box on the video display screen.

EXAMPLE 8–21

```
                              ;A program that displays a green box on the video
                              ;screen using video mode 13H.
                              ;
                                       .MODEL TINY
0000                                   .CODE
                                       .STARTUP
0100 FC                                CLD                    ;select increment

0101 B8 0013                           MOV AX,13H             ;select mode 13H
0104 CD 10                             INT 10H                ;and clear screen

0106 B0 02                             MOV  AL,2              ;use color 02H
0108 B9 0064                           MOV  CX,100            ;starting column
010B BE 000A                           MOV  SI,10             ;starting row
010E BD 004B                           MOV  BP,75             ;size
0111 E8 000D                           CALL BOX               ;display box

0114 B4 01                             MOV  AH,1              ;wait for any key
0116 CD 21                             INT  21H

0118 B8 0003                           MOV AX,3               ;switch to DOS mode
011B CD 10                             INT 10H
                                       .EXIT
                              ;
                              ;The BOX procedure displays a box on the mode 13H
                              ;display.
                              ;***input parameters***
                              ;AL = color number (0-255)
```

```
                        ;CX = starting column number (0-319)
                        ;SI = starting row number (0-199)
                        ;BP = size of box
                        ;
0121                    BOX     PROC NEAR

0121 BB A000                    MOV  BX,0A000H        ;address segment A000
0124 8E C3                      MOV  ES,BX
0126 50                         PUSH AX               ;save color
0127 B8 0140                    MOV  AX,320           ;find starting PEL
012A F7 E6                      MUL  SI
012C 8B F8                      MOV  DI,AX            ;address start of BOX
012E 03 F9                      ADD  DI,CX
0130 58                         POP  AX
0131 57                         PUSH DI               ;save start address
0132 8B CD                      MOV  CX,BP            ;save size in BP
0134                    BOX1:
0134 F3/ AA                     REP  STOSB            ;draw top line
0136 8B CD                      MOV  CX,BP
0138 83 E9 02                   SUB  CX,2             ;adjust CX
013B                    BOX2:
013B 5F                         POP  DI
013C 81 C7 0140                 ADD  DI,320           ;address next row
0140 57                         PUSH DI
0141 AA                         STOSB                 ;draw PEL
0142 03 FD                      ADD  DI,BP
0144 83 EF 02                   SUB  DI,2
0147 AA                         STOSB                 ;draw PEL
0148 E2 F1                      LOOP BOX2

014A 5F                         POP  DI
014B 81 C7 0140                 ADD  DI,320           ;address last row
014F 8B CD                      MOV  CX,BP
0151 F3/ AA                     REP  STOSB
0153 C3                         RET

0154                    BOX     ENDP
                                END
```

8–7   HIGHER RESOLUTION GRAPHICS DISPLAYS

In this section we discuss the use and operation of the 16-color, 640 × 480 high resolution graphics display. Although displays are available with higher resolutions and more colors, they are not standard. We end our discussion of VGA displays with this section. Additional information is available from the manufacturer of video cards for higher resolutions.

Organization of the Video Memory

The **VGA graphics memory** is 64K bytes in size and requires bit-planes to address the 153,600 bytes of memory required for a 16-color, 640 × 480 display. Many video cards contain at least 256K bytes of memory that is addressed in sections **(pages)** through the 64K-byte memory window at locations A0000H–AFFFFH and sometimes B0000H–BFFFFH, depending on the mode and video adapter present. Each time a new bit-plane is selected, the adapter internally addresses a separate 64K-byte section of memory that appears at locations A0000H–AFFFFH.

The 16-color, 640 × 480 VGA graphics mode is mode 12H when specified using the video BIOS. The memory organization for this mode is illustrated in Figure 8–9. Notice that the memory is organized in four **bit-planes.** Each **byte** in a bit-plane represents eight picture elements (PELS) on the video screen. This means that the 640 × 480 display uses 38,400 bytes of

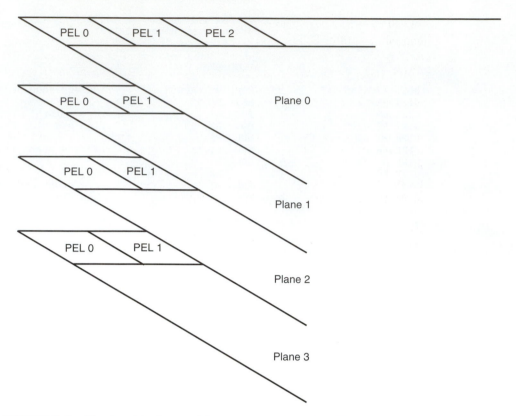

FIGURE 8–9   The four bit-planes of the 640 × 480, 16-color VGA display.

memory in each bit-plane to address the 307,200 PELS found on this display. The **first scanning line** of 640 bits is stored in the first 80 bytes of a video memory plane. The second scanning line is stored in the next 80 bytes, and so forth. The combination of the 4 bit-planes is used to specify one of 16 colors for each of the PELS. To change one PEL on the video display, four bits are changed, one in each bit-plane, to represent the new color for the PEL. Table 8–5, earlier in this chapter, lists the color codes used for a standard VGA display using 4 bit-planes. If all 4 bit-planes are cleared, a black is displayed for the PEL, and so forth.

Programming the 16-Color, 640 × 480 Mode

Programming the 256-color, 320 × 200 mode was fairly easy. The 16-color mode requires more effort because of the way the memory is organized in bit-planes. To plot a single dot:

1. Read the memory locations that must be changed to load the bit-plane information into the video card.
2. Select and address a single bit (PEL) through the **graphics address register** (GAR) and **bit mask register** (BMR).
3. Address and set the **map mask register** (MMR) to 0FH and write 0 (black) to the address containing the PEL to clear the old color from the PEL.
4. Set the desired PEL color through the MMR.
5. Write the PEL by changing the memory location that contains the video information.

The Bit Mask Register.   The bit mask register (BMR) selects the bit or bits that are modified when a byte is written to the display adapter memory. Each bit position of the display adapter memory represents a PEL in the 16-color, 640 × 480 mode. Memory from locations A0000H– A95FFH store

38,400 bytes that represent the 307,200 PELs. (It is interesting to note that commercial television transmits 211,000 PELs to form a video image, but the number of colors displayed are infinite instead of 16.) Memory location A0000H holds the first 8 PELs, location A0001H holds the next 8 PELS, and so forth. If only the leftmost PEL is changed, the bit mask register is programmed with an 80H before any information is written to memory location A0000H. Likewise, other bits or multiple bits can be modified by changing the bit mask register contents.

The BMR is accessed by programming the **graphics address register** (GAR) with an index of 8 to select the BMR. The bit mask is then programmed. The I/O address of the GAR is 03CEH and the BMR uses I/O address 03CFH. Example 8–22 shows how the BMR is programmed so all 8 bits (8 PELs) change together. The first sequence of three instructions selects the bit mask register and the second set of three instructions selects all 8 bits. This setting stays in effect until it is changed by the same sequence of instructions.

EXAMPLE 8–22

```
0000 BA 03CE              MOV   DX,3CEH         ;graphics address
0003 B0 08                MOV   AL,8            ;select bit mask
0005 EE                   OUT   DX,AL

0006 BA 03CF              MOV   DX,03CFH        ;bit mask register
0009 B0 FF                MOV   AL,0FFH         ;select all 8 bits
000B EE                   OUT   DX,AL
```

The Map Mask Register.   The **map mask register** (MMR) selects the bit-planes that are enabled for a write operation. If all bit-planes are enabled, the color F (bright white by default) is written. Note that enabling a bit only allows it to be set. The only color that cannot be written in this manner is black (color 0). To write a new color to a PEL or up to 8 PELs, if the bit mask is FFH, set all four bit-planes (0FH) and write a 00H to memory. This clears it to the color black. Next, select the color to be written by placing the binary color number (0–F) into the MMR, and write the new PELs to memory.

Access to the MMR is provided by using index 2 for sequence address register (port 3C4H). Once the MMR is accessed, select bit-planes by writing to the MMR at I/O port 3C5H. Example 8–23 shows how to write the first (leftmost) PEL on the video screen and set it to color 2 (green). The video memory location must be read before it is changed. If the location is not read, the result is unpredictable. The read operation loads the video byte into an internal latch so it can be changed before it is rewritten to the video memory by the display adapter.

EXAMPLE 8–23

```
0000 B4 00                MOV   AH,0            ;set mode to 12H
0002 B0 12                MOV   AL,12H
0004 CD 10                INT   10H

0006 BA 03CE              MOV   DX,3CEH         ;graphics address
0009 B0 08                MOV   AL,8            ;select bit mask
000B EE                   OUT   DX,AL

000C BA 03CF              MOV   DX,3CFH         ;bit mask register
000F B0 80                MOV   AL,080H         ;select leftmost bit
0011 EE                   OUT   DX,AL

0012 BA 03C4              MOV   DX,3C4H         ;sequence address
0015 B0 02                MOV   AL,2            ;select map mask
0017 EE                   OUT   DX,AL

0018 BA 03C5              MOV   DX,3C5H         ;map mask register
001B B0 0F                MOV   AL,0FH          ;enable all planes
001D EE                   OUT   DX,AL
```

```
001E B8 A000                MOV    AX,0A000H      ;address video memory
0021 8E D8                  MOV    DS,AX
0023 BF 0000                MOV    DI,0
0026 8A 05                  MOV    AL,[DI]        ;must read first
0028 C6 05 00              MOV    BYTE PTR [DI],0 ;clear old color

002B B0 02                  MOV    AL,02H         ;select color 2
002D EE                     OUT    DX,AL

002E C6 05 FF              MOV    BYTE PTR [DI],0FFH  ;write memory
```

Suppose a procedure is required that plots any PEL on the video display. Example 8–24 shows such a procedure and a program that displays some dots at various points and in various colors on the video screen. Look very closely for the single red dot below and to the right of the short cyan line when you execute this program.

The heart of this program is the DOT procedure that displays one dot of any color at any location on the video display. Parameters are passed to DOT through BX (row address), SI (column address), and DL (color). The row address is multiplied by 80 to find the memory byte that corresponds to the row on the video display. Remember that each byte contains 8 PELs. Next the column address is divided by 8. This provides a quotient that is added to the row address byte to locate the memory byte that corresponds to the desired PEL and a remainder. The remainder is used as a shift count to shift an 80H right to locate the correct bit in the bit mask register. Once the proper bit and memory location have been calculated, the remainder of the procedure sets the bit mask register and displays the correct color.

EXAMPLE 8–24

```
                            ;A program that displays a short cyan line that is
                            ;10 PELs wide with a red dot below and to the right
                            ;of the cyan line.
                            ;
                                   .MODEL TINY
0000                               .CODE
                                   .STARTUP
0100 B8 A000                MOV    AX,0A000H      ;address video RAM
0103 8E D8                  MOV    DS,AX
0105 FC                     CLD                   ;select increment

0106 B8 0012                MOV    AX,12H         ;set mode to 12H
0109 CD 10                  INT    10H            ;and clear screen

010B B9 000A                MOV    CX,10          ;set dot count to 10
010E BB 000A                MOV    BX,10          ;row address
0111 BE 0064                MOV    SI,100         ;column address
0114 B2 03                  MOV    DL,3           ;color 3 (cyan)
0116              MAIN1:                           ;plot 10 dots
0116 E8 001B                CALL   DOT            ;display one dot
0119 46                     INC    SI
011A E2 FA                  LOOP   MAIN1          ;repeat 10 times

011C BB 0028                MOV    BX,40          ;row address
011F BE 00C8                MOV    SI,200         ;column address
0122 B2 04                  MOV    DL,4           ;color 4 (red)
0124 E8 000D                CALL   DOT            ;display one red dot

0127 B4 01                  MOV    AH,1           ;wait for key
0129 CD 21                  INT    21H

012B B8 0003                MOV    AX,3
012E CD 10                  INT    10H            ;return to DOS mode
                                   .EXIT
                            ;
```

```
                              ;The DOT procedure displays one dot or PEL on the
                              ;video display.
                              ;BX = row address (0 to 479)
                              ;SI = column address (0 to 639)
                              ;DL = color (0 to 15)
                              ;
0134                DOT       PROC NEAR

0134 51                       PUSH CX
0135 52                       PUSH DX                 ;save color

0136 B8 0050                  MOV  AX,80              ;find row address
0139 F7 E3                    MUL  BX
013B 8B F8                    MOV  DI,AX              ;save it
013D 8B C6                    MOV  AX,SI              ;find column address
013F B6 08                    MOV  DH,8
0141 F6 F6                    DIV  DH
0143 8A CC                    MOV  CL,AH              ;get shift count
0145 B4 00                    MOV  AH,0
0147 03 F8                    ADD  DI,AX              ;form address of PEL
0149 B0 80                    MOV  AL,80H
014B D2 E8                    SHR  AL,CL              ;find bit in bit mask
014D 50                       PUSH AX                 ;save bit mask

014E BA 03CE                  MOV  DX,3CEH            ;graphics address
0151 B0 08                    MOV  AL,8               ;select bit mask
0153 EE                       OUT  DX,AL

0154 BA 03CF                  MOV  DX,3CFH            ;bit mask register
0157 58                       POP  AX                 ;get bit mask
0158 EE                       OUT  DX,AL

0159 BA 03C4                  MOV  DX,3C4H            ;sequence address
015C B0 02                    MOV  AL,2               ;select map mask
015E EE                       OUT  DX,AL

015F BA 03C5                  MOV  DX,3C5H            ;map mask register
0162 B0 0F                    MOV  AL,0FH             ;enable all planes
0164 EE                       OUT  DX,AL

0165 8A 05                    MOV  AL,[DI]            ;must read first
0167 C6 05 00                 MOV  BYTE PTR [DI],0    ;clear old color
016A 58                       POP  AX                 ;get color from stack
016B 50                       PUSH AX
016C EE                       OUT  DX,AL
016D C6 05 FF                 MOV  BYTE PTR [DI],0FFH ;write memory

0170 5A                       POP  DX                 ;restore registers
0171 59                       POP  CX
0172 C3                       RET

0173                DOT       ENDP
                              END
```

Video displays are often organized into text areas Example 8–25 is another procedure (BLOCK) that plots a block that corresponds to text. This procedure breaks the 640×480 display into a series of blocks that are 8 PELs wide and 9 PELs high for a display that is 80× slightly more than 53 (3 raster lines are lost at the bottom of the display). The BLOCK procedure uses BX (column number), SI (row number), and DL for the color of the block. The BLOCK procedure is useful for filling larger areas of the video display at a higher speed.

Example 8–25 displays a cyan line across the top of the screen with a white background on the remainder of the display. This is accomplished by using the BLOCK procedure for filling the display.

EXAMPLE 8–25

```
                    ;A program that displays a cyan bar across the top
                    ;of a white screen.
                    ;
                            .MODEL TINY
0000                        .CODE
                            .STARTUP
0100 B8 A000                MOV   AX,0A000H         ;address video RAM
0103 8E D8                  MOV   DS,AX
0105 FC                     CLD                     ;select increment

0106 B8 0012                MOV   AX,12H            ;set mode to 12H
0109 CD 10                  INT   10H               ;and clear screen

010B B9 0050                MOV   CX,80             ;block count
010E BB 0000                MOV   BX,0              ;row address
0111 BE 0000                MOV   SI,0              ;column address
0114 B2 03                  MOV   DL,3              ;color 3 (cyan)
0116                MAIN1:                          ;plot 80 blocks
0116 E8 0028                CALL  BLOCK             ;display a block
0119 46                     INC   SI                ;address next column
011A E2 FA                  LOOP  MAIN1             ;repeat 80 times

011C BB 0001                MOV   BX,1              ;row address
011F B2 07                  MOV   DL,7              ;color 7 (white)
0121 B6 34                  MOV   DH,52             ;row count
0123                MAIN2:
0123 BE 0000                MOV   SI,0              ;column address
0126 B9 0050                MOV   CX,80             ;column count
0129                MAIN3:
0129 E8 0015                CALL  BLOCK             ;display a block
012C 46                     INC   SI                ;address next column
012D E2 FA                  LOOP  MAIN3             ;repeat 80 times
012F 43                     INC   BX                ;increment row
0130 FE CE                  DEC   DH
0132 75 EF                  JNZ   MAIN2             ;repeat 52 times

0134 B4 01                  MOV   AH,1              ;wait for key
0136 CD 21                  INT   21H

0138 B8 0003                MOV   AX,3
013B CD 10                  INT   10H               ;return to DOS mode
                            .EXIT
                    ;
                    ;The BLOCK procedure displays one block that is 8
                    ;PELs wide by 9 PELs high.
                    ;BX = row address (0 to 52)
                    ;SI = column address (0 to 79)
                    ;DL = block color (0 to 15)
                    ;
0141                BLOCK   PROC NEAR

0141 51                     PUSH  CX
0142 52                     PUSH  DX                ;save color

0143 BA 03CE                MOV   DX,3CEH           ;graphics address
0146 B0 08                  MOV   AL,8              ;select bit mask
0148 EE                     OUT   DX,AL
0149 BA 03CF                MOV   DX,3CFH           ;bit mask register
014C B0 FF                  MOV   AL,0FFH           ;enable all 8 bits
014E EE                     OUT   DX,AL

014F BA 03C4                MOV   DX,3C4H           ;sequence address
0152 B0 02                  MOV   AL,2              ;select map mask
0154 EE                     OUT   DX,AL
```

```
0155 B8 02D0                MOV   AX,80*9          ;find row address
0158 F7 E3                  MUL   BX
015A 8B F8                  MOV   DI,AX            ;save it
015C 03 FE                  ADD   DI,SI            ;form address of PEL

015E B9 0009                MOV   CX,9             ;byte count
0161 BA 03C5                MOV   DX,3C5H          ;map mask register
0164 58                     POP   AX               ;get color
0165 50                     PUSH  AX
0166 8A E0                  MOV   AH,AL
0168              BLOCK1:
0168 B0 0F                  MOV   AL,0FH           ;enable all planes
016A EE                     OUT   DX,AL
016B 8A 05                  MOV   AL,[DI]          ;must read first
016D C6 05 00               MOV   BYTE PTR [DI],0  ;clear old color
0170 8A C4                  MOV   AL,AH
0172 EE                     OUT   DX,AL
0173 C6 05 FF               MOV   BYTE PTR [DI],0FFH ;write memory
0176 83 C7 50               ADD   DI,80
0179 E2 ED                  LOOP  BLOCK1

017B 5A                     POP   DX
017C 59                     POP   CX
017D C3                     RET

017E             BLOCK      ENDP
                            END
```

Text with Graphics.   Displaying text in a graphics display mode such as mode 12H (16 color, 640 × 480) is also difficult. Sometimes the DOS function call 21H is used to place text on the screen, but only if the video information below the text can change to both the text and its background color selected by the video attribute byte. A better way to display text in a graphics mode is to use a lookup table that contains the graphics display characters so they appear without a different background color.

The **video BIOS ROM** contains 8 × 8, 8 × 14, 9 × 14, 8 × 16, and 9 × 16 character sets. These character sets, or your own, may be used as a basis for displaying text in a graphics display mode. Figure 8–10 shows the format of the 8 × 8 character set and a few example characters. The 8 × 8 character set is most often used for the 640 × 480 resolution display. The character set is obtained from the video BIOS ROM via the INT 10H function, which returns the address of the character set in ES:BP.

Example 8–26 gets the 8 × 8 standard character set copied from the video BIOS ROM into the data segment at memory location CHAR. Because each ASCII character requires 8 bytes of memory and because the ASCII code contains 128 characters, the size of the character table is 1,024 bytes in length. Note that BH = 03H to obtain the 8 × 8 standard character set. The extended character set (code 128–255) is obtained is the same fashion except BH = 04H. Table 8–6 lists the character sets available in most video card ROMs.

FIGURE 8–10   Some 8 × 8 characters for the VGA graphics display.

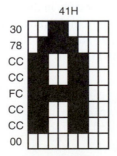
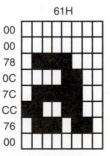
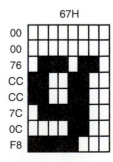

TABLE 8–6 Contents
of register BH for ROM
character sets

BH	Character Set
02H	8 (wide) × 14 (high)
03H	8 × 8 (standard)
04H	8 × 8 (extended) (ASCII 80H–0FFH)
05H	9 × 14 (alternate)
06H	8 × 16
07H	8 × 16 (alternate)

EXAMPLE 8–26

```
0000                    GETC    PROC    FAR

0000 FC                         CLD                     ;select increment

0001 B4 11                      MOV     AH,11H          ;get character set
0003 B0 30                      MOV     AL,30H
0005 B7 03                      MOV     BH,03H          ;select 8 x 8
0007 CD 10                      INT     10H

0009 8C C0                      MOV     AX,ES           ;address memory
000B 8E D8                      MOV     DS,AX
000D B8 ---- R                  MOV     AX,DATA
0010 8E C0                      MOV     ES,AX
0012 BF 0000 R                  MOV     DI,OFFSET CHAR
0015 8B F5                      MOV     SI,BP
0017 B9 0400                    MOV     CX,1024         ;load count
001A F3/ A4                     REP     MOVSB           ;copy character set

001C CB                         RET

001D                    GETC    ENDP
```

Once the character set is fetched from the video BIOS, it is used to display text on the graphics display. The procedure in Example 8–27 uses the techniques learned in Example 8–26 to obtain the character table used to display ASCII text on a graphics display. This procedure assumes that the display uses nine scanning lines for each character line and 80 characters appear across the screen on a character line. This allows 53 text lines, each containing 80 characters, to be displayed on the graphics mode screen. Before the procedure is called, AL = ASCII to be displayed; DH = character line (0–52); DL = character column (0–79); BL = text color (0–F). This procedure does not display a background color for the text. Instead, it superimposes the text on top of whatever graphics presentation appears on the video display. This same procedure can display any character set as long as it is an 8 × 8.

Example 8–27 uses the CHAR procedure to display a cyan 'A' at row 5, column 0, and a bright red 'B' at row 0, column 0. The character table in the ROM is accessed each time that the CHAR procedure is called, which is different from the procedure in Example 8–26. The CHAR procedure uses the bit mask register to change only the bits in the display that correspond to the ASCII character obtained from the video BIOS ROM.

EXAMPLE 8–27

```
                        ;A program that displays a bright red B at row 0,
                        ;column 0, and a cyan A at row 5, column 0.
                        ;
                                .MODEL TINY
0000                            .CODE
```

```
                                .STARTUP
0100 B8 A000            MOV   AX,0A000H        ;address video RAM
0103 8E D8              MOV   DS,AX
0105 FC                 CLD                    ;select increment

0106 B8 0012            MOV   AX,12H           ;set mode to 12H
0109 CD 10              INT   10H              ;and clear screen

010B B0 41              MOV   AL,'A'           ;display 'A'
010D B2 03              MOV   DL,3             ;cyan
010F BB 0005            MOV   BX,5             ;row 5
0112 BE 0000            MOV   SI,0             ;column 0
0115 E8 001A            CALL  CHAR             ;display cyan 'A'

0118 B0 42              MOV   AL,'B'           ;display 'B'
011A B2 0C              MOV   DL,12            ;bright red
011C BB 0000            MOV   BX,0             ;row 0
011F BE 0000            MOV   SI,0             ;column 0
0122 E8 000D            CALL  CHAR             ;bright red 'B'

0125 B4 01              MOV   AH,1             ;wait for key
0127 CD 21              INT   21H

0129 B8 0003            MOV   AX,3
012C CD 10              INT   10H              ;return to DOS mode
                                .EXIT
                        ;
                        ;The CHAR procedure displays a character (8 x 8) on
                        ;the mode 12H display without changing the
                        ;background color.
                        ;AL = ASCII code
                        ;DL = color (0 to 15)
                        ;BX = row (0 to 52)
                        ;SI = column (0 to 79)
                        ;
0132            CHAR     PROC NEAR

0132 51                 PUSH CX
0133 52                 PUSH DX
0134 53                 PUSH BX                ;save row address
0135 50                 PUSH AX                ;save ASCII
0136 B8 1130            MOV   AX,1130H         ;get 8 x 8 set
0139 B7 03              MOV   BH,3
013B CD 10              INT   10H              ;segment is in ES
013D 58                 POP   AX               ;get ASCII code
013E B4 00              MOV   AH,0
0140 D1 E0              SHL   AX,1             ;multiply by 8
0142 D1 E0              SHL   AX,1
0144 D1 E0              SHL   AX,1
0146 03 E8              ADD   BP,AX            ;index character
0148 5B                 POP   BX               ;get row address
0149 B8 02D0            MOV   AX,80*9          ;find row address
014C F7 E3              MUL   BX
014E 8B F8              MOV   DI,AX
0150 03 FE              ADD   DI,SI            ;add in column
0152 B9 0008            MOV   CX,8             ;set count to 8 rows

0155 BA 03CE    C1:     MOV   DX,3CEH          ;address bit mask
0158 B0 08              MOV   AL,8             ;load index 8
015A 26: 8A 66 00       MOV   AH,ES:[BP]       ;get character row
015E 45                 INC   BP               ;point to next row
015F EF                 OUT   DX,AX            ;modify bit mask
0160 BA 03C4            MOV   DX,3C4H          ;address map mask
0163 B8 0F02            MOV   AX,0F02H
0166 EF                 OUT   DX,AX            ;select all planes
0167 42                 INC   DX
0168 8A 05              MOV   AL,[DI]          ;read data
016A C6 05 00           MOV   BYTE PTR [DI],0  ;write black
```

```
016D 58                        POP  AX                 ;get color
016E 50                        PUSH AX
016F EE                        OUT  DX,AL              ;write color
0170 C6 05 FF                  MOV  BYTE PTR [DI],0FFH
0173 83 C7 50                  ADD  DI,80              ;address next row
0176 E2 DD                     LOOP C1                 ;repeat 8 times

0178 5A                        POP  DX
0179 59                        POP  CX
017A C3                        RET

017B               CHAR        ENDP
                               END
```

Example 8–28 illustrates how to display a graphics screen and place text on top of the graphics. This program selects VGA mode 12H and then displays a cyan screen. Next it places two lines of text in two different colors on top of the cyan screen. This program uses the BLOCK procedure to display a cyan screen by filling all the blocks with cyan. It then uses the CHAR procedure to place text on top of the cyan background. A new procedure (LINE) is used to display null strings that represent test text data displayed by this program.

EXAMPLE 8-28

```
                               ;A program that displays two test lines of text
                               ;on a cyan graphics background screen.
                               ;
                                       .MODEL SMALL
0000                                   .DATA
0000 54 68 69 73 20   MES1      DB     'This is test line 1.',0
     69 73 20 74 65
     73 74 20 6C 69
     6E 65 20 31 2E
     00
0015 54 68 69 73 20   MES2      DB     'This is test line 2.',0
     69 73 20 74 65
     73 74 20 6C 69
     6E 65 20 32 2E
     00
0000                                   .CODE
                                       .STARTUP
0017 B8 A000                   MOV  AX,0A000H       ;address video RAM
001A 8E D8                     MOV  DS,AX
001C FC                        CLD                  ;select increment

001D B8 0012                   MOV  AX,12H          ;set mode to 12H
0020 CD 10                     INT  10H             ;and clear screen

0022 B2 03                     MOV  DL,3            ;color cyan
0024 B6 35                     MOV  DH,53           ;row counter
0026 BB 0000                   MOV  BX,0            ;row 0
0029                 MAIN1:
0029 B9 0050                   MOV  CX,80           ;column counter
002C BE 0000                   MOV  SI,0            ;column 0
002F                 MAIN2:
002F E8 0094                   CALL BLOCK           ;display cyan block
0032 46                        INC  SI              ;address column
0033 E2 FA                     LOOP MAIN2           ;repeat 80 times
0035 43                        INC  BX              ;address next row
0036 FE CE                     DEC  DH              ;decrement row
0038 75 EF                     JNZ  MAIN1           ;repeat for 53 rows

003A B8 ---- R                 MOV  AX,@DATA        ;address data
003D 8E C0                     MOV  ES,AX           ;with ES

003F B2 09                     MOV  DL,9            ;bright blue text
0041 BB 0005                   MOV  BX,5            ;row 5
```

```
0044 BE 0000                    MOV  SI,0            ;column 0
0047 BF 0000 R                  MOV  DI,OFFSET MES1  ;address MES1
004A E8 001B                    CALL LINE            ;bright blue MES1

004D B2 0C                      MOV  DL,12           ;bright red
004F BB 000F                    MOV  BX,15           ;row 15
0052 BE 0000                    MOV  SI,0            ;column 0
0055 BF 0015 R                  MOV  DI,OFFSET MES2  ;address MES2
0058 E8 000D                    CALL LINE            ;bright red MES2

005B B4 01                      MOV  AH,1            ;wait for key
005D CD 21                      INT  21H

005F B8 0003                    MOV  AX,3
0062 CD 10                      INT  10H             ;return to DOS mode
                                .EXIT
                        ;
                        ;The line procedure displays the line of text
                        ;addressed by ES:DI.
                        ;DL = color of text (0 to 15).
                        ;The text must be stored as a null string.
                        ;BX = row
                        ;SI = column
                        ;
0068                    LINE     PROC NEAR

0068 26: 8A 05                  MOV  AL,ES:[DI]      ;get character
006B 0A C0                      OR   AL,AL           ;test for null
006D 74 0D                      JZ   LINE1           ;if null
006F 06                         PUSH ES              ;save registers
0070 57                         PUSH DI
0071 56                         PUSH SI
0072 E8 0008                    CALL CHAR            ;display characters
0075 5E                         POP  SI              ;restore registers
0076 5F                         POP  DI
0077 07                         POP  ES
0078 46                         INC  SI              ;address column
0079 47                         INC  DI              ;address character
007A EB EC                      JMP  LINE            ;repeat until null
007C                    LINE1:
007C C3                         RET

007D                    LINE     ENDP
                        ;
                        ;The CHAR procedure displays a character
                        ;(8 x 8) on the mode 12H display without
                        ;changing the background color.
                        ;AL = ASCII code
                        ;DL = color (0 to 15)
                        ;BX = row (0 to 52)
                        ;SI = column (0 to 79)
                        ;
007D                    CHAR     PROC NEAR

007D 51                         PUSH CX
007E 52                         PUSH DX
007F 53                         PUSH BX              ;save row address
0080 50                         PUSH AX              ;save ASCII
0081 B8 1130                    MOV  AX,1130H        ;get 8 x 8 set
0084 B7 03                      MOV  BH,3
0086 CD 10                      INT  10H
0088 58                         POP  AX              ;get ASCII code
0089 B4 00                      MOV  AH,0
008B D1 E0                      SHL  AX,1            ;multiply by 8
008D D1 E0                      SHL  AX,1
008F D1 E0                      SHL  AX,1
0091 03 E8                      ADD  BP,AX           ;index character
0093 5B                         POP  BX              ;get row address
0094 B8 02D0                    MOV  AX,80*9         ;find row address
```

```
0097 F7 E3                              MUL  BX
0099 8B F8                              MOV  DI,AX
009B 03 FE                              ADD  DI,SI            ;add column address
009D B9 0008              C1:           MOV  CX,8             ;set count to 8
00A0 BA 03CE                            MOV  DX,3CEH          ;address bit mask
00A3 B0 08                              MOV  AL,8             ;load index 8
00A5 26: 8A 66 00                       MOV  AH,ES:[BP]       ;get character row
00A9 45                                 INC  BP               ;point to next row
00AA EF                                 OUT  DX,AX
00AB BA 03C4                            MOV  DX,3C4H          ;address map mask
00AE B8 0F02                            MOV  AX,0F02H
00B1 EF                                 OUT  DX,AX            ;select all planes
00B2 42                                 INC  DX
00B3 8A 05                              MOV  AL,[DI]          ;read data
00B5 C6 05 00                           MOV  BYTE PTR [DI],0  ;write black
00B8 58                                 POP  AX               ;get color
00B9 50                                 PUSH AX
00BA EE                                 OUT  DX,AL            ;write color
00BB C6 05 FF                           MOV  BYTE PTR [DI],0FFH
00BE 83 C7 50                           ADD  DI,80            ;address next row
00C1 E2 DD                              LOOP C1               ;repeat 8 times
00C3 5A                                 POP  DX
00C4 59                                 POP  CX
00C5 C3                                 RET

00C6                      CHAR     ENDP
                         ;
                         ;The BLOCK procedure displays one block that is
                         ;8 PELs wide by 9 PELs high.
                         ;BX = row address (0 to 52)
                         ;SI = column address (0 to 79)
                         ;DL = block color (0 to 15)
                         ;
00C6                      BLOCK    PROC NEAR

00C6 51                                 PUSH CX
00C7 52                                 PUSH DX               ;save color

00C8 BA 03CE                            MOV  DX,3CEH          ;graphics address
00CB B0 08                              MOV  AL,8             ;select bit mask
00CD EE                                 OUT  DX,AL
00CE BA 03CF                            MOV  DX,3CFH          ;bit mask register
00D1 B0 FF                              MOV  AL,0FFH          ;enable all 8 bits
00D3 EE                                 OUT  DX,AL

00D4 BA 03C4                            MOV  DX,3C4H          ;sequence address
00D7 B0 02                              MOV  AL,2             ;select map mask
00D9 EE                                 OUT  DX,AL

00DA B8 02D0                            MOV  AX,80*9          ;find row address
00DD F7 E3                              MUL  BX
00DF 8B F8                              MOV  DI,AX            ;save it
00E1 03 FE                              ADD  DI,SI            ;form PEL address
00E3 B9 0009                            MOV  CX,9             ;byte count
00E6 BA 03C5                            MOV  DX,3C5H          ;map mask register
00E9 58                                 POP  AX               ;get color
00EA 50                                 PUSH AX
00EB 8A E0                              MOV  AH,AL
00ED                      BLOCK1:
00ED B0 0F                              MOV  AL,0FH           ;enable all planes
00EF EE                                 OUT  DX,AL
00F0 8A 05                              MOV  AL,[DI]          ;must read first
00F2 C6 05 00                           MOV  BYTE PTR [DI],0  ;clear old color
00F5 8A C4                              MOV  AL,AH
00F7 EE                                 OUT  DX,AL
00F8 C6 05 FF                           MOV  BYTE PTR [DI],0FFH  ;write memory
00FB 83 C7 50                           ADD  DI,80
00FE E2 ED                              LOOP BLOCK1
```

```
0100  5A                        POP   DX
0101  59                        POP   CX
0102  C3

0103                   BLOCK    ENDP
                                END
```

Although this has not been a complete coverage of graphics displays, you should now have a foundation in video display modes 12H and 13H and understand how text and graphics can be mixed in the graphics display modes.

8–8   SUMMARY

1. The mouse is accessed via mouse interrupt INT 33H. Through this interrupt, the position, type, and various other information about the mouse is obtained. Refer to Appendix A for a complete list of the INT 33H mouse functions.

2. Data is stored in tracks, sectors, clusters, and cylinders on a disk. A track is a concentric ring of data; a sector is a section of a track; a cluster is a group of sectors; a cylinder is a group of tracks.

3. The disk is formatted with four main storage areas: the boot sector, the file allocation table (FAT), the root directory, and the file storage area. The boot sector is 512 bytes in length and contains a program the boots the system or loads DOS into memory. The FAT is a table that indicates which clusters are free, which are occupied, and which are corrupt. The root directory is the main directory on the disk that stores filenames and directory names. The file storage area is where the data and programs are stored.

4. File and directory names are up to eight characters in length with an extension of up to three characters. Data is organized in directories and subdirectories and stored as files and programs. In Windows 95, a file or directory name can be up to 255 characters in length.

5. Files are accessed by first opening or creating them using DOS function 21H calls. Once a file is opened or created, it can be read or written using other DOS function 21H calls. Before a program that uses a file ends, the file must be closed or problems can result. Files are accessed by a filename stored in an ASCII-Z string. The ASCII-Z string is an ASCII character string that ends with a null (00H).

6. Once a file is opened via an ASCII-Z string, it is referred to by a file handle. The file handle allows DOS to process file references more efficiently. File handles can also be used to access the keyboard and video display.

7. The file pointer is a 32-bit number used to refer to any byte in a file. A DOS function 21H call (42H) is provided that allows the file pointer to be moved any number of bytes from the start of the file, the current location, or the end of the file.

8. Random access files are created by using sequential access files and the file pointer. The file pointer can be used to randomly access any portion of the sequential file.

9. The directory entry is accessed in much the same manner as the file except that the 3BH function is used to change to a new directory.

10. The command line begins at offset address 81H in the program segment prefix (PSP). This information is often used within a DOS program to ask for help or enable some feature of the DOS program.

11. The video display is accessed through the video BIOS or directly through the video display memory. For a VGA display adapter, the video display memory exists at locations A0000H–AFFFFH for the graphics modes 12H and 13H. Mode 12H is a 16-color, 640×480 display mode. Mode 13H is a 256-color, 320×200 display mode.

12. The 256-color mode stores one picture element per byte of memory and the 16-color mode is a bit-mapped mode storing 8 PELs per memory byte.

13. The bit mask register selects the PEL to be written in the 16-color, 640×480 video display. The map mask register selects the bit-planes (color) to be written in the 16-color, 640×480 video display.

14. The video BIOS ROM contains character sets that are copied to an application program to display text on the graphics mode VGA screen. This allows the display of standard characters, custom characters or character sets, and even multiple character sets.

8–9   QUESTIONS AND PROBLEMS

1. Which INT instruction is used to access the mouse?
2. Describe how to test for the existence of the mouse in a computer system.
3. How is it determined if the mouse is a serial or a bus mouse?
4. How is it determined if the right mouse button is pressed?
5. Why must the mouse be disabled when data are displayed in the video display?
6. Develop a program that enables the mouse in graphics mode 13H. Describe the appearance of the mouse pointer.
7. How is the velocity of the mouse pointer adjusted? (Refer to Appendix A.)
8. A filename may contain up to _____ characters in a DOS-based system.
9. A filename extension may contain _____ characters in a DOS-based system.
10. Which of the following filenames are valid?
    (a) FRIENDLY.1
    (b) 123.FIL
    (c) A*BC.EXT
    (d) WALL-EYE.FIS
    (e) FROG.LEG
11. What is the root directory?
12. If the line C:\DATA1\DATA2 appears, identify the purpose of each of the following parts:
    (a) C:
    (b) DATA1
    (c) DATA2
13. What is another name for a hard disk drive?
14. What is a track?
15. When a track is divided into parts, each part is called a _____.
16. Define the term cylinder.
17. How many bytes of memory are stored on a $3^{1}/_{2}''$ floppy disk that is formatted as a HD disk?
18. How many bytes of memory are stored on a $3^{1}/_{2}''$ floppy disk that is formatted as a DSDD disk?
19. What is the purpose of the boot sector on a disk memory?
20. What information is stored in the file allocation table on a disk memory?
21. What is a cluster?
22. If a cluster contains eight sectors, the minimum file allocation unit size is _____ bytes.
23. How many directory entries fit in the root directory of a $3^{1}/_{2}''$ DSDD floppy disk?
24. What is fragmentation?
25. What information is found in a directory entry?
26. How many characters can be used in a file extension?
27. How many bytes are used to store a directory name?

28. Develop a short sequence of instructions that opens a file whose name is stored at data offset address FILE1 as a read-only file.

29. Develop a short sequence of instructions that creates a hidden file whose name is stored at data segment offset address FILES.

30. Develop a program that creates a file whose name is FROG.LST and fills it with 512 bytes of 00H. (Don't forget to close the file.)

31. Write a macro instruction sequence called READ. The syntax for this macro is READ BUFFER,COUNT,HANDLE where BUFFER is the offset address of a file buffer; COUNT is the number of bytes read; HANDLE is the address of the file handle.

32. Develop a macro sequence called APPEND. The syntax for this macro is APPEND BUFFER,COUNT,HANDLE where BUFFER is the offset address of the file buffer; COUNT is the number of bytes to append to the file; HANDLE is the address of the file handle. (Don't forget to use the move file pointer function to append the end of the file.)

33. Create a random access file (RAN.FIL) containing 200 records with a record length of 100 bytes. To initialize this random access file, place a 00H in every byte.

34. Using the random access file developed in Question 33, develop a macro sequence that accesses and reads any record. The syntax for this macro is ACCESS BUFFER,RECORD, HANDLE where BUFFER is the offset address of the record buffer; RECORD is the record number; HANDLE is the address of the file handle.

35. Modify the macro in Question 34 so it can read or write a record. The syntax is ACCESS T,BUFFER,RECORD,HANDLE where T is added to specify a read ('R') or a write ('W') using the ASCII characters for the letters R and W.

36. Develop a short program that displays the first 256 bytes of a file on the video screen. Your display must appear as in Example 8–29 showing the actual data in any file. It must ask for the name of the file to be displayed, as in the example, and then display the first 256 bytes. (Use text mode.)

EXAMPLE 8–29

```
FILENAME = TEST.TXT

00:   00 00 00 00 00 00 00 00 00 00 00 00 00 00 00 00
10:   00 00 00 00 00 00 00 00 00 00 00 00 00 00 00 00
20:   00 00 00 00 00 00 00 00 00 00 00 00 00 00 00 00
30:   00 00 00 00 00 00 00 00 00 00 00 00 00 00 00 00
40:   00 00 00 00 00 00 00 00 00 00 00 00 00 00 00 00
50:   00 00 00 00 00 00 00 00 00 00 00 00 00 00 00 00
60:   00 00 00 00 00 00 00 00 00 00 00 00 00 00 00 00
70:   00 00 00 00 00 00 00 00 00 00 00 00 00 00 00 00
80:   00 00 00 00 00 00 00 00 00 00 00 00 00 00 00 00
90:   00 00 00 00 00 00 00 00 00 00 00 00 00 00 00 00
A0:   00 00 00 00 00 00 00 00 00 00 00 00 00 00 00 00
B0:   00 00 00 00 00 00 00 00 00 00 00 00 00 00 00 00
C0:   00 00 00 00 00 00 00 00 00 00 00 00 00 00 00 00
D0:   00 00 00 00 00 00 00 00 00 00 00 00 00 00 00 00
E0:   00 00 00 00 00 00 00 00 00 00 00 00 00 00 00 00
F0:   00 00 00 00 00 00 00 00 00 00 00 00 00 00 00 00
```

37. Develop a program that uses DOS INT 21H function number 0EH to log to disk drive B.

38. Develop a program that displays only the execute files (.EXE) found in the root directory of drive C. You must refer to Appendix A for information on the find first matching file function (4EH), and find subsequent matching file function (4FH). You may use wildcards ????????? to represent the filename for these functions.

39. Develop a program that displays the directory entries found in the root directory of drive C.

40. When the find functions are used in DOS, where is the file or directory name that is found by these functions located?

41. What is the DTA?
42. Use DOS function number 5AH in a program to create a unique file in the root directory of disk drive C.
43. Where is the video data stored when mode 13H is selected for the VGA video display adapter?
44. Where is the video text data stored in the personal computer?
45. How many bytes of memory are required to store a screen of information for the 256-color, 320×200 video display mode?
46. How is the INT 10H instruction used to change to a new video display mode?
47. Write a short program that displays color number 19 at each PEL on the mode 13H video display.
48. Develop a program that places the video adapter into mode 13H and then displays a red bar (color number 4) at the left margin of the screen that is 50 PELs high and 30 PELs wide. Place this bar 20 PELs down from the top of the screen.
49. What is a bit-plane?
50. How is the bit mask register used in the 16-color, 640×480 VGA video display mode?
51. How is the map mask register used in the 16-color, 640×480 VGA video display mode?
52. Develop a program that places the video adapter into mode 12H and displays a thin bright blue line, 1 PEL wide, from the top center of the screen to the bottom center of the screen.
53. Develop a program that places the video adapter into mode 12H and then draws a green line, 2 PELs wide, across the bottom of the screen.
54. Develop a program that places the video adapter into mode 12H and then displays a vertical color bar pattern that displays 16 color bars (each 40 PELs wide) showing the 16 colors available in this mode.
55. Develop a program that places the video adapter into mode 12H and then displays a white screen (color 7) with a cyan horizontal bar across the top that is 10 scanning lines high.
56. Develop your own character set for mode 12H of the VGA display. Your character set must contain characters that are 16 bits wide and 16 bits high. (Luckily you only need the characters N, a, m, e, space, and =.)
57. Using the characters developed in Question 56, develop a program using mode 12H that displays a blue video screen that contains the red character line "Name = " at the upper-left margin of the screen.
58. Use the Internet to find the Western Digital Web site and list the current sizes of hard disk drives available from them.
59. Use the Internet to find as many video card manufacturers as possible and list them.
60. Use the Internet to locate information on video drivers from Diamond and list them.

CHAPTER 9

The 80186 and 80188
Embedded Controllers

INTRODUCTION

The Intel 80186 and 80188 16-bit microprocessors are upward compatible to the 8086/8088. Even the hardware is similar to the earlier versions. This chapter presents an overview of each microprocessor, and points out the differences and enhancements that are present in each version. Both 16-bit embedded controllers are also available in CMOS as the 80C186 and 80C188 in four different models: XL, EA, EB, and EC. The CMOS versions are discussed here because they are new (late 1995) and have supplanted the older MOS versions. The first two models (XL and EA) are similar to much older versions of the 80188 and 80186, while the EB and EC are advanced and provide features such as parallel I/O pins, serial channels, enhanced DMA controllers, and enhanced interrupt controllers.

CHAPTER OBJECTIVES

Upon completion of this chapter, you will be able to:

1. Describe the hardware and software enhancements of the 80186 and 80188.
2. Explain the timing generated by the 16-bit embedded controller.
3. Describe the operation of the peripheral control block (PCB) used to control the 80186 and 80188 enhancements.
4. Develop software to initialize the 80186 and 80188 for system operation.

9–1   80186 AND 80188 ARCHITECTURE

The 80186 and 80188, like the 8086 and 8088, are nearly identical. The only difference between them is the width of their data buses. The 80186 (like the 8086) contains a 16-bit data bus, while the 80188 (like the 8088) contains an 8-bit data bus. The internal register structure of the 80186/80188 is virtually identical to the 8086/8088. About the only difference is that the 80186/80188 contain additional reserved interrupt vectors and some very powerful built-in I/O features as well as some additional instructions. The 80186/80188 are often called embedded controllers because of their application, not as a microprocessor-based computer, but as a controller used to control another system of machine.

80C186XL Block Diagram

Figure 9–1 provides the block diagram of the basic 80C186XL version of the embedded controller and Figure 9–2 shows the pin-out of the 80C186XL Notice that this embedded controller contains a great deal of internal circuitry. The block diagram of the 80186 and 80188 is identical except for the prefetch queue, which is 4 bytes in the 80188 and 6 bytes in the 80186. Like the earlier 8086, the 80186 contains a **bus interface unit** (BIU) and an **execution unit** (EU). The BIU interfaces the microprocessor to its address, data, and controls buses, and the EU executes instructions. Both units function independently of each other unless data are passed between them.

In addition to the BIU and EU, the basic 80186/80188 contain a clock generator, a programmable interrupt controller, programmable timers, a programmable DMA controller, and a programmable chip selection unit. These enhancements, not available in the 8086 microprocessor, greatly increase the utility of the 80186/80188 and reduce the number of peripheral components required to implement a system. Many popular subsystems for the personal computer use the 80186 or 80188 embedded controller as caching disk controllers, local area network (LAN) controllers, etc. The 80186 and 80188 also find application in the cellular telephone network as a switcher. Other versions (EA, EB, and EC) containing additional features are presented in later sections of this chapter. Refer to Table 9–1 for a comparison of the members of the 80186/80188 family.

Software for the 80186/80188 is identical to the internal architecture of the 8086 microprocessor mentioned in Chapter 2, where the internal structure of the microprocessor and embedded

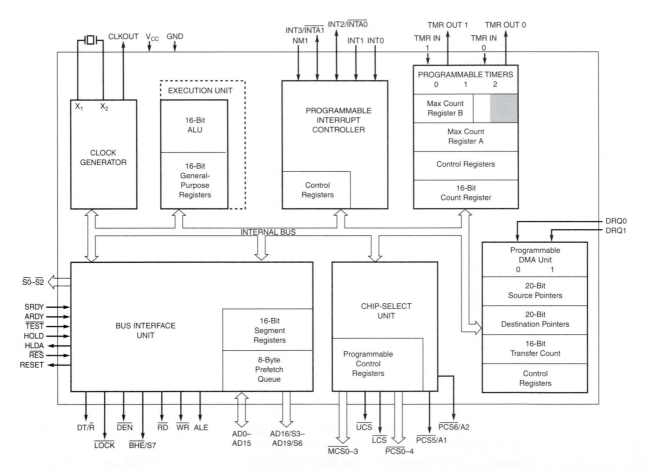

FIGURE 9–1   The block diagram of the 80C186XL microprocessor. Note the block diagram of the 80C188XL is identical, except $\overline{BHE}$/S7 is missing and AD15–AD8 are relabeled A15–A8. (Courtesy of Intel Corporation)

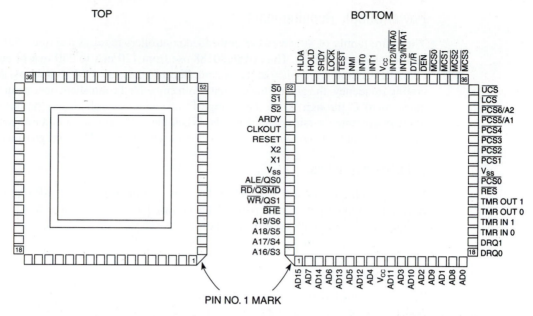

FIGURE 9–2 Pin-out of the 80C186XL microprocessor. (Courtesy of Intel Corporation)

controller is compared. Another difference between the 8086 microprocessor and the 80186/80188 is the following new software features: immediate multiplication; immediate shift counts; string I/O, PUSHA, POPA, BOUND, ENTER, and LEAVE. These new features are also found in the 80286 through the Pentium Pro.

TABLE 9–1 The four versions of 80186/80188 embedded controllers

Feature	80C186XL 80C188XL	80C186EA 80C188EA	80C186EB 80C188EB	80C186EC 80C188EC
80286-like instruction set	✔	✔	✔	✔
Power-save (green mode)	✔	✔		✔
Power-down mode		✔	✔	✔
80C187 interface	✔	✔	✔	✔
ONCE mode	✔	✔	✔	✔
Interrupt controller	✔	✔	✔	✔ 8259-like
Timer unit	✔	✔	✔	✔
Chip selection unit	✔	✔	✔ Enhanced	✔ Enhanced
DMA controller	✔ 2-channel	✔ 2-channel		✔ 4-channel
Serial communications unit			✔	✔
Refresh controller	✔	✔	✔ Enhanced	✔ Enhanced
Watchdog timer				✔
I/O ports			✔ 16 bits	✔ 22 bits

Power Supply Requirements

The entire family of 80186/80188 embedded controllers requires a single +5.0 V power supply with a tolerance of ±5%. The 80186/80188 use from 120 mA to 240 mA of current depending upon the operating frequency of the system. Because these are CMOS devices, the higher the operating frequency, the greater the current. Both embedded controllers operate in ambient temperatures of 0°C through 70°C. This range is not wide enough to use outdoors, but extended temperature range versions are available. When these devices are wired into a system, it is very important to provide good V_{CC} and ground connections on all V_{CC} and ground pins.

DC Characteristics

It is impossible to connect anything to the pins of a microprocessor without first knowing the pins' input and output characteristics. This knowledge allows you to select the proper interface components for use with the embedded controller without fear of damaging anything or causing a system failure later in the life of the product.

Input Characteristics

The input characteristics of these embedded controllers are compatible with all standard logic components. Table 9–2 lists the input voltage levels and input current requirements for any input pin on any version of the 80186/80188. The input current levels are small because the inputs are gate connections on MOSFET transistors and represent only leakage current. Notice that the input voltage levels are slightly different from standard TTL, which uses a maximum logic 0 level of 0.8 V and a minimum logic 1 level of 2.0 V. This might impact upon the noise immunity in a system.

Output Characteristics

Table 9–3 lists the output characteristics of any output pin on the 80186 and 80188 embedded controllers. The logic 1 output level is compatible with most standard logic components, but the logic 0 level is not. Standard logic circuits have a maximum logic 0 output voltage of 0.4 V, and the 80186/80188 have a maximum of 0.45 V.

The difference between standard and embedded controller logic 0 output voltage levels reduces the noise immunity from a standard 400 mV to 350 mV. (**Noise immunity** is the difference between the logic 0 output voltage and the logic 0 input voltage levels.) This reduced noise immunity may result in problems with long wire connections or too many loads. It is therefore recommended that no more than 10 loads be connected to any output pin of the embedded controller. If this loading must be exceeded, then the embedded controller must be buffered as explained later in this chapter.

TABLE 9–2 The input characteristics of the 80186/80188 embedded controllers

Logic Level	Voltage	Current
0	0.7 V maximum	±10 µA maximum
1	1.9 V minimum	±10 µA maximum

TABLE 9–3 The output characteristics of the 80186/80188 embedded controllers

Logic Level	Voltage	Current
0	0.45 V maximum	2.0 mA maximum
1	2.4 V minimum	–2.0 mA maximum

TABLE 9–4 Recommended fan-out from any 80186/80188 output pin

Family	Fan-Out	Sink Current	Source Current
TTL (74XX)	1	−1.6 mA	40 μA
TTL (74LSXX)	5	−0.4 mA	20 μA
TTL (74ASXX)	1	−2.0 mA	50 μA
TTL (74ALSXX)	10	−0.2 mA	20 μA
CMOS (74HCXX)	10	−10 μA	10 μA
CMOS (CD4000)	10	−10 μA	10 μA
NMOS	10	−10 μA	10 μA

Table 9–4 lists some of the more common logic family members and the recommended fan-out from the 80186/80188. The best choice of component types for connection to the 80186/80188 is the 74LS, 74ALS, or 74HC logic family members.

Pin Connections

The following is a list of the pin connections for the basic 80C188XL and 80C186XL embedded controllers. In later sections, the other versions (XA, XB, XC) are explained with the additional pin connections. These pins represent the core for all versions of the 80186/80188 embedded controllers.

AD0–AD7
The address/data connections contain addressing information (A0–A7) during the active phase (logic 1) of ALE and at other times contain data (D0–D7). During a HOLD or DMA action, these pins go to their high-impedance state.

A8–A15 (80188)
Pins contain address bits A8–A15 throughout the microprocessor bus cycle. During a HOLD or DMA action, these pins go to their high-impedance state.

AD8–AD15 (80186)
These connections also contain both address information and data bus information on the 80186 embedded controller. The reason is that the 80186 has a 16-bit wide data bus instead of the 8-bit wide data bus found on the 80188.

A16–A18 and A19/ONCE
These multiplexed address pins must be de-multiplexed with ALE, as are the address bits found on AD0–AD7 (80188) or AD0–AD15 (80186). The A19 pin also acts as an input on reset to enable the ONCE mode of operation. The ONCE mode allows a bed-of-nails tester to test the system connected to the embedded controller because, in the ONCE mode, the embedded controller is completely passive and all output pins are in their high-impedance state.

ALE
The address latch enable output is a signal to the system that the multiplexed address/data connections and A16–A19 contain memory address information. This signal typically connects to an external latch that captures the memory address information.

CLKIN (X1)
The clock input is connected to a crystal or to an external clocking signal. If connected to a crystal as the clock source, the other lead of the crystal connects to the OSCOUT pin. Note that the embedded

	controller operates at one-half the frequency of the crystal or clock input signal.
CLKOUT	The clock output pin provides a TTL compatible signal at one-half the crystal or clock input frequency. The CLKOUT signal is often connected to peripheral components that require a timing signal.
$\overline{DEN}$	The data bus enable signal is an output that enables external data bus buffers in very large systems.
DT/$\overline{R}$	The data transmit/receive output signals an external data bus buffer with the direction of the data transfer.
HLDA	The hold acknowledge indicates that the microprocessor has entered the hold state during a DMA action.
HOLD	An input used to request a DMA action. In response to the HOLD input, the microprocessor stops executing the program; places its address, data, and control buses at their high-impedance states; and activates the HLDA pin.
INT0, INT1, INT4	These maskable interrupt inputs are used to request an interrupt. The INT0 and INT1 inputs can also be used to expand the interrupt structure of the embedded controller by connecting to an external 8259A programmable interrupt controller or similar device. Note that INT4 is not found on all versions of the 80186/80188.
INT2/$\overline{INTA0}$, INT3/$\overline{INTA1}$	These are additional interrupt inputs (INT2 and INT3) or are used as an external interrupt controller to acknowledge an interrupt.
$\overline{LCS}$	The lower chip select is an output that selects the memory device that begins at memory location 00000H.
$\overline{LOCK}$	This output pin is a logic 0 for any instruction that contains the LOCK: prefix. This signal is used in a system that requires an external DMA controller that is locked off the system.
$\overline{MCS3}$–$\overline{MCS0}$	The middle chip selection output pins are used to select memory located between the upper and lower memory areas. These pins are not available on all versions of the 80186/80188. In some versions, these pins are replaced by eight general-purpose chip selection pins labeled $\overline{GCS0}$–$\overline{GCS7}$.
NMI	The non-maskable interrupt input is used to request interrupt type number 2.
OSCOUT (X2)	Used only when a crystal is used to time the operation of the 80186/80188. Never use this pin as an output signal from the oscillator.
$\overline{PCS6}$–$\overline{PCS0}$	The peripheral chip select pins are used to select peripheral I/O devices in a system. These pins are not available on all versions of the 80186/80188.
$\overline{RD}$	The read signal indicates that the microprocessor is ready to receive data through its data bus. This signal goes to its high-impedance state during a HOLD or DMA action.
READY	An input (or inputs ARDY and SRDY) to the microprocessor that indicates that the memory and I/O are ready for the microprocessor to read or write information. This input is used only if it is selected when programming the chip selection unit.

TABLE 9–5 The 80186/80188 status signals

$\overline{S2}$	$\overline{S1}$	$\overline{S0}$	Function
0	0	0	Interrupt acknowledge
0	0	1	I/O read
0	1	0	I/O write
0	1	1	Halt
1	0	0	Instruction fetch
1	0	1	Memory read
1	1	0	Memory write
1	1	1	Passive

$\overline{\text{RESIN}}$	Used to reset the 80186/80188 and begin executing software at location FFFF0H. This pin also causes the RESOUT pin to become a logic 1 for resetting system components.
RESOUT	An output that provides the system with a reset signal.
$\overline{\text{RFSH}}$	The refresh output shows that the 80186/80188 is in a refresh bus cycle.
$\overline{S2}, \overline{S1}, \overline{S0}$	The status output signals are encoded to provide information about the current bus cycle. Table 9–5 lists the function of each code on these three pins. Note that $\overline{S2}$ is often used to indicate a memory (1) or an I/O (0) operation.
$\overline{\text{TEST}}$	Provided for interfacing to the 80187 arithmetic coprocessor.
T0IN and T1IN	Inputs to two of the three internal timers.
T0OUT and T1OUT	Outputs from two of the three internal timers.
$\overline{\text{UCS}}$	The upper chip select is an output that selects the uppermost memory device in the system. This is usually an EPROM or flash memory that contains the operating system.
$\overline{\text{WR}}$	The write output is a signal to the memory or I/O that the microprocessor contains data on its data bus for writing to the memory or I/O system.

9–2

CLOCK GENERATION AND RESET

In this section we discuss the clock generation and reset circuitry required to operate the 80186/80188 embedded controllers.

Clock Generation

The clock generator within the 80186/80188 contains three pins: CLKIN, CLKOUT, and OS-COUT. The CLKIN pin (labeled X1 on some versions) is used to connect the 80186/80188 to an external clocking source or, if a crystal is used as a timing source, to one lead of the crystal. The other lead of the crystal is connected to the OSCOUT pin (labeled X2 on some versions). OS-COUT may be connected to only one side of the crystal and cannot be used as a clocking source in the system. The CLKOUT pin provides a TLL-compatible signal at one-half the crystal (or CLKIN) frequency. This is also the internal operating frequency of the 80186/80188.

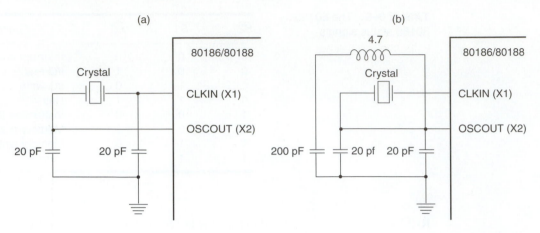

FIGURE 9–3   The crystal circuitry for (a) a fundamental mode crystal (less than 20 MHz) and (b) a third overtone crystal (20 MHz or more).

Figure 9–3 illustrates the clock pins connected to a crystal. Notice that the circuit required for fundamental operation is different for third overtone operation. A crystal operating in the fundamental mode operates at the design frequency and is usually the mode found in crystals operating below 20 MHz, while a crystal operating in the third overtone mode operates at the third harmonic of the fundamental. For example, a crystal that has been cut to operate at 1 MHz will also operate at 3 MHz in the third overtone mode.

In the fundamental mode the capacitors are usually 20 pF, while in the third overtone mode, the capacitors are still 20 pF, but a third is added that has a value of 200 pF in series with the inductor. Table 9–6 lists some values for the inductor when using a third overtone crystal. Remember that the purpose of the series resonant filter is to eliminate the fundamental frequency of the third overtone crystal. A series resonant filter is a short circuit at resonance.

In many applications, the system designer uses a potted or canned oscillator as a timing source connected to the CLKIN pin. This is the most common choice because no capacitors or inductors are required. The output of the potted oscillator connects to the CLKIN pin without the need for any other components. One problem with a crystal oscillator is that its frequency can vary slightly from the stated frequency. This does not happen when a potted oscillator is used as a timing source.

Reset

The reset circuitry within the embedded controller contains two pins: $\overline{\text{RESIN}}$ (reset input) and RESOUT (reset out). The reset input is placed at a logic 0 level to reset the 80186/80188. This is often accomplished with a push-button switch for manual resetting and an RC circuit for automatic resetting whenever power is applied to the system.

TABLE 9–6   Values for the inductor when using a third overtone crystal

Operating Frequency	Crystal Frequency	Inductor Value
10	20	6.8 µH
12	24	5.6 µH
16	32	3.9 µH
20	40	2.2 µH

FIGURE 9–4 The reset circuitry for the 80186/80188 embedded controller.

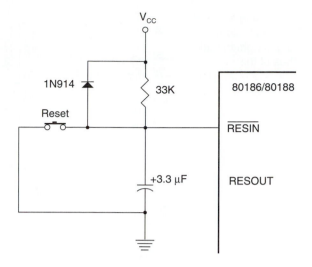

Figure 9–4 illustrates the circuit most often connected to the $\overline{\text{RESIN}}$ pin. If the system is powered down for a period of time, the charge on the capacitor drops to 0.0 V. The instant power is applied to the system, the capacitor applies 0.0 V to the $\overline{\text{RESIN}}$ input. The voltage on the capacitor eventually charges to 5.0 V through the resistor, but this takes time and in the interim allows the embedded controller to reset. The time constant (T = RC) found for the RC circuit is 100 ms for most microprocessor or embedded systems. Never choose a resistor value of greater than 100 KΩ because of the internal resistance of the capacitor. The values chosen in the example are 33 KΩ for the resistor and 3.3 μf for the capacitor. This yields a time constant of 100 ms. The diode is used to clamp the reset input pin to no more than 5.7 V. Because switches are noisy and can have significant inductance, there is always a possibility that oscillations (ringing) could cause the reset input to exceed 7.0 V. In the case of a circuit without the diode, this could destroy the 80186/80188.

The reset function causes the contents of CS to become FFFFH and the contents of IP to become 0000H. This causes the program to start executing at memory location FFFF0H. Because the first instruction is fetched from this location in the memory system, an EPROM or flash memory is placed at the top of the memory system. The memory device is selected with the $\overline{\text{UCS}}$ (upper chip select) pin from the 80186/80188. Note that every other internal function is disabled by the reset operation.

9–3   BUS BUFFERING AND LATCHING

Before the 80186/80188 can be used with memory or I/O interfaces, their multiplexed buses must be de-multiplexed. This section provides the information required to de-multiplex the buses and also illustrates how to buffer the system for connection to a large system. A large system is defined as one where more than 10 external memory or I/O devices are attached to the buses of the embedded controller. A large system is also where one or more devices connected to the bus system require a large amount of current.

De-multiplexing the Buses

The address/data bus on the 80186/80188 is multiplexed (shared) to reduce the number of pins required for the integrated circuit package. Unfortunately, this gives you the task of extracting or de-multiplexing information found on the address/data buses.

FIGURE 9–5 Using a pair of 74AL5373 latches to de-multiplex the address from the address/data bus of the 80C188XL.

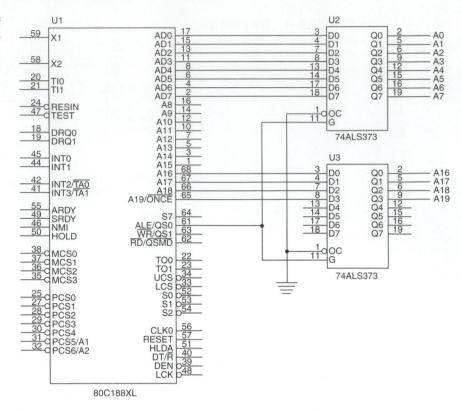

Why not leave the buses multiplexed? Memory and I/O require that the address remains valid and stable during the entire period of time required for a read or write operation. If the address remains multiplexed with data on the address/data bus, the address changes at the memory device, which causes the data to be read or written to the wrong memory or I/O location.

All computer systems have three buses: (1) an address bus that provides the memory and I/O with the memory address or I/O port number, (2) a data bus that transfers data between the microprocessor and the memory and I/O systems, and (3) a control bus that provides the memory and I/O with the read and write control signals that cause a read or write operation. These buses must be present, with all bit positions intact, for the memory or I/O to be interfaced to the microprocessor.

De-multiplexing the 80188.   Figure 9–5 illustrates the 80188 embedded controller and the components required to de-multiplex its buses. In this circuit, two 74ALS373 transparent latches are used to capture the address from the multiplexed address/data bus (AD0–AD7) and the multiplexed address information from A16–A18 and A19/ONCE pins.

The transparent latches, which act as wires that pass information whenever the G input is a logic 1, transfer inputs to the outputs whenever ALE is a logic 1. When ALE returns to a logic 0 level, the transparent latches store the inputs. In the circuit of Figure 9–5, the transparent latches store A0–A7 from address/data lines AD0–AD7 and also A16–A18 and A19/ONCE. With this circuit, the system now has a complete address bus of A0–A19 that is used to address the entire 1M-byte memory system.

In many systems, the latches attached to A16–A18 and A19/ONCE are not used because these address pins do not normally connect to the memory system unless the memory devices are larger than 64K bytes. The 80186/80188 has an on-chip programmable chip selection unit that internally decodes an address to activate a memory chip select pin. For example, if a 64K EPROM is placed at upper memory for the system program, the chip selection logic activates the

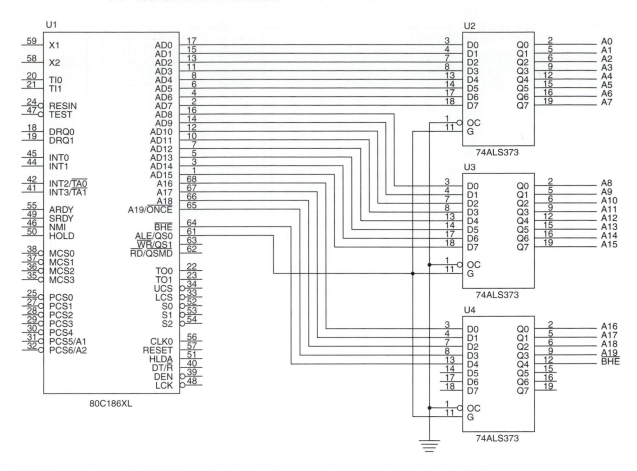

FIGURE 9–6 Using 74ALS373 latches to de-multiplex the address from the address/data bus of the 80C186XL.

memory device's chip select pin whenever address bits A16–A19 are 1111_2 (addresses F0000H–FFFFFH). The 64K memory device has address input pins A0–A15 that are available without latching A16–A19.

De-multiplexing the 80186. As with the 80188, the 80186 requires a separate address, data, and control bus. The difference between these two embedded controllers is that the 80186 has a 16-bit multiplexed address/data bus, while the 80188 has an 8-bit multiplexed address/data bus. The 80186 requires an additional latch to de-multiplex its 16-bit wide address/data bus (AD0–AD15).

The circuit in Figure 9–6 is almost identical to the one in Figure 9–5 except for the additional latch to capture A8–A15 and the $\overline{BHE}$ (bus high enable) signal. The $\overline{BHE}$ signal selects memory that connects to the upper half of the data bus. With a memory and I/O system that is 16 bits wide, the microprocessor needs a way to select any 8-bit or 16-bit memory or I/O location. This is accomplished by the use of an extra control signal called bus high enable.

The Buffered System

If more than ten unit loads are attached to any bus pin, that pin must be buffered and in many cases all the bus pins must be buffered. The address connections that are de-multiplexed by the latches are already buffered, but the data bus and control bus are not. The devices commonly used to buffer a system include the 74ALS373 octal transparent latch, the 74ALS244 octal non-

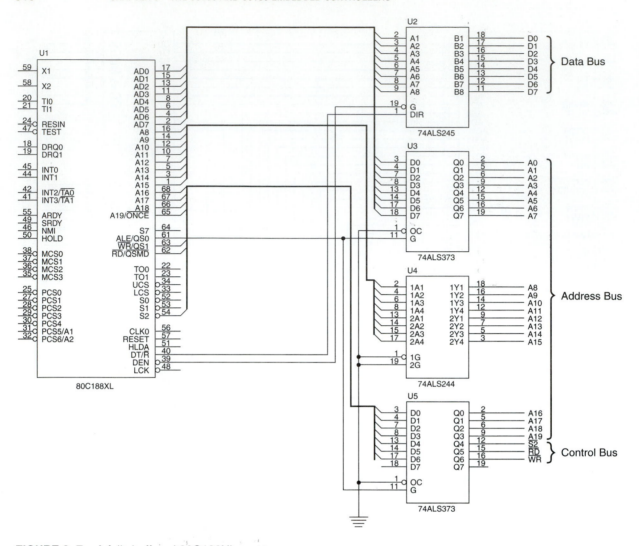

FIGURE 9–7   A fully buffered 80C188XL system.

inverting buffer, and the 74ALS245 bi-directional bus buffer. Each of these devices is designed to drive a fairly large system. The logic 0 output current is 32 mA and the logic 1 output current is –5.2 mA. The logic 0 output current is twice that of standard TTL and the logic 1 output current is many times that of standard TTL. These devices are designed to drive bus systems that contain fairly large capacitive loads.

The fully buffered system introduces a small timing delay into the system. This causes no difficulty unless memory or I/O devices are used that function at near the maximum speed of the bus system. Later in this chapter we will discuss timing and the time delays involved in complete detail.

The Fully Buffered 80188.   Figure 9–7 depicts a fully buffered 80188 embedded controller. Notice that the remaining eight address lines (A8–A15) use a 74ALS244 octal buffer; the eight data bus connections (D0–D7) use a 74ALS245 bi-directional buffer; the control bus signals ($\overline{RD}$, $\overline{WR}$, and $\overline{S2}$) are buffered with a 74ALS373 octal latch that also de-multiplexes and buffers A16–A19. The $\overline{S2}$ line is buffered because it is sometimes used in a large system to select an external memory ($\overline{S2} = 1$) or I/O ($\overline{S2} = 0$) device.

The fully buffered 80188 requires one 74ALS 244 octal buffer, one 74ALS245 bi-directional buffer, and two 74ALS373 octal transparent latches. The direction of data through the 74ALS245 bi-directional buffer is selected by the DT/$\overline{\text{R}}$ signal that is connected to the DIR input. When the DIR input is a logic 1, data flows from the A0–A7 to the B0–B7 pins on the 74ALS245. The $\overline{\text{DEN}}$ signal enables the 74ALS245.

Fully Buffered 80186. Figure 9–8 illustrates the fully buffered 80186 embedded microprocessor. The only additional parts required to fully buffer the 80186 are a pair of 74ALS245 bi-directional bus buffers to buffer the data bus connection. (Compare Figures 9–8 and 9–6.)

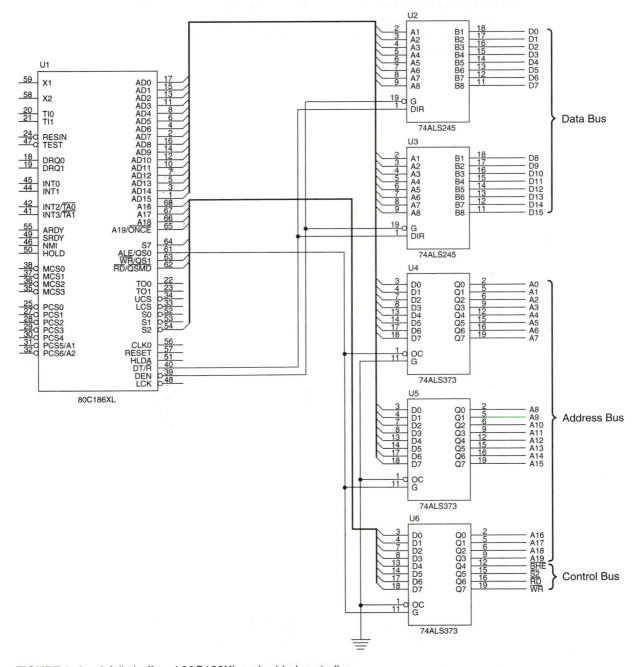

FIGURE 9–8 A fully buffered 80C186XL embedded controller.

9–4   BUS TIMING

You must understand bus timing before choosing a memory or an I/O device for interfacing to the 80186 or 80188 embedded controllers. This section presents the bus timing signals from the embedded controller, which are essential to interfacing memory and I/O.

Basic Bus Operation

The three buses in the 80186/80188—address, data, and control—function in exactly the same manner as those of any other microprocessor. If data is written to the memory (see the simplified timing of Figure 9–9), the microprocessor outputs the memory address on its address bus pins, outputs the data to be written to memory on its data bus pins, and issues a write ($\overline{WR}$) signal. The $\overline{S2}$ pin is also activated as a logic 1 for a memory operation or a logic 0 for an I/O operation. If data is read from memory (see the simplified timing diagram in Figure 9–10), the microprocessor outputs the memory address on its address bus pins, issues a read ($\overline{RD}$) signal, and then accepts the data from the memory device through its data bus pins.

Timing in General

The 80186/80188 use the memory and I/O in periods of time called **bus cycles.** Each of these bus cycles contains four system clocking periods called **T states.** If the crystal frequency is 16 MHz, the system clock is 8 MHz. (Recall that the crystal frequency is divided by a factor of 2 to generate the system clock.) Because a bus cycle requires four clocking periods, the bus cycle is 500 ns (4×125 ns) if the microprocessor is operated with an 8 MHz clock. The microprocessor can write or read data in one bus cycle. The bus cycle at 8 MHz is 500 ns, which means that the microprocessor can read or write 2 million times per second. Because there is an internal queue, the microprocessor can actually execute 4 million bytes per second in bursts. Other versions of the 80186/80188 are available that operate up to clock frequencies of 20 MHz.

T1.   During the first clocking period of a bus cycle, called T1, many things happen. The memory or I/O address is sent to the system through the address and address/data bus connections. The address/data bus is multiplexed and contains memory or I/O address information during T1. Also output during T1 are the control signals ALE, DT/$\overline{R}$, and the status signals $\overline{S2}$–$\overline{S0}$. The status signals indicate the type of bus cycle, as depicted in Table 9–5.

T2.   During T2, the microprocessor issues the $\overline{RD}$ or $\overline{WR}$ signal, $\overline{DEN}$, and, in the case of a write operation, data to be written appears on the data bus. These events cause the memory or I/O

FIGURE 9–9   Simplified 8086/8088 write bus cycle.

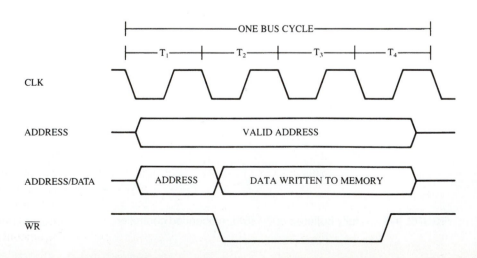

FIGURE 9–10 Simplified 8086/8088 read bus cycle.

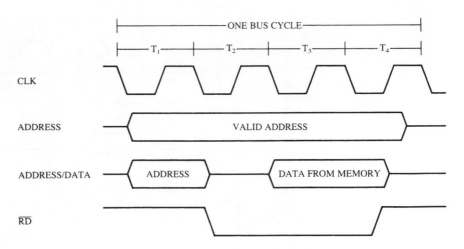

device to perform a read or write operation. The $\overline{DEN}$ signal turns on the data bus buffers, if present in a system, so the memory or I/O device can receive data from the microprocessor for a read operation. If it happens to be a write operation, then the $\overline{DEN}$ signal enables the buffers so data can be sent to the memory or I/O.

The READY input (ARDY or SRDY in some versions of the 80186/80188) is sampled during T2 to determine wait states (Tw) and inserted in the timing to lengthen access time. More about wait states is provided later in this chapter.

T3. This clocking period is provided to allow the access time for the memory or I/O. Note that the data is sampled at the end of T3/beginning of T4 for a read operation. A write operation normally terminates with the deactivation of the $\overline{WR}$ during T4.

T4. In T4, all bus signals are deactivated in preparation for the next bus cycle. This is also the time where the 80186/80188 samples the data bus during a read operation or writes the data during a write operation. The positive edge of the $\overline{WR}$ signal (0-to-1 transition) terminates the write operation.

Read Timing

Figure 9–11(a) shows the read timing for the 80186/80188 microprocessors. A close examination of the timing should allow you to identify the T states and also the major events for each T state.

The most important information gleaned from the read timing is the amount of time allowed for the memory or I/O to accomplish a read operation. Memory is chosen by its **access time,** which is the fixed amount of time required for it to read data. The amount of time is determined by the memory device manufacturer and the type of MOSFET components within the memory device. In order for a given memory device to function in a system, the amount of time allowed by the microprocessor must be known.

The amount of time allowed by the microprocessor for memory access is dependent on the system clock speed. Generally, the higher the clock frequency, the less time allowed to the memory to access data. To find the memory access time in the read timing diagram, we must first locate the point where the microprocessor samples its data bus during a read operation. If you look closely at the read timing diagram (see Figure 9–11), you will notice that the data is input at the end of T3/beginning of T4.

Memory access time starts when the microprocessor sends the address out to the memory or I/O device and ends when the data is sampled at the data bus connections. The time is approximately three clocking periods. (Refer to Figure 9–12 for the times that correspond to the timing diagram of Figure 9–11.) Notice in the read timing diagram that the address appears on the address connection some time after the start of T1. The amount of time is labeled T_{CLAV} time

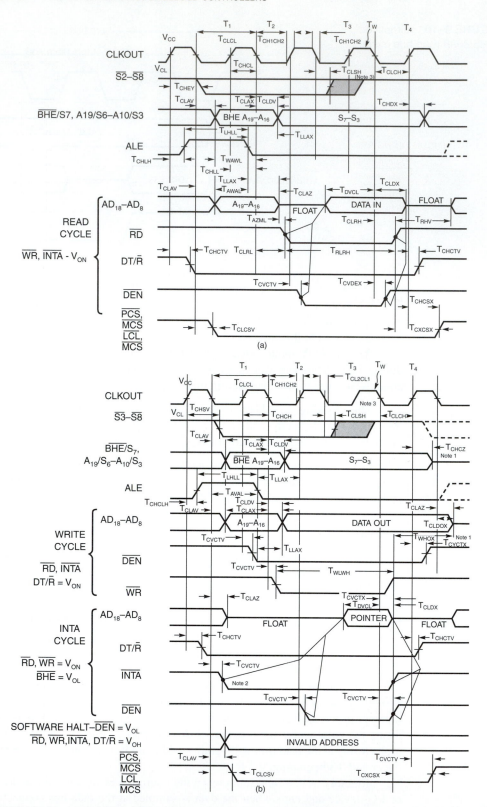

FIGURE 9–11   80C186XL/80C188XL timing. (a) Read cycle timing and (B) write cycle timing. (Courtesy of Intel Corporation)

$T_A = 0°C$ to $+70°C$, $V_{CC} = 5V \pm 10\%$ except $V_{CC} = 5V \pm5\%$ at f > 12.5 MHz

All timings are measured at 1.5V and 100 pF loading on CLKOUT unless otherwise noted.
All output test conditions are with $C_L = 50-200$ pF (10 MHz) and $C_L = 50-100$ pF (12.5–16 MHz).
For A.C. tests, input $V_{IL} = 0.45V$ and $V_{IH} = 2.4V$ except at X1, where $V_{IH} = V_{CC} - 0.5V$.

Symbol	Parameter	80C188		80C188-12		80C188-16		Unit	Test Conditions
		Min	Max	Min	Max	Min	Max		
80C186 TIMING REQUIREMENTS									
T_{DVCL}	Data in Setup (A/D)	15		15		10		ns	
T_{CLDX}	Data in Hold (A/D)	5		5		5		ns	
T_{ARYCH}	ARDY Resolution Transition Setup Time[1]	15		15		15		ns	
T_{ARYLCL}	Asynchronous Ready (ARDY) Setup Time	25		25		25		ns	
T_{CLARX}	ARDY Active Hold Time	15		15		15		ns	
T_{ARYCHL}	ARDY Inactive Hold Time	15		15		15		ns	
T_{SRYCL}	Synchronous Ready (SRDY) Transition Setup Time[2]	15		15		15		ns	
T_{CLSRY}	SRDY Transition Hold Time[2]	15		15		15		ns	
T_{HVCL}	HOLD Setup[1]	15		15		15		ns	
T_{INVCH}	INTR, NMI, TEST, TMR IN Setup Time[1]	15		15		15		ns	
T_{INVCL}	DRQ0, DRQ1, $\overline{RES}$, Setup Time[1]	15		15		15		ns	
80C188 MASTER INTERFACE TIMING RESPONSES									
T_{CLAV}	Address Valid Delay	5	50	5	36	5	33	ns	$C_L = 50$ pF – 200 pF all outputs (except T_{CLTMV}) @ 10 MHz
T_{CLAX}	Address Hold	0		0		0		ns	
T_{CLAZ}	Address Float Delay	T_{CLAX}	30	T_{CLAX}	25	T_{CLAX}	20	ns	
T_{CHCZ}	Command Lines Float Delay		40		33		28	ns	
T_{CHCV}	Command Lines Valid Delay (After Float)		45		37		32	ns	
T_{LHLL}	ALE Width (min)	$T_{CLCL} - 30$		$T_{CLCL} - 30$		$T_{CLCL} - 30$		ns	$C_L = 50$ pF – 100 pF all outputs @ 12.5 & 16 MHz
T_{CHLH}	ALE Active Delay		30		25		20	ns	
T_{CHLL}	ALE Inactive Delay		30		25		20	ns	
T_{LLAX}	Address Hold to ALE Inactive (min)	$T_{CHCL} - 20$		$T_{CHCL} - 15$		$T_{CHCL} - 15$		ns	
T_{CLDV}	Data Valid Delay	5	40	5	36	5	33	ns	
T_{CLDOX}	Data Hold Time	3		3		3		ns	
T_{WHDX}	Data Hold After $\overline{WR}$ (min)	$T_{CLCL} - 34$		$T_{CLCL} - 20$		$T_{CLCL} - 20$		ns	
T_{CVCTV}	Control Active Delay 1	3	56	3	47	3	31	ns	
T_{CHCTV}	Control Active Delay 2	5	44	5	37	5	31	ns	
T_{CVCTX}	Control Inactive Delay	3	44	3	37	3	31	ns	
T_{CVDEX}	$\overline{DEN}$ Inactive Delay (Non-write Cycle)	5	56	5	47	5	35	ns	

Notes:
1. To guarantee recognition at next clock.
2. To guarantee proper operation.

FIGURE 9–12   80C186XL/80C188XL A.C. characteristics. (Courtesy of Intel Corporation)

(continued on next page)

$T_A = 0°C$ to $+70°C$, $V_{CC} = 5V \pm 10\%$ except $V_{CC} = 5V \pm 5\%$ at $f > 12.5$ MHz

All timings are measured at 1.5V and 100 pF loading on CLKOUT unless otherwise noted.
All output test conditions are with $C_L = 50-200$ pF (10 MHz) and $C_L = 50-100$ pF (12.5–16 MHz).
For A.C. tests, input $V_{IL} = 0.45V$ and $V_{IH} = 2.4V$ except at X1, where $V_{IH} = V_{CC} - 0.5V$.

Symbol	Parameter	80C186		80C186-12		80C186-16		Unit	Test Conditions
		Min	Max	Min	Max	Min	Max		
80C186 MASTER INTERFACE REQUIREMENTS									
T_{AZRL}	Address Float to $\overline{RD}$ Active	0		0		0		ns	$C_L = 50 - 200$ pF
T_{CLRL}	$\overline{RD}$ Active Delay	5	44	5	37	5	31	ns	All outputs
T_{CLRH}	$\overline{RD}$ Inactive Delay	5	44	5	37	5	31	ns	(except T_{CLTMV}) @ 10 MHz
T_{RHAV}	$\overline{RD}$ Inactive to Address Active (min)	$T_{CLCL} - 40$		$T_{CLCL} - 20$		$T_{CLCL} - 20$		ns	$C_L = 50 - 100$ pF
T_{CLHAV}	HLDA Valid Delay	3	40	3	33	3	25	ns	All outputs @ 12.5 & 16 MHz
T_{RLRH}	$\overline{RD}$ Pulse Width (min)	$2T_{CLCL} - 46$		$2T_{CLCL} - 40$		$2T_{CLCL} - 30$		ns	
T_{WLWH}	$\overline{WR}$ Pulse Width (min)	$2T_{CLCL} - 34$		$2T_{CLCL} - 30$		$2T_{CLCL} - 25$		ns	
T_{AVLL}	Address Valid to ALE Low (min)	$T_{CLCH} - 19$		$T_{CLCH} - 15$		$T_{CLCH} - 15$		ns	Equal Loading
T_{CHSV}	Status Active Delay	5	45	5	35	5	31	ns	
T_{CLSH}	Status Inactive Delay	5	50	5	35	5	30	ns	
T_{CLTMV}	Timer Output Delay		48		40		30	ns	100 pF max @ 10 MHz
T_{CLRO}	Reset Delay		48		40		30	ns	$C_L = 50 - 200$ pF
T_{CHQSV}	Queue Status Delay		28		28		25	ns	All outputs (except T_{CLTMV}) @10MHz
T_{CHDX}	Status Hold Time	5		5		5		ns	
T_{AVCH}	Address Valid to Clock High	0		0		0		ns	$C_L = 50 - 100$ pF All outputs @ 12.5 & 16 MHz
T_{CLLV}	$\overline{LOCK}$ Valid/Invalid Delay	3	45	3	40	3	35	ns	
T_{DXDL}	$\overline{DEN}$ Inactive to DT/$\overline{R}$ Low	0		0		0		ns	Equal Loading
80C186 CHIP-SELECT TIMING RESPONSES									
T_{CLCSV}	Chip-Select Active Delay		45		33		30	ns	
T_{CXCSX}	Chip-Select Hold from Command Inactive	$T_{CLCH} - 10$		$T_{CLCH} - 10$		$T_{CLCH} - 10$		ns	Equal Loading
T_{CHCSX}	Chip-Select Inactive Delay	5	32	5	28	5	23	ns	

FIGURE 9–12   *(continued)*

on the diagram. (T_{CLAV} is 33 ns for the 16 MHz version.) Because the timing starts with T1 and ends at the beginning of T4, T_{CLAV} is subtracted from the time it takes for three clocking periods. For the 16 MHz version, this is 187.5 ns (3 × 62.5 ns) minus 33 ns or 154.5 ns. One other time must be considered when finding the access time allowed by the microprocessor. Notice in the timing diagram that there is a setup time before the data is sampled. The setup time is labeled T_{DVCL} and for the 16 MHz version is given as 10 ns. This time also must be subtracted to determine access time (see Equation 9–1). This means that the amount of access time allowed the memory or I/O when the microprocessor is operated with a 16 MHz clock is 187.5 ns –33 ns –10 ns or 144.4 ns.

EQUATION 9–1

$$Tacc = 3 \bullet CLK - T_{CLAV} - T_{DVCL}$$

The memory or I/O device chosen for the system must operate within this time. In the case of the 16 MHz version, a memory device with an access time of less than 144.5 ns would be

$T_A = 0°C$ to $+70°C$, $V_{CC} = 5V \pm 10\%$ except $V_{CC} = 5V \pm 5\%$ at f > 12.5 MHz

All timings are measured at 1.5V and 100 pF loading on CLKOUT unless otherwise noted.
All output test conditions are with $C_L = 50$–200 pF (10 MHz) and $C_L = 50$–100 pF (12.5–16 MHz).
For A.C. tests, input $V_{IL} = 0.45V$ and $V_{IH} = 2.4V$ except at X1, where $V_{IH} = V_{CC} - 0.5V$.

Symbol	Parameter	80C188		80C188-12		80C188-16		Unit	Test Conditions
		Min	Max	Min	Max	Min	Max		
80C188 CLKIN REQUIREMENTS Measurements taken with following conditions: external clock input to X1 and X2 not connected (float)									
T_{CKIN}	CLKIN Period	50	1000	40	1000	31.25	1000	ns	
T_{CKHL}	CLKIN Fall Time		5		5		5	ns	3.5 to 1.0V
T_{CKLH}	CLKIN Rise Time		5		5		5	ns	1.0 to 3.5V
T_{CLCK}	CLKIN Low Time	20		16		13		ns	$1.5V^2$
T_{CHCK}	CLKIN High Time	20		16		13		ns	$1.5V^2$
80C188 CLKOUT TIMING 200 pF load maximum for 10 MHz or less, 100 pF load maximum above 10 MHz									
T_{CICO}	CLKIN to CLKOUT Skew		25		21		17	ns	
T_{CLCL}	CLKOUT Period	100	2000	80	2000	62.5	2000	ns	
T_{CLCH}	CLKOUT	$0.5 T_{CLCL} - 8$		$0.5 T_{CLCL} - 7$		$0.5 T_{CLCL} - 7$		ns	$C_L = 100$ pF^2
	Low Time (min)	$0.5 T_{CLCL} - 6$		$0.5 T_{CLCL} - 5$		$0.5 T_{CLCL} - 5$		ns	$C_L = 50$ pF^3
T_{CHCL}	CLKOUT	$0.5 T_{CLCL} - 8$		$0.5 T_{CLCL} - 7$		$0.5 T_{CLCL} - 7$		ns	$C_L = 100$ pF^4
	High Time (min)	$0.5 T_{CLCL} - 6$		$0.5 T_{CLCL} - 5$		$0.5 T_{CLCL} - 5$		ns	$C_L = 50$ pF^3
T_{CH1CH2}	CLKOUT Rise Time		10		10		8	ns	1.0 to 3.5V
T_{CL2CL1}	CLKOUT Fall Time		10		10		8	ns	3.5 to 1.0V

Notes:
1. T_{CLCK} and T_{CHCK} (CLKIN Low and High times) should not have a duration less than 40% of T_{CKIN}.
2. Tested under worst case conditions: $V_{CC} = 5.5$ V (5.25 V @ 16 MHz), $T_A = 70°C$.
3. Not tested.
4. Tested under worst case conditions: $V_{CC} = 4.5$ V (4.75 V @ 16 MHz), $T_A = 0°C$.

FIGURE 9–12   *(continued)*

chosen. Because the memory address must be decoded, the amount of time required for the decoder must also be considered in the access time calculation. A typical decoder might require 12 ns to decode the memory address and select the memory device. In this case, 132.5 ns is left for the memory to function. The device chosen would probably have an access time of 120 ns or less. The access time can be lengthened by inserting wait states (Tw) as described later in this chapter.

The only other timing factor that might affect memory performance is the width of the $\overline{RD}$ signal. The read signal width (T_{RLRH}) is given as 2 clocks – 30 ns for the 16 MHz version. In this case the read pulse width is 95 ns.

Write Timing

Figure 9–11(b) illustrates the write timing diagram for the 80186/80188. The difference between the read and write timing are minimal. The $\overline{RD}$ signal is replaced by the $\overline{WR}$ signal; the data bus contains information for the memory instead of from it; DT/$\overline{R}$ is a logic 1 instead of a logic 0.

When interfacing some memory devices, timing may be especially critical between the point where the $\overline{WR}$ signal returns to a logic 1 and the point where the data is removed from the data bus by the microprocessor. This occurs because most memory devices use the 0-to-1 transition on the write input to perform the write operation. According to the timing diagram, this critical period is T_{WHDX}, which is given as clock –20 or 42.5 ns for the 16 MHz version of the microprocessor. In a memory device this time corresponds to write hold time. In most cases, memory has a hold time of 0 ns, but not always, so this time must also be checked when interfacing RAM to the microprocessor.

9–5   READY AND WAIT STATES

As mentioned earlier in this chapter, the READY (ARDY or SRDY on some versions of the 80188/80186) causes wait states for slower memory and I/O components. A **wait state** (Tw) is an extra clocking period inserted in the timing between T3 and T4 to lengthen the access time allowed by the microprocessor for the memory or I/O. If one wait state is inserted, then the memory access time, normally 144.5 ns for a 16 MHz 80186/80188, is lengthened by one clocking period (62.5 ns) to 207 ns. As many wait states as required can be added to the timing. If two waits are added, then the access time is lengthened to 268.5 ns.

The READY (ARDY or SRDY) Input

The READY input is sampled during T2, as are the ARDY or SRDY inputs to some versions of the 80186/80188. Refer to Figure 9–13 for the timing of these signals. There are two methods for using the READY input: **the normally ready system** and **normally not ready system.** In the normally ready system, the READY input is held high and in the normally not ready system the READY input is held low. The timing for ARDY is identical to READY and the timing for SRDY is sampled only with respect to the 1-to-0 transition of the clock in the normally not ready system and available only in the bottom-most timing diagram of Figure 9–13.

In the normally not ready system, the READY signal is a logic 0 (meaning not ready), which causes an infinite number of wait states unless it becomes a logic 1 during T2, T3, or Tw. This is indicated in the top timing diagram of Figure 9–13. In the normally ready system, the

FIGURE 9–13   The generation of wait states using the not ready and ready methods.

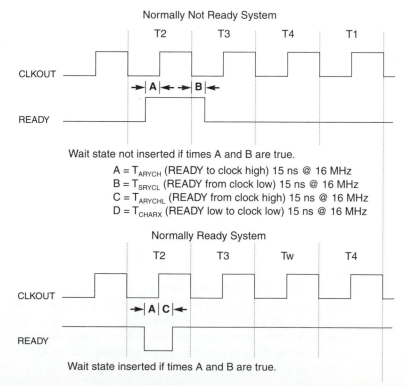

Normally Not Ready System

Wait state not inserted if times A and B are true.

A = T_{ARYCH} (READY to clock high) 15 ns @ 16 MHz
B = T_{SRYCL} (READY from clock low) 15 ns @ 16 MHz
C = T_{ARYCHL} (READY from clock high) 15 ns @ 16 MHz
D = T_{CHARX} (READY low to clock low) 15 ns @ 16 MHz

Normally Ready System

Wait state inserted if times A and B are true.

FIGURE 9–13   *(continued)*

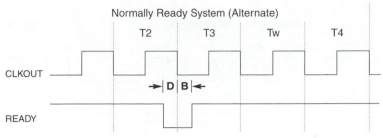

Normally Ready System (Alternate)

Wait state inserted if times A and B are true.

READY pin is held at a logic 1 level until a wait state is needed. Notice that the wait state is in-serted if the READY pin becomes a logic 0 during T2. As we mentioned, this lengthens the ac-cess time for the memory. It is again sampled during T3 and Tw to determine if additional wait states are required. In all cases the setup and hold times are 15 ns before and after the clock to the READY input.

In order to generate wait states, a circuit is needed to apply a signal to the READY input that will cause a variety of wait states. Such a circuit is illustrated in Figure 9–14. This circuit is designed so a different number of wait states can be inserted for a RAM access and a different number of waits for a ROM access. Suppose that a 250-ns EPROM is purchased and a 150-ns RAM. Because each has a different access time, a circuit is needed that can cause different num-bers of wait states for each memory device. In this case if a 16 MHz clock is in use, the RAM re-quires the insertion of 1 wait state (144.5 ns [access time] + 62.5 ns [1 wait] = 209 ns) and the EPROM requires the insertion of 2 wait states (144.5 ns [access time] + 125 ns [2 waits] = 269.5 ns). To insert these wait states, switch S2 is set so pin 3 connects to pin 14, and S1 is set so pin 2 connects to pin 15. This means that the QA output inserts 0 wait states, QB inserts 1 wait, QC in-serts 2 waits, and so forth.

The circuit works in the following manner: As long as the microprocessor is not reading or writing data, there is no ALE pulse. As soon as a read or write operation is started, the ALE

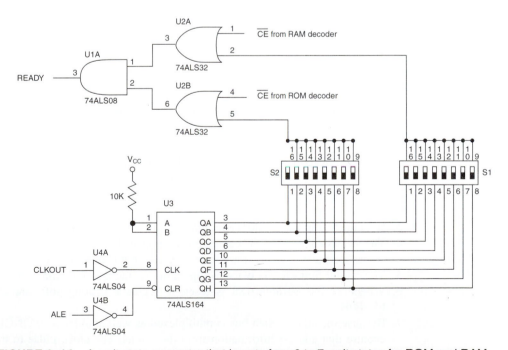

FIGURE 9–14   A wait state generator that inserts from 0 to 7 wait states for ROM and RAM.

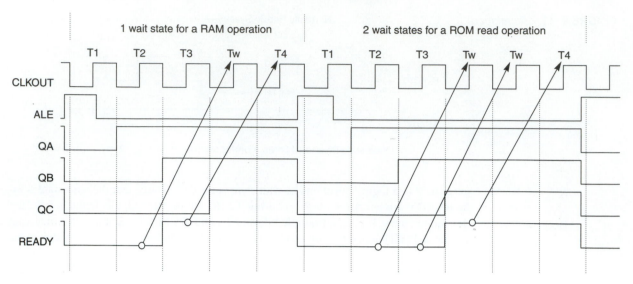

FIGURE 9–15   The timing diagram generated by the circuit of Figure 9–14 for 1 and 2 wait states.

signal becomes a logic 1, which places a logic 0 on the clear input to the shift register clearing all 8 output pins to a logic 0. Once ALE returns to its logic 1 level, before the negative edge of T1, the shift register begins shifting ones into output QA, then QB, and so forth on each clock.

Figure 9–15 shows the wave-forms attained at the QA, QB, and QC outputs of the shift register and the wave-form found on the READY pin for a RAM access and for a ROM access. The other outputs (QD–QH) are not shown in this timing diagram because they do not connect to the READY pin. Notice how the setup and hold times have been met by using a negative edge trigger for the shift clock input.

As we'll see in the chapter on memory interface, the wait state circuit shown here is used only in a system that requires either an external address decoder or many wait states. The XL and EA versions of the 80186/80188 can be programmed to insert between 0 and 3 wait states and the XB and XC versions can be programmed to insert between 0 and 15 wait states.

9–6   SUMMARY

1. The main differences between the 80188 and the 80186 are the width of the data bus and the inclusion of the $\overline{BHE}$ signal on the 80186.
2. The $\overline{S2}$ signal is used to indicate a memory operation (1) or an I/O operation (0) in some systems.
3. Both the 80186 and the 80188 require a + 5.0 V power supply with a tolerance of ±5 percent.
4. The 80186/80188 are TTL-compatible as long as the logic 0 current does not exceed 2.0 mA and as long as a 350-mV noise immunity does not cause problems in the system.
5. The crystal frequency is twice the operating frequency of the microprocessor.
6. Whenever the 80186/80188 is reset, it begins executing software at memory location FFFF0H.
7. Because the address/data bus is multiplexed as well as A16–A19/$\overline{ONCE}$, a latch is used to capture that address information from these pins. The clock pulse to the latch is obtained from the ALE (address latch enable) pin on the microprocessor.

8. In a large system the buses must be buffered because the microprocessor can only drive up to only ten loads.

9. Bus timing is a very important characteristic of the microprocessor. The bus timing is divided into bus cycles of four clocking periods each that accomplish a read or write operation.

10. The amount of time allowed for memory or I/O access is determined by subtracting T_{CLAV} and T_{DVCL} from the time it takes for three clocking periods. In the case of the 16 MHz 80186/80188, the access time is 144.5 ns.

11. Wait states are inserted in the timing to lengthen access time for the memory. A wait state is equal to one clocking period.

12. The READY input is used to insert wait states into the timing.

9–7 QUESTIONS AND PROBLEMS

1. Explain the difference between the 80186 and the 80188.
2. Describe the differences between the 80C188XL and the 80C188EB versions of the 80188 microprocessor.
3. Is the 80186/80188 TTL-compatible? Explain your answer.
4. What is the fan-out from the microprocessor to each of the following logic family members:
   (a) 74XXX
   (b) 74LSXXX
   (c) 74HCXXX
5. What information is available on the address/data bus when ALE is a logic 1?
6. What information is available on the address/data bus when ALE is a logic 0?
7. Describe the purpose of the $\overline{S2}$ status bit.
8. What condition is indicated when $\overline{RD} = 0$?
9. If the 80186 is operated at 8 MHz, which frequency crystal is chosen?
10. If a 22K Ω resister is chosen for the reset circuit, which value capacitor is selected?
11. If the CLLIN signal is at 20 MHz, what frequency is found on the CLKOUT pin?
12. What does the $\overline{WR}$ signal indicate about the operation of the 80186/80188?
13. When DT/$\overline{R}$ is a logic 1, the microprocessor is performing what type of operation?
14. One bus cycle = _____ clocking periods.
15. Which pins must be de-multiplexed in the 80188 microprocessor?
16. Which pins must be de-multiplexed in the 80186 microprocessor?
17. What is a T state?
18. Describe the purpose of each T state: T1, T2, T3, T4, and Tw.
19. If the 80188 microprocessor is operated at 10 MHz, the amount of access time allowed the memory and I/O system is _____ ns.
20. If the 80186 microprocessor is operated at 12 MHz, the amount of access time allowed the memory and I/O system is _____ ns.
21. What is the purpose of the $\overline{DEN}$ signal?
22. If the READY pin is connected to +5.0 V, how many wait states are inserted into the timing?
23. If the READY pin is connected to ground, what happens?
24. If output QE of the circuit in Figure 9–14 is used to generate READY, how many wait states are inserted into the timing?
25. When is the READY line sampled by the microprocessor?
26. Use the Internet to obtain the pin-out of the 80C188EB microprocessor from the Intel Web site.
27. Use the Internet to obtain the block diagram of the 80C188EC microprocessor from the Intel Web site.
28. Download a copy of the Intel ApBuilder program from the Intel Web site.

CHAPTER 10

Memory Interface

INTRODUCTION

Whether simple or complex, every microprocessor-based system has a memory system. The Intel family of microprocessors is no different from others in this respect.

Almost all systems contain two main types of memory: read-only memory (ROM) and random access memory (RAM) or read/write memory. ROM contains system software and permanent system data, while RAM contains temporary data and application software. This chapter explains how to interface both memory types to the Intel family of microprocessors. We demonstrate memory interface to an 8- and a 16-bit data bus as it applies to the 80186/80188 and 80386EX embedded microprocessors using various memory address sizes. This allows virtually any microprocessor to be interfaced to any memory system.

CHAPTER OBJECTIVES

Upon completion of this chapter, you will be able to:

1. Decode the memory address and use the outputs of the decoder to select various memory components.
2. Use programmable logic devices (PLDs) to decode memory addresses.
3. Explain how to interface both RAM and ROM to a microprocessor.
4. Explain how parity can detect memory errors.
5. Interface memory to an 8- and a 16-bit data bus.
6. Interface dynamic RAM to the microprocessor.

10–1 MEMORY DEVICES

Before attempting to interface memory to the microprocessor, you must completely understand the operation of memory components. In this section, we explain the function of the four common types of memory: **read-only memory** (ROM), **flash memory** (EEPROM), **static random access memory** (SRAM), and **dynamic random access memory** (DRAM).

Memory Pin Connections

Pin connections that are common to all memory devices are the address inputs, data outputs or input/outputs, some type of selection input, and at least one control input used to select a read or write operation. See Figure 10–1 for ROM and RAM generic memory devices.

Address Connections.   All memory devices have address inputs that select a memory location within the memory device. Address inputs are almost always labeled from A_0, the least-significant address input, to A_n where subscript n can be any value, but is always labeled as one less than the total number of address pins. For example, a memory device that has 10 address pins has its address pins labeled from A_0 to A_9. The number of address pins found on a memory device is determined by the number of memory locations found within it.

Today, the more common memory devices have between 1K (1,024) and 16M (16,777,216) memory locations, with 256M memory location devices on the horizon. A 1K memory device has 10 address pins (A_0–A_9); therefore, 10 address inputs are required to select any of its 1,024 memory locations. It takes a 10-bit binary number (1,024 different combinations) to select any single location on a 1,024-location device. If a memory device has 11 address connections (A_0–A_{10}), it has 2,048 (2K) internal memory locations. The number of memory locations can thus be extrapolated from the number of address pins. For example, a 4K memory device has 12 address connections, an 8K device has 13, and so forth. A device that contains 1M locations requires a 20-bit address (A_0–A_{19}).

A 400H represents a 1K byte section of the memory system. If a memory device is decoded to begin at memory address 10000H, and it is a 1K device, its last location is at address 103FFH—one location less than 400H. Another important hexadecimal number to remember is a 1000H, because 1000H is 4K. A memory device that contains a starting address of 14000H that is 4K bytes in size, ends at location 14FFFH—one location less than 1000H. A third number is 64K or 10000H. A memory that starts at location 30000H and ends at location 3FFFFH is a 64K-byte memory. Finally, because 1M of memory is not uncommon, a 1M memory contains 100000H memory locations.

Data Connections.   All memory devices have a set of data outputs or input/outputs. The device illustrated in Figure 10–1 has a common set of input/output (I/O) connections. Today many memory devices have bi-directional common I/O pins.

The data connections are the points at which data is entered for storage or extracted for reading. Data pins on memory device are almost always labeled D_0 through D_7 for an 8-bit wide

FIGURE 10–1   A pseudo-memory component illustrating the address, data, and control connections.

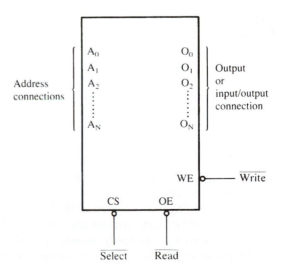

memory device. In this sample memory device, there are eight I/O connections, which means that the memory device stores 8 bits of data in each of its memory locations. An 8-bit-wide memory device is often called a **byte-wide** memory. Although most devices are currently 8 bits wide, not all memory devices are 8 bits wide. Some devices are 16 bits, 4 bits, or just 1 bit wide.

Catalog listings of memory devices often refer to memory locations times bits per location. For example, a memory device with 1K memory locations and 8 bits in each location is often listed as a 1K × 8 by the manufacturer. A 16K × 1 is a memory device containing 16K 1-bit memory locations. Memory devices are also often classified according to total bit capacity. For example, a 1K × 8-bit memory device is sometimes listed as an 8K memory device, or a 64K × 4 memory device is listed as a 256K device. These variations occur from one manufacturer to another.

Selection Connections.   Each memory device has an input—sometimes more than one—that selects or enables the memory device. This kind of input is most often called a **chip select ($\overline{\text{CS}}$)**, **chip enable ($\overline{\text{CE}}$)**, or simply **select ($\overline{\text{S}}$)** input. RAM memory generally has at least one $\overline{\text{CS}}$ or $\overline{\text{S}}$ input, and ROM at least one $\overline{\text{CE}}$. If the $\overline{\text{CE}}$, $\overline{\text{CS}}$, or $\overline{\text{S}}$ input is active (a logic 0 in this case because of the overbar), the memory device performs a read or a write; if it is inactive (a logic 1 in this case), the memory device cannot do a read or a write because it is turned off or disabled. If more than one $\overline{\text{CS}}$ connection is present, all must be activated to read or write data.

Control Connections.   All memory devices have some form of control input or inputs. A ROM usually has only one control input, while a RAM often has one or two control inputs.

The control input most often found on a ROM is the **output enable ($\overline{\text{OE}}$)** or **gate ($\overline{\text{G}}$)** connection, which allows data to flow out of the output data pins of the ROM. If $\overline{\text{OE}}$ and the selection input are both active, then the output is enabled; if $\overline{\text{OE}}$ is inactive, the output is disabled at its high-impedance state. The $\overline{\text{OE}}$ connection enables and disables a set of three state buffers located within the memory device and must be active to read data.

A RAM memory device has either one or two control inputs. If there is one control input, it is often called R/$\overline{\text{W}}$. This pin selects a read or write operation only if the device is selected by the selection input ($\overline{\text{CS}}$). If the RAM has two control inputs, they are usually labeled $\overline{\text{WE}}$ (or $\overline{\text{W}}$) and $\overline{\text{OE}}$ (or $\overline{\text{G}}$). Here $\overline{\text{WE}}$ (**write enable**) must be active to perform a memory write, and $\overline{\text{OE}}$ must be active to perform a memory read operation. When these two controls ($\overline{\text{WE}}$ and $\overline{\text{OE}}$) are present, they must never both be active at the same time. If both control inputs are inactive (logic 1's), then data are neither written nor read and the data connections are at their high-impedance state.

ROM Memory

The **read-only memory** (ROM) permanently stores programs and data that are resident to the system and must not change when power is disconnected. The ROM is permanently programmed so data is always present, even when power is disconnected. This type of memory is often called **nonvolatile memory.**

The ROM is available in many forms today. A device we call a ROM is purchased in mass quantities from a manufacturer and programmed during its fabrication at the factory. The EPROM **(erasable programmable read-only memory),** a type of ROM, is more commonly used when software must be changed often or when too limited a number are in demand to make the ROM economical. For a ROM to be practical, we usually must purchase at least 10,000 devices. An EPROM is programmed in the field on a device called an EPROM programmer. The EPROM is also erasable if exposed to high-intensity ultraviolet light for about 20 minutes or less, depending on the type of EPROM.

PROM memory devices are also available, but they are not as common today. The PROM **(programmable read-only memory)** is also programmed in the field by burning open tiny Nichrome or silicon oxide fuses, but once programmed, it cannot be erased.

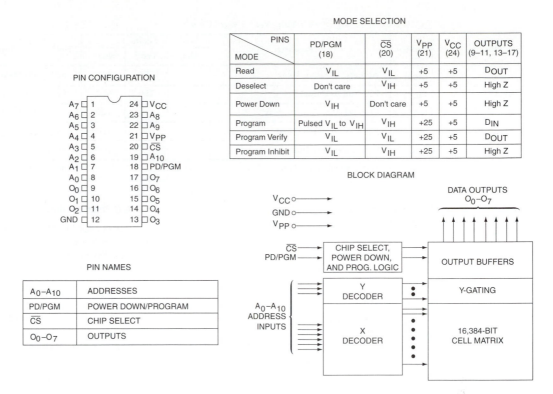

PIN CONFIGURATION

A7	1	24	VCC
A6	2	23	A8
A5	3	22	A9
A4	4	21	VPP
A3	5	20	$\overline{CS}$
A2	6	19	A10
A1	7	18	PD/PGM
A0	8	17	O7
O0	9	16	O6
O1	10	15	O5
O2	11	14	O4
GND	12	13	O3

MODE SELECTION

MODE \ PINS	PD/PGM (18)	$\overline{CS}$ (20)	VPP (21)	VCC (24)	OUTPUTS (9–11, 13–17)
Read	V_{IL}	V_{IL}	+5	+5	DOUT
Deselect	Don't care	V_{IH}	+5	+5	High Z
Power Down	V_{IH}	Don't care	+5	+5	High Z
Program	Pulsed V_{IL} to V_{IH}	V_{IH}	+25	+5	DIN
Program Verify	V_{IL}	V_{IL}	+25	+5	DOUT
Program Inhibit	V_{IL}	V_{IH}	+25	+5	High Z

PIN NAMES

A_0–A_{10}	ADDRESSES
PD/PGM	POWER DOWN/PROGRAM
$\overline{CS}$	CHIP SELECT
O_0–O_7	OUTPUTS

BLOCK DIAGRAM

FIGURE 10–2   The pin-out of the 2716, 2K × 8 EPROM. (Courtesy of Intel Corporation)

Still another, newer type of **read-mostly memory** (RMM) is called the **flash memory.** The flash memory[1] is also often called an EEPROM **(electrically erasable programmable ROM),** EAROM **(electrically alterable ROM),** or a NOVRAM **(nonvolatile RAM).** These memory devices are electrically erasable in the system, but require more time to erase than a normal RAM. The flash memory device is used to store setup information for systems such as the video card in the computer. It may also soon replace the EPROM in the computer for the BIOS memory. Some systems contain a password stored in the flash memory device.

Figure 10–2 illustrates the 2716 EPROM, which is representative of most EPROMs. This device contains 11 address inputs and 8 data outputs. The 2716 is a 2K × 8 memory device. The 27XXX series of EPROMs contains the following part numbers: 2704 (512 × 8), 2708 (1K × 8), 2716 (2K × 8), 2732 (4K × 8), 2764 (8K × 8), 27128 (16K × 8), 27256 (32K × 8), 27512 (64K × 8), and 271024 (128K × 8). Each of these parts contains address pins, eight data connections, one or more chip selection inputs ($\overline{CE}$), and an output enable pin ($\overline{OE}$).

Figure 10–3 illustrates the timing diagram for the 2716 EPROM. Data only appear on the output connections after a logic 0 is placed on both the $\overline{CE}$ and $\overline{OE}$ pin connections. If $\overline{CE}$ and $\overline{OE}$ are not both logic 0's, the data output connections remain at their high-impedance or off states. Note that the V_{PP} pin must be placed at a logic 1 for data to be read from the EPROM. In some cases, the V_{PP} pin is in the same position as the $\overline{WE}$ pin on the SRAM. This can allow a single socket to hold either an EPROM or an SRAM. An example is the 2716 EPROM and the 6116 SRAM, both 2K × 8 devices that have the same pin-out except for V_{PP} on the EPROM and $\overline{WE}$ on the SRAM.

---

[1]Flash memory is a registered trademark of Intel Corporation.

$T_A = 0°C$ to $70°C$, $V_{CC}{}^1 = +5V \pm 5\%$, $V_{PP}{}^2 = V_{CC} \pm 0.6 V^3$

Symbol	Parameter	Limits			Unit	Test Conditions
		Min.	Typ.[4]	Max.		
t_{ACC1}	Address to Output Delay		250	450	ns	PD/PGM = $\overline{CS}$ = V_{IL}
t_{ACC2}	PD/PGM to Output Delay		280	450	ns	$\overline{CS}$ = V_{IL}
t_{CO}	Chip Select to Output Delay			120	ns	PD/PGM = V_{IL}
t_{PF}	PD/PGM to Output Float	0		100	ns	$\overline{CS}$ = V_{IL}
t_{DF}	Chip Deselect to Output Float	0		100	ns	PD/PGM = V_{IL}
t_{OH}	Address to Output Hold	0			ns	PD/PGM = $\overline{CS}$ = V_{IL}

Capacitance[5] $T_A = 25°C$, f = 1 MHz

Symbol	Parameter	Typ.	Max.	Unit	Conditions
C_{IN}	Input Capacitance	4	6	pF	V_{IN} = 0V
C_{OUT}	Output Capacitance	8	12	pF	V_{OUT} = 0V

A.C. Test Conditions:

Output Load: 1 TTL gate and C_L = 100 pF
Input Rise and Fall Times: ≤20 ns
Input Pulse Levels: 0.8 V to 2.2 V
Timing Measurement Reference Level:
  Inputs   1 V and 2 V
  Outputs   0.8 V and 2 V

WAVEFORMS

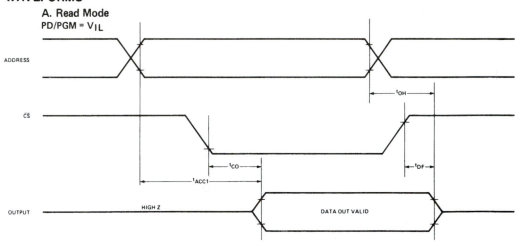

FIGURE 10–3   The timing diagram of AC characteristics of the 2716 EPROM. (Courtesy of Intel Corporation)

One important piece of information provided by the timing diagram and data sheet is the memory access time—the time that it takes the memory to read information. As Figure 10–3 illustrates, memory access time (T_{ACC}) is measured from the appearance of the address at the address inputs until the appearance of the data at the output connections. This is based on the assumption that the $\overline{CE}$ input goes low at the same time that the address inputs become stable. Also, $\overline{OE}$ must be a logic 0 for the output connections to become active. The slowest speed version of this EPROM is 450 ns. (Recall from Chapter 9 that if the 80186/80188 operated with a 16-MHz clock, it allowed the memory 144.5 ns to access data.) This type of memory component requires wait states to operate properly with the 80186/80188 microprocessors because of its rather long access time. If wait states are not desired, higher speed versions of the EPROM are available at an additional cost. Today EPROM memory is available with access times of as little as 100 ns. We discuss flash memory, a type of EEPROM, with I/O devices because it is a programmable memory device.

Static RAM (SRAM) Devices

Static RAM memory devices retain data for as long as DC power is applied. Because no special action (except power) is required to retain stored data, this device is called a **static memory.** It is also called **volatile memory** because it will not retain data without power. The main difference between a ROM and a RAM is that a RAM is written under normal operation and a ROM is programmed outside the computer and is only normally read. The SRAM stores temporary data and is used when the size of the read/write memory is relatively small. Today, a small memory is one that is less than 1M byte.

Figure 10–4 illustrates the 4016 SRAM, which is a 2K × 8 read/write memory. This device has 11 address inputs and 8 data input/output connections. It is representative of all SRAM devices.

The control inputs of this RAM are slightly different from those presented earlier. The $\overline{OE}$ pin is labeled $\overline{G}$, the $\overline{CS}$ pin $\overline{S}$, and the $\overline{WE}$ pin $\overline{W}$. Despite the altered designations, the control pins function exactly the same as those outlined previously. Other manufacturers make this popular SRAM under the part numbers 2016 and 6116.

Figure 10–5 depicts the timing diagram for the 4016 SRAM. As the read cycle timing reveals, the access time is t_a. On the slowest version of the 4016, this time is 250 ns, which is fast enough to connect to an 80186/80188 if operated at lower clock frequencies, without using wait states. Again, it is important to remember that the access time must be checked to determine the compatibility of memory components with the microprocessor and determine if and how many wait states are required for proper operation.

Figure 10–6 illustrates the pin-out of the 62256, 32K × 8 static RAM. This device is packaged in a 28-pin integrated circuit, and is available with access times of 120 ns or 150 ns. Other common SRAM devices are available in 8K × 8 and 128K × 8 sizes with access times of as little as 10 ns for SRAM used in computer cache memory systems.

Dynamic RAM (DRAM) Memory

About the largest static RAM available today is a 128K × 8. Dynamic RAMs, on the other hand, are available in much lager sizes: up to 64M × 1. In all other respects, DRAM is essentially the

FIGURE 10–4 The pin-out of the TMS4016, 2K × 8 static RAM (SRAM). (Courtesy of Texas Instruments Incorporated)

TMS4016 . . . NL PACKAGE
(TOP VIEW)

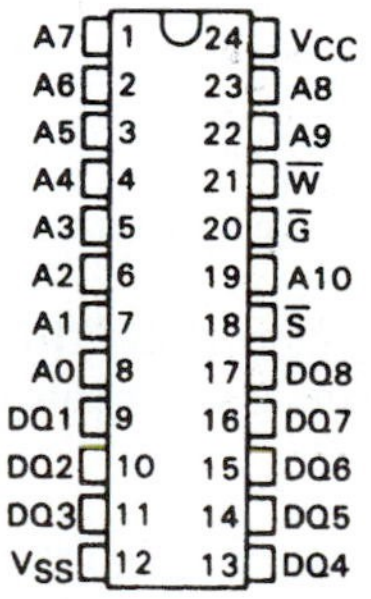

A7	1	24	V_{CC}
A6	2	23	A8
A5	3	22	A9
A4	4	21	$\overline{W}$
A3	5	20	$\overline{G}$
A2	6	19	A10
A1	7	18	$\overline{S}$
A0	8	17	DQ8
DQ1	9	16	DQ7
DQ2	10	15	DQ6
DQ3	11	14	DQ5
V_{SS}	12	13	DQ4

PIN NOMENCLATURE	
A0 – A10	Addresses
DQ1 – DQ8	Data In/Data Out
$\overline{G}$	Output Enable
$\overline{S}$	Chip Select
V_{CC}	+5-V Supply
V_{SS}	Ground
$\overline{W}$	Write Enable

electrical characteristics over recommended operating free-air temperature range (unless otherwise noted)

PARAMETER		TEST CONDITIONS		MIN	TYP†	MAX	UNIT
V_{OH}	High level voltage	$I_{OH} = -1$ mA,	$V_{CC} = 4.5$ V	2.4			V
V_{OL}	Low level voltage	$I_{OL} = 2.1$ mA,	$V_{CC} = 4.5$ V			0.4	V
I_I	Input current	$V_I = 0$ V to 5.5 V				10	μA
I_{OZ}	Off-state output current	$\overline{S}$ or $\overline{G}$ at 2 V or $\overline{W}$ at 0.8 V, $V_O = 0$ V to 5.5 V				10	μA
I_{CC}	Supply current from V_{CC}	$I_O = 0$ mA, $T_A = 0°C$ (worst case)	$V_{CC} = 5.5$ V,		40	70	mA
C_i	Input capacitance	$V_I = 0$ V,	f = 1 MHz			8	pF
C_O	Output capacitance	$V_O = 0$ V,	f = 1 MHz			12	pF

†All typical values are at $V_{CC} = 5$ V, $T_A = 25°C$.

timing requirements over recommended supply voltage range and operating free-air temperature range

PARAMETER		TMS4016-12		TMS4016-15		TMS4016-20		TMS4016-25		UNIT
		MIN	MAX	MIN	MAX	MIN	MAX	MIN	MAX	
$t_{c(rd)}$	Read cycle time	120		150		200		250		ns
$t_{c(wr)}$	Write cycle time	120		150		200		250		ns
$t_{w(W)}$	Write pulse width	60		80		100		120		ns
$t_{su(A)}$	Address setup time	20		20		20		20		ns
$t_{su(S)}$	Chip select setup time	60		80		100		120		ns
$t_{su(D)}$	Data setup time	50		60		80		100		ns
$t_{h(A)}$	Address hold time	0		0		0		0		ns
$t_{h(D)}$	Data hold time	5		10		10		10		ns

switching characteristics over recommended voltage range, $T_A = 0°C$ to $70°C$

PARAMETER		TMS4016-12		TMS4016-15		TMS4016-20		TMS4016-25		UNIT
		MIN	MAX	MIN	MAX	MIN	MAX	MIN	MAX	
$t_{a(A)}$	Access time from address		120		150		200		250	ns
$t_{a(S)}$	Access time from chip select low		60		75		100		120	ns
$t_{a(G)}$	Access time from output enable low		50		60		80		100	ns
$t_{v(A)}$	Output data valid after address change	10		15		15		15		ns
$t_{dis(S)}$	Output disable time after chip select high		40		50		60		80	ns
$t_{dis(G)}$	Output disable time after output enable high		40		50		60		80	ns
$t_{dis(W)}$	Output disable time after write enable low		50		60		60		80	ns
$t_{en(S)}$	Output enable time after chip select low	5		5		10		10		ns
$t_{en(G)}$	Output enable time after output enable low	5		5		10		10		ns
$t_{en(W)}$	Output enable time after write enable high	5		5		10		10		ns

Notes: 3. $C_L = 100$ pF for all measurements except $t_{dis(W)}$ and $t_{en\,(W)}$.
           $C_L = 5$ pF for $t_{dis(W)}$ and $t_{en(W)}$.
        4. t_{dis} and t_{en} parameters are sampled and not 100% tested.

FIGURE 10–5   The AC characteristics of the TMS4016 SRAM and the timing diagrams of the TMS4016 SRAM. (Courtesy of Texas Instruments Incorporated)

same as SRAM except it retains data for only 2 or 4 ms on an integrated capacitor. After 2 or 4 ms, the contents of the DRAM must be completely rewritten (**refreshed**) because the capacitors, which store a logic 1 or logic 0, lose their charges.

Instead of requiring the almost impossible task of reading the contents of each memory location with a program and then rewriting them, the manufacturer has internally constructed the DRAM so, in the 64K × 1 version, the entire contents of the memory are refreshed with 256 reads in a 4-ms interval. Refreshing also occurs during a write, a read, or during a special refresh cycle. Much more information on refreshing DRAMs is provided in a later section that describes DRAM interface to the embedded controller.

timing waveform of read cycle (see note 5)

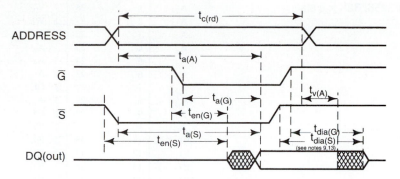

timing waveform of write cycle no. 1 (see note 6)

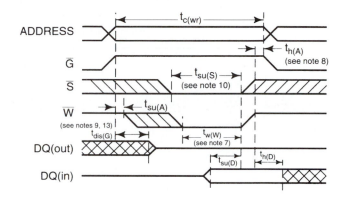

timing waveform of write cycle no. 2 (see notes 6 and 11)

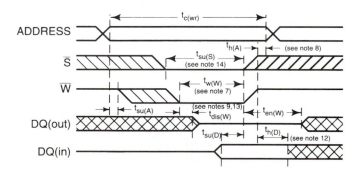

Notes: 5. $\overline{W}$ is high read cycle.

6. $\overline{W}$ must be high during all address transitions.

7. A write occurs during the overlap of a low $\overline{S}$ and a low $\overline{W}$.

8. $t_{h(A)}$ is measured from the earlier of $\overline{S}$ or $\overline{W}$ going high to the end of the write cycle.

9. During this period, I/O pins are in the output state so that the input signals of opposite phase to the outputs must not be applied.

10. If the Slow transition occurs simultaneously with the $\overline{W}$ low transitions or after the $\overline{W}$ transition, output remains in a high-impedance state.

11. G is continuously low ($G = V_{IL}$).

12. If $\overline{S}$ is low during this period, I/O pins are in the output state. Data input signals of opposite phase to the outputs must not be applied.

13. Transition is measured ± 200 mV from steady-state voltage.

14. If the $\overline{S}$ low transition occurs before the W low transition, then the data input signals of opposite phase to the outputs must not be applied for the duration of $t_{dis(W)}$ after the $\overline{W}$ low transition.

FIGURE 10–5 *(continued)*

FIGURE 10–6   Pin diagram of the 62256, 32K × 8 static RAM.

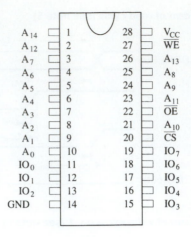

A_{14}	1		28	V_{CC}
A_{12}	2		27	$\overline{WE}$
A_7	3		26	A_{13}
A_6	4		25	A_8
A_5	5		24	A_9
A_4	6		23	A_{11}
A_3	7		22	$\overline{OE}$
A_2	8		21	A_{10}
A_1	9		20	$\overline{CS}$
A_0	10		19	IO_7
IO_0	11		18	IO_6
IO_1	12		17	IO_5
IO_2	13		16	IO_4
GND	14		15	IO_3

PIN FUNCTION

$A_0 - A_{14}$	Addresses
$IO_0 - IO_7$	Data connections
$\overline{CS}$	Chip select
$\overline{OE}$	Output enable
$\overline{WE}$	Write enable
V_{CC}	+5V Supply
GND	Ground

Another disadvantage of DRAM memory is that it requires so many address pins that the manufacturers have multiplexed the address inputs. Figure 10–7 illustrates a 64K × 4 DRAM, the TMS4464, which stores 256K bits of data. Notice that it contains only 8 address inputs where it should contain 16—the number required to address 64K memory locations. The only way that

FIGURE 10–7   The pin-out of the TMS4464, 64K × 4 dynamic RAM (DRAM). (Courtesy of Texas Instruments Incorporated)

TMS4464 . . . JL OR NL PACKAGE
(TOP VIEW)

$\overline{G}$	1		18	V_{SS}
DQ1	2		17	DQ4
DQ2	3		16	$\overline{CAS}$
$\overline{W}$	4		15	DQ3
$\overline{RAS}$	5		14	A0
A6	6		13	A1
A5	7		12	A2
A4	8		11	A3
V_{DD}	9		10	A7

(a)

PIN NOMENCLATURE	
A0-A7	Address Inputs
$\overline{CAS}$	Column Address Strobe
DQ1-DQ4	Data-In/Data-Out
$\overline{G}$	Output Enable
$\overline{RAS}$	Row Address Strobe
V_{DD}	+ 5-V Supply
V_{SS}	Ground
$\overline{W}$	Write Enable

(b)

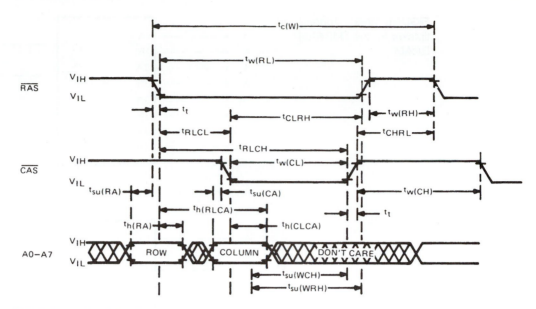

FIGURE 10–8 RAS, CAS, and address input timing for the TMS4464 DRAM. (Courtesy of Texas Instruments Incorporated)

16 address bits can be forced into 8 address pins is in two 8-bit increments. This operation requires two special pins called **column address strobe** ($\overline{\text{CAS}}$) and **row address strobe** ($\overline{\text{RAS}}$). First, A_0–A_7 are placed on the address pins and strobed into an internal row latch by $\overline{\text{RAS}}$ as the row address. Next, the address bits A_8–A_{15} are placed on the same eight address inputs and strobed into an internal column latch by $\overline{\text{CAS}}$ as the column address (see Figure 10-8 for this timing). The 16-bit address held in these internal latches addresses the contents of one of the 4-bit memory locations. Note that $\overline{\text{CAS}}$ also performs the function of the chip selection input to the DRAM.

Figure 10–9 illustrates a set of multiplexers used to strobe the column and row addresses into the eight address inputs of a pair of TMS4464 DRAMs. Here the $\overline{\text{RAS}}$ not only strobes the row address into the DRAMs, but it also changes the address applied to the address inputs. This is possible due to the long propagation delay time of the multiplexers. When $\overline{\text{RAS}}$ is a logic 1, the B inputs are connected to the Y outputs of the multiplexers, and when the $\overline{\text{RAS}}$ input goes to a logic 0, the A inputs connect to the Y outputs. Because the internal row address latch is edge-triggered, it captures the row address before the address at the inputs change to the column address.

As with the SRAM, the R/$\overline{\text{W}}$ pin writes data to the DRAM when a logic zero, but there is no pin labeled $\overline{\text{G}}$ (enable). There also is no $\overline{\text{S}}$ (select) input to the DRAM. As mentioned, the $\overline{\text{CAS}}$ input selects the DRAM. If selected, the DRAM is written if R/$\overline{\text{W}}$ = 0 and read if R/$\overline{\text{W}}$ = 1.

Figure 10–10 shows the pin-out of the 41256 dynamic RAM. This device is organized as a 256K × 1 memory requiring as little as 70 ns to access data.

More recently, larger DRAMs have become available that are organized as a 1M × 1 memory, 4M × 1, and 16M × 1. On the horizon is the 256M × 1 memory, which is in the planning stages. Because DRAM memory is usually placed on small circuit boards called SIMMs, Figure 10–11 shows the pin-outs of the two most common SIMMs (single in-line memory modules). The 30-pin SIMM is organized most often as 1M × 8 or 1M × 9 and 4M × 8 or 4M × 9. (Illustrated in Figure 10–11 is a 4M × 9). The ninth bit is the parity bit. Also shown is the newer 72-pin SIMM. The 72-pin SIMMs are often organized as 1M x 32 or 1M × 36 (with parity). Other sizes are 2M × 32, 4M × 32, or 8M × 32. These are also available with parity.

FIGURE 10–9   Address multiplexer for the TMS4464 DRAM.

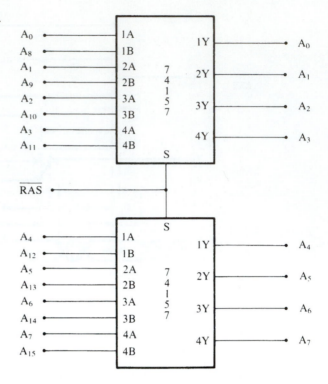

Illustrated in Figure 10–11 is a 4M × 36 SIMM, which has 16M bytes of memory. Not illustrated are the new 168-pin DIMMs (dual in-line memory modules), which are organized as 64-bit wide memory because they are not applicable to the embedded controller at this time.

FIGURE 10–10   The 41256 dynamic RAM organized as a 256K × 1 memory device.

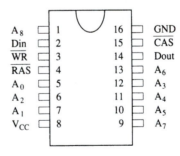

PIN FUNCTIONS

A_0 - A_8	Addresses
Din	Data in
Dout	Data out
$\overline{CAS}$	Column Address Strobe
$\overline{RAS}$	Row Address Strobe
WR	Write enable
V_{CC}	+5V Supply
GND	Ground

(a)

(TOP VIEW)

(b)

(TOP VIEW)

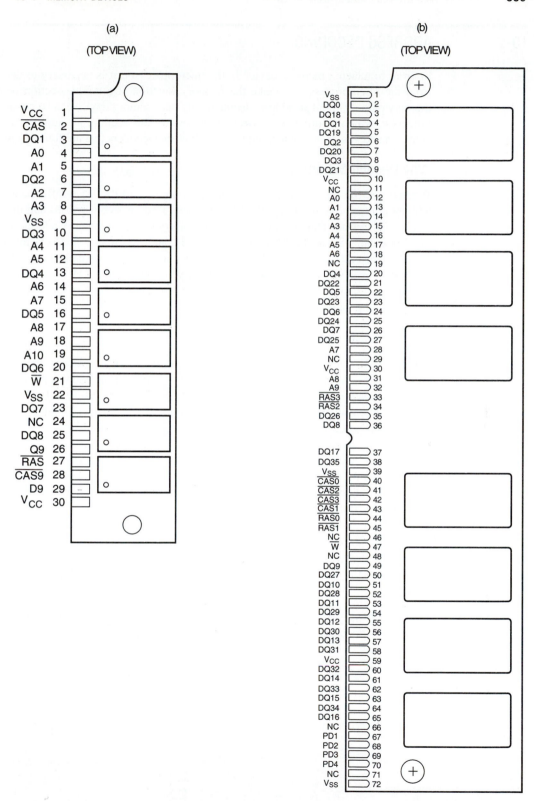

FIGURE 10–11 The pin-outs of the (a) 4M × 9 SIMM and the (b) 1M × 36 SIMM.

10–2   ADDRESS DECODING

In order to attach a memory device to the microprocessor, it is necessary to decode the address from the microprocessor to make the memory function at a unique section or partition of the memory map. Without an address decoder, only one memory device can be connected to a microprocessor, which would make it virtually useless. In this section, we describe a few of the more common address-decoding techniques as well as the decoders that are found in many systems.

Why Decode Memory?

When the 80188 microprocessor is compared to the 2716 EPROM, a difference in the number of address connections surfaces—the EPROM has 11 address connections and the microprocessor has 20. This means that the microprocessor sends out a 20-bit memory address whenever it reads or writes data. Since the EPROM has only 11 address inputs, this is a mismatch that must somehow be corrected. If only 11 of the 80188's address pins are connected to the memory, then the 80188 will see only 2K bytes of memory instead of the 1M bytes that it "expects" the memory to contain. The decoder corrects the mismatch by decoding the address pins that do not connect to the memory component.

Simple NAND Gate Decoder

When the $2K \times 8$ EPROM is used, address connections $A1_{10}$–A_0 of the 80188 are connected to address inputs A_{10}–A_0 of the EPROM, and the remaining nine address pins (A_{19}–A_{11}) are connected to the inputs of a NAND gate decoder (see Figure 10–12). The decoder selects the EPROM from one of the many 2K-byte sections of the entire 1M-byte address range of the 80188 microprocessor.

In this circuit, a single NAND gate decodes the memory address. The output of the NAND gate is a logic 0 whenever the 80188 address pins attached to its inputs (A_{19}–A_{11}) are all logic 1's at the same time that $\overline{S2}$ is a logic 1. (Recall that $\overline{S2}$ is a logic 1 for any memory operation.) The active low (logic 0) output of the NAND gate decoder is connected to the $\overline{CE}$ input, which selects (**enables**) the EPROM. (Recall that whenever $\overline{CE}$ is a logic 0, data will be read from the EPROM only if $\overline{OE}$ is also a logic 0). The $\overline{OE}$ pin is activated by the 80188 $\overline{RD}$ signal.

If the 20-bit binary address, decoded by the NAND gate, is written so the leftmost 9 bits are 1's and the rightmost 11 bits are don't cares (X), the actual address range of the EPROM can be determined. (A don't care is a logic 1 or a logic 0, whichever is appropriate.)

FIGURE 10–12   A simple NAND gate decoder used to select a 2616 EPROM for memory locations FF800H–FFFFFH for the 80188 microprocessor.

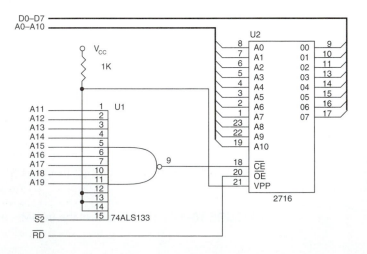

Example 10–1 illustrates how the address range for this EPROM is determined by writing down the externally decoded address bits (A_{19}–A_{11}) and the address bits decided by the EPROM (A_{10}–A_0) as don't cares. As the example illustrates, the don't cares are first written as 0's to locate the lowest address and then as 1's to find the highest address. Example 10–1 also shows these binary boundaries as hexadecimal addresses. Here the 2K EPROM is decoded at memory address locations FF800H–FFFFFH. Notice that this is a 2K-byte section of the memory and is also located at the reset location for the 80186/80188, the most likely place for an EPROM.

EXAMPLE 10–1

```
1111 1111 1XXX XXXX XXXX

             or

1111 1111 1000 0000 0000 = FF800H
            to
1111 1111 1111 1111 1111 = FFFFFH
```

Although this example serves to illustrate decoding, NAND gates are rarely used to decode memory because each memory device requires its own NAND gate decoder. Because of the excessive cost of the NAND gate decoder and inverters that are often required, this option requires that an alternate be found.

The 3-to-8 Line Decoder (74LS138)

One of the more common, although not only, integrated circuit decoders found in many microprocessor-based systems is the 74LS138 3-to-8 line decoder. Figure 10–13 illustrates this decoder and its truth table.

FIGURE 10–13 The 74LS138, 3-to-8 line decoder and function table.

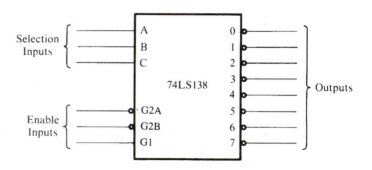

Inputs						Outputs							
Enable		Select											
G2A	G2B	G1	C	B	A	0	1	2	3	4	5	6	7
1	X	X	X	X	X	1	1	1	1	1	1	1	1
X	1	X	X	X	X	1	1	1	1	1	1	1	1
X	X	0	X	X	X	1	1	1	1	1	1	1	1
0	0	1	0	0	0	0	1	1	1	1	1	1	1
0	0	1	0	0	1	1	0	1	1	1	1	1	1
0	0	1	0	1	0	1	1	0	1	1	1	1	1
0	0	1	0	1	1	1	1	1	0	1	1	1	1
0	0	1	1	0	0	1	1	1	1	0	1	1	1
0	0	1	1	0	1	1	1	1	1	1	0	1	1
0	0	1	1	1	0	1	1	1	1	1	1	0	1
0	0	1	1	1	1	1	1	1	1	1	1	1	0

The truth table shows that only one of the eight outputs ever goes low at any time. For any of the decoder's outputs to go low, the three enable inputs ($\overline{G2A}$, $\overline{G2B}$, and G1) must all be active. To be active, the $\overline{G2A}$ and $\overline{G2B}$ inputs must both be low (logic 0), and G1 must be high (logic 1). Once the 74LS138 is enabled, the address inputs (C, B, and A) select which output pin goes low. Imagine eight EPROM $\overline{CE}$ inputs connected to the eight outputs of the decoder! This is a very powerful device because it selects eight different memory devices at the same time.

Sample Decoder Circuit. Notice that the outputs of the decoder illustrated in Figure 10–14 are connected to eight different 2764 EPROM memory devices. Here the decoder selects eight, 8K-byte blocks of memory for a total of 64K bytes of memory. This figure also illustrates the address range of each memory device and the common connections to the memory devices. Notice that all the address connections from the 80188 are connected to this circuit. Also notice that the decoder's outputs are connected to the $\overline{CE}$ inputs of the EPROMs, and the $\overline{RD}$ signal from the 80188 is connected to the $\overline{OE}$ inputs of the EPROMs. This allows only the selected EPROM to be enabled and to send its data to the microprocessor through the data bus whenever $\overline{RD}$ becomes a logic 0.

In this circuit, a 4-input NAND gate is connected to address bits A_{19}–A_{17} and $\overline{S2}$ (a logic 1 for memory). When all three address inputs are high, the output of this NAND gate goes low and enables input $\overline{G2A}$ and $\overline{G2B}$ of the 74LS138. Input G1 is connected directly to A16. In other words, in order to enable this decoder, the first four address connections (A_{19}–A_{16}) must all be high.

The address inputs C, B, and A connect to microprocessor address pins A_{15}–A_{13}. These three address inputs determine which output pin goes low and which EPROM is selected whenever the 80188 outputs a memory address within this range to the memory system.

Example 10–2 shows how the address range of the entire decoder is determined. Notice that the range is location F0000H–FFFFFH. This is a 64K-byte span of the memory.

EXAMPLE 10–2

```
1111 XXXX XXXX XXXX XXXX

          or

1111 0000 0000 0000 0000 = F0000H
        to
1111 1111 1111 1111 1111 = FFFFFH
```

How is it possible to determine the address range of each memory device attached to the decoder's outputs? Again, the binary bit pattern is written down, and this time the C, B, and A address inputs are not don't cares. Example 10–3 shows how output 0 of the decoder is made to go low to select the EPROM attached to that pin. Here C, B, and A are shown as logic 0's.

EXAMPLE 10–3

```
     CBA
1111 000X XXXX XXXX XXXX

            or

1111 0000 0000 0000 0000 = F0000H
            to
1111 0001 1111 1111 1111 = F1FFFH
```

If the address range of the EPROM connected to output 1 of the decoder is required, it is determined in exactly the same way as that of output 0. The only difference is that now the C, B, and A inputs contain a 001 instead of a 000 (see Example 10–4). The remaining output address ranges are determined in the same manner by substituting the binary address of the output pin into C, B, and A.

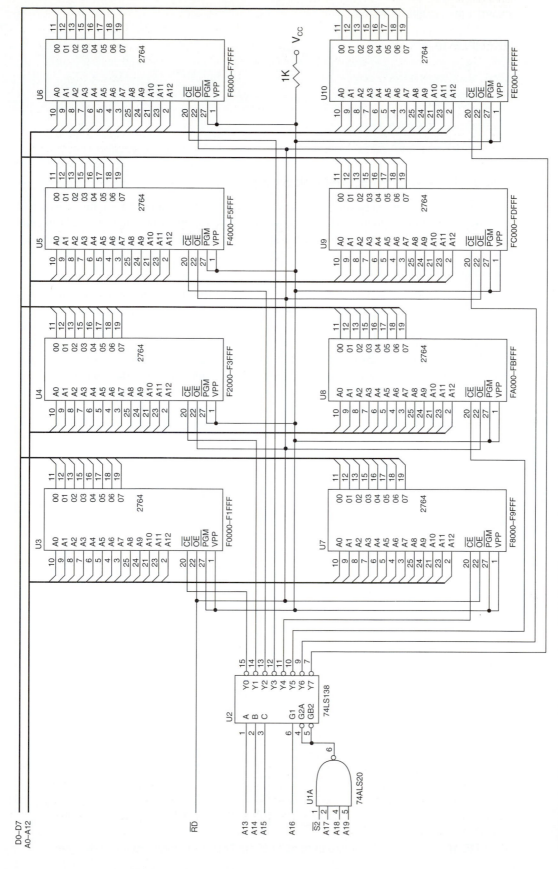

FIGURE 10–14 A circuit that uses eight 2764 EPROMs for a 64K × 8 section of the 80188 memory. The addresses selected are F0000H–FFFFFH.

EXAMPLE 10–4

```
     CBA
1111 001X XXXX XXXX XXXX

            or

1111 0010 0000 0000 0000 = F2000H
           to
1111 0011 1111 1111 1111 = F3FFFH
```

The Dual 2-to-4 Line Decoder (74LS139)

Another decoder that finds some application is the 74LS139 dual 2-to-4 line decoder. Figure 10–15 illustrates both the pin-out and the truth table for this decoder. The 74LS139 contains two separate 2-to-4 line decoders—each with its own address, enable, and output connections.

PROM Address Decoder

Another, once common, address decoder is the bipolar PROM, used because of its larger number of input connections, which reduce the number of other circuits required in a system memory address decoder. The 74LS138 decoder has six inputs used for address connections. The PROM decoder may have many more inputs for address decoding.

An example is the 82S147 (512×8) PROM used as an address decoder. It has 10 input connections and 8 output connections and can replace the circuit in Figure 10–14 without the extra 4-input NAND gate. This saves space on the printed circuit board and reduces the cost of a system.

Figure 10–16 illustrates this address decoder with the PROM in place. The PROM is a memory device that must be programmed with the correct binary bit pattern to select the eight

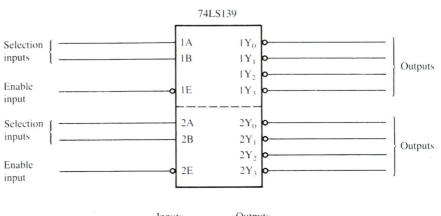

Inputs			Outputs			
$\overline{E}$	A	B	$\overline{Y_0}$	$\overline{Y_1}$	$\overline{Y_2}$	$\overline{Y_3}$
0	0	0	0	1	1	1
0	0	1	1	0	1	1
0	1	0	1	1	0	1
0	1	1	1	1	1	0
1	X	X	1	1	1	1

FIGURE 10–15   The pin-out and truth table of the 74LS139, dual 2-to-4 line decoder.

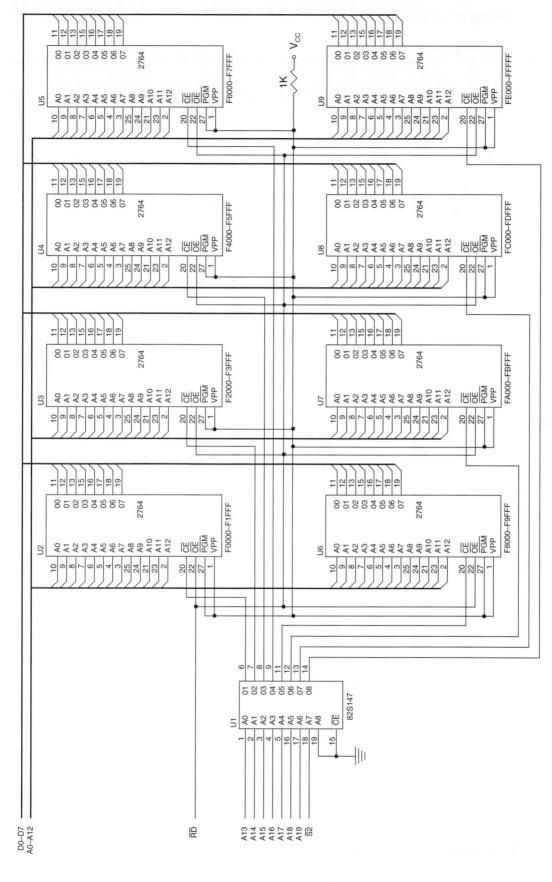

FIGURE 10–16 A memory system using the 82S147 PROM decoder to select eight 2764 EPROMs.

345

TABLE 10–1   The 82S147 PROM programming pattern for the circuit of Figure 10–16

Inputs										Outputs							
$\overline{OE}$	A_8	A_7	A_6	A_5	A_4	A_3	A_2	A_1	A_0	O_0	O_1	O_2	O_3	O_4	O_5	O_6	O_7
0	0	1	1	1	1	1	0	0	0	0	1	1	1	1	1	1	1
0	0	1	1	1	1	1	0	0	1	1	0	1	1	1	1	1	1
0	0	1	1	1	1	1	0	1	0	1	1	0	1	1	1	1	1
0	0	1	1	1	1	1	0	1	1	1	1	1	0	1	1	1	1
0	0	1	1	1	1	1	1	0	0	1	1	1	1	0	1	1	1
0	0	1	1	1	1	1	1	0	1	1	1	1	1	1	0	1	1
0	0	1	1	1	1	1	1	1	0	1	1	1	1	1	1	0	1
0	0	1	1	1	1	1	1	1	1	1	1	1	1	1	1	1	0
all other combinations										1	1	1	1	1	1	1	1

EPROM memory devices. The PROM itself has nine address inputs that select one of the 512 internal 8-bit memory locations. The remaining input ($\overline{CE}$) must be grounded because if this PROM's outputs float to their high-impedance state, then one or more of the EPROMs might be selected by noise impulses in the system.

Table 10–1 lists the binary bit pattern programmed into each PROM location in order to select the eight different EPROMs. The main advantage to using a PROM is that the address map is easily changed in the field. Because the PROM comes with all the locations programmed as logic 1's, only 8 of the 512 locations must be programmed. This saves valuable time for the manufacturer.

PLD Programmable Decoders

In this section, we explain the use of the programmable logic device or PLD as a decoder. Recently, the PAL has replaced PROM address decoders in the latest memory interfaces. There are three PLD devices that function in basically the same manner, but have different names: PLA **(programmable logic array),** PAL **(programmable array logic),** and GAL **(gated array logic).** Although these devices have been in existence since the mid-1970s, they have only recently appeared, in the late 1980s, in memory system and digital designs. The PAL and the PLA are fuse programmed as is the PROM, while some of PLD devices are erasable devices as are EPROMs. Still other PLD devices are electrically erasable. In essence, all three devices are arrays of logic elements that are programmable.

Combinatorial Programmable Logic Arrays.   One of the two basic types of PALs is the combinatorial programmable logic array. This device is internally structured as a programmable array of combinational logic circuits. Figure 10–17 illustrates internal structure of the PAL16L8 that is constructed with AND/OR gate logic. This device, which is very common, has 10 fixed inputs, two fixed outputs, and six pins that are programmable as inputs or outputs. Each output pin is generated from a 7-input OR gate that has an AND gate attached to each input. The outputs of the OR gates pass through a three-state inverter that defines each out as an AND/NOR function. Initially all of the fuses connect all of the vertical/horizontal connections illustrated in Figure 10–17. Programming is accomplished by blowing fuses to connect various inputs to the OR gate array. The wired-AND function is performed at each input connection that allows a product term of up to 16 inputs. A logic expression using the PAL16L8 can have 7 product terms with 16 inputs NORed together to generate the output expression. This device is ideal as a memory address decoder because of its structure. It is also ideal because the outputs are active low.

Fortunately, we don't have to choose the fuses by number as was customary in the early days of this device. Today we program the PAL using a software package such as PALASM, the PAL assembler program. The PALASM program and its syntax are an industry standard for programming PAL devices. Example 10–5 decodes the same areas of memory as decoded in Figure 10–16. This program was developed using a text editor such as EDIT, available with Microsoft DOS version 7.0, or Notepad in Windows 95. The program can also be developed using an editor that

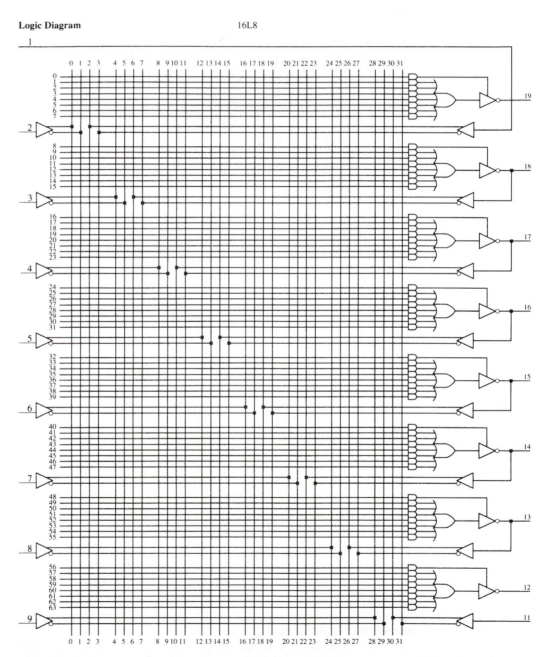

FIGURE 10–17 The PAL16L8. (Copyright Advanced Micro Devices, Inc., 1988. Reprinted with permission of copyright owner. All rights reserved.)

comes with the PALASM package or any other PAL assembler program. Various editors attempt to ease the task of defining the pins, but we believe it is easier to use EDIT and the listing as shown.

EXAMPLE 10–5

```
TITLE        Address Decoder
PATTERN      Test 1
REVISION     A
AUTHOR       Barry B. Brey
COMPANY      BreyCo
DATE         6/6/97
CHIP         DECODER1  PAL16L8

;pins 1   2   3   4   5   6   7   8  9  10
      A19 A18 A17 A16 A15 A14 A13 S2 NC GND

;pins 11 12 13 14 15 16 17 18 19  20
      NC O8 O7 O6 O5 O4 O3 O2 O1 VCC
EQUATIONS
/O1 = A19 * A18 * A17 * A16 * /A15 * /A14 * /A13 * S2
/O2 = A19 * A18 * A17 * A16 * /A15 * /A14 * A13 * S2
/O3 = A19 * A18 * A17 * A16 * /A15 * A14 * /A13 * S2
/O4 = A19 * A18 * A17 * A16 * /A15 * /A14 * A13 * S2
/O5 = A19 * A18 * A17 * A16 * A15 * /A14 * /A13 * S2
/O6 = A19 * A18 * A17 * A16 * A15 * A14 * A13 * S2
/O7 = A19 * A18 * A17 * A16 * A15 * /A14 * /A13 * S2
/O8 = A19 * A18 * A17 * A16 * A15 * /A14 * A13 * S2
```

The first eight lines of the program illustrated in Example 10–5 identify the program title, pattern, revision, author, company, date, and chip type with the program name. Although it is normal to find entries in each heading, the only entry that is absolutely necessary is the CHIP statement. In this example, the chip type is a PAL16L8 and the program is called DE-CODER1. After the program is identified, a comment statement (;pins) identifies the pin numbers. Below this comment state appear the pins as defined for this application. Once all the pins are defined, we use the EQUATIONS statement to indicate that the equations for this application follow. In this example, the equations define the eight chip enable outputs for the eight EPROM memory devices. Refer to Figure 10–18 for the complete schematic diagram of this PAL decoder.

Note that each equation specifies one of the active low output pins as defined by the / in front of the pin name. For example, /O1 is used in place of $\overline{O1}$. We normally place an overbar on top of an active low output, but that is not possible when typing, so the slash in front of a pin name is used to indicate active low outputs. In this example all outputs are active low because of the PAL 16L8, which has only active low outputs. Other PAL devices are available with active high outputs, if needed. Normally all pins are defined before the equation statement as active high.

Logic symbols used in PAL equations include the * for the AND operation and the + for the OR operation. This example illustrates only AND operations. If a single input is inverted, the / is placed in front of the pin name. If a group must be inverted, a slash is placed in front of the group that is surrounded by parentheses. For example, /(A + B) is the same as $\overline{A + B}$, the NOR function. A NAND function would then be /(A * B) which is the same as $\overline{A} * \overline{B}$.

Let us examine the very first equation in Example 10–5. The output pin /O1 is active low, which means that it becomes a logic 0, enabling the EPROM in Figure 10–18 when the equation on the other side of the equal sign is true. The equation for this output contains a number of inputs that are ANDed together. In this example, when S2 (note this is not inverted by the PAL so when the $\overline{S2}$ pin goes to a logic 1 so does S2), A_{19}, A_{18}, A_{17}, and A_{16} are all ones while A_{15}, A_{14}, and A_{13} are all low, the output pin /O1 becomes a logic 0. This binary number corresponds to memory locations F0000H–F1FFFH.

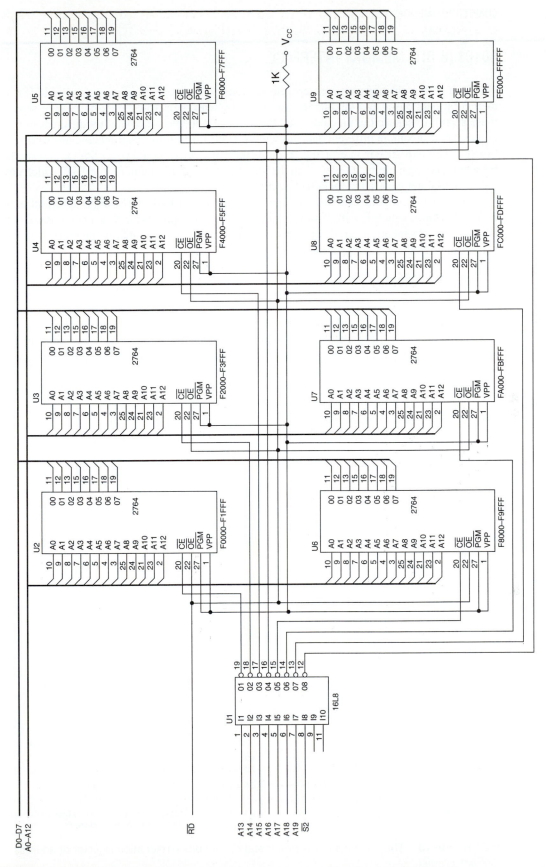

FIGURE 10-18 A memory system using the PAL16L8 decoder to select eight 2764 EPROMs.

349

10–3   80188 (8-BIT) MEMORY INTERFACE

This part of the chapter contains separate sections on memory interfacing for the 80188 with their 8-bit data buses and the 80186 and 80386EX with their 16-bit data buses. We separated these sections because the methods used to address the memory are slightly different in microprocessors that contain different data bus widths. Hardware engineers or technicians who wish to broaden their expertise in interfacing 16-bit memory interface should cover both sections. The 8-bit section is much more complete than the section on the 16-bit memory interface, which covers only material not covered in the 80188 section.

In this section, we examine the memory interface to both RAM and ROM and explain parity checking, which is still commonplace in many microprocessor-based computer systems. We also briefly mention error correction schemes currently available to memory system designers on the Pentium Pro processor.

Basic 80188 Memory Interface

The 80188 microprocessor has an 8-bit data bus, which makes it ideal to connect to the common 8-bit memory devices available today. The 8-bit memory size makes the 80188 ideal as a simple embedded controller. For the 80188 to function correctly with the memory, however, the memory system must decode the address to select a memory component, and it must use the $\overline{RD}$, $\overline{WR}$, and $\overline{S2}$ control signals provided by the 80188 to control the memory system.

In the last section we looked at decoding, which is still used with the 80188 occasionally. This section details the programmable memory chip selection logic pins found on the 80188. There are two main versions of the 80188 that require explanation; the 80C188XL and 80C188EA versions are programmed differently than the 80C188EB and 80C188 EC versions.

The 80C186XL/80C188XL and 80C186XA/80C188EA Chip Selection Unit

The chip selection unit simplifies the interface of memory and I/O to the 80186/80188. This unit contains programmable chip selection logic. In small- and medium-sized systems no external decoder is required to select memory and I/O. Large systems, however, may still require external decoders, as we saw in the last section. The chip selection unit for the XL and EA versions uses a register stored in an area of memory called the **peripheral control block** or PCB. The peripheral control block is initially at I/O addresses FF00H–FFFFH and contains a register called the PCB control register that can relocate the PCB to any other memory or I/O block that is 256 bytes in length. Figure 10–19 illustrates the PCB relocation register, which by default is initially at I/O address FFFEH and FFFFH.

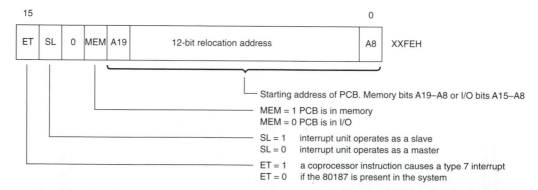

FIGURE 10–19   The contents of the peripheral control block relocation register at address XXFEH.

Suppose that the location of the peripheral control block must be placed in memory at address 20000H–200FFH. This is accomplished as illustrated in Example 10–6, but in most causes it remains at its default I/O addresses of FF00H–FFFFH as assumed throughout the remainder of our discussion.

EXAMPLE 10–6

```
0100  BA FFFE     MOV    DX,0FFFEH     ;address of PCB relocation register
0103  B8 1200     MOV    AX,1200H      ;memory address 20000H-200FFH
0106  EE          OUT    DX,AL         ;you only need an 8-bit output
```

Memory Chip Selects.   Six pins (XL and EA versions) are used to select different external memory components in a small- or medium-sized 80186/80188–based system. The $\overline{UCS}$ **(upper chip select)** pin enables the memory device located in the upper portion of the memory map most often populated with ROM. This programmable pin allows the size of the ROM to be specified and also the number of wait states required. Note that the ending address of the ROM is FFFFFH. The $\overline{LCS}$ **(lower chip select)** pin selects the memory device (usually a RAM) that begins at memory location 00000H. As with the $\overline{UCS}$ pin, the memory size and number of wait states are programmable. The remaining four or eight pins select **middle memory** devices. The four pins in the XL and EA version ($\overline{MCS1}$–$\overline{MCS0}$) are programmed for both the starting (base) address and memory size. All devices must be the same size.

Programming the Chip Selection Unit for XL and EA Versions.   The number of wait states in each section of the memory and the I/O are programmable. The 80186/80188 (versions XL and EA) have a built-in wait state generator that can introduce between 0 and 3 wait states. Table 10–2 lists the logic levels required on bits R2–R0 in each programmable register to select various numbers of wait states. These three lines also select if an external READY signal is required to generate wait states. If READY is selected, the external READY signal is active.

Suppose that a 64K-byte EPROM is located at the top of the memory system and requires two wait states for proper operation. To select this device for this section of memory, the $\overline{UCS}$ pin is programmed for a memory range of F0000H–FFFFFH with two wait states. Figure 10–20 shows the control registers for all memory and I/O selection in the peripheral control block at offset addresses XXA0H–XXA9H. Notice that the rightmost 3 bits of these control registers are from Table 10–2. The control register for the upper memory area is location at PCB offset address A0H. This 16-bit register is programmed with the starting address of the memory area (F0000H in this case) and the number of wait states. Please note that the upper two bits of the address must be programmed as 00 and that only address bits A_{17}–A_{10} are programmed into the control register. Refer to Table 10–3 for the codes for various memory sizes. Because our example requires two wait states, the basic address is the same as in the table for a 64K device, except the rightmost three bits are 110 instead of 100. The data sent to the upper memory control register is 3006H.

Suppose that a 32K-byte SRAM that requires no waits and no READY input is located at the bottom of the memory system. To program the $\overline{LCS}$ pin to select this device, register A2 is loaded in exactly the same manner as register A0H. In this example, a 07FCH is sent to register A2H. Table 10–4 lists the programming values for the lower chip selection output.

TABLE 10–2   Wait state control bits R2, R1, and R0 (XL and EA versions)

R2	R1	R0	Number of Waits	READY Required
0	X	X	—	Yes
1	0	0	0	No
1	0	1	1	No
1	1	0	2	No
1	1	1	3	No

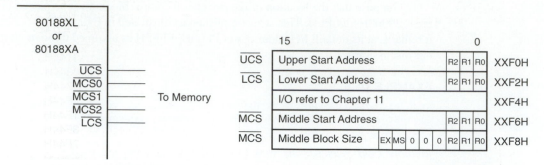

FIGURE 10–20 The control register for programming the memory selection pins. Note that R2, R1, and R0 appear in Table 10–2, and EX and MS are defined in Chapter 11 with I/O.

TABLE 10–3 Upper memory programming for register A0H (XL and EA versions)

Start Address	Block Size	Value for No Waits, No READY
FFC00H	1K	3FC4H
FF800H	2K	3F84H
FF000H	4K	3F04H
FE000H	8K	3E04H
FC000H	16K	3C04H
F8000H	32K	3804H
F0000H	64K	3004H
E0000H	128K	1004H
C0000H	256K	0004H

TABLE 10–4 Lower memory programming for register A2H. (XL and EA versions)

Ending Address	Block Size	Value for No Waits, No READY
003FFH	1K	0004H
007FFH	2K	0044H
00FFFH	4K	00C4H
01FFFH	8K	01C4H
03FFFH	16K	03C4H
07FFFH	32K	07C4H
0FFFFH	64K	0FC4H
1FFFFH	128K	1FC4H
3FFFFH	256K	3FC4H

The central part of the memory is programmed via two registers: A6H and A8H. Register A6H programs the beginning or base address of the middle memory select lines ($\overline{MCS3}$–$\overline{MCS0}$) and number of waits. Register A8H defines the size of the block of memory and the individual memory device size (refer to Table 10–5). In addition to block size, the number of peripheral wait states are programmed as with other areas of memory. The EX (bit 7) and MS (bit 6) specify the peripheral selection lines and will be discussed in Chapter 11.

TABLE 10–5   Middle memory programming for register A8H.(XL and EA versions)

Block Size	Chip Size	Value for No Waits, No READY, and EX = 0 MS = 1
8K	2K	0144H
16K	4K	0344H
32K	8K	0744H
64K	16K	0F44H
128K	32K	1F44H
256K	64K	3F44H
512K	128K	7F44H

For example, suppose that four 32K-byte SRAMs are added to the middle memory area beginning at location 80000H and ending at location 9FFFFH with no wait states. To program the middle memory selection lines for this area of memory, we place the leftmost seven address bits in register A6H with bits 8 through 3 containing logic 0's and the rightmost three bits containing the ready control bits. For this example register A6H is loaded with 8004H. Register A8H is programmed with a 1F44H, assuming that EX = 0 and MS = 1 and no wait states and no READY are required for the peripherals.

Figure 10–21 illustrates a 27256 (32K × 8) EPROM interfaced at upper memory, a 62256 (32K × 8) SRAM interfaced at lower memory, and two 62256 (32K × 8) SRAM interfaced at middle memory devices 0 and 1. The address range of upper memory begins at location F8000H and ends at FFFFFH and the address range for lower memory begins at 00000H and ends at 07FFFH. The middle memory devices are placed at locations 08000H–0FFFFH and 10000H–17FFFH.

Example 10–7 illustrates the program required to initialize the 80C188XL to function with the memory illustrated in Figure 10–21. In this example, two wait states are inserted for the EPROM and zero wait states for the SRAM. Depending on the clock frequency of the microprocessor, the number of wait states may need to be changed for proper operation. Notice how the reset offset address and bottom of the EPROM offset address are forced by using ORG statements.

EXAMPLE 10–7

```
                        .MODEL SMALL
0000                    .CODE
                                ORG 8000H
8000                    TOP:
8000 BA FFA2                    MOV DX,0FFA2H     ;address LCS
8003 B8 07F4                    MOV AX,07C4H      ;set start address at 00000H
8006 EE                         OUT DX,AL

8007 BA FFA6                    MOV DX,0FFA6H     ;address MCS base address
800A B8 0804                    MOV AX,0804H      ;set start address at 08000H
800D EE                         OUT DX,AL

800E BA FFA8                    MOV DX,0FFA8H     ;address MCS block size
8011 B8 9034                    MOV AX,1F44H      ;set block size to 128K
8014 EE                         OUT DX,AL

;
;System software is placed here
;

                        .STARTUP

                                ORG 0FFF0H        ;reset location
```

```
FFF0 BA FFA0                MOV  DX,0FFA0H    ;address UCS
FFF3 B8 F834                MOV  AX,03806H    ;set start address at 0F8000H
FFF6 EE                     OUT  DX,AL
FFF7 E9 8006                JMP  TOP

              END
```

Programming the Chip Selection Unit for EB and EC Versions.   As we mentioned earlier, the EB and EC versions have a different chip selection unit. These newer versions of the 80186/80188 contain an upper and lower memory chip selection pin as do earlier versions, but they do not contain middle selection and peripheral selection pins. In place of the middle and peripheral chip selection pins, the EB and EC versions contain eight general-purpose chip selection pins (GCS7–GCS0) that select either a memory device or an I/O device.

Programming is also different because each of the chip selection pins contains a starting address register and an ending address register. Refer to Figure 10–22 for the offset address of each pin and also the contents of the start and end registers.

Notice that programming for the EB and EC versions of the 80186/80188 is much easier than for the earlier XL and XA versions. For example, to program the UCS pin for an address

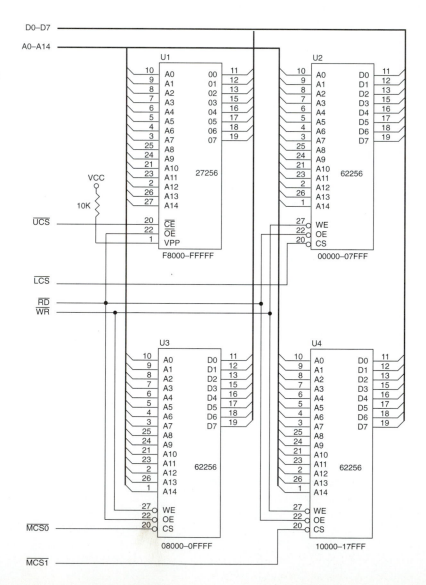

FIGURE 10–21   A small memory system interfaced to the 80C186XL embedded controller. (Notice there is no decoder.)

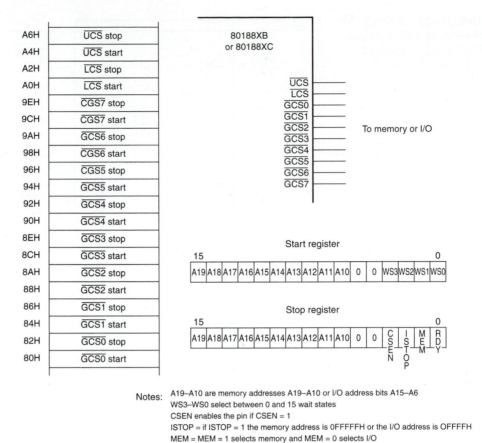

A6H	$\overline{\text{UCS}}$ stop
A4H	$\overline{\text{UCS}}$ start
A2H	$\overline{\text{LCS}}$ stop
A0H	$\overline{\text{LCS}}$ start
9EH	$\overline{\text{CGS7}}$ stop
9CH	$\overline{\text{CGS7}}$ start
9AH	$\overline{\text{GCS6}}$ stop
98H	$\overline{\text{CGS6}}$ start
96H	$\overline{\text{CGS5}}$ stop
94H	$\overline{\text{GCS5}}$ start
92H	$\overline{\text{GCS4}}$ stop
90H	$\overline{\text{GCS4}}$ start
8EH	$\overline{\text{GCS3}}$ stop
8CH	$\overline{\text{GCS3}}$ start
8AH	$\overline{\text{GCS2}}$ stop
88H	$\overline{\text{GCS2}}$ start
86H	$\overline{\text{GCS1}}$ stop
84H	$\overline{\text{GCS1}}$ start
82H	$\overline{\text{GCS0}}$ stop
80H	$\overline{\text{GCS0}}$ start

80188XB or 80188XC

$\overline{\text{UCS}}$, $\overline{\text{LCS}}$, $\overline{\text{GCS0}}$, $\overline{\text{GCS1}}$, $\overline{\text{GCS2}}$, $\overline{\text{GCS3}}$, $\overline{\text{GCS4}}$, $\overline{\text{GCS5}}$, $\overline{\text{GCS6}}$, $\overline{\text{GCS7}}$ To memory or I/O

Start register

| 15 | | | | | | | | | | | | | | | 0 |
| A19 | A18 | A17 | A16 | A15 | A14 | A13 | A12 | A11 | A10 | 0 | 0 | WS3 | WS2 | WS1 | WS0 |

Stop register

| 15 | | | | | | | | | | | | | | | 0 |
| A19 | A18 | A17 | A16 | A15 | A14 | A13 | A12 | A11 | A10 | 0 | 0 | CSEN | ISTOP | MEM | RDY |

Notes: A19–A10 are memory addresses A19–A10 or I/O address bits A15–A6
WS3–WS0 select between 0 and 15 wait states
CSEN enables the pin if CSEN = 1
ISTOP = if ISTOP = 1 the memory address is 0FFFFFH or the I/O address is 0FFFFH
MEM = MEM = 1 selects memory and MEM = 0 selects I/O
RDY = enables external ready if RDY = 1 for more than 15 wait states

FIGURE 10–22 The chip selection unit in the EB and EC versions of the 80186/80188.

that begins at location F0000H and ends at location FFFFFH (64K bytes), the starting address register (offset = A4H) is programmed with F002H for a starting address of F0000H with two wait states. The ending address register (offset = A6H) is programmed with 000EH for an ending address of FFFFFH for memory with no external ready synchronization. The other chip selection pins are programmed in a likewise fashion.

Figure 10–23 shows a memory interface for a 16 MHz 80186CEB microprocessor. This interface contains two 27512 (64K-byte EPROM) and a single 62256 SRAM. The EPROM is located at the top 128K bytes (E0000H–FFFFFH) of the memory system and the SRAM is located at the bottom 32K bytes of memory (00000H–07FFFH). Notice that $\overline{\text{UCS}}$ selects one EPROM and $\overline{\text{GCS0}}$ selects the other EPROM. The $\overline{\text{LCS}}$ pin selects the SRAM. Because the EPROM used in this example requires 450 ns for access, wait states are required because the microprocessor operating at 16 MHz allows only 144.5 ns for memory access. In this system six wait states are inserted ($6 \times 62.5 + 144.5 = 519.5$ ns) to allow enough time for the memory to function. Note that five wait states would allow 457 ns, too close to allow for proper memory operation. A margin of about 20 ns should be allowed when finding wait states. The SRAM requires 250 ns for access, which means that it requires two wait states (269.5 ns), which allows 19.5 ns for a margin.

Because there is no decoder in Figure 10–23 and the 80C188EB has an internal programmable decoder, a short sequence of instructions (see Example 10–8) are required to program the $\overline{\text{UCS}}$, $\overline{\text{LCS}}$, and $\overline{\text{GCS0}}$ pins. Notice that the stop address is one byte above the actual ending address of a memory device.

FIGURE 10–23 A circuit that interfaces 128K bytes of EPROM and 32K bytes of SRAM to the 80C188EB.

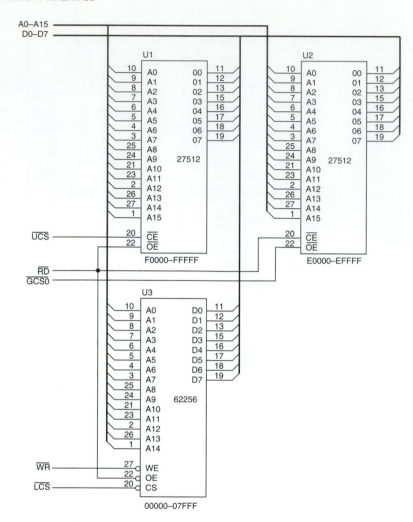

EXAMPLE 10–8

```
                .model small
0000            .code

0000            TOP:
0000 BA FFA6        MOV DX,0FFA6H    ;address UCS stop of FFFFFH
0003 B8 000E        MOV AX,000EH
0006 EE             OUT DX,AL

0007 BA FFA0        MOV DX,0FFA0H    ;address LCS
000A B8 0002        MOV AX,0002H     ;Set start address at 00000H
000D EE             OUT DX,AL        ;with 2 waits.

000E BA FFA2        MOV DX,0FFA2H    ;address LCS stop at 07FFFH
0011 B8 080A        MOV AX,080AH
0014 EE             OUT DX,AL

0015 BA FF80        MOV DX,0FF80H    ;address GCS0
0018 B8 E006        MOV AX,0E006H    ;Set start address at E0000H
001B EE             OUT DX,AL        ;with 6 waits.

001C BA FF82        MOV DX,0FF82H    ;address GCS0 stop at EFFFFH
001F B8 F00A        MOV AX,0F00AH
0022 EE             OUT DX,AL
```

```
;
;System software is placed at this point in the upper EPROM.
;
                    .STARTUP

                    ORG 0FFF0H        ;reset location

FFF0 BA FFA4        MOV DX,0FFA4H     ;address UCS
FFF3 B8 F006        MOV AX,0F006H     ;Set start address at 0F0000H
FFF6 EE             OUT DX,AL         ;with 6 waits.
FFF7 E9 0006        JMP TOP

                    END
```

Interfacing Flash Memory

Flash memory (EEPROM) is becoming commonplace for storing setup information on video cards as well as for storing the system BIOS in the personal computer. Flash memory is also found in many other applications to store information that is changed only occasionally.

The only difference between a flash memory device and SRAM is that the flash memory device requires a 12 V programming voltage to erase and write new data. The 12 V can be available either at the power supply or a 5-V-to-12-V converter designed for use with flash memory can be obtained.

Figure 10–24 illustrates a 28F400 Intel flash memory device interfaced to the 80C188EB microprocessor. The 28F400 can be used as either a $512K \times 8$ memory device or as a $256K \times 16$ memory device. Because it is interfaced to the 80C188EB, its configuration is $512K \times 8$. Notice that the control connections on this device are identical to that of an SRAM: $\overline{CE}$, $\overline{OE}$, and $\overline{WE}$. The only new pins are V_{PP}, which is connected to 12 V for erase and programming; $\overline{PWD}$, which selects the power-down mode when it is a logic 0 and is also used for programming; and $\overline{BYTE}$, which selects a byte (0) or word (1) operation. The pin DQ15 functions as the least-significant address input when operated in the byte mode. Another difference is the amount of time required to accomplish a write operation. The SRAM can perform a write operation in as little as 10 ns, but the flash memory requires approximately 0.4 second to erase a byte. The next chapter covers the topic of programming the flash memory device along with I/O devices. The flash memory device has some internal registers and is programmed using I/O techniques. This chapter concentrates on its interface to the microprocessor.

Notice in Figure 10–24, that the $\overline{UCS}$ is chosen to select the flash memory. The software, not shown for this illustration, would place the flash memory at locations 80000H–FFFFFH.

FIGURE 10–24 The 28F400 flash memory device interfaced to the 80C186XB microprocessor.

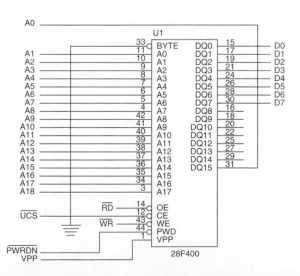

Parity for Memory Error Detection

Because such large memories are available in today's systems and because circuit costs are minimal, many memory board manufacturers have added parity checking to their RAM memory boards, although recently there seems to be a trend away from parity. Parity checking counts the number of 1's in data and indicates whether there is an even or odd number. If all data is stored with even parity (with an even number of 1 bits), a 1-bit error can be detected. Memory that contains parity is 9 bits wide or 36 bits wide for the newer 72-pin SIMM (**single-in-line memory module**) components found in computer systems.

Figure 10–25 illustrates the 74AS280 parity generator/detector integrated circuit. This circuit has nine inputs and generates even or odd parity for the 9-bit number placed on its inputs. It also checks the parity of a 9-bit number connected to its inputs.

Figure 10–26 illustrates a 64K × 8 static RAM system using two 62556 32K × 8 SRAM devices for data storage that has parity generation and detection. Notice that a 74AS280 generates a parity bit stored in a 6287 64K × 1 SRAM. This circuit decodes the memory at locations 80000H–8FFFFH using the $\overline{GCS2}$ and $\overline{GCS3}$ pins on an 80C188XB microprocessor. Here the eight data bus connections are attached to the parity generator's (U5) inputs A–H. Input I is grounded, so if an even number of 1's appear on the data bus, a 1 (at the even output) is stored in the parity RAM. If an odd number of 1's appear, a 0 is stored in the parity RAM. Here odd parity is stored for each byte of data, including the parity bit written to the memory.

When data is read from the memory, each datum is connected to another 74AS280 (U6) to check its parity. In this case, all the inputs to the checker are connected. Inputs A–H are connected to the data RAM's outputs, and input I is connected to the parity RAM. Note that the parity RAM control pins are different. This SRAM reads data from its output pin when selected and writes data if selected with $\overline{WE}$ = 0. It does not have an $\overline{OE}$ connection for $\overline{RD}$. If parity is odd, as it is if everything is correct, the even parity output of the 74AS280 (U6) is a logic 0. If a bit of the information read from the memory changes for any reason, then the even output pin of the 74AS280 will become a logic 1. The parity output pin is connected to a special input of the 80C188XB called the non-maskable interrupt (NMI) input. The NMI input can never be turned off. If it is placed at its logic 1 level, the program being executed is interrupted, and a special subroutine indicates that a parity error has been detected by the memory system. (More detail on interrupts is provided in Chapter 12.)

The application of the parity error is timed so the data read from the memory is settled to its final state before an NMI input occurs. The operation is timed by a D-type flip-flop that latches the output of the parity checker at the end of an $\overline{RD}$ cycle from this section of the memory. In this way, the memory has enough time to read the information and pass it through the generator before the output of the generator is sampled by the NMI input.

Error Correction

Error correction schemes have been around for a long time, but integrated circuit manufacturers have only recently started to produce error correcting circuits. One such circuit is the 74LS636,

FIGURE 10–25 A parity generator/detector. Generates the ninth parity bit and detects parity on a 9-bit code.

74LS280

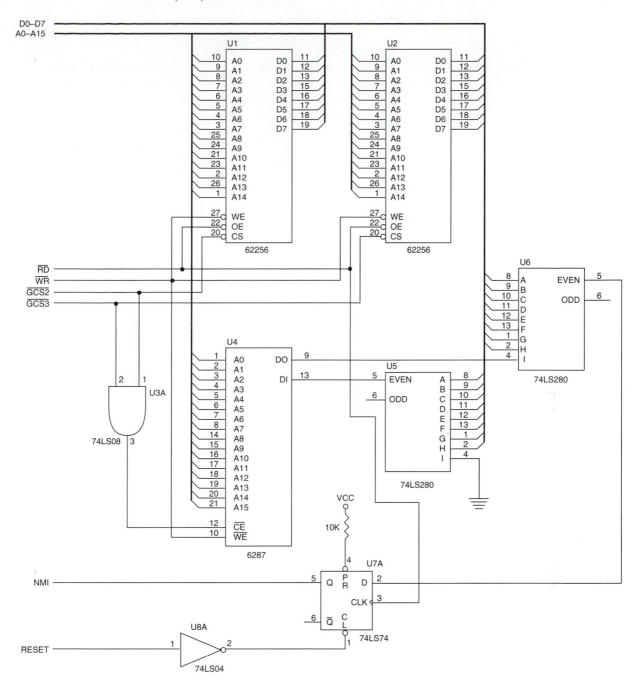

FIGURE 10–26 A 64K-byte memory system that contains a parity error detection circuit.

an 8-bit error correction and detection circuit that corrects any single-bit memory read error and flags any 2-bit error. This device is found in very high-end computer systems because of the cost of implementing a system that uses error correction.

The 74LS636 corrects errors by storing five parity bits with each byte of memory data. This does increase the amount of memory required, but it also provides automatic error correction for single-bit errors. If more than two bits are in error, this circuit may not detect it. Fortunately, this is rare, and the extra effort required to correct more than a single-bit error is very

pin assignments

	J, N PACKAGES		
1	DEF	11	CB4
2	DB0	12	nc
3	DB1	13	CB3
4	DB2	14	CB2
5	DB3	15	CB1
6	DB4	16	CB0
7	DB5	17	S0
8	DB6	18	S1
9	DB7	19	SEF
10	GND	20	V_{CC}

(a)

functional block diagram

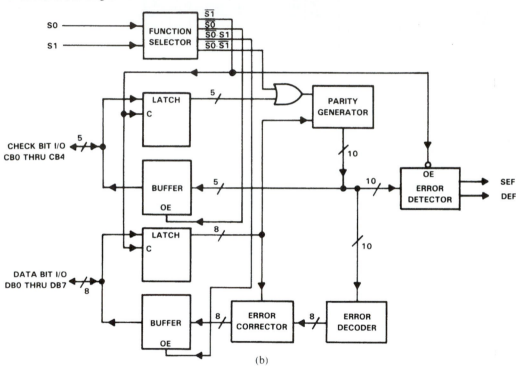

(b)

FIGURE 10–27   (a) The pin connections of the 74LS636. (b) The block diagram of the 74LS636. (Courtesy of Texas Instruments Incorporated)

expensive and not worth the effort. Whenever a memory component fails completely, its bits are all high or all low. In this case, the circuit flags the processor with a multiple-bit error indication.

Figure 10–27 depicts the pin-out of the 74LS636. Notice that it has eight data I/O pins, five check bit I/O pins, two control inputs (S0 and S1), and two error outputs: single error flag (SEF) and double error flag (DEF). The control inputs select the type of operation to be performed and are listed in the truth table shown in Table 10–6.

When a single error is detected, the 74LS636 goes through an error correction cycle: it places a 01 on S0 and S1 by causing a wait and then a read following error correction.

Figure 10–28 illustrates a circuit used to correct single-bit errors with the 74LS636 and to interrupt the processor through the NMI pin for double-bit errors. To simplify the illustration, we depict only one 2K × 8 RAM and a second 2K × 8 RAM to store the 5-bit check code.

TABLE 10–6 Control bits
S0 and S1

S0	S1	Function	SEF	DEF
0	0	Write check word	0	0
0	1	Correct data word	*	*
1	0	Read data	0	0
1	1	Latch data	*	*

Note: These levels are determined by the type of error.

The connection of this memory component is different from that of the previous example. Notice that the $\overline{S}$ or $\overline{CS}$ pin is grounded, and data bus buffers control the flow to the system bus. This is necessary if the data is to be accessed from the memory before the $\overline{RD}$ strobe goes low.

On the next negative edge of the clock after an $\overline{RD}$, the 74LS636 checks the single-error flag (SEF) to determine whether an error has occurred. If so, then a correction cycle causes the single-error defect to be corrected. If a double error occurs, then an interrupt request is generated by the double-error flag (DEF) output, which is connected to the NMI pin of the microprocessor.

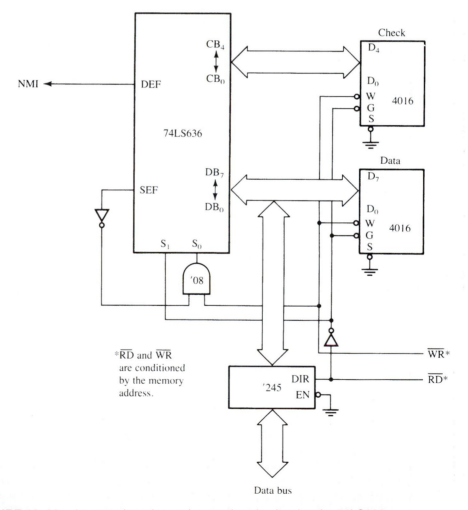

FIGURE 10–28 An error detection and correction circuit using the 74LS636.

10–4   80186 AND 80386EX (16-BIT) MEMORY INTERFACE

The 80186 and 80386EX embedded microprocessors differ from the 80188 in two ways: (1) the data bus is 16 bits wide instead of 8 bits wide as in the 80188, and (2) there is a new control signal, bus high enable ($\overline{\text{BHE}}$). The address bit A_0 or $\overline{\text{BLE}}$ is also used differently. Because this section is based on information provided in Section 10–3, it is extremely important that you read the previous section first. A few other differences exists between the 80186 and the 80386EX. For example, the 80386EX contains a 26-bit address bus (A_{25}–A_1) instead of the 20-bit address bus (A_{19}–A_0) of the 80186. Note that the A_0 address bit for the 80386EX is encoded as the $\overline{\text{BLE}}$ signal.

16-Bit Bus Control

The data bus of the 80186 and 80386EX is twice as wide as the bus for the 80188. This wider data bus presents us with a unique set of problems that have not been encountered before. The 80186 and 80386EX must be able to write data to any 16-bit location or any 8-bit location. This means that the 16-bit data bus must be divided into two separate sections (**banks**) that are 8 bits in width so the microprocessor can write to either half (8-bit) or both halves (16-bit). Figure 10–29 illustrates the two banks of the memory. One bank (**low bank**) holds all the even-numbered memory locations, and the other bank (**high bank**) holds all the odd-numbered memory locations.

The 8086, 80186, 80286, and 80386SX use the $\overline{\text{BHE}}$ signal (high bank) and the A_0 address bit or $\overline{\text{BLE}}$ (bus low enable) to select one or both banks of memory used for the data transfer. Table 10–7 depicts the logic levels on these two pins and the bank or banks selected.

Bank selection is accomplished in two ways: (1) a separate write signal is developed to select a write to each bank of the memory, or (2) separate decoders are used for each bank. As a careful comparison reveals, the first technique is by far the least costly approach to memory interface for the 80186 and 80386EX embedded microprocessors. Because the 80186 and 80386EX have programmable chip selection pins, the pins are used to select most memory devices. In very large systems, external decoders are still used to select memory. Next we'll discuss external decoders for 16-bit memory systems. Then we'll program the chip select pins of the 80186 and 80386EX for memory selection.

Separate Bank Decoders.   The use of separate bank decoders is often the least effective way to decode memory addresses for the 80186 and 80386EX. This method is sometimes used, but it is

FIGURE 10–29   The high (odd) and low (even) memory banks of the 80186 and 80386EX microprocessors.

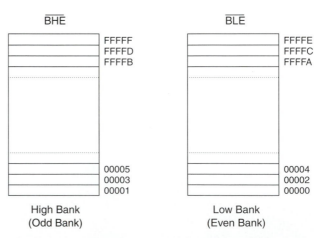

Note that A0 is labeled BLE (bus low enable) on the 80386EX.

TABLE 10–7   Memory bank selection using $\overline{BHE}$ and $\overline{BLE}$ (A0)

$\overline{BHE}$	$\overline{BLE}$ (A0)	Function
0	0	Both banks enabled for a 16-bit transfer
0	1	High bank enabled for an 8-bit transfer
1	0	Low bank enabled for an 8-bit transfer
1	1	No banks enabled

difficult to understand why in most cases. One reason may be to conserve energy, because only the bank or banks selected are enabled. Later we'll discuss separate bank read and write signals as another method of interfacing 16-bit wide memory.

Figure 10–30 illustrates two 74LS138 decoders used to select 64K RAM memory components for the 80386EX microprocessor (26-bit address). Here decoder U3 has the $\overline{BLE}$ pin (A_0) attached to $\overline{G_2B}$, and decoder U2 has the $\overline{BHE}$ signal attached to its $\overline{G_2B}$ input. Because the decoder will not activate until all its enable inputs are active, decoder U2 activates for a 16-bit

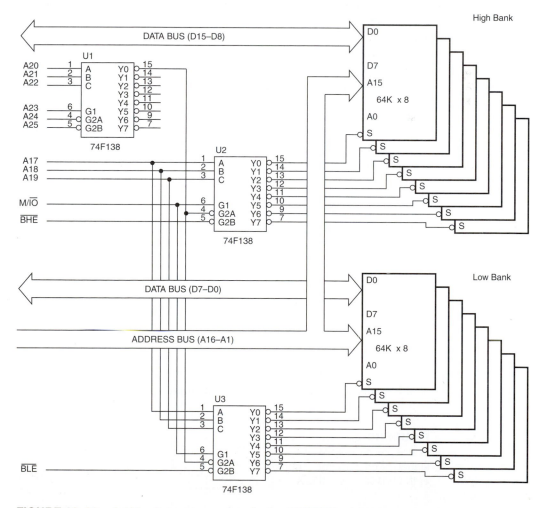

FIGURE 10–30   A 1M × 8 memory system for the 80386EX microprocessor using separate bank decoders.

operation or an 8-bit operation from the high bank, and decoder U3 activates for a 16-bit operation or an 8-bit operation to the low bank. These two decoders and the 16 64K-byte RAMs they control represent a 1M range of the 80386EX memory system. Decoder U1 enables U2 and U3 for the memory address range 0800000H–08FFFFFH.

Notice from Figure 10–30 that the A_0 address pin does not connect to the memory; in fact, it does not even exist on the 80386EX microprocessor. Also notice that address bus bit position A_1 is connected to the memory chip's address input A_0; A_2 is connected to A_1, and so forth. The reason is that A_0 from the 80186 (or $\overline{BLE}$ from the 80386EX) is already connected to decoder U2 and does not need to be connected again to the memory. If A_0 or $\overline{BLE}$ is attached to the A_0 address pin of memory, every other memory location in each bank of memory would be used. This means that half of the memory is wasted if A_0 or $\overline{BLE}$ is connected to A_0.

Separate Bank Write Strobes. The most effective way to handle bank selection is to develop a separate write strobe for each memory bank. This technique requires only one decoder to select a 16-bit wide memory. This often saves money and reduces the number of components in a system.

Why not also generate separate read strobes for each memory bank? This is usually unnecessary, because the 80186 and 80386EX read only the byte of data that they need at any given time from half of the data bus. If 16-bit sections of data are always presented to the data bus during a read, the microprocessor ignores the 8-bit section it doesn't need, without any conflicts or special problems.

Figure 10–31 depicts the generation of separate 80386EX write strobes for the memory. Here a 74LS32 OR gate combines $\overline{BLE}$ with $\overline{WR}$ for the low bank selection signal ($\overline{LWR}$) and $\overline{BHE}$ with $\overline{WR}$ for the high bank selection signal ($\overline{HWR}$). The 80186 uses the same circuit, except that A_0 replaces the $\overline{BLE}$ signal.

A memory system that uses separate write strobes is constructed differently from either the 8-bit system (80188) or the system using separate memory banks. Memory in a system that uses separate write strobes is decoded as 16-bit wide memory. For example, suppose that a memory system will contain 64K bytes of SRAM memory. This memory requires two 32K-byte memory devices (62256) so a 16-bit wide memory can be constructed. Note that 16-bit wide memory is always populated with pairs of 8-bit wide memory devices. Because the memory is 16 bits wide and another circuit generates the bank write signals, address bit A_0 becomes a don't care. In fact, A_0 is not even a pin on the 80386EX microprocessor.

EXAMPLE 10–9

```
00 0000 0110 0000 0000 0000 0000 = 0060000H
              to
00 0000 0110 1111 1111 1111 1111 = 006FFFFH


00 0000 0110 XXXX XXXX XXXX XXXX = 006XXXXH
```

Example 10–9 shows how a 16-bit wide memory stored at locations 0060000H–006FFFFH is decoded for the 80386EX microprocessor. Memory in this example is decoded so

FIGURE 10–31 Generation of a separate write strobe signal for each memory bank of the 80186 or 80386EX.

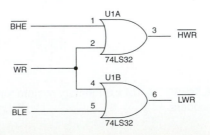

FIGURE 10–32   A 16-bit memory interfaced to the 80386EX at memory locations 060000H–06FFFFH.

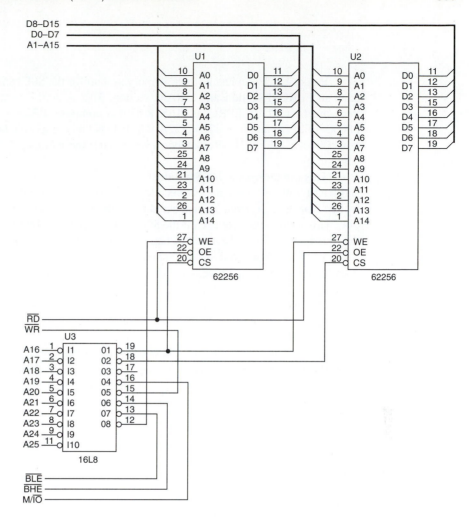

bit A_0 is a don't care for the decoder. Bit positions A_1–A_{15} are connected to memory component address pins A_0–A_{14}. The decoder (PAL16L8) enables both memory devices by using address connection A_{25}–A_{15} to select memory whenever address 006XXXXH appears on the address bus.

Figure 10–32 illustrates this simple circuit using a PAL16L8 to both decode memory and generate the separate write strobe. The program for the PAL16L8 decoder is illustrated in Example 10–10. Notice that not only is the memory selected, but both the lower and upper write strobes are also generated by the PAL.

EXAMPLE 10–10

```
TITLE      Address Decoder
PATTERN    Test 2
REVISION   A
AUTHOR     Barry B. Brey
COMPANY    BreyCo
DATE       6/7/96
CHIP       DECODER2 PAL16L8

;pins 1    2    3    4    5    6    7    8    9    10
     A16  A17  A18  A19  A20  A21  A22  A23  A24  GND

;pins 11  12   13   14   15  16   17  18   19   20
      A25 LWR  BLE  BHE  WR  MIO  NC  HWR  SEL  VCC
EQUATIONS
```

```
/SEL = /A25*/A24*/A23*/A22*/A21*/A20*/A19*A18*A17*/A16*MIO
/LWR = /WR*/BLE
/HWR = /WR*/BHE
```

Figure 10–33 depicts a small memory system for the 80C186EB microprocessor that contains an EPROM section and a RAM section. This system uses conventional decoding. Here there are four 27128 EPROMs (16K × 8) that compose a 32K × 16-bit memory at locations F0000–FFFFFH and four 62256 (32K × 8) RAMs that compose a 64K × 16-bit memory at locations 00000H–1FFFFH. (Remember, even though the memory is 16 bits wide, it is still numbered in bytes.)

This circuit uses a 74LS139 dual 2-to-4 line decoder that selects EPROM with one half and RAM with the other half. It decodes memory that is 16 bits wide and not 8 bits as before. Notice that the $\overline{RD}$ strobe is connected to all the EPROM $\overline{OE}$ inputs and all the RAM $\overline{G}$ input pins. This is done because even if the 80C186EB is only reading 8 bits of data, the application of the remaining 8 bits to the data bus has no effect on the operation of the 80C186EB.

The $\overline{LWR}$ and $\overline{HWR}$ strobes are connected to different banks of the RAM memory. Here it does matter if the microprocessor is doing a 16- or an 8-bit write. If the 80C186EB writes a 16-bit number to memory, both $\overline{LWR}$ and $\overline{HWR}$ go low and enable the $\overline{W}$ pins both memory banks. But, if the 80C186EB does an 8-bit write, then only one of the write strobes goes low, writing to only one memory bank. Again the only time that the banks make a difference is for a memory write operation.

Notice that an EPROM decoder signal is sent to the 80C186EB wait state generator because EPROM memory usually requires a wait state. The signal comes from the NAND gate (U1A) used to select the EPROM decoder section, so that if EPROM is selected, a wait state is requested.

Figure 10–34 illustrates a memory system connected to the 80C186XB microprocessor using a PAL16L8 as a decoder. This interface contains 256K bytes of EPROM in the form of four 27512 (64K × 8) EPROMs and 128K bytes of SRAM memory found in four 62256 (32K × 8) SRAMs. Note that the $\overline{UCS}$ signal selects two of the EPROMs for the upper 128K bytes of memory (E0000H–FFFFFH) and the $\overline{LCS}$ signal selects the bottom 64K bytes of SRAM (00000H–0FFFFH). The remaining memory is selected using conventional decoding techniques.

Notice from Figure 10–34 that the PAL also generates the memory bank write signals $\overline{LWR}$ and $\overline{HWR}$. As can be gleaned from this circuit, the number of components required to interface memory has been reduced to just one in most cases—the PAL. Example 10–11 is the program listing for the PAL. The PAL decodes the 16-bit wide memory addresses at locations 10000H–1FFFFH for the SRAM and locations C0000H–DFFFFH for the remaining EPROM.

EXAMPLE 10–11

```
TITLE       Address Decoder
PATTERN     Test 3
REVISION    A
AUTHOR      Barry B. Brey
COMPANY     BreyCo
DATE        6/8/97
CHIP        DECODER3 PAL16L8
;pins 1  2    3    4    5    6   7   8  9   10
      A19 A18 A17 A16 BLE BHE NC WR NC GND

;pins 11 12 13 14 15 16  17  18  19  20
      S2 NC NC NC NC LWR HWR RAM ROM VCC
EQUATIONS

/LWR = /WR * /BLE
/HWR = /WR * /BHE
/ROM = S2 * A19 * A18 * /A17
/RAM = S2 * /A19 * /A18 * /A17 * A16
```

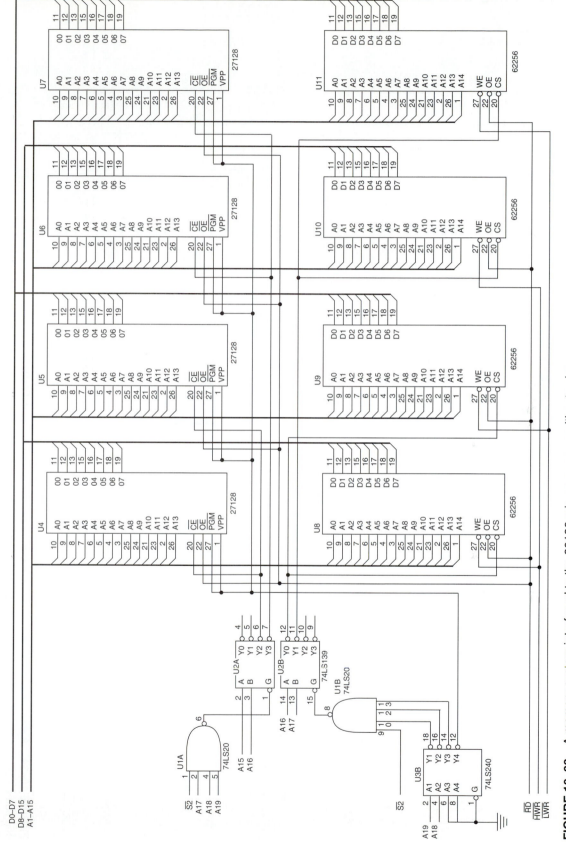

FIGURE 10–33 A memory system interfaced to the 80186 microprocessor without using the chip select pins.

367

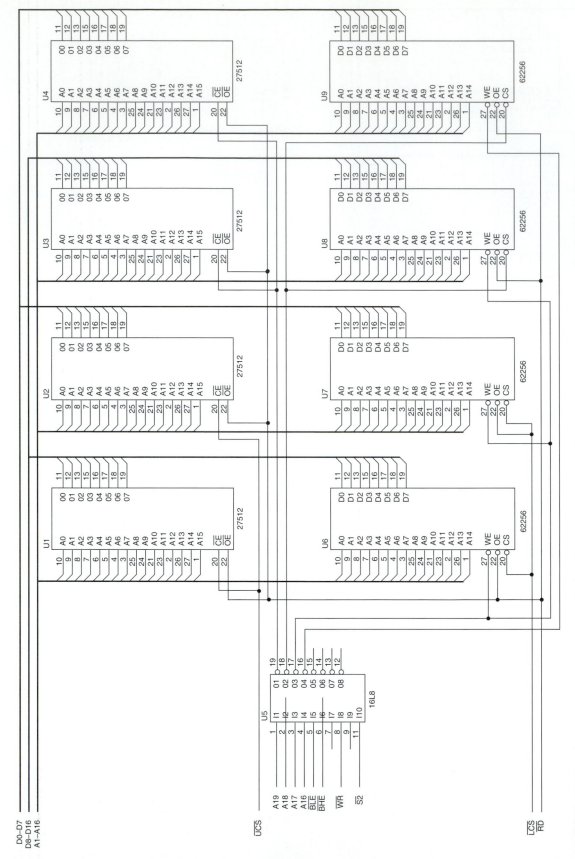

FIGURE 10–34 A memory system connected to the 80186 using some chip select pins and some conventional decoding.

Example 10–12 depicts the software required to program the $\overline{\text{LCS}}$ and the $\overline{\text{UCS}}$ pins to select the upper EPROM and the lower SRAM of the memory in Figure 10–34. The EPROM is active for addresses E0000H–FFFFFH and the SRAM is active for addresses 00000H–0FFFFH.

EXAMPLE 10–12

```
                    .MODEL SMALL
0000                .CODE

0000            TOP:
0000 BA FFA6        MOV     DX,0FFA6H    ;address UCS stop register
0003 B8 000E        MOV     AX,000EH     ;stop address = FFFFFH
0006 EE             OUT     DX,AL

0007 BA FFA0        MOV     DX,0FFA0H    ;address LCS start register
000A B8 0000        MOV     AX,0000H     ;start address = 0000H
000D EE             OUT     DX,AL        ;with no waits

000E BA FFA2        MOV     DX,0FFA2H    ;address LCS stop register
0011 B8 100A        MOV     AX,100AH     ;stop address = 0FFFFH
0014 EE             OUT     DX,AL
;
;System software is placed at this point.
;
                    .STARTUP

                    ORG     0FFF0H       ;reset address

FFF0 BA FFA4        MOV     DX,0FFA4H    ;address UCS start register
FFF4 B8 E000        MOV     AX,0E000H    ;start address = E0000H
FFF6 EE             OUT     DX,AL        ;with no waits
FFF7 E9 0006        JMP     TOP

                    END
```

80386EX Memory Interface

As with the 80186/80188, the 80386EX also contains a programmable chip selection unit. In fact, the 80386EX looks much like the EB and EC versions of the 80186/80188 when the chip selection unit is examined. The main difference is that the 80386EX has only seven general chip select pins instead of the eight found on the 80186/80188 EB and EC versions and there is no $\overline{\text{LCS}}$ pin.

Figure 10–35 illustrates the chip selection pins on the 80386EX and also the register locations and contents used for programming the chip selection unit. As with the 80186/80188 EB and EC versions, the 80386EX also shares the function of the chip select pins with other signals.

Another major advantage the 80386EX has is the way that the general and upper chip select pins are programmed. Note that the only pin that is not shared with other functions is the $\overline{\text{UCS}}$ pin. The $\overline{\text{GCS0}}$–$\overline{\text{GCS6}}$ pins are shared with either a parallel port (port 2), $\overline{\text{DACK0}}$, or $\overline{\text{REFRESH}}$. Because of this, the configuration register must be programmed before a pin can be used to select memory or I/O. Programming the configuration register should be done by reading the register, setting the bits that require activation, and then writing the register. The reason that this technique is required is that the configuration register is also used for other functions in the system and this prevents the other bits from being modified.

The address programmed into the chip selection unit is 15 bits wide for the 80386EX. The lower address register is programmed with the least-significant address 5 bits, and the high address register with the most-significant 10 bits. The address register provides the most-significant 15 bits of the 26-bit memory address and the most-significant 15 bits of the I/O port address. For example, if a memory device is to start at address 02000H, the address programmed into the two address registers is a 0000100000 00000$_2$.

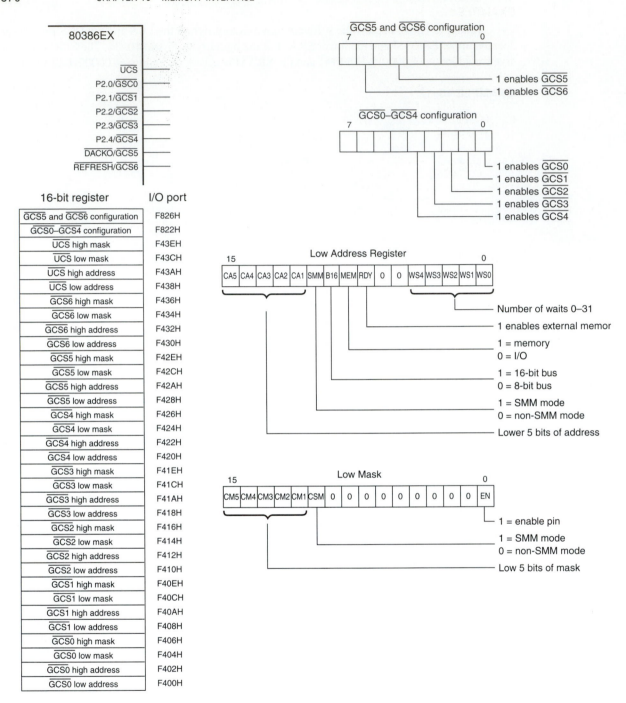

FIGURE 10–35   The chip selection unit within the 80386EX embedded microprocessor. Note that the high address and high mask registers are not illustrated, but contain the most-significant 10 bits of the mask or address.

The mask register is used to disable address bits used for memory selection, which determines the size of the memory device. For example, if the memory device starts at address 02000H and ends at address 02FFFFH, the mask register is programmed with a $0000001111\ 11111_2$, which uses only the most-significant 6 bits of the address for memory selection. This is illustrated in Example 10–13.

EXAMPLE 10–13

```
Address register = 00 0010 0000 0000 00
   Mask register = 00 0000 1111 1111 11
Selected address = 00 0010 XXXX XXXX XX = 02XXXXH or 020000H-02FFFFH
```

Another example assumes that the memory device is to exist at locations 0E0000H–0FFFFFH. This is accomplished by programming the address as a $0011100000\ 00000_2$ with a mask of $0000011111\ 11111_2$, which causes the microprocessor to select memory as indicated in Example 10–14. Note that a 1 in a mask bit causes the corresponding address bit to become a don't care (X).

EXAMPLE 10–14

```
Address register = 00 1110 0000 0000 00
   Mask register = 00 0001 1111 1111 11
Selected address = 00 111X XXXX XXXX XX - 0E0000H-0FFFFFH
```

Figure 10–36 illustrates a 16-bit wide memory system interfaced to the 80386EX microprocessor. The system illustrated contains six 27512 EPROM memory devices interfaced at memory locations 3A0000H–3FFFFFH, the top of the system memory. The reset location for the 80386EX is at memory address 3FFFF0H, which is 16 bytes from the top of the system memory. The system also contains a pair of SRAM devices (62256) interfaced at addresses 000000H–00FFFFH.

The software required to program the $\overline{UCS}$ pin and general chip seection pins used in Figure 10–36 appears in Example 10–15. Note that the EPROM in this system requires 10 wait states and the SRAM requires three wait states.

EXAMPLE 10–15

```
0017 BA F43E      MOV   DX,0F43EH   ;address UCS high address
001A B8 03E0      MOV   AX,3E0H     ;start address = 3E0000H
001D EF           OUT   DX,AX

001E BA F43C      MOV   DX,0F43CH   ;address UCS low address
0021 B8 030A      MOV   AX,30AH     ;10 waits
0024 EF           OUT   DX,AX

0025 BA F43A      MOV   DX,0F43AH   ;address UCS high mask
0028 B8 03E0      MOV   AX,3E0H     ;address = 3E0000H-3FFFFFH
002B EF           OUT   DX,AX

002C BA F438      MOV   DX,0F438H   ;address UCS low mask
002F B8 0001      MOV   AX,1        ;enable pin
0032 EF           OUT   DX,AX

0033 BA F406      MOV   DX,0F406H   ;address GCS0 high address
0036 B8 03C0      MOV   AX,3C0H     ;start address = 3C0000H
0039 EF           OUT   DX,AX

003A BA F404      MOV   DX,0F404H   ;address GCS0 low address
003D B8 030A      MOV   AX,30AH     ;10 waits
0040 EF           OUT   DX,AX

0041 BA F402      MOV   DX,0F402H   ;address GCS0 high mask
0044 B8 03E0      MOV   AX,3E0H     ;address = 3C0000H-3DFFFFH
0047 EF           OUT   DX,AX

0048 BA F        MOV   DX,0F400H   ;address GCS0 low mask
004B B8 0001      MOV   AX,1        ;enable pin
004E EF           OUT   DX,AX

004F BA F40E      MOV   DX,0F40EH   ;address GCS1 high address
0052 B8 03A0      MOV   AX,3A0H     ;start address = 3A0000H
0055 EF           OUT   DX,AX
```

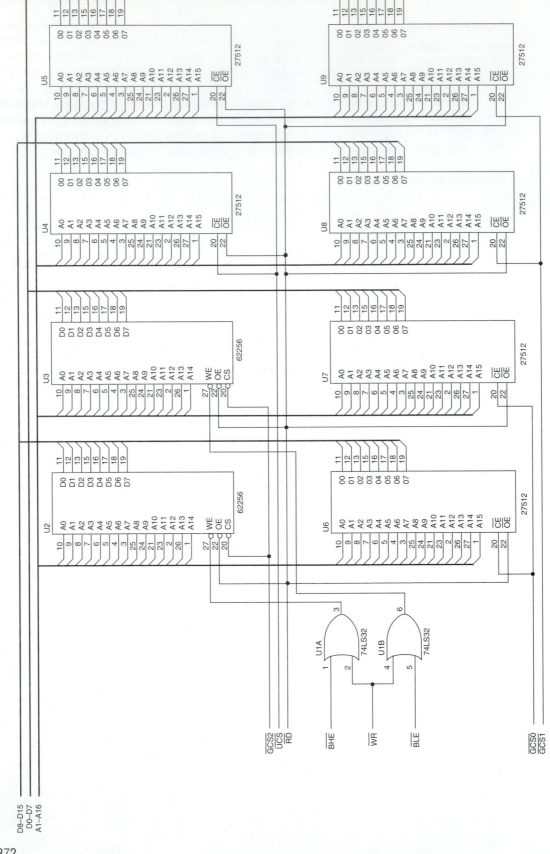

FIGURE 10–36 A small memory system interfaced to the 80386EX embedded microprocessor.

372

```
0056 BA F40C      MOV   DX,0F40CH    ;address GCS₁ low address
0059 B8 030A      MOV   AX,30AH      ;10 waits
005C EF           OUT   DX,AX

005D BA F40A      MOV   DX,0F40AH    ;address GCS₁ high mask
0060 B8 03E0      MOV   AX,3E0H      ;address = 3A0000H-3BFFFFH
0063 EF           OUT   DX,AX

0064 BA F408      MOV   DX,0F408H    ;address GCS₁ low mask
0067 B8 0001      MOV   AX,1         ;enable pin
0069 EF           OUT   DX,AX

006A BA F416      MOV   DX,0F416H    ;address GCS₂ ligh address
006D B8 0000      MOV   AX,0         ;start address = 000000H
0070 EF           OUT   DX,AX

0071 BA F414      MOV   DX,0F414H    ;address GCS₂ low address
0074 B8 0303      MOV   AX,303H      ;3 waits
0077 EF           OUT   DX,AX

0078 BA F412      MOV   DX,0F412H    ;address GCS₂ high mask
007B B8 03F0      MOV   AX,3F0H      ;address = 000000H-00FFFFH
007E EF           OUT   DX,AX

007F BA F410      MOV   DX,0F410H    ;address GCS₂ low mask
0082 B8 0001      MOV   AX,1         ;enable pin
0085 EF           OUT   DX,AX

0086 BA F822      MOV   DX,0F822H    ;address GCS₀-GCS₄ config
0089 EC           IN    AL,DX        ;read register

008A 0C 07        OR    AL,7         ;enable GCS₀-GCS₂ pins
008C EF           OUT   DX,AX
```

10–5 DYNAMIC RAM

Because RAM is often very large, it requires many SRAM devices at a great cost or just a few DRAMs (dynamic RAMs) at a much reduced cost. The DRAM memory, as discussed briefly in Section 10–1, is fairly complex because it requires address multiplexing and refreshing. Luckily, the integrated circuit manufacturers have provided a dynamic RAM controller that includes the address multiplexers and all the timing circuitry necessary for refreshing.

This section of the text covers the DRAM memory device in much more detail than Section 10–1 and provides information on the use of a dynamic controller in a memory system.

DRAM Revisited

As mentioned in Section 10–1, a DRAM retains data for only 2–4 ms and requires the multiplexing of address inputs. We have already covered address multiplexers in Section 10–1, but we will examine the operation of the DRAM during refresh in detail here.

As previously mentioned, a DRAM must be refreshed periodically because it stores data internally on capacitors that lose their charge in a short period of time. In order to refresh a DRAM, the contents of a section of the memory must periodically be read or written. Any read or write automatically refreshes an entire section of the DRAM. The number of bits refreshed depends on the size of the memory component and its internal organization.

Refresh cycles are accomplished by doing a read, a write, or a special refresh cycle that doesn't read or write data. The refresh cycle is totally internal to the DRAM and is accomplished while other memory components in the system operate. This type of refresh is called either hidden refresh, transparent refresh, or sometimes cycle stealing.

In order to accomplish a hidden refresh while other memory components are functioning, an $\overline{RAS}$-only cycle strobes a row address into the DRAM to select a row of bits to be refreshed. The $\overline{RAS}$ input also causes the selected row to be read out internally and rewritten into the selected bits. This recharges the internal capacitors that store the data. This type of refresh is hidden from the system because it occurs while the microprocessor is reading or writing to other sections of the memory.

The DRAM's internal organization contains a series of rows and columns. A 256K × 1 DRAM has 256 columns, each containing 256 bits, or rows organized into four sections of 64K bits each. Whenever a memory location is addressed, the column address selects a column (or internal memory word) of 1,024 bits (one per section of the DRAM). Refer to Figure 10–37 for the internal structure of a 256K × 1 DRAM. Note that larger memory devices are structured similarly to the 256K × 1 device. The difference usually lies in either the size of each section or the number of sections in parallel.

Figure 10–38 illustrates the timing for an $\overline{RAS}$-only refresh cycle. The main difference between the $\overline{RAS}$ and a read or write is that it applies only a refresh address, which is usually obtained from a 7- or an 8-bit binary counter. The size of the counter is determined by the type of DRAM being refreshed. The refresh counter is incremented at the end of each refresh cycle so all the rows are refreshed in 2 or 4 ms, depending on the type of DRAM.

If there are 256 rows to be refreshed within 4 ms, as in a 256K × 1 DRAM, then the refresh cycle must be activated at least once every 15.6 μs in order to meet the refresh specification. For example, it takes the 80186/80188, running at a 16 MHz clock rate, 250 ns to do a read or a write. Because the DRAM must have a refresh cycle every 15.6 μs, this means that for every 62 or so memory reads or writes, the memory system must run a refresh cycle or memory data will be lost. This represents a loss of 1.6 percent of the computer's time, a small price to pay for the savings represented by using the dynamic RAM.

EDO Memory

A slight modification to the structure of the DRAM changes the device into an EDO (**extended data output**) DRAM device. In the EDO memory, any memory access, including a refresh, stores the 256 bits selected by $\overline{RAS}$ into latches. These latches hold the next 256 bits of information so in most programs, which are sequentially executed, the data is available without any wait states. This slight modification to the internal structure of the DRAM increases system performance by about 25 to 33 percent.

DRAM Controllers

In most systems, a DRAM controller integrated circuit performs the task of address multiplexing and the generation of the DRAM control signals. Some newer embedded microprocessor such as the 80186/80188 and 80386EX include the refresh circuitry as a part of the microprocessor.

Figure 10–39 illustrates the pins on the 80186/80188 that are used with the dynamic RAM and also the register that programs the DRAM refresh function. The 80186/80188 do not provide the address multiplexer for the DRAM or the control signals $\overline{RAS}$ or $\overline{CAS}$. Also note from Figure 10–39 that the locations of the refresh control registers differ for the XL and EA versions when compared to the EB and EC versions. The EB and EC versions contain a refresh address register, while the XL and EA versions do not.

Operation of the Refresh Section of the 80186/80188.   The refresh interval counter counts down the system CLKOUT signal until the counter reaches zero, which triggers the refresh cycle. The refresh interval counter is programmed to cause a refresh as illustrated Example 10–16. Note that the refresh period (R_{PERIOD}) for most DRAM memory is either 2,000 μs or 4,000 μs. The frequency of

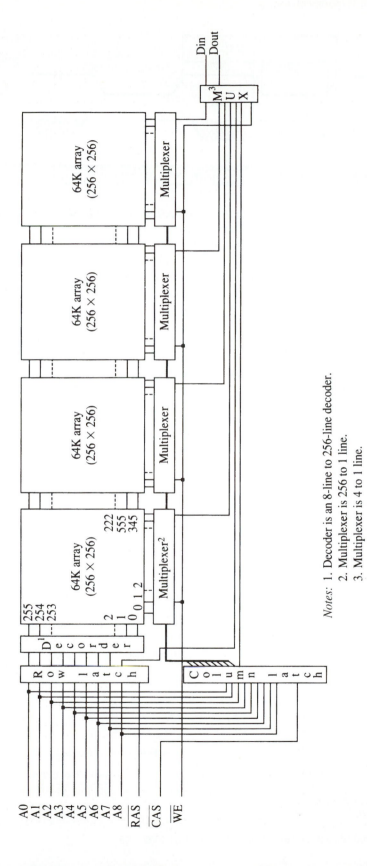

FIGURE 10–37 The internal structure of a 256K × 1 DRAM. Note that each of the internal 256 words are 1,024 bits wide.

Notes: 1. Decoder is an 8-line to 256-line decoder.
2. Multiplexer is 256 to 1 line.
3. Multiplexer is 4 to 1 line.

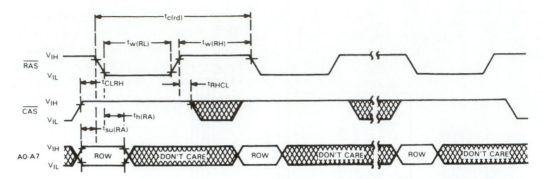

FIGURE 10–38 The timing diagram of the RAS refresh cycle for the TMS4464 DRAM. (Courtesy of Texas Instruments Corporation)

the CLKOUT pin (F_{CPU}) is determined by the crystal or clock frequency. The number of rows is given in the manufacturer's data sheet and is usually 128 or 256. The 5 percent is a scaling factor that accounts for bus holds and so forth that might occur in a system to ensure a refresh within the refresh period.

EXAMPLE 10–16

$$\text{Refresh Interval} = \frac{R_{PERIOD} \times F_{CPU}}{ROWS + (ROWS \times 5\%)}$$

Suppose that a DRAM has 128 rows, and a refresh requirement of 2,000 μs is attached to an 80186/80188 operating at 8 MHz. The refresh interval, programmed into the refresh interval counter, is calculated as 119 in Example 10–17. This means that a refresh cycle will occur every 119 clocks or once per 14.875 μs in this example.

EXAMPLE 10–17

$$\text{Refresh Interval} = \frac{2,000 \times 8}{128 + 6.4} = \frac{16,000}{134.4} \approx 119$$

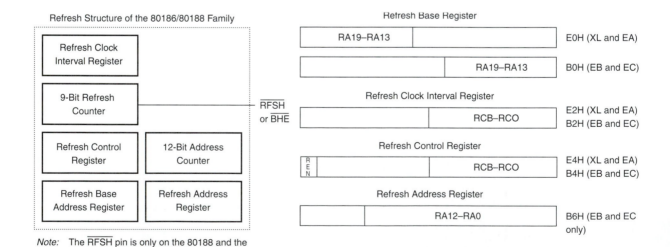

Note: The $\overline{\text{RFSH}}$ pin is only on the 80188 and the $\overline{\text{BHE}}$ pin is only on the 80186.

FIGURE 10–39 The internal refresh control structure of the 80186/80188.

Each refresh cycle causes the microprocessor to execute a dummy read cycle where it sends out the read signal along with the refresh address on the address bus. This causes the memory systems to refresh one row inside the DRAM. The refresh row address is generated by the microprocessor, as dictated by the refresh base address register and in the 80186/80188 EB and EC versions by the refresh address register. The 80186/80188 XL and EA versions specify only the base address (A_{19}–A_{13}) of the DRAM for a refresh operation, while the EB and EC versions also specify the base address (A_{19}–A_{13}) and a refresh address (A_{12}–A_0).

Note that after each refresh, the contents of a 12-bit refresh address register are incremented to the next row address. This causes the next sequential row to be refreshed on the next refresh cycle.

Programming the 80186/80188 Refresh Control Unit. Once the refresh interval is determined, the refresh control unit is ready for programming. Using Example 10–17, the interval of 119 is programmed into the refresh interval counter. To program the DRAM refresh controller:

1. Program the refresh base address register.
2. Program the refresh interval register.
3. Program the refresh control register.

Example 10–18 illustrates a program that assumes the DRAM base address is 00000H, using the refresh interval of 119 for the EB or EC version of the 80186/80188. In this program we exercise the DRAM by performing a write operation before ending the initialization of the internal refresh unit. This set of dummy writes is usually required for a DRAM before it can be used for memory.

EXAMPLE 10–18

```
0017 BA FFB0        MOV   DX,0FFB0H   ;address base register
001A B8 0000        MOV   AX,0        ;base address = 00000H
001D EE             OUT   DX,AL

001E BA FFB2        MOV   DX,0FFB2H   ;address interval counter
0021 B8 0077        MOV   AX,119      ;count = 119
0024 EE             OUT   DX,AL

0025 BA FFB4        MOV   DX,0FFB4H   ;address refresh control
0028 B8 8000        MOV   AX,8000H    ;enable refresh unit
002B EE             OUT   DX,AL

002C B9 0008        MOV   CX,8        ;load loop counter
002F B8 0000        MOV   AX,0
0032 8E D8          MOV   DS,AX       ;segment = 0000H
0034 BF 0000        MOV   DI,0        ;offset = 0000H

0037        REPS:

0037 89 05          MOV   [DI],AX     ;exercise DRAM
0039 E2 FC          LOOP  REPS
```

In the EB and EC versions the refresh address register is also available for programming the starting address of the refresh using bits A_{12}–A_0, which is not normally required unless the DRAM is very small.

Also note that the $\overline{RFSH}$ signal, which appears on the 80188, is missing from the pin-out of the 80186. The 80186 indicates a refresh cycle by placing a logic 1 on both the A_0 and $\overline{BHE}$ pins. In an 80186 system, these two signals are combined to generate a refresh signal for the memory system.

DRAM Interfaced to the 80188

Figure 10–40 illustrates a small system that contains a pair of 256K DRAMs (41256 SIMMs) interfaced at locations 00000H–3FFFFH and 40000H–7FFFFH. It also shows a 27256 EPROM

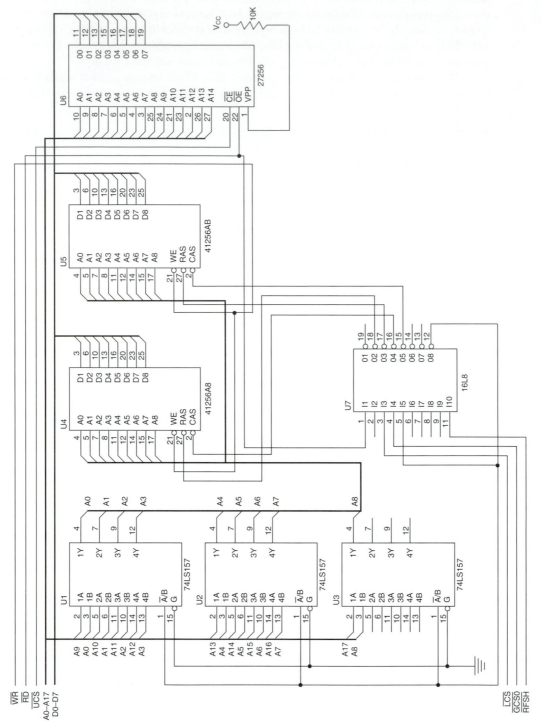

FIGURE 10–40 An 80188EB system that contains 512K of DRAM and 32K of EPROM.

interfaced at locations F8000H–FFFFFH. Note the extra circuitry required to interface the DRAM memory to the 80188EB microprocessor. Extra circuitry is required to multiplex the address for the DRAM and to generate the control signals.

This system uses three 74LS157 quad 2-to-1 line multiplexers to generate the multiplexed address signals for the DRAM. It also uses a single PAL16L8 to generate the control signals for the DRAM. The program for the PAL16L8 is listed in Example 10–19.

EXAMPLE 10–19

```
TITLE       DRAM signal generator U7
PATTERN     Test 4
REVISION    A
AUTHOR      Barry B. Brey
COMPANY     BreyCo
DATA        6/8/97
CHIP        GEN6    PAL16L8

;pins 1   2   3    4     5   6   7   8   9   10
      RD  NC  LCS  GCS0  AB  NC  NC  NC  NC  GND

;pins 11    12  13  14  15    16    17    18    19  20
      RFSH  BA  NC  NC  CAS2  CAS1  RAS2  RAS1  NC  VCC

EQUATIONS

/BA   = /LCS + /GCS0 + /RFSH
/RAS1 = /LCS + /RFSH
/RAS2 = /GCS0 + /FRSH
/CAS1 = /AB * /LCS * RFSH
/CAS2 = /AB * /GCS0 * RFSH
```

Notice from the PAL equations that the $\overline{A}/B$ inputs to the multiplexers are a logic 1 until either $\overline{LCS}$, $\overline{GCS0}$, or $\overline{RFSH}$ becomes a logic 0. This presents the DRAM with row address A_0–A_8, when $\overline{A}/B = 1$, and a column address A_9–A_{17}, when $\overline{A}/B = 0$. Also notice that both DRAM $\overline{RAS}$ inputs are active during a refresh cycle ($\overline{RFSH}$), which refreshes the DRAMs together. A small time delay is obtained for the $\overline{CAS}$ inputs of the DRAMs by using the PAL time delay from the AB input to the BA output. (A DRAM is selected by the activation of its $\overline{CAS}$ input). The $\overline{CAS}$ signal is passed to only one DRAM because of the steering provided by the PAL. (The program required to initialize this system is not shown here.)

Refresh Control for the 80386EX

The refresh control circuitry within the 80386EX is almost identical to that of the 80186/80188. One notable difference is the address of the refresh control registers. The refresh interval register is located at address F4A2H and is programmed in the same manner as that of the 80186/80188. Another difference is in the refresh base address register, which is located at I/O address F4A0H and contains address bits A_{25}–A_{14} for programming the base address. The refresh control register is located at I/O address F44A4H and is enabled by writing an 8000H to the register just as it is in the 80186/80188.

10–6   SUMMARY

1. All memory devices have address inputs; data inputs and outputs, or just outputs; a pin for selection; and one or more pins that control the operation of the memory.

2. Address connections on a memory component are used to select one of the memory locations within the device. Ten address pins have 1,024 combinations and therefore are able to address 1,024 different memory locations.

3. Data connections on a memory are used to enter information to be stored in a memory location and to retrieve information read from a memory location. Manufacturers list their memory as, for example, $4K \times 4$, which means that the device has 4K memory locations (4,096) and 4 bits are stored in each location.

4. Memory selection is accomplished via a chip selection pin ($\overline{CS}$) on many RAMs or a chip enable pin ($\overline{CE}$) on many EPROM or ROM memories.

5. Memory function is selected by an output enable pin ($\overline{OE}$) for reading data, which normally connects to the system read signal ($\overline{RD}$). The write enable pin ($\overline{WE}$) for writing data normally connects to the system write signal ($\overline{WR}$).

6. EPROM memory is programmed by an EPROM programmer and can be erased if exposed to ultraviolet light. Today EPROMs are available in sizes from $1K \times 8$ to $128K \times 8$ and larger.

7. The flash memory (EEPROM) is programmed in the system by using a 12-V programming pulse.

8. Static RAM (SRAM) retains data for as long as the system power supply is attached. These memory types are available in sizes up to $128K \times 8$.

9. Dynamic RAM (DRAM) retains data for only a short period, usually 2–4 ms. This creates problems for the memory system designer because the DRAM must be refreshed periodically. DRAMs also have multiplexed address inputs that require an external multiplexer to provide each half of the address at the appropriate time.

10. Memory address decoders select an EPROM or RAM at a particular area of the memory. Commonly found address decoders include the 74LS138 3-to-8 line decoder, the 74LS139 2-to-4 line decoder, and programmed selection logic in the form of a PROM or PLD.

11. The PROM and PLD address decoders for microprocessors like the 8088 through the Pentium Pro reduce the number of integrated circuits required to complete a functioning memory system.

12. The 8088 minimum mode memory interface contains 20 address lines, 8 data lines, and 3 control lines: $\overline{RD}$, $\overline{WR}$, and $\overline{S2}$. The 80188 memory functions correctly only when all these lines are used for memory interface.

13. The access speed of the EPROM must be compatible with the microprocessor to which it is interfaced. Many EPROMs available today have an access time of 450 ns, which is too slow for most speed versions of the 80186/80188 or the 80386EX. In order to circumvent this problem, a wait state is inserted to increase memory access time to at least 480 ns.

14. The 80186/80188 XL and EA versions contain six memory chip selection signals and can insert from 0 to 3 wait states for each memory device. The 80186/80188 EB and EC versions contain 10 chip selection signals that can be programmed for 0 to 15 wait states.

15. The 80386EX embedded controller has nine chip selection pins that can be programmed to insert from 0 to 31 wait states for each memory component.

16. Parity checkers are becoming commonplace today in many microprocessor-based microcomputer systems. An extra bit is stored with each byte of memory, making the memory 9 bits wide instead of 8.

17. Error correction features are also available for memory systems, but they require the storage of many more bits. If an 8-bit number is stored with an error correction circuit, it actually takes 13 bits of memory: 5 for an error checking code and 8 for the data. Most error correction integrated circuits are able to correct only a single-bit error.

18. The 80186 and 80386EX memory interface has a 16-bit data bus and contains an $\overline{S2}$ control pin, whereas the 80188 has an 8-bit data bus. In addition to these changes, there is an extra control signal called bush high enable ($\overline{BHE}$).

19. The 80186 and 80386EX memory system is organized into two 8-bit banks: high bank and low bank. The high bank memory is enabled by the $\overline{\text{BHE}}$ control signal and the low bank by the $\overline{\text{BLE}}$ signal on the 80386EX or by the A_0 address line on the 80186.
20. Two common schemes for selecting the banks in an 80186 and 80386EX-based system include (1) a separate decoder for each bank, and (2) separate $\overline{\text{WR}}$ control signals for each bank with a common decoder.
21. Dynamic RAM controllers are designed to control DRAM memory components. Many DRAM controllers today contain address multiplexers, refresh counters, and the circuitry required to do a periodic DRAM memory refresh.
22. The refresh control unit within the 80186/80188 is programmed to generate a refresh cycle through its refresh interval register. The base address register controls the starting address of the external DRAM memory system for the refresh operation.

10–7   QUESTIONS AND PROBLEMS

1. Which types of connections are common to all memory devices?
2. List the number of words found in each memory device for the following numbers of address connections:
   (a) 8
   (b) 11
   (c) 12
   (d) 13
3. List the number of data items stored in each of the following memory devices and the number of bits in each datum:
   (a) $2K \times 4$
   (b) $1K \times 1$
   (c) $4K \times 8$
   (d) $16K \times 1$
   (e) $64K \times 4$
4. What is the purpose of the $\overline{\text{CS}}$ or $\overline{\text{CE}}$ pin on a memory component?
5. What is the purpose of the $\overline{\text{OE}}$ pin on a memory device?
6. What is the purpose of the $\overline{\text{WE}}$ pin on a RAM?
7. How many bytes of storage do the following EPROM memory devices contain?
   (a) 2708
   (b) 2716
   (c) 2732
   (d) 2764
   (e) 27128
8. Why won't a 450-ns EPROM work directly with an 8-MHz 80188?
9. What can be stated about the amount of time it takes to erase and write a location in a flash memory device?
10. SRAM is an acronym for which type of device?
11. The 4016 memory has a $\overline{\text{G}}$ pin, an $\overline{\text{S}}$ pin, and a $\overline{\text{W}}$ pin. What are these pins used for in this RAM?
12. How much memory access time is required by the slowest 4016?
13. DRAM is an acronym for which type of device?
14. The TMS4464 has eight address inputs, yet it is a 64K DRAM. Explain how a 16-bit memory address is forced into eight address inputs.

15. What are the purposes of the $\overline{CAS}$ and $\overline{RAS}$ inputs of a DRAM?
16. How much time is required to refresh the typical DRAM?
17. Why are memory address decoders important?
18. Modify the NAND gate decoder of Figure 10–12 to select the memory for address range DF800H–DFFFFH.
19. Modify the NAND gate decoder in Figure 10–12 to select the memory for address range 40000H–407FFH.
20. When the G1 input is high and $\overline{G2A}$ and $\overline{G2B}$ are both low, what happens to the outputs of the 74LS138 3-to-8 line decoder?
21. Modify the circuit of Figure 10–14 to address memory range 70000H–7FFFFH.
22. Modify the circuit of Figure 10–14 to address memory range 40000H–4FFFFH.
23. Describe the 74LS139 decoder.
24. Why is a PROM address decoder often found in a memory system?
25. Reprogram the PROM in Table 10–1 to decode memory address range 80000H–8FFFFH.
26. Reprogram the PROM in Table 10–1 to decode memory address range 30000H–3FFFFH.
27. Modify the circuit of Figure 10–18 by rewriting the PAL program to address memory at locations 40000H–4FFFFH.
28. Modify the circuit of Figure 10–18 by rewriting the PAL program to address memory at locations B0000H–BFFFFH.
29. Develop a small memory interface for the 80C188XL microprocessor that contains a single 27512 EPROM at locations F0000H–FFFFFH and a single 62256 SRAM at locations 00000H–07FFFH. Make sure that you include software required to program the microprocessor and include three wait states for the EPROM and no wait states for the SRAM.
30. Develop a memory system for the 80C188XL microprocessor that contains a 26256 EPROM at locations F8000H–FFFFFH and three 62256 SRAMs at locations 00000H–07FFFH, 80000H–87FFFH, and 88000H–8FFFFH. Make sure that you include the software required to program the microprocessor and include two wait states for the EPROM and one wait state for each of the SRAM devices.
31. Develop a memory system for the 80C188EB that contains two 27256 EPROMs at locations F0000H–F7FFFH and F8000H–FFFFFH. Also include SRAM (62256) at locations 00000H–07FFFH and 08000H–0FFFFH. Make sure that you include the software required to program the microprocessor and include six wait states for the EPROM and two wait states for the SRAM.
32. Develop a memory system for the 80C188EB that uses a pair of 27512 EPROM located at addresses E0000H–FFFFFH and four 62256 SRAMs at locations 00000H–1FFFFH. Make sure that you include the required software for programming the microprocessor and include five wait states for the EPROM and three wait states for the SRAM.
33. Explain how odd parity is stored in a memory system and how it is checked.
34. The 74LS636 error correction and detection circuit stores a check code with each byte of data. How many bits are stored for the check code?
35. What is the purpose of the SEF pin on the 74LS636?
36. The 74LS636 will correct _____ bits that are in error.
37. Explain the major difference between the buses of the 80186 and 80188 microprocessors.
38. What is the purpose of the $\overline{BHE}$ and A_0 pins on the 80186 microprocessor?
39. What is the $\overline{BLE}$ pin and which other pin has it replaced?
40. Which two methods are used to select the memory in the 80186 microprocessor?
41. If $\overline{BHE}$ is a logic 0, then the _____ memory bank is selected.
42. If A_0 is a logic 0, then the _____ memory bank is selected.
43. Why don't separate bank read ($\overline{RD}$) strobes need to be developed when interfacing memory to the 80186?

44. Develop a 16-bit-wide memory system for the 80C186XL microprocessor that contains 64K of EPROM (two waits) and 64K of SRAM (no waits). Include the software required to place the EPROM at locations F0000H–FFFFFH and the SRAM at locations 00000H–0FFFFH.

45. Develop a 16-bit-wide memory system for the 80C186EC that contains 128K bytes of EPROM (three waits) and 64K bytes of SRAM (one wait). Include the software required to place the EPROM at locations E0000H–FFFFFH and the SRAM at locations 00000H–0FFFFH.

46. Develop a 16-bit-wide memory system for the 80386EX that contains 256K bytes of EPROM (nine waits) and 128K bytes of SRAM (six waits). Include the software required to place the EPROM at locations 3C0000H–3FFFFFH and the SRAM at locations 000000H–007FFFH and 080000H–080FFFH.

47. What is an $\overline{RAS}$-only cycle?

48. When DRAM is refreshed, can it be done while other sections of the memory operate?

49. If a 1M × 1 DRAM requires 4 ms for a refresh and has 256 rows to be refreshed, no more than _____ ms must pass before another row is refreshed.

50. What count is programmed into the refresh interval counter of the 80C188EC microprocessor if the clock frequency is 16 MHz, the DRAM refresh period is 2,000 µs, and the DRAM contains 256 rows?

51. Assuming that a DRAM is interfaced to a 12-MHz 80C186EC microprocessor at locations 80000H–BFFFFH and the DRAM has a refresh period of 4,000 µs with 128 rows, develop the program required for the refresh unit.

52. Use the Internet to locate data on the 27256 EPROM and print the data sheet.

53. Use the Internet to locate information on DRAM memory devices and print at least two data sheets of different size devices.

54. Use the Internet to locate information on EDO memory and write a report describing its operation.

CHAPTER 11

Basic Input/Output Interfacing

INTRODUCTION

A microprocessor or embedded controller is great at solving problems, but if it can't communicate with the outside world, it is of little worth. This chapter outlines some of the basic methods of communications, both serial and parallel, between humans or machines and the microprocessor.

First we introduce the basic I/O interface and discuss decoding for I/O devices. Then we provide detail on parallel and serial interfacing, both of which have a wide variety of applications. As applications, we connect analog-to-digital and digital-to-analog converters as well as both DC and stepper motors to the microprocessor.

CHAPTER OBJECTIVES

Upon completion of this chapter, you will be able to:

1. Explain the operation of the basic input and output interfaces.
2. Decode 8-, 16-, and 32-bit I/O devices so they can be used at any I/O port address.
3. Define handshaking and explain how to use it with I/O devices.
4. Interface and program the 82C55 programmable parallel interface.
5. Interface and program the 8279 programmable keyboard/display controller.
6. Interface and program the 16550 serial communications interface adapter.
7. Interface and program the 8254 programmable interval timer.
8. Interface an analog-to-digital converter and a digital-to-analog converter to the microprocessor.
9. Interface both DC and stepper motors to the microprocessor.
10. Interface and use a flash memory device with the microprocessor.

11–1 INTRODUCTION TO INPUT/OUTPUT INTERFACING

In this section we explain the operation of the I/O instructions: IN, INS, OUT, and OUTS. We also explain the concept of isolated (sometimes called direct or I/O mapped I/O) and memory-mapped I/O, the basic input and output interfaces, and handshaking. A working knowledge of these topics will make it easier to understand the connection and operation of the programmable interface components and I/O techniques presented in the remainder of this chapter.

Input/Output Instructions

The instruction set contains one type of instruction that transfers information to an I/O device (OUT) and another to read information from an I/O device (IN). We also look at instructions (INS and OUTS) that transfer character strings of data between the memory and an I/O device. Table 11–1 lists all versions of each instruction found in the microprocessor's instruction set.

Both the IN and OUT instructions transfer data between an I/O device and the microprocessor's accumulator (AL, AX, or EAX). The I/O address is stored in register DX as a 16-bit I/O address or in the byte (p8) immediately following the opcode as an 8-bit I/O address. Intel calls the 8-bit form (p8) a **fixed address** because it is stored with the instruction, usually in a

TABLE 11–1   Input/output instructions

Instruction	Data Width	Function
IN AL,p8	8	A byte is input from port p8 into AL
IN AX,p8	16	A word is input from port p8 into AX
IN EAX, p8	32	A doubleword is input from port p8 into EAX
IN AL,DX	8	A byte is input from the port addressed by DX into AL
IN AX,DX	16	A word is input from the port addressed by DX into AX
IN EAX,DX	32	A word is input from the port addressed by DX into EAX
INSB	8	A byte is input from the port addressed by DX into the extra segment memory location addressed by DI, then DI = DI ± 1
INSW	16	A word is input from the port addressed by DX into the extra segment memory location addressed by DI, then DI = DI ± 2
INSD	32	A doubleword is input from the port addressed by DX into the extra segment memory location addressed by DI, then DI ± 4
OUT p8,AL	8	A byte is output from AL to port p8
OUT p8,AX	16	A word is output from AX to port p8
OUT p8,EAX	32	A doubleword is output from EAX to port p8
OUT DX,AL	8	A byte is output from AL to the port addressed by DX
OUT DX,AX	16	A word is output from AX to the port addressed by DX
OUT DX,EAX	32	A doubleword is output from EAX to the port addressed by DX
OUTSB	8	A byte is output from the data segment memory location addressed by SI to the port addressed by DX, then SI = SI ± 1
OUTSW	16	A word is output from the data segment memory locations addressed by SI to the port addressed by DX, then SI = SI ± 2
OUTSD	32	A doubleword is output from the data segment memory locations addressed by SI to the port addressed by DX, then SI = SI ± 4

ROM. The 16-bit I/O address in DX is called a **variable address** because it is stored in a DX and then increments/decrements SI by 4 registers, which can be varied. Both INS and OUTS use a variable I/O address contained in the DX register.

Whenever data is transferred using the IN or OUT instruction, the I/O address (often called a **port number** or **port address**) appears on the address bus. The external I/O interface decodes it the same way that it decodes a memory address. The 8-bit fixed port number (p8) appears on address bus connections A_7–A_0 with bits A_{15}–A_8 equal to 00000000_2. The address connections above A_{15} are undefined for an I/O instruction. The 16-bit variable port number (DX) appears on address connections A_{15}–A_0. This means that the first 256 I/O port addresses (00H–FFH) are accessed by both the fixed and variable I/O instructions, but any I/O address from 0100H–FFFFH is only accessed by the variable I/O address. In many dedicated task systems, such as those based upon the embedded controller or microprocessor, only the rightmost 8 bits of the address are decoded, thus reducing the amount of circuitry required for decoding. In a PC computer, all 16 address bus bits are decoded with locations 00XXH–03XXH, which are the I/O addresses used for I/O inside the PC.

The INS and OUTS instructions address the I/O device using the DX register, but do not transfer data between the accumulator and the I/O device as IN and OUT. Instead, these instructions transfer data between memory and the I/O device. The memory address is located by ES:DI for the INS instruction and DS:SI for the OUTS instruction. As with other string instructions, the contents of the pointers are incremented or decremented as dictated by the state of the direction flag (DF). Both INS and OUTS can also be prefixed with the REP prefix allowing more than one byte, word, or doubleword to be transferred between I/O and memory.

Isolated and Memory-Mapped I/O

There are two completely different methods of interfacing I/O to the microprocessor: **isolated I/O** and **memory-mapped I/O.** In the isolated I/O method, the IN, INS, OUT, and OUTS instructions transfer data between the microprocessor accumulator or memory and the I/O device. In the memory-mapped I/O method, any instruction that references memory can accomplish the transfer. Both isolated and memory-mapped I/O are in use, so both are discussed here.

Isolated I/O.   The most common I/O transfer technique used in the Intel microprocessor-based system is isolated I/O. The term *isolated* describes how the I/O locations are isolated from the memory system in a separate I/O address space. (Figure 11–1 illustrates both the isolated and memory-mapped address spaces for any Intel 80X86, Pentium, or Pentium Pro microprocessor.) The addresses for isolated I/O devices, called *ports,* are separate from the memory. As a result, you can expand the memory to its full size without using any of this space for I/O devices. A disadvantage of isolated I/O is that the data transferred between I/O and the microprocessor must be accessed by the IN, INS, OUT, and OUTS instructions. Separate control signals for the I/O space are developed (using $\overline{S2}$, $\overline{RD}$, and $\overline{WR}$) that indicate an I/O read (sometimes using the $\overline{IORC}$ control signal) or an I/O write (sometimes using the $\overline{IOWC}$ signal) operation. These signals indicate that an I/O port address appears on the address bus that is used to select the I/O device. In the personal computer, isolated I/O ports are used for controlling peripheral devices. As a rule, an 8-bit port address is used to access devices located on the system board, such as the timer and keyboard interface, and a 16-bit port is used to access serial and parallel ports as well as video and disk drive systems.

Memory-Mapped I/O.   Unlike isolated I/O, memory-mapped I/O does not use IN, INS, OUT, and OUTS instructions. Instead, it uses any instruction that transfers data between the microprocessor and memory. A memory-mapped I/O device is treated as a memory location in the memory map. The main advantage of memory-mapped I/O is that any memory transfer instruction can be used to access the I/O device. The main disadvantage is that a portion of the memory

FIGURE 11–1 The memory and I/O maps for the 8086/8088 microprocessors. (a) Isolated I/O. (b) Memory-mapped I/O.

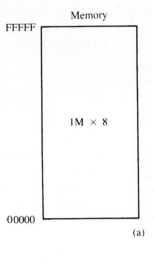

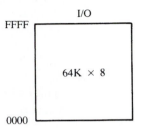

(a)

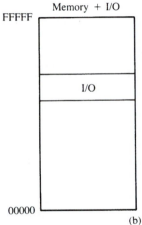

(b)

system is used as the I/O map. This reduces the amount of memory available to applications. Another advantage is that the $\overline{\text{IORC}}$ and $\overline{\text{IOWC}}$ signals have no function in a memory-mapped I/O system, which may often result in a reduction of circuitry required for decoding.

Personal Computer I/O Map

The personal computer uses part of the I/O map for dedicated functions. Figure 11–2 shows the I/O map for the PC. Because I/O space between ports 0000H and 03FFH are normally reserved for the computer system. The I/O ports located at 0400H–FFFFH are generally available for user applications. Note that the 80287 arithmetic coprocessor uses I/O address 00F8H–00FFH for communications, Intel reserves I/O ports 00F0H–00FFH. The 80386–Pentium Pro use I/O ports 800000F8–800000FFH for their arithmetic coprocessors. The I/O ports located between 0000H and 00FFH are accessed via the fixed port I/O instructions and the ports located above 00FFH are accessed via the variable I/O port instructions.

Basic Input and Output Interfaces

The basic input device is a set of three-state buffers. The basic output device is a set of data latches. The term IN refers to moving data from the I/O device into the microprocessor and the term OUT refers to moving data out of the microprocessor to the I/O device.

FIGURE 11–2 The I/O map of a personal computer illustrating many of the fixed I/O areas.

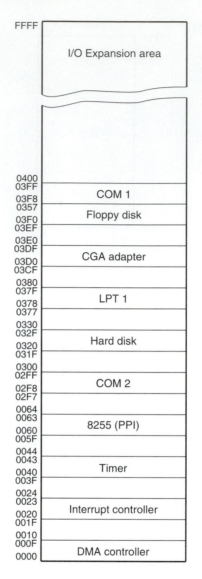

FFFF	I/O Expansion area
0400 / 03FF – 03F8	COM 1
0357 – 03F0	Floppy disk
03EF – 03E0	
03DF – 03D0	CGA adapter
03CF – 0380	
037F – 0378	LPT 1
0377 – 0330	
032F – 0320	Hard disk
031F – 0300	
02FF – 02F8	COM 2
02F7 – 0064	
0063 – 0060	8255 (PPI)
005F – 0044	
0043 – 0040	Timer
003F – 0024	
0023 – 0020	Interrupt controller
001F – 0010	
000F – 0000	DMA controller

The Basic Input Interface.   Three-state buffers are used to construct the 8-bit input port depicted in Figure 11–3. Notice that the external TTL data (simple toggle switches in this example) are connected to the inputs of the buffers. The outputs of the buffers connect to the data bus. The exact data bus connections depend on the version of the microprocessor. For example, the 80188 has data bus connections D_7–D_0, while the 80186 and 80386EX have D_{15}–D_0. The circuit of Figure 11–3 allows the microprocessor to read the contents of the eight switches that connect to any 8-bit section of the data bus whenever the select signal's $\overline{SEL}$ becomes a logic 0.

When the microprocessor executes an IN instruction, the I/O port address is decoded to generate the logic 0 on $\overline{SEL}$. A zero placed on the output control inputs ($\overline{1G}$ and $\overline{2G}$) of the 74ALS244 buffer causes the data input connections (A) to be connected to the data output (Y) connections. Notice that the Y output connects to the system data bus. The IN instruction fetches the contents of the data bus into the microprocessor's accumulator register, in this case from the switches. If a logic 1 is placed on the output control inputs of the 74ALS244 buffer, the device enters the three-state high-impedance mode that effectively disconnects the switches from the data bus.

FIGURE 11–3 The basic input interface illustrating the connection of eight switches. Note that the 74ALS244 is a three-state that controls the application of the switch data to the data bus.

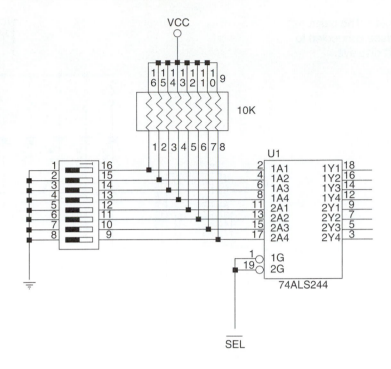

This basic input circuit is not optional and must appear any time that input data is interfaced to the microprocessor. Sometimes it appears as a discrete part of the circuit as in Figure 11–3 and sometimes it is built into a programmable I/O device.

Sixteen-bit data can also be interfaced to various versions of the 80186 and 80386EX, but not nearly as commonly as 8-bit data. To interface 16-bit data, the circuit in Figure 11–3 is modified to include two 74ALS244 buffers that connect 16 bits of input data to the 16-bit data bus.

The Basic Output Interface. The basic output interface receives data from the microprocessor and must usually hold it for some external device. Its latches, like the buffers found in the input device, are often built into the I/O device.

Figure 11–4 shows how eight simple light-emitting diodes (LEDs) connect to the microprocessor through a set of eight data latches. The latch stores the number output by the microprocessor from the data bus so the LEDs can be lit with any 8-bit binary number. Latches are needed to hold the data because when the microprocessor executes an OUT instruction, the data is present on the data bus for less than 1.0 μs. Without a latch, the viewer would never see the LEDs illuminate.

When the OUT instruction executes, the data from AL, AX, and EAX are transferred to the latch via the data bus. Here the D inputs of a 74ALS374 octal latch are connected to the data bus to capture the output data, and the Q outputs of the latch are attached to the LEDs. When a Q output becomes a logic 0, the LED lights. Each time that the OUT instruction executes, the $\overline{SEL}$ signal to the latch activates, capturing the data output to the latch from any 8-bit section of the data bus. The data is held until the next OUT instruction executes.

Handshaking

Many I/O devices accept or release information at a much slower rate than the microprocessor. Another method of I/O control, called **handshaking** or **polling,** synchronizes the I/O device with the microprocessor. An example of a device that requires handshaking is a parallel printer that prints 100 characters per second (CPS). It is obvious that the microprocessor can definitely send

FIGURE 11–4 The basic output interface connected to a set of LED displays.

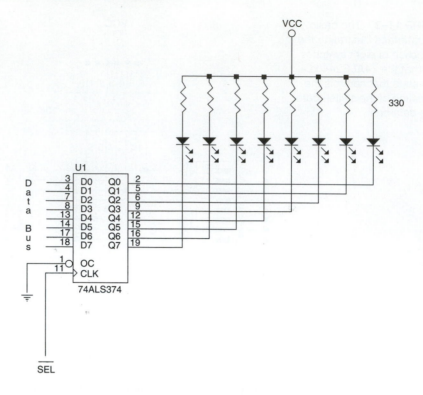

more than 100 CPS to the printer, so we need a way to slow the down microprocessor to match speeds with the printer.

Figure 11–5 illustrates the typical input and output connections found on a printer. Here data is transferred through a series of data connections (D_7–D_0); BUSY indicates that the printer is busy; STB is a clock pulse used to send data into the printer for printing.

The ASCII data to be printed by the printer is placed on D_7–D_0 and a pulse is applied to the $\overline{STB}$ connection. The strobe signal sends the data into the printer so it can be printed. As soon as the printer receives the data, it places a logic 1 on the BUSY pin, indicating that it is busy printing data. The microprocessor polls or tests the BUSY pin to decide if the printer is busy. If the printer is busy, the microprocessor waits; if it is not busy, the microprocessor sends another ASCII character to the printer. The process of interrogating the printer is called **handshaking** or **polling.** Example 11–1 illustrates a simple procedure that tests the printer BUSY flag and sends data to the printer if it is not busy. The PRINT procedure prints the ASCII-coded contents of BL only if the BUSY flag is a logic 0, indicating the printer is not busy.

EXAMPLE 11–1

```
                           ;procedure that prints the ASCII contents of BL
                           ;
0000                       PRINT   PROC    NEAR

0000 E4 4B                         IN      AL,BUSY         ;get BUSY flag
0002 A8 04                         TEST    AL,BUSY_BIT     ;test BUSY bit
0004 75 FA                         JNE     PRINT           ;if printer busy
0006 8A C3                         MOV     AL,BL           ;get data from BL
0008 E6 4A                         OUT     PRINTER,AL      ;send data to
printer
000A CB                            RET                     ;return from
                                                            procedure

000B                       PRINT   ENDP
```

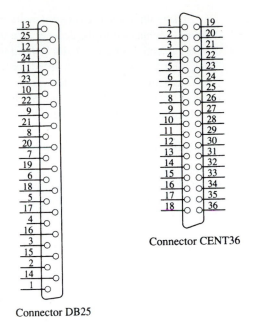

Connector CENT36

Connector DB25

DB25 Pin number	CENT36 Pin number	Function	DB25 Pin number	CENT36 Pin number	Function
1	1	$\overline{\text{Data Strobe}}$	12	12	Paper empty
2	2	Data 0 (D0)	13	13	Select
3	3	Data 1 (D1)	14	14	Afd
4	4	Data 2 (D2)	15	32	$\overline{\text{Error}}$
5	5	Data 3 (D3)	16	—	$\overline{\text{RESET}}$
6	6	Data 4 (D4)	17	31	Select in
7	7	Data 5 (D5)	18–25	19–30	Ground
8	8	Data 6 (D6)	—	17	Frame ground
9	9	Data 7 (D7)	—	16	Ground
10	10	$\overline{\text{Ack}}$	—	33	Ground
11	11	Busy			

FIGURE 11–5 The DB25 connector found on computers and the Centronics 36-pin connector found on printers for the Centronics parallel printer interface.

11–2 INPUT/OUTPUT PORT ADDRESS DECODING

I/O port address decoding is very similar to memory address decoding, especially for memory-mapped I/O devices. In fact, we do not discuss memory-mapped I/O decoding because it is treated exactly the same as memory, except that the $\overline{\text{IORC}}$ and $\overline{\text{IOWC}}$ are not used, since there is no IN or OUT instruction. The decision to use memory-mapped I/O is often determined by the size of the memory system and the placement of the I/O devices in the system. Decoding is illustrated for both the internal chip selection units of the 80186/80188 and 80386EX (as introduced in Chapter 10) with memory interface and also using traditional decoding techniques.

The main difference between memory decoding and isolated I/O decoding is the number of address pins connected to the decoder. We decode A_{25}–A_0 (80396EX) or A_{19}–A_0 (80186/80188) for memory selection and A_{15}–A_0 for isolated I/O device selection. Sometimes, if the I/O devices use only fixed, 8-bit I/O addressing, we decode only A_7–A_0. Another difference is that we use the $\overline{IORC}$ and $\overline{IOWC}$ to activate I/O devices for a read or write operation. On earlier versions of the microprocessor $IO/\overline{M} = 1$ and $\overline{RD}$ or $\overline{WR}$ are used to activate I/O devices. On some versions these signals are developed by combining the $\overline{S2}$ signal with the $\overline{RD}$ and $\overline{WR}$ signals whenever external decoding is in use.

Decoding 8-Bit I/O Addresses

As mentioned, the fixed I/O instruction uses an 8-bit I/O port address that appears on A_{15}–A_0 as 0000H–00FFH. If a system never contains more than 256 I/O devices, we often decode only address connections A_7–A_0 for an 8-bit I/O port address. Thus we ignore address connection A_{15}–A_8. (You *cannot* ignore these address connections in a personal computer—all 16 bits must be used.) Please note that the DX register can also address I/O ports 00H–FFH. Also note that if the address is decoded as an 8-bit address, then we can never include I/O devices that use a 16-bit I/O address.

Figure 11–6 illustrates a 74ALS138 decoder that decodes 8-bit I/O ports F0H through F7H. (We assume that the system will use only I/O ports 00H–FFH for this decoder.) This decoder is identical to a memory address decoder except we only connect address bits A_7–A_0 to the inputs of the decoder. Figure 11–7 shows the PAL version of this decoder. Notice that the PAL is a better decoder circuit because the number of integrated circuits has been reduced to one device. The program for the PAL appears in Example 11–2.

EXAMPLE 11–2

```
AUTHOR      Barry B. Brey
COMPANY     BreyCo
DATE        7/1/97
CHIP        DECODER8 PAL16L8

;pins  1  2  3  4  5  6  7  8  9   10
       A0 A1 A2 A3 A4 A5 A6 A7 NC  GND

;pins  11 12 13 14 15 16 17 18 19  20
       NC F7 F6 F5 F4 F3 F2 F1 F0  VCC

EQUATIONS

/F0 = A7 * A6 * A5 * A4 * A3 * /A2 * /A1 * /A0
/F1 = A7 * A6 * A5 * A4 * A3 * /A2 * /A1 * A0
/F2 = A7 * A6 * A5 * A4 * A3 * /A2 * A1  * /A0
/F3 = A7 * A6 * A5 * A4 * A3 * /A2 * A1  * A0
/F4 = A7 * A6 * A5 * A4 * A3 * A2  * /A1 * /A0
/F5 = A7 * A6 * A5 * A4 * A3 * A2  * /A1 * A0
/F6 = A7 * A6 * A5 * A4 * A3 * A2  * A1  * /A0
/F7 = A7 * A6 * A5 * A4 * A3 * A2  * A1  * A0
```

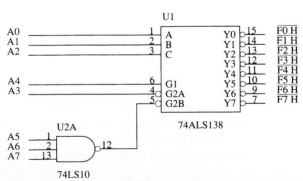

FIGURE 11–6 A port decoder that decodes 8-bit I/O ports. This decoder generates active low outputs for ports F0H–F7H.

FIGURE 11–7 A PAL16L8 decoder that generates I/O port signals for port F0H–F7H.

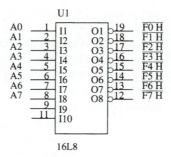

Decoding 16-Bit I/O Addresses

We also decode 16-bit I/O addresses, especially in a personal computer system. The main difference between decoding an 8-bit I/O address and a 16-bit I/O address is that eight additional address lines (A_{15}–A_8) must be decoded. Figure 11–8 illustrates a circuit that contains a PAL16L8 and an 8-input NAND gate used to decode I/O ports EFF8H–EFFFH. These are common I/O port assignments in a PC used for the serial communications port.

The NAND gate decodes the first 8 bits of the I/O port address (A_{15}–A_8) so it generates a signal to enable the PAL16L8 for any I/O address between EF00H and $\overline{\text{EFFFH}}$. The PAL16L8 further decodes the I/O address to produce eight active low-output strobes $\overline{\text{EFF8H}}$–$\overline{\text{EFFFH}}$. The program for the PAL16L8 appears in Example 11–3.

EXAMPLE 11–3

```
AUTHOR     Barry B. Brey
COMPANY    BreyCo
DATE       7/2/97
CHIP       DECODER9 PAL16L8

;pins  1  2  3  4  5  6  7  8  9    10
       A0 A1 A2 A3 A4 A5 A6 A7 NAND GND

;pins 11    12     13     14     15     16     17     18     19   20
      NC  EFFFH  EFFEH  EFFDH  EFFCH  EFFBH  EFFAH  EFF9H  EFF8H VCC

EQUATIONS

/EFF8H = A7 * A6 * A5 * A4 * A3 * /A2 * /A1 * /A0 * /NAND
/EFF9H = A7 * A6 * A5 * A4 * A3 * /A2 * /A1 *  A0 * /NAND
/EFFAH = A7 * A6 * A5 * A4 * A3 * /A2 *  A1 * /A0 * /NAND
/EFFBH = A7 * A6 * A5 * A4 * A3 * /A2 *  A1 *  A0 * /NAND
/EFFCH = A7 * A6 * A5 * A4 * A3 *  A2 * /A1 * /A0 * /NAND
/EFFDH = A7 * A6 * A5 * A4 * A3 *  A2 * /A1 *  A0 * /NAND
/EFFEH = A7 * A6 * A5 * A4 * A3 *  A2 *  A1 * /A0 * /NAND
/EFFF7 = A7 * A6 * A5 * A4 * A3 *  A2 *  A1 *  A0 * /NAND
```

FIGURE 11–8 A PAL16L8 decoder that decodes 16-bit address EFF8H–EFFFH.

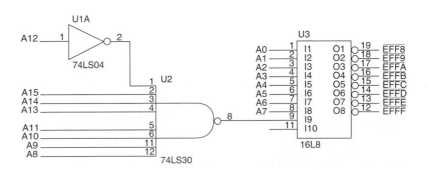

8- and 16-Bit I/O Ports

Now that we understand that decoding the I/O port address is probably simpler than decoding a memory address (because of the number of bits), we explain how data is transferred between the microprocessor and 8- or 16-bit I/O devices. Data transferred to an 8-bit I/O device exists in one of the I/O banks in a 16-bit microprocessor such as the 80186 or 80386EX. The I/O system contains two 8-bit memory banks just as memory does. This is illustrated in Figure 11–9, which shows the separate I/O banks for a 16-bit system such as the 80386SX.

Because two I/O banks exist, any 8-bit I/O write requires a separate write strobe to function correctly. I/O reads do not require separate read strobes because, as with memory, the microprocessor reads only the byte it expects and ignores the other byte. The only time that a read can cause problems is when the I/O device responds incorrectly to a read operation. In the case of an I/O device that responds to a read from the wrong bank, we may need to include separate read signals. This is discussed later in this chapter.

Figure 11–10 illustrates a system that contains two different 8-bit output devices located at 8-bit I/O addresses 40H and 41H. Because these are 8-bit devices and because they appear in different I/O banks, we generate separate I/O write signals. The program for the PAL16L8 decoder used in Figure 11–10 is illustrated in Example 11–4.

EXAMPLE 11–4

```
AUTHOR     Barry B. Brey
COMPANY    BreyCo
DATE       7/3/97
CHIP       DECODERA PAL16L8

;pins  1     2   3    4  5  6  7  8  9  10
       BHE IOWC BLE A1 A2 A3 A4 A5 A6 GND

;pins 11 12 13 14 15 16 17 18 19  20
      A7 NC NC NC NC NC NC 40 41 VCC

EQUATIONS

/40 = /BLE * /IOWC * /A7 * A6 * /A5 * /A4 * /A3 * /A2 * /A1
/41 = /BHE * /IOWC * /A7 * A6 * /A5 * /A4 * /A3 * /A2 * /A1
```

When selecting 16-bit wide I/O devices, the $\overline{BLE}$ (A_0) and $\overline{BHE}$ pins have no function because both I/O banks are selected together. Although 16-bit I/O devices are relatively rare, a few

FIGURE 11–9   The I/O banks found in the 80186 and 80386EX microprocessor-based systems.

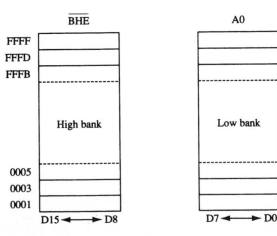

FIGURE 11–10 An I/O port decoder that selects ports 40H and 41H for output data.

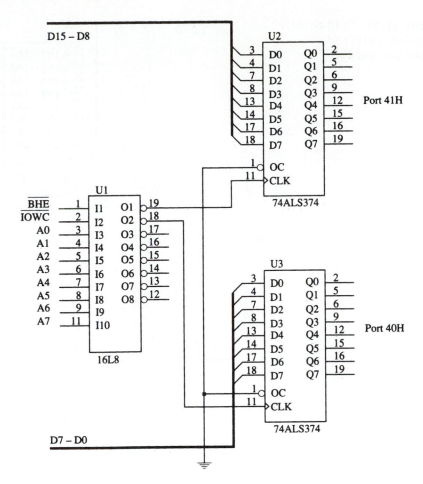

do exist for analog-to-digital and digital-to-analog converters as well as for some video and disk memory interfaces.

Figure 11–11 illustrates a 16-bit input device connected to function at 8-bit I/O addresses 64H and 65H. Notice that the PAL16L8 decoder does not have a connection for address bit $\overline{BLE}$ (A$_0$) and $\overline{BHE}$ because these signals do not apply to 16-bit-wide I/O devices. The program for the PAL16L8 is illustrated in Example 11–5 to show how the enable signals are generated for the three-state buffers (74ALS244) used as input devices.

EXAMPLE 11–5

```
AUTHOR      Barry B. Brey
COMPANY     BreyCo
DATE        7/5/97
CHIP        DECODERB PAL16L8

;pins   1    2  3  4  5  6  7  8  9  10
        IORC A1 A2 A3 A4 A5 A6 A7 NC GND

;pins  11 12 13 14 15 16 17 18 19  20
       NC NC NC NC NC NC NC NC O6X VCC

EQUATIONS

/O6X = /IORC * /A7 * A6 * A5 * /A4 * /A3 * /A2 * /A1
```

FIGURE 11–11   A 16-bit I/O port decoded at I/O addresses 64H and 65H.

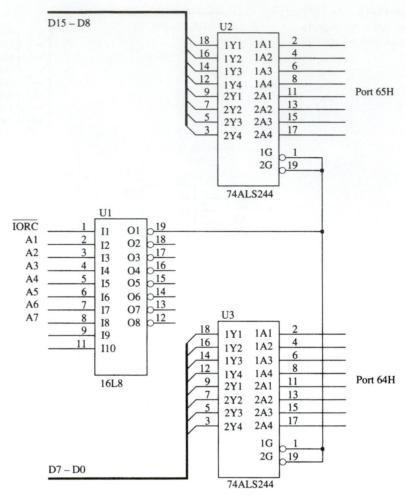

I/O Ports and the 80186/80188

As with the memory chip selection, I/O chip selection is most often accomplished by using the internal chip selection unit of the 80186 or 80188. There are two basic versions of the chip selection interface available to the 80186/80188. One version, the XL and EA, uses a series of seven peripheral chip selection pins to select I/O devices, and the other version, EB and EC, uses the general chip selection pins to select either a memory device or an I/O device.

Programming the XL and EA Version I/O Ports.   The XL and EA versions of the 80186 and 80188 contain seven pins that are dedicated to selecting peripheral devices ($\overline{PCS0}$–$\overline{PCS6}$). The pin-out and control register structure for the XL and EA versions are shown in Figure 11–12.

The peripheral control pins are programmed in a block using the base address (U19–U13 bits in register A4H) to program the starting address of the I/O block. Note that the number of wait states programmed into register A4H sets the number of wait states for all peripheral selection pins. Table 11–2 illustrates the location of the I/O ports in relation to the base address. Note that the ports are spaced by 80H bytes. The function of the peripheral bits is programmed in register A8H, which also programs the block size of the middle memory selection pins defined in Chapter 10.

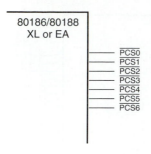

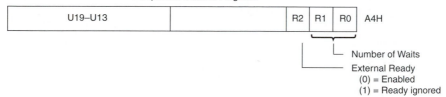

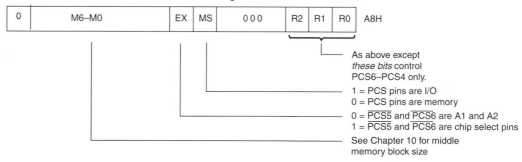

FIGURE 11–12 The peripheral chip select pins and registers for the 80186/80188 XL and EA versions.

TABLE 11–2 The address ranges of the $\overline{PCS}$ pins

Pin	Starting Address	Ending Address
$\overline{PCS}$	Base	Base + 7FH
$\overline{PC1}$	Base + 80H	Base + FFH
$\overline{PC2}$	Base + 100H	Base + 17FH
$\overline{PC3}$	Base + 180H	Base + 1FFH
$\overline{PC4}$	Base + 200H	Base + 27FH
$\overline{PC5}$	Base + 280H	Base + 2FFH
$\overline{PC6}$	Base + 300H	Base + 37FH

Figure 11–13 illustrates a pair of 74LS374 latches interfaced to the 80C188XL microprocessor using the $\overline{PCS0}$ and $\overline{PCS1}$ outputs to provide a clock pulse to capture data from the microprocessor. The clock pulse is generated by combining the peripheral signal with the $\overline{WR}$ signal from the microprocessor. Notice that one latch is labeled port 8000H and the other 8080H. The program to enable and set these port addresses appears in Example 11–6.

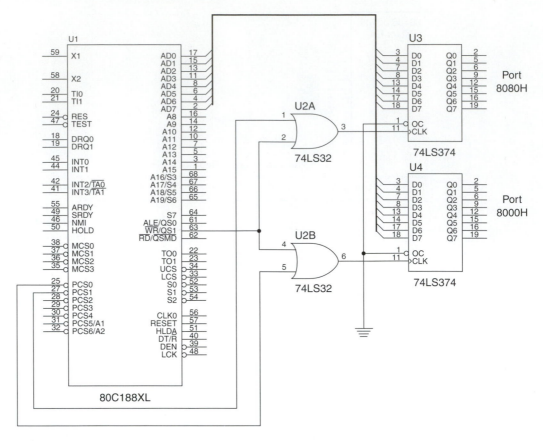

FIGURE 11–13 A pair of 74LS374 latches interfaced to the 80C188XL microprocessor.

EXAMPLE 11–6

```
0000 BA FFA4              MOV     DX,0FFA4H       ;address peripheral control
0003 B8 8004              MOV     AX,8004H        ;base = 8000H, no waits
0006 EE                   OUT     DX,AL

0007 BA FFA8              MOV     DX,0FFA8H       ;address alternate control
000A B8 0040              MOV     AX,0040H        ;PCS pins = I/O
000D EE                   OUT     DX,AL
```

Programming the EB and EC Version I/O Ports.   The EB and EC versions of the 80186/80188 I/O selection are programmed in much the same manner as memory selection (see Chapter 10). The main difference is that one bit is changed in the stopping address register to select the I/O function for the general chip selection pins. An example interface to the 80C188EB appears in Figure 11–14. This example circuit interfaces a latch at I/O port 1000H for output data and a buffer at port 2000H for input data. As with memory, each general chip select output is individually programmed so the addresses do not need to be placed in a block as with the XL and EA versions.

The software required to program the $\overline{\text{GCS1}}$ and $\overline{\text{GCS2}}$ pins to select the latches appears in Example 11–7. Notice that the starting address for $\overline{\text{GCS1}}$ is programmed at 1000H and the ending address is set for 103FH. This is the smallest block of I/O port addresses that can be assigned to an I/O device.

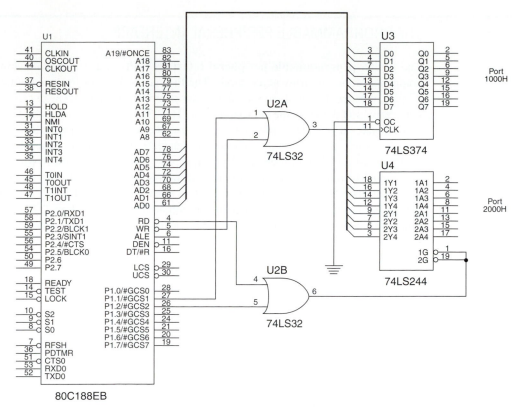

FIGURE 11–14 An input port interfaced at location 2000H and an output port interfaced at location 1000H.

EXAMPLE 11–7

```
0000  BA FF84              MOV    DX,0FF84H      ;address GCS1 start
0003  B8 1000              MOV    AX,1000H       ;start = 1000H, no waits
0006  EE                   OUT    DX,AL

0007  BA FF86              MOV    DX,0FF86H      ;address GCS1 stop
000A  B8 1048              MOV    AX,1048H       ;stop = 103FH, I/O
000D  EE                   OUT    DX,AL

000E  BA FF88              MOV    DX,0FF88H      ;address GCS2 start
0011  B8 2000              MOV    AX,2000H       ;start = 2000H, no waits
0014  EE                   OUT    DX,AL

0015  BA FF8A              MOV    DX,0FF8AH      ;address GCS2 stop
0018  B8 2048              MOV    AX,2048H       ;stop = 203FH, I/O
001B  EE                   OUT    DX,AL
```

I/O Ports and the 80386EX

As with the 80186/80188 EB and EC versions, the 80386EX contains a series of general chip selection pins that are programmed for either a memory device or an I/O device. The main difference is that the 80386EX contains only seven general chip selection pins instead of eight. Another difference is the way the chip selection pins are programmed. Refer to Chapter 10 for details on programming the chip selection pins of the 80386EX.

11–3   THE PROGRAMMABLE PERIPHERAL INTERFACE

The 82C55 **programmable peripheral interface** (PPI) is a very popular low-cost interfacing component found in many applications. The PPI has 24 pins for I/O, programmable in groups of 12 pins, that are used in three separate modes of operation. The 82CC55 can interface any TTL-compatible I/O device to the microprocessor. The 82C55A (CMOS version) requires the insertion of wait states if operated with a microprocessor that reads data in less than 120 ns after the $\overline{RD}$ signal becomes a logic 0. As we learned in Chapter 9, the amount of time allowed for read access is 205 ns for an 8 MHz processor. This requires no wait states. If the 16 MHz processor is used, the amount of time allowed for the read operation is about 85 ns. To ensure that the 82C55 functions with the 16 MHz version, one wait state must be inserted. The 82C55 provides at least 2.5 mA of sink (logic 0) current at each output with a maximum of 4.0 mA. Because I/O devices are inherently slow, wait states used during I/O transfers do not impact significantly upon the speed of the system. The 82C55 still finds application (compatible for programming, although it may not appear in the system as a discrete 82C55) even in the latest 80486- or Pentium Pro–based computer system. The 82C55 is used for interface to the keyboard and the parallel printer port in many of these personal computers. It also controls the timer and reads data from the keyboard interface.

Basic Description of the 8255

Figure 11–15 illustrates the pin-out diagram of the 82C55. Its three I/O ports (labeled A, B, and C) are programmed in groups of 12 pins. Group A connections consist of port A (PA_7–PA_0) and the upper half of port C (PC_7–PC_4) and group B consists of port B (PB_7–PB_0) and the lower half of port C (PC_3–PC_0). The 82C55 is selected by its $\overline{CS}$ pin for programming and for reading or writing to a port. Register selection is accomplished through the A_1 and A_0 input pins that select an internal register for programming or operation. Table 11–3 lists the I/O port assignments used for programming and access to the I/O ports. In the personal computer, an 82C55 or its equivalent is decoded at I/O ports 60H–63H.

The 82C55 is a fairly simple device to program and interface to the microprocessor. For the 82C55 to be read or written, the $\overline{CS}$ input must be a logic 0 and the correct I/O address must

FIGURE 11–15 The pin-out of the 82C55 peripheral interface adapter (PPI).

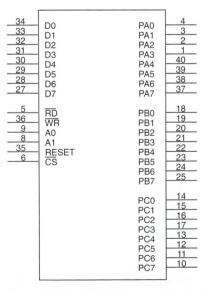

82C55

TABLE 11–3 I/O port assignments for the 82C55

A_1	A_0	Function
0	0	Port A
0	1	Port B
1	0	Port C
1	1	Command Register

be applied to the A_1 and A_0 pins. The remaining port address pins are don't cares as far as the 82C55 is concerned and are externally decoded to select the 82C55.

Figure 11–16 shows an 82C55 connected to the 80C188EB so it functions at I/O port addresses 00C0H (port A), 00C1H (port B), 00C2H (port C), and 00C3H (command register). Notice from this interface that all the 82C55 pins are direct connections to the 80C188EB except for the $\overline{CS}$ pin. The $\overline{CS}$ pin is decoded and selected by the $\overline{GCS2}$ output pin from the microprocessor.

The RESET input to the 82C55 initializes the device whenever the microprocessor is reset. A RESET input to the 82C55 causes all ports to be set up as simple input ports using mode 0 operation. Because the port pins are internally programmed as input pins on a reset, this prevents damage when the power is first applied to the system. After a RESET no other commands are needed to program the 82C55 as long as it used as an input device at all three ports. An 82C55 is interfaced to the personal computer at port addresses 60H–63H for keyboard control and also for

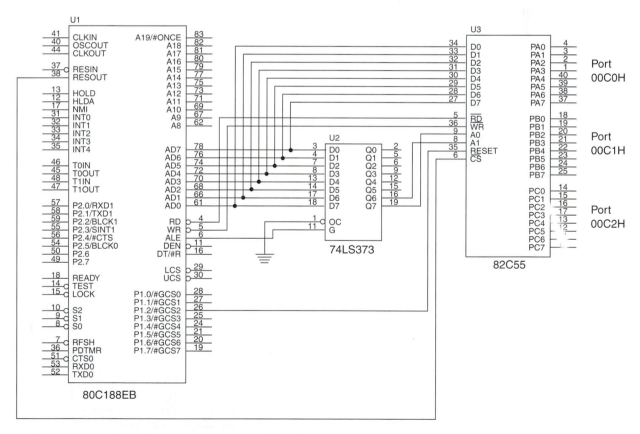

FIGURE 11–16 The 82C55 interfaced to the 80C188EB microprocessor.

controlling the speaker, timer, and other internal devices such as memory expansion. This is true for any AT or earlier style personal computer system.

Programming the 82C55

The 82C55 is easy to program because it contains only two internal command registers as illustrated in Figure 11–17. Notice that bit position 7 selects either command byte A or command byte B. Command byte A programs the function of groups A and B, while command byte B sets (1) or resets (0) bits of port C only if the 82C55 is programmed in mode 1 or 2.

FIGURE 11–17 The command byte of the command register in the 82C55. (a) Programs ports A, B, and C. (b) Sets or resets the bit indicated in the select a bit field.

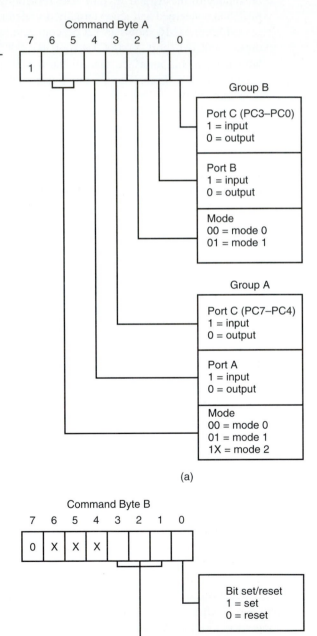

Group B pins (port B and the lower part of port C) are programmed as either input or output pins. Group B can operate in either mode 0 or mode 1. Mode 0 is the basic input/output mode that allows the pins of group B to be programmed as simple input and latched output connections. Mode 1 operation is the strobed operation for group B connections where data is transferred through port B and port C provides handshaking signals.

Group A pins (port A and the upper part of port C) are also programmed as either input or output pins. The difference is that group A can operate in modes 0, 1, and 2. Mode 2 operation is a bi-directional mode of operation for port A.

If a 0 is placed in bit position 7 of the command byte, command byte B is selected. This command allows any bit of port C to be set (1) or reset (0) if the 82C55 is operated in either mode 1 or 2. Otherwise this command byte is not used for programming. We often use the bit set/reset function in control systems to set or clear a control bit at port C.

Mode 0 Operation

Mode 0 operation causes the 82C55 to function as either a buffered input device or as a latched output device. These are the same as the basic input and output circuits discussed in the first section of this chapter.

Figure 11–18 shows the 82C55 connected to a set of 8 seven-segment LED displays. In this circuit, both ports A and B are programmed as (mode 0) simple latched output ports. Port A provides the segment data inputs to the display and port B provides a means of selecting a display position at a time for multiplexing the displays. The 82C55 is interfaced to an 80C188EB microprocessor through a PAL16L8 so it functions at I/O port numbers 0700H–0703H. The program for the PAL16L8 is listed in Example 11–8. The PAL decodes the I/O address and also develops the lower write strobe for the $\overline{\text{WR}}$ pin of the 82C55.

EXAMPLE 11–8

```
AUTHOR      Barry B. Brey
COMPANY     BreyCo
DATE        7/6/96
CHIP        DECODERD PAL16L8

;pins   1   2   3   4   5   6   7   8   9   10
        A2  A3  A4  A5  A6  A7  A8  A9  A10 GND

;pins  11   12   13  14   15   16   17  18  19   20
       A11  NC   S2  A12  A13  A14  A15 NC  CS   VCC

EQUATIONS

/CS = /A15 * /A14 * /A13 * /A12 * /A11 * A10 * A9 * A8 * /A6 * /A5 * /A4 * /A3 * /A2 * /S2
```

The resistor values are chosen in Figure 11–18 so the segment current is 80 mA. This current is required to produce an average current of 10 mA per segment as the displays are multiplexed. A six-digit display would use a segment current of 60 mA for an average of 10 mA per segment. In this type of display system, only one of the eight display positions is on at any given instant. The peak anode current in an eight-digit display is 560 mA (7 segments × 80 mA), but the average anode current is 80 mA. In a six-digit display the peak current would be 420 mA (7 segments × 60 mA. Whenever displays are multiplexed, we increase the segment current from 10 mA (for a display that uses 10 mA per segment as the nominal current) to a value equal to the number of display positions times 10 mA. This means that a four-digit display uses 40 mA per segment, a 5-digit display uses 50 mA, etc.

In this display the segment load resistor passes 80 mA of current and has a voltage of approximately 3.0 V across it. The LED is 1.65 V nominally and a few tenths are dropped across

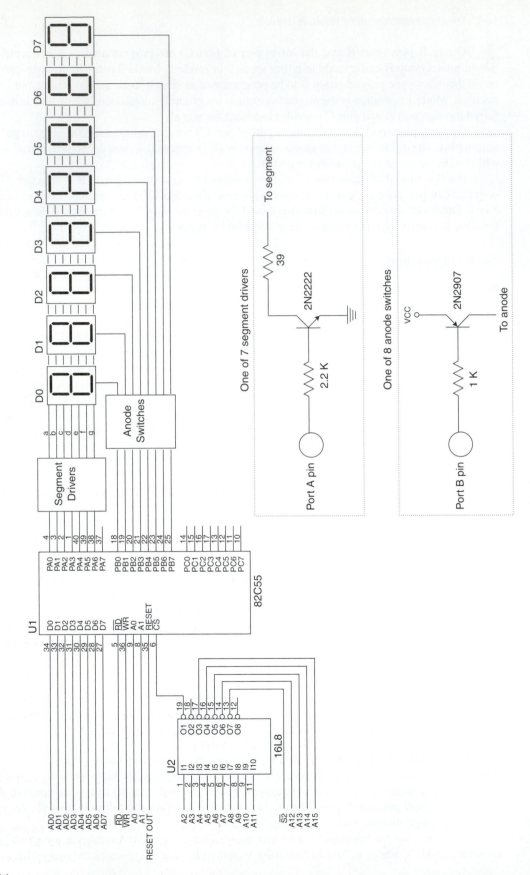

FIGURE 11–18 An eight-digit LED display interfaced to the 80188 microprocessor through an 82C55 PPI.

the anode switch, and the segment switch, hence a voltage of 3.0 V across the segment load resistor. The value of the resistor is $\frac{3.0\,V}{80\,mA}$ = 37.5 Ω. The closest standard resistor value of 39 Ω is used in Figure 11–18 for the segment load.

The resistor in series with the base of the segment switch assumes that the minimum gain of the transistor is 100. The base current is therefore $\frac{80\,mA}{100}$ = 0.8 mA. The voltage across the base resistor is approximately 3.0 V (the minimum logic 1 voltage level of the 82C55) minus the drop across the emitter-base junction (0.7 V) or 2.3 V. The value of the base resistor is then $\frac{2.3\,V}{0.8\,mA}$ = 2.875 KΩ. The closest standard resistor value is 2.7 KΩ, but a 2.2 KΩ is chosen for this circuit.

The anode switch has a single resistor on its base. The current through the resistor is $\frac{560\,mA}{100}$ = 5.6 mA because the minimum gain of the transistor is 100. This exceeds the maximum current of 4.0 mA from the 82C55, but this is small enough that it will work without a problem. The maximum current assumes that you are using the port pin as a TTL input to another circuit. If the amount of current was over 8.0–10.0 mA, then appropriate circuitry in the form of either a Darlington pair or another transistor switch would be required. Here the voltage across the base resistor is 5.0 V minus the drop across the emitter-base junction (0.7 V) minus the voltage at the port pin (0.4 V) for a logic 0 level. The value of the resistor is $\frac{3.9\,V}{5.6\,mA}$ = 696 Ω. The closest standard resistor value is 690 Ω, but a 1 KΩ is chosen in this example.

Before we examine software to operate the display, we must program the 82C55 using the short sequence of instructions listed in Example 11–9. Here ports A and B are both programmed as outputs.

EXAMPLE 11–9

```
                          ;programming the 82C55 PIA
                          ;
0000 B0 80                          MOV       AL,10000000B
0002 BA 0703                        MOV       DX,703H          ;address command
0005 EE                             OUT       DX,AL            ;program 82C55
```

The procedure to drive these displays is listed in Example 11–10. For this display system to function correctly, we must call this procedure often. Notice that the procedure calls another procedure (DELAY) that causes a 1-ms time delay. This time delay is not illustrated in this example, but is used to allow time for each display position to turn on. It is recommended by the manufacturers of LED displays that the display flash be between 100 Hz and 1500 Hz. Using a 1-ms time delay, we light each digit for 1 ms for a total display flash rate of 1000 Hz/8 displays or a flash rate of 125.

EXAMPLE 11–10

```
                          ;Procedure that scans the 8-digit LED display.
                          ;This procedure must be called from a program
                          ;whenever possible to display 7-segment
                          ;coded data from memory.
                          ;
0006                      DISP    PROC      NEAR USES AX BX DX SI

000A 9C                           PUSHF                ;save registers

                          ;setup registers for display

000B BB 0008                      MOV       BX,8           ;load count
000E B4 7F                        MOV       AH,7FH         ;load selection
                                                            pattern
0010 BE 00FF R                    MOV       SI,OFFSET MEM-1 ;address data
0013 BA 0701                      MOV       DX,701H        ;address Port B

                          ;display 8 digits
```

```
0016                         DISP1:
0016 8A C4                        MOV       AL,AH          ;select a digit
0018 EE                           OUT       DX,AL
0019 4A                           DEC       DX             ;address Port A
001A 8A 00                        MOV       AL,[BX+SI]     ;get 7-segment
                                                            data
001C EE                           OUT       DX,AL
001D E8 029A R                    CALL      DELAY          ;wait one
                                                            millisecond
0020 D0 CC                        ROR       AH,1           ;address next
                                                            digit
0022 42                           INC       DX             ;address Port B
0023 4B                           DEC       BX             ;adjust count
0024 75 F0                        JNZ       DISP1          ;repeat 8 times

0026 9D                           POPF
                                  RET

002C              DISP            ENDP
```

The display procedure (DISP) addresses an area of memory where the data, in 7-segment code, is stored for the 8 display digits. The AH register is loaded with a code (7FH) that initially addresses the most-significant display position. Once this position is selected, the contents of memory location MEM + 7 is addressed and sent to the most-significant digit. The selection code is then adjusted to select the next display digit as the address. This process repeats 8 times to display the contents of location MEM through MEM + 7 on the 8 display digits. Note that the time delay procedure is not illustrated and neither is the area of memory, which must be set up in the data segment.

A Stepper Motor Interfaced to the 82C55.   Another device often interfaced to a computer system is the stepper motor. A stepper motor is a digital motor because it is moved in discrete steps as it traverses through 360°. A common stepper motor is geared to move perhaps 15° per step in an inexpensive stepper motor, to 1° per step on a more costly high-precision stepper motor. In all cases these steps are gained through many magnetic poles and/or gearing. Notice that two coils are energized in Figure 11–19. If less power is required, one coil may be energized at a time causing the motor to step at 45°, 135°, 225°, and 315°.

Figure 11–19 shows a four-coil stepper motor that uses an armature with a single pole. Notice in this illustration, the stepper motor is shown four times with the armature (permanent magnetic) rotated to four discrete places. This is accomplished by energizing the coils as shown. This is an illustration of full stepping. The stepper motor is driven using NPN Darlington amplifier pairs to provide a large current to each coil.

A circuit that can drive this stepper motor is illustrated in Figure 11–20 with the four coils shown in place. This circuit uses the 82C55 to provide it with the drive signals used to rotate the armature of the motor in either the right-hand or left-hand direction.

A simple procedure that drives the motor (assuming port A is programmed in Mode 0 as an output device) is listed in Example 11–11. This subroutine is called with CX holding the number of steps and direction of the rotation. If CX > 8000H, the motor spins in the right-hand direction and if CX < 8000H, it spins in the left-hand direction. The leftmost bit of CX is removed and the remaining 15 bits contain the number of steps. Notice that the procedure uses a time delay (not illustrated) that causes a 1-ms time delay. This time delay is required to allow the stepper motor armature time to move to its next position.

EXAMPLE 11–11

```
= 0040                            PORT      EQU   40H      ;assign Port A
                                  ;
                                  ;Procedure to control stepper motor.
                                  ;
0000                              STEP      PROC  NEAR
```

```
0000 A0 0000 R            MOV    AL,POS        ;get position
0003 81 F9 8000           CMP    CX,8000H
0007 77 10                JA     RH            ;if right-hand
                                                direction
0009 83 F9 00             CMP    CX,0
000C 74 14                JE     STEP_OUT      ;if no steps
000E              STEP1:
000E D0 C0                ROL    AL,1          ;step left
0010 E6 40                OUT    PORT,AL
0012 E8 0011              CALL   DELAY         ;wait one
                                                millisecond
0015 E2 F7                LOOP   STEP1         ;repeat until CX = 0
0017 EB 09                JMP    STEP_OUT
0019            RH:
0019 81 E1 7FFF           AND    CX,7FFFH      ;clear bit 15
001D            RH1:
001D D0 C8                ROR    AL,1          ;step right
001F E6 40                OUT    PORT,AL
0021 E8 0006              CALL   DELAY         ;wait one
                                                millisecond
0024 E2 F7                LOOP   RH1           ;repeat until CX = 0
0026        STEP_OUT:
0026 A2 0000              MOV    POS,AL        ;save position
0029 C3                   RET

0029                   STEP    ENDP
```

The current position is stored in memory location POS, which must be initialized with 33H, 66H, 0CCH, or 99H. This allows a simple ROR (step right) or ROL (step left) instruction to rotate the binary bit pattern for the next step.

FIGURE 11–19 The stepper motor showing full-step operation. (a) 45°, (b) 135°, (c) 225°, and (d) 317°.

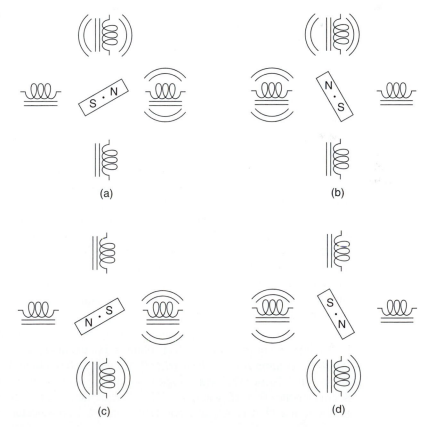

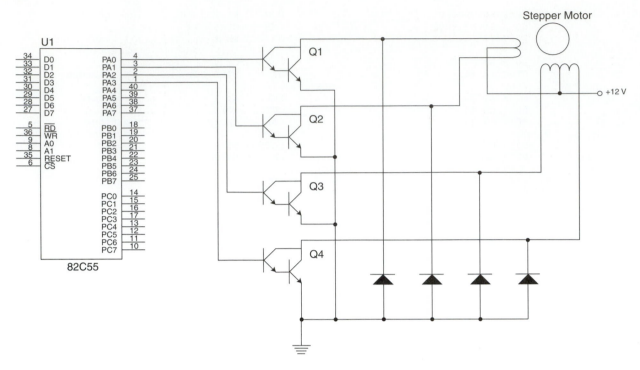

FIGURE 11–20   A stepper motor interfaced to the 82C55. Note that no decoder is shown in this illustration.

Stepper motors can also be operated in the half-step mode, which allows eight steps per sequence. This is accomplished by using the full step sequence described with a half step obtained by energizing one coil interspersed between the full steps. Half stepping allows the armature to be positioned at 0°, 90°, 180°, and 270°. The half-step position codes are 11H, 22H, 44H, and 88H. A complete sequence of eight steps would follow as: 11H, 33H, 22H, 66H, 44H, 0CCH, 88H, and 99H. This sequence could be either output from a lookup table or generated with software.

Key Matrix Interface.   Keyboards come in a vast variety of sizes from the standard 101-key QWERTY keyboards interfaced to the microprocessor to small specialized keyboards that may contain only 4 to 16 keys. This section concentrates on the smaller keyboards that may be purchased pre-assembled or may be constructed out of individual key switches.

Figure 11–21 illustrates a small key matrix that contains 16 switches interfaced to ports A and B of an 82C55. In this example the switches are formed into a 4 × 4 matrix, but any matrix could be used, such as a 2 × 8. Notice that the keys are all simple, single-pole, single-throw, push-button switches organized into four rows (ROW0–ROW3) and four columns (COL0–COL3). Also notice that each row is connected to 5.0 V through a 10 KΩ pull-up resistor to ensure that the row is pulled high when no push-button switch is closed.

The 82C55 is decoded (no decoder illustrated) at I/O ports 0050H–0053H for an 80188 microprocessor. Port A is programmed as an input port to read the rows and port B is programmed as an output port to select columns. For example, if 1110 is output from the port B pins PB_3–PB_0, column zero (with switches 0, 1, 2, and 3) has a logic 1 so the four keys in column zero are selected. Notice that with a logic 0 on PB0, the only switches that can place a logic 0 on port A are switches 0–3. If switches 4–F are closed, the corresponding port A pins remain a logic 1. Likewise if a 1101 is output to port B, switches 4–7 are selected, and so forth.

FIGURE 11–21 A 16-keypad connected to the 82C55 PPI.

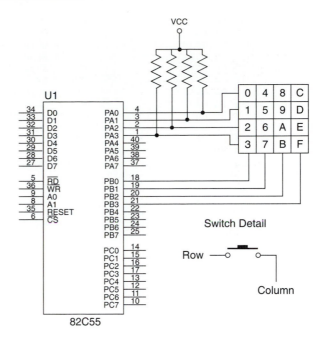

Figure 11–22 shows a flowchart of the software required to read a key from the keyboard matrix and de-bounce the key. Keys must be de-bounced and that is normally accomplished with a short time delay of from 10–20 ms. The flowchart contains three main sections. The first waits for the release of a key. This seems awkward, but software executes very quickly in a microprocessor and there is a possibility that the program will return to the top of this program before the key is released so we must wait for a release first. Next the flowchart shows that we wait for a keystroke. Once the keystroke is detected, the position of the key is calculated in the final part of the flowchart.

The software uses a procedure called SCAN to scan the keys, and another procedure called DELAY to waste 10 ms of time for de-bouncing. The main keyboard procedure is called KEY and it appears with the other procedures in Example 11–12. The SCAN procedure is generic so it can handle any configuration of keyboard from a 2×2 matrix to an 8×8. Changing the two equates at the start of the program (ROW and COL) will change the configuration of the software for any size keyboard. Not shown the steps required to initialize the 82C55 so port A is an input port and port B is an output port.

EXAMPLE 11–12

```
                      ;Keyboard procedure that scans the keyboard and
                      ;returns with the numeric code of the key in AL.
                      ;
= 0004                ROWS    EQU     4          ;number of rows
= 0004                COLS    EQU     4          ;number of columns
= 0050                PORTA   EQU     50H        ;port A address
= 0051                PORTB   EQU     51H        ;port B address

0000                  KEY     PROC    NEAR USES CX

0001 E8 002F                  CALL    SCAN       ;test all keys
0004 75 FA                    JNZ     KEY        ;if key closed
0006 E8 0048                  CALL    DELAY      ;wait for about 10 ms
0009 E8 0027                  CALL    SCAN       ;test all keys
000C 75 F2                    JNZ     KEY        ;if key closed
000E                  KEY1:
```

```
000E E8 0022                CALL    SCAN            ;test all keys
0011 74 FB                  JZ      KEY1            ;if no key closed
0013 E8 003B                CALL    DELAY           ;wait for about 10 ms
0016 E8 001A                CALL    SCAN            ;test all keys
0019 74 F3                  JZ      KEY1            ;if no key closed
001B 50                     PUSH    AX              ;save row codes
001C B0 04                  MOV     AL,COLS         ;calculate starting row key
001E 2A C1                  SUB     AL,CL
0020 B5 04                  MOV     CH,ROWS
0022 F6 E5                  MUL     CH
0024 8A C8                  MOV     CL,AL
0026 FE C9                  DEC     CL
0028 58                     POP     AX
0029            KEY2:
0029 D0 C8                  ROR     AL,1            ;find row position
002B FE C1                  INC     CL
002D 72 FA                  JC      KEY2
002F 8A C1                  MOV     AL,CL           ;mode code to AL
                            RET

0033            KEY     ENDP
0033            SCAN    PROC    NEAR USES BX

0034 B1 04                  MOV     CL,ROWS         ;form row mask
0036 B7 FF                  SHL     BH,CL
003A B9 0004                MOV     CX,COLS         ;load column count
003D B3 FE                  MOV     BL,0FEH         ;get selection code
003F           SCAN1:
003F 8A C3                  MOV     AL,BL           ;select column
0041 E6 51                  OUT     PORTB,AL
0043 D0 C3                  ROL     BL,1
0045 E4 50                  IN      AL,PORTA        ;read rows
0047 0A C7                  OR      AL,BH
0049 3C FF                  CMP     AL,0FF  H       ;test for a key
004B 75 04                  JNZ     SCAN2
004D FE C9                  DEC     CL
004F 75 EE                  JNZ     SCAN1
0053           SCAN2:
                            RET

0053           SCAN            ENDP

0053           DELAY PROC       NEAR USES CX

0055 B9 1388                MOV     CX,5000         ;10 ms (8 MHz clock)
0058           DELAY1:
0058 E2 FE                  LOOP    DELAY1
                            RET

005C           DELAY           ENDP
```

A note about the SCAN procedure: The time between where the keyboard column is selected and where the rows are read is very short. In a very high speed system, a small time delay must be placed between these two points to allow for the data at port A to settle to its final state. In most cases this is not needed. Note that in the SCAN procedure, a DEC CL instruction is used. This could be replaced with LOOP, except that LOOP does not affect the zero flag, which would cause problems with the software.

Mode 1 Strobed Input

Mode 1 operation causes port A and/or port B to function as latching input devices. This allows external data to be stored into the port until the microprocessor is ready to retrieve it. Port C is also used in mode 1 operation, not for data, but for control or handshaking signals that help

FIGURE 11–22 The flow-chart used for keyboard scanning software and de-bounce.

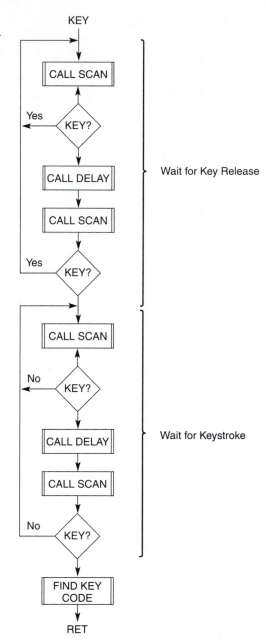

operate ports A and/or B as strobed input ports. Figure 11–23 shows the timing diagram and how both ports are structured for mode 1 strobed input operation.

The strobed input port captures data from the port pins when the strobe ($\overline{\text{STB}}$) is activated. Strobe captures the port data on the 0-to-1 transition. The $\overline{\text{STB}}$ signal causes data to be captured in the port and it also activates the IBF **(input buffer full)** and INTR **(interrupt request)** signals. Once the microprocessor, through software (IBF) or hardware (INTR), notices that data is strobed into the port, it executes an IN instruction to read the port ($\overline{\text{RD}}$). The act of reading the port restores both IBF and INTR to their inactive states until the next datum is strobed into the port.

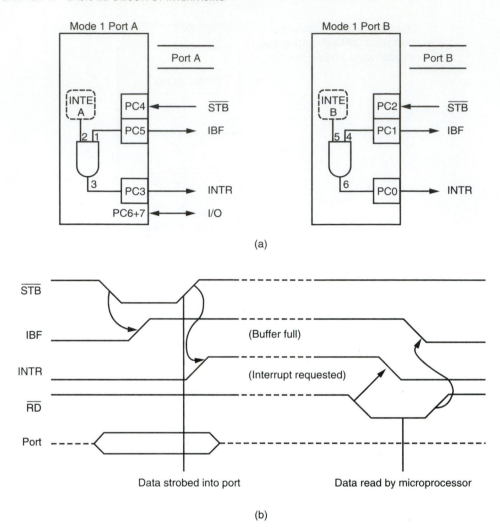

FIGURE 11–23   Strobed input operation (mode 1) of the 82C55. (a) Internal structure. (b) Timing diagram.

Signal Definitions for Mode 1 Strobed Input.

$\overline{\text{STB}}$	The **strobe** input loads data into the port latch, which holds the information until it is input to the microprocessor via the IN instruction.
IBF	**Input buffer full** is an output that indicates the input latch contains information.
INTR	**Interrupt request** is an output that requests an interrupt. The INTR pin becomes a logic 1, when the $\overline{\text{STB}}$ input returns to a logic 1, and is cleared when the data are input from the port by the microprocessor.
INTE	The **interrupt enable** signal is neither an input nor an output, but an internal bit programmed via the port PC4 (port A) or PC2 (port B) bit position.
PC_7, PC_6	The port C pins 7 and 6 are general purpose I/O pins that are available for any purpose.

Strobed Input Example.   An excellent example of a strobed input device is a keyboard. The keyboard encoder de-bounces the key-switches and provides a strobe signal whenever a key is de-

FIGURE 11–24 Using the 82C55 for strobed input operation of a keyboard.

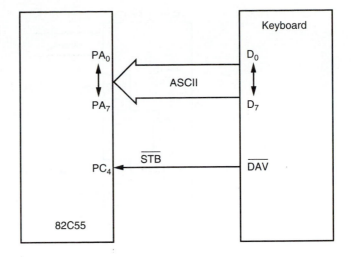

pressed and the data output contains the ASCII-coded key-code. Figure 11–24 illustrates a keyboard connected to strobed input port A. Here $\overline{DAV}$ **(data available)** is activated for 1 μs each time a key is typed on the keyboard. This causes data to be strobed into port A because $\overline{DAV}$ is connected to the $\overline{STB}$ input of port A. So each time a key is typed, it is stored in port A of the 82C55. The $\overline{STB}$ input also activates the IBF signal indicating that data is in port A.

Example 11–13 reads data from the keyboard each time a key is typed. This procedure reads the key from port A and returns with the ASCII code in AL. To detect a key, port C is read and the IBF bit (bit position PC_5) is tested to see if the buffer is full. If the buffer is empty (IBF = 0), then the procedure keeps testing this bit waiting for a character to be typed on the keyboard.

EXAMPLE 11–13

```
                        ;Procedure that reads the keyboard encoder
                        ;and returns the ASCII character in AL.
                        ;
= 0020                  BIT5    EQU     20H
= 0022                  PORTC   EQU     22H
= 0020                  PORTA   EQU     20H

0000                    READ    PROC    NEAR

0000 E4 22                      IN      AL,PORTC    ;read Port C
0002 A8 20                      TEST    AL,BIT5     ;test IBF
0004 74 FA                      JZ      READ        ;if IBF = 0
0006 E4 20                      IN      AL,PORTA    ;read data
0008 C3                         RET

0009                    READ    ENDP
```

Mode 1 Strobed Output

Figure 11–25 illustrates the internal configuration and timing diagram of the 82C55 when it is operated as a strobed output device under mode 1. Strobed output operation is similar to mode 0 output except that control signals are included to provide handshaking.

Whenever data is written to a port programmed as a strobed output port, the $\overline{OBF}$ **(output buffer full)** signal becomes a logic 0 to indicate that data is present in the port latch. This signal indicates that data is available to an external I/O device that removes the data by strobing the $\overline{ACK}$ **(acknowledge)** input to the port. The $\overline{ACK}$ signal returns the $\overline{OBF}$ signal to a logic 1, indicating that the buffer is not full.

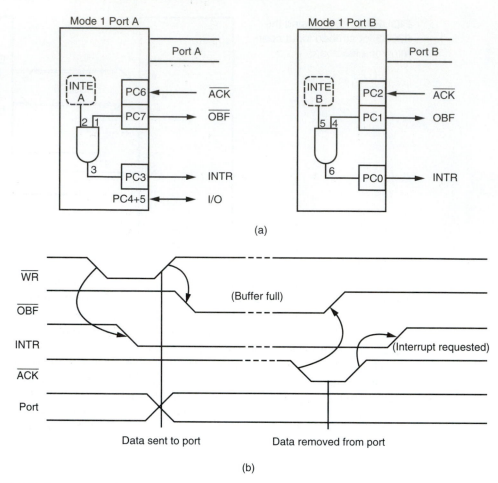

FIGURE 11–25 Strobed output operation (mode 1) of the 82C55. (a) Internal structure. (b) Timing diagram.

Signal Definitions for Mode 1 Strobed Output.

$\overline{OBF}$ **Output buffer full** is an output that goes low whenever data are output (OUT) to the port A or port B latch. This signal is set to a logic 1 whenever the $\overline{ACK}$ pulse returns from the external device.

$\overline{ACK}$ The **acknowledge** signal causes the $\overline{OBF}$ pin to return to a logic 1 level. The $\overline{ACK}$ is a response from an external device that indicates it has received the data from the 82C55 port.

INTR   **Interrupt request** is a signal that interrupts the microprocessor when the external device receives the data via the $\overline{ACK}$ signal. This pin is qualified by the internal INTE (**interrupt enable**) bit.

INTE   **Interrupt enable** is neither an input nor an output, but an internal bit programmed to enable or disable the INTR pin. The INTE A bit is programmed as PC_6 and INTE B as PC_2.

PC_5, PC_4 Port C bits 5 and 4 are general-purpose I/O pins. The bit set and reset commands may be used to set or reset these two pins.

FIGURE 11–26 The 82C55 connected to a parallel printer interface that illustrates the strobed output mode of operation for the 82C55.

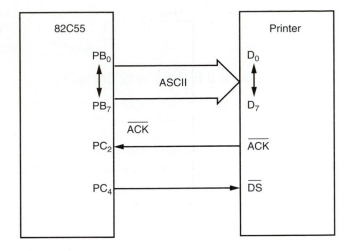

Strobed Output Example. The printer interface discussed in Section 11–1 is used here to demonstrate how to achieve strobed output synchronization between the printer and the 82C55. Figure 11–26 illustrates port B connected to a parallel printer with eight data inputs for receiving ASCII-coded data, a $\overline{\text{DS}}$ (**data strobe**) input to strobe data into the printer, and an $\overline{\text{ACK}}$ output to acknowledge the receipt of the ASCII character.

In this circuit, there is no signal to generate the $\overline{\text{DS}}$ signal to the printer, so PC_4 is used with software that generates the $\overline{\text{DS}}$ signal. The $\overline{\text{ACK}}$ signal that is returned from the printer acknowledges the receipt of the data and is connected to the $\overline{\text{ACK}}$ input of the 82C55.

Example 11–14 lists the software that sends the ASCII-coded character in AH to the printer. The procedure first tests $\overline{\text{OBF}}$ to decide if the printer has removed the data from port B. If not, the procedure waits for the $\overline{\text{ACK}}$ signal to return from the printer. If $\overline{\text{OBF}} = 1$, then the procedure sends the contents of AH to the printer through port B and also sends the $\overline{\text{DS}}$ signal.

EXAMPLE 11–14

```
                              ;Procedure that transfers the ASCII character
                              ;from AH to the printer via port B.
                              ;
= 0002                        BIT1    EQU     2
= 0062                        PORTC   EQU     62H
= 0061                        PORTB   EQU     61H
= 0063                        CMD     EQU     63H

0000                          PRINT   PROC    NEAR

                              ;check if printer is ready

0000 E4 62                            IN      AL,PORTC        ;get OBF
0002 A8 02                            TEST    AL,BIT1         ;test OBF
0004 74 FA                            JZ      PRINT           ;if OBF = 0

                              ;send character to printer

0006 8A C4                            MOV     AL,AH           ;get data
0008 E6 61                            OUT     PORTB,AL        ;print data

                              ;send data strobe to printer

000A B0 08                            MOV     AL,8            ;clear DS
000C E6 63                            OUT     CMD,AL
000E B0 09                            MOV     AL,9            ;set DS
```

```
0010 E6                                    OUT        CMD,AL
0012 C3                                    RET

0013                        PRINT          ENDP
```

Mode 2 Bi-directional Operation

In mode 2, which is allowed with group A only, port A becomes bi-directional, allowing data to be transmitted and received over the same eight wires. Bi-directional bused data is useful when interfacing two computers. It is also used for the IEEE-488 parallel high-speed GPIB (**general purpose instrumentation bus**) interface standard. Figure 11–27 shows the internal structure and timing diagram for mode 2 bi-directional operation.

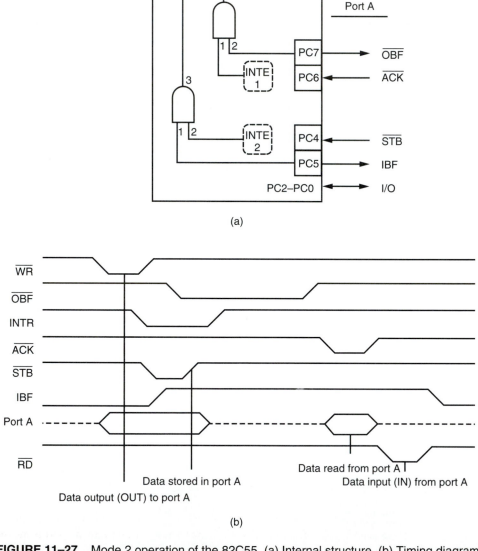

(a)

(b)

FIGURE 11–27   Mode 2 operation of the 82C55. (a) Internal structure. (b) Timing diagram.

Signal Definitions for Bi-directional Mode 2.

INTR	**Interrupt request** is an output used to interrupt the microprocessor for both input and output conditions.
$\overline{\text{OBF}}$	**Output buffer full** is an output that indicates that the output buffer contains data for the bi-directional bus.
$\overline{\text{ACK}}$	**Acknowledge** is an input that enables the three-state buffers so data can appear on port A. If $\overline{\text{ACK}}$ is a logic 1, the output buffers of port A are at their high-impedance state.
$\overline{\text{STB}}$	The **strobe** input loads the port A input latch with external data from the bi-directional port A bus.
IBF	**Input buffer full** is an output used to signal that the input buffer contains data for the external bi-directional bus.
INTE	**Interrupt enable** are internal bits ($INTE_1$ and $INTE_2$) that enable the INTR pin. The state of the INTR pin is controlled through port C bits PC_6 ($INTE_1$) and PC_4 ($INTE_2$).
PC_2, PC_1, and PC_0	These bits are general-purpose I/O pins in mode 2 controlled by the bit set and reset commands.

The Bi-directional Bus. The bi-directional bus is used by referencing port A with the IN and OUT instructions. To transmit data through the bi-directional bus, the program first tests the $\overline{\text{OBF}}$ signal to determine whether the output buffer is empty. If it is, then data is sent to the output buffer via the OUT instruction. The external circuitry also monitors the $\overline{\text{OBF}}$ signal to decide if the microprocessor has sent data to the bus. As soon as the output circuitry sees a logic 0 on $\overline{\text{OBF}}$, it sends back the $\overline{\text{ACK}}$ signal to remove it from the output buffer. The $\overline{\text{ACK}}$ signal sets the $\overline{\text{OBF}}$ bit and also enables the three-state output buffers so data may be read. Example 11–15 transmits the contents of the AH register through bi-directional port A.

EXAMPLE 11–15

```
                            ;Procedure that transmits AH through the bi-
                            ;directional bus of port A.
                            ;
= 0080                      BIT7    EQU     80H
= 0062                      PORTC   EQU     62H
= 0060                      PORTA   EQU     60H

0000                        TRANS   PROC    NEAR

0000 E4 62                          IN      AL,PORTC    ;get OBF
0002 A8 80                          TEST    AL,BIT7     ;test OBF
0004 74 FA                          JZ      TRANS       ;if OBF = 1

0006 8A C4                          MOV     AL,AH       ;get data
0008 E6 60                          OUT     PORTA,AL    ;send data
000A C3                             RET

000B                        TRANS   ENDP
```

To receive data through the bi-directional port A bus, the IBF bit is tested with software to decide if data have been strobed into the port. If IBF = 1, then data is input using the IN instruction. The external interface sends data into the port using the $\overline{\text{STB}}$ signal. When $\overline{\text{STB}}$ is activated, the IBF signal becomes a logic 1 and the data at port A is held inside the port in a latch. When the IN instruction executes, the IBF bit is cleared and the data in the port is moved into AL. Example 11–16 is a procedure that reads data from the port.

FIGURE 11–28   A summary of the port connections for the 82C55 PIA.

	Mode 0		Mode 1		Mode 2
Port A	IN	OUT	IN	OUT	I/O
Port B	IN	OUT	IN	OUT	Not used
Port C 0			INTR$_B$	INTR$_B$	I/O
1			IBF$_B$	$\overline{\text{OBF}}_B$	I/O
2			$\overline{\text{STB}}_B$	$\overline{\text{ACK}}_B$	I/O
Port C 3	IN	OUT	INTR$_A$	INTR$_A$	INTR
4			$\overline{\text{STB}}_A$	I/O	$\overline{\text{STB}}$
5			IBF$_A$	I/O	IBF
6			I/O	$\overline{\text{ACK}}_A$	$\overline{\text{ACK}}$
7			I/O	$\overline{\text{OBF}}_A$	$\overline{\text{OBF}}$

EXAMPLE 11–16

```
                                ;Procedure that reads data from the bi-
                                ;direction port A and returns it in AL.
                                ;
= 0020                          BIT5    EQU     20H
= 0062                          PORTC   EQU     62H
= 0060                          PORTA   EQU     60H
0000                            READ    PROC    NEAR

0000 E4 62                              IN      AL,PORTC        ;get IBF
0002 A8 20                              TEST    AL,BIT5         ;test IBF
0004 74 FA                              JZ      READ            ;if IBF = 0
0006 E4 60                              IN      AL,PORTA        ;get data
0008 C3                                 RET

0009                            READ    ENDP
```

The INTR **(interrupt request)** pin can be activated from both directions of data flow through the bus. If INTR is enabled by both INTE bits, then the output and input buffers cause interrupt requests. This occurs when data is strobed into the buffer using $\overline{\text{STB}}$ or when data is written using OUT.

82C55 Mode Summary

Figure 11–28 is a graphical summary of the three modes of operation for the 82C55: Mode 0 provides simple I/O, mode 1 provides strobed I/O, and mode 2 provides bi-directional I/O. As we mentioned, these modes are selected through the command register of the 82C55.

11–4   80186/80188 AND 80386EX I/O PORTS

As mentioned earlier, some versions of the 80186, 80188, and 80386EX contain I/O ports. In this section we explain how to program and use these I/O pins.

80186/80188 EB and EC Versions

Both the 80186/80188 EB and EC versions contain I/O pins. Port 1 is shared between I/O and chip selection and port 2 is shared with other functions that apply to the serial interfaces. The EC version contains 22 I/O pins, while the EB version contains 16 I/O pins. Figure 11–29 shows the ports available to both the EB and EC versions and the registers used to control the I/O port pins.

FIGURE 11–29   The I/O ports on the 80186/80188 EB and EC versions.

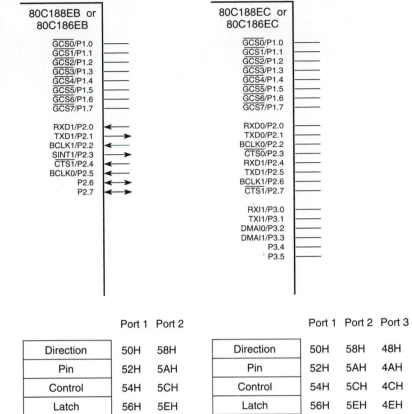

Port 1	Port 2	
Direction	50H	58H
Pin	52H	5AH
Control	54H	5CH
Latch	56H	5EH

Port 1	Port 2	Port 3
Direction	50H	58H	48H
Pin	52H	5AH	4AH
Control	54H	5CH	4CH
Latch	56H	5EH	4EH

Each port is controlled by a series of four registers that is 16 bits in width, but only the rightmost 8 bits are used to program the port pins. The registers for a port function as follows:

Control Register Programs the pins on the port as port pins for I/O or as special function pins. If a 1 is placed in a bit, the corresponding port pin is programmed as a special function pin. For example, if a 0001H is output to the port 1 control register, the P1.0 pin is programmed as $\overline{GCS0}$, while the remaining pins of port 1 are programmed as I/O pins. A hardware reset operation programs all port pins as special function pins.

Direction Register Controls the direction of the I/O pins (in or out). If a pin is programmed for I/O operation, the direction register selects input (1) or output (0) operation. For example, if port 2 is programmed for I/O operation and the direction register is programmed with 00FEH, all of the port 2 pins are input pins except for P2.0.

Latch Register Holds output data for a port that is programmed as an output port. The latch can also be read, but a read from the latch returns the contents of the latch, not the port pin.

Pin Register Contains data read from the port that is programmed as an input port.

Controlling a Printer Through the Parallel Port Pins. A Centronics parallel printer port can be created by using the parallel ports on the 80186/80188 EB or EC version of the microprocessor. Because port 1 is used for chip selection, port 2 is selected in this example for the parallel data

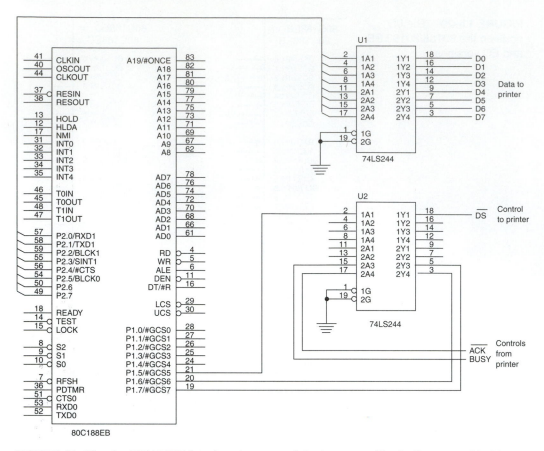

FIGURE 11–30   An 80C188EB interfaced to a parallel printer port. The buffers are added for protection.

connections for the printer interface. Some of the port 1 pins are used for control signals between the microprocessor and the printer. No attempt has been made to connect the error lines from the printer, although this would be the normal practice. Refer to Figure 11–30 for the fully buffered printer interface.

Two procedures are written to control this interface. One initializes the port pins, and the other transmits data to the printer for printing. Example 11–17 illustrates the software (INIT) to initialize ports 1 and 2 as simple output ports for transmitting data to the printer. The INIT procedure also programs the pins of port 2 for controlling the printer interface. One of the pins is programmed as an output pin and the other two are programmed as input pins.

EXAMPLE 11–17

```
0000                      INIT    PROC    NEAR    USES AX DX

0002 BA 0054                      MOV     DX,0054H    ;address port 1 control
0005 B8 0007                      MOV     AX,0007H    ;make bits 3-7 I/O
0008 EE                           OUT     DX,AL

0009 BA 005C                      MOV     DX,005CH    ;address port 2 control
000C B8 0000                      MOV     AX,0000H    ;make all bits I/O
010F EE                           OUT     DX,AL

0010 BA 0058                      MOV     DX,0058H    ;address port 3 direction
```

```
0013 B8 00CA                    MOV    AX,00CAH        ;program direction
0016 EE                         OUT    DX,AL

0017 BA 0056                    MOV    DX,0056H        ;address port 1 latch
001A B8 00FF                    MOV    AX,00FFH        ;data strobe = 1
001D EE                         OUT    DX,AL
                                RET

0021                     INIT   ENDP
```

Controlling the interface to the printer can be accomplished by using the BUSY or $\overline{ACK}$ signal. The $\overline{ACK}$ line pulses low for at least 500 ns when the printer completes printing a character and also after power is applied to the printer. The BUSY pin is used in Example 11–18 to test for a busy condition. If the printer is busy, no printing can ensue. If the printer is not busy, the software sends a character from AL to output ports 1 and 2 and then sends a pulse to the printer on the $\overline{DS}$ (data strobe) line. The data strobe signal informs the printer that data is available on D_0–D_7 for printing.

EXAMPLE 11–18

```
0000                     SEND   PROC   NEAR    USES AX DX

0002 50                         PUSH   AX              ;save ASCII code
0003 BA FF5A                    MOV    DX,0FF5AH       ;address port 2 pins
                                .REPEAT
0006 EC                                IN     AL,DX    ;get port 2 pins
0007 A8 10                             TEST   AL,10H   ;test BUSY
                                .UNTIL ZERO?           ;until printer not busy
000B 58                         POP    AX              ;get ASCII code
000C 50                         PUSH   AX
000D 8A E0                      MOV    AH,AL
000F 24 F0                      AND    AL,0F0H         ;assemble bits 7-4
0011 C0 EC 03                   SHR    AH,3
0014 80 E4 60                   AND    AH,060H
0017 80 C4 04                   ADD    AH,4
001A 80 E4 0A                   AND    AH,0AH
001D 80 F4 08                   XOR    AH,8
0020 02 C4                      ADD    AL,AH
0022 EE                         OUT    DX,AL           ;send bits 7-4
0023 58                         POP    AX              ;get ASCII
0024 C0 E0 04                   SHL    AL,4
0027 0C 04                      OR     AL,4
0029 BA FF56                    MOV    DX,0FF56H       ;address port 1 latch
002C EE                         OUT    DX,AL           ;send bits 3-0
002D 24 F0                      AND    AL,0F0H
002F EE                         OUT    DX,AL           ;data strobe = 0
0030 0C F0                      OR     AL,0F0H
0032 EE                         OUT    DX,AL           ;data strobe = 1
                                RET

0036                     SEND   ENDP
```

80386EX I/O Structure

The 80386EX is much like the 80186/80188 EC versions because it also contains three I/O ports. The difference is that the 80386EX contains three 8-bit ports or 24 pins of I/O, while the 80186/80188 contains 22 pins of I/O. Refer to Figure 11–31 for the I/O port pins and register structure of the 80386EX microprocessor.

Note that the control structure for the 80386EX is identical to that of the 80186/80188 EB and EC versions. The operation of the control register is similar to the 80186/80188 except as noted.

FIGURE 11–31   The pin configuration of the 80386EX I/O ports and the register structure.

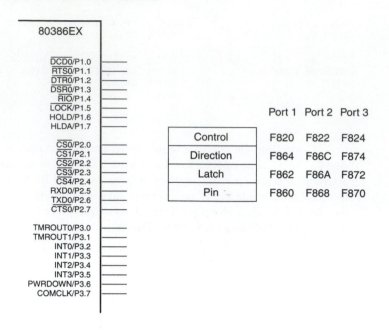

	Port 1	Port 2	Port 3
Control	F820	F822	F824
Direction	F864	F86C	F874
Latch	F862	F86A	F872
Pin	F860	F868	F870

Control Register

The control register selects I/O operation (0) for a port pin or peripheral or special function operation (1), which is identical to the 80186/80188.

Direction Register

The port pins of the 80386EX are programmed as input, output, or open-drain output pins. A logic 0 in the direction register programs a corresponding port pin as an output pin. A logic 1 in the direction register programs a port pin to function as an input and at the same time as an open-drain output pin. For input operation, the latch register must be programmed with a logic 1. If a logic 0 is placed in the latch register, while the pin is programmed as an open-drain output, the port pin becomes a logic 0.

Latch Register

Places a logic 1 or 0 on a port pin that is programmed as an output. If a port pin is programmed as an open-drain output, a logic 0 in the latch register places a logic 0 on the port pin; a logic 1 in the latch register places the port pin at its high-impedance state (open-drain).

Pin Register

This is where data read from a port is found. To read data from a port pin, the port direction register must be programmed with a 1 and the latch must also be programmed with a 1.

11–5   THE 8279 PROGRAMMABLE KEYBOARD/DISPLAY INTERFACE

The 8279 is a programmable keyboard and display interfacing component that scans and encodes up to a 64-key keyboard and controls up to a 16-digit numerical display. The keyboard interface has a built-in first-in, first-out (FIFO) buffer that allows it to store up to eight keystrokes before the microprocessor must retrieve a character. The display section controls up to 16 numeric displays from an internal 16×8 RAM that stores the coded display information.

Basic Description of the 8279

As we shall see, the 8279 is designed for ease of interfacing with any microprocessor. Figure 11–32 illustrates the pin-out of this device and the definition of each pin connection follows.

Pin Definitions for the 8279.

A_0	The A_0 address input selects data or control for reads and writes between the microprocessor and the 8279. A logic 0 selects data and a logic 1 selects control or status register.
$\overline{BD}$	**Blank** is an output used to blank the displays.
CLK Clock	**Clock** is an input that generates the internal timing for the 8279. The maximum allowable frequency on the CLK pin is 3.125 MHz for the 8279-5 and 2.0 MHz for the 8279. Other timings require wait states in microprocessors executing at above 5 MHz.
CN/ST	**Control/strobe** is an input normally connected to the control key on a keyboard.
$\overline{CS}$	**Chip select** is an input that enables the 8279 for programming, reading the keyboard and status information, and writing control and display data.
$DB_7–DB_0$	The **data bus** is consists of bi-directional pins that connect to the data bus on the microprocessor.
IRQ	**Interrupt request** is an output that becomes a logic 1 whenever a key is pressed on the keyboard. This signal indicates that keyboard data is available for the microprocessor.
OUTA3–OUTA0	Outputs that send data to the displays (most-significant).
OUTB3–OUTB0	Outputs that send data to the displays (least-significant).

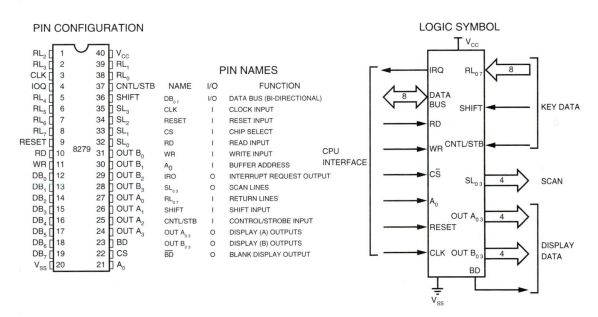

FIGURE 11–32 The pin-out and logic symbol of the 8279 programmable keyboard/display interface. (Courtesy of Intel Corporation)

$\overline{\text{RD}}$	The **read** input is directly connected to the $\overline{\text{IORC}}$ or $\overline{\text{RD}}$ signal from the system. When $\overline{\text{CS}}$ is a logic 0, the $\overline{\text{RD}}$ input causes a read from the data registers or status register.
RESET	The **reset** input that connects to the system RESET signal.
RL7–RL0	**Return lines** are inputs used to sense any key depression in the keyboard matrix.
SHIFT	The **shift** input normally connects to the shift key on a keyboard.
SL3–SL0	The **scan line** outputs scan both the keyboard and the displays.
$\overline{\text{WR}}$	**Write** is an input that connects to write strobe signal that is developed with external logic. The $\overline{\text{WR}}$ input causes data to be written to a data registers or control registers within the 8279.
VCC	A power supply pin connected to the system +5.0-V bus.
VSS	The ground pin connects to the system ground.

Interfacing the 8279 to the Microprocessor

Figure 11–33 shows the 8279 connected to the 80C188EB microprocessor. The 8279 is decoded to function at 8-bit I/O addresses 10H and 11H where port 10H is the data port and 11H is the control port. This circuit uses a PAL16L8 (see Example 11–19) to decode the I/O address for the 8279. Address bus bit A0 selects either the data or control port. Notice that the $\overline{\text{CS}}$ signal selects the 8279 and also provides a signal called $\overline{\text{WAIT2}}$ that is used to cause two wait states so that this device functions with an 8MHz 80C188EB. The maximum clock input frequency is 3.125 MHz, which means that either a timer within the 80C188EB must generate the clock or it can be generated by dividing the clock output signal by a factor of 4 to provide a 2.0 MHz clock to the 8279.

The only signal not connected to the microprocessor is the IRQ output. This is an interrupt request pin and is beyond the scope of our discussion. The next chapter explains interrupts and where they operate and function in a system.

FIGURE 11–33 The 8279 interfaced to the 80188 microprocessor.

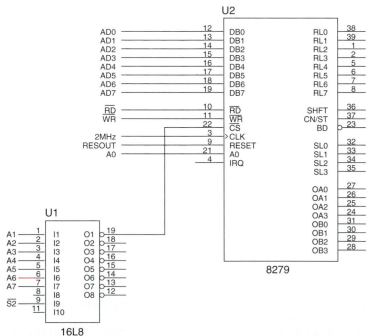

EXAMPLE 11–19

```
TITLE        Address Decoder
PATTERN      Test 14
REVISION     A
AUTHOR       Barry B. Brey
COMPANY      BreyCo
DATE         7/10/97
CHIP         DECODERE PAL16L8

;pins  1   2   3   4   5   6   7   8   9   10
       A1  A2  A3  A4  A5  A6  A7  NC  S2  GND

;pins  11  12  13  14  15  16  17  18  19  20
       NC  NC  NC  NC  NC  NC  NC  NC  CS  VCC

EQUATIONS

/CS = /A7 * /A6 * /A5 * A4 * /A3 * /A2 * /A1 * /S2
```

Keyboard Interface

Suppose that a 64-key keyboard (with no numeric displays) is connected through the 8279 to the 8088 microprocessor. Figure 11–34 shows this connection as well as the keyboard. With the 8279, the keyboard matrix is any size from a 2×2 matrix (4 keys) to an 8×8 matrix (64 keys). (Each crossover point in the matrix contains a normally open push-button switch that connects one vertical column with one horizontal row when a key is pressed.)

The I/O port number decoded is the same as that decoded in Figure 11–33. The I/O port number is 10H for the data port and 11H for the control port in this circuit.

The 74ALS138 decoder generates eight active low-column strobe signals for the keyboard. The selection pins SL_2–SL_0 sequentially scan each column of the keyboard, and the internal circuitry of the 8279 scans the RL pins searching for a key-switch closure. Pull-up resistors, normally found on input lines of a keyboard, are not required because the 8279 contains its own internal pull-ups on the RL inputs.

Programming the Keyboard Interface.   Before any keystroke is detected, the 8279 must be programmed—a more involved procedure than with the 82C55. The 8279 has eight control words to

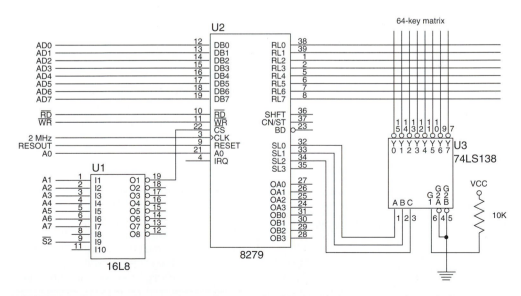

FIGURE 11–34   A 64-key matrix interfaced to the 80188 microprocessor through the 8279.

TABLE 11–4   The 8279 control word summary

D_7	D_6	D_5	Function	Purpose
0	0	0	Mode set	Selects the number of display positions, left or right entry, and type of keyboard scan
0	0	1	Clock	Programs the internal clock and sets the scan and de-bounce times
0	1	0	Read FIFO	Selects the type of FIFO read and the address of the read
0	1	1	Read display	Selects the type of display read and the address of the read
1	0	0	Write display	Selects the type of write and the address of the write
1	0	1	Display write inhibit	Allows half-bytes to be blanked
1	1	0	Clear	Clears the display or FIFO
1	1	1	End interrupt	Clears the IRQ signal to the microprocessor

consider before it is programmed. The first three bits of the number sent to the control port (11H in this example) select one of the eight different control words. Table 11–4 lists all eight control words and briefly describes them.

Control Word Descriptions.   The following is a list of the control words that program the 8279. Note that the first three bits are the control register number from Table 11–4, which are followed by other binary bits of information as they apply to each control.

000DDMMM   The mode set is a command with an opcode of 000 and two fields programmed to select the mode of operation for the 8279. The DD field selects the mode of operation for the displays (see Table 11–5) and the MMM field selects the mode of operation (see Table 11–6) for the keyboard.

The DD field selects either an 8- or a 16-digit display and determines whether new data is entered to the rightmost or leftmost display position. The MMM field is quite a bit more complex. It provides encoded, decoded, or strobed keyboard operation.

In encoded mode, the SL outputs are active-high and follow the binary bit pattern 0 through 7 or 0 through 15 depending whether 8- or 16-digit displays are selected. In decoded mode, the SL outputs are active-low, and only one of the four outputs is low at any given instant. The decoded outputs repeat the pattern: 1110, 1101, 1011, and 0111. In strobed mode, an active-high pulse on the CN/ST input pin strobes data from the RL pins into an internal FIFO where they are held for the microprocessor.

TABLE 11–5   Binary bit assignment for DD of the mode set control word

DD	Function
00	8-digit display with left entry
01	16-digit display with left entry
10	8-digit display with right entry
11	16-digit display with right entry

TABLE 11–6 Binary bit assignment for MMM of the mode set control word

MMM	Function
000	Encoded keyboard with 2-key lockout
001	Decoded keyboard with 2-key lockout
010	Encoded keyboard with N-key rollover
011	Decoded keyboard with N-key rollover
100	Encoded sensor matrix
101	Decoded sensor matrix
110	Strobed keyboard, encoded display scan
111	Strobed keyboard, decoded display scan

It is also possible to select either 2-key lockout or N-key rollover. Two-key lockout prevents two keys from being recognized if pressed simultaneously. N-key rollover will accept all keys pressed simultaneously, from first to last.

001PPPPP The **clock command** word programs the internal clock divider. The code PPPPP is a prescaler that divides the clock input pin (CLK) to achieve the desired operating frequency or approximately 100 KHz. An input clock of 1 MHz thus requires a prescaler of 01010_2 for PPPPP.

010Z0AAA The **read FIFO** control word selects the address of a keystroke from the internal FIFO buffer. Bit positions AAA select the desired FIFO location from 000 to 111, and Z selects auto-increment for the address. Under normal operation, this control word is used only with the sensor matrix operation of the 8279.

011ZAAAA The **display read** control word selects the read address of one of the display RAM positions for reading through the data port. AAAA is the address of the position to be read and Z selects auto-increment mode. This command is used if the information stored in the display RAM must be read.

100ZAAAA The **write display** control word selects the write address of one of the displays. AAAA addresses the position to be written to through the data port and Z selects auto-increment so subsequent writes through the data port are to subsequent display positions.

1010WWBB The **display write inhibit** control word inhibits writing to either half of each display RAM location. The leftmost W inhibits writing to the leftmost four bits of the display RAM location, and the rightmost W inhibits the rightmost four bits. The BB field functions in a like manner, except it blanks (turns off) either half of the output pins.

1100CCFA The **clear** control word clears the display, the FIFO, or both the display and FIFO. Bit F clears the FIFO and the display RAM status, and sets the address pointer to 000. If the CC bits are 00 or 01, the all display RAM locations become 0000000; if CC = 10, all locations become 00100000; if CC = 11, all locations become 11111111.

111E0000 The **end of interrupt** control word is issued to clear the IRQ pin to zero in the sensor matrix mode. If E is a 1, the special error mode is used. In the special error mode, the status register indicates if multiple key closures have occurred.

The large number of control words make programming the keyboard interface appear complex. Before anything is programmed, the clock divider rate must be determined. In the circuit illustrated in Figure 11–34, we use a 2.0 MHz clock input signal. To program the prescaler to generate a 100 KHz internal rate, we program PPPPP of the clock control word with a 20 or 10100_2.

The next step involves programming the keyboard type. In the example keyboard of Figure 11–36, we have an encoded keyboard. Notice that the circuit includes an external decoder that converts the encoded data from the SL pins into decoded column selection signals. We are free in this example to choose either 2-key lockout or N-key rollover, but most applications use 2-key lockout.

Finally, we program the operation of the FIFO. Once the FIFO is programmed, it never needs to be reprogrammed unless we need to read prior keyboard codes. Each time a key is typed, the data is stored in the FIFO; if it is read from the FIFO before the FIFO is full (8 characters), then the data from the FIFO follows the same order as the typed data. Example 11–20 initializes the 8279 to control the keyboard illustrated in Figure 11–36.

EXAMPLE 11–20

```
                                  ;Initialization dialog for the keyboard interface
                                  ;of Figure 11-36.
                                  ;
0000 B0 34                        MOV    AL,00110100B        ;program clock
0002 E6 11                        OUT    11H,AL

0004 B0 00                        MOV    AL,0
0006 E6 11                        OUT    11H,AL              ;program mode

0008 B0 50                        MOV    AL,01010000B
000A E6 11                        OUT    11H,AL              ;program FIFO
```

Once the 8279 is initialized, a procedure is required to read data from the keyboard. We determine if a character is typed in the keyboard by looking at the FIFO status register. Whenever the control port is addressed by the IN instruction, the contents of the FIFO status word is copied into the AL register. Figure 11–35 shows the contents of the FIFO status register and defines the purpose of each status bit.

Example 11–21 first tests the FIFO status register to see if it contains any data. If NNN = 000, the FIFO is empty. Upon determining that the FIFO is not empty, the procedure inputs data to AL and returns with the keyboard code in AL.

EXAMPLE 11-21

```
                                  ;Procedure that reads data from the FIFO and
                                  ;returns it in AL.
                                  ;
= 0007                            MASKS  EQU    7

0000                              READ   PROC   NEAR

0000 E4 11                               IN     AL,11H       ;read status
0002 A8 07                               TEST   AL,MASKS     ;test NNN
0004 74 FA                               JZ     READ         ;if NNN = 0

0006 E4 10                               IN     AL,10H       ;read FIFO data
0008 C3                                  RET

0009                              READ   ENDP
```

The data found in AL upon returning from the subroutine contains raw data from the keyboard. Figure 11–36 shows the format of this data for both the scanned and strobed modes of operation. The scanned code is returned from our keyboard interface and is converted to ASCII

FIGURE 11–35 The 8279-5 FIFO status register.

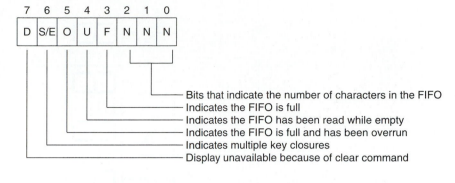

FIGURE 11–36 The (a) scanned keyboard code and (b) strobed keyboard code for the 8279-5 FIFO.

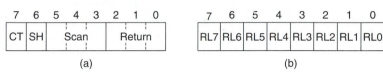

code by using the XLAT instruction with an ASCII-code lookup table. The scanned code is returned with the row and column number occupying the rightmost six bits.

The SH bit shows the state of the shift pin and the CT bit shows the state of the control pin. In the strobed mode, the contents of the eight RL inputs appear as they are sampled by placing a logic 1 on the strobe input pin to the 8279.

Six-Digit Display Interface

Figure 11–37 depicts the 8279 connected to the 80C188 microprocessor and a six-digit numeric display. This interface uses a PAL16L8 (program not shown) to decode the 8279 at I/O ports 20H (data) and 21H (control/status). The segment data is supplied to the displays through the OUTA and OUTB pins of the 8279. These bits are buffered by a segment driver (ULN2003A) to drive the segment inputs to the display.

A 74ALS138 3-to-8 line decoder enables the anode switches of each display position. The SL_2–SL_0 pins supply the decoder with the encoded display position from the 8279. Notice that the left-hand display is at position 0101 and the right-hand display is at position 0000. These are the addresses of the display positions as indicated in control words for the 8279.

It is necessary to choose resistor values that allow 60 mA of current flow per segment. In this circuit we use 47 Ω resisters. If we allow 60 mA of segment current, then the average segment current is 10 mA, or one-sixth of 60 mA because current only flows for one-sixth of the time through a segment. The anode switches must supply the current for all seven segments plus the decimal point. Here the total anode current is 8×60 mA or 480 mA.

Example 11–22 lists the initialization dialog for programming the 8279 to function with this six-digit display. This software programs the display and clears the display RAM.

EXAMPLE 11–22

```
                              ;Initialization dialog for the 6-digit display of
                              ;Figure 11-37.
                              ;
0000 B0 34              MOV     AL,00110100B        ;program clock
0002 E6 21              OUT     21H,AL

0004 B0 00              MOV     AL,0                ;program mode set
0006 E6 21              OUT     21H,AL

0008 B0 C1              MOV     AL,11000001B        ;clear display
000A E6 21              OUT     21H,AL
```

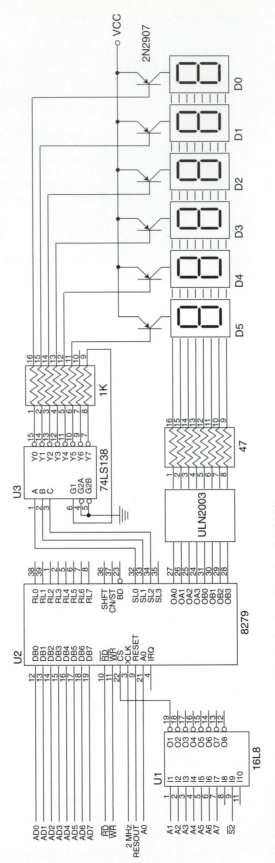

FIGURE 11–37 A six-digit numeric display interfaced to the 8279.

Example 11–23 lists a procedure for displaying information on the displays. Data is transferred to the procedure through the AX register. AH contains the seven-segment display code and AL contains the address of the displayed digit.

EXAMPLE 11–23

```
                        ;Procedure that displays AH on the display
                        ;position addressed by AL.
                        ;
= 0080                  MASKS   EQU     80H

0000                    DISP    PROC    NEAR

0000 50                         PUSH    AX          ;save data
0001 0C 80                      OR      AL,MASKS    ;select digit
0003 E6 21                      OUT     21H,AL

0005 8A C4                      MOV     AL,AH       ;display data
0007 E6 20                      OUT     20H,AL
0009 58                         POP     AX          ;restore data
000A C3                         RET

000B                    DISP    ENDP
```

11–6 PROGRAMMABLE TIMERS

This section describes the 8254 programmable interval timer and the timers built into the 80186/80188 and 80386EX microprocessors. The 8254 programmable interval timer consists of three independent 16-bit programmable counters (**timers**). Each counter is capable of counting in binary or binary-coded decimal (BCD). The maximum allowable input frequency to any counter is 10 MHz. This device is useful wherever the microprocessor must control real-time events. Some examples of usage include: real time clock, events counter, and motor speed and direction control.

This timer also appears in the personal computer decoded at ports 40H–43H to (1) generate a basic timer interrupt that occurs at approximately 18.2 Hz, (2) cause the DRAM memory system to be refreshed, and (3) provide a timing source to the internal speaker and other devices. The timer in the personal computer is an 8253 instead of an 8254.

8254 Functional Description

Figure 11–38 shows the pin-out of the 8254, which is a higher-speed version of the 8253, and a diagram of one of the three counters. Each timer contains a CLK input, a gate input, and an output (OUT) connection. The CLK input provides the basic operating frequency to the timer, the gate pin controls the timer in some modes, and the OUT pin is where we obtain the output of the timer.

The signals that connect to the microprocessor are the data bus pins (D_7–D_0), $\overline{RD}$, $\overline{WR}$, $\overline{CS}$, and address inputs A_1 and A_0. The address inputs are present to select any of the four internal registers used for programming, reading, or writing to a counter. The personal computer contains an 8253 timer or its equivalent decoded at I/O ports 40H–43H. Timer zero is programmed to generate an 18.2 Hz signal that interrupts the microprocessor at interrupt vector 8 for a clock tick. The tick is often used to time programs and events. Timer 1 is programmed for a 15 µs output that is used on the PC/XT personal computer to request a DMA action used to refresh the dynamic RAM. Timer 2 is programmed to generate tone on the personal computer speaker.

FIGURE 11–38 The 8254
programmable interval
timer. (a) Internal structure.
(b) Pin-out. (Courtesy of Intel
Corporation)

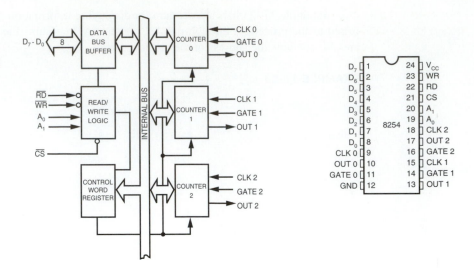

Pin Definitions.

A_1, A_0	The **address inputs** select one of four internal registers within the 8254. See Table 11–7 for the function of the A_1 and A_0 address bits.
CLK	The **clock** input is the timing source for each of the internal counters. This input is often connected to the CLKOUT signal from the microprocessor or an external timing source.
$\overline{CS}$	**Chip select** enables the 8254 for programming and reading or writing a counter.
G	The **gate** input controls the operation of the counter in some modes of operation.
GND	**Ground** connects to the system ground bus.
OUT	A **counter output** is where the waveform generated by the timer is available.
$\overline{RD}$	**Read** causes data to be read from the 8254 and often connects to the $\overline{IORC}$ signal, found on some microprocessors or to the $\overline{RD}$ pin of the 80186/80188 or 80386EX.
VCC	**Power** connects to the +5.0 V power supply.
$\overline{WR}$	**Write** causes data to be written to the 8254 and often connects to the write strobe ($\overline{WR}$).

Programming the 8254

Each counter is individually programmed by writing a control word followed by the initial count. Figure 11–39 shows the program control word structure of the 8254. The **control word** allows the programmer to select the counter, mode of operation, and type of operation (read/write). The

TABLE 11–7 Address
selection inputs to the 8254

A_1	A_0	Function
0	0	Counter 0
0	1	Counter 1
1	0	Counter 2
1	1	Control word

FIGURE 11–39 The control word for the 8254-2 timer.

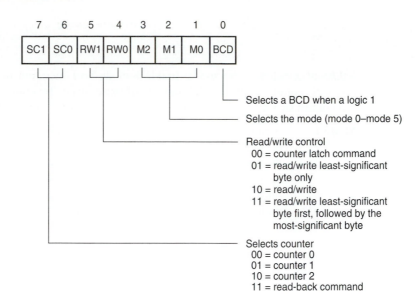

control word also selects either a binary or BCD count. Each counter may be programmed with a count of 1 to FFFFH. A count of 0 is equal to FFFFH + 1 (65,536) or 10,000 in BCD. The minimum count of 1 applies to all modes of operation except modes 2 and 3, which have a minimum count of 2. Timer 0 is used in the personal computer with a divide by count of 64K (FFFFH) to generate the 18.2 Hz (18.196 Hz) interrupt clock tick. Timer 0 has a clock input frequency of 4.77 MHz ÷ 4 or 1.1925 MHz.

The control word uses the BCD bit to select a BCD count (BCD = 1) or a binary count (BCD = 0). The M2, M1, and M0 bits select one of the six different modes of operation (000–101) for the counter. The RW1 and RW0 bits determine how the data is read from or written to the counter. The SC1 and SC0 bits select a counter or the special read-back mode of operation discussed later in this section.

Each counter has a program control word used to select the way the counter operates. If two bytes are programmed into a counter, then the first byte (LSB) will stop the count, and the second byte (MSB) will start the counter with the new count. The order of programming is important for each counter, but programming of different counters may be interleaved for better control. For example, the control word may be sent to each counter before the counts for individual programming. Example 11–24 shows a few ways to program counters 1 and 2. The first method programs both control words, then the LSB of the count for each counter, which stops them from counting. Finally the MSB portion of the count is programmed, starting both counters with the new count. The second example shows one counter programmed before the other.

EXAMPLE 11–24

```
PROGRAM CONTROL WORD 1        ;set up counter 1
PROGRAM CONTROL WORD 2        ;set up counter 2
PROGRAM LSB 1                 ;stop counter 1 and program LSB
PROGRAM LSB 2                 ;stop counter 2 and program LSB
PROGRAM MSB 1                 ;program MSB of counter 1 and start it
PROGRAM MSB 2                 ;program MSB of counter 2 and start it

        or

PROGRAM CONTROL WORD 1        ;set up counter 1
PROGRAM LSB 1                 ;stop counter 1 and program LSB
PROGRAM MSB 1                 ;program MSB of counter 1 and start it
```

```
PROGRAM CONTROL WORD 2          ;set up counter 2
PROGRAM LSB 2                   ;stop counter 2 and program LSB
PROGRAM MSB 2                   ;program MSB of counter 2 and start it
```

Modes of Operation.   Six modes (mode 0–mode 5) of operation are available to each of the 8254 counters. Figure 11–40 shows how each of these modes functions with the CLK input, the gate (G) control signal, and OUT signal. A description of each mode follows.

Mode 1   Allows the 8254 counter to be used as an events counter. In this mode, the output becomes a logic 0 when the control word is written and remains the until N plus the number of programmed counts. For example, if a count of 5 is programmed, the output will remain a logic 0 for 6 counts beginning with N. The gate (G) input must be a logic 1 to allow the counter to count. If G becomes a logic 0 in the middle of the count, the counter will stop until G again becomes a logic 1.

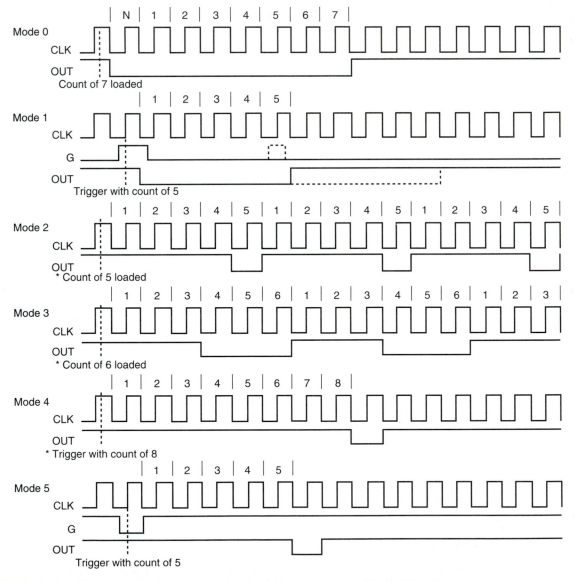

FIGURE 11–40   The six modes of operation for the 8454-2 programmable interval timer.
Note: The G input stops the count when 0 in modes 2, 3, and 4.

Mode 1   Causes the counter to function as a retriggerable monostable multivibrator (one-shot). In this mode, the G input triggers the counter so that it develops a pulse at the OUT connection that becomes a logic 0 for the duration of the count. If the count is 10, then the OUT connection goes low for 10 clocking periods when triggered. If the G input occurs within the duration of the output pulse, the counter is again reloaded with the count and the OUT connection continues for the total length of the count.

Mode 2   Allows the counter to generate a series of continuous pulses that are one clock pulse in width. The separation between pulses is determined by the count. For example, for a count of 10, the output is a logic 1 for nine clock periods and low for one clock period. This cycle is repeated until the counter is programmed with a new count or until the G pin is placed at a logic 0 level. The G input must be a logic 1 for this mode to generate a continuous series of pulses.

Mode 3   Generates a continuous square-wave at the OUT connection provided the G pin is a logic 1. If the count is even, the output is high for one-half the count and low for one-half the count. If the count is odd, the output is high for one clock period longer than it is low. For example, if the counter is programmed for a count of 5, the output is high for three clocks and low for two.

Mode 4   Allows the counter to produce a single pulse at the output. If the count is programmed as a 10, the output is high for 10 clocking periods and then low for one clocking period. The cycle does not begin until the counter is loaded with its complete count. This mode operates as a software triggered one-shot. As with modes 2 and 3, this mode also uses the G input to enable the counter. The G input must be a logic 1 for the counter to operate for these three modes.

Mode 5   A hardware triggered one-shot that functions as mode 4 except it is started by a trigger pulse on the G pin instead of by software. This mode is also similar to mode 1 because it is retriggerable.

Generating a Wave-Form with the 8254.   Figure 11–41 shows an 8254 connected to function at I/O ports 0700H, 0701H, 0702H, and 0703H of an 8 MHz, 80188 microprocessor. The addresses are decoded using a PAL16L8 that also generates a write strobe signal for the 8254, which is connected to the low-order data bus connections. The program for the PAL is not illustrated here because it is basically the same as many of the prior examples.

Example 11–25 generates a 100 KHz square-wave at OUT_0 and a 200 KHz continuous pulses at OUT_1. We use mode 3 for counter 0 and mode 2 for counter 1. The count programmed into counter 0 is 80 and the count for counter 1 is 40. These counts generate the desired output frequencies with an 8 MHz input clock.

EXAMPLE 11–25

```
                        ;Procedure that programs the 8254 timer to function
                        ;as illustrated in Figure 11-41.
                        ;
0000                    TIME    PROC    NEAR

0000 50                         PUSH    AX              ;save registers
0001 52                         PUSH    DX

0002 BA 0703                    MOV     DX,703H         ;address control word
0005 B0 36                      MOV     AL,00110110B    ;program counter 0
0007 EE                         OUT     DX,AL           ;for mode 3

0008 B0 74                      MOV     AL,01110100B    ;program counter 1
000A EE                         OUT     DX,AL           ;for mode 2
```

```
000B BA 0700              MOV    DX,700H      ;address counter 0
000E B0 50                MOV    AL,80        ;load count of 80
0010 EE                   OUT    DX,AL
0011 32 C0                XOR    AL,AL
0013 EE                   OUT    DX,AL

0014 BA 0701              MOV    DX,701H      ;address counter 1
0017 B0 28                MOV    AL,40        ;load count of 40
0019 EE                   OUT    DX,AL
001A 32 C0                XOR    AL,AL
001C EE                   OUT    DX,AL

001D 5A                   POP    DX           ;restore registers
001E 58                   POP    AX
001F C3                   RET

0020              TIME    ENDP
```

Reading a Counter.   Each counter has an internal latch that is read with the read counter port operation. These latches will normally follow the count. If the contents of the counter are needed at a particular time, then the latch can remember the count by programming the counter latch control word (see Figure 11–42), which causes the contents of the counter to be held in a latch until it is read. Whenever a read from the latch or the counter is programmed, the latch tracks the contents of the counter.

When it is necessary for the contents of more than one counter to be read at the same time, we use the read-back control word illustrated in Figure 11–43. With the read-back control word, the $\overline{CNT}$ bit is a logic 0 to cause the counters selected by CNT0, CNT1, and CNT2 to be latched. If the status register is to be latched, then the $\overline{ST}$ bit is placed at a logic 0. Figure 11–44 shows the status register, which shows the state of the output pin, whether the counter is at its null state (0), and how the counter is programmed.

FIGURE 11–41 The 8254 interfaced to the 80188 to produce two different waveforms.

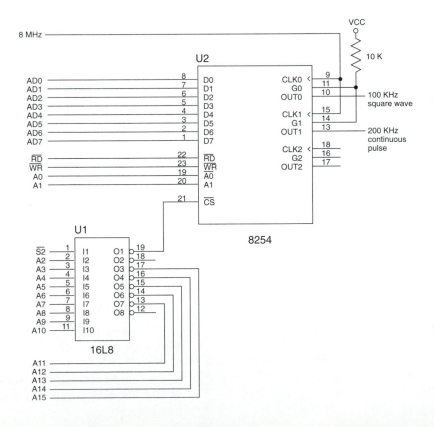

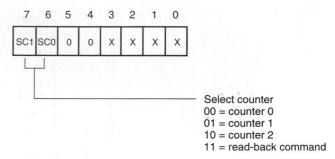

FIGURE 11–42 The 8254-2 counter latch control word.

FIGURE 11–43 The 8254-2 read-back control word.

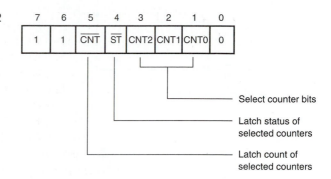

FIGURE 11–44 The 8254-2 status register.

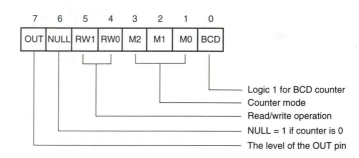

DC Motor Speed and Direction Control

One application of the 8254 timer is as a motor speed controller for a DC motor. Figure 11–45 shows the schematic diagram of the motor and its associated driver circuitry. It also illustrates the interconnection of the 8254, a flip-flop, and the motor and its driver.

The operation of the motor driver circuitry is fairly straightforward. If the Q output of the 74ALS112 is a logic 1, the base Q2 is pulled up to +12 V through the base pull-up resistor and the base of Q2 is open-circuited. This means that Q1 is off and Q2 is on, with ground applied to the positive lead of the motor. The bases of both Q3 and Q4 are pulled low to ground through the inverters. This causes Q3 to conduct or turn on and Q4 to turn off applying ground to the negative lead of the motor. The logic 1 at the Q output of the flip-flop therefore connects +12 V to the positive lead of the motor and ground to the negative lead. This connection causes the motor to spin in its forward direction. If the state of the Q output of the flip-flop becomes a logic 0, then the conditions of the transistors are reversed and +12 V is attached to the negative lead of the motor with ground attached to the positive lead. This causes the motor to spin in the reverse direction.

If the output of the flip-flop is alternated between a logic 1 and 0, the motor spins in either direction at various speeds. If the duty cycle of the Q output is 50 percent the motor will not spin

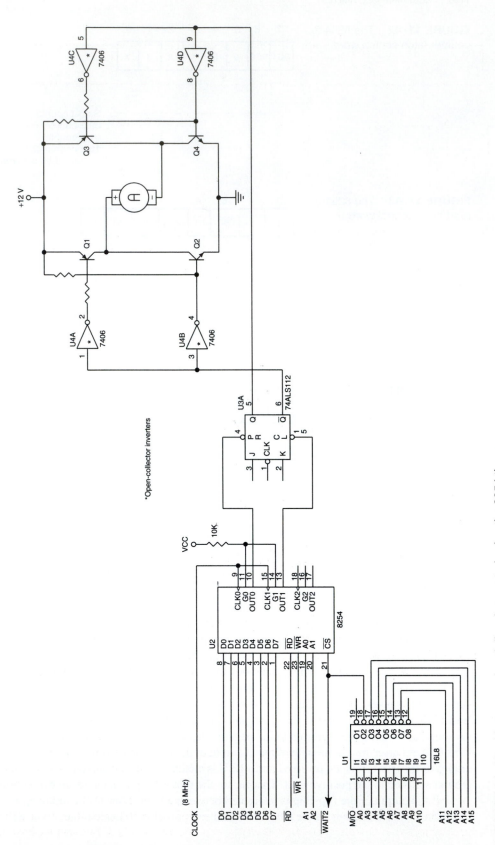

FIGURE 11–45 Motor speed and direction control using the 8254 timer.

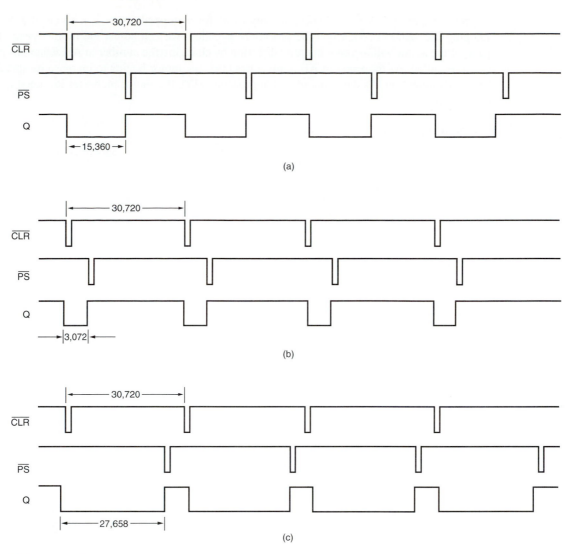

FIGURE 11–46   Timing for the motor speed and direction control circuit of Figure 11–45. (a) High-speed rotation. (b) High-speed rotation in the reverse direction. (c) High-speed rotation in the forward direction.

at all and exhibits some holding torque because current flows through it. Figure 11–46 shows some timing diagrams and their effects on the speed and direction of the motor. Notice how each counter generates pulses at different positions to vary the duty cycle at the Q output of the flip-flop. This output is also called *pulse width modulation.*

To generate these wave-forms, counters 0 and 1 are both programmed to divide the input clock (PCLK) by 30,720. We change the duty cycle of Q by changing the point at which counter 1 is started in relationship to counter 0. This changes the direction and speed of the motor. But why divide the 8 MHz clock by 30,720? The divide rate of 30,720 is divisible by 256 so we can develop a short program that allows 256 different speeds. This also produces a basic operating frequency for the motor of about 260 Hz, which is low enough in frequency to power the motor. It is important to keep this operating frequency below 1,000 Hz but above 60 Hz.

Example 11–26 is a procedure that controls the speed and direction of the motor. The speed is controlled by the value of AH when this procedure is called. Because we have an 8-bit number to represent speed, a 50 percent duty cycle for a stopped motor is a count of 128. By changing the value in AH, when the procedure is called, we can adjust the motor speed. The speed of the motor will increase in either direction by changing the number in AH when this procedure is called. As the value in AH approaches 00H, the motor begins to increase its speed in the reverse direction. As the value of AH approaches FFH, the motor increases its speed in the forward direction.

EXAMPLE 11–26

```
                        ;Procedure that controls the speed and direction
                        ;of the motor in Figure 11-45.
                        ;
                        ;When this procedure is called, the contents of
                        ;AH determine the speed and direction of the
                        ;motor where AH is between 00H and FFH.
                        ;
= 0703                  CNTR    EQU     703H
= 0700                  CNT0    EQU     700H
= 0701                  CNT0    EQU     701H
= 7800                  COUNT   EQU     30720

0000                    SPEED   PROC    NEAR

0000 50                         PUSH    AX              ;save registers
0001 51                         PUSH    DX
0002 53                         PUSH    BX

0003 8A CD                      MOV     BL,AL           ;calculate count
0005 B8 0078                    MOV     AX,120
0008 F6 E3                      MUL     BL
000A 8B D8                      MOV     BX,AX
000C B8 7800                    MOV     AX,COUNT
000F 2B C3                      SUB     AX,BX
0011 8B D8                      MOV     BX,AX

0013 BA 0703                    MOV     DX,CNTR         ;program control words
0016 B0 34                      MOV     AL,00110100B
0018 EE                         OUT     DX,AL
0019 B0 74                      MOV     AL,01110100B
001B EE                         OUT     DX,AL

001C BA 0701                    MOV     DX,CNT1         ;program counter 1
001F B8 7800                    MOV     AX,COUNT        ;to generate a clear
0022 EE                         OUT     DX,AL
0023 8A C4                      MOV     AL,AH
0025 EE                         OUT     DX,AL
0026            SPE:
0026 EC                         IN      AL,DX           ;wait for counter 1
0027 86 C4                      XCHG    AL,AH           ;to reach calculated
0029 EC                         IN      AL,DX           ;count
002A 86 C4                      XCHG    AL,AH
002C 3B C3                      CMP     AX,BX
002E 72 F6                      JB      SPE

0030 BA 0700                    MOV     DX,CNT0         ;program counter 0
0033 B8 7800                    MOV     AX,COUNT        ;to generate a set
0036 EE                         OUT     DX,AL
0037 8A C4                      MOV     AL,AH
0039 EE                         OUT     DX,AL

003A 5B                         POP     BX              ;restore registers
003B 5A                         POP     DX
003C 58                         POP     AX
003D C3                         RET

003E                    SPEED   ENDP
```

The procedure adjusts the wave-form at Q by first calculating the count that counter 0 is to start in relationship to counter 1. This is accomplished by multiplying AH by 120 and then subtracting it from 30,720. This is required because the counters are down-counters that count from the programmed count to 0, before restarting. Next, counter 1 is programmed with a count of 30,720 and started to generate the clear wave-form for the flip-flop. After counter 1 is started, it is read and compared with the calculated count. Once it reaches this count, counter 0 is started with a count of 30,720. From this point forward, both counters continue generating the clear and set wave-forms until the procedure is again called to adjust the speed and direction of the motor.

The Timer Structure of the 80186/80188

There are three timers located within the 80186/80188 microprocessors. All versions contain the same timer configuration (refer to Figure 11–47), although the I/O port addresses vary in different versions. Each timer is a 16-bit programmable modulus counter. Timer 2 is clocked from the output of a divide-by-4 counter, which internally divides the system CLLOUT signal by a factor of 4. For example, in a system that uses a 32-MHz crystal, the CLKOUT and microprocessor operating frequency is 16 MHz. The signal presented to timer 2 is 4 MHz in this system. No output signal from timer 2 is available on the microprocessor; its output is used internally to cause a DMA action, an interrupt, or as a clocking source for the other two timers.

Timers 0 and 1 can be clocked from timer 2, CLKOUT/4, or from the external timer input pin. In Figure 11–47, the switches are programmed via software, and bit positions in the timer control register for each timer. The external timer input pin can provide a clock to the timer or it can be used to reset the count or to enable the count. The output of timers 0 and 1 can be used to cause an interrupt and are also available on the timer output pins.

Timer Control. The timers are controlled by a group of 4 registers for each timer. Figure 11–48 shows the structure of the timer control register and lists the I/O port addresses of all versions of the 80186/80188 timers. Each timer contains a control register, a count register, and one or two compare registers. The control register programs the operation of the timer. The count register holds the current count. The compare registers contain the maximum count value used to program the counter sequence.

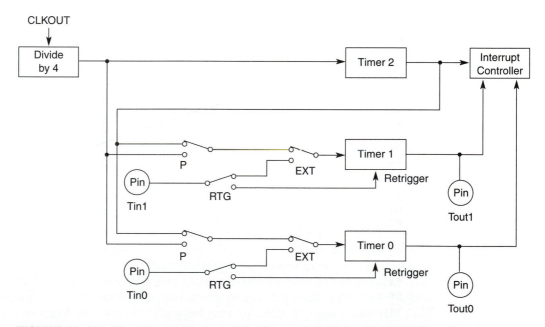

FIGURE 11–47 The internal structure of the timer unit within the 80186/80188 microprocessors.

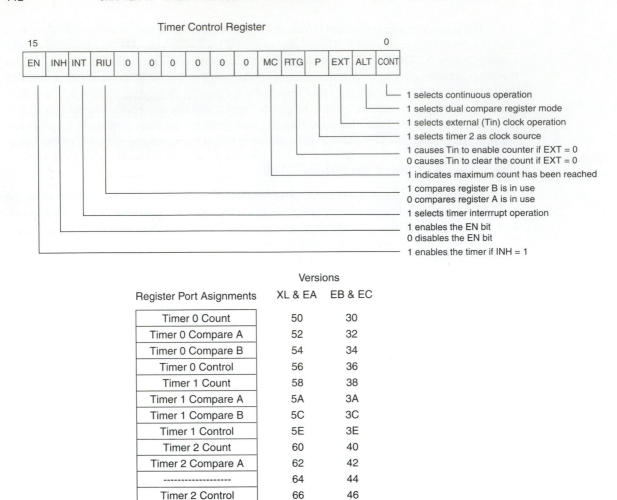

FIGURE 11–48   The 80186/80188 timer control register and port assignments.

The timer count register contains a 16-bit number that is incremented by the selected input clock. For example, if the CLKOUT/4 is the clocking source at 4 MHz, then the timer counts at a 4 MHz rate. The compare register determines the counter reset and begins counting from zero. For example, if single compare register operation is selected (compare register A) and it is programmed with an FFH (255 decimal), the counter will count from zero until it reaches FFH, where it is cleared to zero on the next input clock pulse. This counter counts from 0000H to 00FFH before resetting and is a modulus or divide by 100H counter. If the Tout pin is monitored, the frequency will be 4 MHz/100H or 15,625 Hz.

Timers 0 and 1 can operate in the alternate compare register mode (ALT = 1). In this mode, the Tout pin is a logic 1 for the duration of the timer dictated by compare register A and a logic 0 for the duration of compare register B. This allows many possible output wave-forms. If a single compare register (A) is used, the output is a logic 1 for the duration of the count in compare register A and then becomes a logic 0 for one count. Figure 11–49 shows the relationship of the clock input to the various outputs in the timer unit.

Suppose that the microprocessor system needs a wave-form that is a logic 1 for 100 µs and a logic 0 for 500 µs. Timer 0 is used to generate this wave-form. Because the input clock to timer 0 is 4 MHz (assuming a 16-MHz operating frequency), the compare A register is programmed for

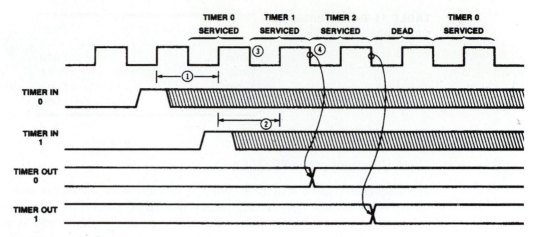

1. Timer in 0 resolution time
2. Timer in 1 resolution time
3. Modified count value written into 80186 timer 0 count register
4. Modified count value written into 80186 time

FIGURE 11–49 Timing for the 80186/80188 timers. (Courtesy of Intel Corporation)

a count of 400 and the compare B register is programmed for a count of 20,000. Example 11–27 programs timer 0 to produce this result. When programming a timer, use the following steps:

1. Clear the count register.
2. Program the compare registers.
3. Program the timer control register.

EXAMPLE 11–27

```
            ;Procedure that generates an output that is a logic 1 for 100 µs
            ;and a logic 0 for 500 µs. (for the 80186/80188 versions XL/EA)
            ;
= FF50      T0_COUNT  EQU 0FF50H          ;port assignments
= FF52      T0_CMPA   EQU 0FF52H
= FF54      T0_CMPB   EQU 0FF54H
= FF56      T0_CONT   EQU 0FF56H


0010        TIMER  PROC  NEAR    USES DX AX

0012 BA FF50       MOV   DX,T0_COUNT    ;address timer 0 count
0015 B8 0000       MOV   AX,0           ;clear to 0000H
0018 EE            OUT   DX,AL

0019 BA FF52       MOV   DX,T0_CMPA     ;address timer 0 compare A
001C B8 0190       MOV   AX,400         ;compare A = 400
001F EE            OUT   DX,AL

0020 BA FF54       MOV   DX,T0_CMPB     ;address timer 0 compare B
0023 B8 4E20       MOV   AX,20000       ;compare B = 20000
0026 EE            OUT   DX,AL

0027 BA FF56       MOV   DX,T0_CONT     ;address timer 0 control
002A B8 C003       MOV   AX,0C003H      ;continuous, alternate
002D EE            OUT   DX,AL

                   RET

0031        TIMER  ENDP
```

TABLE 11–8   The 80386EX
timer ports

Function	Port
Timer 0	F040
Timer 1	F041
Timer 2	F042
Control register	F043

Timer Structure of the 80386EX

The timer control unit within the 80386EX is identical to the 8254/8254 covered earlier in this section of the chapter. The only difference is that the I/O port assignments are in the peripheral control block of the 80386EX and are not defined by an external decoder as with the 8254 we saw earlier. Table 11–8 lists the port assignments in the 80386EX assuming that they have not been reassigned by changing the peripheral control block location.

The timer clock input can be driven for the CLK pin on the microprocessor for each timer or from an internal prescaler. The prescaler programming is discussed in Chapter 15 with the 80386EX embedded controller.

11–7   PROGRAMMABLE COMMUNICATIONS INTERFACE

This section explains the operation of the 16550 programmable communications interface and the operation of the serial ports available on some versions of the 80186/80188. The first part explains the 16550 and the last explains the 80186/80188 communications interface.

The National Semiconductor Corporation's PC16550D is a programmable communications interface designed to connect to virtually any type of serial interface. The 16550 is a universal asynchronous receiver/transmitter (UART) that is fully compatible with the Intel microprocessors. The 16550 is capable of operating at 0–1.5 M Baud. Baud rate is the number of bits transferred per second, including start, stop, data, and parity. The 16550 also includes a programmable Baud rate generator and separate FIFOs for input and output data to ease the load on the microprocessor. Each FIFO contains 16 bytes of storage. This is the most common communications interface found in modern microprocessor-based equipment including the personal computer and many modems.

Asynchronous Serial Data

Asynchronous serial data information is transmitted and received without a clock or timing signal. Figure 11–50 illustrates two frames of asynchronous serial data. Each frame contains a start bit, seven data bits, parity, and one stop bit. In this figure, a frame, which contains one ASCII character, has 10 bits. Most dial-up communications systems, such as CompuServe, Prodigy, and America Online use 10 bits for asynchronous serial data with even parity. Most

FIGURE 11–50   Asynchronous serial data.

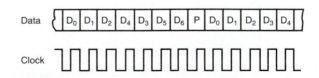

Internet and bulletin board services also use 10 bits, but they normally do not use parity. Instead, 8 data bits are transferred replacing parity with a data bit. This makes byte transfers of non-ASCII data much easier to accomplish.

16550 Functional Description

Figure 11–51 illustrates the pin-out of the 16550 UART. This device is available as a 40-pin DIP **(dual in-line package)** or as a 44-pin PLCC **(plastic lead-less chip carrier).** Two completely separate sections are responsible for data communications: the receiver and the transmitter. Because each of these sections are independent of each other, the 16550 is able to function in simplex, half-duplex, or full-duplex modes. One of the main features of the 16550 is its internal receiver and transmitter FIFO (first-in, first-out) memories. Because each is 16 bytes deep, the UART only requires attention from the microprocessor after receiving 16 bytes of data. It also holds 16 bytes before the microprocessor must wait for the transmitter. The FIFO makes this UART ideal when interfacing to high-speed systems because less time is required to service it.

An example of a **simplex** system is one where the transmitter or receiver is used by itself such as in an FM **(frequency modulation)** radio station. An example of a **half-duplex** system is a CB **(citizens band)** radio where we transmit and receive, but not both at the same time. The **full-duplex** system allows transmission and reception in both directions simultaneously. An example of a full-duplex system is the telephone.

The 16550 can control a **modem** (modulator/demodulator), which is a device that converts TTL levels of serial data into audio tones that can pass through the telephone system. Five pins on the 16650 are devoted to modem control: $\overline{DSR}$ (data set ready), $\overline{DTR}$ (data terminal ready), $\overline{CTS}$ (clear-to-send), $\overline{RTS}$ (request-to-send), $\overline{RI}$ (ring indicator), and $\overline{DCD}$ (data carrier detect). The modem is referred to as the **data set** and the 16550 is referred to as the **data terminal.**

16550 Pin Functions.

A_0, A_1, A_2	The address inputs are used to select an internal register for programming and also data transfer. Refer to Table 11–9 for a list of each combination of the address inputs and the registers selected.
$\overline{ADS}$	The **address strobe** input is used to latch the address lines and chip select lines. If not needed (as in the Intel system), connect this pin to ground. The $\overline{ADS}$ pin is designed for use with Motorola microprocessors.

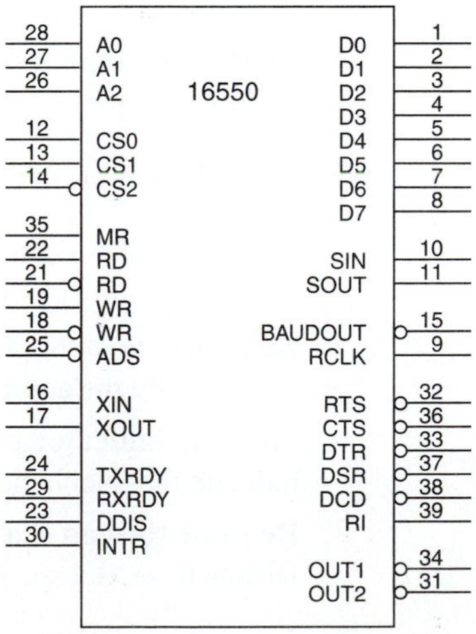

FIGURE 11–51 The pin-out of the 16550 UART.

TABLE 11–9 The registers selected by A_0, A_1, and A_2

A_2	A_1	A_0	Register
0	0	0	Receiver buffer (read) and transmitter holding (write)
0	0	1	Interrupt enable
0	1	0	Interrupt identification (read) and FIFO control (write)
0	1	1	Line control
1	0	0	Modem control
1	0	1	Line status
1	1	0	Modem status
1	1	1	Scratch

$\overline{\text{BAUDOUT}}$	The **Baud out** pin is where the clock signal generated by the Baud rate generator from the transmitter section is made available. It is most often connected to the RCLK input to generate a receiver clock that is equal to the transmitter clock.
CS0, CS1, $\overline{\text{CS2}}$	The **chip select** inputs must all be active to enable the 16550 UART.
$\overline{\text{CTS}}$	The **clear to send** (if low) indicates that the modem or data set is ready to exchange information. This pin is often used in a half-duplex system to turn the line around.
D_7–D_0	The **data bus** pins are connected to the microprocessor data bus.
$\overline{\text{DCD}}$	The **data carrier detect** input is used by the modem to signal the 16550 that a carrier is present.
DDIS	The **disable driver** output becomes a logic 0 to indicate that the microprocessor is reading data from the UART. DDIS can be used to change the direction of data flow through a buffer.
$\overline{\text{DSR}}$	**Data set ready** is an input to the 16550 that indicates the modem or data set is ready to operate.
$\overline{\text{DTR}}$	**Data terminal ready** is an output that indicates that the data terminal (16550) is ready to function.
INTR	I**nterrupt request** is an output to the microprocessor used to request an interrupt (INTR = 1) whenever the 16550 has a receiver error, has received data, or if the transmitter is empty.
MR	**Master reset** initializes the 16550 and should be connected to the system RESOUT signal.
$\overline{\text{OUT1}}$, $\overline{\text{OUT2}}$	These are user-defined output pins that can provide signals to a modem or any other device as needed in a system.
RCLK	**Receiver clock** is the clock input to the receiver section of the UART. This input is always 16 times the desired receiver Baud rate.
RD, $\overline{\text{RD}}$	**Read inputs** (either may be used) causes data to be read from the register specified by the address inputs to the UART.
$\overline{\text{RI}}$	The **ring indicator** input is placed at the logic 0 level by the modem to indicate the telephone is ringing.
$\overline{\text{RTS}}$	**Request-to-send** is a signal to the modem indicating that the UART wishes to send data.

SIN, SOUT	These are the serial data pin; SIN accepts serial data and SOUT transmits serial data.
$\overline{\text{RXRDY}}$	**Receiver ready** is a signal used to transfer received data via DMA techniques.
$\overline{\text{TXRDY}}$	**Transmitter ready** is a signal used to transfer transmitter data via DMA techniques.
WR, $\overline{\text{WR}}$	**Write** either connects to the microprocessor write signal to transfer commands and data to the 16550.
XIN, XOUT	These are the main clock connections. A crystal is connected across these pins to form a crystal oscillator or XIN is connected to an external timing source.

Programming the 16550

Programming the 16550 is fairly simple, although slightly more involved, compared to some of the other programmable interfaces described in this chapter. Programming is a two-part process that includes initialization dialog and operational dialog.

Initializing the 16550. Initialization dialog, which occurs after a hardware or software reset, consists of two parts: programming the line control register and the Baud rate generator. The line control register selects the number of data bits, number of stop bits, and parity (whether it's even or odd or if parity is sent as a one or a zero). The Baud rate generator is programmed with a divisor that determines the Baud rate of the transmitter section.

Figure 11–52 illustrates the line control register. The line control register is programmed by outputting information to I/O port 011 (A_2, A_1, A_0). The rightmost two bits of the line control register select the number of transmitted data bits (5, 6, 7, or 8). The number of stop bits is

FIGURE 11–52 The contents of the 16550 line control register.

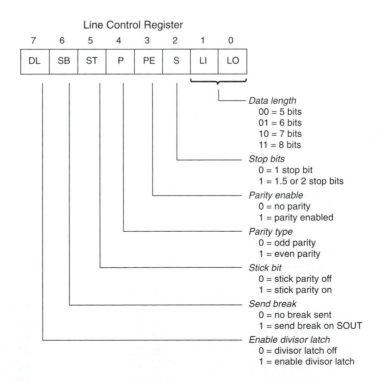

TABLE 11–10 The operation of the ST and parity bits

ST	P	PE	Function
0	0	0	No parity
0	0	1	Odd parity
0	1	0	No parity
0	1	1	Even parity
1	0	0	Undefined
1	0	1	Send/receive 1
1	1	0	Undefined
1	1	1	Send/receive 0

selected by S in the line control register. If S = 0, one stop bit is used. If S = 1, 1.5 stop bits are used for 5 data bits and 2 stop bits are used with 6, 7, or 8 data bits.

The next three bits are used together to send even or odd parity, to send no parity, or to send a 1 or a 0 in the parity bit position. To send even or odd parity, the ST (**stick**) bit must be placed at a logic 0 level and parity enable must be a logic 1. The value of the parity bit then determines even or odd parity. To send no parity (common in Internet connections), ST = 0 as well as the parity enable bit. This sends and receives data without parity. Finally if a 1 or a 0 must be sent and received in the parity bit position for all data, ST = 1, with a 1 in parity enable. To send a 1 in the parity bit position, place a 0 in the parity bit; to send a 0, place a 1 in the parity bit. (Refer to Table 11–10 for the operation of the parity and stick bits.)

The remaining bits in the line control register are used to send a break and to select programming for the Baud rate divisor. If bit position 6 of the line control register is a logic 1, a break is transmitted. As long as this bit is a 1, the break is sent from the SOUT pin. A break by definition is at least two frames of logic 0 data. The software in the system is responsible for timing the transmission of the break. To end the break, bit position 6 or the line control register is returned to a logic 0 level. The Baud rate divisor is programmable only when bit position 7 of the line control register is a logic 1.

Programming the Baud Rate.   The Baud rate generator is programmed at I/O addresses 000 and 001 (A_2, A_1, A_0). Port 000 is used to hold the least-significant part of the 16-bit divisor and port 001 the most-significant part. The value used for the divisor depends on the external clock or crystal frequency. Table 11–11 illustrates common Baud rates obtainable if a 18.432 MHz crystal is used as a timing source. It also shows the divisor values programmed into the Baud rate generator to obtain these Baud rates. The actual number programmed into the Baud rate generator causes it to produce a clock that is 16 times the desired Baud rate. For example, if 240 is programmed into the Baud rate divisor, the Baud rate is $\frac{18.432\ \text{MHz}}{16 \times 240} = 4{,}800$ Baud.

TABLE 11–11 The divisor used with the Baud rate generator for an 18.432 MHz crystal illustrating common Baud rates

Baud Rate	Divisor Value
110	10,473
300	3,840
1,200	920
2,400	480
4,800	240
9,600	120
19,200	60
38,400	30
57,600	20
115,200	10

FIGURE 11–53 The 16550 interfaced to the 80188 microprocessor at ports 00F0H–00F7H.

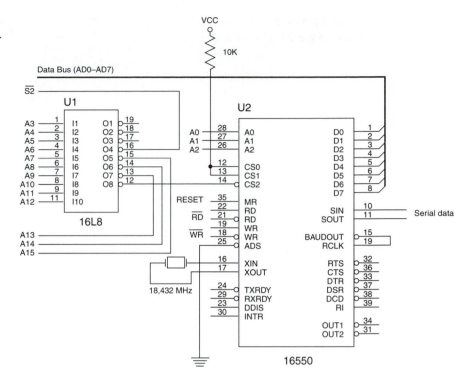

Sample Initialization. Suppose that an asynchronous system requires seven data bits, odd parity, a Baud rate of 9600, and one stop bit. Example 11-28 lists a procedure that initializes the 16550 to function in this manner. Figure 11–53 shows the interface to the 80188 microprocessor using a PAL16L8 to decode the 8-bit port addresses F0H and F7H. (The PAL program is not shown.) Here port F3H accesses the line control register, and F0H and F1H access the Baud rate divisor registers. The last part of Example 11–28 is then described with the function of the FIFO control register.

EXAMPLE 11–28

```
                        ;Initialization dialog for Figure 11-53.
                        ;Baud rate 9600, 7 data, odd parity, one stop
                        ;
= 00F3                  LINE    EQU    0F3H
= 00F0                  LSB     EQU    0F0H
= 00F1                  MSB     EQU    0F1H
= 00F2                  FIFO    EQU    0F2H

0000                    START   PROC   NEAR

0000 B0 8A                      MOV    AL,10001010B    ;enable Baud divisor
0002 E6 F3                      OUT    LINE,AL

0004 B0 78                      MOV    AL,120          ;program Baud rate
0006 E6 F0                      OUT    LSB,AL
0008 B0 00                      MOV    AL,0
000A E6 F1                      OUT    MSB,AL

000C B0 0A                      MOV    AL,00001010B    ;program 7 data, odd
000E E6 F3                      OUT    LINE,AL         ;parity, one stop

0010 B0 07                      MOV    AL,00000111B    ;enable transmitter and
0012 E6 F2                      OUT    FIFO,AL         ;and receiver

0014 C3                         RET

0015                    START   ENDP
```

FIGURE 11–54 The FIFO control register of the 16550 UART.

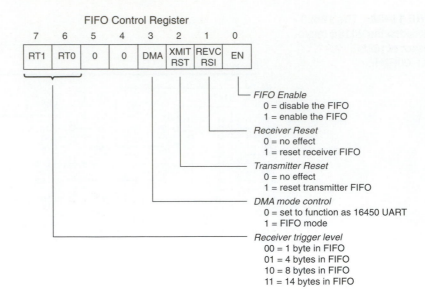

After the line control register and Baud rate divisor are programmed into the 16550, it is still not ready to function. After programming the line control register and Baud rate, we still must program the FIFO control register, which is at port F2H in the circuit of Figure 11–53. Figure 11–54 illustrates the FIFO control register for the 16550. This register enables the transmitter and receiver (bit 0 = 1) and clears the transmitter and receiver FIFOs. It also provides control for the 16550 interrupts, which are discussed in the next chapter. Notice that the last section of Example 10-24 places a 7 into the FIFO control register. This enables the transmitter and receiver, and clears both FIFOs. The 16550 is now ready to operate, but without interrupts. Interrupts are automatically disabled when the MR (master reset) input is placed at a logic 1 by the system RESET signal.

Sending Serial Data.   Before serial data can be sent or received through the 16550, we need to know the function of the line status register (see Figure 11–55). The line status register contains information about error conditions and the state of the transmitter and receiver. This register is tested before a byte is transmitted or received.

Suppose that a procedure (see [Example 11–29]) is written to transmit the contents of AH to the 16550 and out through its serial data pin (SOUT). The TH bit is polled by software to determine if the transmitter is ready to receive data. This procedure uses the circuit of Figure 11–53.

EXAMPLE 11–29

```
                         ;Procedure that transmits AH via the 16550 UART.
                         ;
= 00F5                   LSTAT   EQU     0F5H            ;line status port
= 00F0                   DATA    EQU     0F0H            ;data port

0000                     SEND    PROC    NEAR

0000 50                          PUSH    AX              ;save AX
0001 E4 F5                       IN      AL,LSTAT        ;get line status register
0003 A8 20                       TEST    AL,20H          ;test TH bit
0005 74 FA                       JZ      SEND            ;if transmitter not ready

0007 8A C4                       MOV     AL,AH           ;get data
0009 E6 F0                       OUT     DATA,AL         ;transmit data
000B 58                          POP     AX              ;restore AX
000C C3                          RET

000D                     SEND    ENDP
```

FIGURE 11–55 The contents of the line status register of the 16550 UART.

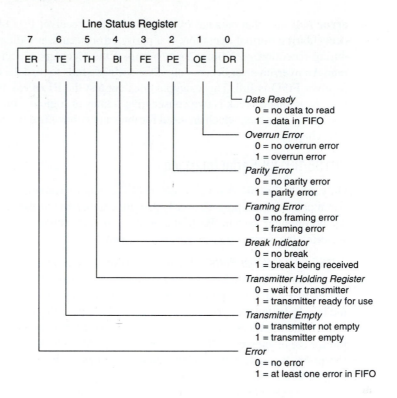

Receiving Serial Data. To read received information from the 16550, we test the DR bit of the line status register. Example 11–30 tests the DR bit to decide if the 16550 has received any data. Upon receiving data, the procedure tests for errors. If an error is detected, the procedure returns with AL equal to an ASCII '?'. If no error has occurred, then the procedure returns with AL equal to the received character.

EXAMPLE 11–30

```
                        ;Procedure that receives data from the 16550 UART
                        ;and returns it in AL.
                        ;
= 00F5          LSTAT   EQU     0F5H            ;line status port
= 00F0          DATA    EQU     0F0H            ;data port

0000            RECV    PROC    NEAR

0000 E4 F5              IN      AL,LSTAT        ;get line status register
0002 A8 01              TEST    AL,1            ;test DR bit
0004 74 FA              JZ      RECV            ;if no data in receiver

0006 A8 0E              TEST    AL,0EH          ;test all 3 error bits
0008 75 03              JNZ     ERR             ;for an error

000A E4 F0              IN      AL,DATA         ;read data from 16550
000C C3                 RET
000D            ERR:
000D B0 3F              MOV     AL,'?'          ;get question mark
000F C3                 RET

0010            REVC    ENDP
```

UART Errors. The types of errors detected by the 16550 are: parity error, framing error, and overrun error. A **parity error** indicates that the received data contains the wrong parity. A **framing error** indicates that the start and stop bits are not in their proper places. An **overrun**

error indicates that data has overrun the internal receiver FIFO buffer. These errors should not occur during normal operation. If a parity error occurs, it indicates that noise was encountered during reception. A framing error occurs if the receiver is receiving data at an incorrect Baud rate. An overrun error occurs only if the software fails to read the data from the UART before the receiver FIFO is full. This example does not test the BI (break indicator bit) for a break condition. Note that a break is two consecutive frames of logic 0's on the SIN pin of the UART. The remaining registers, which are used for interrupt control and modem control, are discussed in the next chapter.

80186/80188 Serial Interface

Only the 80186/80188 versions EB and EC contain serial interfaces. These microprocessors contain a pair of serial ports (0 and 1) that are comparable to the communications ports on the personal computer system. Each microprocessor also contains an internal Baud rate generator for selecting the desired transmit/receive frequency.

80186/80188 Serial Ports.   Figure 11–56 shows the internal structure of the serial ports found on the 80186/80188 EB and EC versions. Notice that both the transmitter and receiver are buffered. Each serial port can operate in four asynchronous modes and in one synchronous mode. Mode 0 is the synchronous mode, and modes 1 through 4 are asynchronous.

Synchronous operation uses the RXD pin for transmitted and received data and the TXD pin for the clock signal. Note that the clock signal is also provided on the TXD pin even for data reception. Because both transmit and receive pins are used for synchronous communication, the serial port can operate only in the half duplex mode.

Asynchronous modes contain a frame that includes one start bit and at least one stop bit. Modes 1, 3, and 4 function in the same manner except for the number of data bits. Mode 2 transmits and receives with 9 data bits; mode 1 uses 8 data bits; mode 4 uses 7 data bits. In all three cases if parity is enabled, one of the bit times is used for parity. For example, in mode 1 there are 8 data bits unless parity is enabled, then there are 7 data bits with parity as the eighth bit. The same applies to modes 3 and 4.

Mode 2 is a special asynchronous mode that is used in conjunction with mode 3. This special mode is used to communicate between systems using one serial link with as many serial connects as required. This is discussed later in this section.

Figure 11–57 shows the Baud rate control registers for the 80186/80188. The Baud rate generator contains two registers per channel. One is the count and the other is a compare register. The combination of the counter and compare registers function in much the same manner as the timer. The input to the Baud rate generator is either from the microprocessor clock (same frequency as clock out) or from the extern BCLK input pin. Any Baud rate is possible, but the standard Baud rates range from 1,200 to 19,200 Baud. Table 11–12 lists some common Baud rates and the value programmed into the compare register. Note that a tolerance of ±5 percent is

TABLE 11–12   Compare register values for typical Baud rates

Baud Rate	20 MHz CLKOUT		16 MHz CLKOUT		8 MHz CLKOUT	
	Modes 1–4	*Mode 0*	*Modes 1–4*	*Mode 0*	*Modes 1–4*	*Mode 0*
19,200	8081H	8410H	8067H	8340H	8033H	81A0H
9,600	8103H	8822H	80CFH	8682H	8067H	8340H
4,800	8208H	9045H	81A0H	8D04H	80CFH	8682H
2,400	8411H	A08CH	8340H	9A0AH	81A0H	8D04H
1,200	8822H	C119H	8682H	B414H	8340H	9A0AH

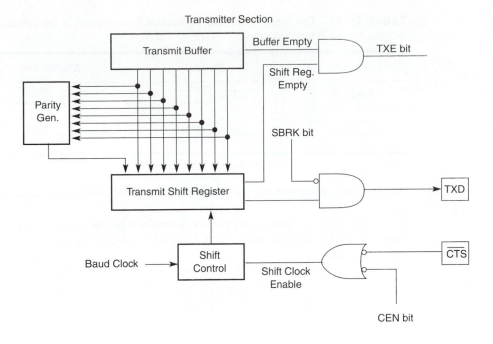

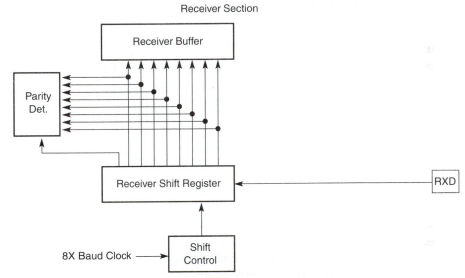

FIGURE 11–56 The internal structure of an 80186/80188 serial port.

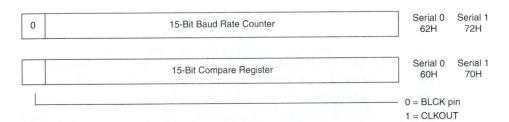

FIGURE 11–57 The Baud rate register of the 80186/80188.

TABLE 11–13   The equations used to calculate the contents of the compare register for the Baud rate clock

Baud Clock Source	Mode 0	Modes 1–4
CLKOUT	$CMP = \dfrac{CLKOUT}{Baud\ Rate} - 1$	$CMP = \dfrac{CLKOUT}{Baud\ Rate \times 8} - 1$
BCLK	$CMP = \dfrac{CLKOUT}{Baud\ Rate}$	$CMP = \dfrac{BCLK}{Baud\ Rate \times 8}$

standard when programming the Baud rate and that the Baud rate compare count is different for mode 0.

The compare register contents are determined by the equations listed in Table 11–13 for either the BCLK input or the CLKOUT signal as selected by the leftmost bit of the Baud rate compare register.

Example 11–31 shows how to program the 80186/80188 serial Baud rate generator for a Baud rate of 10,000 for mode 0 operation. Here the microprocessor crystal frequency is assumed to be 8 MHz, which means CLKOUT is 4 MHz. The value programmed into the Baud rate compare register is 818FH.

EXAMPLE 11–31

```
= FF62              S0_BCNT    EQU     0FF62H
= FF60              50_BCMP    EQU     0FF60H

0010                P_PORT PROC    NEAR   USES DX AX

0012 BA FF62                   MOV     DX,S0_BCNT      ;address Baud rate counter
0015 B8 0000                   MOV     AX,0            ;clear counter
0018 EE                        OUT     DX,AL

0019 BA FF60                   MOV     DX,S0_BCMP      ;address Baud rate compare
001C B8 818F                   MOV     AX,818FH        ;'set mode 0 rate to 10,000
001F EE                        OUT     DX,AL

                               RET

0023                P_PORT ENDP
```

Figure 11–58 illustrates the serial port register structure for the 80186/80188. There are four registers associated with each port; two are control registers and two are data registers. The transmit and receive data registers are used to convey data to and from the port, while the control and status registers are used to control and operate the port.

The control register selects the mode of operation and enables the receiver. It also selects the type of parity and the operation of the clear-to-send pin as well as the eighth bit in a mode 2 or 3 transmission. For example, suppose that a dial-up service is selected for connection to serial port 0. The service operates with even parity and 7 data bits. Mode 1 is selected because it contains ten bits, one start, one stop, seven data bits, and a parity bit. The software required to program serial port 0 for operation with this service appears in Example 11–32. Note that the clear-to-send pin is not enabled. If clear-to-send is enabled, the $\overline{CTS}$ pin must be at a logic 0 for the port to transmit data.

EXAMPLE 11–32

```
= FF64              S0_CONT    EQU      0FF64H

0010                S_PORT PROC    NEAR    USES DX AX
```

```
0012 BA FF64          MOV     DX,S0_CONT      ;address
0015 B8 0039          MOV     AX,0039H        ;10 bits
0018 EE               OUT     DX,AL
                      RET

001C          S_PORT  ENDP
```

The status register provides information about the operation of the seri...
been programmed. Error information is available as well as information about the con...
data registers. The RI bit becomes a logic 1 each time that data is received and available. The
bit becomes a logic 1 each time that the transmitter buffer is ready to accept additional data. The

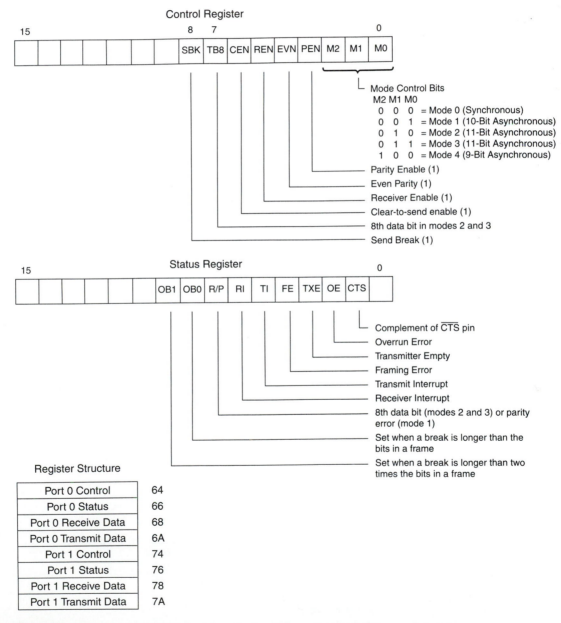

FIGURE 11–58 The register structure and contents of the control and status registers for the serial ports of the 80186 and 80188.

TXE bit becomes a logic 1 whenever the transmit buffer and transmitter shift register are empty. Example 11–33 transmits a string of serial data through serial port 0. The only parameter transferred to the procedure is the address of the data (DS:SI). Transmission continues until a carriage return (0DH)/line feed (0AH) combination occurs.

EXAMPLE 11–33

```
= FF6A              S0_TD  EQU    0FF6AH
= FF66              S0_STS EQU    0FF66H

0010                T_PORT PROC   NEAR    USES SI DX AX

                         .REPEAT
                           .REPEAT

0013 BA FF66                     MOV   DX,S0_STS  ;address serial port 0 status
0016 ED                          IN    AX,DX
0017 A8 20                       TEST  AL,32      ;test RI

                           .UNTIL !ZERO?          ;until RI = 1

001B AC                    LODSB                   ;get data
001C BA FF6A               MOV   DX,S0_TD          ;address serial port 0 transmit
001F EE                    OUT   DX,AL             ;send data

                         .UNTIL BYTE PTR [SI-2]== 0DH && BYTE PTR [SI-1]==0AH
                         RET

0030                T_PORT ENDP
```

Another procedure is required to read information from the serial port. As with the transmit procedure of Example 11–33, the receive procedure (in Example 11–34) returns received data in AL, and must poll the status register to determine if the receiver is ready. The RI bit indicates that the receiver has information in its buffer. One other difference exists in the receiver procedure: a test of the receiver errors bits is done to determine if an error occurred during reception. Parity errors are normally detected only in mode 1 operation, which is the standard for many serial communications systems.

EXAMPLE 11–34

```
= FF66              S0_STAT    EQU    0FF66H
= FF68              S0_RDAT    EQU    0FF68H

0010                R_PORT PROC   NEAR    USES DX AX

0012 BA FF66               MOV   DX,S0_STAT  ;address port 0 status
                             .REPEAT
0015 EC                        IN    AL,DX   ;get status
0016 A8 40                     TEST  AL,40H  ;test RI
                             .UNTIL !ZERO?

001A A8 94                 TEST  AL,94H       ;test for errors
                             .IF    !ZERO?
001E B0 3F                     MOV    AL,'?'
                             .ELSE
0022 BA FF68                   MOV    DX,S0_RDAT;address data register
0025 EC                        IN     AL,DX    ;get data
                             .ENDIF

                           RET

0029                R_PORT   ENDP
```

FIGURE 11–59 Mode 2 operation showing the interconnection between master and slave microprocessors.

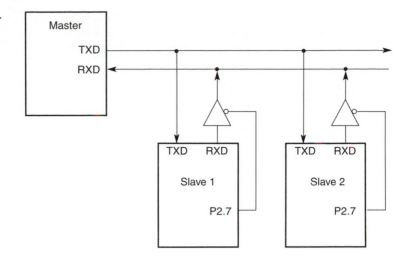

Mode 2 operation is a special mode used in conjunction with mode 3 operation for multiple drop serial interfaces similar to the Universal Serial Bus (USB) in the latest personal computer systems. This mode is often called the master/slave mode as illustrated in the simple schematic of Figure 11–59. The master addresses and communicates with one slave at a time. The master initializes communications by transmitting an address to a slave using mode 2 operation with the ninth bit set to a logic 1. All slaves receive the address, but only the one programmed to accept the address responds by switching to mode 3 operation, while all other slaves remain in mode 2 operation. Once communications between the master and the selected slave are established, data is transferred with the ninth bit equal to a logic 0, which causes the other slaves to ignore the serial data stream.

The software used to communicate between the master and a slave appears in Example 11–35. This main procedure sends a serial data stream from the memory of the master to one of the slave units. When this procedure is called, the contents of BL contain the address of the slave (01H for slave 1 and 02H for slave 2) and DS:SI addresses and area of memory in the master to be serially transferred to the slave. The number of bytes transferred between the master and slave are held in the CX register when the procedure is called. Notice that two commands are sent to the slave with the COMD procedure: one is the slave address and the other is the disconnect command (02H). The data is also sent by the COMD procedure.

EXAMPLE 11–35

```
               ;This procedure transfers the master block of data addressed by DS:SI to
               ;the slave addressed by BL. The number of bytes transferred are held
               ;in CX upon calling this procedure.
               ;
               ;Addresses =  01H = address of slave 1
               ;             02H = address of slave 2
               ;Commands  =  01H = accept block from master
               ;             02H = disconnect from slave
               ;
0010           SEND    PROC  NEAR  USES BX CX SI

0013 53                PUSH  BX      ;push slave address on stack
0014 E8 0039           CALL  SELECT  ;select slave
0017 83 C4 02          ADD   SP,2    ;dump slave address from stack
001A 74 30             JZ    SEND1   ;if slave not available set Z flag
001C 6A 01             PUSH  1       ;send command to slave
001E E8 0059           CALL  COMD
0021 83 C4 02          ADD   SP,2    ;dump command from stack
```

```
0024 51                         PUSH    CX      ;send count to slave
0025 E8 0052                    CALL    COMD
0028 83 C4 02                   ADD     SP,2
002B 51                         PUSH    CX
002C 8A CD                      MOV     CL,CH
002E 51                         PUSH    CX
002F E8 0048                    CALL    COMD
0032 83 C4 02                   ADD     SP,2
0035 59                         POP     CX


                                .REPEAT         ;send data to slave

0036 AC                           LODSB
0037 50                           PUSH    AX
0038 E8 003F                      CALL    COMD    ;send data
003B 83 C4 02                     ADD     SP,2
                                .UNTILCXZ

0040 6A 02                       PUSH    2
0042 E8 0035                      CALL    COMD    ;send slave disconnect
0045 83 C4 02                     ADD     SP,2
0048 B0 01                        MOV     AL,1
004A 0A C0                        OR      AL,AL   ;clear Z flag for valid operation
004C               SEND1:
                                RET

0050               SEND    ENDP

= FF64             S0_CON  EQU   0FF64H ;serial port 0 control
= FF6A             S0_SEN  EQU   0FF6AH ;serial port 0 send data
= FF66             S0_STAT EQU   0FF66H ;serial port 0 status
= FF68             S0_BUF  EQU   0FF68H ;serial port 0 read buffer

0050               SELECT PROC  NEAR    USES BP DX

0052 8B EC                        MOV     BP,SP
0054 BA FF64                      MOV     DX,S0_CON   ;address control register
0057 B8 0083                      MOV     AX,83H      ;mode 3, bit 9 = 1
005A EF                           OUT     DX,AX
005B BA FF6A                      MOV     DX,S0_SEN   ;address send data
005E 8A 46 06                     MOV     AL,[BP+6]   ;get slave address from stack
0061 EE                           OUT     DX,AL
0062 BA FF64                      MOV     DX,S0_CON   ;address control register
0065 B8 0023                      MOV     AX,23H      ;enable receiver
0068 EF                           OUT     DX,AX
0069 BA FF66                      MOV     DX,S0_STAT
                                .REPEAT

006C EC                           IN      AL,DX   ;get status
006D A8 40                        TEST    AL,40H  ;test RI bit

                                .UNTIL !ZERO?

0071 BA FF68                      MOV     DX,S0_BUF
0074 EC                           IN      AL,DX       ;get slave response
0075 22 C0                        AND     AL,AL       ;see if slave responded
                                RET

007A               SELECT ENDP

007A               COMD    PROC  NEAR    USES BP DX

007C 8B EC                        MOV     BP,SP
007E BA FF64                      MOV     DX,S0_CON   ;address control register
0081 B8 0003                      MOV     AX,3        ;mode 3, bit 9 = 0
0084 EF                           OUT     DX,AX
```

```
0085 BA FF6A                    MOV    DX,S0_SEN   ;address transmitter
0088 8A 46 06                   MOV    AL,[BP+6]
008B EE                         OUT    DX,AL       ;send data/command to slave
                                RET

008F            COMD    ENDP
```

The master required software for communications and so does the slave, as shown in Example 11-36. Note that port P2.7 is used to enable the three-state buffer illustrated in Figure 11–59. A logic 0 enables the buffer for transmission and a logic 1 disables the buffer. Again note that the RXD line is used for transmission to the master.

EXAMPLE 11–36

```
                ;This procedure transfer is used by the slave unit to accept
                ;a block of data sent by the master.  Note that the slave stores the
                ;data in an area of memory called the buffer.
                ;
                ;Slave commands = 01H accept block from master
                ;               = 02H disconnect from the master
                ;
= FF5E          P2_LAT  EQU     0FF5EH          ;port 2 latch register
= FF66          S0_STAT EQU     0FF66H          ;serial port 0 status
= FF64          S0_CON  EQU     0FF64H          ;serial port 0 control
= FF68          S0_DAT  EQU     0FF68H          ;serial port zero receive data
= FF6A          S0_TRAN EQU     0FF6AH          ;serial port 0 transmit data
= 0001          SLAVE_1 EQU     01H             ;slave 1 address
= 0002          DISC    EQU     02H             ;disconnect command
= 0001          BLOCK   EQU     01H             ;receive block command
                ;
0010            SLAVE   PROC    NEAR

0010 BA FF66    TOP:    MOV     DX,S0_STAT
0013 EC                 IN      AL,DX           ;clear status register
0014 BA FF5E            MOV     DX,P2_LAT       ;address P2 latch
0017 B0 80             MOV     AL,80H          ;disable three-state buffer
0019 EE                 OUT     DX,AL
001A BA FF64            MOV     DX,S0_CON       ;address control register
001D B8 0022            MOV     AX,22H          ;mode 2, receive mode
0020 EF                 OUT     DX,AX
                        .REPEAT
0021 E8 003D              CALL   RECV          ;get address from master
                        .UNTIL    AL==SLAVE_1
0028  BA FF64           MOV     DX,S0_CON       ;address control register
002B  B8 0003           MOV     AX,3            ;mode 3, transmit
002E  EF                OUT     DX,AX
002F  BA FF5E           MOV     DX,P2_LAT       ;address port 2 latch
0032  B0 00             MOV     AL,0            ;enable buffer
0034  EE                OUT     DX,AL
0035  BA FF6A           MOV     DX,S0_TRAN      ;address transmit data
0038  B0 01             MOV     AL,SLAVE_1      ;echo slave address to master
003A  EE                OUT     DX,AL
003B  BA FF64           MOV     DX,S0_CON       ;address control register
003E  B8 0023           MOV     AX,23H          ;mode 3, receive
0041  EF                OUT     DX,AX
0042  E8 001C           CALL    RECV            ;get command
0045  3C 02             CMP     AL,DISC
0047  74 C7             JE      TOP             ;disconnect

                        .IF AL==BLOCK           ;test for block command
004D E8 0011              CALL   RECV           ;get low count
0050 8A C8                MOV    CL,AL
0052 E8 000C              CALL   RECV           ;get high count
0055 8A E8                MOV    CH,AL
0057 BE 0000 R            MOV    SI,OFFSET BUFFER
                          .REPEAT
005A E8 0004                CALL   RECV         ;get data
```

```
005D AA                     STOSB
                        .UNTILCXZ
                      .ENDIF
005F EB AE              JMP    TOP          ;go disconnect and wait

0061              SLAVE ENDP

0061              RECV   PROC   NEAR        ;receive one byte

0061 BA FF66           MOV    DX,S0_STAT    ;address status register
                      .REPEAT
0064 EC                    IN     AL,DX     ;read status
0065 A8 40                 TEST   AL,40H    ;test RI
                      .UNTIL !ZERO?
0069 BA FF68           MOV DX,S0_DAT        ;address receiver buffer
006C EC                IN     AL,DX         ;get data
006D C3                RET

006E              RECV   ENDP
```

11–8   ANALOG-TO-DIGITAL (ADC) AND DIGITAL-TO-ANALOG CONVERTERS (DAC)

Analog-to digital (ADC) and digital-to-analog (DAC) converters are used to interface the microprocessor to the analog world. Many events that are monitored and controlled by the microprocessor are analog events. These often include monitoring all forms of events, even speech, to controlling motors and like devices. In order to interface the microprocessor to these events, we understand the interface and control of the ADC and DAC, which convert between analog and digital data.

The DAC0830 Digital-to-Analog Converter

A fairly common and low-cost digital-to-analog converter is the DAC0830 (from National Semi-conductor Corporation). This device is an 8-bit converter that transforms an 8-bit binary number into an analog voltage. Other converters are available that convert from 10, 12, or 16 bits into analog voltages. The number of voltage steps generated by the converter is equal to the number of binary input combinations. Therefore, an 8-bit converter generates 256 different voltage levels, a 10-bit converter generates 1,024 levels, and so forth. The DAC0830 is a medium speed converter that transforms a digital input to an analog output in approximately 1 μs. It is also available in other versions such as the DAC0831.

Figure 11–60 illustrates the pin-out of the DAC0830. This device has a set of eight data bus connections for the application of the digital input code and a pair of analog outputs labeled

FIGURE 11–60   The pin-out of the DAC0830 digital-to-analog converter.

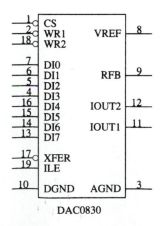

DAC0830

FIGURE 11–61 The internal
structure of the DAC0830.

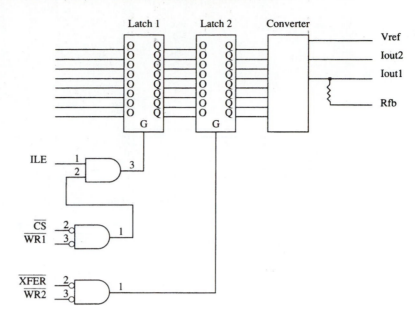

Iout1 and Iout2 that are designed as inputs to an external operational amplifier. Because this is an 8-bit converter, its output step voltage is defined as $-V_{REF}$ (**reference voltage**) divided by 255. For example, if the reference voltage is -5.0 V, its output step voltage is $+.0196$ V. Note that the output voltage is the opposite polarity of the reference voltage. If an input of $1001\ 0010_2$ is applied to the device, the output voltage will be the step voltage times $1001\ 0010_2$, or in this case $+2.862$ V. By changing the reference voltage to -5.1 V, the step voltage becomes $+.02$ V. The step voltage is also often called the **resolution** of the converter.

Internal Structure of the DAC0830.   Figure 11–61 illustrates the internal structure of the DAC0830. Notice that this device contains two internal registers. The first is a holding register, while the second connects to the R–2R internal ladder converter. The two latches allow one byte to be held while another is converted. In many cases we disable the first latch and only use the second for entering data into the converter. This is accomplished by connecting a logic 1 to ILE and a logic 0 to $\overline{\text{CS}}$ (**chip select**).

Both latches within the DAC0830 are transparent latches. That is, when the G input to the latch is a logic 1, data passes through the latch, but when the G input becomes a logic 0, data is latched or held. The converter has a reference input pin (V_{REF}) that establishes the full-scale output voltage. If -10 V is placed on V_{REF}, the full-scale (11111111_2) output voltage is $+10$ V. The output of the R–2R ladder within the converter appears at Iout1 and Iout2. These outputs are designed to be applied to an operational amplifier such as a 741 or similar device.

Connecting the DAC0830 to the Microprocessor.   The DAC0830 is connected to the microprocessor as illustrated in Figure 11–62. Here a PAL16L8 is used to decode the DAC0830 at 8-bit I/O port address 20H. Whenever an OUT 20H,AL instruction is executed, the contents of data bus connections AD_0–AD_7 are passed to the converter within the DAC0830. The 741 operational amplifier along with the -12 V zener reference voltage causes the full-scale output voltage to equal $+12$ V. The output of the operational amplifier feeds a driver that powers a 12 V DC motor. This driver is a Darlington amplifier for large motors. This example shows the converter driving a motor, but other devices could also be used as an output.

The ADC080X Analog-to-Digital Converter

A common lost-cost ADC is the ADC0804, which belongs to a family of converters that are identical except for accuracy. This device is compatible with a wide range of microprocessors

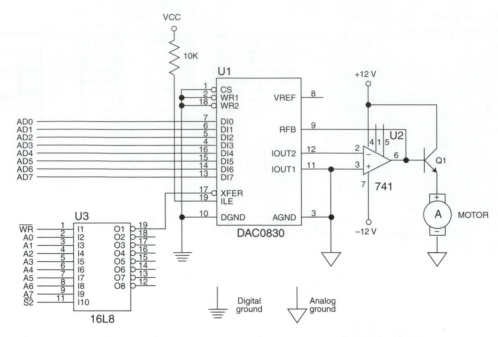

FIGURE 11–62   A DAC0830 interfaced to the 80186 microprocessor at 8-bit I/O address 20H.

such as the Intel family. There are faster ADCs available and some with more resolution than 8 bits, but this device is ideal for many applications that do not require a high degree of accuracy. The ADC0804 requires up to 100 μs to convert an analog input voltage into a digital output code.

Figure 11–63 shows the pin-out of the ADC0804 (from National Semiconductor Corporation) converter. To operate the converter, the $\overline{WR}$ pin is pulsed with $\overline{CS}$ grounded to start the conversion process. Because this converter requires a considerable amount of time for the conversion, a pin labeled INTR signals the end of the conversion. Refer to Figure 11–64 for a timing diagram that shows the interaction of the control signals. As we can see, we start the converter with the $\overline{WR}$ pulse, wait for INTR to return to a logic 0 level, and then read the data from the converter. If a time delay is used that allows at least 100 μs of time, then we don't need to test the INTR pin. Another option is to connect the INTR pin to an interrupt input so that when the conversion is complete, an interrupt occurs.

The Analog Input Signal.   Before the ADC0804 can be connected to the microprocessor, we must understand its analog inputs. There are two analog inputs to the ADC0804: V_{IN} (+) and V_{IN} (–). These inputs are connected to an internal operational amplifier and are differential inputs, as

FIGURE 11–63   The pin-out of the ADC0804 analog-to-digital converter.

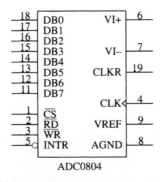

ADC0804

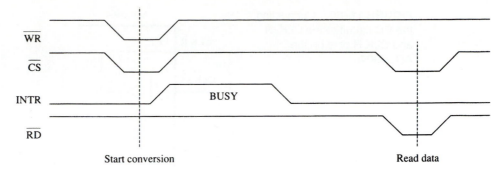

FIGURE 11–64   The timing for ADC0804 analog-to-digital converter.

shown in Figure 11–65. The differential inputs are summed by the operational amplifier to produce a signal for the internal analog-to-digital converter. Figure 11–65 shows a few ways to use these differential inputs. The first way (see Figure 11–65a) uses a single input that can vary between 0 V and +5.0 V. The second (see Figure 11–65b) shows a variable voltage applied to the V_{IN} (−) pin so the zero reference for V_{IN} (+) can be adjusted.

Generating the Clock Signal.   The ADC0804 requires a clock source for operation. The clock can be an external clock applied to the CLK IN pin or it can be generated with an RC circuit. The permissible range of clock frequencies is between 100 KHz and 1460 KHz. It is desirable to use a frequency that is as close as possible to 1460 KHz so conversion time is kept to a minimum.

If the clock is generated with an RC circuit, we use the CLK IN and CLK R pins connected to an RC circuit as illustrated in Figure 11–66. When this connection is in use, the clock frequency is calculated by the equation in Example 11–37.

EXAMPLE 11–37

$$\text{Fclk} = \frac{1}{1.1\,\text{RC}}$$

Connecting the ADC0804 to the Microprocessor.   The ADC0804 is interfaced to the 80186 or 80386EX microprocessor as illustrated in Figure 11–67. Notice that the V_{REF} signal is not attached to anything and this is normal. Suppose that the ADC0804 is decoded at 8-bit I/O port address 40H for the data and port address 42H for the INTR signal and a procedure is required to

FIGURE 11–65   The analog inputs to the ADC0804 converter. (a) To sense a 0- to +5.0-V input. (b) To sense an input offset from ground.

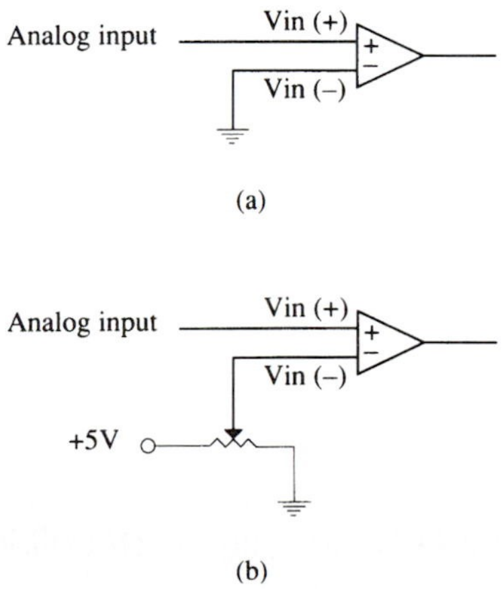

Analog input ─── V_{in} (+)

V_{in} (−)

(a)

Analog input ─── V_{in} (+)

V_{in} (−)

+5V

(b)

FIGURE 11–66   Connecting the RC circuit to the CLK IN and CLK R pins on the ADC0804.

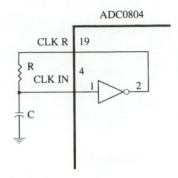

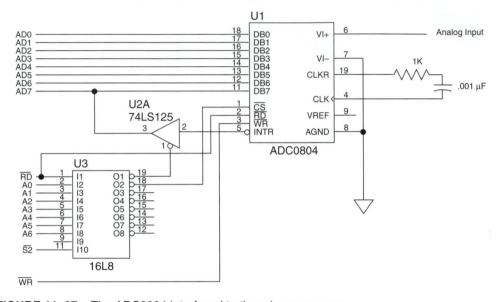

FIGURE 11–67   The ADC0804 interfaced to the microprocessor.

start and read the data from the ADC. This procedure is listed in Example 11–38. Notice that the INTR bit is polled and if it becomes a logic 0, the procedure ends with AL containing the converted digital code.

EXAMPLE 11–38

```
                        ;Procedure that reads data from the ADC and returns
                        ;it in AL.
                        ;
0000                    ADCX    PROC    NEAR
0000 E6 40                      OUT     40H,AL          ;start conversion
0002                    ADCX1:
0002 E4 42                      IN      AL,42H          ;read INTR
0004 A8 80                      TEST    AL,80H          ;test INTR
0006 75 FA                      JNZ     ADCX1           ;repeat until INTR = 0
0008 E4 40                      IN      AL,40H          ;get ADC data
000A C3                         RET

000B                    ADCX    ENDP
```

Using the ADC0804 and the DAC0830

This section gives an example using both the ADC0804 and the DAC0830 to capture and replay audio signals or speech. In the past, we often used a speech synthesizer to generate speech, but the quality of the speech was poor. For human-quality speech we can use the ADC0804 to capture an audio signal and store it in memory for later playback through the DAC0830.

Figure 11–68 illustrates the circuitry required to connect the ADC0804 at I/O ports 0700H and 0702H. The DAC0830 is interfaced at I/O port 704H. These I/O ports are in the low bank of a 16-bit microprocessor such as the 80186 or 80386EX. The software used to run these converters appears in Example 11–39. This software reads a 1-second burst of speech and then plays it back 10 times. This process repeats until the system is turned off. In this example, speech is

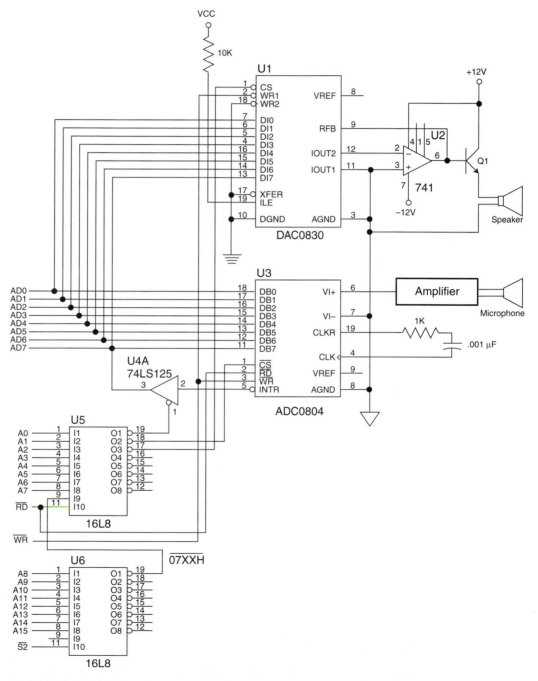

FIGURE 11–68 A circuit to store speech and play it back through a speaker.

sampled and stored in a section of memory called WORDS. The sample rate is chosen at 2048 samples per second, which renders acceptable-sounding speech.

EXAMPLE 11–39

```
                          ;Software that records a 1-second passage of speech
                          ;and plays it back 10 times before recording the
                          ;next 1-second passage of speech.
                          ;
                          ;assumes a clock of 8 MHz (8086) for the time delay.
                          ;
                          .MODEL SMALL
                          .DATA
0000 0500 [               WORDS   DB   2048 DUP (?)      ;space for speech
          0000
                ]
                          .CODE
                          .STARTUP
0018              AGAIN:
0018 E8 000A              CALL    READ                   ;read speech
001B B9 000A              MOV     CX,10                  ;set count to 10
001E             LOOP1:
001E E8 0023              CALL    WRITE                  ;playback speech
0021 E2 FB                LOOP    LOOP1                  ;repeat 10 times
0023 EB EA                JMP     AGAIN                  ;repeat forever

0025              READ    PROC    NEAR

0025 BF 0000 R            MOV     DI,OFFSET WORDS        ;address data area
0028 B9 0500              MOV     CX,2048                ;load count
002B BA 0700              MOV     DX,0700H               ;address port
002E             READ1:
002E EE                   OUT     DX,AL                  ;start converter
002F 83 C2 C0             ADD     DX,2                   ;address status port
0032             READ2:
0032 EC                   IN      AL,DX                  ;get INTR
0033 A8 80                TEST    AL,80H                 ;test INTR
0035 75 FB                JNZ     READ2                  ;wait for INTR = 0
0037 83 EA 02             SUB     DX,2                   ;address data port
003A EC                   IN      AL,DX                  ;get data from ADC
003B 88 05                MOV     [DI],AL                ;store data in array
003D 47                   INC     DI                     ;address next element
003E E8 0018              CALL    DELAY                  ;wait for 1/2048 second
0041 E2 EB                LOOP    READ1                  ;repeat 2,048 times
0043 C3                   RET

0044              READ    ENDP

0044              WRITE PROC    NEAR

0044 51                   PUSH    CX
0045 BF 0000 R            MOV     DI,OFFSET WORDS        ;address data
0048 B9 0500              MOV     CX,2048                ;load count
004B BA 0704              MOV     DX,0704H               ;address DAC
004E             WRITE1:
004E 8A 05                MOV     AL,[DI]                ;get data from array
0050 EE                   OUT     DX,AL                  ;send data to DAC
0051 47                   INC     DI                     ;address next element
0052 E8 0004              CALL    DELAY                  ;wait 1/2048 second
0055 E2 F7                LOOP    WRITE1                 ;repeat 2,048 times
0057 59                   POP     CX
0058 C3                   RET

0059              WRITE ENDP

0059              DELAY PROC    NEAR
```

```
0059 51                    PUSH    CX
005A B9 00E1               MOV     CX,225              ;approximately 1/2048 sec.
005D           DELAY1:
005D E2 FE                 LOOP    DELAY1
005F 59                    POP     CX
0060 C3                    RET

0061           DELAY ENDP
               .END
```

11-9   SUMMARY

1. The 8086–Pentium Pro microprocessors have two basic types of I/O instructions: IN and OUT. The IN instruction inputs data from an external I/O device into either the AL (8-bit) or AX (16-bit) register. The IN instruction is available as a fixed port instruction, a variable port instruction, or as a string instruction (80186–Pentium Pro) INSB or INSW. The OUT instruction outputs data from AL or AX to an external I/O device and is also available as a fixed, variable, or string instruction OUTSB or OUTSW. The fixed port instruction uses an 8-bit I/O port address, while the variable and string I/O instructions use a 16-bit port number found in the DX register.

2. Isolated I/O, sometimes called direct I/O, uses a separate map for the I/O space, freeing the entire memory for use by the program. Isolated I/O uses the IN and OUT instructions to transfer data between the I/O device and the microprocessor. The control structure of the I/O map uses $\overline{IORC}$ (I/O read control) and $\overline{IOWC}$ (I/O write control) plus the bank selection signals $\overline{BHE}$ and $\overline{BLE}$ (A_0 on the 80186) to effect the I/O transfer. The 80186/80188 uses the $\overline{S2}$ signal with $\overline{RD}$ and $\overline{WR}$ to generate the I/O control signals.

3. Memory-mapped I/O uses a portion of the memory space for I/O transfers. This reduces the amount of memory available, but it negates the need to use the $\overline{IORC}$ and $\overline{IOWC}$ signals for I/O transfers. In addition, any instruction that addresses a memory location using any addressing mode can be used to transfer data between the microprocessor and the I/O device using memory-mapped I/O.

4. All input devices are buffered so the I/O data is connected to the data bus only during the execution of the IN instruction. The buffer is either built into a programmable peripheral or located separately.

5. All output devices use a latch to capture output data during the execution of the OUT instruction. This is necessary because data appears on the data bus for less than 100 ns for an OUT instruction and most output devices require the data for a longer time. In many cases the latch is built into the peripheral.

6. Handshaking or polling is the act of two independent devices synchronizing with a few control lines. For example, the computer asks a printer if it is busy by inputting the BUSY signal from the printer. If it isn't busy, the computer outputs data to the printer and informs the printer that data is available with a data strobe ($\overline{DS}$) signal. This communication between the computer and the printer is a handshake or poll.

7. The I/O port number appears on address bus connections A_7–A_0 (note that A_{15}–A_8 contain zeros for an 8-bit port) for a fixed port I/O instruction and on A_{15}–A_0 for a variable port I/O instruction. In both cases, address bits above A_{15} are undefined.

8. Because the 80186 and 80386EX contain a 16-bit data bus and the I/O addresses reference byte-sized I/O locations, the I/O space is also organized in banks as is the memory system. In order to interface an 8-bit I/O device to the 16-bit data bus, we often require separate write strobes, an upper and a lower, for I/O write operations. Likewise, the 80486 and Pentium/ Pentium Pro also have I/O arranged in banks.

9. The I/O port decoder is much like the memory address decoder except instead of decoding the entire address, the I/O port decoder decodes only a 16-bit address for variable port instructions and often an 8-bit port number for fixed I/O instructions.

10. I/O ports are programmed in the 80186/80188 and the 80386EX as were the memory addresses we discussed in the previous chapter. The difference is that in the 80186/80188, the first 11 bits of the I/O port are specified, which means that the smallest block of I/O is 32 locations.

11. The 82C55 is a programmable peripheral interface (PIA) that has 24 I/O pins that are programmable in two groups of 12 pins each (group A and group B). The 82C55 operates in three modes: simple I/O (mode 0), strobed I/O (mode 1), and bi-directional I/O (mode 2). When the 82C55 is interfaced to the 8086 operating at 8 MHz, we insert two wait states because the speed of the microprocessor is faster than the 82C55 can handle.

12. The 8279 is a programmable keyboard/display controller that can control a 64-key keyboard and a 16-digit numeric display.

13. The 8254 is a programmable interval timer that contains three 16-bit counters that count in binary or binary-coded decimal (BCD). Counters are independent of each other, and operate in six different modes: (1) events counter, (2) retriggerable monostable multivibrator, (3) pulse generator, (4) square-wave generator, (5) software-triggered pulse generator, and (6) hardware-triggered pulse generator.

14. The 16550 is a programmable communications interface capable of receiving and transmitting asynchronous serial data.

15. The 80186/80188 serial interface contains a programmable Baud rate generator and operates in five different modes. Mode 0 is a synchronous mode, while modes 1 and 4 are asynchronous. Modes 2 and 3 are used to implement a serial bus between microprocessors and other devices.

16. The DAC0830 is an 8-bit digital-to-analog converter that converts a digital signal to an analog voltage within 1 μs.

17. The ADC0804 is an 8-bit analog-to-digital converter that converts an analog signal into a digital signal within 100 μs.

11–10   QUESTIONS AND PROBLEMS

1. Explain which way the data flows for an IN and an OUT instruction.
2. Where is the I/O port number stored for a fixed I/O instruction?
3. Where is the I/O port number stored for a variable I/O instruction?
4. Where is the I/O port number stored for a string I/O instruction?
5. To which register is data input by the 16-bit IN instruction?
6. Describe the operation of the OUTSB instruction.
7. Describe the operation of the INSW instruction.
8. Contrast a memory-mapped I/O system with an isolated I/O system.
9. What is the basic input interface?
10. What is the basic output interface?
11. Explain the term handshaking as it applies to computer I/O systems.
12. An even-numbered I/O port address is found in the _____ I/O bank in the 80186 microprocessor.
13. Show the circuitry required to generate the upper and lower I/O write strobes.
14. Develop an I/O port decoder, using a 74ALS138, that generates low-bank I/O strobes for the 8-bit I/O port addresses: 10H, 12H, 14H, 16H, 18H, 1AH, 1CH, and 1EH.
15. Develop an I/O port decoder, using a 74ALS138, that generates high-bank I/O strobes for the 8-bit I/O port addresses: 11H, 13H, 15H, 17H, 19H, 1BH, 1DH, and 1FH.

16. Develop an I/O port decoder, using a PAL16L8, that generates 16-bit I/O strobes for the 16-bit I/O port addresses: 1000H–1001H, 1002H–1003H, 1004H–1005H, 1006H–1007H, 1008H–1009H, 100AH–100BH, 100CH–100DH, and 100EH–100FH.

17. Develop an I/O port decoder, using the PAL16L8, that generates the following low-bank I/O strobes: 00A8H, 00B6H, and 00EEH.

18. Develop an I/O port decoder, using the PAL16L8, that generates the following high-bank I/O strobes: 300DH, 300BH, 1005H, and 1007H.

19. Why are both $\overline{BHE}$ and $\overline{BLE}$ (A0) ignored in a 16-bit port address decoder?

20. An 8-bit I/O device, located at I/O port address 0010H, is connected to which data bus connections?

21. An 8-bit I/O device located at I/O port address 100DH is connected to which data bus connections?

22. Write the software required to program the $\overline{PCS}$ pins of an 80C188XL microprocessor for I/O ports beginning at I/O address 3000H with 2 wait states.

23. Write the software required to program the $\overline{GCS3}$ pin of the 80C188EB microprocessor for I/O port 1000H, with 3 wait states.

24. Write the software required to program the $\overline{GCS2}$ pin of the 80386EX microprocessor for I/O port 4000H with 30 wait states.

25. The 82C55 has how many programmable I/O pin connections?

26. List the pins that belong to group A and to group B in the 82C55.

27. Which two 82C55 pins accomplish internal I/O port address selection?

28. The $\overline{RD}$ connection on the 82C55 is attached to which 80186 system control bus connection?

29. Using a PAL16L8, interface an 82C55 to the 80186 microprocessor so it functions at I/O locations 0380H, 0382H, 0384H, and 0386H.

30. When the 82C55 is reset, its I/O ports are all initialized as _____.

31. Which three modes of operation are available to the 82C55?

32. What is the purpose of the $\overline{STB}$ signal in strobed input operation of the 82C55?

33. Explain the operation of simple four-coil stepper motor.

34. What sets the IBF pin in strobed input operation of the 82C55?

35. Write the software required to place a logic 1 on the PC7 pin of the 82C55 during strobed input operation.

36. How is the interrupt request pin (INTR) enabled in the strobed input mode of operation of the 82C55?

37. In strobed output operation of the 82C55, what is the purpose of the $\overline{ACK}$ signal?

38. What clears the $\overline{OBF}$ signal in strobed output operation of the 82C55?

39. Write the software required to decide if PC4 is a logic 1 when the 82C55 is operated in the strobed output mode.

40. Which group of pins are used during bi-directional operation of the 82C55?

41. Which pins are general-purpose I/O pins during mode 2 operation of the 82C55?

42. What changes must be made to Figure 11–21 so it functions with a keyboard matrix that contains three rows and five columns?

43. What time is usually used to de-bounce a keyboard?

44. What is normally connected to the CLK pin of the 8279?

45. How many wait states are required to interface the 8279 to the 80186 microprocessor operating with an 8 MHz clock?

46. If the 8279 CLK pin is connected to a 3.0 MHz clock, program the internal clock.

47. What is an overrun error in the 8279?

48. What is the difference between encoded and decoded as defined for the 8279?

49. Interface the 8279 so it functions at 8-bit I/O ports 40H–7FH. Use the 74ALS138 as a decoder and use either the upper or lower data bus.

50. Interface a 16-key keyboard and an 8-digit numeric display to the 8279.

51. The 8254 interval timer functions from DC to _____ Hz.

52. Each counter in the 8254 functions in how many different modes?

53. Interface an 8254 to function at I/O port addresses: XX10H, XX12H, XX14H, and XX16H. Write the software required to cause counter 2 to generate an 80 KHz square-wave if the CLK input to counter 2 is 8 MHz.

54. What number is programmed in an 8254 counter to count 300 events?

55. If a 16-bit count is programmed into the 8254, which byte of the count is programmed first?

56. Explain how the read-back control word functions in the 8254.

57. Program counter 1 of the 8254 so it generates a continuous series of pulses that have a high time of 100 μs and low time of 1 μs. Make sure to indicate the CLK frequency required to accomplish this task.

58. Why does a 50 percent duty cycle cause the motor to stand still in the motor speed and direction control circuit presented in this chapter?

59. Write the software required to program timer 1 in the 80188 microprocessor, so it generates an output wave-form that is a logic 1 for 10 ms and a logic 0 for 150 μs. Assume that the CLLOUT frequency is used as a timing source at a frequency of 1 MHz.

60. Using timer 2 in the 80188 as a prescaler for timer 0, develop the software required to program timer 0 so it generates a 1-second square-wave if the crystal frequency is 8 MHz.

61. What is the purpose of the T_{IN} pin on timer 0 and timer 1 in the 80186 microprocessor?

62. What is asynchronous serial data?

63. What is Baud rate?

64. Program the 16550 for operation using six data bits, even parity, one stop bit, and a Baud rate of 19,200 using a 18.432 MHz clock. (Assume that the I/O ports are numbered 20H and 22H.)

65. If the 16550 is to generate a serial signal at a Baud rate of 2400 Baud, and the Baud rate divisor is programmed for 16, what is the frequency of the signal?

66. Describe the following terms: simplex, half-duplex, and full-duplex.

67. How is the 16550 reset?

68. Write a procedure for the 16550 that transmits 16 bytes from a small buffer in the data segment (DS is loaded externally) address by SI (SI is loaded externally).

69. Develop the software required to program serial port 0 (80C188EB) for a mode 0 transmit/receive Baud rate of 1200 Baud. Use the internal clock at a CLKOUT frequency of 4 MHz.

70. Develop the software required to program the mode 1 operation of serial port 0 so it functions at 9600 Baud with 7 data bits and even parity. The CLKOUT frequency is 2 MHz.

71. The DAC0830 converts an 8-bit digital input to an analog output in approximately _____ μs.

72. What is the step voltage at the output of the DAC0830 if the reference voltage is –2.55 V?

73. Interface a DAC0830 to the 80186 so it operates at I/O port 400H.

74. Develop a program for the interface of Question number 73 so the DAC0830 generates a triangular voltage wave-form. The frequency of this wave-form must be approximately 100 Hz.

75. The ADC080X requires approximately _____ to convert an analog voltage into a digital code.

76. What is the purpose of the INTR pin on the ADC080X?

77. The $\overline{WR}$ pin on the ADC080X is used for what purpose?

78. Interface an ADC080X at I/O port 0260H for data and 0270H to test the INTR pin.

79. Develop a program for the ADC080X in Question number 78 so it reads an input voltage once per 100 ms and stores the results in a memory array that is 100H bytes in length.

80. Use the Internet to search for "I/O." Visit at least five sites that have information about I/O and write a paragraph describing the contents of these sites.

81. Use the Internet to search for "keyboard switch." Find at least two keyboard switches and list the bounce times from the technical data.

82. Use the Internet to search for "LCD displays." Write a short report detailing the history of the LCD.

83. Use the Internet to locate technical data on at least two different stepper motors.

CHAPTER 12

Interrupts

INTRODUCTION

In this chapter, we expand our coverage of basic I/O and programmable peripheral interfaces by examining a technique called interrupt-processed I/O. An interrupt is a hardware-initiated procedural call that interrupts the currently executing program.

We provide examples and a detailed explanation of the interrupt structure of the entire Intel family of microprocessors and in particular the 80186 and 80188.

CHAPTER OBJECTIVES

Upon completion of this chapter, you will be able to:

1. Explain the interrupt structure of the Intel family of microprocessors.
2. Explain the operation of software interrupt instructions: INT, INTO, INT 3, and BOUND.
3. Explain how the interrupt enable flag bit (IF) modifies the interrupt structure.
4. Describe the function of the trap interrupt flag bit (TF) and the operation of trap-generated tracing.
5. Detail and program the interrupt structure of the 80186/80188.
6. Develop interrupt service procedures that control lower-speed external peripheral devices.
7. Expand the interrupt structure of the microprocessor using the 8259A programmable interrupt controller and other techniques.
8. Explain the purpose and operation of a real-time clock.

12–1 BASIC INTERRUPT PROCESSING

In this section, we discuss the function of an interrupt in a microprocessor-based system. We also discuss the structure and features of interrupts available to the Intel family of microprocessors.

The Purpose of Interrupts

Interrupts are particularly useful when interfacing I/O devices that provide or require data at relatively low data transfer rates. In Chapter 11 we saw a keyboard example using strobed input operation of the 82C55. In that example, software polled the 82C55 and its IBF bit to decide if data

FIGURE 12–1 A time line that indicates interrupt usage in a typical system.

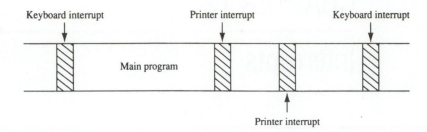

was available from the keyboard. If the person using the keyboard types one character per second, the software for the 82C55 waits an entire second between each keystroke for the person to type another key. This process is such a tremendous waste of time that designers have developed another technique called interrupt processing to handle it.

Unlike the polling technique, interrupt processing allows the microprocessor to execute other software while the keyboard operator is thinking about what key to type next. As soon as a key is pressed, the keyboard encoder de-bounces the switch and sends a pulse to an interrupt input pin, which interrupts the microprocessor. In this way, the microprocessor executes other software until the key is actually pressed. The interrupt procedure reads the key, stores the key code in a buffer, and then returns to the program that was interrupted. As a result, the microprocessor can print reports or complete a multitude of other tasks while the operator is typing a document or thinking about what to type next.

Figure 12–1 shows a time line that indicates a typist typing data on a keyboard, a printer removing data from the memory, and a program executing. The program is the main program that is interrupted for each keystroke from the keyboard and each character that is printed on the printer. The keyboard interrupt service procedure, called by the keyboard interrupt, and the printer interrupt service procedure each take little time to execute. This frees the microprocessor to perform other tasks while no key is typed and no character is printed.

Interrupts

The interrupts of the entire Intel family of microprocessors include two hardware pins that request interrupts (INTR and NMI) and one hardware pin ($\overline{\text{INTA}}$) that acknowledges the interrupt requested through INTR. In addition to the pins, the microprocessor also has software interrupts: INT, INTO, INT 3, and BOUND. Two flag bits, IF (interrupt flag) and TF (trap flag), are also used with the interrupt structure and a special return instruction IRET (IRETD in the 80386, 80486, or Pentium/Pentium Pro). More interrupt inputs are available on the 80186/80188 and the 80386EX microprocessors through their expanded interrupt capabilities.

Interrupt Vectors.   The interrupt vectors and the interrupt vector table are crucial to an understanding of hardware and software interrupts. The **interrupt vector table** is located in the first 1,024 bytes of memory at addresses 000000H–0003FFH. The vector table contains 256 different four-byte interrupt vectors. An **interrupt vector** contains the address (segment and offset) of the interrupt service procedure.

Table 12–1 illustrates the interrupt vector table for the microprocessor. The first 5 interrupt vectors are identical in all Intel microprocessor family members, from the 8086 through the Pentium Pro. Other interrupt vectors exist for the 80286 that are upward compatible to the 80386, 80486, and Pentium/Pentium Pro, but not downward compatible to the 8086 or 8088. Notice that the 80186/80188 and the 80386EX contain many different vectors compared to the other Intel microprocessors. Intel reserves the first 32 interrupt vectors to use in various microprocessor family members. The last 224 vectors are available as user interrupt vectors. Each vector is 4 bytes long and contains the **starting address** of the interrupt service procedure. The first two bytes of the vector contain the offset address and the last two bytes contain the segment address.

TABLE 12–1 The interrupts found in the 8086–Pentium Pro microprocessors

Type	Vector Address	Microprocessor	Name
0	000–003	8086–Pentium Pro	Divide error
1	004–007	8086–Pentium Pro	Single-step
2	008–00B	8086–Pentium Pro	NMI pin
3	00C–00F	8086–Pentium Pro	1-byte breakpoint
4	010–013	8086–Pentium Pro	INTO instruction
5	014–017	80186–Pentium Pro	Bound
6	018–01B	80186–Pentium Pro	Unidentified opcode
7	01C–01F	80186–Pentium Pro	Coprocessor not available
8	020–023	80286–Pentium Pro 80186/80188	Double-fault Timer 0
9	024–027	80386	Coprocessor segment overrun
A (10)	028–02B	80286–Pentium Pro	Invalid task state segment
B (11)	02C–02F	80286–Pentium Pro	Segment not present
C (12)	030–033	80286–Pentium Pro 80186/80188	Stack segment overrun INT0
D (13)	034–037	80286–Pentium Pro 80186/80188	General protection fault INT1
E (14)	038–03B	80386–Pentium Pro 80186/80188	Page fault INT2
F (15)	03C–03F	80186/80188	INT3
10 (16)	040–043	80186EB–Pentium Pro	Coprocessor error
11 (17)	044–047	80486SX–Pentium Pro 80186/80188	Alignment check INT4
12 (18)	048–04B	Pentium–Pentium Pro 80186/80188	Machine check Timer 1
13 (19)	04C–04F	80186/80188	Timer 2
14 (20)	050–053	80186/80188	Serial 0 receive (DMA0)
15 (21)	054–057	80186/80188	Serial 0 transmit (DMA1)
16–1F	058–07F	8086–Pentium Pro	Reserved
20–FF	080–3FF	8086–Pentium Pro	User interrupts

The following list describes the function of each dedicated interrupt vector.

Type 0   The **divide error** interrupt occurs whenever the result of a division causes an overflow or whenever a division by zero is attempted.

Type 1   A **single-step** interrupt occurs after the execution of each instruction if the trap flag bit is set. Upon accepting this interrupt, the trap flag bit is cleared so execution within the interrupt service procedure proceeds at full speed. More detail is provided about this interrupt later in this section.

Type 2   The **non-maskable** hardware interrupt occurs as a result of placing a logic 1 on the NMI input pin to the microprocessor. This input is non-maskable, which means that it cannot be disabled.

Type 3   A **one-byte** interrupt is a special one-byte instruction (INT 3) that uses this vector to access its interrupt service procedure. The INT 3 instruction is often used to store a breakpoint for debugging a program.

Type 4   The **overflow** interrupt is a special vector used with the INTO instruction. The INTO instruction interrupts the program if an overflow condition exists as reflected by the overflow flag (OF).

Type 5   A **BOUND** instruction compares a register with boundaries stored in the memory and may cause an interrupt. If the contents of the register are greater than or equal to the first word in memory and less than or equal to the second word, no interrupt occurs because the contents of the register are within bounds. If the contents of the register are out of bounds, a type 5 interrupt ensues.

Type 6   The **invalid opcode** interrupt occurs whenever an undefined opcode is encountered in a program.

Type 7   A **coprocessor not available** interrupt occurs when a coprocessor is not found in the system as dictated by the machine status word (MSW) in the coprocessor control bits. If an ESC or WAIT instruction executes and the coprocessor is not found, a type 7 exception or interrupt occurs.

Type 8   The **double fault** interrupt is activated whenever two separate interrupts occur during the same instruction.

Type 9   A **coprocessor segment overrun** interrupt (80386 only) occurs if the ESC instruction (coprocessor opcode) memory operand extends beyond offset address FFFFH.

Type A   The **invalid task state segment** interrupt occurs if the TSS is invalid because the segment limit field is not 002BH or higher. In most cases this occurs because the TSS is not initialized.

Type B   A **segment not present** interrupt occurs when the P bit (P = 0) in a descriptor indicates that the segment is not present or not valid.

Type C   The **stack segment overrun** interrupt occurs if the stack segment is not present (P = 0) or if the limit of the stack segment is exceeded. In the 80186/80188 versions XL, EA, and EB, this interrupt vector is in response to the INT0 pin.

Type D   A **general protection** interrupt occurs for most protection violation in the 80286–Pentium Pro protected-mode system. (These errors occur in Windows as **general protection faults**.) A list of these protection violations follows:
a. descriptor table limit exceeded
b. privilege rules violated
c. invalid descriptor segment type loaded
d. write to protected code segment
e. read from execute-only code segment
f. write to read-only data segment
g. segment limit exceeded
h. CPL = 0 when executing CTS, HLT, LGDT, LIDT, LLDT, LMSW, and LTR
i. CPL > IOPL when executing CLI, IN, INS, LOCK, OUT, OUTS, and STI

The 808186/80188 use this interrupt vector in response to the INT1 pin.

Type E   The **page fault** interrupt occurs for any page fault memory or code access in the 80386, 80486, and Pentium/Pentium Pro microprocessors. In the 80186/80188 this vector is in response to the INT2 input.

Type F   This interrupt occurs in response to the INT3 input pin on the 80186/80188 microprocessors.

Type 10   A **coprocessor error** takes effect whenever a coprocessor error ($\overline{\text{ERROR}}$ = 0) occurs for the ESCape or WAIT instructions for the 80386, 80486, and Pentium/Pentium Pro microprocessors only. In the 80186EB, this interrupt is in response to the numeric coprocessor.

Type 11 An **alignment check** interrupt indicates that word data is addressed at an odd memory location or doubleword data at an incorrect location. This interrupt is active in the 80486 and Pentium/Pentium Pro microprocessors. The 80186/80186-EB versions also use this vector for the INT4 input.

Type 12 The **machine check** interrupt activates a system memory management mode interrupt in the Pentium and Pentium Pro microprocessors. In the 80186/80188, this vector is used by the Timer 1 interrupt.

Type 13 A **timer 2** interrupt occurs in the 80186/80188 in response to a timer 2 rollover.

Type 14 This 80186/80188 interrupt vector occurs for the serial port 0 receiver in the EB version and for DMA channel 0 in the XL and EA versions.

Type 15 The serial port 0 transmitter causes this interrupt in the 80186/80188EB versions or DMA channel 1 in the XL and EA versions.

Interrupt Instructions: BOUND, INTO, INT, INT 3, and IRET

Of the five software interrupt instructions available to the microprocessor, INT and INT 3 are very similar, BOUND and INTO are conditional, and IRET is a special interrupt return instruction.

The BOUND instruction, which has two operands, compares a register with two words of memory data. For example, if the instruction BOUND AX,DATA is executed, AX is compared with the contents of DATA and DATA + 1 and also with DATA + 2 and DATA + 3. If AX is less than the word contents of DATA and DATA + 1, a type 5 interrupt occurs. If AX is greater than the word contents of DATA + 2 and DATA + 3, a type 5 interrupt occurs. If AX is within the bounds of these two memory words, no interrupt occurs.

The INTO instruction checks the overflow flag (OF). If OF = 1, the INTO instruction calls the procedure whose address is stored in interrupt vector type number 4. If OF = 0, then the INTO instruction performs no operation and the next sequential instruction in the program executes.

The INT n instruction calls the interrupt service procedure that begins at the address represented in vector number n. For example, an INT 80H or INT 128 calls the interrupt service procedure whose address is stored in vector type number 80H (000200H–00203H). To determine the vector address, multiply the vector type number (n) by 4. This yields the beginning address of the four-byte long interrupt vector. For example, an INT 5 = 4 × 5 or 20 (14H). The vector for INT 5 begins at address 000014H and continues to 000017H. Each INT instruction is stored in two bytes of memory: the first byte contains the opcode and the second the interrupt type number. The only exception to this is the INT 3 instruction, a one-byte instruction. The INT 3 instruction is often used as a breakpoint interrupt because it is easy to insert a one-byte instruction into a program. Breakpoints are often used to debug faulty software.

The IRET instruction is a special return instruction used to return from both software and hardware interrupts. The IRET instruction is much like a normal far RET, because it retrieves the return address from the stack. It is unlike the normal return because it also retrieves a copy of the flag register from the stack. An IRET instruction removes 6 bytes from the stack: 2 for the IP, 2 for the CS, and 2 for the flags.

In the 80386–Pentium Pro there is also an IRETD instruction because these microprocessors can push the EFLAG register (32 bits) on the stack as well as the 32-bit EIP in the protected mode. If operated in the real mode we use the IRET instruction with the 80386–Pentium Pro/80486 microprocessors.

The Operation of a Real-Mode Interrupt

When the microprocessor completes executing the current instruction, it determines whether an interrupt is active by checking: (1) instruction executions, (2) single-step, (3) NMI, (4) coprocessor segment overrun, (5) INTR, and (6) INT instructions in the order presented. If one or more of these interrupt conditions are present, the following sequence of events occurs:

1. The contents of the flag register are pushed onto the stack.
2. Both the interrupt (IF) and trap (TF) flags are cleared. This disables the INTR pin and also the trap or single-step feature.
3. The contents of the code segment register (CS) are pushed onto the stack.
4. The contents of the instruction pointer (IP) are pushed onto the stack.
5. The interrupt vector contents are fetched and placed into both IP and CS so the next instruction executes at the interrupt service procedure addressed by the vector.

Whenever an interrupt is accepted, the microprocessor (1) places the contents of the flag register, CS and IP, onto the stack; (2) clears both IF and TF; (3) jumps to the procedure addressed by the interrupt vector. After the flags are pushed onto the stack, IF and TF are cleared. The IF and TF flags are returned to the state prior to the interrupt when the IRET instruction is encountered at the end of the interrupt service procedure. Therefore, if interrupts were enabled prior to the interrupt service procedure, they are automatically re-enabled by the IRET instruction at the end of the interrupt procedure.

The return address (in CS and IP) is pushed onto the stack during the interrupt. Sometimes the return address points to the next instruction in the program and sometimes it points to the instruction or point in the program where the interrupt occurred. Interrupt type numbers 0, 5, 6, 7, 8, 10, 11, 12, and 13 push a return address that points to the offending instruction, instead of the next instruction in the program. This allows the interrupt service procedure to possibly retry the instruction in certain error cases.

Some of the protected-mode interrupts (types 8, 10, 11, 12, and 13) place an error code on the stack following the return address. The error code identifies the selector that caused the interrupt. In cases where no selector is involved, the error code is a 0.

Operation of a Protected-Mode Interrupt

In the protected mode, interrupts have exactly the same assignments as in real mode, but the interrupt vector table is different. In place of interrupt vectors, the protected mode uses a set of 256 interrupt descriptors stored in an interrupt descriptor table (IDT). The interrupt descriptor table is 256×8 (2K) bytes in length, with each descriptor containing 8 bytes. The interrupt descriptor table is located at any memory location in the system by the interrupt descriptor table address register (IDTR).

Each entry in the IDT contains the address of the interrupt service procedure in the form of a segment selector and a 32-bit offset address. It also contains the P bit (present) and DPL bits to describe the privilege level of the interrupt. Figure 12–2 shows the contents of the interrupt descriptor.

Real-mode interrupt vectors can be converted into protected-mode interrupts by copying the interrupt procedure addresses from the interrupt vector table and converting them to 32-bit offset addresses that are stored in the interrupt descriptors. A single selector and segment descriptor can be placed in the global descriptor table that identifies the first 1 M byte of memory as the interrupt segment.

FIGURE 12–2 The protected-mode interrupt descriptor.

Interrupt Descriptor (Protected Mode)

7	Offset (A31–A24)	Offset (A23–A16)	6
5	P	DPL	0	1110	00000000	4
3	Selector (S15–S8)	Selector (S7–S0)	2
1	Offset (A15–A8)	Offset (A7–A0)	0

FIGURE 12–3 The flag register.

Other than the IDT and interrupt descriptors, the protected-mode interrupts function like the real-mode interrupts. Both types of interrupt procedures are returned using the IRET (real mode) or IRETD (protected mode) instruction. The only difference is that in protected mode the microprocessor accesses the IDT instead of the interrupt vector table.

Interrupt Flag Bits

The interrupt flag (IF) and the trap flag (TF) are both cleared after the contents of the flag register are stacked during an interrupt. Figure 12–3 illustrates the contents of the flag register and the location of IF and TF. When the IF bit is set, it allows the INTR pin to cause an interrupt; when the IF bit is cleared, it prevents the INTR pin from causing an interrupt. When TF = 1, it causes a trap interrupt (type number 1) to occur after each instruction executes. This is why we often call a trap a single-step. When TF = 0, normal program execution occurs. This flag bit allows debugging, as explained in later chapters that detail the 80386–Pentium Pro.

The interrupt flag is set and cleared by the STI and CLI instructions, respectively. There are no special instructions that set or clear the trap flag. Example 12–1 shows an interrupt service procedure that turns tracing on by setting the trap flag bit on the stack from inside the procedure. Example 12–2 shows an interrupt service procedure that turns tracing off by clearing the trap flag on the stack from within the procedure.

EXAMPLE 12–1

```
                        ;Procedure that sets TF to enable trap.
                        ;
0000                    TRON    PROC    NEAR
0000 50                         PUSH    AX              ;save registers
0001 55                         PUSH    BP
0002 8B EC                      MOV     BP,SP           ;get SP
0004 8B 46 08                   MOV     AX,[BP+8]       ;get flags from stack
0007 80 CC 01                   OR      AH,1            ;set TF
000A 89 46 08                   MOV     [BP+8],AX       ;save flags
000D 5D                         POP     BP              ;restore registers
000E 58                         POP     AX
000F CF                         IRET

0010                    TRON    ENDP
```

EXAMPLE 12–2

```
                        ;Procedure that clears TF to disable trap.
                        ;
0000                    TROFF   PROC    NEAR

0000 50                         PUSH    AX              ;save registers
0001 55                         PUSH    BP
0002 8B EC                      MOV     BP,SP           ;get SP
0004 8B 46 08                   MOV     AX,[BP+8]       ;get TF
0007 80 E4 FE                   AND     AH,0FEH         ;clear TF
000A 89 46 08                   MOV     [BP+8],AX       ;save flags
000D 5D                         POP     BP              ;restore registers
000E 58                         POP     AX
000F CF                         IRET

0010                    TROFF   ENDP
```

In both examples, the flag register is retrieved from the stack by using the BP register, which by default addresses the stack segment. After the flags are retrieved, the TF bit is either set (TRON) or cleared (TROFF) before returning from the interrupt service procedure. The IRET instruction restores the flag register with the new state of the trap flag.

Trace Procedure.   Assuming that TRON is accessed by an INT 40H instruction and TROFF is accessed by an INT 41H instruction, Example 12–3 traces through a program immediately following the INT 40H instruction. The interrupt service procedure illustrated in Example 12–3 responds to interrupt type number 1 or a trap interrupt. Each time the trap interrupt occurs—after each instruction executes following INT 40H—the TRACE procedure displays the contents of all the 16-bit microprocessor registers on the CRT screen. This provides a register trace of all the instructions between the INT 40H (TRON) and INT 41H (TROFF).

EXAMPLE 12–3

```
                                  .MODEL TINY
0000                              .CODE
0000 41 58 20 3D 20 42   RNAME    DB      'AX = ','BX = ','CX = ','DX = '
     58 20 3D 20 43 58
     20 3D 20 44 58 20
     3D 20
0014 53 50 20 3D 20 42            DB      'SP = ','BP = ','SI = ','DI = '
     50 20 3D 20 53 49
     20 3D 20 44 49 20
     3D 20
0028 49 50 20 3D 20 46            DB      'IP = ','FL = ','CS = ','DS = '
     4C 20 3D 20 43 53
     20 3D 20 44 53 20
     3D 20
003C 45 53 20 3D 20 53            DB      'ES = ','SS = '
     53 20 3D 20

                         DISP     MACRO   PAR1
                                  PUSH    AX
                                  PUSH    DX
                                  MOV     DL,PAR1
                                  MOV     AH,6
                                  INT     21H
                                  POP     DX
                                  POP     AX
                                  ENDM

                         CRLF     MACRO
                                  DISP    13
                                  DISP    10
                                  ENDM

0046                     TRACE    PROC    FAR USES AX BP BX

0049 BB 0000 R                    MOV     BX,OFFSET RNAME     ;address names
                                  CRLF
0060 E8 004D                      CALL    DREG               ;display AX
0063 58                           POP     AX                 ;get BX
0064 50                           PUSH    AX
0065 E8 0048                      CALL    DREG               ;display BX
0068 8B C1                        MOV     AX,CX
006A E8 0043                      CALL    DREG               ;display CX
006D 8B C2                        MOV     AX,DX
006F E8 003E                      CALL    DREG               ;display DX
0072 8B C4                        MOV     AX,SP
0074 83 C0 0C                     ADD     AX,12
0077 E8 0036                      CALL    DREG               ;display SP
007A 8B C5                        MOV     AX,BP
007C E8 0031                      CALL    DREG               ;display BP
```

```
007F 8B C6                          MOV     AX,SI
0081 E8 002C                        CALL    DREG            ;display SI
0084 8B C7                          MOV     AX,DI
0086 E8 0027                        CALL    DREG            ;display DI
0089 8B EC                          MOV     BP,SP
008B 8B 46 06                       MOV     AX,[BP+6]
008E E8 001F                        CALL    DREG            ;display IP
0091 8B 46 0A                       MOV     AX,[BP+10]
0094 E8 0019                        CALL    DREG            ;display flags
0097 8B 46 08                       MOV     AX,[BP+8]
009A E8 0013                        CALL    DREG            ;display CX
009D 8C D8                          MOV     AX,DS
009F E8 000E                        CALL    DREG            ;display DS
00A2 8C C0                          MOV     AX,ES
00A4 E8 0009                        CALL    DREG            ;display ES
00A7 8C D0                          MOV     AX,SS
00A9 E8 0004                        CALL    DREG            ;display SS
                                    IRET


00B0                     TRACE      ENDP


00B0                     DREG       PROC    NEAR USES CX

00B1 B9 0005                        MOV     CX,5            ;load count
00B4              DREG1:
                                    DISP    CS:[BX]         ;display character
00BF 43                             INC     BX              ;address next
00C0 E2 F2                          LOOP    DREG1           ;repeat 5 times
00C2 B9 0004                        MOV     CX,4            ;load count
00C5              DREG2:
00C5 D3 C8                          ROL     AX,1            ;position digit
00C7 D3 C8                          ROL     AX,1
00C9 D3 C8                          ROL     AX,1
00CB D3 C8                          ROL     AX,1
00CD 50                             PUSH    AX
00CE 24 0F                          AND     AL,0FH          ;convert to ASCII
                                    .IF AL > 9
00D4 04 07                             ADD AL,7
                                    .ENDIF
00D6 04 30                          ADD     AL,30H
                                    DISP    AL
00E2 58                             POP     AX
00E3 E2 E0                          LOOP    DREG2           ;repeat 4 times
                                    DISP    ' '
                                    RET

00F1                     DREG       ENDP
                                    END
```

Storing an Interrupt Vector in the Vector Table

In order to install an interrupt vector—sometimes called a **hook**—the assembler must address absolute memory. Example 12–4 shows how a new vector is added to the interrupt vector table by using the assembler and a DOS function call. Here INT 21H function call number 25H initializes the interrupt vector. Notice that the first thing done in this procedure is to save the old interrupt vector number using DOS INT 21H function call number 35H to read the current vector. Refer to Appendix A for more detail on DOS INT 21H function calls.

EXAMPLE 12–4

```
                        .MODEL TINY
                        .CODE
                        ;Program that installs NEW40 at INT 40H.
                        ;
                        .STARTUP
```

```
0100 EB 05                              JMP    START
0102 00000000              OLD    DD    ?
                           ;
                           ;new interrupt procedure
                           ;
0106                       NEW40  PROC   FAR

0106 CF                           IRET

0107                       NEW40  ENDP

0107              START:
0107 8C C8                        MOV    AX,CS      ;get data segment
0109 8E D8                        MOV    DS,AX
010B B4 35                        MOV    AH,35H     ;get old interrupt vector
010D B0 40                        MOV    AL,40H
010F CD 21                        INT    21H
0111 89 1E 0102 R                 MOV    WORD PTR OLD,BX
0115 8C 06 0104 R                 MOV    WORD PTR OLD+2,ES
                           ;
                           ;install new interrupt vector 40H
                           ;
0119 BA 0106 R                    MOV    DX,OFFSET NEW40
011C B4 25                        MOV    AH,25H
011E B0 40                        MOV    AL,40H
0120 CD 21                        INT    21H
                           ;
                           ;leave NEW40 in memory
                           ;
0122 BA 0107 R                    MOV    DX,OFFSET START
0125 D1 EA                        SHR    DX,1
0127 D1 EA                        SHR    DX,1
0129 D1 EA                        SHR    DX,1
012B D1 EA                        SHR    DX,1
012D 42                           INC    DX
012E B8 3100                      MOV    AX,3100H
0131 CD 21                        INT    21H
                                  END
```

12–2   HARDWARE INTERRUPTS

The microprocessor (except for the 80188/80186 and 80386EX) has two hardware interrupt inputs: non-maskable interrupt (NMI) and interrupt request (INTR). Whenever the NMI input is activated, a type 2 interrupt occurs because NMI is internally decoded. The INTR input must be externally decoded to select a vector. Any interrupt vector can be chosen for the INTR pin, but we usually use an interrupt type number between 20H and FFH. The $\overline{\text{INTA}}$ signal is also an interrupt pin on the microprocessor, but it is an output that is used in response to the INTR input to apply a vector type number to the data bus connections D_7–D_0. Figure 12–4 shows the three user interrupt connections on the microprocessor.

The non-maskable interrupt (NMI) is an edge-triggered input that requests an interrupt on the positive edge (0-to-1 transition). After a positive edge, the NMI pin must remain a logic 1 until it is recognized by the microprocessor. Before the positive edge is recognized, the NMI pin must be a logic 0 for at least two clocking periods.

The NMI input is often used for parity errors and other major system faults such as power failures. Power failures are easily detected by monitoring the AC power line and causing an NMI interrupt whenever AC power drops out. In response to this type of interrupt, the microprocessor stores all the internal register in a battery backed-up-memory or an EEPROM. Figure 12–5 shows a power failure detection circuit that provides a logic 1 to the NMI input whenever AC power is interrupted.

FIGURE 12–4 The interrupt pins.

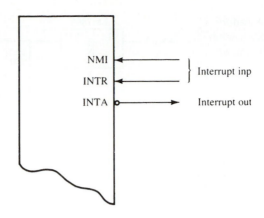

In this circuit, an optical isolator provides isolation from the AC power line. The output of the isolator is shaped by a Schmitt-trigger inverter that provides a 60-Hz pulse to the trigger input of the 74LS122 retriggerable monostable multivibrator. The values of R and C are chosen so the 74LS122 has an active pulse width of 33 ms or two AC input periods. Because the 74LS122 is retriggerable, as long as AC power is applied, the Q output remains triggered at a logic 1 and $\overline{Q}$ remains a logic 0.

If the AC power fails, the 74LS122 no longer receives trigger pulses from the 74ALS14, which means that Q returns to a logic 0 and $\overline{Q}$ returns to a logic 1, interrupting the microprocessor through the NMI pin. The interrupt service procedure, not shown here, stores the contents of all internal registers and other data into a battery-backed-up memory. This system assumes that the power supply has a large enough filter capacitor to provide energy for at least 75 ms after the AC power ceases.

Figure 12–6 shows a circuit that supplies power to a memory after the DC power fails. Here diodes are used to switch supply voltages from the DC power supply to the battery. The diodes used are standard silicon diodes because the power supply to this memory circuit is elevated above +5.0 V to +5.7 V. Notice that the resistor is used to trickle-charge the battery, which is either NiCad, lithium, or a gel cell.

When DC power fails, the battery provides a reduced voltage to the V_{CC} connection on the memory device. Most memory devices will retain data with V_{CC} voltages as low as 1.5 V, so the battery voltage does not need to be +5.0 V. The $\overline{WR}$ pin is pulled to V_{CC} during a power outage, so no data will be written to the memory.

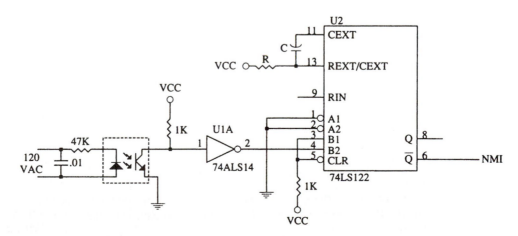

FIGURE 12–5 A power failure detection circuit.

FIGURE 12–6   A battery-backed-up memory system using a NiCad, lithium, or gel cell.

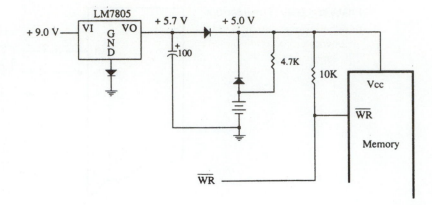

INTR and $\overline{\text{INTA}}$

The interrupt request input (INTR) is level-sensitive, which means that it must be held at a logic 1 level until it is recognized. The INTR pin is set by an external event and cleared inside the interrupt service procedure. This input is automatically disabled once it is accepted by the microprocessor and re-enabled by the IRET instruction at the end of the interrupt service procedure. The 80386–Pentium Pro use the IRETD instruction in the protected mode of operation.

The microprocessor responds to the INTR input by pulsing the $\overline{\text{INTA}}$ output in anticipation of receiving an interrupt vector type number on data bus connection D_7–D_0. Figure 12–7 shows the timing diagram for the INTR and $\overline{\text{INTA}}$ pins of the microprocessor. There are two $\overline{\text{INTA}}$ pulses generated by the system that are used to insert the vector type number on the data bus.

Figure 12–8 illustrates a simple circuit that applies interrupt vector type number FFH to the data bus in response to an INTR. Notice that the $\overline{\text{INTA}}$ pin is not connected in this circuit. Because resistors are used to pull the data bus connections (D_0–D_7) high, the microprocessor automatically sees vector type number FFH in response to the INTR input. This is possibly the least-expensive way to implement the INTR pin on the microprocessor.

Using a Three-State Buffer for $\overline{\text{INTA}}$.   Figure 12–9 shows how interrupt vector type number 80H is applied to the data bus (D_0–D_7) in response to an INTR. In response to the INTR, the micro-

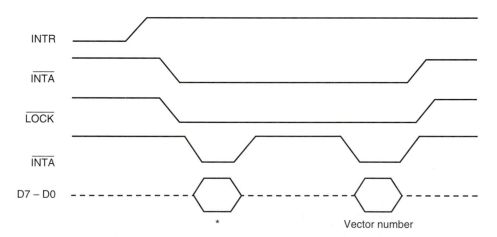

FIGURE 12–7   The timing for the INTR and $\overline{\text{INTA}}$ pulses for the microprocessor.
Note: This portion of the data bus is ignored and usually contains the vector number.

FIGURE 12–8 A simple method for generating inter-rupt vector type number FFH in response to INTR.

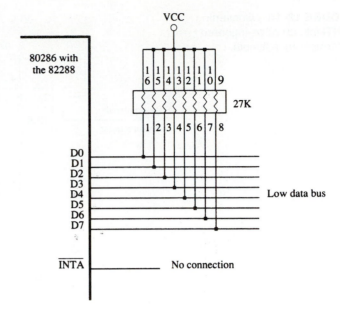

FIGURE 12–9 A circuit that applies any interrupt vector type number in response to INTA. Here the circuit is ap-plying type number 80H.

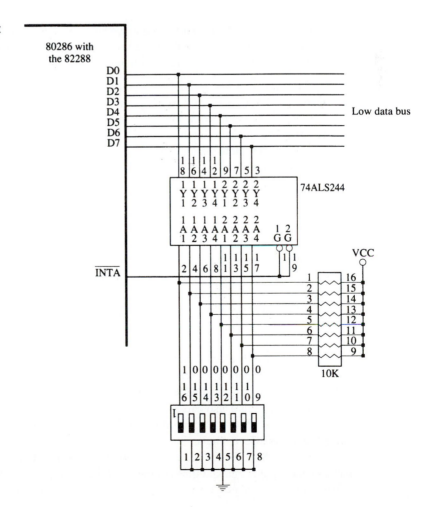

FIGURE 12–10 Converting INTR into an edge-triggered interrupt request input.

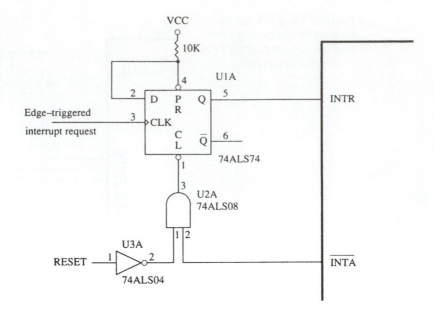

processor outputs the $\overline{\text{INTA}}$ that is used to enable a 74ALS244 three-state octal buffer. The octal buffer applies the interrupt vector type number to the data bus in response to the $\overline{\text{INTA}}$ pulse. The vector type number is easily changed with the DIP switches that are shown in this illustration.

Making the INTR Input Edge-Triggered.   Often an edge-triggered input is needed instead of a level-sensitive input. The INTR input is converted to an edge-triggered input by using a D-type flip-flop as illustrated in Figure 12–10. Here the clock input becomes an edge-triggered interrupt request input and the clear input is used to clear the request when the $\overline{\text{INTA}}$ signal is output by the microprocessor. Note that the RESET signal initially clears the flip-flop so no interrupt is requested when the system is first powered.

The 82C55 Keyboard Interrupt

The keyboard example presented in the last chapter provides a simple example of the operation of the INTR input and an interrupt. Figure 12–11 illustrates the interconnection of the 82C55 with the microprocessor and the keyboard. It also shows how a 74ALS244 octal buffer is used to provide the microprocessor with interrupt vector type number 40H in response to the keyboard interrupt during the $\overline{\text{INTA}}$ pulse.

The 82C55 is decoded at 80186 I/O port addresses 0500H, 0502H, 0504H, and 0506H by a PAL16L8 (the program is not illustrated). The 82C55 is operated in mode 1 (strobed input mode), so whenever a key is typed, the INTR output (PC3) becomes a logic 1, requesting an interrupt through the INTR pin on the microprocessor. The INTR pin remains high until the ASCII data are read from port A. In other words, every time a key is typed, the 82C55 requests a type 40H interrupt through the INTR pin. The $\overline{\text{DAV}}$ signal from the keyboard causes data to be latched into port A and also causes INTR to become a logic 1.

Example 12–5 lists the interrupt service procedure for the keyboard. It is very important that all registers affected by an interrupt are saved before they are used. In the software required to initialize the 82C55 (not shown here), the FIFO is initialized so both pointers are equal, the INTR request pin is enabled through the INTE bit inside the 82C55, and the mode of operation is programmed.

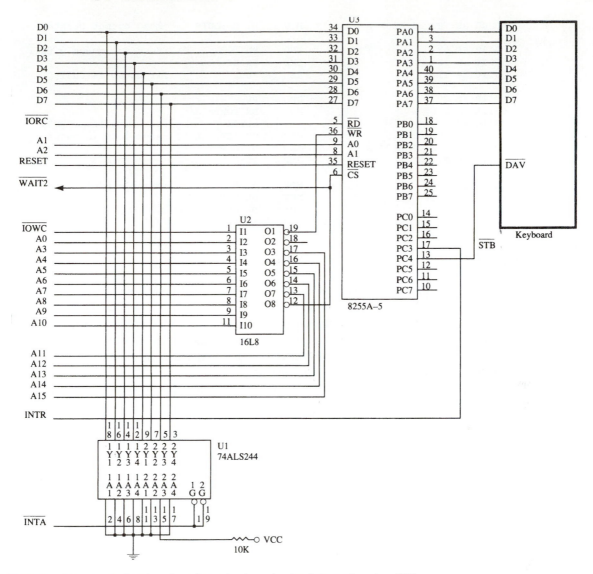

FIGURE 12–11 An 82C55 interfaced to a keyboard using interrupt vector 40H.

EXAMPLE 12–5

```
                      ;Interrupt service procedure that reads a key
                      ;from the keyboard in Figure 12-11.
                      ;
= 0500                PORTA   EQU     500H
= 0506                CNTR    EQU     506H

0000 0100 [           FIFO    DB   256 DUP (?)         ;queue
     00
     ]
0100 0000             INP     DW      ?                ;input pointer
0102 0000             OUTP    DW      ?                ;output pointer

0104                  KEY     PROC    FAR USES AX BX DI DX

0108 2E: 8B 1E 0100 R         MOV     BX,CS:INP        ;load input pointer
```

```
010D 2E: 8B 3E 0102 R          MOV    DI,CS:OUTP     ;load output pointer

0112 FE C3                     INC    BL             ;test for queue = full
0114 3B DF                     CMP    BX,DI
0116 74 11                     JE     FULL           ;if queue is full

0118 FE CB                     DEC    BL
011A BA 0500                   MOV    DX,PORTA
011D EC                        IN     AL,DX          ;get data from 82C55
011E 2E: 88 07                 MOV    CS:[BX],AL     ;save data in queue
0121 2E: FE 06 0100 R          INC    BYTE PTR INP
0126 EB 07 90                  JMP    DONE
0129                    FULL:
0129 B0 08                     MOV    AL,8           ;disable 82C55 interrupt
012B BA 0506                   MOV    DX,CNTR
012E EE                        OUT    DX,AL
012F                   DONE:
012F                           IRET

0134                   KEY    ENDP
```

The procedure is fairly short because the 80186 already knows that keyboard data is available when the procedure is called. Data is input from the keyboard and then stored in the FIFO (first-in, first-out) buffer. Most keyboard interfaces contain a FIFO that is at least 16 bytes in depth. The FIFO in this example is 256 bytes, which is more than adequate for a keyboard interface. Notice how the INC BYTE PTR INP is used to add one to the input pointer and also make sure that it always addresses data in the queue.

This procedure first checks to see if the FIFO is full. A full condition is indicated when the input point (INP) is one byte below the output pointer (OUTP). If the FIFO is full, the interrupt is disabled with a bit set/reset command to the 82C55, and a return from the interrupt occurs. If the FIFO is not full, the data is input from port A, and the input pointer is incremented before a return occurs.

Example 12–6 removes data from the FIFO. This procedure first determines whether the FIFO is empty by comparing the two pointers. If the pointers are equal, the FIFO is empty and the software waits at the EMPTY loop where it continuously tests the pointers. The EMPTY loop is interrupted by the keyboard interrupt, which stores data into the FIFO so it is no longer empty. This procedure returns with the character in register AH.

EXAMPLE 12–6

```
                       ;Procedure that reads data from the queue of
                       ;Example 12-5 and returns with it in AH.
                       ;
0134                   READ   PROC   FAR USES BX DI DX

0137                   EMPTY:
0137 2E: 8B 1E 0100 R          MOV    BX,CS:INP      ;load input pointer
013D 2E: 8B 3E 0102 R          MOV    DI,CS:OUTP     ;load output pointer
0142 3B DF                     CMP    BX,DI
0144 74 F2                     JE     EMPTY          ;if queue is empty

0146 2E: 8A 25                 MOV    AH,CS:[DI]     ;get data
0149 B0 09                     MOV    AL,9           ;enable 82C55 interrupt
014B BA 0506                   MOV    DX,CNTR
014E EE                        OUT    DX,AL
014F 2E: FE 06 0102 R          INC    BYTE PTR CS:OUTP
                               RET

0157                   READ   ENDP
```

FIGURE 12–12 Interrupt pins on the 80186/80188 microprocessors.

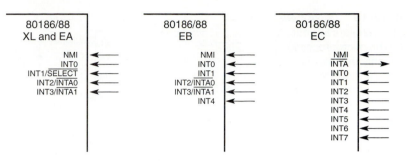

80186/80188 Interrupts

The interrupt structure of the 80186/80188 microprocessors is enhanced when compared to the other Intel microprocessors. In addition to the NMI input, the 80186/80188 contain additional dedicated interrupt input pins and the ability to cause interrupts from the internal devices such as the timers. Figure 12–12 illustrates the interrupt connections on each of the 80186/80188 family members.

Notice that the EB version contains an extra interrupt input (INT4) not found on the XL and EA versions. It also contains interrupts for the two serial ports. Although the EC version contains many additional interrupt inputs, it is programmed differently from the XL, EA, and EB versions and explained in its own section. All versions contain internal and external interrupts. The timers, serial port, and DMA channels are examples of internally activated interrupts, while the INT 2 pin is an example of an externally activated interrupt.

Table 12–2 lists the interrupt registers and their functions in controlling the interrupt structure of the 80186/80188. Also listed are the I/O ports in the peripheral control block, assuming the block is at I/O ports FFXXH, and the version where the registers are used.

Operation of the Interrupt Unit. The interrupt controller operates in two modes: master and slave. The mode of operation is selected by a bit in the interrupt control register called the CAS

TABLE 12–2 Interrupt control registers for the 80186/80188

Register	XL and EA Port Addresses	EB Port Address
INT0 Control	FF38H	FF18H
INT1 Control	FF3AH	FF1AH
INT2 Control	FF3CH	FF1CH
INT3 Control	FF3EH	FF1EH
INT4 Control	—	FF16H
Serial Control	—	FF14H
DMA Control	FF34H	—
Timer Control	FF32H	FF12H
Interrupt Status	FF30H	FF10H
Interrupt Request	FF2EH	FF0EH
In-Service	FF2CH	FF0CH
Priority Mask	FF2AH	FF0AH
Interrupt Mask	FF28H	FF08H
Poll Status	FF26H	FF06H
Poll	FF24H	FF04H
End-of-Interrupt	FF22H	FF02H

FIGURE 12–13 Two 8259A interrupt controllers interfaced to the 80186 microprocessor. Note that only the interrupt control connections are illustrated.

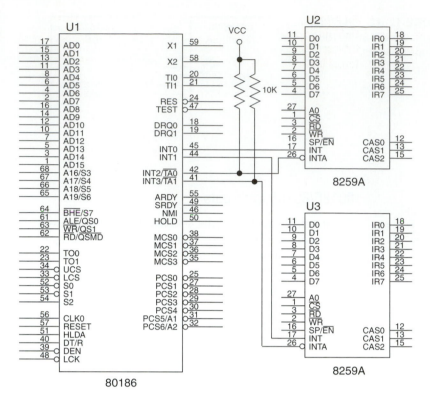

bit. If the CAS bit is a logic 1, the interrupt controller connects to external 8259A programmable interrupt controllers; if CAS is a logic 0, the internal interrupt controller is selected. In many cases there are enough interrupts within the 80186/80188 so the slave mode is not normally used.

When the interrupt controller operates in the slave mode, it uses up to two external 8259A programmable interrupt controllers for interrupt input expansion. Figure 12–13 shows how the external interrupt controllers connect to the 80186/80188 interrupt input pins for slave operation. Here the INT0 and INT1 inputs are used as external connections to the interrupt request outputs of the 8259s and $\overline{INTA0}$ (INT2) and $\overline{INTA1}$ (INT3) are used as interrupt acknowledge signals to the external controllers.

Figure 12–14 illustrates the interrupt controller registers. Notice from Table 12–2 that each interrupt input pin and each internal device has its own control register. Also notice that the control register for the cascade pins (slave mode) are different from the other control registers.

In the interrupt control register, the mask bit enables (0) or disables (1) the interrupt input represented by the control word, and the priority bits set the priority level of the interrupt source. The highest priority level is 000 and the lowest is 111. The CAS bit is used to enable slave or cascade mode (0 enables slave mode) and the SFNM bit selects the special fully nested mode. The SFNM allows the priority structure of the 8259A to be maintained. (The 8258A is discussed in detail later in this chapter.)

The interrupt request register contains an image of the interrupt sources in each mode of operation. When an interrupt is requested, the corresponding interrupt request bit becomes a logic 1 even if the interrupt is masked. The request is cleared when the 80186/80188 acknowledges the interrupt. Figure 12–15 illustrates the binary bit pattern of the interrupt request register for both the master and slave modes.

The interrupt mask register has the same format as the interrupt request register illustrated in Figure 12–15. If a source is masked (disabled), the corresponding bit of the interrupt mask

FIGURE 12–14 The structure of the control registers for internal interrupts and external interrupt pins.

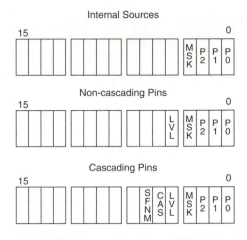

P2, P1, P0 = Interrupt Priority Level (000 = highest)
MSK = 0 enables interrupt
LVL = 0 enables edge triggering and 1 enables level triggering
CAS = 1 enables cascade mode
SFNM = 1 enables special fully nested mode

FIGURE 12–15 The interrupt request and mask register.

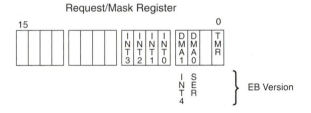

register contains a logic 1; if enabled, it contains a logic 0. The interrupt mask register is read to determine which interrupt sources are masked and which are enabled. A source is masked by setting the source's mask bit in its interrupt control register. Both the mask and request registers are used in some systems to determine if an interrupt is pending (request register) and if it is unmasked (mask register).

The priority mask register, illustrated in Figure 12–16, shows the priority of the interrupt currently being serviced by the 80186/80188. The level of the interrupt is indicated by priority bits P2–P0. Internally, these bits prevent an interrupt by a lower priority source. These bits are automatically set to the next lower level at the end of an interrupt as issued by the 80186/80188. These bits are set (111) to enable all priority levels after a reset, or if no other interrupts are pending.

The in-service register has the same binary bit pattern as the request register of Figure 12–15. The bit that corresponds to the interrupt source is currently in control of the microprocessor. The bit is reset at the end of the interrupt service procedure.

Both the interrupt poll and interrupt poll status registers share the same binary bit patterns, as illustrated in Figure 12–17. These registers have a bit (INT REQ) that indicates an interrupt is pending. This bit is set if an interrupt is received with sufficient priority and cleared when an interrupt is acknowledged. The S bits indicate the interrupt vector type number of the highest priority pending interrupt.

FIGURE 12–16 The priority mask register.

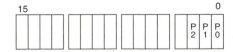

FIGURE 12–17   The poll and poll status registers.

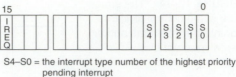

S4–S0 = the interrupt type number of the highest priority pending interrupt
IREQ = 1 if the interrupt is pending

FIGURE 12–18   The contents of the end-of-interrupt register.

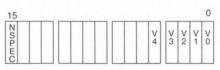

V4–V0 = vector type for a specific end-of-interrupt
NSPEC = 1 for a nonspecific end-of-interrupt

Although these two registers may appear to be identical because they contain the same information, they differ in function. When the interrupt poll register is read, the interrupt is acknowledged. When the interrupt poll status register is read, no acknowledge is sent. These registers are used only in the master mode, not in the slave mode. Reading the interrupt poll register does not update any of the internal registers. If this technique is used to acknowledge an interrupt, it is up to the program to read the interrupt type and execute the proper interrupt service procedure. In most cases, the end-of-interrupt register is used to acknowledge and end an interrupt procedure.

The end-of-interrupt (EOI) register causes the termination of an interrupt when written by a program. Figure 12–18 shows the contents of the EOI register for both master and slave modes.

In the master mode, writing to the EOI register ends either a specific interrupt level or whatever level is currently active (nonspecific). In the nonspecific mode, the NSPEC bit must be set (8000H) before the EOI register is written to end a nonspecific interrupt. The nonspecific EOI clears the highest-level interrupt bit in the in-service register. The specific EOI clears the selected bit in the in-service register. The nonspecific EOI is used in all cases unless an interrupt procedure can be interrupted by another interrupt procedure.

In the slave mode, the level of the interrupt to be terminated is written to the EOI register. The slave mode does not allow a nonspecific EOI.

The format of interrupt status register is depicted in Figure 12–19. In the master mode, T2–T0 indicate which timer (timer 0, timer 1, or timer 2) is causing an interrupt. This is necessary because all three timers have the same interrupt priority level. These bits are set when the timer requests an interrupt and cleared when the interrupt is acknowledged. The DHLT (DMA halt) bit is used only in the master mode and when set, it stops a DMA action. Notice that the interrupt status register is different for the EB version.

The interrupt vector register is present only in the slave mode and only in the XL and EA versions at offset address 20H. It is used to specify the most-significant five bits of the interrupt vector type number. Figure 12–20 illustrates the format of this register.

FIGURE 12–19   The interrupt status register.

NMI = 1 if an NMI interrupt is pending (EB version)
    = 1 suspends DMA actions (XL and EA versions)
STX = 1 for a pending serial transmit interrupt (EB version)
SRX = 1 for a pending serial receiver interrupt (EB version)
T2–T0 = 1 for a pending timer interrupt from T0, T1, and T2

FIGURE 12–20 The slave mode interrupt vector register for the XL and EA versions of the 80186/80188.

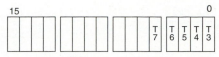

T7–T3 = the five most-significant bits of the vector type number

An Interrupt Example. The interrupt control is very complicated, so let's look at an example. Suppose an application requires a 16-bit counter that increments once per millisecond. In this application, the word-sized memory location COUNT is used to hold a number incremented once per millisecond. The contents of memory location COUNT can be read and saved, and then read again at a later time to determine the amount of time between the reads in milliseconds. To accomplish this millisecond counter task, timer 2 is used to divide the system clock down to 1 millisecond. At the end of each counting sequence, an interrupt is generated that increments the contents of memory location COUNT.

Example 12–7 shows the software required to program timer 2 and to program the interrupt control unit so timer 2 generates an interrupt. The system clock is assumed to be running with a 32-MHz crystal, which generates a system clock frequency of 16 MHz. Because the clock input to timer 2 is one-quarter of the system clock frequency, timer 2 counts once every 250 ns (4 MHz). In order to accumulate a count of 1 millisecond, timer 2 is programmed to divide by 4,000. Recall that a programmable counter multiplies time and divides frequency; 250 ns $\times$ 4,000 = 1 ms. The STI instruction is required to enable the interrupt structure once the interrupt controller is programmed. The STI instruction would appear in the software that calls SETUP.

EXAMPLE 12–7

```
= FF60              T2_CNT   EQU  0FF60H    ;timer 2 count register
= FF5E              T2_CON   EQU  0FF5EH    ;timer 2 control register
= FF62              T2_CMP   EQU  0FF62H    ;timer 2 compare register
= FF32              IT2_CN   EQU  0FF32H    ;timer interrupt control
= FF22              EOI      EQU  0FF22H    ;end-of-interrupt

;Procedure that initializes timer 2 to cause an interrupt once per
;millisecond and update COUNT. Note that this software is written for
;either the XL or EA version of the 80186/80188.

0010                SETUP    PROC   NEAR USES DX DS

0012 BA FF60                 MOV    DX,T2_CNT ;clear timer 2 count
0015 B8 0000                 MOV    AX,0
0018 EF                      OUT    DX,AX

0019 BA FF62                 MOV    DX,T2_CMP ;set count to 4,000
001C B8 0FA0                 MOV    AX,4000
001F EF                      OUT    DX,AX

0020 BA FF5E                 MOV    DX,T2_CON ;program timer 2
0023 B8 E001                 MOV    AX,0E001H ;enable interrupt and count
0026 EF                      OUT    DX,AX

0027 B8 0000                 MOV    AX,0
002A 8E D8                   MOV    DS,AX     ;address segment 0000H
002C 8C C8                   MOV    AX,CS     ;get code segment address
                    ;
                    ;install interrupt vector
                    ;
002E C7 06 004C 0041  MOV    WORD PTR DS:[4CH],TIME2
0034 A3 004E          MOV    WORD PTR DS:[4EH],AX
```

```
0037 BA FF32          MOV    DX,IT2_CN   ;address timer control
003A B8 0007          MOV    AX,7        ;enable timer interrupt
003D EF               OUT    DX,AX

                      RET

0041            SETUP ENDP
```

Example 12–8 shows the interrupt service procedure for the timer 2 interrupt. Note that this is a far procedure that is installed in Example 12–7 at vector address 4CH–4FH. All this procedure does is increment COUNT, once per millisecond, and then acknowledges the timer 2 interrupt with a nonspecific end-of-interrupt instruction. All registers used by the interrupt are saved and the IRET instruction is used to return from the interrupt service procedure. As before, this software is written for the XL and EA versions of the 80186/80188.

EXAMPLE 12–8

```
0041               TIME2  PROC   FAR USES DX AX    ;interrupt procedure

0043 FF 06 0000 R         INC    COUNT             ;increment count
0047 BA FF22              MOV    DX,EOI            ;end of interrupt
004A B8 8000              MOV    AX,8000H          ;nonspecific interrupt
004D EF                   OUT    DX,AX             ;clear timer interrupt
                          IRET                     ;interrupt return

0051               TIME2  ENDP
```

12–3   EXPANDING THE INTERRUPT STRUCTURE

In this section, we discuss three of the more common methods of expanding the interrupt structure of the microprocessor. We explain how, with software and some hardware, it is possible to expand the INTR input (INT0 or INT1 on the 80186/80188) so it accepts 7 interrupt inputs. The $\overline{INTA}$ referred to here is available on the INT2 and INT3 pins of the 80186/80188 if programmed in the slave mode. We also explain how to "daisy-chain" interrupts by software polling. In the next section, we see another technique, which allows up to 63 interrupting inputs that are added by means of the 8259A programmable interrupt controller.

Using the 74ALS244 to Expand

The circuit shown in Figure 12–21 allows the microprocessor to expand its interrupt structure by up to seven additional interrupt inputs. The only hardware change is the addition of an 8-input NAND gate, which provides the INTR signal to the microprocessor when any of the $\overline{IR}$ inputs become active.

Operation.   If any of the $\overline{IR}$ inputs become a logic 0, the output of the NAND gate generates a logic 1, which requests an interrupt through the INTR input. The interrupt vector fetched during the $\overline{INTA}$ (interrupt acknowledge) pulse depends on which of the interrupt request lines is active. Table 12–3 shows the interrupt vectors used by a single interrupt request input.

If two or more interrupt request inputs are simultaneously active, a new interrupt vector is generated. For example, if $\overline{IR1}$ and $\overline{IR0}$ are both active, the interrupt vector generated is FCH (252). Priority is resolved at this vector location (FCH). If the $\overline{IR0}$ input is to have the higher priority, the vector address for $\overline{IR0}$ is stored at vector location FCH. The entire top half of the vector table and its 128 interrupt vectors must be used to accommodate all possible conditions of these seven interrupt request inputs. This seems wasteful, but in many dedicated applications it is a cost-effective approach to interrupt expansion. In the 80186/80188 this allows seven interrupt inputs in addition to the inputs already present. Of course to enable this mode, two of the interrupt inputs are lost.

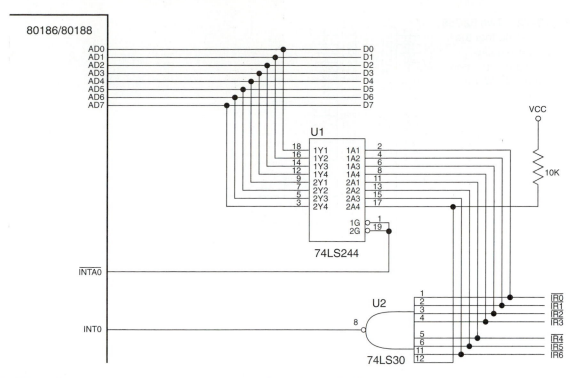

FIGURE 12–21 Expanding the interrupt structure of the 80186/80188.

TABLE 12–3 Single interrupt request for the circuit of Figure 12–21

IR6	IR5	IR4	IR3	IR2	IR1	IR0	Vector
1	1	1	1	1	1	0	FEH
1	1	1	1	1	0	1	FDH
1	1	1	1	0	1	1	FBH
1	1	1	0	1	1	1	F7H
1	1	0	1	1	1	1	EFH
1	0	1	1	1	1	1	DFH
0	1	1	1	1	1	1	BFH

Note: Although not illustrated, the IR inputs are all active-low.

Daisy-Chained Interrupt

Expansion by means of a daisy-chained interrupt is better than using the 74ALS244 interrupt expansion because it requires only one interrupt vector. The task of determining priority is left to the interrupt service procedure. Setting priority for a daisy-chain does require additional software execution time, but in general this is a much better approach to expanding the interrupt structure of the microprocessor.

Figure 12–22 illustrates a set of two 82C55 peripheral interfaces with their four INTR outputs daisy-chained and connected to the single INTR input of the microprocessor. If any interrupt output becomes a logic 1 from either 82C55, the INTR input to the microprocessor becomes a logic 1, causing an interrupt.

When a daisy-chain is used to request an interrupt, it is better to pull the data bus connections (D_0–D_7) high using pull-up resisters so interrupt vector FFH is generated and used for the

FIGURE 12–22   Two 82C55 PIAs connected so that the INTR outputs are daisy-chained to produce an interrupt request at INT0 for the 80186/80188.

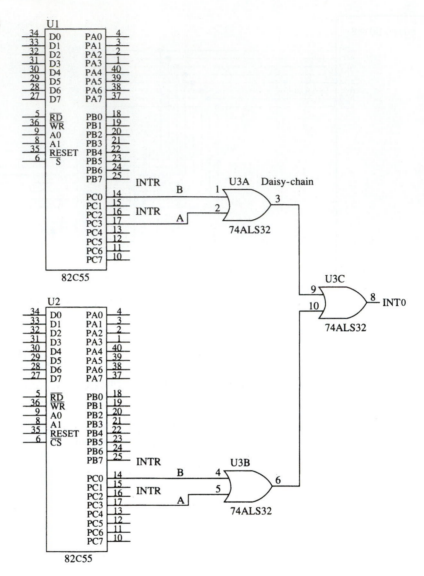

chain. Actually any interrupt vector can be used to respond to a daisy-chain. In the circuit, any of the four INTR outputs from the two 82C55s will cause the INTR pin on the microprocessor to go high requesting an interrupt.

When the INTR pin does go high with a daisy-chain, the hardware gives no direct indication as to which 82C55 or which INTR output caused the interrupt. The task of locating the INTR output that became active is up to the interrupt service procedure, which must poll the 82C55s to determine what caused the interrupt.

Example 12–9 lists the interrupt service procedure that responds to the daisy-chain interrupt request. This procedure polls each 82C55 and each INTR output to decide which interrupt service procedure to utilize.

EXAMPLE 12–9

```
                              ;Procedure that services the daisy-chain interrupt
                              ;of Figure 12-22.
                              ;
= 0504                        C1      EQU     504H       ;first 82C55
```

```
= 0604                      C2      EQU     604H        ;second 82C55
= 0001                      MASK1   EQU     1           ;INTRB
= 0008                      MASK2   EQU     8           ;INTRA

0000                        POLL    PROC    FAR USES AX DX

0002 BA 0504                        MOV     DX,C1       ;address first 82C55
0005 EC                             IN      AL,DX       ;get port C
0006 A8 01                          TEST    AL,MASK1
0008 75 0F                          JNZ     LEVEL_0     ;if INTRB is set
000A A8 08                          TEST    AL,MASK2
000C 75 13                          JNZ     LEVEL_1     ;if INTRA is set

000E BA 0604                        MOV     DX,C2       ;address second 82C55
0011 EC                             IN      AL,DX       ;get port C
0012 A8 01                          TEST    AL,MASK1
0014 75 1B                          JNZ     LEVEL_2     ;if INTRB is set
0016 EB 29 00                       JMP     LEVEL_3     ;for INTRA

0019                        POLL    ENDP
```

12–4   8259A PROGRAMMABLE INTERRUPT CONTROLLER

The 8259A programmable interrupt controller (PIC) adds eight vectored priority-encoded interrupts to the microprocessor. This controller can be expanded without additional hardware to accept up to 64 interrupt request inputs. This expansion requires a master 8259A and eight 8259A slaves. The 8259A is used in this section to explain the interrupt structure of the 80186/80188 microprocessor. We also describe the 80186/80188 version EC and the 80386EX, which incorporate the function of the 8259A interrupt controller.

General Description of the 8259A

Figure 12–23 shows the pin-out of the 8259A. The 8259A is easy to connect to the microprocessor because all of its pins are direct connections except the $\overline{CS}$ pin, which must be decoded for programming, and the $\overline{WR}$ pin, which must receive an I/O bank write pulse from the lower bank in a 16-bit system. The following list describes each pin on the 8259A.

D_7–D_0 The **bi-directional data connections** are normally connected to the lower data bus on the 80186 and 80386EX microprocessors.

IR_7–IR_0 **Interrupt request** inputs are used to request an interrupt and to connect to a slave in a system with multiple 8259As.

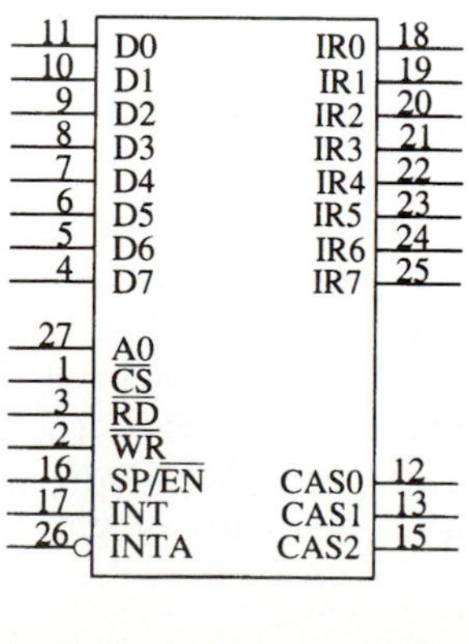

FIGURE 12–23   The pin-out of the 8259A programmable interrupt controller (PIC).

$\overline{\text{WR}}$	The **write** input connects to the lower write strobe signal in a 16-bit system or any other bus write strobe in any size system. The write strobe is used only for programming.
$\overline{\text{RD}}$	The **read** input connects to the microprocessor read signal.
INT	The **interrupt** output connects to the INTR pin on the microprocessor from the master, and connects to a master IR pin on a slave. This is the INT_0 or INT_1 pin on the 80186/80188.
$\overline{\text{INTA}}$	**The interrupt acknowledge** input connects to the $\overline{\text{INTA}}$ signal on the system. In a system with a master and slaves, only the master $\overline{\text{INTA}}$ signal is connected. This is the INT_2 or INT_3 pin on the 80186/80188.
A_0	The A_0 address input selects different command words within the 8259A. In a 16-bit system, this is often connected to A_1 from the microprocessor.
$\overline{\text{CS}}$	**Chip select** enables the 8259A for programming and control.
$\text{SP}/\overline{\text{EN}}$	**The slave program/enable buffer** is a dual-function pin. When the 8259A is in buffered mode, this is an output that controls the data bus transceivers in a large microprocessor-based system. When the 8259A is not in the buffered mode, this pin programs the device as a master (1) or a slave (0).
$\text{CAS}_2\text{--}\text{CAS}_0$	The **cascade** lines are used as outputs from the master to the slaves for cascading multiple 8259As in a system.

Connecting a Single 8259A

Figure 12–24 shows a single 8259A connected to the 80186 microprocessor. Here the $\text{SP}/\overline{\text{EN}}$ pin is pulled high to indicate that it is a master. Notice that the 8259A is decoded at I/O ports 0400H and 0402H by the PAL16L8 (no program shown). Like other peripherals discussed in Chapter 11, the 8259A requires four wait states for it to function properly with a 16-MHz 80386EX, and more wait states for some other versions of the Intel microprocessor family.

Cascading Multiple 8259As

Figure 12–25 shows two 8259As connected to the 80386EX microprocessor in a way that is often found in the AT-style computer, which has two 8259As for interrupts. The XT- or PC-style computer uses one 8259A at interrupt vectors 08H–0FH. The AT-style computer uses interrupt vector 0AH as a cascade input from a second 8259A located at vectors 70H–77H. Appendix A contains a table that lists the functions of all the interrupt vectors used in the PC-, XT-, and AT-style computers.

This circuit uses vectors 08H–0FH and I/O ports 0300H and 0302H for U1, the master; vectors 70H–77H and I/O ports 0304H and 0306H for U2, the slave. Notice that we also include data bus buffers to illustrate the use of the $\text{SP}/\overline{\text{EN}}$ pin on the 8259A. These buffers are used only in very large systems that have many devices connected to their data bus connections. In practice, we seldom find these buffers.

Programming the 8259A

The 8259A is programmed by initialization and operation command words. **Initialization command words** (ICW) are programmed before the 8259A is able to function in the system and dictate the basic operation of the 8259A. **Operation command words** (OCW) are programmed during the normal course of operation and allow the 8259A to function properly.

Initialization Command Words.   There are four initialization command words (ICW) for the 8259A that are selected when the A0 pin is a logic 1. When the 8259A is first powered up, it must be programmed with ICW1, ICW2, and ICW4. If the 8259A is programmed in cascade

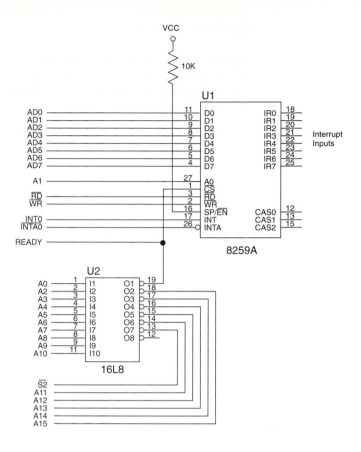

FIGURE 12–24 The 8259A interfaced to an 80186 microprocessor at I/O ports 4000H and 402H.

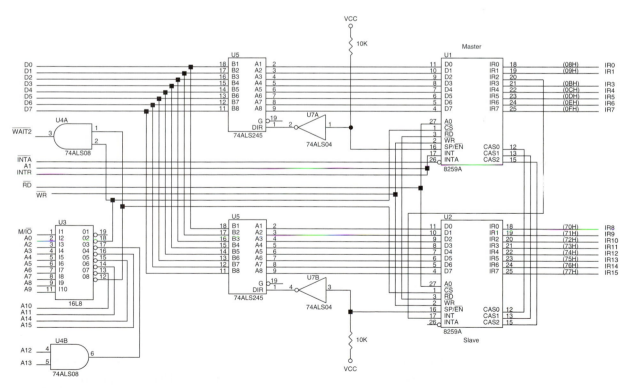

FIGURE 12–25 Two 8259As interfaced to the 8259A at I/O ports 0300H and 0302H for the master, and 0304H and 0306H for the slave.

mode by ICW1, then we must also program ICW3. So, if a single 8259A is used in a system, ICW1, ICW2, and ICW4 must be programmed. If cascade mode is used in a system, then all four ICWs must be programmed. Refer to Figure 12–26 for the format of all four ICWs. The following list describes each ICW.

ICW1   Programs the basic operation of the 8259A. To program ICW1 for 8086–Pentium Pro operation, place a logic 1 in bit IC4. Bits ADI, A7, A6, and A5 are don't cares for microprocessor operation and apply to the 8259A only when used with an 8-bit 8085 microprocessor (not covered in this book). This ICW selects single or cascade operation by programming the SNGL bit. If cascade operation is selected, we must also program ICW3. The LTIM bit determines whether the interrupt request inputs are positive edge-triggered or level triggered.

ICW2   Selects the vector number used with the interrupt request inputs. For example, if we decide to program the 8259A so it functions at vector locations 08H–0FH, we place a 08H into this command word. Likewise, if we decide to program the 8259A for vectors 70H–77H, we place a 70H in this ICW.

ICW3   Used only when ICW1 indicates that the system is operated in cascade mode. This ICW indicates where the slave is connected to the master. For example, in Figure 12–25 we connected a slave to IR2. To program ICW3 for this connection, in both master and slave, we place a 04H in ICW3. Suppose we have two slaves connected to a master using IR0 and IR1. The master is programmed with an ICW3 of 03H; one slave is programmed with an ICW3 of 01H; the other slave is programmed with an ICW3 of 02H.

ICW4   Programmed for use with the 8086–Pentium Pro microprocessors. This ICW is not programmed in a system that functions with the 8085 microprocessor. The rightmost bit must be a logic 1 to select operation with the 8086–Pentium Pro microprocessors and the remaining bits are programmed as follows:

SFNM—Delects the special fully nested mode of operation for the 8259A if a logic 1 is placed in this bit. This allows the highest priority interrupt request from a slave to be recognized by the master while it is processing another interrupt from a slave. Normally only one interrupt request is processed at a time and others are ignored until the process is complete.

BUF and **M/S**—Buffer and master slave are used together to select buffered operation or non-buffered operation for the 8259A as a master or a slave.

AEOI—Selects automatic or normal end-of-interrupt (discussed more fully under operation command words). The EOI commands of OCW2 are used only if the AEOI mode is not selected by ICW4. If AEOI is selected, the interrupt automatically resets the interrupt request bit and does not modify priority. This is the preferred mode of operation for the 8259A and reduces the length of the interrupt service procedure.

Operation Command Words.   The operation command words (OCW) are used to direct the operation of the 8259A once it is programmed with the initialization command words. The OCW are selected when the A_0 pin is at a logic 0 level, except for OCW1, which is selected when A_0 is a logic 1. Figure 12–27 shows the binary bit patterns for all three operation command words of the 8259A. The following list describes the function of each OCW.

OCW1   Used to set and read the interrupt mask register. When a mask bit is set, it will **turn off** (mask) the corresponding interrupt input. The mask register is read when OCW1 is read. Because the state of the mask bits is unknown when the 8259A is first initialized, OCW1 must be programmed after programming the ICW upon initialization.

FIGURE 12–26 The 8259A initialization command words (ICWs). (Courtesy of Intel Corporation)

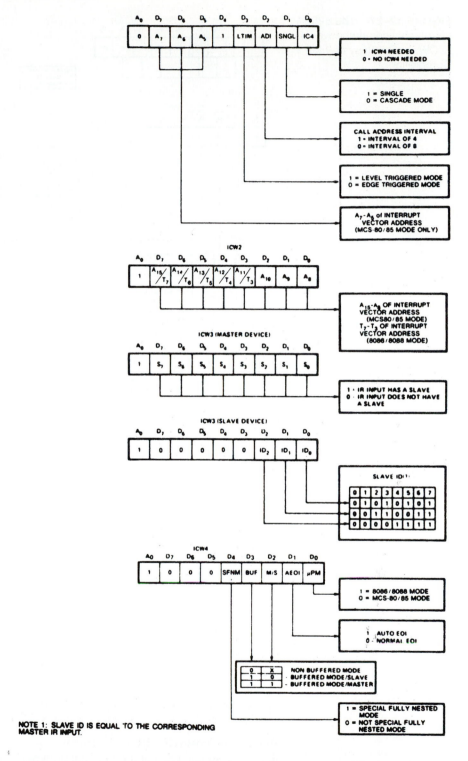

NOTE 1: SLAVE ID IS EQUAL TO THE CORRESPONDING MASTER IR INPUT.

OCW2   Programmed only when the AEOI mode is not selected for the 8259A. In this case, this OCW selects how the 8259A responds to an interrupt. The modes are listed as follows:

Nonspecific End-of-Interrupt—Sent by the interrupt service procedure to signal the end of the interrupt. The 8259A automatically determines which interrupt

FIGURE 12–27 The 8259A operational command words (OCWs). (Courtesy of Intel Corporation)

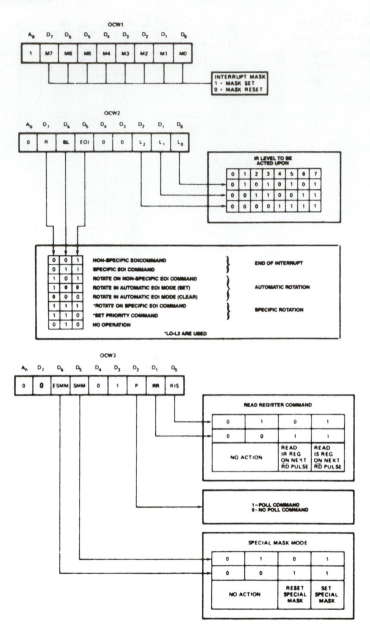

level was active and resets the correct bit of the interrupt status register. Resetting the status bit allows the interrupt to take action again or a lower priority interrupt to take effect.

Specific End-of-Interrupt—Allows a specific interrupt request to be reset. The exact position is determined with bits L2–L0 of OCW2.

Rotate-on-Nonspecific EOI—Functions exactly like the nonspecific end-of-interrupt command except it rotates interrupt priorities after resetting the interrupt status register bit. The level reset by this command becomes the lowest priority interrupt. For example, if IR4 was just serviced by this command, it becomes the lowest priority interrupt input and IR5 becomes the highest priority.

Rotate-on-Automatic EOI—Selects automatic EOI with rotating priority. This command must be sent to the 8259A only once if this mode is desired. If this mode must be turned off, use the clear command.

Rotate-on-Specific EOI—Functions as the specific EOI, except that it selects rotating priority.

Set priority—Allows you to set the lowest priority interrupt input using the L2–L0 bits.

OCW3 Selects the register to be read, the operation of the special mask register, and the poll command. If polling is selected, the P bit must be set and then output to the 8259A. The next read operation will read the poll word. The rightmost three bits of the poll word indicate the active interrupt request with the highest priority. The leftmost bit indicates whether there is an interrupt, and must be checked to determine whether the rightmost three bits contain valid information.

Status Register. Three status registers are readable in the 8259A: interrupt request register (IRR), in-service register (ISR), and interrupt mask register (IMR). (Refer to Figure 12–28 for all three status registers; they all have the same bit configuration.) The IRR is an 8-bit register that indicates which interrupt request inputs are active. The ISR is an 8-bit register that contains the level of the interrupt being serviced. The IMR is an 8-bit register that holds the interrupt mask bits and indicates which interrupts are masked off.

Both the IRR and ISR are read by programming OCW3, and IMR is read through OCW1. To read the IMR, A0 = 1, and to read either IRR or ISR, A0 = 0. Bit positions D0 and D1 of OCW3 select which register (IRR or ISR) is read when A0 = 0.

8259A Programming Example

Figure 12–29 illustrates the 8259A programmable interrupt controller connected to a 16550 programmable communications controller. In this circuit, the INTR pin from the 16550 is connected to the programmable interrupt controller's interrupt request input IR0. An IR0 occurs when: (1) the transmitter is ready to send another character, (2) when the receiver has received a character, (3) when an error is detected while receiving data, and (4) a modem interrupt occurs. Notice that the 16550 is decoded at I/O ports 40H and 47H, and the 8259A is decoded at 8-bit I/O ports 48H and 49H. Both devices are interfaced to the data bus of an 80188 microprocessor.

Initialization Software. The first portion of the software for this system must program both the 16550 and the 8259A and then enable the INTR pin on the 80188 so interrupts can take effect. Example 12–10 lists the software required to program both devices and enable INTR. This software uses two memory FIFOs that hold data for the transmitter and for the receiver. Each memory FIFO is 16K bytes in length and is addressed by a pair of pointers (input and output).

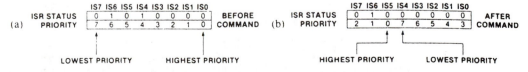

FIGURE 12–28 The 8259A in-service register (ISR). (a) Before IR$_4$ is accepted, and (b) after IR$_4$ is accepted. (Courtesy of Intel Corporation)

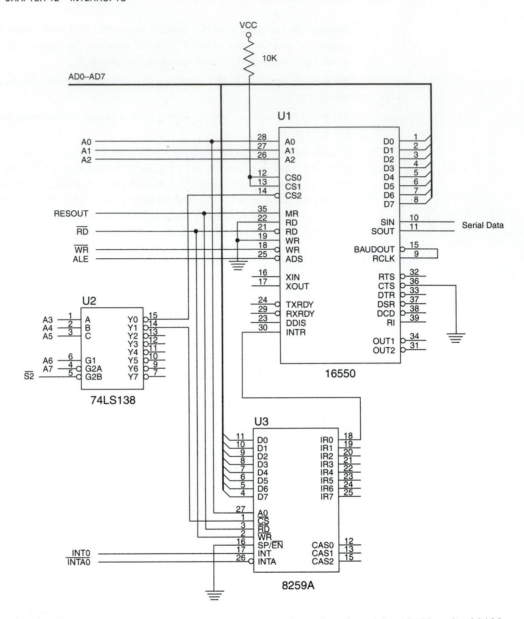

FIGURE 12–29   A communications circuit interfacing the 8259A and the 16550 to the 80188.

EXAMPLE 12–10

```
                         ;Initialization software for the 16550 and 8259A
                         ;of the circuit in Figure 12-29.
                         ;
= 0048          PIC1     EQU    48H      ;8259A control A0 = 0
= 0049          PIC1     EQU    49H      ;8259A control A0 = 1
= 001B          ICW1     EQU    1bH      ;8259A ICW1
= 0080          ICW2     EQU    80H      ;8259A ICW2
= 0003          ICW4     EQU    3        ;8259A ICW4
= 00FE          OCW1     EQU    0FEH     ;8259A OCW1
= 0043          LINE     EQU    43H      ;16550 line register
= 0040          LSB      EQU    40H      ;16550 Baud divisor LSB
= 0041          MSB      EQU    41H      ;16550 Baud divisor MSB
= 0042          FIFO     EQU    42H      ;16550 FIFO register
```

```
= 0041                          ITR     EQU     41H        ;16550 interrupt register
                                ;
0000                    START   PROC    NEAR
                                ;
                                ;Program 16550, but do not enable interrupts yet.
                                ;
0000 B0 8A                              MOV     AL,10001010B   ;enable Baud divisor
0002 E6 43                              OUT     LINE,AL
0004 B0 78                              MOV     AL,120         ;program Baud rate
0006 E6 40                              OUT     LSB,AL         ;9600 Baud rate
0008 B0 00                              MOV     AL,0
000A E6 41                              OUT     MSB,AL
000C B0 0A                              MOV     AL,00001010B   ;program 7-data, odd
000E E6 43                              OUT     LINE,AL        ;parity, one stop
0010 B0 07                              MOV     AL,00000111B   ;enable transmitter
0012 E6 42                              OUT     FIFO,AL        ;and receiver
;
                                ;Program 8259A
                                ;
0014 B0 1B                              MOV     AL,ICW1        ;program ICW1
0016 E6 48                              OUT     PIC1,AL

0018 B0 80                              MOV     AL,ICW2        ;program ICW2
001A E6 49                              OUT     PIC2,AL

001C B0 03                              MOV     AL,ICW4        ;program ICW4
001E E6 49                              OUT     PIC2,AL

0020 B0 FE                              MOV     AL,OCW1        ;program OCW1
0022 E6 49                              OUT     PIC2,AL

0024 FB                                 STI                    ;enable system INTR pin
                                ;
                                ;enable 16550 interrupts
                                ;
0025 B0 07                              MOV     AL,5           ;enable receiver and
0027 E6 41                              OUT     ITR,AL         ;error interrupts
0029 C3                                 RET

002A                    START   ENDP
```

The first portion of the procedure (START) programs the 16550 UART for operation with seven data bits, odd parity, one stop bit, and a Baud rate clock of 9600. The FIFO control register also enables both the transmitter and receiver.

The second part of the procedure programs the 8259A with its three ICWs and its one OCW. The 8259A is set up so it functions at interrupt vectors 80H–87H and operates with automatic EOI. The ICW enables the interrupt for the 16550 UART. Also enabled is the INTR pin of the microprocessor by using the STI instruction.

The final part of the software enables the receiver and error interrupts of the 16550 UART through the interrupt control register. The transmitter interrupt is not enabled until data is available for transmission. Refer to Figure 12–30 for the contents of the interrupt control register of the 16550 UART. Notice that the control register can enable or disable the receiver, transmitter, line status (error), and modem interrupts.

FIGURE 12–30 The interrupt control register for the 16550.

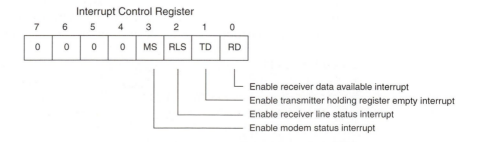

FIGURE 12–31   The interrupt identification register for the 16550.

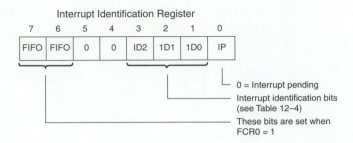

TABLE 12–4   The interrupt control bits of the 16550

Bit 3	Bit 2	Bit 1	Bit 0	Priority	Type	Reset Control
0	0	0	1	—	No interrupt	—
0	1	1	0	1	Receiver error (parity, framing, overrun, or break)	Reset by reading the line register
0	1	0	0	2	Receiver data available	Reset by reading the data
1	1	0	0	2	Character time-out, nothing has been removed from the receiver FIFO for at least 4 character times	Reset by reading the data
0	0	1	0	3	Transmitter empty	Reset by writing to the transmitter
0	0	0	0	4	Modem status	Reset by reading the modem status

Note: 1 is the highest priority and 4 the lowest.

Handling the 16550 UART Interrupt Request.   Because the 16550 generates only one interrupt request for various interrupts, the interrupt handler must poll the 16550 to determine which type of interrupt has occurred. This is accomplished by examining the interrupt identification register (see Figure 12–31). Note that the interrupt identification register (read-only) shares the same I/O port as the FIFO control register (write-only).

The interrupt identification register indicates if an interrupt is pending and the type of interrupt. It also shows if the transmitter and receiver FIFO memories are enabled. Refer to Table 12–4 for the contents of the interrupt control bits.

The interrupt service procedure must examine the contents of the interrupt identification register to determine which event caused the interrupt and pass control to the appropriate procedure for the event. Example 12–11 shows the first part of an interrupt handler that passes control to RECV for a receiver data interrupt, to TRANS for a transmitter data interrupt, and to ERR for a line status error interrupt. Note that the modem status is not tested in this example.

EXAMPLE 12–11

```
                  ;Interrupt handler for the 16550 UART of
                  ;Figure 12-29.
                  ;
0000              INT80   PROC    FAR

0000 50                   PUSH    AX
```

```
0001 E4 42                      IN     AL,42H          ;input interrupt ID reg
0003 3C 06                      CMP    AL,6            ;test for error
0005 74 20                      JE     ERR             ;for receiver error

0007 3C 02                      CMP    AL,2            ;test for transmitter
0009 74 55                      JE     TRANS           ;for transmitter ready

000B 3C 04                      CMP    AL,4            ;test for receiver
000D 74 11                      JE     RECV            ;for receiver ready
```

Receiving Data from the 16550. The data received by the 16550 is stored, not only in the FIFO within the UART, but also in a FIFO memory until the software in the main program can use them. The FIFO memory used for received data is 16K bytes in length, so many characters can easily be stored and received before any intervention from the microprocessor is required to empty the receiver's memory FIFO. The receiver memory FIFO is stored in the extra segment so string instructions, using the DI register, can be used to access it.

Receiving data from the 16550 requires two procedures: one reads the data register of the 16550 each time that the INTR pin requests an interrupt, and stores it into the memory FIFO; the other reads data from the memory FIFO from the main program.

Example 12–12 reads data from the memory FIFO from the main program. This procedure assumes that the pointers (IIN and IOUT) are initialized in the initialization dialog for the system (not shown). The READ procedure returns with AL containing a character read from the memory FIFO. If the memory FIFO is empty, the procedure returns with the carry flag bit set to a logic 1. If AL contains a valid character, the carry flag bit is cleared upon return from READ.

Notice how the FIFO is reused by changing the address from the top of the FIFO to the bottom whenever it exceeds the start of the FIFO plus 16K. This is located at the CMP instruction at offset address 0015. Also notice that interrupts are enabled at the end of this procedure in case they are disabled by a full-memory FIFO condition by the RECV interrupt procedure.

EXAMPLE 12–12

```
                        ;Procedure that reads one character from the memory
                        ;FIFO and returns with it in AL.
                        ;If the FIFO is empty, the return occurs with carry = 1.
                        ;
0000                    READ    PROC    NEAR USES BX DI

0002 26: 8B 3E 4002 R          MOV    DI,IOUT          ;get output pointer
0007 26: 8B 1E 4000 R          MOV    BX,IIN           ;get input pointer

000C 3B DF                     CMP    BX,DI            ;compare pointers
000E F9                        STC                     ;set carry flag
000F 74 16                     JE     DONE1            ;if empty

0011 26: BA 06                 MOV    AL,ES:[DI]       ;get data from FIFO
0014 47                        INC    DI               ;address next byte
0015 81 FF 4000 R              CMP    DI,OFFSET FIFO+16*1024
0019 26: 89 3E 4002 R          MOV    IOUT,DI          ;save pointer
001E 76 07                     JBE    DONE             ;if within bounds
0020 26: C7 06 4002 R          MOV    IOUT,OFFSET FIFO
     0000 R
0027           DONE:
0027 F8                        CLC                     ;clear carry flag
0028           DONE1:
0028 9C                        PUSHF                   ;save carry flag
0029 E4 41                     IN     AL,41H           ;read interrupt control
002B 06 05                     OR     AL,5             ;enable receiver interrupts
002D E6 41                     OUT    41H,AL
002F 9D                        POPF
                               RET

0033           READ    ENDP
```

Example 12–13 lists the RECV interrupt service procedure that is called each time the 16550 receives a character for the microprocessor. In this example, this interrupt uses vector type number 80H, which must address the interrupt handler of Example 12–11. Each time this interrupt occurs, the REVC procedure is accessed by the interrupt handler's reading a character from the 16550. The RECV procedure stores the character into the memory FIFO. If the memory FIFO is full, the receiver interrupt is disabled by the interrupt control register within the 16550. This may result in lost data, but at least it will not cause the interrupt to overrun valid data already stored in the memory FIFO. Any error conditions detected by the 8251A store a ? (3FH) in the memory FIFO. Note that errors are detected by the ERR portion of the interrupt handler (not shown).

EXAMPLE 12–13

```
                          ;RECV portion of the interrupt handler in
                          ;Example 12-11.
                          ;
0020                      RECV:                        ;continues from Example 12-11
0020 53                           PUSH    BX           ;save registers
0021 57                           PUSH    DI
0022 56                           PUSH    SI
0023 26: 8B 1E 4002 R             MOV     BX,IOUT      ;load output pointer
0028 26: 8B 36 4000 R             MOV     SI,IIN       ;load input pointer
002D 8B FE                        MOV     DI,SI
002F 46                           INC     SI
0030 81 FE 4000 R                 CMP     SI,OFFSET FIFO+16*1024
0034 76 03                        JBE     NEXT
0036 BE 0000 R                    MOV     SI,OFFSET FIFO
0039             NEXT:
0039 3B DE                        CMP     BX,SI        ;is FIFO full?
003B 74 0B                        JE      FULL         ;if it is full
003D E4 40                        IN      AL,40H       ;read 16550 receiver
003F AA                           STOSB                ;save it in FIFO
0040 26: 89 36 4000 R             MOV     IIN,SI       ;save input pointer
0045 EB 06 90                     JMP     DONE         ;end up
0048             FULL:
0048 E4 41                        IN      AL,41H       ;read interrupt control
004A 24 FA                        AND     AL,0FAH      ;disable receiver
004C E6 41                        OUT     41H,AL
004E             DONE:
004E B0 20                        MOV     AL,20H       ;signal 8259A EOI
0050 E6 49                        OUT     49H,AL
0052 5E                           POP     SI           ;restore registers
0053 5F                           POP     DI
0054 5B                           POP     BX
0055 58                           POP     AX
0056 CF                           IRET
```

Transmitting Data to the 16550.   Data is transmitted to the 16550 in much the same manner as it is received, except the interrupt service procedure removes transmit data from a second 16K-byte memory FIFO.

Example 12–14 fills the output FIFO. It is similar to the procedure listed in Example 12–12, except it determines whether the FIFO is full rather than empty.

EXAMPLE 12–14

```
                          ;Procedure that places data into the memory FIFO for
                          ;transmission by the transmitter interrupt.
                          ;AL = character to be transmitted.
                          ;
0000                      SAVE    PROC    NEAR USES BX DI SI
0003 26: 8B 36 8004 R             MOV     SI,OIN       ;get input pointer
0008 26: 8B 1E 8006 R             MOV     BX,OOUT      ;get output pointer
000D 8B FE                        MOV     DI,SI
```

```
000F 46                                   INC   SI
0010 81 FE 8004 R                          CMP   SI,OFFSET OFIFO+16*1024
0014 76 03                                 JBE   NEXT
0016 BE 4004 R                             MOV   SI,OFFSET OFIFO
0019                            NEXT:
0019 3B DE                                 CMP   BX,SI
001B 74 06                                 JE    DONE       ;if full
001D AA                                    STOSB            ;save data in OFIFO
001E 26: 89 36 8004 R                      MOV   OIN,SI
0023                            DONE:
0023 E4 41                                 IN    AL,41H     ;read interrupt control
0025 06 01                                 OR    AL,1       ;enable transmitter
0027 E6 41                                 OUT   41H,AL
                                           RET
002D                            SAVE  ENDP
```

Example 12–15 lists the interrupt service subroutine for the 16550 UART transmitter. This procedure is a continuation of the interrupt handler presented in Example 12–11 and is similar to the RECV procedure of Example 12–13, except it determines whether the FIFO is empty rather than full. Note that we do not include an interrupt service procedure for the break interrupt or any errors.

EXAMPLE 12–15

```
                               ;Interrupt service procedure for the 16550
                               ;transmitter.
                               ;
0060                           TRANS:
0060 53                                 PUSH  BX              ;save registers
0061 57                                 PUSH  DI
0062 26: 8B 1E 8004 R                   MOV   BX,OIN          ;load input pointer
0068 26: 8B 3E 8006 R                   MOV   DI,OOUT         ;load output pointer
006D 3B DF                              CMP   BX,DI
006F 74 17                              JE    EMPTY           ;if empty
0071 26: 8A 05                          MOV   AL,ES:[DI]      ;get character
0074 E6 40                              OUT   40H,AL          ;send it to UART
0076 47                                 INC   DI
0077 81 FF 8004 R                       CMP   DI,OFFSET OFIFO+16*1024
007B 76 03                              JBE   NEXT1
007D BF 4004 R                          MOV   DI,OFFSET OFIFO
0080                           NEXT1:
0080 26: 89 3E 8006 R                   MOV   OOUT,DI
0085 EB 07 90                           JMP   DONES
0088                           EMPTY:
0088 E4 41                              IN    AL,41H          ;read interrupt control
008A 24 FD                              AND   AL,0FDH         ;disable transmitter
008C E6 41                              OUT   41H,AL
008E                           DONES:
008E B0 20                              MOV   AL,20H          ;signal 8259A EOI
0090 E6 49                              OUT   49H,AL
0092 5F                                 POP   DI
0093 5B                                 POP   BX
0094 58                                 POP   AX
0095 CF                                 IRET
```

The 16550 also contains a scratch register, which is a general-purpose register that can be used in any way you deem necessary. Also contained within the 16550 are a modem control register and a modem status register. These registers allow the modem to cause interrupts and control the operation of the 16550 with a modem. Refer to Figure 12–32 for the contents of both the modem status register and the modem control register.

The modem control register uses bit positions 0–3 to control various pins on the 16550. Bit position 4 enables the internal loop-back test for testing purposes. The modem status register allows the status of the modem pins to be tested; it also allows the modem pins to be checked for a change, or in the case of $\overline{RI}$, a trailing edge.

FIGURE 12–32   The modem status and control register of the 16550.

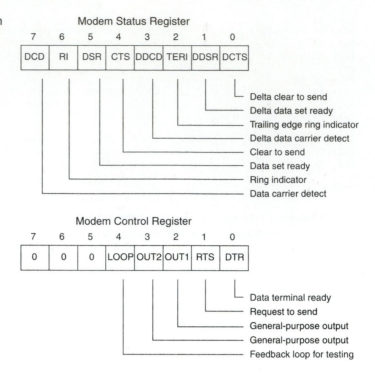

Figure 12–33 illustrates the 16550 UART connected to an RS-232C interface that is often used to control a modem. Included in this interface are line driver and receiver circuits used to convert between TTL levels on the 16550 to RS–232C levels found on the interface. Note that RS–232C levels are usually +12 V for a logic 0 and –12 V for a logic 1 level.

In order to transmit or receive data through the modem, the $\overline{\text{DTR}}$ pin is activated (logic 0) and the UART then waits for the $\overline{\text{DSR}}$ pin to become a logic 0 from the modem indicating that

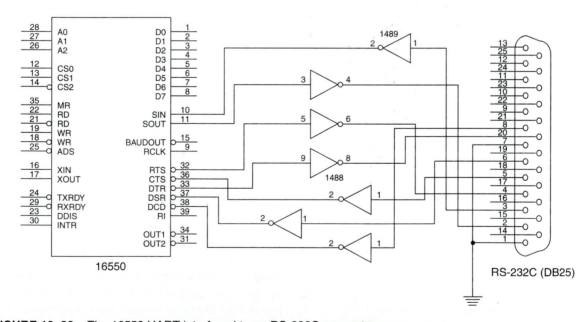

FIGURE 12–33   The 16550 UART interfaced to an RS-232C connector.

the modem is ready. Once this handshake is complete, the UART sends the modem a logic 0 on the $\overline{RTS}$ pin. When the modem is ready, it returns the $\overline{CTS}$ signal (logic 0) to the UART. Communications can now begin. The $\overline{DCD}$ signal from the modem is an indication that the modem has detected a carrier. This signal must also be tested before communications can begin.

12–5   REAL-TIME CLOCK

In this section, we present a real-time clock as an example use of an interrupt. A **real-time clock** keeps time in real time, that is, in hours and minutes. The example illustrated here keeps time in hours, minutes, seconds, tenths of seconds, and milliseconds using five memory locations to hold the BCD time of day.

A method for producing a periodic interrupt is to use one of the internal timers in the 80186/80188. Suppose that timer 2 is programmed to cause an interrupt once per millisecond. This provides a basis for counting time. Example 12–16 is a procedure that programs timer 2 to generate a 1-ms interrupt. Note that the clock frequency is assumed to be 16 MHz (32-MHz crystal) in this example.

EXAMPLE 12–16

```
                      ;Software that initializes timer 2 of the 80186/80188 version EB
                      ;and loads the interrupt vector with procedure address CLOCK.
                      ;
= FF40                T2_CNT EQU 0FF40H      ;timer 2 count register
= FF42                T2_CMP EQU 0FF42H      ;timer 2 compare register
= FF46                T2_CON EQU 0FF46H      ;timer 2 control register
                      ;
0100                  SETUP  PROC   NEAR  USES AX DX DS

0103 B8 0000                 MOV    AX,0          ;address segment 0000
0106 8E D8                   MOV    DS,AX

0108 8C C8                   MOV    AX,CS         ;set timer 2 interrupt vector
010A A3 004E                 MOV    WORD PTR DS:[4EH],AX
010D B8 012D R               MOV    AX,OFFSET CLOCK
0110 A3 004C                 MOV    WORD PTR DS:[4CH],AX
0113 BA FF40                 MOV    DX,T2_CNT  ;address timer 2 count
0116 B8 0000                 MOV    AX,0
0119 EE                      OUT    DX,AL         ;clear timer 2 count
011A BA FF42                 MOV    DX,T2_CMP  ;address timer 2 compare
011D B8 0FA0                 MOV    AX,4000       ;divide by 4,000
0120 EE                      OUT    DX,AL         ;to generate a 1 ms interrupt

0121 BA FF46                 MOV    DX,T2_CON  ;address timer 2 control
0124 B8 E001                 MOV    AX,0E001H  ;enable timer 2 interrupt
0127 EE                      OUT    DX,AL
                      ;
                      ;program interrupt controller here
                      ;
0128 FB                      STI                  ;enable interrupts
                             RET

012D                  SETUP  ENDP
```

The software for the real-time clock contains an interrupt service procedure that is called once per millisecond and a procedure that updates the count located in five memory locations. Example 12–17 lists both procedures. Note that in this example the time is stored at absolute memory location 00400H–00404H.

EXAMPLE 12–17

```
                      ;Interrupt service procedure for the real-time clock.
                      ;
012D                  CLOCK  PROC   FAR     USES AX SI DS

0130 B8 0000                 MOV    AX,0
0133 8E D8                   MOV    DS,AX      ;address time in segment 0000H
0135 BE 0400 R               MOV    SI,OFFSET TIME

0138 B4 00                   MOV    AH,0       ;divide by 100
013A E8 0025                 CALL   UPDATE     ;adjust 1/1000 counter
013D 75 18                   JNZ    DONE       ;if no rollover from 99 to 00
013F B4 10                   MOV    AH,10H     ;divide by 10
0141 E8 001E                 CALL   UPDATE     ;adjust divide by 1/10 second counter
0144 75 11                   JNZ    DONE       ;if no rollover from 09 to 00
0146 B4 60                   MOV    AH,60H     ;divide by 60
0148 E8 0017                 CALL   UPDATE     ;adjust seconds counter
014B 75 0A                   JNZ    DONE       ;if no rollover from 59 to 00
014D E8 0012                 CALL   UPDATE     ;adjust minutes counter
0150 75 05                   JNZ    DONE       ;if no rollover from 59 to 00
0152 B4 24                   MOV    AH,24H     ;divide by 24
0154 E8 000B                 CALL   UPDATE     ;adjust hours counter
0157                 DONE:
0157 BA FF02                 MOV    DX,0FF02H  ;end of interrupt command
015A B8 8000                 MOV    AX,8000H
015D EE                      OUT    DX,AL
                             IRET

0162                  CLOCK  ENDP

0162                  UPDATE PROC   NEAR

0162 AC                      LODSB             ;get counter
0163 04 01                   ADD    AL,1       ;add one (increment will not work)
0165 27                      DAA               ;make count BCD
0166 38 C4                   CMP    AH,AL      ;test modulus
0168 75 02                   JNZ    UPDATE1    ;if no rollover
016A B0 00                   MOV    AL,0
016C                 UPDATE1:
016C 88 44 FF                MOV    [SI-1],AL  ;save new count
016F C3                      RET

0170                  UPDATE ENDP
```

12–6   SUMMARY

1. An interrupt is a hardware- or software-initiated call that interrupts the currently executing program at any point and calls a procedure. The procedure is called by the interrupt handler or an interrupt service procedure.
2. Interrupts are useful when an I/O device needs to be serviced only occasionally at low data transfer rates.
3. The microprocessor has five instructions that apply to interrupts: BOUND, INT, INT3, INTO, and IRET. The INT and INT3 instructions call procedures with addresses stored in an interrupt vector whose type is indicated by the instruction. The BOUND instruction is a conditional interrupt that uses interrupt vector type number 5. The INTO instruction is a conditional interrupt that interrupts a program only if the overflow flag is set. Finally, the IRET instruction is used to return from interrupt service procedures.

4. The microprocessor has many pins that apply to its hardware interrupt structure: NMI, INT0, INT1, INT2, INT3, and in some versions INT4. The interrupts inputs are used to request interrupts. Internal interrupts are caused by the timers, serial ports, and DMA circuitry in various versions of the 80186/80188.

5. Real-mode interrupts are referenced through a vector table that occupies memory locations 00000H–003FFH. Each interrupt vector is four bytes long and contains the offset and segment addresses of the interrupt service procedure. In protected mode, the interrupts reference the interrupt descriptor table (IDT), which contains 256 interrupt descriptors. Each interrupt descriptor contains a segment selector and a 32-bit offset address.

6. Two flag bits are used with the interrupt structure of the microprocessor: trap (TF) and interrupt enable (IF). The IF flag bit enables the INTR interrupt input, and the TF flag bit causes interrupts to occur after the execution of each instruction as long as TF is active.

7. The first 32 interrupt vector locations are reserved for Intel use with many predefined in the microprocessor. The last 224 interrupt vectors are for programmer use and can perform any function desired.

8. Whenever an interrupt is detected, the following events occur: (1) the flags are pushed onto the stack, (2) the IF and TF flag bits are both cleared, (3) the IP and CS registers are both pushed onto the stack, and (4) the interrupt vector is fetched from the interrupt vector table and the interrupt service subroutine is accessed through the vector address.

9. Tracing or single-stepping is accomplished by setting the TF flag bit. This causes an interrupt to occur after the execution of each instruction for debugging.

10. The nonmaskable interrupt input (NMI) calls the procedure whose address is stored at interrupt vector type number 2. This input is positive-edge triggered.

11. The INTR pin is not internally decoded as is the NMI pin. Instead, $\overline{\text{INTA}}$ is used to apply the interrupt vector type number to data bus connections D_0–D_7 during the $\overline{\text{INTA}}$ pulse.

12. Methods of applying the interrupt vector type number to the data bus during $\overline{\text{INTA}}$ vary widely. One method uses resisters to apply interrupt type number FFH to the data bus, while another uses a three-state buffer to apply any vector type number.

13. The 8259A programmable interrupt controller (PIC) adds at least eight interrupt inputs to the microprocessor. If more interrupts are needed, this device can be cascaded to provide up to 64 interrupt inputs.

14. Programming the 8259A is a two-step process: First a series of initialization command words (ICWs) are sent to the 8259A, then a series of operation command words (OCWs) are sent.

15. The 8259A contains three status registers: IMR (interrupt mask register), ISR (in-service register), and IRR (interrupt request register).

16. A real-time clock is used to keep time in real time. In most cases time is stored in either binary or BCD form in several memory locations.

12–7   QUESTIONS AND PROBLEMS

1. What is interrupted by an interrupt?
2. Define the term interrupt.
3. What is called by an interrupt?
4. Why do interrupts free-up time for the microprocessor?
5. List the interrupt pins found on the microprocessor.
6. List the five interrupt instructions for the microprocessor.
7. What is an interrupt vector?

8. Where are the interrupt vectors located in the microprocessor's memory?
9. How many different interrupt vectors are found in the interrupt vector table?
10. Which interrupt vectors are reserved by Intel?
11. Explain how a type 0 interrupt occurs.
12. Where is the interrupt descriptor table located for protected-mode operation?
13. Each protected-mode interrupt descriptor contains what information?
14. Describe the differences between a protected- and real-mode interrupt.
15. Describe the operation of the BOUND instruction.
16. Describe the operation of the INTO instruction.
17. Which memory locations contain the vector for an INT 44H instruction?
18. Explain the operation of the IRET instruction.
19. What is the purpose of interrupt vector type number 7?
20. List the events that occur when an interrupt becomes active.
21. Explain the purpose of the interrupt flag (IF).
22. Explain the purpose of the trap flag (TF).
23. How is IF cleared and set?
24. How is TF cleared and set?
25. The NMI interrupt input automatically vectors through which vector type number?
26. Does the $\overline{\text{INTA}}$ signal activate for the NMI pin?
27. The INT3 input is programmed with which 80186/80188 PCB ports?
28. The NMI input is _____ -sensitive.
29. When the $\overline{\text{INTA}}$ signal becomes a logic 0, it indicates that the microprocessor is waiting for an interrupt _____ number to be placed on the data bus (D_0–D_7).
30. Which 80186/80188 interrupt vector responds to a timer 2 interrupt request?
31. Which 80186/80188 interrupt vector responds to the INT2 interrupt input?
32. What is a FIFO?
33. Develop a circuit that places interrupt type number 86H on the data bus in response to the INTR input.
34. Develop a circuit that places interrupt type number CCH on the data bus in response to the INTR input.
35. Explain why pull-up resistors on D_0–D_7 cause the microprocessor to respond with interrupt vector type number FFH for the $\overline{\text{INTA}}$ pulse.
36. What is a daisy-chain?
37. Why must interrupting devices be polled in a daisy-chained interrupt system?
38. What is the 8259A?
39. How many 8259As are required to have 64 interrupt inputs?
40. What is the purpose of the IR_0–IR_7 pins on the 8259A?
41. When are the CAS_2–CAS_0 pins used on the 8259A?
42. Where is a slave INT pin connected on the master 8259A in a cascaded system?
43. What is an ICW?
44. What is an OCW?
45. How many ICWs are needed to program the 8259A when operated as a single master in a system?
46. Where is the vector type number stored in the 8259A?
47. Where is the sensitivity of the IR pins programmed in the 8259A?
48. What is the purpose of ICW1?
49. What is a nonspecific EOI?
50. Explain priority rotation in the 8259A.
51. What is the purpose of IRR in the 8259A?
52. At which I/O ports is the master 8259A PIC found in the personal computer?
53. At which I/O ports is the slave 8259A found in the personal computer?

54. Which bit is set in the interrupt control register of the 16550 to enable the receiver data interrupt?
55. Which bit is set in the interrupt control register of the 16550 to enable receiver error interrupts?
56. Use to the Internet to locate the National Semiconductor Web site and download the data sheet for the 16550.
57. Use the Internet to locate the Motorola Web site and list the communications interfaces available from Motorola.
58. Use the Internet to locate the Zilog Web site and list the communications interfaces available from Zilog.

CHAPTER 13

Direct Memory Access and DMA-Controlled I/O

INTRODUCTION

In previous chapters, we discussed basic and interrupt-processed I/O. Now we turn to the final form of I/O called **direct memory access** (DMA). The DMA I/O technique provides direct access to the memory while the microprocessor is temporarily disabled. This allows data to be transferred between memory and the I/O device at a rate that is limited only by the speed of the memory components in the system or the DMA controller. The DMA transfer speed can approach 15–20M-byte transfer rates with many of today's high-speed RAM memory components and DMA controllers.

DMA transfers are used for many purposes, but more common are DRAM refresh (in some systems), video displays for refreshing the screen memory, sound cards, and disk memory system reads and writes. The DMA transfer is also used to do high-speed memory-to-memory transfers.

This chapter also explains the operation of disk memory systems and video systems that are often DMA processed. Disk memory includes floppy, fixed, and optical disk storage. Video systems include digital (rarely used today) and analog monitors.

CHAPTER OBJECTIVES

Upon completion of this chapter, you will be able to:

1. Describe a DMA transfer.
2. Explain the operation of the HOLD and HLDA direct memory access control signals.
3. Explain the function of the 8237 DMA controller when used for DMA transfers.
4. Program the 8237 to accomplish DMA transfers.
5. Describe the operation of the 80186/80188 DMA controller.
6. Program the 80186/80188 DMA controllers for DMA transfers.
7. Describe the disk standards found in personal computer systems.
8. Describe the various video interface standards that are found in the personal computer.

13–1 BASIC DMA OPERATION

Two control signals are used to request and acknowledge a direct memory access transfer in the microprocessor-based system. The HOLD pin is an input used to request a DMA action; the HLDA pin is an output that acknowledges the DMA action. In addition to the HOLD and HLDA

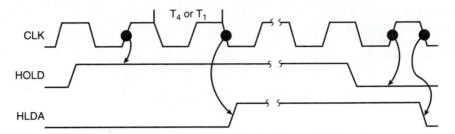

FIGURE 13–1   HOLD and HLDA timing for the 80186/80188 microprocessors.

pins, some versions of the 80186/80188 microprocessors also contain DRQ pins to request a DMA action. Figure 13–1 shows the timing that is typically found on the HOLD and HLDA pins. The DRQ inputs are detailed later in Section 13–2.

When the HOLD input is placed at a logic 1 level, a DMA action (sometimes called a hold) is requested. The microprocessor responds, within a few clocks, by suspending the execution of the program and by placing its address, data, and control bus at their high-impedance states. The high-impedance state causes the microprocessor to appear as if it has been removed from its socket. This state allows external I/O devices, controllers, or other microprocessors to gain access to the system buses so memory can be accessed directly.

As Figure 13–1 indicates, HOLD is sampled in the middle of any clocking cycle. Thus the hold can take effect at any time during the operation of any instruction in the microprocessor. As soon as the microprocessor recognizes the hold, it stops executing software and enters hold cycles. The HOLD input has a higher priority than the INTR or NMI interrupt inputs. Interrupts take effect at the end of an instruction, while a HOLD can take effect even in the middle of an instruction. The only microprocessor pin that has a higher priority than a HOLD is the RESET pin. The HOLD input may not be active during a RESET or the reset is not guaranteed and the microprocessor could lock up.

The HLDA signal becomes active to indicate that the microprocessor has indeed placed its buses at their high-impedance state as can be seen in the timing diagram of Figure 13–1. Note that there are one and sometimes two clock cycles between the time that HOLD changes until the time that HLDA changes. The HLDA output is a signal to the external requesting device that the microprocessor has relinquished control of its memory and I/O space. You could call the HOLD input a DMA request input and the HLDA output a DMA grant signal. In some versions of the 80186/80188, the HOLD input is labeled DREQ or DMA request.

Basic DMA Definitions

Direct memory accesses normally occur between an I/O device and memory without the use of the microprocessor. A *DMA read* transfers data directly from the memory to the I/O device. A *DMA write* transfers data directly from an I/O device to memory. In both operations, the memory and I/O are controlled simultaneously; that is why many systems contain separate memory and I/O control signals. This special control bus structure of the microprocessor allows DMA transfers. A DMA read causes the $\overline{MRDC}$ and $\overline{IOWC}$ signals to both activate, transferring data from the memory to the I/O device. A DMA write causes the $\overline{IORC}$ and signals to both activate. These control bus signals are available to all microprocessors in the Intel family except the 80186/80188 systems. If the 80186/80188 use the HOLD and HLDA signals (EB version), the generation of separate memory and I/O control signals is required such as those provided by a system controller or by a circuit such as the one illustrated in Figure 13–2. A DMA controller provides the memory with its address and a signal ($\overline{DACK}$) that selects the I/O device during the DMA transfer.

FIGURE 13–2 A circuit that develops control bus signals in a system that uses the 80186/80188 microprocessors.

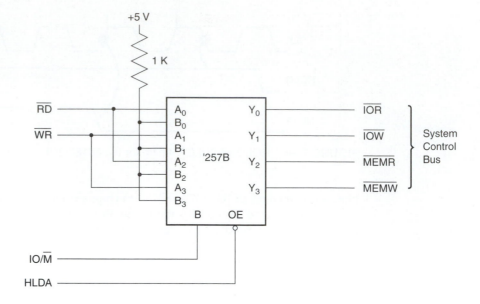

The data transfer speed is determined by the speed of the memory device or a DMA controller that often controls DMA transfers. If the memory speed is 100 ns, DMA transfers occur at rates of up to $\frac{1}{100\,\text{ns}}$ or 10M bytes per second. If the DMA controller in a system functions at a maximum rate of 5 MHz and we still use 100 ns of memory, the maximum transfer rate is 5 MHz because the DMA controller is slower than the memory. In many cases the DMA controller slows the speed of the system when DMA transfers occur.

13–2   THE 8237 DMA CONTROLLER

The 8237 DMA controller supplies the memory and I/O with control signals and furnishes the memory address information during the DMA transfer. The 8237 is a special-purpose microprocessor whose job is the high-speed data transfer between memory and the I/O. Figure 13–3 shows the pin-out and block diagram of the 8237 programmable DMA controller. Although this device may not appear as a discrete component in modern microprocessor-based systems, it does appear within system controller chip sets found in most newer systems. Although not described because of its complexity, the chip set (82357 ISP, or integrated peripheral controller), and its integral set of two DMA controllers are programmed exactly as the 8237. The ISP also provides a pair of 8259A programmable interrupt controllers for the system. Another chip set that finds application is the 80430, which also provides a pair of DMA controllers to the system.

Some of the 80186/80188 family members (XL, EA, and EC versions) contain their own built-in DMA controller and do not require the use of the 8237. These internal DMA controllers are explained later in this chapter.

The 8237 is a four-channel device that is compatible with all versions of the 80186/80188 microprocessors that do not contain a built-in DMA controller. The 8237 is also compatible with all personal computer systems. The 8237 can be expanded to include any number of DMA channel inputs, although four channels seem to be adequate for many small systems. The 8237 is capable of DMA transfers at rates of up to 1.6M bytes per second. Each channel is capable of addressing a full 64K-byte section of memory and can transfer up to 64K bytes with a single programming.

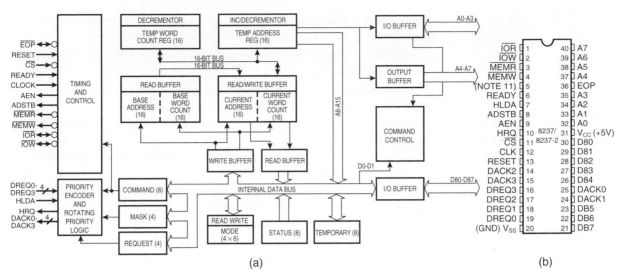

FIGURE 13–3 The 8237A-5 programmable DMA controller. (a) Block diagram and (b) pin-out. (Courtesy of Intel Corporation)

Pin Definitions

CLK	The **clock** input is connected to the system clock signal as long as that signal is 5 MHz or less. In the 80186/80188 systems, the clock must be inverted for the proper operation of the 8237.
$\overline{CS}$	**Chip select** enables the 8237 for programming. The $\overline{CS}$ pin is normally connected to the output of a decoder. The decoder does not use the 80186/80188 control signal $IO/\overline{M}$ ($M/\overline{IO}$) because it contains the new memory and I/O control signals ($\overline{MEMR}$, $\overline{MEMW}$, $\overline{IOR}$, and $\overline{IOW}$).
RESET	The **reset** pin clears the command, status, request, and temporary registers. It also clears the first/last flip-flop and sets the mask register. This input primes the 8237 so it is disabled until programmed otherwise.
READY	A logic 0 on the **ready** input causes the 8237 to enter wait states for slower memory and/or I/O components.
HLDA	**Hold acknowledge** signals the 8237 that the microprocessor has relinquished control of the address, data, and control buses.
$DREQ_3$– $DREQ_0$	The **DMA request** inputs are used to request a DMA transfer for each of the four DMA channels. Because the polarity of these inputs is programmable, they are either active-high or active-low inputs.
DB_7–DB_0	The **data bus** pins are connected to the microprocessor data bus connections and are used during the programming of the DMA controller.
$\overline{IOR}$	**I/O read** is a bi-directional pin used during programming and during a DMA write cycle.
$\overline{IOW}$	**I/O write** is a bi-directional pin used during programming and during a DMA read cycle.
$\overline{EOP}$	**End-of-process** is a bi-directional signal that is used as an input to terminate a DMA process or as an output to signal the end of the DMA transfer. This input is often used to interrupt a DMA transfer at the end of a DMA cycle.

A_3–A_0 These **address pins** select an internal register during programming and also provide part of the DMA transfer address during a DMA action.

A_7–A_4 These **address pins** are outputs that provide part of the DMA transfer address during a DMA action.

HRQ   **Hold request** is an output that connects to the HOLD input of the microprocessor to request a DMA transfer.

$DACK_3$–   **DMA channel acknowledge** outputs acknowledge a channel DMA request.
$DACK_0$ These outputs are programmable as either active-high or active-low signals. The DACK outputs are often used to select the DMA controlled I/O device during the DMA transfer.

AEN   The **address enable** signal enables the DMA address latch connected to the DB_7–DB_0 pins on the 8237. It is also used to disable any buffers in the system connected to the microprocessor.

ADSTB   **Address strobe** functions as ALE, except that it is used by the DMA controller to latch address bits A_{15}–A_8 during the DMA transfer.

$\overline{MEMR}$ **Memory read** is an output that causes memory to read data during a DMA read cycle.

$\overline{MEMW}$ **Memory write** is an output that causes memory to write data during a DMA write cycle.

Internal Registers

CAR   The **current address register** is used to hold the 16-bit memory address used for the DMA transfer. Each channel has its own current address register for this purpose. When a byte of data is transferred during a DMA operation, the CAR is either incremented or decremented, depending on how it is programmed.

CWCR   The **current word count register** programs a channel for the number of bytes (up to 64K) transferred during a DMA action. The number loaded into this register is one less than the number of bytes transferred. For example, if a 10 is loaded into the CWCR, then 11 bytes are transferred during the DMA action.

BA and   The **base address** (BA) and **base word count** (BWC) registers are used when
BWC   auto-initialization is selected for a channel. In the auto-initialization mode, these registers are used to reload both the CAR and CWCR after the DMA action is completed. This allows the same count and address to be used to transfer data from the same memory area.

CR   The **command register** programs the operation of the 8237 DMA controller. Figure 13–4 depicts the function of the command register.

The command register uses bit position 0 to select the memory-to-memory DMA transfer mode. Memory-to-memory DMA transfers use DMA channel 0 to hold the source address and DMA channel 1 to hold the destination address. (This is similar to the operation of a MOVSB instruction.) A byte is read from the address accessed by channel 0 and saved within the 8237 in a temporary holding register. Next, the 8237 initiates a memory write cycle where the contents of the temporary holding register are written into the addresses selected by DMA channel 1. The number of bytes transferred are determined by the channel 1 count register.

The channel 0 address hold enable bit (bit position 1) programs channel 0 for memory-to-memory transfers. For example, suppose that you must fill an area of memory with data, channel 0 can be held at the same address while channel 1 changes for memory-to-memory transfer. This copies the contents of the address accessed by channel 0 into a block of memory accessed by channel 1.

FIGURE 13–4 8237A-5 command register. (Courtesy of Intel Corporation)

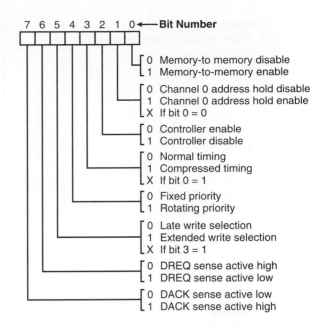

7 6 5 4 3 2 1 0 ←— **Bit Number**

- 0 Memory-to memory disable
- 1 Memory-to-memory enable

- 0 Channel 0 address hold disable
- 1 Channel 0 address hold enable
- X If bit 0 = 0

- 0 Controller enable
- 1 Controller disable

- 0 Normal timing
- 1 Compressed timing
- X If bit 0 = 1

- 0 Fixed priority
- 1 Rotating priority

- 0 Late write selection
- 1 Extended write selection
- X If bit 3 = 1

- 0 DREQ sense active high
- 1 DREQ sense active low

- 0 DACK sense active low
- 1 DACK sense active high

The controller enable/disable bit (bit position 2) turns the entire controller on and off. The normal and compressed bit (bit position 3) determine whether a DMA cycle contains 2 (compressed) or 4 (normal) clocking periods. Bit position 5 is used in normal timing to extend the write pulse so it appears one clock earlier in the timing for I/I devices that require a wider write pulse.

Bit position 4 selects priority for the four DMA channel DREQ inputs. In the fixed priority scheme, channel 0 has the highest priority and channel 3 has the lowest. In the rotating priority scheme, the most recently serviced channel assumes the lowest priority. For example, if channel 2 just had access to a DMA transfer, it assumes the lowest priority and channel 3 assumes the highest priority position. Rotating priority is an attempt to give all channels equal priority.

The remaining two bits (bit positions 6 and 7) program the polarities of the DREQ inputs and the DACK outputs.

MR   The **mode register** programs the mode of operation for a channel. Note that each channel has its own mode register (see Figure 13–5) as selected by bit positions 1 and 0. The remaining bits of the mode register select the operation, auto-initialization, increment/ decrement, and mode for the channel. Verification operations generate the DMA addresses without generating the DMA memory and I/O control signals.

The modes of operation include demand, single, block, and cascade. Demand mode transfers data until an external $\overline{EOP}$ is input or until the DREQ input becomes inactive. Single mode releases the HOLD after each byte of data is transferred. If the DREQ pin is held active, the 8237 again requests a DMA transfer through the DRQ line to the microprocessor's HOLD input. Block mode automatically transfers the number of bytes indicated by the count register for the channel. DREQ need not be held active through the block mode transfer. Cascade mode is used when more than one 8237 is present in a system.

RR   The **request register** is used to request a DMA transfer via software (see Figure 13–6). This is very useful in memory-to-memory transfers where an external signal is not available to begin the DMA transfer.

FIGURE 13–5 8237A-5 mode register. (Courtesy of Intel Corporation)

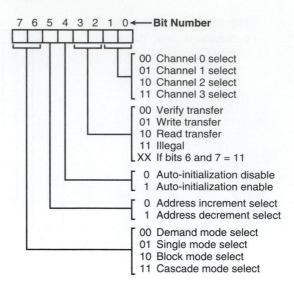

FIGURE 13–6 8237A-5 request register. (Courtesy of Intel Corporation)

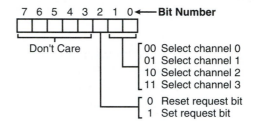

MRSR The **mask register set/reset** sets or clears the channel mask as illustrated in Figure 13–7. If the mask is set, the channel is disabled. Recall that the RESET signal sets all channel masks to disable them.

MSR The **mask register** (see Figure 13–8) clears or sets all of the masks with one command instead of individual channels as with the MRSR.

FIGURE 13–7 8237A-5 mask register set/reset mode. (Courtesy of Intel Corporation)

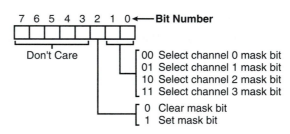

FIGURE 13–8 8237A-5 mask register. (Courtesy of Intel Corporation)

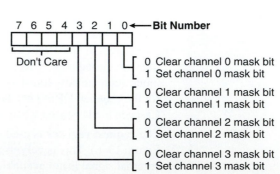

FIGURE 13–9 8237A-5
status register. (Courtesy of
Intel Corporation)

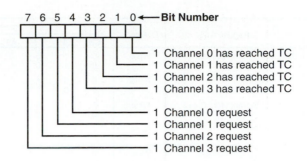

SR   The **status register** shows the status of each DMA channel (see Figure 13–9).
The TC bits indicate if the channel has reached its terminal count (transferred
all its bytes). Whenever the terminal count is reached, the DMA transfer is
terminated for most modes of operation. The request bits indicate whether the
DREQ input for a given channel is active.

Software Commands

Three software commands are used to control the operation of the 8237. These commands do not
have a binary bit pattern as do the various control registers within the 8237. A simple output to
the correct port number enables the software command. Figure 13–10 shows the I/O port assignments that access all registers and the software commands.

The functions of the software commands are explained in the following list.

1. **Clear the first/last flip-flop**—Clears the first/last (F/L) flip-flop within the 8237. The F/L flip-
   flop selects which byte (low or high order) is read/written in the current address and current
   count registers. If F/L = 0, the low-order byte is selected; if F/L = 1, the high-order byte is selected. Any read or write to the address or count register automatically toggles the F/L flip-flop.
2. **Master clear**—Acts exactly the same as the RESET signal to the 8237. As with the RESET
   signal, this command disables all channels.
3. **Clear mask register**—Enables all four DMA channels.

Programming the Address and Count Registers

Figure 13–11 illustrates the I/O port locations for programming the count and address registers
for each channel. Notice that the state of the F/L flip-flop determines whether the LSB or MSB is

FIGURE 13–10 8237A-5
command and control port assignments. (Courtesy of Intel
Corporation)

Signals						Operation
A3	A2	A1	A0	$\overline{IOR}$	$\overline{IOW}$	
1	0	0	0	0	1	Read Status Register
1	0	0	0	1	0	Write Command Register
1	0	0	1	0	1	Illegal
1	0	0	1	1	0	Write Request Register
1	0	1	0	0	1	Illegal
1	0	1	0	1	0	Write Single Mask Register Bit
1	0	1	1	0	1	Illegal
1	0	1	1	1	0	Write Mode Register
1	1	0	0	0	1	Illegal
1	1	0	0	1	0	Clear Byte Pointer Flip/Flop
1	1	0	1	0	1	Read Temporary Register
1	1	0	1	1	0	Master Clear
1	1	1	0	0	1	Illegal
1	1	1	0	1	0	Clear Mask Register
1	1	1	1	0	1	Illegal
1	1	1	1	1	0	Write All Mask Register Bits

Channel	Register	Operation	Signals							Internal Flip-Flop	Data Bus DB0-DB7
			$\overline{CS}$	$\overline{IOR}$	$\overline{IOW}$	A3	A2	A1	A0		
0	Base and Current Address	Write	0	1	0	0	0	0	0	0	A0-A7
			0	1	0	0	0	0	0	1	A8-A15
	Current Address	Read	0	0	1	0	0	0	0	0	A0-A7
			0	0	1	0	0	0	0	1	A8-A15
	Base and Current Word Count	Write	0	1	0	0	0	0	1	0	W0-W7
			0	1	0	0	0	0	1	1	W8-W15
	Current Word Count	Read	0	0	1	0	0	0	1	0	W0-W7
			0	0	1	0	0	0	1	1	W8-W15
1	Base and Current Address	Write	0	1	0	0	0	1	0	0	A0-A7
			0	1	0	0	0	1	0	1	A8-A15
	Current Address	Read	0	0	1	0	0	1	0	0	A0-A7
			0	0	1	0	0	1	0	1	A8-A15
	Base and Current Word Count	Write	0	1	0	0	0	1	1	0	W0-W7
			0	1	0	0	0	1	1	1	W8-W15
	Current Word Count	Read	0	0	1	0	0	1	1	0	W0-W7
			0	0	1	0	0	1	1	1	W8-W15
2	Base and Current Address	Write	0	1	0	0	1	0	0	0	A0-A7
			0	1	0	0	1	0	0	1	A8-A15
	Current Address	Read	0	0	1	0	1	0	0	0	A0-A7
			0	0	1	0	1	0	0	1	A8-A15
	Base and Current Word Count	Write	0	1	0	0	1	0	1	0	W0-W7
			0	1	0	0	1	0	1	1	W8-W15
	Current Word Count	Read	0	0	1	0	1	0	1	0	W0-W7
			0	0	1	0	1	0	1	1	W8-W15
3	Base and Current Address	Write	0	1	0	0	1	1	0	0	A0-A7
			0	1	0	0	1	1	0	1	A8-A15
	Current Address	Read	0	0	1	0	1	1	0	0	A0-A7
			0	0	1	0	1	1	0	1	A8-A15
	Base and Current Word Count	Write	0	1	0	0	1	1	1	0	W0-W7
			0	1	0	0	1	1	1	1	W8-W15
	Current Word Count	Read	0	0	1	0	1	1	1	0	W0-W7
			0	0	1	0	1	1	1	1	W8-W15

FIGURE 13–11   8237A-5 DMA channel I/O port addresses. (Courtesy of Intel Corporation)

programmed. If the state of the F/L flip-flop is unknown, the count and address could be programmed incorrectly. A system RESET clears the state of the F/L flip-flop as does the clear command. It is also important that the DMA channel be disabled before its address and count are programmed.

There are four steps required to program the 8237: (1) the F/L flip-flop is cleared using a clear F/L command, (2) the channel is disabled, (3) the LSB and then MSB of the address are programmed, and (4) the LSB and MSB of the count are programmed. Once these four operations are performed, the channel is programmed and ready to use. Additional programming is required to select the mode of operation before the channel is enabled and the DMA transfer is started.

The 8237 Connected to the 80X86 Microprocessor

Figure 13–12 shows an 80188-based (EB version) system that contains the 8237 DMA controller.

The address enable (AEN) output of the 8237 controls the output pins of the latches and the outputs of the 74LS257 (U8). During normal 80C188EB operation (AEN = 0), latches U4 and U6, and the multiplexer (U8) provide address bus bits A_{19}–A_{16} and A_7–A_0. Recall that address connections A_{19}–A_{16} are multiplexed and must be de-multiplexed for use in the system. The multiplexer provides the system control signals as long as the 80C188EB is in control of the system. During a DMA action (AEN = 1), latches U4 and U6 are disabled along with the multiplexer (U8). Latches U3 and U7 now provide address bits A_{19}–A_{16} and A_{15}–A_8. Address bus bits A_7–A_0 are provided directly by the 8237 and contain a part of the DMA transfer address. The control signals $\overline{MEMR}$, $\overline{MEMW}$, $\overline{IOR}$, and $\overline{IOW}$ are also provided by the DMA controller during a DMA action.

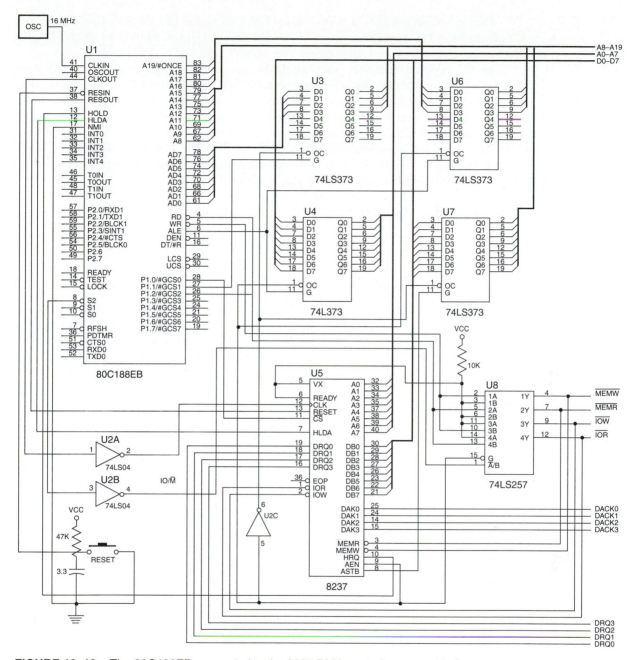

FIGURE 13–12 The 80C188EB connected to the 8237 DMA controller to provide four DMA channels.

The address strobe output (ASTB) from the 8237 clocks the address (A_{15}–A_8) into latch U7 during the DMA action so the entire DMA transfer address becomes available on the address bus. Address bus bits A_{19}–A_{16} are provided by latch U3, which must be programmed with these four address bits before the controller is enabled for the DMA transfer. The DMA operation of the 8237 is limited to a transfer of not more than 64K bytes within the same 64K-byte section of the memory.

The general chip selection pins select the 8237 for programming and also the 4-bit latch (U3) for the uppermost four address bits. The microprocessor is programmed (not shown) so it enables the 8237 for I/O port addresses 0040H–007FH and the I/O latch (U3) for ports 0000H–003FH.

During normal 80C188EB operation, the DMA controller and integrated circuits U3 and U7 are disabled. During a DMA action, integrated circuits U4, U5, and U8 are disabled so the 8237 can take control of the system through the address, data, and control buses.

In the personal computer, the two DMA controllers are programmed at I/O ports 0000H–000FH for DMA channels 0–3 and ports 00C0H–00DFH for DMA channels 4–7. Note that the second controller is programmed at only even addresses so channel 4 base and current address is programmed at I/O port 00C0H and the channel 4 base and current count is programmed at port 00C2H. The page register, which holds address bits A23–A16 of the DMA address, are located at I/O ports 0087H (CH-0), 0083H (CH-1), 0081H (CH-2), 0082H (CH-3), (no channel 4), 008BH (CH-5), 0089H (CH-6), and 008AH (CH-7). The page register functions as the address latch described with the examples in this text.

Memory-to-Memory Transfer with the 8237

The memory-to-memory transfer is much more powerful than even the automatically repeated MOVSB instruction. While the repeated MOVSB instruction tables the 80188 4.2 μs per byte, the 8237 requires only 2.0 μs per byte. This is over twice as fast as a software data transfer. This is not true if an 80386, 80846, or Pentium/Pentium Pro is in use in the system.

Sample Memory-to-Memory DMA Transfer.   Suppose that the contents of memory locations 10000H–13FFFH are to be transferred into memory locations 14000H–17FFFH. This is accomplished with a repeated string move instruction, or, at a much faster rate, with the DMA controller.

Example 13–1 illustrates the software required to initialize the 8237 and program latch U3 in Figure 13–12 for this DMA transfer.

EXAMPLE 13–1

```
                          ;Procedure that transfers a block of data using the
                          ;8237A DMA controller in Figure 13-12.  This is a
                          ;memory-to-memory block transfer.
                          ;
                          ;Calling parameters:
                          ;      SI = source address
                          ;      DI = destination address
                          ;      CX = count
                          ;      ES = segment of source and destination
                          ;
= 0010                    LATCH   EQU   10H             ;latch U3
= 007C                    CLEAR_F EQU   7CH             ;F/L flip flop
= 0070                    CH0_A   EQU   70H             ;channel 0 address
= 0072                    CH1_A   EQU   72H             ;channel 1 address
= 0073                    CH1_C   EQU   73H             ;channel 1 count
= 007B                    MODE    EQU   7BH             ;mode
= 0078                    CMMD    EQU   78H             ;command
= 007F                    MASKS   EQU   7FH             ;masks
= 0079                    REQ     EQU   79H             ;request register
= 0078                    STATUS  EQU   78H             ;status register

0000                      TRANS   PROC  FAR USES AX

0001 8C C0                        MOV   AX,ES           ;program latch U3
0003 8A C4                        MOV   AL,AH
0005 C0 E8 04                     SHR   AL,4
0008 E6 10                        OUT   LATCHB,AL

000A E6 7C                        OUT   CLEAR_F,AL      ;clear F/L flip-flop

000C 8C C0                        MOV   AX,ES           ;program source
000E C1 E0 04                     SHL   AX,4
0011 03 C6                        ADD   AX,SI           ;form source offset
0013 E6 70                        OUT   CH0_A,AL
0015 8A C4                        MOV   AL,AH
0017 E6 70                        OUT   CH0_A,AL
```

```
0019 8C C0                      MOV     AX,ES            ;program destination
001B C1 E0 04                   SHL     AX,4
001E 03 C7                      ADD     AX,DI            ;form destination
0020 E6 72                      OUT     CH1_A,AL
0022 8A C4                      MOV     AL,AH
0024 E6 72                      OUT     CH1_A,AL

0026 8B C1                      MOV     AX,CX            ;program count
0028 48                         DEC     AX               ;adjust count
0029 E6 73                      OUT     CH1_C,AL
002B 8A C4                      MOV     AL,AH
002D E6 73                      OUT     CH1_C,AL

002F B0 88                      MOV     AL,88H           ;program mode
0031 E6 7B                      OUT     MODE,AL
0033 B0 85                      MOV     AL,85H
0035 E6 7B                      OUT     MODE,AL

0037 B0 01                      MOV     AL,1             ;enable transfer
0039 E6 78                      OUT     CMMD,AL

003B B0 0E                      MOV     AL,0EH           ;unmask channel 0
003D E6 7F                      OUT     MASKS,AL

003F B0 04                      MOV     AL,4             ;start DMA transfer
0041 E6 79                      OUT     REQ,AL

                                .REPEAT                  ;wait until complete
0043 E4 78                      IN      AL,STATUS
                                .UNTIL AL & 1

                                RET

              TRANS ENDP
```

Programming the DMA controller requires a few steps, as illustrated in Example 13–1. The leftmost digit of the 5-digit hexadecimal address is sent to latch U3. Next, the channels are programmed after the F/L flip-flop is cleared. We use channel 0 as the source and channel 1 as the destination for a memory-to-memory transfer. The count is next programmed with a value that is one less than the number of bytes to be transferred. Next, the mode register of each channel is programmed, the command register selects a block move, channel 0 is enabled, and a software DMA request is initiated. Before return is made from the procedure, the status register is tested for a terminal count. Recall that the terminal count flag indicates that the DMA transfer is complete. The TC also disables the channel preventing additional transfers.

Sample Memory Fill Using the 8237.   In order to fill an area of memory with the same data, the channel 0 source register is programmed to point to the same address throughout the transfer. This is accomplished with the channel 0 hold mode. The controller copies the contents of this single memory location to an entire block of memory addressed by channel 1. This has many useful applications.

For example, suppose that a video display must be cleared. This operation can be performed using the DMA controller with the channel 0 hold mode and a memory-to-memory transfer. If the video display contains 80 columns and 25 lines, it has 2,000 display positions that must be set to 20H (an ASCII space) to clear the screen.

Example 13–2 clears an area of memory addressed by ES:DI. The CX register transfers the number of bytes to be cleared to the CLEAR procedure. Notice that this procedure is nearly identical to Example 13–1, except the command register is programmed so the channel 0 address is held. The source address is programmed as the same address as ES:DI and then the destination is programmed as one location beyond ES:DI. Also note that this program is designed to function with the hardware in Figure 13–12 and will not function in the personal computer unless you have the same hardware.

EXAMPLE 13–2

```
                            ;Procedure that clears an area of memory using the
                            ;8237A DMA controller in Figure 13-12.  This is a
                            ;memory-to-memory block transfer with a channel 0 hold.
                            ;
                            ;Calling parameters:
                            ;       DI = offset address of area cleared
                            ;       ES = segment address of area cleared
                            ;       CX = number of bytes cleared
                            ;
= 0010                      LATCHB  EQU   10H         ;latch U3
= 007C                      CLEAR_F EQU   7CH         ;F/L flip flop
= 0070                      CH0_A   EQU   70H         ;channel 0 address
= 0072                      CH1_A   EQU   72H         ;channel 1 address
= 0073                      CH1_C   EQU   73H         ;channel 1 count
= 007B                      MODE    EQU   7BH         ;mode
= 0078                      CMMD    EQU   78H         ;command
= 007F                      MASKS   EQU   7FH         ;masks
= 0079                      REQ     EQU   79H         ;request register
= 0078                      STATUS  EQU   78H         ;status register
= 0000                      ZERO    EQU   0H          ;zero

0000                        CLEAR   PROC  FAR USES AX

0001 8C C0                          MOV   AX,ES       ;program latch U3
0003 8A C4                          MOV   AL,AH
0005 C0 E8 04                       SHR   AL,4
0008 E6 10                          OUT   LATCHB,AL

000A E6 7C                          OUT   CLEAR_F,AL  ;clear F/L flip-flop

000C 2E: A0 0000                    MOV   AL,CS:ZERO
0010 26: 88 05                      MOV   ES:[DI],AL  ;clear first byte

0013 8C C0                          MOV   AX,ES       ;program source
0015 C1 E0 04                       SHL   AX,4
0018 03 C7                          ADD   AX,DI       ;form source offset
001A E6 70                          OUT   CH0_A,AL
001C 8A C4                          MOV   AL,AH
001E E6 70                          OUT   CH0_A,AL

0020 8C C0                          MOV   AX,ES       ;program destination
0022 C1 E0 04                       SHL   AX,4
0025 03 C7                          ADD   AX,DI       ;form destination offset
0027 48                             INC   AX
0028 E6 72                          OUT   CH1_A,AL
002A 8A C4                          MOV   AL,AH
002C E6 72                          OUT   CH1_A,AL

002E 8B C1                          MOV   AX,CX       ;program count
0030 48                             DEC   AX          ;adjust count
0031 48                             DEC   AX
0032 E6 73                          OUT   CH1_C,AL
0034 8A C4                          MOV   AL,AH
0036 E6 73                          OUT   CH1_C,AL

0038 B0 88                          MOV   AL,88H      ;program mode
003A E6 7B                          OUT   MODE,AL
003C B0 85                          MOV   AL,85H
003E E6 7B                          OUT   MODE,AL

0040 B0 03                          MOV   AL,3        ;enable block hold transfer
0042 E6 78                          OUT   CMMD,AL

0044 B0 0E                          MOV   AL,0EH      ;unmask channel 0
0046 E6 7F                          OUT   MASKS,AL

0048 B0 04                          MOV   AL,4        ;start DMA transfer
004A E6 79                          OUT   REQ,AL
```

```
                                  .REPEAT                   ;wait until DMA complete
004C E4 78                        IN            AL,STATUS
                                  .UNTIL AL &1
                                  RET

0054                     CLEAR    ENDP
```

DMA in the 80186/80188

All versions of the 80186 and 80188 are capable of doing DMA operations. The difference is that some of these devices do not contain DMA controllers, while others contain extensive DMA control capabilities. The XL and EA versions contain a two-channel DMA controller, while the EC version contains a four-channel DMA controller. The EB version does not contain a DMA controller, but it does have HOLD and HLDA pins so that the 8237 can be used with the EB version.

The XL and EA Versions of the DMA Controller. As mentioned, the XL and EA versions of the 80186/80188 both contain a two-channel, fully programmable DMA controller. Each channel has its own set of 20-bit address registers so any memory or I/O location is accessible for a DMA transfer without the need for an external 4-bit latch. Also notice that the DMA request is activated by a DREQ input for each channel. In addition, each channel is programmable for auto-increment or auto-decrement to either source or destination registers.

Figure 13–13 illustrates the internal register structure of the DMA controller found within the XL and EA versions of the 80186/80188. These registers are located in the peripheral control block at offset addresses C0H–DBH.

Notice that both DMA channel register sets are identical. Each channel contains a control word, a 20-bit wide source and destination pointer, and a transfer count. The transfer count is 16 bits and allows unattended DMA transfers of bytes (80188/80186) and words (80186 only). Each time that a byte or word is transferred, the count is decremented by 1 until it reaches 0000H—the terminal count.

The source and destination pointers are each 20 bits wide so DMA transfers can occur to any memory location or I/O address without concern for segment and offset addresses. If the

FIGURE 13–13 The DMA controller register structure of the 80186/80188 (versions XL and EA).

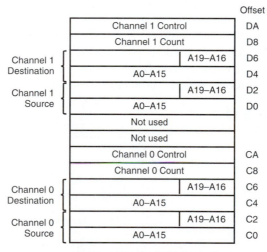

I/O Port Map for DMA Controller

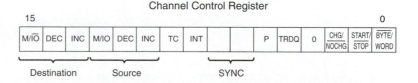

Channel Control Register

source or destination address is an I/O port, bits A19–A16 must be 0000 or a malfunction may occur. Recall that the 8237 contained only a 16-bit address register, which required a 4-bit external latch to accomplish DMA transfers.

Each DMA channel contains its own channel control register (refer to Figure 13–13), which defines its operation. The leftmost 6 bits specify the operation of both the source and destination registers. The M/$\overline{\text{IO}}$ bit selects a memory or I/O location; DEC causes the pointer to be decremented; INC causes the pointer to be incremented. If both the INC and DEC bits are 1, then the pointer is unchanged after each DMA transfer. Notice that memory-to-memory transfers are possible with this DMA controller if both DEC and INC are set.

The TC (terminal count) bit causes the DMA channel to stop transfers when the channel count register is decremented to 0000H. If this bit is a logic 1, the DMA controller continues to transfer data even after the terminal count is reached.

The INT bit enables interrupts to the interrupt controller. If set, this bit causes an interrupt to be issued when the terminal count of the channel is reached. (Refer to Chapter 12, which details the interrupt structure of the 80186/80188, for more information about a DMA interrupt.)

The SYNC bits select the type of synchronization for the channel: 00 = no synchronization, 01 = source synchronization, and 10 = destination synchronization. When either unsynchronized or source synchronization is selected, data is transferred at the rate of 2M bytes per second. These two types of synchronization allow transfers to occur without interruption. If destination synchronization is selected, the transfer rate is slower (1.3M bytes per second), and the controller relinquishes control to the 80186/80188 after each DMA transfer. This is required in some applications so that the DMA controller does not use all of the CPU time for DMA transfers.

The P bit selects the channel priority. If P = 1, the channel has the highest priority. If both channels have the same priority, and both channels are actively transferring data, the DMA controller alternates transfers between channels.

The TRDQ bit enables DMA transfers from timer 2. If this bit is a logic 1, the DMA request originates from timer 2. This can also be used to prevent the DMA transfers from using all of the microprocessor's time for the transfer.

The CHG/$\overline{\text{NOCHG}}$ bit determines whether START/$\overline{\text{STOP}}$ changes for a write to the control register. The START/$\overline{\text{STOP}}$ bit starts or stops the DMA transfer. To start a DMA transfer, both CHG/$\overline{\text{NOCHG}}$ and START/$\overline{\text{STOP}}$ are placed at a logic 1 level.

The $\overline{\text{BYTE}}$/WORD selects whether the transfer is byte- or word-sized.

The following is a description and the program required to accomplish a memory-to-memory DMA transfer. The built-in DMA controller is capable of performing memory-to-memory transfers. The procedure used to program the controller and start the transfer is listed in Example 13–3.

EXAMPLE 13–3

```
                        ,MODEL SMALL
                        .186
0000                    .CODE


                        ;Memory-to-memory DMA transfer procedure.
                        ;
                        ;Calling parameters:
                        ;
                        ;      DS:SI = source address
                        ;      ES:DI = destination address
                        ;      CX = count
                        ;
                        GETA    MACRO   SEGA,OFFA,DMAA
                                MOV     AX,SEGA            ;;get segment
                                SHL     AX,4               ;shift left 4 places
                                ADD     AX,OFFA            ;;add in offset
                                MOV     DX,DMAA            ;;address controller
                                OUT     DX,AX              ;;rightmost 16 bits
```

```
                              PUSHF                         ;;save carry
                              MOV     AX,SEGA               ;;get segment
                              SHR     AX,12                 ;;leftmost 4 bits
                              POPF
                              ADC     AX,0                  ;;add in carry
                              ADD     DX,2
                              OUT     DX,AX
                              ENDM

0000                  MOVES   PROC    FAR

                              GETA    DS,SI,0FFC0H          ;source address
                              GETA    ES,DI,0FFC4H          ;destination address

0032 BA FFC8                  MOV     DX,0FFC8H             ;program count
0035 8B C1                    MOV     AX,CX
0037 EE                       OUT     DX,AL

0038 BA FFCA                  MOV     DX,0FFCAH             ;program control
003B B8 B606                  MOV     AX,0B606H
003E EE                       OUT     DX,AL                 ;start transfer

003F CB                       RET

0040                  MOVES   ENDP
                              END
```

The procedure in Example 13–3 transfers data from the data segment location address by SI into the extra segment location addressed by DI. The number of bytes transferred is held in register CX. This operation is identical to the REP MOVSB instruction, but execution occurs at a much higher speed. Notice that the GETA macro converts the segment and offset addresses into a 20-bit physical address for the DMA controller.

The EC Version of the DMA Controller.   As mentioned, the EC version of the 80186/80188 is a four-channel DMA controller, which is much more powerful than the other versions. The EC version contains a pair of DMA controllers that are very similar to the single two-channel DMA controller found in the XL and EA versions. Unlike the XL and EA versions, the EC version's DMA controller can also be internally activated by the serial port receiver and transmitter (refer to Figure 13–14 for

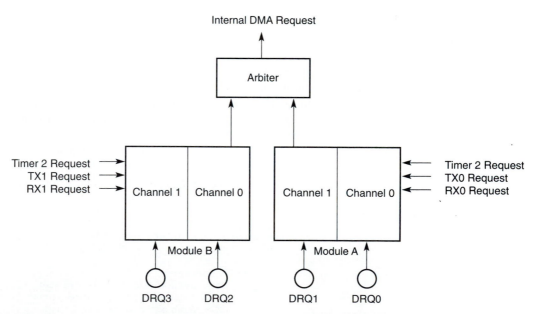

FIGURE 13–14   The internal structure of the EC version's DMA controller.

the internal structure of the EC version's DMA controller). Recall that the DMA controller in the XL and EA versions could only be internally activated by the timer 2 signals.

The EC version of the DMA controller contains four DRQ lines that are organized into a pair of DMA modules (A and B). The A module is accessed via DRQ0 and DRQ1 or from a request generated by timer 2 or serial port 0. The B module is accessed via DRQ2 and DRQ3 or from a request generated by timer 2 or serial port 1. Priority between modules is rotated when the priority is equal.

Figure 13–15 illustrates the programming model for the DMA controllers within the EC version of the 80186/80188. Note that the model is essentially the same as the model for the XL and EA versions presented in Figure 13–13. The main difference is that the EC version contains four channels, while the XL and EA versions contain only two channels. Also notice that two additional registers have been added to the DMA controller architecture. One sets the DMA module priority and the other allows (halt register) software to suspend a DMA action.

The DMA module priority register contains two bits (DMPA and DMPB) that set the priority of the module. If the DMPA bit is set, module A has the higher priority so DRQ0 and DRQ1 take precedence over DRQ2 and DRQ3. If both modules are set to the same priority, then priority rotates between the modules. The IRA and IRB bits, in the DMA module priority register, determine whether timer 2 is a source for a DMA request when the IRA or IRB bit is cleared. Module A, IRA bit selects the serial port 0 as a DMA requesting source if set and module B, IRB bit selects the serial port 1 as a DMA requesting source if set.

The DMA halt register is also new to the EC version of the 80186/80188 microprocessors. The rightmost two bits (HA and HB) are set to suspend DMA transfers for modules A and B. The HNI bit (bit position 7) is automatically set whenever a NMI interrupt request is presented to the microprocessor. The HMI (bit position 15) must be set to clear the HMI bit during a write operation. Finally, the HMA and HMB bits must be set to modify the HA and HB bits, respectively. The DMA halt register is used by software to stop a DMA action.

DMA Processed Printer Interface

Figure 13–16 illustrates the hardware required to process a DMA-controlled printer interface. In this system, the 80C188XL microprocessor is used to interface to a Centronics-type parallel printer. The write pulse that is passed through to the latch (U4) during the DMA action also generates the data strobe ($\overline{DS}$) signal to the printer through the single-shot or monostable multivibrator (U5). The $\overline{ACK}$ signal returns from the printer each time it is ready for additional data. In this circuit $\overline{ACK}$ is used to request a DMA action through a flip-flop (U3).

Notice that the I/O device is not selected by decoding the address on the address bus. Instead of a decoder, the $\overline{PCS0}$ signal is combined with the $\overline{WR}$ strobe from the microprocessor to clear the DMA request, latch data for the printer, and generate the data strobe signal for the printer.

Software that controls this interface is simple because only the address of the data and the number of characters to be printed are programmed. Once programmed, the channel is enabled, and the DMA action transfers a byte at a time to the printer interface each time that the interface receives the $\overline{ACK}$ signal from the printer.

The procedure that prints data from the current data segment is illustrated in Example 13–4. This procedure programs the DMA controller of the 80C188Xl, but doesn't actually print anything. Printing is accomplished by the DMA controller and the printer interface.

EXAMPLE 13–4

```
;Procedure that prints data via the printer interface
;in Figure 13-16.
;
;Calling parameters:
;        BX = offset address of printer data
;        DS = segment address of printer data
;        CX = number of bytes to print
;
```

FIGURE 13–15   The DMA controller register structure of the 80186/80188 (version EC).

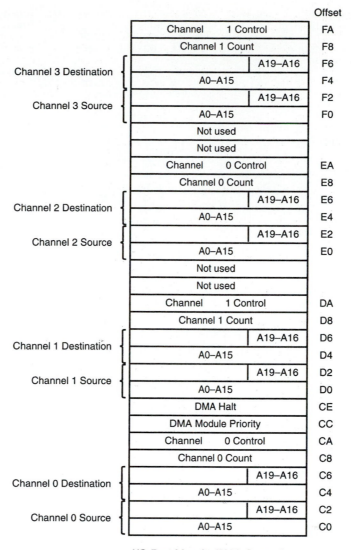

	Offset
Channel 1 Control	FA
Channel 1 Count	F8
A19–A16	F6
A0–A15	F4
A19–A16	F2
A0–A15	F0
Not used	
Not used	
Channel 0 Control	EA
Channel 0 Count	E8
A19–A16	E6
A0–A15	E4
A19–A16	E2
A0–A15	E0
Not used	
Not used	
Channel 1 Control	DA
Channel 1 Count	D8
A19–A16	D6
A0–A15	D4
A19–A16	D2
A0–A15	D0
DMA Halt	CE
DMA Module Priority	CC
Channel 0 Control	CA
Channel 0 Count	C8
A19–A16	C6
A0–A15	C4
A19–A16	C2
A0–A15	C0

Channel 3 Destination
Channel 3 Source
Channel 2 Destination
Channel 2 Source
Channel 1 Destination
Channel 1 Source
Channel 0 Destination
Channel 0 Source

I/O Port Map for DMA Controller

Channel Control Register

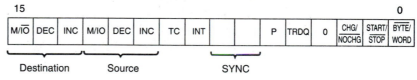

15															0
M/IO	DEC	INC	M/IO	DEC	INC	TC	INT			P	TRDQ	0	CHG/ NOCHG	START/ STOP	BYTE/ WORD

Destination   Source   SYNC

DMA Module Priority Register

				10		8						2		0
				IRB		IRA						DMPB		DMPA

DMA Halt Register

15							9	8	7					1	0
HMI							HMB	HMA	HNI					HB	HA

```
= FFC0                    C0_S    EQU    0FFC0H        ;channel 0 source address
= FFC4                    C0_D    EQU    0FFC4H        ;channel 0 destination address
= FFC8                    C0_CN   EQU    0FFC8H        ;channel 0 count
= FFCA                    C0_CTL  EQU    0FFCAH        ;channel 0 control

0100                      PRINT   PROC   DX
0101   8C D8                      MOV    AX,DS         ;translate address
0103   C1 E0 04                   SHL    AX,4
0106   03 C3                      ADD    AX,BX
0108   9C                         PUSHF                ;save possible carry
0109   BA FFC0                    MOV    DX,C0_S       ;address source
010C   EF                         OUT    DX,AX         ;program A0-A15
010D   8C D8                      MOV    AX,DS
010F   C1 E8 0C                   SHR    AX,12
0112   9D                         POPF
0113   83 D0 00                   ADC    AX,0          ;add in carry
0116   83 C2 02                   ADD    DX,2
0119   EF                         OUT    DX,AX         ;program A16-A19
011A   BA FFC4                    MOV    DX,C0_D       ;address destination
011D   B8 2000                    MOV    AX,2000H      ;port 2000H for PCS0
0120   EF                         OUT    DX,AX
0121   B8 0000                    MOV    AX,0
0124   83 C2 02                   ADD    DX,2
0127   EF                         OUT    DX,AX
0128   BA FFC8                    MOV    DX,C0_CNT     ;address count
012B   8B C1                      MOV    AX,CX
```

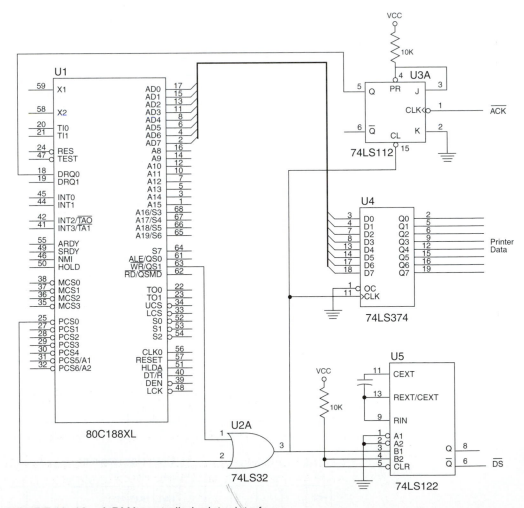

FIGURE 13–16   A DMA controlled printer interface.

```
012D  EF                      OUT   DX,AX               ;program count
012E  B8 1686                 MOV   AX,1686H            ;start DMA transfer
                              RET

0133                    PRINT ENDP
```

A secondary procedure is needed to determine if the DMA action has been completed. Example 13–5 tests the DMA controller to see if the DMA transfer is complete. The TEST_P procedure is called before programming the DMA controller to see if the prior transfer is complete.

EXAMPLE 13–5

```
                        ;Procedure that tests for a complete DMA action.

= FFCA                  C0_CTL EQU   0FFCAH            ;control register

0000                    TEST_P PROC   NEAR USES AX DX

0102  BA FFCA                  MOV   DX,C0_C           ;address control register
                              .REPEAT
0105  ED                       IN    AX,DX             ;get control register
0106  25 0200                  AND AX,200H             ;test TC bit
                              .UNTIL !ZERO?             ;until not zero
                              RET

010E                    TEST_P ENDP
```

Printed data can be double-buffered by first loading buffer 1 with data to be printed. Next, the PRINT procedure is called to begin printing buffer 1. Because it takes very little time to program the DMA controller, a second buffer (buffer 2) can be filled with new printer data while the first buffer (buffer 1) is printed by the printer interface and DMA controller. This process is repeated until all data is printed.

13–3   DISK MEMORY SYSTEMS

Disk memory is used to store long-term data. Many types of disk storage systems are available today. All disk memory systems use magnetic media except the optical disk memory, which stores data on a plastic disk. Optical disk memory is either a **CD-ROM** (compact disk/read only memory) that is read, but never written, or a **WORM** (write once/read mostly) that is read most of the time, but can be written once by a laser beam. Also becoming available is optical disk memory that can be read and written many times, but there is still a limitation on the number of write operations allowed. This section provides an introduction to disk memory systems and detail of their operation.

Floppy Disk Memory

The most common, basic form of disk memory is the floppy or flexible disk. This magnetic recording media is available in three sizes: the 8″ **standard,** 5^1/$_4$″ **mini-floppy,** and the 3^1/$_2$″ **micro-floppy.** Today the 8″ standard version has been replaced by the mini- and micro-floppy disks. The micro-floppy disk is quickly replacing the mini-floppy in newer systems because of its reduced size, ease of storage, and durability. Even so, many systems are still marketed with both the mini- and micro-floppy disk drives. In fact, one vendor markets a single disk drive that accepts both the 5^1/$_4$″ and 3^1/$_2$″ floppy disks.

All disks have several things in common. They are all organized so data is stored in tracks. A **track** is a concentric ring of data that is stored on a surface of a disk. Figure 13–17 illustrates the surface of a 5^1/$_4$″ mini-floppy disk showing a track that is divided into sectors. A **sector** is a common subdivision of a track that is designed to hold a reasonable amount of data. In many

FIGURE 13–17 The format of a 5$^1/_4$″ floppy disk.

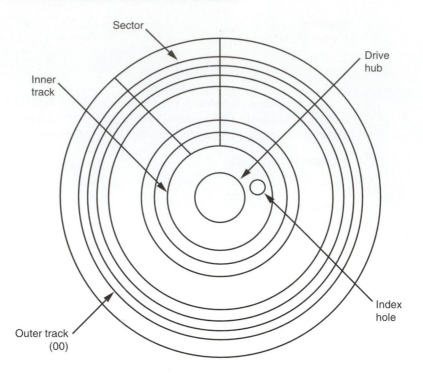

systems a sector holds either 512 or 1,024 bytes of data. The size of a sector can vary from 128 bytes to the length of one entire track.

Notice in Figure 13–17 that there is a hole through the disk that is labeled an index hole. The **index hole** is designed so the electronic system that reads the disk is able to find the beginning of a track and its first sector (00). Tracks are numbered from track 00, the outermost track, in increasing value toward the center or innermost track. Sectors are often numbered from sector 00 on the outermost track, to whatever value is required to reach the innermost track and its last sector.

The 5$^1/_4$″ Mini-Floppy Disk.   Today, the 5$^1/_4$″ floppy is probably the most popular disk size used with older microcomputer systems. Figure 13–18 illustrates this mini-floppy disk. The floppy disk is rotated at 300 RPM inside its semi-rigid plastic jacket. The head mechanism in a floppy disk drive makes physical contact with the surface of the disk; this eventually causes wear and tear to the disk.

Today most mini-floppy disks are double-sided. This means that data is written on both the top and bottom surfaces of the disk. A set of tracks is called a **cylinder** and consists of one top and one bottom track. Cylinder 00, for example, consists of the outermost top and bottom tracks.

Floppy disk data is stored in the double-density format, which uses a recording technique called **MFM** (modified frequency modulation) to store the information. Double-density, double-sided (**DS-DD**) disks are normally organized with 40 tracks of data on each side of the disk. A double-density disk track is typically divided into 9 sectors with each sector containing 512 bytes of information. This means that the total capacity of a double-density, double-sided disk is 40 tracks per side × 2 sides × 9 sectors per track × 512 bytes per sector or 368,640 (360K) bytes of information.

Earlier disk memory systems used single-density and **FM** (frequency modulation) to store information in 40 tracks on one or two sides of the disk. Each of the 8 or 9 sectors on the single-density disk stored 256 bytes of data. This means that a single-density disk stored 90K bytes of data per side. A single-density, double-sided disk stored 180K bytes of data.

Also common today are **high-density** (HD) mini-floppy disks. A high-density mini-floppy disk contains 80 tracks of information per side with 8 sectors per track. Each sector contains

FIGURE 13–18 The 5¹/₄″ mini-floppy disk.

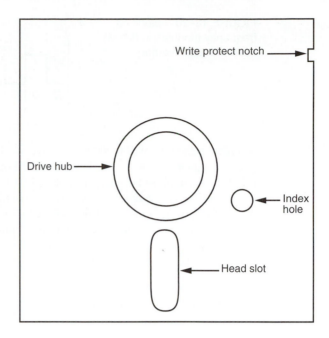

1,024 bytes of information. This gives the 5¹/₄″ high-density mini-floppy disk a total capacity of 80 tracks per side × 2 sides × 15 sectors per track × 512 bytes per sector or 1,228,800 (approximately 1.2M) bytes of information.

The magnetic recording technique used to store data on the surface of the disk is called **non-return to zero** (NRZ) recording. With NRZ recording, magnetic flux placed on the surface of the disk never returns to zero. Figure 13–19 illustrates the information stored in a portion of a track. It also shows how the magnetic field encodes the data. Arrows are used in this illustration to show polarity of the magnetic field stored on the surface of the disk.

The main reason that this form of magnetic encoding was chosen is that it automatically erases old information when new information is recorded. If another technique were used, a separate erase head would be required. The mechanical alignment of a separate erase head and a separate read/write head is virtually impossible. The magnetic flux density of the NRZ signal is so intense that it completely saturates (magnetizes) the surface of the disk erasing all prior data. It also ensures that information will not be affected by noise because the amplitude of the magnetic field contains no information. The information is stored in the placement of the changes of magnetic field.

Data is stored in the form of MFM (modified frequency modulation) in modern floppy disk systems. The MFM recording technique stores data in the form illustrated in Figure 13–20. Notice that each bit time is 2 μs in width on a double-density disk. This means that data is

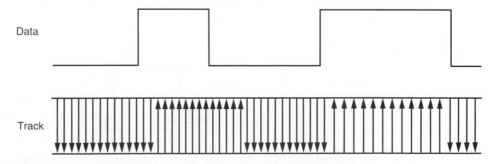

FIGURE 13–19 The non-return to zero (NRZ) recording technique.

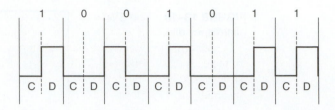

FIGURE 13–20 Modified frequency modulation (MFM) used with disk memory.

recorded at the rate of 500,000 bits per second. Each 2 μs bit time is divided into two parts. One part is designated to hold a clock pulse and the other holds a data pulse. If a clock pulse is present, it is 1 μs in width, as is a data pulse. Clock and data pulses are never present at the same time in one bit period. (High-density disk drives halve these times so that a bit time is 1 μs and a clock or data pulse is 0.5 μs in width. This also doubles the transfer rate to 1 million bits per second.)

If a data pulse is present, the bit time represents a logic 1. If no data or no clock is present, the bit time represents a logic 0. If a clock pulse is present with no data pulse, the bit time also represents a logic 0. The following rules apply when data is stored using MFM:

1. A data pulse is always stored for a logic 1.
2. No data and no clock are stored for the first logic 0 in a string of logic 0's.
3. The second and subsequent logic 0's in a row contain a clock pulse, but no data pulse.

A clock is inserted as the second and subsequent zero in a row to maintain synchronization as data is read from the disk. The electronics used to recapture the data from the disk drive uses a phase-locked loop to generate a clock and a data window. The phase-locked loop needs a clock or data to maintain synchronized operation.

The 3 1/2" Micro-Floppy Disk.   Another very popular disk size is the $3^{1}/2''$ micro-floppy disk. Recently this size has begun to sell very well and in the future promises to be the dominant size floppy disk. The micro-floppy disk is a much improved version of the mini-floppy disk described earlier. Figure 13–21 illustrates the $3^{1}/2''$ micro-floppy disk.

Soon after it was released, disk designers noticed several shortcomings of the mini-floppy, a scaled down version of the 8" standard floppy. Probably one of the biggest problems with the mini-floppy is that it is packaged in a semi-rigid plastic cover that bends easily. The micro-

FIGURE 13–21 The $3^{1}/2''$ micro-floppy disk.

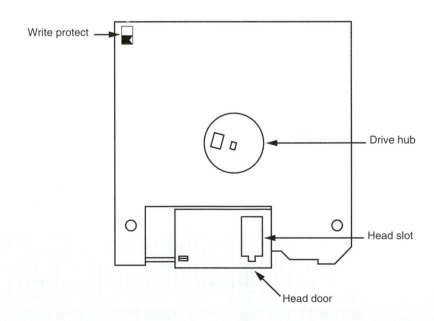

floppy is packaged in a rigid plastic jacket that will not bend easily. This provides a much greater degree of protection to the disk inside the jacket.

Another problem with the mini-floppy is the head slot that continually exposes the surface of the disk to contaminants. This problem is also corrected on the micro-floppy because it is constructed with a spring-loaded sliding head door. The head door remains closed until the disk is inserted into the drive. Once inside the drive, the drive mechanism slides open the door, exposing the surface of the disk to the read/write heads. This provides a great deal of protection to the surface of the micro-floppy disk.

Yet another improvement is the sliding plastic write protection mechanism on the micro-floppy disk. On the mini-floppy disk a piece of tape was placed over a notch on the side of the jacket to prevent writing. This plastic tape easily became dislodged inside disk drives, causing problems. On the micro-floppy, an integrated plastic slide has replaced the tape write protection mechanism. To write-protect (prevent writing) the micro-floppy disk, the plastic slide is moved to **open** the hole though the disk jacket. This allows light to strike a sensor, which inhibits writing.

Still another improvement is the replacement of the index hole with a different drive mechanism. The drive mechanism on the mini-floppy allowed the disk drive to grab the disk at any point. This required the index hole so that the electronics could find the beginning of a track. The index hole was another trouble spot because it collected dirt and dust. The micro-floppy has a drive mechanism that is keyed so that it only fits one way inside the disk drive. The index hole is no longer required because of this keyed drive mechanism. Because of the sliding head mechanism and the fact that no index hole exists, the micro-floppy disk has no place to catch dust or dirt.

Two types of micro-floppy disks are widely available: the double-sided, double-density (DS-DD) and the high-density (HD). The double-sided, double-density micro-floppy disk has 80 tracks per side with each track containing 9 sectors. Each sector contains 512 bytes of information. This allows 80 tracks per side × 2 sides × 9 sectors × 512 bytes per sector or 737,280 (720K) bytes of data to be stored on a double-density, double-sided floppy disk.

The high-density, double-sided micro-floppy disk stores even more information. The high-density version has 80 tracks per side, but the number of sectors is doubled to 18 per track. This format still uses 512 bytes per sector as did the double-density format. The total number of bytes on a high-density, double-sided micro-floppy disk is 80 tracks per side × 2 sides × 18 sectors per track × 512 bytes per sector or 1,474,560 (1.44M) bytes of information.

Recently a new size $3^1/_2''$ floppy disk has been introduced, the EHD (extended high-density) floppy disk. This new format stores 2.88M bytes of data on a single floppy disk. At this time the format is expensive and not yet common. Also available is the floptical disk that stores data magnetically using an optical tracking system. The floptical disk stores 21M bytes of data.

Hard Disk Memory

Larger disk memory is available in the form of the hard disk drive. The hard disk drive is often called a **fixed or rigid disk** because it is not removable like the floppy disk. The term **Winchester drive** is also used to describe a hard disk drive, but less commonly today. Hard disk memory has a much larger capacity than the floppy disk memory. Hard disk memory is available in sizes exceeding 1G byte of data. Common, low-cost (less than $0.25 per megabyte), sizes are presently 1.2G bytes or 1.6G bytes.

There are several differences between the floppy disk and the hard disk memory. The hard disk memory uses a flying head to store and read data from the surface of the disk. A flying head, which is very small and light, does not touch the surface of the disk. It flies above the surface on a film of air that is carried with the surface of the disk as it spins. The hard disk typically spins at 3,000 to 5,400 RPM, which is more than 10 times faster than the floppy disk. This higher rotational speed allows the head to fly (just as an airplane flies) just over the top of the surface of the disk. This is an important feature because there is no wear on the surface as there is with the floppy disk.

Problems can arise because of flying heads. One problem is a head crash. If the power is abruptly interrupted or the hard disk drive is jarred, the head can crash onto the disk surface. This can damage the disk surface or the head. To help prevent crashes, some drive manufacturers have included a system that automatically parks the head when power is interrupted. This type of disk drive has auto-parking heads. When the heads are parked, they are moved to a safe landing zone (unused track) when the power is disconnected. Some drives do not have auto-parking. This type of drive usually requires a program that parks the heads on the innermost track before power is disconnected. The innermost track is a safe landing area because it is the very last track filled by the disk drive. Parking is the responsibility of the operator in this type of disk drive.

Another difference between a floppy disk drive and a hard disk drive is the number of heads and disk surfaces. A floppy disk drive has two heads, one for the upper surface and one for the lower surface. The hard disk drive may have up to eight disk surfaces (four platters) with up to two heads per surface. Each time that a new cylinder is obtained by moving the head assembly, 16 new tracks are available under the heads. Refer to Figure 13–22, which illustrates a hard disk system.

Heads are moved from track to track using either a stepper motor or a voice coil. The stepper motor is slow and noisy, while the voice coil mechanism is quiet and quick. Moving the head assembly requires one step per cylinder in a system that uses a stepper motor to position the heads. In a system that uses a voice coil, the heads can be moved many cylinders with one sweeping motion. This makes the disk drive faster when seeking new cylinders.

Another advantage of the voice coil system is that a servo mechanism can monitor the amplitude of the signal as it comes from the read head and make slight adjustments in the position of the heads. This is not possible with a stepper motor, which relies strictly on mechanics to position the head. Stepper motor type head positioning mechanisms can often become misaligned with use, while the voice coil mechanism corrects for any misalignment.

Hard disk drives often store information in sectors that are 512 bytes in length. Data is addressed in **clusters** of eight or more sectors that contain 4,096 or more bytes on most hard disk drives. Hard disk drives use either MFM or RLL to store information. MFM is described with floppy disk drives. **Run-length limited** (RLL) is described here.

A typical older MFM hard disk drive uses 18 sectors per track so that 18K bytes of data are stored per track. If a hard disk drive has a capacity of 40M bytes, it contains approximately 2,280 tracks. If the disk drive has two heads, it contains 1,140 cylinders. If it contains four heads, then it has 570 cylinders. These specifications vary from disk drive to disk drive.

RLL Storage.   Run-length limited (RLL) disk drives use a different method for encoding the data than MFM. The term RLL means that the run of zeros (zeros in a row) is limited. A common

FIGURE 13–22   A hard disk drive that uses four heads per platter.

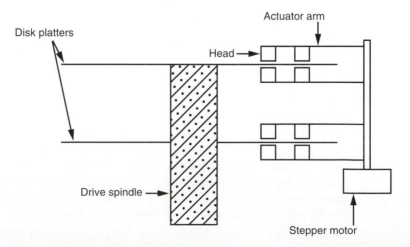

TABLE 13–1 Standard
RLL 2,7 coding

Input Data Stream	RLL Output
000	000100
10	0100
010	100100
0010	00100100
11	1000
011	001000
0011	00001000

RLL encoding scheme in use today is RLL 2,7. This means that the run of zeros is always between two and seven. Table 13–1 illustrates the coding used with standard RLL.

Data is first encoded using the values in Table 13–1 before being sent to the drive electronics for storage on the disk surface. Because of this encoding technique, it is possible to achieve a 50 percent increase in data storage on a disk drive when compared to MFM. The main difference is that the RLL drive often contains 27 tracks instead of the 18 found on the MFM drive. (Some RLL drives also use 35 sectors per track.)

It is interesting to note that in most cases RLL encoding requires no change to the drive electronics or surface of the disk. The only difference is a slight decrease in the pulse width using RLL, which may require slightly finer oxide particles on the surface of the disk. Disk manufacturers test the surface of the disk and grade the disk drive as either an MFM-certified or an RLL-certified drive. Other than grading, there is no difference in the construction of the disk drive or the magnetic material that coats the surface of the disks.

Figure 13–23 shows a comparison of MFM and RLL data. The amount of time (space) required to store RLL data is reduced when compared to MFM. Here a 101001011 is coded in both MFM and RLL so that these two standards can be compared. Notice that the width of the RLL signal has been reduced so that 3 pulses fit in the same space as a clock and a data pulse for MFM. A 40M-byte MFM disk can hold 60M bytes of RLL-encoded data. Besides holding more information, the RLL drive can be written and read at a higher rate.

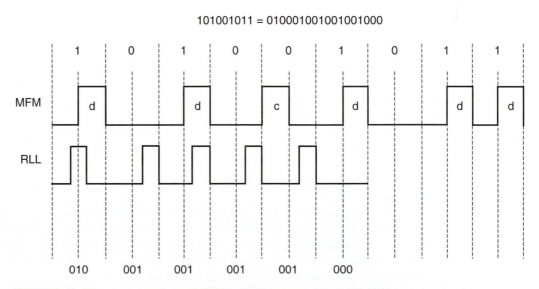

101001011 = 010001001001001000

FIGURE 13–23 A comparison of MFM with RLL using data 101001011.

All hard disk drives use either MFM or RLL encoding. There are a number of disk drive interfaces in use today. The oldest is the ST-506 interface, which uses either MFM or RLL data. A disk system using this interface is called either an MFM or RLL disk system. Newer standards are also in use today. These include ESDI, SCSI, and IDE. All of these newer standards use RLL, even though they normally do not call attention to it. The main difference is the interface between the computer and the disk drive. The IDE system is becoming the standard hard disk memory interface.

The **enhanced small disk interface** (ESDI) system, which has disappeared, is capable of transferring data between itself and the computer at rates approaching 10M bytes per second. An ST-506 interface can approach a transfer rate of 860K bytes per second.

The **small computer system interface** (SCSI) system is also used because it allows up to seven different disk or other interfaces to be connected to the computer through the same interface controller. SCSI is found in some PC-type computers and also in the Apple Macintosh system. An improved version, SCSI-II, has started to appear in some systems. In the future this interface may be replaced with IDE in most applications.

The newest system is **integrated drive electronics** (IDE), which incorporates the disk controller in the disk drive and attaches the disk drive to the host system through a small interface cable. This allows many disk drives to be connected to a system without worrying about bus or controller conflicts. IDE drives are found in newer IBM PS-2 systems and many clones. Even Apple computer systems are starting to be found with IDE drives in place of the SCSI drives found in older Apple computers. The IDE interface is also capable of driving other I/O devices besides the hard disk. This interface usually contains at least a 32K-byte cache memory for disk data. The cache speeds disk transfers. Common access times for an IDE drive are often less than 10 ms, whereas the access time for a floppy disk is about 200 ms. Even the size advantage of the SCSI drive has all but disappeared with IDE drives of over 3G bytes.

Optical Disk Memory

Optical disk memory (see Figure 13–24) is commonly available in two forms: the CD-ROM (compact disk/read only memory) and the WORM (write once/read mostly). The CD-ROM is the lowest cost optical disk, but it suffers from lack of speed. Access times for a CD-ROM are typically 300 ms or longer, about the same as a floppy disk. [Note that slower CD-ROM devices (less than 4×) are on the market and should be avoided.] Hard disk magnetic memory can have access times as little as 11 ms. A CD-ROM stores 660M bytes of data or a combination of data and musical passages. As systems develop and become more visually active, the use of the CD-ROM drive will become even more common.

The WORM drive sees far more commercial application than the CD-ROM. The problem is that its application is very specialized due to the nature of the WORM. Because data may only be written once, the main application is in the banking industry, insurance industry, and other massive data-storing organizations. The WORM is normally used to form an audit trail of transactions that are spooled onto the WORM and retrieved only during an audit. You might call the WORM an archiving device.

Many WORM and read/write optical disk memory systems are interfaced to the microprocessor using the SCSI or ESDI interface standards used with hard disk memory. The difference is that the current optical disk drives are no faster than most floppy drives. Some CD-ROM drives are interfaced to the microprocessor through proprietary interfaces that are not compatible with other disk drives.

The main advantage of the optical disk is its durability. Because a solid-state laser beam is used to read the data from the disk, and the focus point is below a protective plastic coating, the surface of the disk may contain small scratches and dirt particles and still be read correctly. This feature allows less care of the optical disk than a comparable floppy disk. About the only way to destroy data on an optical disk is to break it or deeply scar it.

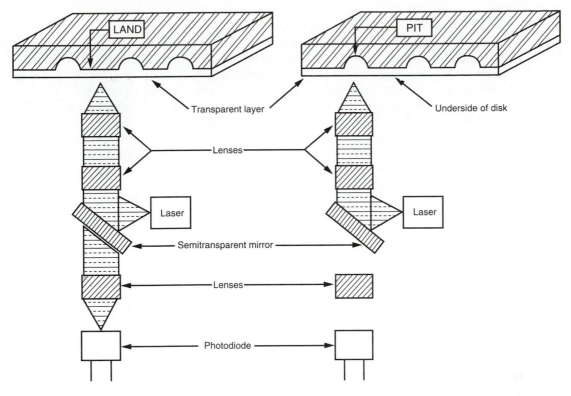

FIGURE 13–24   The optical CD-ROM memory system.

The read/write CD-ROM drive is here and its cost is dropping rapidly. In the near future we should start seeing the read/write CD-ROM replacing floppy disk drives. The main advantage is the vast storage available on the read/write CD-ROM. Soon the format will change so many gigabytes of data will be available. The new versatile read/write CD-ROM, called a DVD (digital video disk), should become available sometime in 1997.

13–4   VIDEO DISPLAYS

Modern video displays are OEM (**original equipment manufacturer**) devices that are usually purchased and incorporated into a system. Today there are many different types of video displays available, either color or monochrome.

Monochrome versions usually display information using amber, green, or paper-white displays. The paper-white display is becoming extremely popular for many applications. The most common of these applications are desktop publishing and computer-aided drafting (CAD). Also available are LCDs (liquid crystal displays) found in small laptop or notebook computer systems.

The color displays are more diverse. Color display systems are available that accept information as a composite video signal much as your home television, as TTL voltage level signals (0 or 5 V), and as analog signals (0–0.7 V). Composite video displays are disappearing because the resolution available is too low. Today many applications require high-resolution graphics that cannot be displayed on a composite display such as a home television receiver. Early composite video displays were found with Commodore 64, Apple 2, and similar computer systems.

Video Signals

Figure 13–25 illustrates the signal sent to a composite video display. This signal is composed of several parts that are required for this type of display. The signals illustrated represent the signals sent to a color composite video monitor. Notice that these signals include not only video, but sync pulses, sync pedestals, and a color burst. Also notice that no audio signal is illustrated because one often does not exist. Rather than include audio with the composite video signal, audio is developed in the computer and output from a speaker inside the computer cabinet. It can also be developed by a sound system and output in stereo to external speakers. The major disadvantages of the composite video display are the resolution and color limitations. Composite video signals were designed to emulate television video signals so a home television receiver could function as a video monitor.

Most modern video systems use direct video signals that are generated with separate sync signals. In a direct video system, video information is passed to the monitor through a cable that uses separate lines for video and also synchronization pulses. Recall that these signals were combined in a composite video signal.

A monochrome (one color) monitor uses one wire for video, one for horizontal sync, and one for vertical sync. Often these are the only signal wires found. A color video monitor uses three video signals. One signal represents red, another green, and the third blue. These monitors are often called RGB monitors for the video primary colors of light: red (R), green (G), and blue (B).

The TTL RGB Monitor

The RGB monitor is available as either an analog or TTL monitor. The RGB video TTL monitor uses TTL level signals (0 or 5 V) as video inputs and a fourth line, called intensity, to allow a change in intensity. The RGB video TTL display can display a total of 16 different colors. The TTL RGB monitor is used primarily in the CGA (**color graphics adapter**) system found in older computer systems.

Table 13–2 lists these 16 colors and the TTL signals present to generate them. Eight of the 16 colors are generated at high intensity and the other eight at low intensity. The three video colors are red, green, and blue, which are the primary colors of light. The secondary colors are cyan, magenta, and yellow. Cyan is a combination of blue and green video signals and is blue-

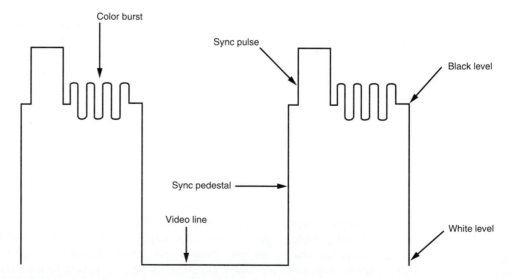

FIGURE 13–25   The composite video signal.

TABLE 13–2 The 16 colors found in the CGA display

Intensity	Red	Green	Blue	Color
0	0	0	0	Black
0	0	0	1	Blue
0	0	1	0	Green
0	0	1	1	Cyan
0	1	0	0	Red
0	1	0	1	Magenta
0	1	1	0	Brown
0	1	1	1	White
1	0	0	0	Gray
1	0	0	1	Bright Blue
1	0	1	0	Bright Green
1	0	1	1	Bright Cyan
1	1	0	0	Bright Red
1	1	0	1	Bright Magenta
1	1	1	0	Yellow
1	1	1	1	Bright White

green in color. Magenta is a combination of blue and red video signals and is a purple color. Yellow (high intensity) and brown (low intensity) are both a combination of red and green video signals. If additional colors are desired, TTL video is not normally used. A scheme was developed using low and medium color TTL video signals, which provided 32 colors, but it proved of little application and never found widespread use in the field.

Figure 13–26 illustrates the 9-pin connector most often found on the TTL RGB monitor or a TTL monochrome monitor. Two of the connections are used for ground, three for video, two for synchronization or retrace signals, and one for intensity. Notice that pin 7 is labeled normal video. This is the pin used on a monochrome monitor for the luminance or brightness signal. Monochrome TTL monitors use the same 9-pin connector as RGB TTL monitors.

The Analog RGB Monitor

In order to display more than 16 colors, an analog video display is required. These are often called analog RGB monitors. Analog RGB monitors still have three video input signals, but don't have the intensity input. Because the video signals are analog signals instead of two-level TTL signals, they are any voltage level between 0.0 V and 0.7 V. This allows an infinite number

FIGURE 13–26 The 9-pin connector found on a TTL monitor.

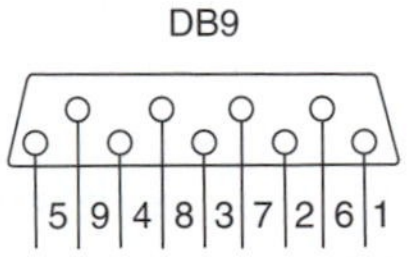

DB9

5 9 4 8 3 7 2 6 1

Pin	Function
1	Ground
2	Ground
3	Red video
4	Green video
5	Blue video
6	Intensity
7	Normal video
8	Horizontal retrace
9	Vertical retrace

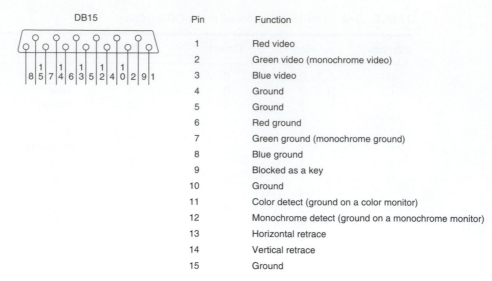

DB15	Pin	Function
	1	Red video
	2	Green video (monochrome video)
	3	Blue video
	4	Ground
	5	Ground
	6	Red ground
	7	Green ground (monochrome ground)
	8	Blue ground
	9	Blocked as a key
	10	Ground
	11	Color detect (ground on a color monitor)
	12	Monochrome detect (ground on a monochrome monitor)
	13	Horizontal retrace
	14	Vertical retrace
	15	Ground

FIGURE 13–27   The 15-pin connector found on an analog monitor.

of colors to be displayed because an infinite number of voltage levels between the minimum and maximum could be generated. In practice, a finite number of levels are generated. This is usually either 256K, 16M, or 24M colors depending on the standard.

Figure 13–27 illustrates the connector used for an analog RGB or analog monochrome monitor. Notice that the connector has 15 pins and supports both RGB and monochrome analog displays. The way data is displayed on an analog RGB monitor depends upon the interface standard used with the monitor. Pin 9 is a key, which means that no hole exists on the female connector for this pin.

Most analog displays use a digital-to-analog converter (DAC) to generate each color video voltage. This includes the color LCD display found in portable applications. A common standard uses a 6-bit DAC for each video signal to generate 64 different voltage levels between 0 V and 0.7 V. There are 64 different red video levels, 64 different green video levels, and 64 different blue video levels. This allows $64 \times 64 \times 64$ different colors to be displayed or 262,144 (256K) colors.

Other arrangements are possible, but the speed of the DAC is critical. Most modern displays require an operating conversion time of 25 ns to 40 ns maximum. When converter technology advances, additional resolution at a reasonable price will become available. If 7-bit converters are used for generating video, $128 \times 128 \times 128$ or 2,097,152 (2M) colors are displayed. In this system a 21-bit color code is needed so that a 7-bit code is applied to each DAC. Eight-bit converters allow $256 \times 256 \times 256$ or 16,777,216 (16M) colors. In most cases, modern video cards use 8-bit DACs for generating 16M colors for a standard called true color.

Figure 13–28 illustrates the video generation circuit used in many common video standards such as the short-lived EGA (enhanced graphics adapter) and VGA (variable graphics array) used with an IBM PC. This circuit is used to generate VGA video. Notice that each color is generated with an 18-bit digital code. Six of the 18 bits are used to generate each video color voltage when applied to the inputs of a 6-bit DAC.

A high-speed palette SRAM (access time of less than 40 ns) is used to store 256 different 18-bit codes that represent 256 different hues. This 18-bit code is applied to the digital-to-analog converters. The address input to the SRAM selects one of the 256 colors stored as 18-bit binary codes. This system allows 256 colors out of a possible 256K colors to be displayed at one time. In order to select any of 256 colors, an 8-bit code that is stored in the computer's video display RAM is used to specify the color of a picture element. If more colors are used in a system, the code must be wider. For example, a system that displays 1,024 colors out of 256K colors requires a 10-bit

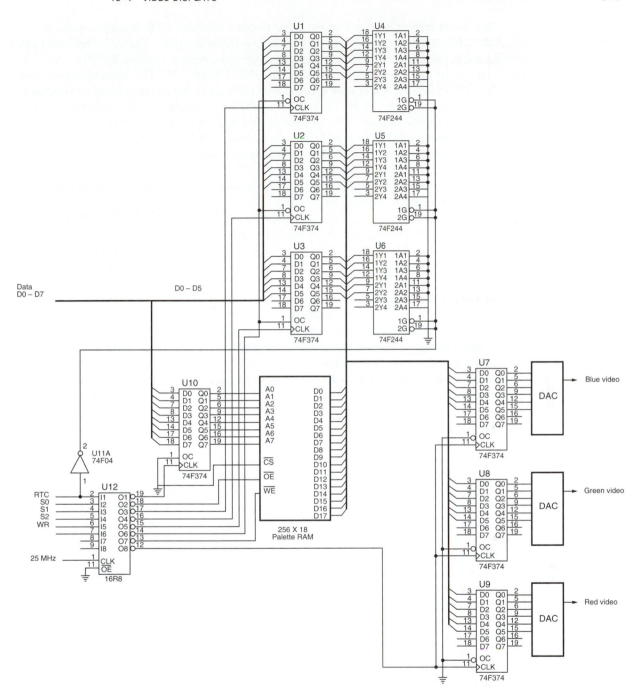

FIGURE 13–28 Generation of VGA video signals.

code to address the SRAM, which contains 1,024 locations each containing an 18-bit color code. Some newer systems use a larger palette SRAM to store up to 64K different colors codes.

The Apple Macintosh IIci uses a 24-bit binary code to specify each color in its color video adapter. Each DAC is 8 bits wide. This means that each converter can generate 256 different video voltage levels. There are $256 \times 256 \times 256$ or 16,777,216 different possible colors. As with the IBM VGA standard, only 256 colors are displayed at a time. The SRAM in the Apple interface is 256×24 instead of 256×18.

Whenever a color is placed on the video display, provided RTC is a logic 0, the system sends the 8-bit code that represents a color to the D0–D7 connections. The PAL 16R8 then generates a clock pulse for U10, which latches the color code. After 40 ns (one 25-MHz clock) the PAL generates a clock pulse for the DAC latches (U7, U8, and U9). This amount of time is required for the palette SRAM to look up the 18-bit contents of the memory location selected by U10. Once the color code (18 bits) is latched into U7–U9, the three DACs convert it to three video voltages for the monitor. This process is repeated for each 40-ns-wide picture element (pixel) that is displayed. The pixel is 40 ns wide because a 25-MHz clock is used in this system. Higher resolution is attainable if a higher clock frequency is used with the system.

If the color codes (18 bits) stored in the SRAM must be changed, this is always accomplished during retrace when RTC is a logic 1. This prevents any video noise from disrupting the image displayed on the monitor.

In order to change a color, the system uses the S0, S1, and S2 inputs of the PAL to select U1, U2, U3, or U10. First the address of the color to be changed is sent to latch U10. This addresses a location in the palette SRAM. Next, each new video color is loaded into U1, U2, and U3. Finally, the PAL generates a write pulse for the $\overline{\text{WE}}$ input to the SRAM to write the new color code into the palette SRAM.

Retrace occurs 70.1 times per second in the vertical direction and 31,500 times per second in the horizontal direction for a 640×480 display. During retrace, the video signal voltage sent to the display must be 0 V. This causes black to be displayed during the retrace. Retrace itself is used to move the electron beam to the upper left-hand corner for vertical retrace and to the left margin of the screen for horizontal retrace.

The circuit illustrated causes U4–U6 buffers be enabled so they apply 00000 each to the DAC latch for retrace. The DAC latches capture this code and generate 0 V for each video color signal to blank the screen. By definition, 0 V is considered the black level for video and 0.7 V is considered full intensity on a video color signal.

The resolution of the display, for example 640×480, determines the amount of memory required for the video interface card. If this resolution is used with a 256-color display (8 bits per pixel), then 640×480 bytes of memory (307,200) are required to store all of the pixels for the display. Higher resolution displays are possible, but as you can imagine, even more memory is required. A 640×480 display has 480 video raster lines and 640 pixels per line. A **raster line** is the horizontal line of video information that is displayed on the monitor. A pixel is the smallest subdivision of this horizontal line.

Figure 13–29 illustrates the video display showing the video lines and retrace. The slant of each video line is greatly exaggerated, as is the spacing between lines. This illustration shows retrace in both the vertical and horizontal directions. In the case of a VGA display, as described, the vertical retrace occurs exactly 70.1 times per second and the horizontal retrace occurs exactly 31,500 times per second. (The Apple Macintosh IIci uses a vertical rate of 66.67 Hz and a horizontal rate of 35 KHz to generate a 640×480 color display.)

In order to generate 640 pixels across one line, it takes 40 ns $\times$ 640 or 25.6 μs. A horizontal time of 31,500 Hz allows a horizontal line time of 1/31,500 or 31.746 μs. The difference between these two times is the retrace time allowed to the monitor. (The Apple Macintosh IIci has a horizontal line time of 28.57 μs.)

Because the vertical retrace repetition rate is 70.1 Hz, the number of lines generated is determined by dividing the vertical time into the horizontal time. In the case of a 640×400 VGA display, this is 449.358 lines. Only 400 of these lines are used to display information; the rest are lost during the retrace. Since 49.358 lines are lost during the retrace, the retrace time is 49.358 $\times$ 31.766 μs or 1,568 μs. It is during this relatively large amount of time that the color palette SRAM is changed or the display memory system is updated for a new video display. In the Apple Macintosh IIci computer (640×480) the number of lines generated is 525. Of these total number of lines, 45 are lost during vertical retrace.

FIGURE 13–29 A video screen illustrating the raster lines and retrace.

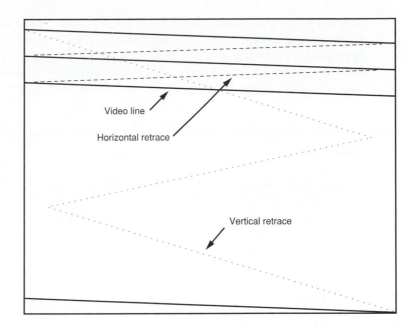

Other display resolutions are 800×600 and $1{,}024 \times 768$. The 800×600 SVGA (super VGA) display is ideal for a 14″ color monitor, while the 1024×768 EVGA or XVGA (extended VGA) is ideal for a 21″ or 25″ monitor used in CAD systems. These resolutions sound like just another set of numbers, but realize that an average home television receiver has a resolution of approximately 400×300. The high-resolution display available on computer systems is much clearer than that available on home television. A resolution of $1{,}024 \times 768$ approaches that found in 35-mm film. The only disadvantage of the video display on a computer screen is the number of colors displayed at a time, but as time passes this will surely improve. Additional colors allow the image to appear more realistic because of subtle shadings that are required for a true high-quality lifelike image.

If a display system operates with a 60-Hz vertical time and a 15,600-Hz horizontal time, the number of lines generated is 15,600/60 or 260 lines. The number of usable lines in this system is most likely 240, where 20 are lost during vertical retrace. It is clear that the number of scanning lines is adjustable by changing the vertical and horizontal scanning rates. The vertical scanning rate must be greater than or equal to 50 Hz or flickering will occur. The vertical rate must not be higher than about 75 Hz or problems with the vertical deflection coil may occur. The electron beam in a monitor is positioned by an electrical magnetic field generated by coils in a yoke that surrounds the neck of the picture tube. Since the magnetic field is generated by coils, the frequency of the signal applied to the coil is limited.

The horizontal scanning rate is also limited by the physical design of the coils in the yoke. Because of this, it is normal to find the frequency applied to the horizontal coils within a narrow range. This is usually 30,000 Hz–37,000 Hz or 15,000 Hz–17,000 Hz. Some newer monitors are called multi-sync monitors because the deflection coil is taped so that it can be driven with different deflection frequencies. Sometimes both the vertical and horizontal coils are taped for different vertical and horizontal scanning rates.

High-resolution displays use either interlaced or non-interlaced scanning. The non-interlaced scanning system is used in all standards except the highest. In the interlaced system, the video image is displayed by drawing half the image first with all of the odd scanning lines, then the other half with the even scanning lines. Obviously this system is more complex and is more efficient only because the scanning frequencies are reduced by 50 percent in an interlaced system. For example, a video system that uses 60 Hz for the vertical scanning frequency and 15,720 Hz for the horizontal frequency generates 262 (15,720/60) lines of video at the rate of 60 full frames per second. If the horizontal frequency is changed slightly to 15,750 Hz, 262.5 (15,750/60) lines are generated so two full sweeps are required to draw one complete picture of 525 video lines. Notice how just a slight change in horizontal frequency doubled the number of raster lines.

13–5   SUMMARY

1. The HOLD input is used to request a DMA action, and the HLDA output signals that the hold is in effect. When a logic 1 is placed on the HOLD input, the microprocessor (1) stops executing the program, (2) places its address, data, and control bus at their high-impedance state, and (3) signals that the hold is in effect by placing a logic 1 on the HLDA pin.

2. A DMA read operation transfers data from a memory location to an external I/O device. A DMA write operation transfers data from an I/O device into the memory. Also available is a memory-to-memory transfer that allows data to be transferred between two memory locations using DMA techniques.

3. The 8237 direct memory access (DMA) controller is a four-channel device that can be expanded to include additional channels of DMA.

4. The 80186/80188 XL and EA versions each contain a programmable two-channel DMA controller. Two pins on these microprocessors are used to gain access to a DMA transfer and are labeled DRQ_0 and DRQ_1.

5. The 80186/80188 EC versions contain a four-channel DMA controller that is similar to the two-channel controller in the XL and EA versions. The main difference is that the serial channels in the EC versions can cause a DMA transfer.

6. Disk memory comes in the form of floppy disk storage, either the $5^{1}/4''$ mini-floppy disk or $3^{1}/2''$ micro-floppy disk. Both disks are found as double-sided, double-density (DS-DD) or as high-density (HD) storage devices. The DS-DD $5^{1}/4''$ disk stores 360K bytes of data and the HD $5^{1}/4''$ disk stores 1.2M bytes of data. The DS-DD $3^{1}/2''$ disk stores 720K bytes of data and the HD $3^{1}/2''$ disk stores 1.44M bytes of data.

7. Floppy disk memory data is stored using NRZ (non-return to zero) recording. This method saturates the disk with one polarity of magnetic energy for a logic 1 and the opposite polarity for a logic 0. In either case, the magnetic field never returns to zero. This technique eliminates the need for a separate erase head.

8. Data is recorded on disks by using either modified frequency modulation (MFM) or by run-length limited (RLL) encoding schemes. The MFM scheme records a data pulse for a logic 1, no data or clock for the first logic 0 of a string of zeros, and a clock pulse for the second and subsequent logic 0 in a string of zeros. The RLL scheme encodes data so 50 percent more information can be packed onto the same disk area. Most modern disk memory systems use the RLL encoding scheme.

9. Video monitors are either TTL or analog. The TTL monitor uses two discrete voltage levels of 0 V and 5.0 V. The analog monitor uses an infinite number of voltage levels between 0.0 V and 0.7 V. The analog monitor can display an infinite number of video levels, while the TTL monitor is limited to two video levels.

10. The color TTL monitor displays 16 different colors. This is accomplished through three video signals (red, green, and blue) and an intensity input. The analog color monitor can display an infinite number of colors through its three video inputs. In practice, the most common form of color analog display system (VGA) can display 256K different colors.

11. The video standards found today include: VGA (640 × 480), SVGA (800 × 600), and EVGA or XVGA (1,024 × 768). In all three cases, the video information can be 256 colors out of a total possible 256K colors.

13–6 QUESTIONS AND PROBLEMS

1. Which microprocessor pins are used to request and acknowledge a DMA transfer?
2. Explain what happens when a logic 1 is placed on the HOLD input pin.
3. A DMA read transfers data from _____ to _____.
4. A DMA write transfers data from _____ to _____.
5. The DMA controller selects the memory location used for a DMA transfer through which bus signals?
6. The DMA controller selects the I/O device used during a DMA transfer by which pin?
7. What is a memory-to-memory DMA transfer?
8. Describe the effect on the microprocessor and DMA controller when the HOLD and HLDA pins are at their logic 1 levels.
9. Describe the effect on the microprocessor and DMA controller when the HOLD and HLDA pins are at their logic 0 levels.
10. The 8237 DMA controller is a _____-channel DMA controller.
11. If the 8237 DMA controller is decoded at I/O ports 2000H–200FH, which ports are used to program channel 1?
12. Which 8237 DMA controller register is programmed to initialize the controller?
13. How many bytes can be transferred by the 8237 DMA controller?
14. Write a sequence of instructions that transfers data from memory locations 21000H–210FFH to 20000H–200FFH using channel 2 of the 8237 DMA controller. You must initialize the 8237 and use the latch described in Section 13–1 to hold A19–A16.
15. Write a sequence of instructions that transfers data from memory to an external I/O device using channel 3 of the 8237. The memory area to be transferred is at locations 20000H–20FFFH.
16. The 80188 XL microprocessor contains a _____-channel DMA controller.
17. Which I/O port addresses are used to program the channel 1 source address into the 80186 EA version of the microprocessor?
18. What is the purpose of the INC bit in the control register for a DMA channel in the 80186/80188 microprocessors?
19. What is the purpose of the TC bit in a DMA channel of the 80186/80188 microprocessors?
20. How many DMA channels are available in the EB version of the 80186/80188?
21. How many DMA channels are available in the EC version of the 80186/80188?
22. What is the purpose of the halt register in the EC version of the 80186/80188?
23. Which I/O port accesses the halt register in the EC version of the 80186/80188?
24. Develop a short sequence of instructions that program the 80186/80188, channel 0, destination memory address with a 20000H.
25. The 5¹/₄″ disk is known as a _____-floppy disk.
26. The 3¹/₂″ disk is known as a _____-floppy disk.
27. Data is recorded in concentric rings on the surface of a disk known as a _____.
28. A track is divided into sections of data called _____.
29. On a double-sided disk, the upper and lower tracks together are called a _____.
30. Why is NRZ recording used on a disk memory system?
31. Draw the timing diagram generated to write a 1001010000 using MFM encoding.
32. Draw the timing diagram generated to write a 1001010000 using RLL encoding.
33. What is a flying head?
34. Why must the heads on a hard disk be parked?
35. What is the difference between a voice coil head position mechanism and a stepper motor head positioning mechanism?
36. What is a WORM?
37. What is a CD-ROM?
38. What is the difference between a TTL monitor and an analog monitor?

39. What are the three primary colors of light?
40. What are the three secondary colors of light?
41. What is a pixel?
42. A video display with a resolution of 800×600 contains _____ lines of video information with each line divided into _____ pixels.
43. Explain how a TTL RGB monitor can display 16 different colors.
44. Explain how an analog RGB monitor can display an infinite number of colors.
45. If an analog RGB video system uses 7-bit DACs, it can generate _____ different colors.
46. Why does standard VGA allow only 256 different colors out of 256K colors to be displayed at one time?
47. If a video system uses a vertical frequency of 60 Hz and a horizontal frequency of 32,400 Hz, how many raster lines are generated?
48. Use the Internet to locate the Sony Web page and write a short report that describes DVD.
49. Use the Internet to locate information on Diamond video cards and write a short report describing the types available.
50. Use the Internet to locate information on hard disk drives from Western Digital and list the currently available sizes of IDE drives.

CHAPTER 14

The Arithmetic Coprocessor

INTRODUCTION

The Intel family of arithmetic coprocessors includes the 8087, 80187, 80287, 80387SX, 80387DX, and the 80487SX for use with the 80486SX microprocessor. The 80486DX and Pentium microprocessors contain their own built-in arithmetic coprocessors. Some of the cloned 80486 microprocessors (from IBM and Cyrix) do not contain arithmetic coprocessors. The instruction sets and programming for all devices are almost identical; the main difference is that each coprocessor is designed to function with a different Intel microprocessor. This chapter describes the entire family of arithmetic coprocessors. Because the coprocessor is a part of the 80486DX, Pentium, and Pentium Pro, and because these microprocessors are commonplace, many programs now require (or at least benefit from) a coprocessor.

The family of coprocessors labeled 80X87 is able to multiply, divide, add, subtract, find the square root, partial tangent, partial arctangent, and logarithms. Data types include: 16-, 32-, and 64-bit signed-integers; 18-digit BCD data; 32-, 64-, and 80-bit floating-point numbers. The operations performed by the 80X87 generally execute many times faster than equivalent operations written with the most efficient programs using the microprocessor's normal instruction set. With the improved Pentium coprocessor, operations execute at about five times faster than those performed by the 80486 microprocessor with an equal clock frequency. The Pentium can often execute a coprocessor instruction and two integer instructions simultaneously. The Pentium Pro coprocessor is similar in performance to the Pentium coprocessor except that new instructions have been added: FMOV and FCOMI.

CHAPTER OBJECTIVES

Upon completion of this chapter, you will be able to:

1. Convert between decimal data and signed-integer, BCD, and floating-point data for use by the arithmetic coprocessor.
2. Explain the operation of the 80X87 arithmetic coprocessor.
3. Connect the 80187 coprocessor to the 80186/80188 microprocessors.
4. Explain the operation and addressing modes of each arithmetic coprocessor instruction.
5. Develop programs that solve complex arithmetic problems using the arithmetic coprocessor.

14–1   DATA FORMATS FOR THE ARITHMETIC COPROCESSOR

In this section we discuss the types of data used with all arithmetic coprocessor family members. (Refer to Table 14–1 for a list of all Intel microprocessors and their companion coprocessors.) These data types include: signed-integer, BCD, and floating-point. Each has a specific use in a system, and many systems require all three data types. Note that assembly language programming with the coprocessor is often limited to modifying the coding generated by a high-level language such as C/C++. In order to accomplish any such modification, the instruction set and some basic programming concepts are required.

Signed-Integers

The signed-integers used with the coprocessor are basically the same as those described in Chapter 1. When used with the arithmetic coprocessor, signed-integers are 16 (word), 32 (short integer), or 64 bits (long integer) in width. Conversion between decimal and signed-integer format is handled in exactly the same manner as for the signed-integers described in Chapter 1. As you recall, positive numbers are stored in true form with a leftmost sign-bit of 0, and negative numbers are stored in two's complement form with a leftmost sign-bit of 1.

The word integers range in value from –32,768 to +32,767; the short integer range is $\pm 2 \times 10^{+9}$; the long integer range is $\pm 9 \times 10^{+18}$. Integer data types are found in some applications that use the arithmetic coprocessor. Figure 14–1 shows these three forms of signed-integer data.

Data is stored in memory using the same assembler directives described and used in earlier chapters. The DW directive defines words, DD defines short integers, and DQ defines long integers. Example 14–1 shows how several different sizes of signed-integers are defined for use by the assembler and arithmetic coprocessor.

EXAMPLE 14–1

```
0000 0002                DATA1   DW    +2          ;16-bit integer
0002 FFDE                DATA2   DW    -34         ;16-bit integer
0004 000004D2            DATA3   DD    +1234       ;short integer
0008 FFFFFF9C            DATA4   DD    -100        ;short integer
000C 0000000000005BA0    DATA5   DQ    +23456      ;long integer
0014 FFFFFFFFFFFFFF86    DATA6   DQ    -122        ;long integer
```

TABLE 14–1
Microprocessor and Intel coprocessor compatibility

Microprocessor	Coprocessor
8086	8087
8088	8087
80186	80187
80188	80187
80286	80287
80386SX	80387SX
80386DX	80387DX
80386EX	80387SX
80486SX	80487SX
80486DX	Built into microprocessor
Pentium	Built into microprocessor

FIGURE 14–1 Integer forms of data for the 80X87 family of arithmetic coprocessors: (a) word, (b) short, and (c) long.

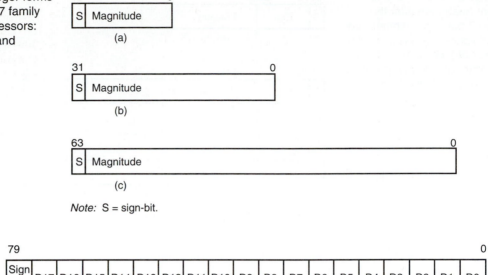

Note: S = sign-bit.

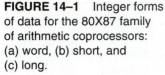

FIGURE 14–2 BCD data format for the 80X87 family of arithmetic coprocessors.

Binary-Coded Decimal (BCD)

The binary-coded decimal (BCD) form requires 80 bits of memory. Each number is stored as an 18-digit packed integer in 9 bytes of memory as 2 digits per byte. The tenth byte contains only a sign bit for the 18-digit signed BCD number. Figure 14–2 shows the format of the BCD number used with the arithmetic coprocessor. Note that both positive and negative numbers are stored in true form and never in 10's complement form. The DT directive stores BCD data in the memory as illustrated in Example 14–2.

EXAMPLE 14–2

```
0000                    DATA1   DT      200             ;200 decimal stored as BCD
            00000000000000000200

000A                    DATA2   DT      -10             ;-10 decimal stored as BCD
            80000000000000000010

0014                    DATA3   DT      10020           ;10,020 1 decimal stored as BCD
            00000000000000010020
```

Floating-Point Numbers

Floating-point numbers are often called real numbers because they hold signed-integers, fractions, and mixed numbers. A floating-point number has three parts: a sign-bit, a biased exponent, and a significand. Floating-point numbers are written in scientific binary notation. The Intel family of arithmetic coprocessors supports three types of floating-point numbers: short (32 bits), long (64 bits), and temporary (80 bits). Refer to Figure 14–3 for the three forms of the floating-point number. The short form is also called a single-precision number and the long form is also called a double-precision number. Sometimes the 80-bit temporary form is called an extended-precision number. The floating-point numbers, and the operations performed by the arithmetic coprocessor,

FIGURE 14–3 Floating-point (real) format for the 80X87 family of arithmetic co-processors. (a) Short (single-precision) with a bias of 7FH, (b) long (double-precision) with a bias of 3FFH, and (c) temporary (extended-precision) with a bias of 3FFFH.

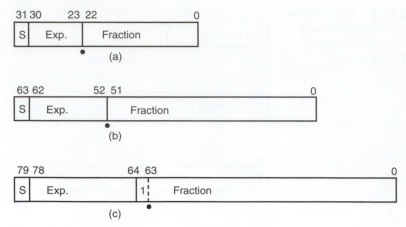

Note: S = sign-bit and Exp. = exponent.

conform to the IEEE-754 standard as adopted by all major personal computer software producers. This includes Microsoft, which has recently stopped supporting the Microsoft floating-point format and the ANSI floating-point standard popular in mainframe computer systems.

Converting to Floating-Point Form. Converting from decimal to floating-point form is a simple task accomplished by the following steps:

1. Convert the decimal number to binary.
2. Normalize the binary number.
3. Calculate the biased exponent.
4. Store the number in floating-point format.

These four steps are illustrated for the decimal number 100.25_{10} in Example 14–3. Here the decimal number is converted to a single-precision (32-bit) floating-point number.

EXAMPLE 14–3

```
Step      Result

1         100.25 = 1100100.01

2         1100100.01 = 1.10010001 x 2⁶

3         110 + 01111111 = 10000101

4         Sign = 0
          Exponent = 10000101
          Significand = 10010001000000000000000
```

In step 3 of Example 14–3, the biased exponent is the exponent, a 2^6 or 110, plus a bias of 01111111 (7FH) or 10000101 (85H). All single-precision numbers use a bias of 7FH, double-precision numbers use a bias of 3FFH, and extended-precision numbers use a bias of 3FFFH.

In step 4 of Example 14–3, the information found in prior steps is combined to form the floating-point number. The leftmost bit is the sign-bit of the number. In this case it is a 0 because the number is $+100.25_{10}$. The biased exponent follows the sign-bit. The significand is a 23-bit number with an implied one-bit. Note that the significand of a number 1.XXXX is the XXXX portion. The 1. is an **implied one-bit** that is only stored in the extended precision form of the floating-point number as an explicit one-bit.

Some special rules apply to a few numbers. The number 0, for example, is stored as all zeros except for the sign-bit, which can be a logic 1 to represent a negative zero. The plus and minus infinity is stored as logic 1's in the exponent with a significand of all zeros and the sign-bit that represents plus or minus. A NAN (not-a-number) is an invalid floating-point result that has all ones in the exponent with a significand that is not all zeros.

Converting from Floating-Point Form.   Conversion to a decimal number from a floating-point number is summarized in the following steps:

1. Separate the sign-bit, biased exponent, and significand.
2. Convert the biased exponent to a true exponent by subtracting the bias.
3. Write the number as a normalized binary number.
4. Convert it to a de-normalized binary number.
5. Convert the de-normalized binary number to decimal.

These five steps convert a single-precision floating-point number to decimal in Example 14–4. Notice how the sign-bit of 1 makes the decimal result negative. Also notice that the implied one-bit is added to the normalized binary result in step 3.

EXAMPLE 14–4

```
Step      Result

1         Sign = 1
          Exponent = 10000011
          Significand = 10010010000000000000000

2         100 = 10000011 - 01111111

3         1.1001001 x 2⁴

4         11001.001

5         -25.125
```

Storing Floating-Point Data in Memory.   Floating-point numbers are stored with the assembler using the DD directive for single precision, DQ for double precision, and DT for extended precision. Some examples of floating-point data storage are shown in Example 14–5. We discovered that the Microsoft version 6.0 macro assembler contains an error that does not allow a plus sign to be used with positive floating-point numbers. For example, a +92.45 must be defined as 92.45 for the assembler to function correctly. Microsoft has assured us that this error has been corrected in version 6.11 of MASM if the REAL4, REAL8, or REAL10 directives are used in place of DD, DQ, and DT to specify floating-point data. The assembler provides access 8087 emulator if your system does not contain a Pentium Pro, Pentium, 80486, or any other microprocessor with a coprocessor. The emulator comes with all Microsoft high-level languages or as shareware programs such as EM87. The emulator is accessed by including the OPTION EMULATOR statement immediately following the .MODEL statement in a program. Be aware that the emulator does not emulate some of the coprocessor instructions. Do not use this option if your system contains a coprocessor. In all cases you must include the .8087, .80187, .80287, .80387, .80487, and .80587 switches to enable the generation of coprocessor instructions. At this time there is no switch for the Pentium Pro, but it will most likely be .80687 when Microsoft produces the next version of the assembler program.

EXAMPLE 14–5

```
0000 C377999A        DATA7    DD      -247.6      ;define single precision
0004 40000000        DATA8    DD      2.0         ;define single precision
0008 486F4200        DATA9    REAL4   2.45E+5     ;define single precision
000C                 DATA10   DQ      100.25      ;define double precision
     4059100000000000
0014                 DATA11   REAL8   0.001235    ;define double precision
     3F543BF727136A40
001C                 DATA12   REAL10  33.9876     ;define extended precision
     400487F34D6A161E4F76
```

14–2   THE 80X87 ARCHITECTURE

The 80X87 is designed to operate concurrently with the microprocessor. The 80486DX, Pentium, and Pentium Pro microprocessors contain their own internal and fully compatible versions of the 80387. With other family members, the coprocessor is an external integrated circuit that parallels most of the connections on the microprocessor. The 80186/80188 family uses the 80187 coprocessor and the 80386EX uses the 80387SX coprocessor. The 80X87 executes 68 different instructions. The microprocessor executes all normal instructions and the 80X87 executes arithmetic coprocessor instructions. Both the microprocessor and coprocessor can execute their respective instructions simultaneously or concurrently. The numeric or arithmetic coprocessor is a special-purpose microprocessor that is specially designed to efficiently execute arithmetic and transcendental operations.

The microprocessor intercepts and executes the normal instruction set; the coprocessor intercepts and executes only the coprocessor instructions. Recall that the coprocessor instructions are actually escape (ESC) instructions. These instructions are used by the microprocessor to generate a memory address for the coprocessor so the coprocessor can execute a coprocessor instruction.

Internal Structure of the 80X87

Figure 14–4 shows the internal structure of the arithmetic coprocessor. Notice that this device is divided into two major sections: the control unit and the numeric execution unit.

The control unit interfaces the coprocessor to the microprocessor system data bus. Both devices monitor the instruction stream. If the instruction is an ESCape (coprocessor) instruction, the coprocessor executes it; if not, the microprocessor executes it.

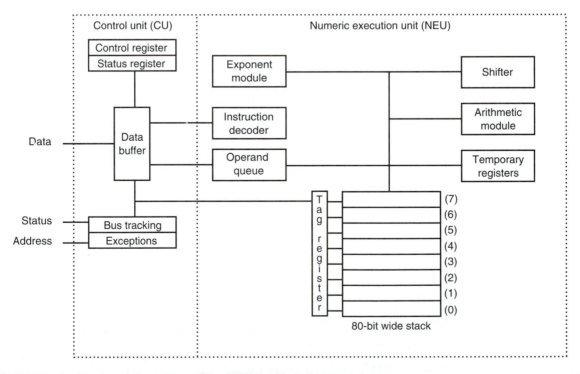

FIGURE 14–4   The internal structure of the 80X87 arithmetic coprocessor.

The numeric execution unit (NEU) is responsible for executing all coprocessor instructions. The NEU has an eight-register stack that holds operands for arithmetic instructions and the results of arithmetic instructions. Instructions either address data in specific stack data registers or use a push and pop mechanism to store and retrieve data on the top of the stack. Other registers in the NEU are status, control, tag, and exception pointers. A few instructions transfer data between the coprocessor and the AX register in the microprocessor. The FSTSW AX instruction is the only instruction available to the coprocessor that allows direct communications to the microprocessor through the AX register. Note that the 8087 does not contain the FSTSW AX instruction.

The stack within the coprocessor contains eight registers that are each 80 bits in width. These stack registers always contain an 80-bit extended-precision floating-point number. The only time data appears as any other form is when it resides in the memory system. The coprocessor converts from signed-integer, BCD, single-precision, or double-precision form as the data is moved between the memory and the coprocessor register stack.

Status Register. The status register (see Figure 14–5) reflects the overall operation of the coprocessor. The status register is accessed by executing the instruction FSTSW, which stores the contents of the status register into a word of memory. The FSTSW AX instruction copies the status register directly to the microprocessor's AX register on the 80287 or above coprocessor. Once status is stored in memory or the AX register, the bit positions of the status register can be examined by normal software. The coprocessor/microprocessor communications are carried out through the I/O ports 00FAH–00FFH on the 80287 and I/O ports 800000FAH–800000FFH on the 80386 through the Pentium Pro. Never use these I/O ports for interfacing I/O devices to the microprocessor.

The newer coprocessors (80287 and above) use status bit position 6 (SF) to indicate a stack overflow or underflow error. The following is a list of the status bits, except for SF, and their application.

B   The **busy bit** indicates that the coprocessor is busy executing a task. Busy can be tested by examining the status register or by using the FWAIT instruction. Newer coprocessors automatically synchronize with the microprocessor so the busy flag need not be tested before performing additional coprocessor tasks.

C3–C0   The **condition code bits** (refer to Table 14–2 for a complete listing of each combination of these bits and their function) indicate conditions about the coprocessor. These bits have different meanings for different instructions as indicated in the table. The top of the stack is denoted as ST in Table 14–2.

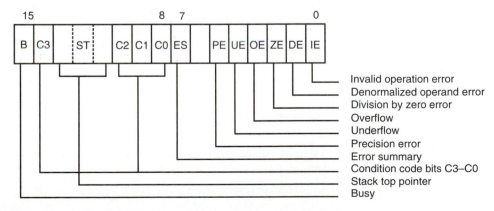

FIGURE 14–5 The 80X87 arithmetic coprocessor status register.

TABLE 14–2   The 80X87 status register condition code bits

Instruction	C3	C2	C1	C0	Indication
FTST, FCOM	0	0	X	0	ST > operand
	0	0	X	1	ST < operand
	1	0	X	1	ST = operand
	1	1	X	1	ST is not comparable
FPREM	Q1	0	Q0	Q2	Rightmost 3 bits of quotient
	?	1	?	?	Incomplete
FXAM	0	0	0	0	+ unnormal
	0	0	0	1	+ NAN
	0	0	1	0	− unnormal
	0	0	1	1	− NAN
	0	1	0	0	+ normal
	0	1	0	1	+ ∞
	0	1	1	0	− normal
	0	1	1	1	− ∞
	1	0	0	0	+ 0
	1	0	0	1	Empty
	1	0	1	0	− 0
	1	0	1	1	Empty
	1	1	0	0	+ denormal
	1	1	0	1	Empty
	1	1	1	0	− denormal
	1	1	1	1	Empty

Notes: Unnormal = leading bits of the significand are zero; denormal = exponent at its most negative value; normal = standard floating-point form; NAN (not-a-number) = an exponent of all ones and a significand not equal to zero and the operand for FTST is zero.

TOP   The **top-of-stack (ST)** bit indicates the current register addressed as the top-of-the-stack (ST). This is normally register 0.

ES   The **error summary** bit is set if any unmasked error bit (PE, UE, OE, ZE, DE, or IE) is set. In the 80X87 coprocessor, the error summary also caused a coprocessor interrupt. Since the advent of 80187, the coprocessor interrupt has been absent from the family.

PE   The **precision error** indicates that the result or operands exceed selected precision.

UE   An **underflow error** indicates a non-zero result that is too small to represent with the current precision selected by the control word.

OE   An **overflow error** indicates a result that is too large to be represented. If this error is masked, the coprocessor generates infinity for an overflow error.

ZE   A **zero error indicates** the divisor was zero while the dividend is a non-infinity or non-zero number.

DE   A **denormalized error** indicates that at least one of the operands is denormalized.

IE   An **invalid error** indicates a stack overflow or underflow, indeterminate form (0 ÷ 0, + ∞, − ∞, etc.), or the use of a NAN as an operand. This flag indicates errors such as those produced by taking the square root of a negative number.

There are two ways to test the bits of the status register once they are moved into the AX register with the FSTSW AX instruction. One method uses the TEST instruction to test individuals bits of the status register. The other uses the SAHF instruction to transfer the leftmost 8 bits of the status register into the microprocessor's flag register. Both methods are illustrated in Example 14–6. This example uses the DIV instruction to divide the top of the stack by the contents of DATA1 and the FSQRT instruction to find the square root of the top of the stack. The example also uses the FCOM instruction to compare the contents of the stack top with DATA1. Note that the conditional jump instructions are used with the SAHF instruction to test for the condition listed in Table 14–3. Although SAHF and conditional jumps cannot test all possible operating conditions of the coprocessor, they can help to reduce the complexity of certain tested conditions. The SAHF places C0 into the carry flag, C2 into the parity flag, and C3 into the zero flag.

EXAMPLE 14–6

```
                              ;using TEST to isolate the divide by zero error bit

0000 67& D8 35 00000000 R          FDIV  DATA1
0007 9B DF E                        FSTSW AX                 ;copy status register to AX
       000A A9 000                            TEST  AX,4            ;test bit position 2
000D 75 18                          JNZ   DIVIDE_ERROR

                              ;using TEST to isolate the invalid operation error bit
                              ;after an FSQRT instruction

000F D9 FA                          FSQRT
0011 9B DF E0                        FSTSW AX                 ;copy status register to AX
0014 A9 0001                        TEST  AX,1               ;test bit position 1
0017 75 0E                          JNZ   FSQRT_ERROR

                              ;using the SAHF instruction and conditional jumps to
                              ;test for the conditions in Table 14-3 after an FCOM

0019 67& D8 15 00000000             RFCOM DATA1
0020 9B DF E0                        FSTSW AX                 ;copy status register to AX
0023 9E                             SAHF                     ;copy status bits to flags
0024 74 04                          JE    ST_EQUAL
0026 72 02                          JB    ST_BELOW
0028 77 00                          JA    ST_ABOVE
```

When the FXAM instruction and FSTSW AX are executed and followed by the SAHF instruction, the zero flag will contain C3. Notice from Table 14–2 that C3 indicates the value +0 when set along with other errors. The FXAM instruction could be used to test a divisor before a division for a zero value by using the JZ instruction following FXAM, FSTSW AX, and SAHF.

Control Register.   The control register, illustrated in Figure 14–6, selects precision, rounding control, and infinity control. It also masks and unmasks the exception bits that correspond to the rightmost 6 bits of the status register. The FLDCW instruction is used to load a value into the control register.

TABLE 14–3   Coprocessor conditions tested with the conditional jump instructions and SAHF after FCOM or FTST as illustrated in Example 14–6

C3	C2	C0	Condition	Jump Instruction
0	0	0	ST > Operand	JA (jump if ST above)
0	0	1	ST < Operand	JB (jump if ST below)
1	0	0	ST = Operand	JE (jump if ST equal)

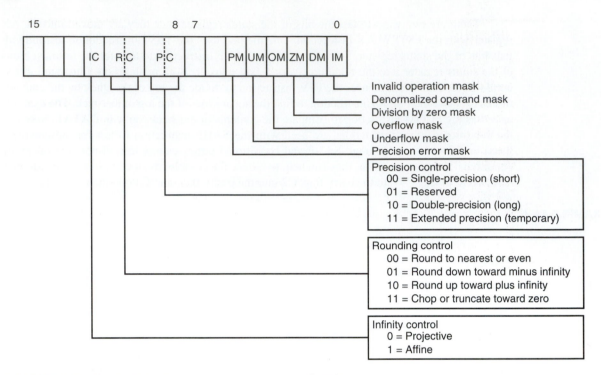

FIGURE 14–6   The 80X87 arithmetic coprocessor control register.

The following is a description of each bit or grouping of bits found in the control register.

IC	**Infinity control** selects either affine or projective infinity. Affine allows positive and negative infinity, while projective assumes infinity is unsigned.
RC	**Rounding control** determines the type of rounding as defined in Figure 14–6.
PC	**Precision control** sets the precision of the result as defined in Figure 14–6.
Exception Masks	Determine whether the error indicated by the exception affects the error bit in the status register. If a logic 1 is placed in one of the exception control bits, the corresponding status register bit is masked off.

Tag Register.   The tag register indicates the contents of each location in the coprocessor stack. Figure 14–7 illustrates the tag register and the status indicated by each tag. The tag indicates whether a register is valid, zero, invalid or infinity, or empty. The only way that a program can view the tag register is by storing the coprocessor environment using the FSTENV, FSAVE, or FRSTOR instructions. Each of these instructions stores the tag register along with other coprocessor data.

FIGURE 14–7   The 80X87 arithmetic coprocessor tag register.

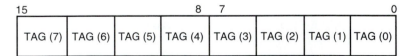

14–3 INSTRUCTION SET

The arithmetic coprocessor executes over 68 different instructions. When a coprocessor instruction references memory, the microprocessor automatically generates the memory address for the instruction. The coprocessor uses the data bus for data transfers during coprocessor instructions, and the microprocessor uses it during normal instructions. The 80287 and the 80186/80188 use the Intel reserved I/O ports 00F8H–00FFH for communications between the coprocessor and the microprocessor (even though the coprocessor uses only ports 00FCH–00FFH). These ports are used mainly for the FSTSW AX instruction. The 80387–Pentium Pro use I/O ports 800000F8H–800000FFH for this communication. The instruction set for the 80187 used with the 80186/80188 microprocessor is identical to the 80387SX used with the 80386EX microprocessor. This update occurred after the release of the latest members of the 80186/80188 family.

In this section we describe the function of each instruction and list its assembly language form. Because the coprocessor uses the microprocessor memory addressing modes, not all possible forms of each instruction are illustrated. Each time that the assembler encounters one of the coprocessor mnemonic opcodes, it converts it into a machine language ESC instruction. The ESC instruction represents an opcode to the coprocessor.

Data Transfer Instructions

There are three basic data transfers: floating-point, signed-integer, and BCD. The only time that data ever appears in the signed-integer or BCD form is in the memory. Inside the coprocessor, data is always stored as an 80-bit extended-precision floating-point number.

Floating-Point Data Transfers. There are four traditional floating-point data transfer instructions in the coprocessor instruction set: FLD (load real), FST (store real), FSTP (store real and pop), and FXCH (exchange). A new instruction is added to the Pentium Pro: a condition floating-point move instruction that uses the opcode FCMOV with a floating-point condition.

The FLD instruction loads floating-point memory data to the top of the internal stack, referred to as ST (stack top). This instruction stores the data on the top of the stack and then decrements the stack pointer by one. Data loaded to the top of the stack is from any memory location or from another coprocessor register. For example, an FLD ST(2) instruction copies the contents of register 2 to the stack top, which is ST. The top of the stack is register 0 when the coprocessor is reset or initialized. Another example is the FLD DATA7 instruction, which copies the contents of memory location DATA7 to the top of the stack. The size of the transfer is automatically determined by the assembler through the directives DD or REAL4 for single precision, DQ or REAL 8 for double precision, and DT or REAL10 for extended precision.

The FST instruction stores a copy of the top of the stack into the memory location or coprocessor register indicated by the operand. At the time of storage, the internal, extended-precision floating-point number is rounded to the size of the floating-point number indicated by the control register.

The FSTP (floating-point store and pop) instruction stores a copy of the top of the stack into memory or any coprocessor register and then pops the data from the top of the stack. You might think of FST as a copy instruction and FSTP as a removal instruction.

The FXCH instruction exchanges the register indicated by the operand with the top of the stack. For example, the FXCH ST(2) instruction exchanges the top of the stack with register 2.

Integer Data Transfer Instructions. The coprocessor supports three integer data transfer instructions: FILD (load integer), FIST (store integer), and FISTP (store integer and pop). These three instructions function as did FLD, FST, and FSTP, except the data transferred are integer data.

The coprocessor automatically converts the internal extended-precision floating-point data to integer data. The size of the data is determined by the way that the label is defined with DW, DD, or DQ in the assembly language program.

BCD Data Transfer Instructions.   Two instructions load or store BCD signed integer data: The FBLD instruction loads the top of the stack with BCD memory data, and the FBSTP stores the top of the stack and does a pop.

Example 14–7 shows how the assembler automatically adjusts the FLD, FILD, and FBLD instructions for different size operands. Look closely at the machine-coded forms of the instructions. Note in this example, that it begins with the .386 and .387 directives that identify the microprocessor as an 80386EX and the coprocessor, as an 80387SX. If the 80186/80188 microprocessors are in use with their coprocessors, the directives .186 and .187 appear. By default the assembler assumes that the software is assembled for an 8086/8088 with an 8087 coprocessor. The .486, .487, .586, and .587 switches are also available for use with the 80486 and Pentium microprocessors. Even though the program in Example 14–7 executes, CodeView or some other debugging tool must be used to view any changes to the coprocessor stack.

EXAMPLE 14–7

```
                              .MODEL   SMALL
                              .386                      ;select 80386 microprocessor
                              .387                      ;select 80387 coprocessor
0000                          .DATA
0000  41F00000      DATA1     DD      30.0              ;single precision
0004                DATA2     DQ      100.25            ;double precision
      4059100000000000
000C                DATA3     DT      33.9876           ;extended precision
      400487F34D6A161E4F76
0016  001E          DATA4     DW      30                ;16-bit integer
0018  0000001E      DATA5     DD      30                ;32-bit integer
001C                DATA6     DQ      30                ;64-bit integer
      000000000000001E
0024                DATA7     DT      30H               ;BCD 30
      00000000000000000030
0000                          .CODE
                              .STARTUP

0010  D9 06 0000 R            FLD     DATA1
0014  DD 06 0004 R            FLD     DATA2
0018  DB 2E 000C R            FLD     DATA3

001C  DF 06 0016 R            FILD    DATA4
0020  DB 06 0018 R            FILD    DATA5
0024  DF 2E 001C R            FILD    DATA6

0028  DF 26 0024 R            FBLD    DATA7
                              .EXIT
                              END
```

The Pentium Pro FCMOV Instruction.   The Pentium Pro contains a new instruction called FCMOV that also contains a condition. If the condition is true, the FCMOV instruction copies the source to the destination. The conditions tested by FCMOV and the opcodes used with FCMOV appear in Table 14–4. Notice that these conditions check for either an ordered or unordered number. The testing for NAN and denormalized numbers are not checked with FCMOV.

Example 14–8 shows how the FCMOVB (move if below) instruction is used to copy the contents of ST(2) to the stack top (ST) if the contents of ST(2) is below ST. Notice that the FCOM instruction must be used to perform the compare and the contents of the status register must still be copied to the flags for this instruction to function. More about the FCMOV instruction appears in our discussion of the FCOMI instruction, which is also new to the Pentium Pro processor.

TABLE 14–4 The variation of the FCMOV instruction and conditions tested

Instruction	Condition
FCMOVB	Move if below
FCMOVE	Move if equal
FCMOVBE	Move if below or equal
FCMOVU	Move if unordered
FCMOVNB	Move if not below
FCMOVNE	Move if not equal
FCMOVNBE	Move if not below or equal
FCMOVNU	Move if not ordered

EXAMPLE 14–8

```
FCOM    ST(2)           ;compare ST and ST(2)
FNSTSW  AX              ;copy floating flags to AX
SAHF                    ;copy floating flags to flags
FCMOVB  ST(2)           ;copy ST(2) to ST if below
```

Arithmetic Instructions

Arithmetic instructions for the coprocessor include: addition, subtraction, multiplication, division, and square root. The arithmetic-related instructions are: scaling, rounding, absolute value, and changing the sign.

Table 14–5 lists the basic addressing modes allowed for the arithmetic operations. Each addressing mode is shown with an example using the FADD (real addition) instruction. All arithmetic operations are floating-point except in some cases when memory data is referenced as an operand.

The classic stack form of addressing operand data (stack addressing) uses the top of the stack as the source operand and the next to the top of the stack as the destination operand. Afterwards a pop removes the source datum from the stack and only the result in the destination register remains at the top of the stack. To use this addressing mode, the instruction is placed in the program without any operands such as FADD or FSUB. The FADD instruction adds ST to ST(1) and stores the answer at the top of the stack; it also removes the original two data from the stack by popping. FSUB subtracts ST from ST(1) and leaves the difference at ST. Therefore a reverse subtraction (FSUBR) subtracts ST(1) from ST and leaves the difference at ST. (An error exists in Intel documentation, including the Pentium data book, that describes the operation of some reverse instructions.) Another use for reverse operations is finding a reciprocal (1/X). This is

TABLE 14–5 Arithmetic addressing modes

Mode	Form	Example
Stack	ST(1),ST	FADD
Register	ST,ST(n)	FADD ST,ST(2)
	ST(n),ST	FADD ST(2),ST
Register Pop	ST(n),ST	FADDP ST(3),ST
Memory	Operand	FADD DATA2

Note: Stack addressing is fixed as ST(1),ST and includes a pop so only the result remains at the top of the stack; n = register number 0–7; register addressing for any instruction can use a destination of ST or ST(n) as illustrated.

accomplished, if X is at the top of the stack, by loading a 1.0 to ST (FLD1) followed by the FDIVR instruction. The FDIVR instruction divides ST(1) into ST, or X into 1, and leaves the reciprocal (1/X) at ST.

The register-addressing mode uses ST for the top of the stack and ST(n) for another location where n is the register number. With this form, one operand must be ST and the other is ST(n). To double the top of the stack, the FADD ST,ST(0) instruction is used where ST(0) also addresses the top of the stack. One of the two operands in the register-addressing mode must be ST, while the other must be in the form ST(n), where n is a stack register 0–7. For many instructions, either ST or ST(n) can be the destination. It is fairly important that the top of the stack is ST(0). This is accomplished by resetting or initializing the coprocessor before using it in a program. Another example of register addressing is FADD ST(1),ST, where the contents of ST are added to ST(1) and the result is placed into ST(1).

The top of the stack is always used as the destination for the memory-addressing mode because the coprocessor is a stack-oriented machine. For example, the FADD DATA instruction adds the real-number contents of memory location DATA to the top of the stack.

Arithmetic Operations.   The letter P in an opcode specifies a register pop after the operation (FADDP compared to FADD). The letter R in an opcode (subtraction and division only) indicates reverse mode. The reverse mode is useful for memory data because normally memory data subtracts from the top of the stack. A reverse subtract instruction subtracts the top of the stack from memory and stores the result in the top of the stack. For example, if the top of the stack contains a 10 and memory location DATA1 contains a 1, the FSUB DATA1 instruction results in a +9 on the stack top and the FSUBR instruction results in a –9. Another example is FSUBR ST,ST(1), which will subtract ST from ST(1) and store the result on ST. A variant is FSUBR ST(1),ST, which will subtract ST(1) from ST and store the result on ST(1).

The letter I as a second letter in an opcode indicates that the memory operand is an integer. For example, the FADD DATA instruction is a floating-point addition, while the FIADD DATA is an integer addition that adds the integer at memory location DATA to the floating-point number at the top of the stack. The same rules apply to FADD, FSUB, FMUL, and FDIV instructions.

Arithmetic-Related Operations.   Other operations that are arithmetic in nature include: FSQRT (square root), FSCALE (scale a number), FPREM/FPREM1 (find partial remainder), FRNDINT (round to integer), FXTRACT (extract exponent and significand), FABS (find absolute value), and FCHG (change sign). The following is a list of these instructions and the functions they perform.

FSQRT	Finds the square root of the top of the stack and leaves the resultant square root at the top of the stack. An invalid error occurs for the square root of a negative number. For this reason, the IE bit of the status register should be tested whenever an invalid result can occur. The IE bit can be tested by loading the status register to AX with the FSTSW AX instruction, followed by TEST AX,1 to test the IE status bit.
FSCALE	Adds the contents of ST(1) (interpreted as an integer) to the exponent at the top of the stack. FSCALE multiplies or divides rapidly by powers of 2. The value in ST(1) must be between 2^{-15} and 2^{+15}.
FPREM/ FPREM1	Performs modulo division of ST by ST(1). The resultant remainder is found in the top of the stack and has the same sign as the original dividend. A modulo division results in a remainder without a quotient. FPREM is supported for the 8086 and 80287 and FPREM1 should be used in newer coprocessors.
FRNDINT	Rounds the top of the stack to an integer.
FXTRACT	Decomposes the number at the top of the stack into two separate parts that represent the value of the unbiased exponent and the value of the significand.

The extracted significand is found at the top of the stack and the unbiased exponent at ST(1). This instruction is often used to convert a floating-point number to a form that can be printed as a mixed number.

FABS   Changes the sign of the top of the stack to positive.

FCHS   Changes the sign from positive to negative or negative to positive.

Comparison Instructions

The comparison instructions examine data at the top of the stack in relation to another element and return the result of the comparison in the status register condition code bits C3–C0. Comparisons that are allowed by the coprocessor are: FCOM (floating-point compare), FCOMP (floating-point compare with a pop), FCOMPP (floating-point compare with two pops), FICOM (integer compare), FICOMP (integer compare and pop), FSTS (test), and FXAM (examine). New with the introduction of the Pentium Pro is the floating compare that moves results to flags or to the FCOMI instruction. The following is a list of these instructions and a description of their function.

FCOM   Compares the floating-point data at the top of the stack with an operand, which may be any register or any memory operand. If the operand is not coded with the instruction, the next stack element ST(1) is compared with the stack top ST.

FCOMP/   Both instructions perform as FCOM, but they also pop one or two data
FCOMPP   from the stack.

FICOM/   The top of the stack is compared with the integer stored at a memory operand.
FICOMP   In addition to the compare, FICOMP also pops the top of the stack.

FTST   Tests the contents of the top of the stack against a zero. The result of the comparison is coded in the status register condition code bits as illustrated in Table 14–2 with the status register. Also see Table 14–3 for a way of using SAHF and the conditional jump instruction with FTST.

FXAM   Examines the stack top and modifies the condition code bits to indicate whether the contents are positive, negative, normalized, etc. Refer to the status register in Table 14–2.

FCOMI/   New to the Pentium Pro, this instruction compares in exactly the same
FUCOMI   manner as the FCOM instruction with one additional feature: it moves the floating-point flags into the flag register just as the FNSTSW AX and SAHF instruction do back in Example 14–8. Intel has combined the FCOM, FNSTSW AX, and SAHF instructions to form FCOMI. Also available is the unordered compare or FUCOMI. Each is also available with a pop by appending the opcode with a P.

Transcendental Operations

The transcendental instructions include: FPTAN (partial tangent), FPATAN (partial arctangent), FSIN (sine), FCOS (cosine), FSINCOS (sine and cosine), F2XM1 ($2^x - 1$), FYL2X (Y $\log_2$ X), and FYL2XP1 [Y $\log_2$ (X + 1)]. A list of these operations follows with a description of each transcendental operation.

FPTAN   Finds the partial tangent of Y/X = tan θ. The value of θ is at the top of the stack and must be between 0 and $\pi/4$ radians for the 8087 and 80287 and less than 2^{63} for the 80187, 80387, 80486/7, Pentium, and Pentium Pro microprocessors. The result is a ratio is found as ST = X and ST(1) = Y. If the value is outside the allowable range, an invalid error occurs as indicated by the status register IE bit. ST(7) must be empty for this instruction to function properly.

TABLE 14–6 Exponential functions

Function	Equation
10^Y	$2^Y \times \log_2 10$
ϵ^Y	$2^Y \times \log_2 \epsilon$
X^Y	$2^Y \times \log_2 X$

FPATAN   Finds the partial arctangent as $\theta = \text{ARCTAN } X/Y$. The value of X is at the top of the stack and Y is at ST(1). The values of X and Y must be as follows: $0 \leq Y < X < \infty$. The instruction pops the stack and leaves θ at the top of the stack.

F2XM1   Finds the function $2^x - 1$. The value of X is taken from the top of the stack and the result if returned to the top of the stack. To obtain 2^x add one to the result at the top of the stack. The value of X must be in the range of -1 and $+1$. The F2XM1 instruction is used to derive the functions listed in Table 14–6. Note that the constants $\log_2 10$ and $\log_2 \epsilon$ are built-in as standard values for the coprocessor.

FSIN/FCOS   Finds the sine or cosine of the argument located in ST expressed in radians ($360° = 2\pi$ radians) with the result found in ST. The values of ST must be less than 2^{63}.

FSINCOS   Finds the sine and cosine of ST, expressed in radians, and leaves the results as ST = sine and ST(1) = cosine. As with FSIN or FCOS, the initial value of ST must be less than 2^{63}.

FYL2X   Finds Y $\log_2$ X. The value X is taken from the stack top and Y is taken from ST(1). The result is found at the top of the stack after a pop. The value of X must range between 0 and ∞, and the value of Y must be between $-\infty$ and $+\infty$. A logarithm with any positive base (b) is found by the equation

$$\text{LOG}_b X = (\text{LOG}_2 b)^{-1} \times \text{LOG}_2 X$$

FYL2XP1   Finds Y $\log_2 (X + 1)$. The value of X is taken from the stack top and Y is taken from ST(1). The result is found at the top of the stack after a pop. The value of X must range between 0 and $1 - \sqrt{2}/2$ and the value of Y must be between $-\infty$ and $+\infty$.

Constant Operations

The coprocessor instruction set includes opcodes that return constants to the top of the stack. A list of these instructions appears in Table 14–7.

TABLE 14–7 Constant operations

Instruction	Constant Pushed to ST
FLDZ	+0.0
FLD1	+1.0
FLDPI	π
FLDL2T	$\log_2 10$
FLDL2E	$\log_2 \epsilon$
FLDLG2	$\log_{10} 2$
FLDLN2	$\log_\epsilon 2$

Coprocessor Control Instructions

The coprocessor has control instructions for initialization, exception handling, and task switching. The control instructions have two forms: for example, FINIT initializes the coprocessor and so does FNINIT. The difference is that FNINIT does not cause any wait states, while FINIT does cause waits. The microprocessor waits for the FINIT instruction by testing the BUSY pin on the coprocessor. All control instructions have these two forms. The following is a list of each control instruction and its function.

FINIT/FNINIT	Performs a reset operation on the arithmetic coprocessor (refer to Table 14–8 for the reset conditions). The coprocessor operates with a closure of projective (unsigned infinity), rounds to the nearest or even, and uses extended precision when reset or initialized. It also sets register 0 as the top of the stack.
FSETPM	Changes the addressing mode of the coprocessor to the protected addressing mode. This mode is used when the microprocessor is also operated in the protected mode. As with the microprocessor, protected mode can only be exited by a hardware reset or in the case of the 80386 through the Pentium Pro with a change to the control register.
FLDCW	Loads the control register with the word addressed by the operand.
FSTCW/FNSTCW	Stores the control register into the word-sized memory operand.
FSTSW AX/ FNSTSW AX	Copies the contents of the control register to the AX register. This instruction is not available to the 8087 coprocessor.
FCLEX/FNCLEX	Clears the error flags in the status register and also the busy flag.
FSAVE/FNSAVE	Writes the entire state of the machine to memory. Figure 14–8 shows the memory contents for this instruction.
FRSTOR	Restores the state of the machine from memory. This instruction is used to restore the information saved by FSAVE/FNSAVE.
FSTENV/FNSTENV	Stores the environment of the coprocessor as shown in Figure 14–9.
FLDENV	Reloads the environment saved by FSTENV/FNSTENV.
FINCST	Increments the stack pointer.
FDECSTP	Decrements the stack pointer.

TABLE 14–8 Coprocessor state after a reset or an initialization

Field	Value	Condition
Infinity	0	Projective
Rounding	00	Round to nearest
Precision	11	Extended precision
Error masks	11111	Error bits disabled
Busy	0	Not busy
C3–C0	????	Unknown
TOP	000	Register 000 or ST(0)
ES	0	No errors
Error bits	00000	No errors
All tags	11	Empty
Registers	ST(0)–ST(7)	Not changed

FIGURE 14–8 Memory format when the 80X87 registers are saved with the FSAVE instruction.

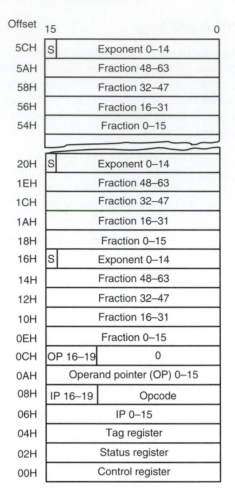

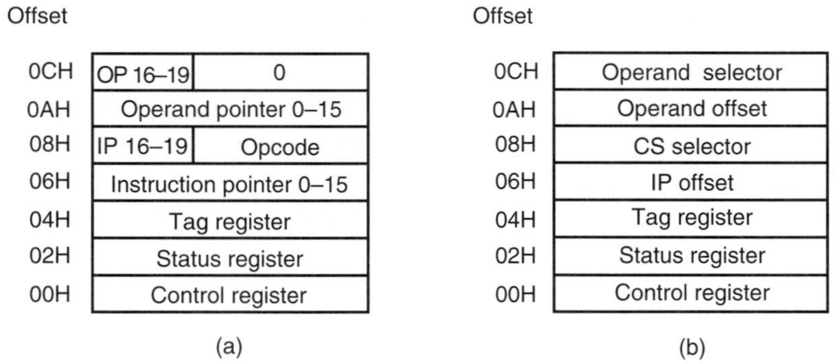

FIGURE 14–9 Memory format for the FSTENV instruction: (a) real mode and (b) protected mode.

Offset

Offset		
0CH	OP 16–19	0
0AH	Operand pointer 0–15	
08H	IP 16–19	Opcode
06H	Instruction pointer 0–15	
04H	Tag register	
02H	Status register	
00H	Control register	

(a)

Offset	
0CH	Operand selector
0AH	Operand offset
08H	CS selector
06H	IP offset
04H	Tag register
02H	Status register
00H	Control register

(b)

FFREE	Frees a register by changing the destination register's tag to empty. It does not affect the contents of the register.
FNOP	Floating-point coprocessor NOP.
FWAIT	Causes the microprocessor to wait for the coprocessor to finish an operation. FWAIT should be used before the microprocessor accesses memory data that is affected by the coprocessor.

Coprocessor Instructions

Although we have not discussed the microprocessor circuitry, we will discuss the instruction sets of these coprocessors and their differences with the other versions of the coprocessor. These newer coprocessors contain the same basic instructions provided by the earlier versions with a few additional instructions.

The 80187, 80387, 80486, 80487SX, and Pentium contain the following additional instructions: FCOS (cosine), FPREM1 (partial remainder), FSIN (sine), FSINCOS (sine and cosine), and FUCOM/FUCOMP/FUCOMPP (unordered compare). The sine and cosine instructions are the most significant addition to the instruction set. In the earlier versions of the coprocessor, the sine and cosine are calculated from the tangent. The Pentium Pro contains two new floating-point instructions: FCMOV (a conditional move) and FCOMI (a compare and move to flags).

Table 14–9 lists the instruction sets for all versions of the coprocessor. It also lists the number of clocking periods required to execute each instructions. Execution times are listed for the 8087, 80187, 80287, 80387, 80486, 80487, Pentium, and Pentium Pro. To determine the execution time of an instruction, the clock time is multiplied by the listed execution time. The FADD instruction requires 70–143 clocks for the 80287. Suppose that an 8-MHz clock is used with the 80287. The clocking period is $1/8$ MHz or 125 ns. The FADD instruction requires between 8.75 μs and 17.875 μs to execute. Using a 33-MHz (33 ns) 80486DX2, this instruction requires between 0.264 μs and 0.66 μs to execute. On the Pentium the FADD instruction requires 1–7 clocks so if operated at 133 MHz (7.52 ns), the FADD requires between 0.00752 μs and 0.05264 μs. The Pentium Pro is even faster than the Pentium.

Table 14–9 uses some shorthand notations to represent the displacement that may or may not be required for an instruction that uses a memory-addressing mode. It uses the abbreviation to represent the mode, mmm, to represent a register/memory addressing mode, and rrr to represent one of the floating-point coprocessor registers ST(0)–ST(7). The d (destination) bit that appears in some instruction opcodes defines the direction of the data flow as in FADD ST,ST(2) or FADD ST(2),ST. The d bit is a logic 0 for flow towards ST as in FADD ST,ST(2) where ST holds the sum after the addition, and a logic 1 for FADD ST(2),ST where ST(2) holds the sum.

Also note that some instructions allow a choice of whether a wait is inserted. For example, the FSTSW AX instruction copies the status register into AX. The FNSTSW AX instruction also copies the status register to AX, but without a wait.

TABLE 14–9   The instruction set of the arithmetic coprocessor (pp. 569–587)

F2XM1 $2^{ST} - 1$		
11011001 11110000		
Example		Clocks
F2XM1	8087	310–630
	80187	213–473
	80287	310–630
	80387	211–476
	80486/7	140–279
	Pentium	13–57
	Pentium Pro	

FABS   Absolute value of ST

11011001 11100001

Example		Clocks
FABS	8087	10–17
	80187	24
	80287	10–17
	80387	22
	80486/7	3
	Pentium	1
	Pentium Pro	

FADD/FADDP/FIADD   Addition

11011000 oo000mmm disp	32-bit memory (FADD)
11011100 oo000mmm disp	64-bit memory (FADD)
11011d00 11000rrr	FADD ST,ST(rrr)
11011110 11000rrr	FADDP ST,ST(rrr)
11011110 oo000mmm disp	16-bit memory (FIADD)
11011010 oo000mmm disp	32-bit memory (FIADD)

Format	Examples		Clocks
FADD	FADD DATA	8087	70–143
FADDP	FADD ST,ST(1)	80187	65–91
FIADD	FADDP	80287	60–143
	FIADD NUMBER	80387	23–72
	FADD ST,ST(3)	80486/7	8–20
	FADDP ST,ST(2)	Pentium	1–7
	FADD ST(2),ST	Pentium Pro	

FCLEX/FNCLEX   Clear errors

11011011 11100010

Example		Clocks
FCLEX	8087	2–8
FNCLEX	80187	13
	80287	2–8
	80387	11
	80486/7	7
	Pentium	9
	Pentium Pro	

FCOM/FCOMP/FCOMPP/FICOM/FICOMP   Compare

```
11011000 oo010mmm disp        32-bit memory (FCOM)
11011100 oo010mmm disp        64-bit memory (FCOM)
11011000 11010rrr             FCOM ST(rrr)
11011000 oo011mmm disp        32-bit memory (FCOMP)
11011100 oo011mmm disp        64-bit memory (FCOMP)
11011000 11011rrr             FCOMP ST(rrr)
11011110 11011001             FCOMPP
11011110 oo010mmm disp        16-bit memory (FICOM)
11011010 oo010mmm disp        32-bit memory (FICOM)
11011110 oo011mmm disp        16-bit memory (FICOMP)
11011010 oo011mmm disp        32-bit memory (FICOMP)
```

Format	Examples		Clocks
FCOM	FCOM ST(2)	8087	40–93
FCOMP	FCOMP DATA	80187	2685
FCOMPP	FCOMPP		
FICOM	FICOM NUMBER	80287	40–93
FICOMP	FICOMP DATA3	80387	24–63
		80486/7	15–20
		Pentium	1–8
		Pentium Pro	

FCOMI/FUCOMI/COMIP/FUCOMIP   Compare and Load Flags

```
11011011 11110rrr             FCOMI ST(rrr)
11011011 11101rrr             FUCOMI ST(rrr)
11011111 11110rrr             FCOMIP ST(rrr)
11011111 11101rrr             FUCOMIP ST(rrr)
```

Format	Examples		Clocks
FCOM	FCOMI ST(2)	8087	—
FUCOMI	FUCOMI ST(4)	80187	—
FCOMIP	FCOMIP ST(0)		
FUCOMIP	FUCOMIP ST(1)	80287	—
		80387	—
		80486/7	—
		Pentium	—
		Pentium Pro	

FCMOVcc   Conditional Move

11011010 11000rrr	FCMOVB ST(rrr)
11011010 11001rrr	FCMOVE ST(rrr)
11011010 11010rrr	FCMOVBE ST(rrr)
11011010 11011rrr	FCMOVU ST(rrr)
11011011 11000rrr	FCMOVNB ST(rrr)
11011011 11001rrr	FCMOVNE ST(rrr)
11011011 11010rrr	FCMOVENBE ST(rrr)
11011011 11011rrr	FCMOVNU ST(rrr)

Format	Examples		Clocks
FCMOVB	FCMOVB ST(2)	8087	—
FCMOVE	FCMOVE ST(3)	80187	—
		80287	—
		80387	—
		80486/7	—
		Pentium	—
		Pentium Pro	

FCOS   Cosine of ST

11011001 11111111

Example		Clocks
FCOS	8087	—
	80187	125–774
	80287	—
	80387	123–772
	80486/7	193–279
	Pentium	18–124
	Pentium Pro	

FDECSTP   Decrement stack pointer

11011001 11110110

Example		Clocks
FDECSTP	8087	6–12
	80187	24
	80287	6–12
	80387	22
	80486/7	3
	Pentium	1
	Pentium Pro	

FDISI/FNDISI   Disable interrupts

11011011 11100001

(ignored on the 80287, 80387, 80486/7, Pentium, and Pentium Pro)

Example

		Clocks
FDISI FNDISI	8087	2–8
	80187	—
	80287	—
	80387	—
	80486/7	—
	Pentium	—
	Pentium Pro	—

FDIV/FDIVP/FIDIV   Division

11011000 oo110mmm disp	32-bit memory (FDIV)
11011100 oo100mmm disp	64-bit memory (FDIV)
11011d00 11111rrr	FDIV ST,ST(rrr)
11011110 11111rrr	FDIVP ST,ST(rrr)
11011110 oo110mmm disp	16-bit memory (FIDIV)
11011010 oo110mmm disp	32-bit memory (FIDIV)

Format	Examples		Clocks
FDIV FDIVP FIDIV	FDIV DATA	8087	191–243
	FDIV ST,ST(3)	80187	90–147
	FDIVP FIDIV NUMBER	80287	191–243
	FDIV ST,ST(5)	80387	88–140
	FDIVP ST,ST(2)	80486/7	78–89
	FDIV ST(2),ST	Pentium	39–42
		Pentium Pro	

FDIVR/FDIVRP/FIDIVR   Division reversed

11011000 oo111mmm disp	32-bit memory (FDIVR)
11011100 oo111mmm disp	64-bit memory (FDIVR)
11011d00 11110rrr	FDIVR ST,ST(rrr)
11011110 11110rrr	FDIVRP ST,ST(rrr)
11011110 oo111mmm disp	16-bit memory (FIDIVR)
11011010 oo111mmm disp	32-bit memory (FIDIVR)

Format	Examples		Clocks
FDIVR FDIVRP FIDIVR	FDIVR DATA	8087	191–243
	FDIVR ST,ST(3)	80187	90–147
	FDIVRP FIDIVR NUMBER	80287	191–243
	FDIVR ST,ST(5)	80387	88–140
	FDIVRP ST,ST(2)	80486/7	8–89
	FDIVR ST(2),ST	Pentium	39–42
		Pentium Pro	

FENI/FNENI   Disable interrupts

11011011 11100000

(ignored on the 80187, 80287, 80387, 80486/7, Pentium, and Pentium Pro)

Example		Clocks
FENI FNENI	8087	2–8
	80187	—
	80287	—
	80387	—
	80486/7	—
	Pentium	—
	Pentium Pro	

FFREE   Free register

11011101 11000rrr

Format	Examples	Clocks	
FFREE	FFREE FFREE ST(1) FFREE ST(2)	8087	9–16
		80187	20
		80287	9–16
		80387	18
		80486/7	3
		Pentium	1
		Pentium Pro	

FINCSTP   Increment stack pointer

11011001 11110111

Example		Clocks
FINCSTP	8087	6–12
	80187	20
	80287	6–12
	80387	21
	80486/7	3
	Pentium	1
	Pentium Pro	

FINIT/FNINIT   Initialize coprocessor

11011001 11110110

Example		Clocks
FINIT FNINIT	8087	2–8
	80187	35
	80287	2–8
	80387	33
	80486/7	17
	Pentium	12–16
	Pentium Pro	

FLD/FILD/FBLD   Load data to ST(0)

11011001 oo000mmm disp	32-bit memory (FLD)	
11011101 oo000mmm disp	64-bit memory (FLD)	
11011011 oo101mmm disp	80-bit memory (FLD)	
11011111 oo000mmm disp	16-bit memory (FILD)	
11011011 oo000mmm disp	32-bit memory (FILD)	
11011111 oo101mmm disp	64-bit memory (FILD)	
11011111 oo100mmm disp	80-bit memory (FBLD)	

Format	Examples		Clocks
FLD FILD FBLD	FLD DATA FILD DATA1 FBLD DEC_DATA	8087	17–310
		80187	16–305
		80287	17–310
		80387	14–275
		80486/7	3–103
		Pentium	1–3
		Pentium Pro	

FLD1   Load +1.0 to ST(0)

11011001 11101000

Example		Clocks
FLD1	8087	15–21
	80187	26
	80287	15–21
	80387	24
	80486/7	4
	Pentium	2
	Pentium Pro	

FLDZ   Load +0.0 to ST(0)

11011001 11101110

Example		Clocks
FLDZ	8087	11–17
	80187	22
	80287	11–17
	80387	20
	80486/7	4
	Pentium	2
	Pentium Pro	

FLDPI   Load π to ST(0)

11011001 11101011

Example		Clocks
FLDPI	8087	16–22
	80187	42
	80287	16–22
	80387	40
	80486/7	8
	Pentium	3–5
	Pentium Pro	

FLDL2E   Load $\log_2 \varepsilon$ to ST(0)

11011001 11101010

Example		Clocks
FLDL2E	8087	15–21
	80187	42
	80287	15–21
	80387	40
	80486/7	8
	Pentium	3–5
	Pentium Pro	

FLDL2T   Load $\log_2 10$ to ST(0)

11011001 11101001

Example		Clocks
FLDL2T	8087	16–22
80187	42
80287	16–22
80387	40
80486/7	8
Pentium	3–5
Pentium Pro

FLDLG2   Load $\log_{10} 2$ to ST(0)

11011001 11101000

Example		Clocks
FLDLG2	8087	18–24
80187	43
80287	18–24
80387	41
80486/7	8
Pentium	3–5
Pentium Pro

FLDLN2   Load $\log_e \varepsilon\, 2$ to ST(0)

11011001 11101101

Example		Clocks
FLDLN2	8087	17–23
80187	43
80287	17–23
80387	41
80486/7	8
Pentium	3–5
Pentium Pro

FLDCW   Load control register

11011001 oo101mmm disp

Format	Examples		Clocks
FLDCW	FLDCW DATA FLDCW STATUS	8087	7–14
		80187	23
		80287	7–14
		80387	19
		80486/7	4
		Pentium	7
		Pentium Pro	

FLDENV   Load environment

11011001 oo100mmm disp

Format	Examples		Clocks
FLDENV	FLDENV ENVIRON FLDENV DATA	8087	35–45
		80187	113
		80287	25–45
		80387	71
		80486/7	34–44
		Pentium	32–37
		Pentium Pro	

FMUL/FMULP/FIMUL   Multiplication

11011000	oo001mmm	disp	32-bit memory (FMUL)
11011100	oo001mmm	disp	64-bit memory (FMUL)
11011d00	11001rrr		FMUL ST,ST(rrr)
11011110	11001rrr		FMULP ST,ST(rrr)
11011110	oo001mmm	disp	16-bit memory (FIMUL)
11011010	oo001mmm	disp	32-bit memory (FIMUL)

Format	Examples		Clocks
FMUL FMULP FIMUL	FMUL DATA FMUL ST,ST(2) FMUL ST(2),ST FMULP FIMUL DATA3	8087	110–168
		80187	31–102
		80287	110–168
		80387	29–82
		80486/7	11–27
		Pentium	1–7
		Pentium Pro	

FNOP   No operation

11011001 11010000

Example		Clocks
FNOP	8087	10–16
80187	14
80287	10–16
80387	12
80486/7	3
Pentium	1
Pentium Pro

FPATAN   Partial arctangent of ST(0)

11011001 11110011

Example		Clocks
FPATAN	8087	250–800
80187	316–489
80287	250–800
80387	314–487
80486/7	218–303
Pentium	17–173
Pentium Pro

FPREM   Partial remainder

11011001 11111000

Example		Clocks
FPREM	8087	15–190
80187	76–157
80287	15–190
80387	74–155
80486/7	70–138
Pentium	16–64
Pentium Pro

FPREM1   Partial remainder (IEEE)

11011001 11110101

Example		Clocks
FPREM1	8087	—
	80187	97–187
	80287	—
	80387	95–185
	80486/7	72–167
	Pentium	20–70
	Pentium Pro	

FPTAN   Partial tangent of ST(0)

11011001 11110010

Example		Clocks
FPTAN	8087	30–450
	80187	193–499
	80287	30–450
	80387	191–497
	80486/7	200–273
	Pentium	17–173
	Pentium Pro	

FRNDINT   Round ST(0) to an integer

11011001 11111100

Example		Clocks
FRNDINT	8087	16–50
	80187	68–82
	80287	16–50
	80387	66–80
	80486/7	21–30
	Pentium	9–20
	Pentium Pro	

FRSTOR Restore state

11011101 oo110mmm disp

Format	Examples		Clocks
FRSTOR	FRSTOR DATA FRSTOR STATE FRSTOR MACHINE	8087	197–207
		80187	482
		80287	197–207
		80387	308
		80486/7	120–131
		Pentium	70–95
		Pentium Pro	

FSAVE/FNSAVE Save machine state

11011101 oo110mmm disp

Format	Examples		Clocks
FSAVE FNSAVE	FSAVE STATE FNSAVE STATUS FSAVE MACHINE	8087	197–207
		80187	550
		80287	197–207
		80387	375
		80486/7	143–154
		Pentium	124–151
		Pentium Pro	

FSCALE Scale ST(0) by ST(1)

11011001 11111101

Example		Clocks
FSCALE	8087	32–38
	80187	69–88
	80287	32–38
	80387	67–86
	80486/7	30–32
	Pentium	20–31
	Pentium Pro	

FSETPM   Set protected mode

11011011 11100100

Example   Clocks

FSETPM	8087	—
	80187	—
	80287	2–18
	80387	12
	80486/7	—
	Pentium	—
	Pentium Pro	

FSIN   Sine of ST(0)

11011001 11111110

Example   Clocks

FSIN	8087	—
	80187	124–773
	80287	—
	80387	122–771
	80486/7	193–279
	Pentium	16–126
	Pentium Pro	

FSINCOS   Find sine and cosine of ST(0)

11011001 11111011

Example   Clocks

FSINCOS	8087	—
	80187	196–811
	80287	—
	80387	194–809
	80486/7	243–329
	Pentium	17–137
	Pentium Pro	

FSQRT   Square root of ST(0)

11011001 11111010

Example	Clocks	
FSQRT	8087	180–186
	80187	124–131
	80287	180–186
	80387	122–129
	80486/7	83–87
	Pentium	70
	Pentium Pro	

FST/FSTP/FIST/FISTP/FBSTP   Store

```
11011001  oo010mmm  disp        32-bit memory (FST)
11011101  oo010mmm  disp        64-bit memory (FST)
11011101  11010rrr              FST ST(rrr)
11011011  oo011mmm  disp        32-bit memory (FSTP)
11011101  oo011mmm  disp        64-bit memory (FSTP)
11011011  oo111mmm  disp        80-bit memory (FSTP)
11011101  11001rrr              FSTP ST(rrr)
11011111  oo010mmm  disp        16-bit memory (FIST)
11011011  oo010mmm  disp        32-bit memory (FIST)
11011111  oo011mmm  disp        16-bit memory (FISTP)
11011011  oo011mmm  disp        32-bit memory (FISTP)
11011111  oo111mmm  disp        64-bit memory (FISTP)
11011111  oo110mmm  disp        80-bit memory (FBSTP)
```

Format	Examples	Clocks	
FST	FST DATA	8087	15–540
FSTP	FST ST(3)	80187	13–536
FIST	FST	80287	15–540
FISTP	FSTP	80387	11–534
FBSTP	FIST DATA2	80486/7	3–176
	FBSTP DATA6	Pentium	1–3
	FISTP DATA9	Pentium Pro	

FSTCW/FNSTCW   Store control register

11011001 oo111mmm disp

Format	Examples	Clocks	
FSTCW	FSTCW CONTROL	8087	12–18
FNSTCW	FNSTCW STATUS	80187	21
	FSTCW MACHINE	80287	12–18
		80387	15
		80486/7	3
		Pentium	2
		Pentium Pro	

FSTENV/FNSTENV   Store environment

11011001 oo110mmm disp

Format	Examples		Clocks
FSTENV	FSTENV CONTROL	8087	40–50
FNSTENV	FNSTENV STATUS	80187	146
	FSTENV MACHINE	80287	40–50
		80387	103–104
		80486/7	58–67
		Pentium	48–50
		Pentium Pro	

FSTSW/FNSTSW   Store status register

11011101 oo111mmm disp

Format	Examples		Clocks
FSTSW	FSTSW CONTROL	8087	12–18
FNSTSW	FNSTSW STATUS	80187	21
	FSTSW MACHINE	80287	12–18
	FSTSW AX	80387	15
		80486/7	3
		Pentium	2–5
		Pentium Pro	

FSUB/FSUBP/FISUB   Subtraction

11011000 oo100mmm disp	32-bit memory (FSUB)	
11011100 oo100mmm disp	64-bit memory (FSUB)	
11011d00 11101rrr	FSUB ST,ST(rrr)	
11011110 11101rrr	FSUBP ST,ST(rrr)	
11011110 oo100mmm disp	16-bit memory (FISUB)	
11011010 oo100mmm disp	32-bit memory (FISUB)	

Format	Examples		Clocks
FSUB	FSUB DATA	8087	70–143
FSUBP	FSUB ST,ST(2)	80187	28–92
FISUB	FSUB ST(2),ST	80287	70–143
FSUBP	FISUB DATA3	80387	29–82
		80486/7	8–35
		Pentium	1–7
		Pentium Pro	

FSUBR/FSUBRP/FISUBR   Reverse subtraction

11011000 oo101mmm disp	32-bit memory (FSUBR)	
11011100 oo101mmm disp	64-bit memory (FSUBR)	
11011d00 11100rrr	FSUBR ST,ST(rrr)	
11011110 11100rrr	FSUBRP ST,ST(rrr)	
11011110 oo101mmm disp	16-bit memory (FISUBR)	
11011010 oo101mmm disp	32-bit memory (FISUBR)	

Format	Examples		Clocks
FSUBR	FSUBR DATA	8087	70–143
FSUBRP	FSUBR ST,ST(2)	80187	28–92
FISUBR	FSUBR ST(2),ST	80287	70–143
FSUBRP	FISUBR DATA3	80387	29–82
		80486/7	8–35
		Pentium	1–7
		Pentium Pro	

FTST   Compare ST(0) with +0.0

11011001 11100100

Example		Clocks
FTST	8087	38–48
	80187	30
	80287	38–48
	80387	28
	80486/7	4
	Pentium	1–4
	Pentium Pro	

FUCOM/FUCOMP/FUCOMPP   Unordered compare

11011101 11100rrr	FUCOM ST,ST(rrr)	
11011101 11101rrr	FUCOMP ST,ST(rrr)	
11011101 11101001	FUCOMPP	

Format	Examples		Clocks
FUCOM	FUCOM ST,ST(2)	8087	—
FUCOMP	FUCOM	80187	26–28
FUCOMPP	FUCOMP ST,ST(3)	80287	—
	FUCOMP	80387	24–26
	FUCOMPP	80486/7	4–5
		Pentium	1–4
		Pentium Pro	

FWAIT   Wait

10011011

Example

FWAIT		Clocks
	8087	4
	80187	6
	80287	3
	80387	6
	80486/7	1–3
	Pentium	1–3
	Pentium Pro	

FXAM   Examine ST(0)

11011001 11100101

Example

FXAM		Clocks
	8087	12–23
	80187	32–40
	80287	12–23
	80387	30–38
	80486/7	8
	Pentium	21
	Pentium Pro	

FXCH   Exchange ST(0) with another register

11011001 11001rrr   FXCH ST,ST(rrr)

Format	Examples		Clocks
FXCH	FXCH ST,ST(1)	8087	10–15
	FXCH	80187	20
	FXCH ST,ST(4)	80287	10–15
		80387	18
		80486/7	4
		Pentium	1
		Pentium Pro	

FXTRACT — Extract components of ST(0)

11011001 11110100

Example

		Clocks
FXTRACT	8087	27–55
	80187	72–78
	80287	27–55
	80387	70–76
	80486/7	16–20
	Pentium	13
	Pentium Pro	

FYL2X — ST(1) x log$_2$ ST(0)

11011001 11110001

Example

		Clocks
FYL2X	8087	900–1100
	80187	122–540
	80287	900–1100
	80387	120–538
	80486/7	196–329
	Pentium	22–111
	Pentium Pro	

FXL2XP1 — ST(1) x log$_2$ [ST(0) + 1.0]

11011001 11111001

Example

		Clocks
FXL2XP1	8087	700–1000
	80187	259–549
	80287	700–1000
	80387	257–547
	80486/7	171–326
	Pentium	22–103
	Pentium Pro	

Notes: d = direction, where d = 0 for ST as the destination, and d = 1 for ST as the source; rrr = floating-point register number; oo = mode; mmm = r/m field; and disp = displacement. Intel has not released clock timings for the Pentium Pro, so they appear blank in the table.

14–4   CONNECTING THE 80187 TO THE 80186

The 80187 coprocessor and the 80186 microprocessor are compatible, provided a latch is added to capture address connections A1 and A2, which are used as command inputs to the 80187. Note that the 80187 cannot be used with the 80188 microprocessor.

Interconnecting the 80187 and 80186

Figure 14–10 illustrates the interconnection between the 80187 and 80C186XL. Notice that peripheral select pins 5 and 6 are used as address connections A1 and A2. The microprocessor generates address signals A1 and A2 for the coprocessor whenever the EX bit in the $\overline{MCS}$ and $\overline{PCS}$ alternate control register is cleared to zero. These latched address signals connect to the CMD0 and CMD1 inputs of the 80187 coprocessor. Also notice that the read and write signals connect to the numeric processor read and write inputs. The clock signal for the coprocessor can come from either the system CLKOUT pin of the microprocessor or from a separate oscillator (not shown in Figure 14–10). The CKM pin is grounded when CLKOUT is the timing source and connected to 5.0 V when an external clock signal is the source. Also note that the signals $\overline{NPS1}$, PEREQ, and $\overline{ERROR}$ are multiplexed with three of the $\overline{MCS}$ pins. The $\overline{MCS}$ can still be used to select memory.

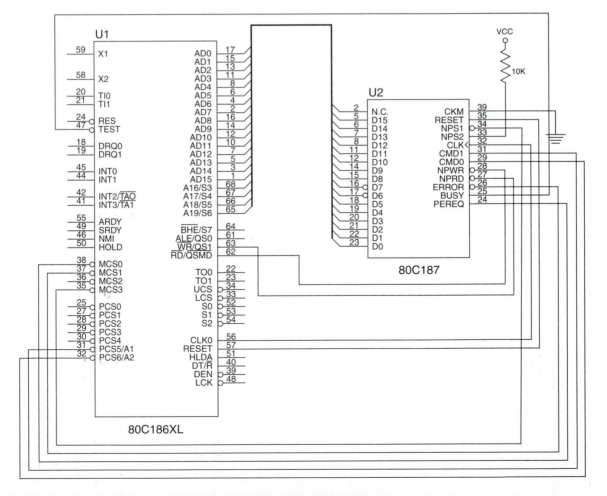

FIGURE 14–10   The 80C187 connected to the 80C186XL microprocessor.

14–5   PROGRAMMING WITH THE ARITHMETIC COPROCESSOR

In this section we look at some examples of programming techniques for the coprocessor.

Calculating the Area of a Circle

This first programming example illustrates a simple method of addressing the coprocessor stack. First recall that the equation for calculating the area of a circle is $A = \pi r^2$. A program that performs this calculation is listed in Example 14–9. This program takes test data from array RAD that contains five sample radii. The five areas are stored in a second array called AREA. No attempt is made in this program to use the data from the AREA array.

EXAMPLE 14–9

```
                        ;A short program that finds the area of five circles whose
                        ;radii are stored in array RAD.
                        ;
                                .MODEL SMALL
                                .386                    ;select 80386
                                .387                    ;select 80387
0000                            .DATA
0000 4015C28F    RAD     DD     2.34,5.66,9.33,234.5,23.4
     40B51EB8
     411547AE
     436A8000
     41BB3333
0014 0005 [      AREA    DD     5 DUP (?)
        00000000
              ]
0000                            .CODE
                                .STARTUP
0010 BE 0000             MOV SI,0                        ;source element 0
0013 BF 0000             MOV DI,0                        ;destination element 0
0016 B9 0005             MOV CX,5                        ;count of 5
0019            MAIN1:
0019 D9 84 0000 R        FLD    RAD [SI]                 ;radius to ST
001D D8 C8               FMUL ST,ST(0)                   ;square radius
001F D9 EB               FLDPI                           ;n to ST
0021 DE C9               FMUL                            ;multiply ST = ST x ST(1)
0023 D9 9D 0014 R        FSTP AREA [DI]                  ;save area
0027 46                  INC  SI
0028 47                  INC  DI
0029 E2 EE               LOOP MAIN1
                                .EXIT
                                END
```

Although this is a simple program, it does illustrate the operation of the tack. To provide a better understanding of the operation of the stack, Figure 14–11 shows the contents of the stack after each instruction of Example 14–9 executes. Note that only one pass through the loop is illustrated, because the program calculates five areas.

The first instruction loads the contents of memory location RAD [SI], one of the elements of the array, to the top of the stack. Next the FMUL ST,ST(0) instruction squares the radius on the top of the stack. The FLDPI instruction loads n to the stack top. The FMUL instruction uses the classic stack addressing mode to multiply ST by ST(1). After the multiplication, the prior values of ST and ST(1) are removed from the stack and the product replaces them at the top of the stack. Finally, the FSTP [DI] instruction copies the top of the stack, the area, to an array memory location AREA and clears the stack.

Notice how care is taken to always remove all stack data. This is important because if data remains on the stack at the end of the procedure, the stack top will no longer be register 0. This could cause problems because software assumes that the top of the stack is register 0. Another

FIGURE 14–11 Operation with the 80X87 stack. Note that the stack is shown after the execution of the indicated instruction.

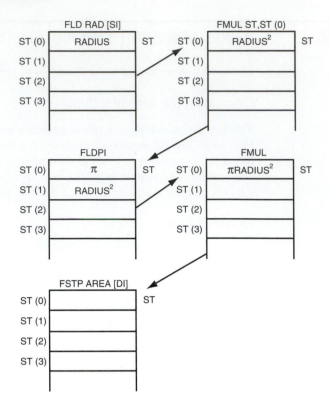

way of ensuring that the coprocessor is initialized is to place the FINIT (initialization) instruction at the start of the program.

Finding the Resonant Frequency

An equation commonly used in electronics is the formula for determining the resonant frequency of an LC circuit. The equation solved by the program in Example 14–10 is

$$Fr = \frac{1}{2\pi\sqrt{LC}}$$

This example uses L1 for the inductance L, C1 for the capacitor C, and RESO for the resultant resonant frequency.

EXAMPLE 14–10

```
                        ;A sample program that finds the resonant frequency of an LC
                        ;tank circuit.
                        ;
                                .MODEL SMALL
                                .386
                                .387
0000                            .DATA
0000 00000000           RESO    DD      ?               ;resonant frequency
0004 358637BD           L1      DD      0.000001        ;inductance
0008 358637BD           C1      DD      0.000001        ;capacitance
000C 40000000           TWO     DD      2.0             ;constant
0000                            .CODE
                                .STARTUP
0010 D9 06 0004 R               FLD     L1              ;get L
0014 D8 0E 0008 R               FMUL    C1              ;find LC

0018 D9 FA                      FSQRT                   ;find √LC
```

```
001A D8 0E 000C R          FMUL TWO                         ;find 2√LC

001E D9 EB                 FLDPI                            ;get π
0020 DE C9                 FMUL                             ;get 2π√LC

0022 D9 E8                 FLD1                             ;get 1
0024 DE F1                 FDIVR                            ;form 1/(2π√LC)

0026 D9 1E 0000 R          FSTP RESO                        ;save frequency
                           .EXIT
                           END
```

Notice the straightforward manner in which the program solves this equation. Very little extra data manipulation is required because of the stack inside the coprocessor. Also notice how the constant TWO is defined for the program and how the DIVRP, using classic stack addressing, is used to form the reciprocal. If you own a reverse-polish entry calculator, such as those produced by Hewlett-Packard, you are familiar with stack addressing. If not, using the coprocessor will increase your experience with this type of entry.

Findng the Roots Using the Quadratic Equation

This example illustrates how to find the roots of a polynomial expression $(ax^2 + bx + c = 0)$ using the quadratic equation. The quadratic equation is

$$b\pm\frac{\sqrt{b^2 - 4ac}}{2a}$$

Example 14–11 illustrates a program that finds the roots (R1 and R2) for the quadratic equation. The constants are stored in memory locations A1, B1, and C1. Note that no attempt is made to determine the roots if they are imaginary. This example tests for imaginary roots and exits to DOS with a zero in the roots (R1 and R2) if it finds them. In practice, imaginary roots could be solved for and stored in a separate set of result memory locations.

EXAMPLE 14–11

```
                    ;A program that finds the roots of a polynomial equation using
                    ;the quadratic equation.  Note imaginary roots are indicated if
                    ;both root 1 (R1) and root 2 (R2) are zero.
                    ;
                           .MODEL SMALL
                           .386
                           .387
0000                       .DATA
0000 40000000       TWO     DD   2.0
0004 40800000       FOUR    DD   4.0
0008 3F800000       A1      DD   1.0
000C 00000000       B1      DD   0.0
0010 C1100000       C1      DD   -9.0
0014 00000000       R1      DD   ?
0018 00000000       R2      DD   ?
0000                       .CODE
                           .STARTUP
0010 D9 EE                 FLDZ
0012 D9 16 0014 R          FST   R1                 ;clear roots
0016 D9 1E 0018 R          FSTP  R2
001A D9 06 0000 R          FLD   TWO
001E D8 0E 0008 R          FMUL  A1                 ;form 2a
0022 D9 06 0004 R          FLD   FOUR
0026 D8 0E 0008 R          FMUL  A1
002A D8 0E 0010 R          FMUL  C1                 ;form 4ac
002E D9 06 000C R          FLD   B1
0032 D8 0E 000C R          FMUL  B1                 ;form b²
0036 DE E1                 FSUBR                    ;form b²-4ac
0038 D9 E4                 FTST                     ;test b²-4ac for zero
```

```
003A 9B DF E0              FSTSW  AX                   ;copy status register to AX
003D 9E                    SAHF                        ;move to flags
003E 74 0E                 JZ    ROOTS1                ;if b²-4ac is zero
0040 D9 FA                 FSQRT                       ;find square root of b²-4ac
0042 9B DF E0              FSTSW  AX
0045 A9 0001               TEST  AX,1                  ;test for invalid error (negative)
0048 74 04                 JZ    ROOTS1
004A DE D9                 FCOMPP                       ;clear stack
004C EB 18                 JMP   ROOTS2                ;end
004E             ROOTS1:
004E D9 06 000C R          FLD   B1
0052 D8 E1                 FSUB  ST,ST(1)
0054 D8 F2                 FDIV  ST,ST(2)
0056 D9 1E 0014 R          FSTP  R1                    ;save root 1
005A D9 06 000C R          FLD   B1
005E DE C1                 FADD
0060 DE F1                 FDIVR
0062 D9 1E 0018 R          FSTP  R2                    ;save root 2
0066             ROOTS2:
                           .EXIT
                           END
```

Using a Memory Array to Store Results

The next programming example illustrates the use of a memory array and the scaled-indexed addressing mode to access the array. Example 14–12 calculates 100 values of inductive reactance. The equation for inductive reactance is XL = 2πFL. In this example the frequency range is from 10 Hz to 1,000 Hz for F and an inductance of 4H. Notice how the instruction FSTP DWORD PTR CS:[EDI+4*ECX] is used to store the reactance for each frequency beginning with the last at 1,000 Hz and ending with the first at 10 Hz. Also notice how the FCOMP instruction is used to clear the stack just before the RET instruction.

EXAMPLE 14–12

```
                           ;A program that calculates the inductive reactance of L
                           ;at a frequency range of 10 Hz to 1000 Hz and stores them
                           ;in array XL.  Note the increment is 10 Hz.
                                  .MODEL SMALL
                                  .386
                                  .387
0000                              .DATA
0000 40800000     L        DD    4.0                   ;4.0H test value
0004 0064 [       XL       DD    100 DUP (?)
          00000000
                  ]
0194 447A0000     F        DD    1000.0                ;start at 1000 Hz
0198 41200000     TEN      DD    10.0                  ;increment of 10 Hz
0000                              .CODE
                                  .STARTUP
0010 66| B9 00000064       MOV   ECX,100               ;load count
0016 66| BF 00000000 R     MOV   EDI,OFFSET XL-4       ;address result
001C D9 EB                 FLDPI                        ;get π
001E D8 C0                 FADD  ST,ST(0)               ;form 2π
0020 D8 0E 0000 R          FMUL  L                      ;form 2πL
0024             L1:
0024 D9 06 0194 R          FLD   F                      ;get F
0028 D8 C9                 FMUL  ST,ST(1)
002A 67& D9 1C 8F          FSTP  DWORD PTR [EDI+4*ECX]
002E D9 06 0194 R          FLD   F
0032 D8 26 0198 R          FSUB  TEN                    ;change frequency
0036 D9 1E 0194 R          FSTP  F
003A E2 E8                 LOOP  L1
003C D8 D9                 FCOMP
                           .EXIT
                           END
```

Displaying a Single-Precision Floating-Point Number

In this section we show how to take the floating-point contents of a 32-bit single-precision floating-point number and display it on the video display. The procedure displays the floating-point number as a mixed number with an integer part and a fractional part separated by a decimal point. In order to simplify the procedure, a limit is placed on the display size of the mixed number so the integer portion is a 32-bit binary number and the fraction is a 24-bit binary number. The procedure will not function properly for larger or smaller numbers.

Example 14–13 calls a procedure for displaying the contents of memory location NUMB on the video display at the current cursor position. The procedure first tests the sign of the number and displays a minus sign for a negative number. After displaying the minus sign, if needed, the number is made positive by the FABS instruction. Next, it is divided into an integer and fractional part and stored at WHOLE and FRACT. Notice how the FRNDINT instruction is used to round (using the chop mode) the top of the stack to form the whole number part of NUMB. The whole number part is then subtracted from the original number to generate the fractional part. This is accomplished with the FSUB instruction that subtracts the contents of ST(1) from ST.

EXAMPLE 14–13

```
                        ;A program that displays the floating-point contents of NUMB
                        ;as a mixed decimal number.
                              .MODEL SMALL
                              .386
                              .387
0000                          .DATA
0000 C50B0200          NUMB   DD    -2224.125              ;test data
0004 0000              TEMP   DW    ?
0006 00000000          WHOLE  DD    ?
000A 00000000          FRACT  DD    ?
0000                          .CODE
                              .STARTUP
0010 E8 000B                  CALL DISP                    ;display NUMB
                              .EXIT
                        ;
                        ;procedure that displays the ASCII code from AL
                        ;
0017                   DISPS PROC NEAR

0017 B4 06                    MOV  AH,6                    ;display AL
0019 8A D0                    MOV  DL,AL
001B CD 21                    INT  21H
001D C3                       RET

001E                   DISPS ENDP
                        ;
                        ;Procedure that displays the floating-point contents of NUMB
                        ;in decimal form.
                        ;
001E                   DISP  PROC  NEAR

001E 9B D9 3E 0004 R          FSTCW TEMP                   ;save current control word
0023 81 0E 0004 R 0C00        OR   TEMP,0C00H              ;set rounding to chop
0029 D9 2E 0004 R             FLDC TEMP
002D D9 06 0000 R             FLD  NUMB                    ;get NUMB
0031 D9 E4                    FTST                         ;test NUMB
0033 9B DF E0                 FSTSW AX                     ;status to AX
0036 25 4500                  AND  AX,4500H                ;get C3, C2, and C0
                              .IF  AX == 0100H
003E B0 2D                      MOV   AL,'-'
0040 E8 FFD4                    CALL     DISPS
0043 D9 E1                      FABS
                              .ENDIF
0045 D9 C0                    FLD  ST
```

```
0047  D9 FC                     FRNDINT                        ;get integer part
0049  DB 16 0006 R              FIST   WHOLE
004D  DE E1                     FSUBR
004F  D9 E1                     FABS
0051  D9 1E 000A R              FSTP   FRACT                   ;save fraction
0055  66| A1 0006 R             MOV    EAX,WHOLE
0059  66| BB 0000000A           MOV    EBX,10
005F  B9 0000                   MOV    CX,0
0062  53                        PUSH   BX
                                .WHILE  1                      ;divide until quotient = 0
0063  66| BA 00000000             MOV    EDX,0
0069  66| F7 F3                   DIV    EBX
006C  80 C2 30                    ADD    DL,30H
006F  52                          PUSH   DX
                                  .BREAK  .IF EAX == 0
0075  41                          INC    CX
                                  .IF    CX == 3
007B  6A 2C                          PUSH   ','
007D  B9 0000                        MOV    CX,0
                                  .ENDIF
                                .ENDW
                                .WHILE  1                      ;display whole number part
0082  5A                          POP    DX
                                  .BREAK .IF  DX == BX
0087  8A C2                        MOV    AL,DL
0089  E8 FF8B                      CALL   DISPS
                                .ENDW
008E  B0 2E                     MOV    AL,'.'                  ;display decimal point
0090  E8 FF84                    CALL DISPS
0093  66| A1 000A R             MOV    EAX,FRACT
0097  9B D9 3E 0004 R           FSTCW  TEMP                    ;save current control word
009C  81 36 0004 R 0C00         XOR    TEMP,0C00H              ;set rounding to nearest
00A2  D9 2E 0004 R              FLDCW  TEMP
00A6  D9 06 000A R              FLD    FRACT
00AA  D9 F4                     FXTRACT
00AC  D9 1E 000A R              FSTP   FRACT
00B0  D9 E1                     FABS
00B2  DB 1E 0006 R              FISTP  WHOLE
00B6  66| 8B 0E 0006 R          MOV    ECX,WHOLE
00BB  66| A1 000A R             MOV    EAX,FRACT
00BF  66| C1 E0 09              SHL    EAX,9
00C3  66| D3 D8                 RCR    EAX,CL
                                .REPEAT
00C6  66| F7 E3                   MUL    EBX
00C9  66| 50                      PUSH   EAX
00CB  66| 92                      XCHG   EAX,EDX
00CD  04 30                       ADD    AL,30H
00CF  E8 FF45                     CALL   DISPS
00D2  66| 58                      POP    EAX
                                .UNTIL  EAX == 0
00D9  C3                        RET

00DA                     DISP  ENDP
                               END
```

The last part of the procedure displays the whole number part followed by the fractional part. The techniques are the same as introduced earlier, dividing a number by 10 and displaying the remainders in reverse order converts and displaying an integer. A multiplication by 10 converts a fraction to decimal for displaying. Note that the fractional part may contain a rounding error for certain values. This occurs because the number has not been adjusted to remove the rounding error that is inherent in floating-point fractional numbers.

Reading a Mixed-Number from the Keyboard

If floating-point arithmetic is used in a program, you must develop a method of reading the number from the keyboard and converting it to floating-point. The procedure in Example 14–14 reads a signed mixed number from the keyboard and converts it to a floating-point number located at memory location NUMB.

EXAMPLE 14–14

```
                        ;A program that reads a mixed number from the keyboard.
                        ;The result is stored at memory location NUMB as a
                        ;double-precision floating-point number.
                        ;
                                .MODEL SMALL
                                .386
                                .387
0000                            .DATA
0000 00         SIGN    DB      ?               ;sign indicator
0001 0000       TEMP1   DW      ?               ;temporary storage
0003 41200000   TEN     DD      10.0            ;10.0
0007 00000000   NUMB    DD      ?               ;result
0000                            .CODE
                GET     MACRO                   ;;read key macro
                        MOV   AH,1
                        INT   21H
                        ENDM
                                .STARTUP
0010 D9 EE              FLDZ                    ;clear ST
                        GET                     ;read a character
                        .IF  AL == '+'          ;test for +
001A C6 06 0000 R 00        MOV   SIGN,0        ;clear sign indicator
                            GET
                        .ENDIF
                        .IF  AL == '-'          ;test for -
0027 C6 06 0000 R 01        MOV   SIGN,1        ;set sign indicator
                            GET
                        .ENDIF
                        .REPEAT
0030 D8 0E 0003 R           FMUL   TEN          ;multiply result by 10
0034 B4 00                  MOV   AH,0
0036 2C 30                  SUB   AL,30H         ;convert from ASCII
0038 A3 0001 R              MOV   TEMP1,AX
003B DE 06 0001 R           FIADD TEMP1          ;add it to result
                            GET                  ;get next character
                        .UNTIL  AL < '0' || AL > '9'
                        .IF AL == '.'            ;do if -
004F D9 E8                  FLD1                 ;get one
                            .WHILE  1
0051 D8 36 0003 R               FDIV   TEN
                                GET
                            .BREAK  .IF  AL < '0' || AL > '9'
0061 B4 00                      MOV   AH,0
0063 2C 30                      SUB   AL,30H      ;convert from ASCII
0065 A3 0001 R                  MOV   TEMP1,AX
0068 DF 06 0001 R               FILD TEMP1
006C D8 C9                      FMUL ST,ST(1)
006E DC C2                      FADD ST(2),ST
0070 D8 D9                      FCOMP
                            .ENDW
0074 D8 D9                  FCOMP               ;clear stack
                        .ENDIF
                        .IF  SIGN == 1
007D D9 E0                  FCHS                ;make negative
                        .ENDIF
007F D9 1E 0007 R       FSTP NUMB               ;save result
                        .EXIT
                        END
```

Unlike other examples in this chapter, Example 14–4 uses some of the high-level language constructs presented in earlier chapters to reduce its size. Here the sign is first read from the keyboard, if present, and saved for later use, as a zero for positive and a one for negative, in adjusting the sign of the resultant floating-point number. Next, the integer portion of the number is read. Notice how the .REPEAT–.UNTIL loop is used to read the number until something other than a number (0–9) is typed. This portion terminates with a period, space, or carriage return. If a period is typed, then the procedure continues and reads a fractional part by using an .IF–.ENDIF construct. If a space or carriage return is entered, the number is converted to floating-point form and stored at NUMB. Notice how a .WHILE–.ENDW loop is used to convert the fractional part of the number. The whole number portion is converted with a multiply by 10 and the fractional portion is converted with a divide by 10.

14–6   SUMMARY

1. The arithmetic coprocessor functions in parallel with the microprocessor. This means that the microprocessor and coprocessor can execute their respective instructions simultaneously.
2. The data types manipulated by the coprocessor include: signed-integer, floating-point, and binary-coded decimal (BCD).
3. There are three forms of integers used with the coprocessor: word (16 bits), short (32 bits), and long (64 bits). Each integer contains a signed number in true magnitude for positive numbers and two's complement form for negative numbers.
4. A BCD number is stored as an 18-digit number in 10 bytes of memory. The most-significant byte contains the sign-bit, and the remaining 9 bytes contain an 18-digit packed BCD number.
5. The coprocessor supports three types of floating-point numbers: single precision (32 bits), double precision (64 bits), and extended precision (80 bits). A floating-point number has three parts: the sign, biased exponent, and significand. In the coprocessor, the exponent is biased with a constant and the integer bit of the normalized number is not stored in the significand except in the extended-precision form.
6. Decimal numbers are converted to floating-point numbers by: (a) converting the number to binary, (b) normalizing the binary number, (c) adding the bias to the exponent, (d) storing the number in floating-point form.
7. Floating-point numbers are converted to decimal by: (a) subtracting the bias from the exponent, (b) unnormalizing the number, and (c) converting it to decimal.
8. The 80287 and 80186/80188 use I/O space for the execution of some of their instructions. This space is invisible to the program and is used internally by the 80286/80287, 80186/80187, or 80188/80187 systems. These 16-bit I/O addresses (00F8H–00FFH) must not be used for I/O data transfers in a system that contains an 80287. The 80387, 80486/7, Pentium, and Pentium Pro use I/O addresses 800000F8H–800000FFH.
9. The coprocessor contains a status register that indicates busy, what conditions follow a compare or test, the location of the top of the stack, and the state of the error bits. The FSTSW AX instruction, followed by SAHF, is often used with conditional jump instructions to test for some coprocessor conditions.
10. The control register of the coprocessor contains control bits that select: infinity, rounding, precision, and error masks.
11. The following directives are often used with the coprocessor for storing data: DW (define word), DD (define doubleword), DQ (define quadword), and DT (define 10 bytes).
12. The coprocessor uses a stack to transfer data between itself and the memory system. Generally, data is loaded to the top of the stack or removed from the top of the stack for storage.
13. Internal coprocessor data is always in the 80-bit extended-precision form. The only time that data is in any other form is when it is stored or loaded from the memory.

14. The coprocessor addressing modes include: the classic stack mode, register, register with a pop, and memory. Stack addressing is implied and the data at ST becomes the source, ST(1) the destination, and the result is found in ST after a pop.

15. The coprocessor's arithmetic operations include: addition, subtraction, multiplication, division, and square root.

16. There are transcendental functions in the coprocessor's instruction set. These functions find the partial tangent or arctangent, $2^x - 1$, Y $\log_2$ X, and Y $\log_2$ (X + 1). The 80387, 80486/7, Pentium, and Pentium Pro also include sine and cosine functions.

17. Constants are stored inside the coprocessor that provide: +0.0, +1.0, π, $\log_2$ 10, $\log_2$ ε, $\log_{10}$ 2, and $\log_\varepsilon$ 2.

18. The 80387 functions with the 80386 microprocessor and the 80487SX functions with the 80486SX microprocessor, but the 80486DX, Pentium, and Pentium Pro contain their own internal arithmetic coprocessors. The instructions performed by the earlier versions are available on these coprocessors. In addition to these instructions, the 80387, 80486/7, Pentium, and Pentium Pro can find the sine and cosine.

19. The Pentium Pro contains two new floating-point instructions: FCMOV and FCOMI. The FCMOV instruction is a conditional move and the FCOMI performs the same task as FCOM, but is also places the floating-point flags into the system flag register.

20. When connecting the 80187 coprocessor to the 80186/80188 microprocessors, address connections A1 and A2 must be latched and applied to CMD0 and CMD1.

14–7 QUESTIONS AND PROBLEMS

1. List the three types of data that are loaded or stored in memory by the coprocessor.
2. List the three integer data types, the range of the integers stored in them, and the number of bits allotted to each.
3. Explain how a BCD number is stored in memory by the coprocessor.
4. List the three types of floating-point numbers used with the coprocessor and the number of binary bits assigned to each.
5. Convert the following decimal numbers into single-precision floating-point numbers:
   (a) 28.75
   (b) 624
   (c) −0.615
   (d) +0.0
   (e) −1000.5
6. Convert the following single-precision floating-point numbers into decimal:
   (a) 11000000 11110000 00000000 00000000
   (b) 00111111 00010000 00000000 00000000
   (c) 01000011 10011001 00000000 00000000
   (d) 01000000 00000000 00000000 00000000
   (e) 01000001 00100000 00000000 00000000
   (f) 00000000 00000000 00000000 00000000
7. Explain what the coprocessor does when a normal microprocessor instruction executes.
8. Explain what the microprocessor does when a coprocessor instruction executes.
9. What is the purpose of the C3–C0 bits in the status register?
10. Which operation is accomplished with the FSTSW AX instruction?
11. What is the purpose of the IE bit in the status register?
12. How can SAHF and a conditional jump instruction be used to determine is the top of the stack (ST) is equal to register ST(2)?

13. How is the rounding mode selected in the 80X87?

14. Which coprocessor instruction uses the microprocessor's AX register?

15. Which I/O ports are reserved for coprocessor use with the 80287?

16. How is data stored inside the coprocessor?

17. What is a NAN?

18. When the coprocessor is reset, the top of the stack register is register number _____.

19. What does the term "chop" mean in the rounding control bits of the control register?

20. What is the difference between affine and projective infinity control?

21. Which microprocessor instruction forms the opcodes for the coprocessor?

22. The FINIT instruction selects _____ precision for all coprocessor operations.

23. Using assembler pseudo-opcodes, write statements that accomplish the following:
    (a) Store a 23.44 into a double-precision floating-point memory location named FROG.
    (b) Store a –123 into a 32-bit signed-integer location named DATA3.
    (c) Store a –23.8 into a single-precision floating-point memory location named DATA1.
    (d) Reserve a double-precision memory location named DATA2.

24. Describe how the FST DATA instruction functions. Assume that DATA is defined as a 64-bit memory location.

25. What does the FILD DATA instruction accomplish?

26. Write an instruction that adds the contents of register 3 to the top of the stack.

27. Describe the operation of the FADD instruction.

28. Choose an instruction that subtracts the contents of register 2 from the top of the stack and stores the result in register 2.

29. What is the function of the FBSTP DATA instruction?

30. What is the difference between a forward and a reverse division?

31. What is the purpose of the Pentium Pro FCOMI instruction?

32. What does a Pentium Pro FCMOVB instruction accomplish?

33. What must occur before executing any FCMOV instruction?

34. Develop a procedure that finds the reciprocal of the single-precision floating-point number. The number is passed to the procedure in EAX and must be returned as a reciprocal in EAX.

35. What is the difference between the FTST and FXAM instructions?

36. Explain what the F2XM1 instruction calculates.

37. Which coprocessor status register bit should be tested after the FSQRT instruction executes?

38. Which coprocessor instruction pushes n onto the top of the stack?

39. Which coprocessor instruction places a 1.0 at the top of the stack?

40. What will FFREE ST(2) accomplish when executed?

41. Which instruction stores the environment?

42. What does the FSAVE instruction save?

43. Develop a procedure that finds the area of a rectangle ($A = L \times W$). Memory locations for this procedure are single-precision floating-point locations A, L, and W.

44. Write a procedure that finds the capacitive reactance

$$\left(XC = \frac{1}{2\pi FC1} \right)$$

Memory locations for this procedure are single-precision floating-point locations XC, F, and C1.

45. Develop a procedure that generates a table of square roots for the integers 2 through 10. The results must be stored as single-precision floating-point numbers in an array called ROOTS.

46. When is the FWAIT instruction used in a program?

47. What is the difference between the FSTSW and FNSTSW instructions?

48. Which two address bits must be latched when connecting the 80187 to the 80186/80188 microprocessors?

FIGURE 14–12 The series/parallel circuit (Question 50).

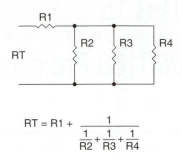

$$RT = R1 + \cfrac{1}{\cfrac{1}{R2} + \cfrac{1}{R3} + \cfrac{1}{R4}}$$

49. Which I/O ports are used by the 80187 coprocessor?

50. Given the series/parallel circuit and equation illustrated in Figure 14–12, develop a program using single-precision values for R1, R2, R3, and R4 that finds the total resistance and stores the result at single-precision location RT.

51. Develop a procedure that finds the cosine of a single-precision floating-point number. The angle, in degrees, is passed to the procedure in EAX and the cosine is returned in EAX. Recall that FCOS finds the cosine of an angle expressed in radians.

52. Given two arrays of double-precision floating-point data (ARRAY1 and ARRAY2) that each contain 100 elements, develop a procedure that finds the product of ARRAY1 times ARRAY2 and stores the double-precision floating-point result in a third array (ARRAY3).

53. Develop a procedure that takes the single-precision contents of register EBX times n and stores the result in register EBX as a single-precision floating-point number. You must use memory to accomplish this task.

54. Write a procedure that raises a single-precision floating-point number X to the power Y. Parameters are passed to the procedure with EAX = X and EBX = Y. The result is passed back to the calling sequence in ECX.

55. Given that $LOG_{10} X = (LOG_2 10)^{-1} \times LOG_2 X$, write a procedure called LOG10 that finds the LOG_{10} of the value at the stack top. Return the LOG_{10} at the stack top at the end of the procedure.

56. Use the procedure developed in Question 55 to solve the equation:

$$\text{Gain in decibels} = 20 \, LOG_{10} \, (Vout/Vin)$$

The program should take arrays of single-precision values for VOUT and VIN and store the decibel gains in a third array called DBG. There are 100 values for VOUT and VIN.

CHAPTER 15

80186/80188 Projects

INTRODUCTION

Although this chapter is not meant to provide new devices or information, it does contain several complete systems that show how to apply the 80186/80188 microprocessors. The software and hardware for these systems are explained in detail so that you can construct similar systems in the industrial environment.

CHAPTER OBJECTIVES

Upon completion of this chapter, you will be able to:

1. Construct simple microprocessor-based systems that use the 80186 or 80188 microprocessor as a core.
2. Write software that initializes the microprocessor for a given application.
3. Develop software that controls not only the microprocessor but also the system connected to the microprocessor.
4. Learn new applications for microprocessor-based systems.

15–1   USING THE 80C188EB MICROPROCESSOR AS AN APPLIANCE TIMER

An appliance timer is a relatively simple device that illustrates some control features and subsystems found in many microprocessor-based systems. One obvious need for the appliance timer is a real-time clock so that appliances can be enabled and disabled at desired times and so that events can be timed. Also required is the ability to control AC current for the appliances.

Hardware

Before the software can be written for this system, the hardware must be developed. Our system contains two 120-VAC electrical outlets for applying power to two appliances. This can be modified to include more outlets if desired (up to eight). The following major components have been chosen for the system:

- Two solid-state relays for switching 120 VAC
- 12-key telephone-style keypad for programming
- 5-digit LED display for displaying time and programming information
- 2716 EPROM for program storage
- 6116 SRAM for parameter storage and for the stack
- 80C188EB microprocessor

Figure 15–1 illustrates the complete schematic diagram of the appliance controller. Note that the segment and anode driver circuitry is not shown in this illustration. For information about these drivers, refer to Chapter 11, Section 11–3. In this system, the memory (U5 and U6) is interfaced to the 80C188EB microprocessor with a 74LS373 (U2) address latch used to supply the memory with address signals A_0–A_7. This provides 2K of memory for program storage (2716) and 2K of storage (6116) for transient data.

There are three I/O devices or ports active in this system. Port 1 from the microprocessor is used to provide the display with segment data and to select the 74LS374 (U4) latch for controlling the solid-state relays. The anode driver connects to port 2 (on the microprocessor) to select a single display at a time. Some of the same port 2 pins that select a display position also select rows of keys on the telephone-style keypad. Other port 2 pins are used to read columns of keys from the keypad. The solid-state relays are controlled from the outputs of a latch (U4) that is connected as a simple output port. Note that the OR gate (U3) is used to provide the latch with a clock pulse whenever $\overline{GCS7}$ is active at the same time that $\overline{WR}$ is active. The outputs of the latch apply a logic 1 to a solid-state relay to enable it, which in turn applies 120 VAC to an AC outlet (P1/P2). Note that six additional pins remain available at the output of the latch so another six solid-state relays could be attached to control six additional AC devices.

It is possible to remove the latch and OR gate if only up to four solid-state relays are needed in the system. This is accomplished (but not shown here) by reconnecting the port 2 pins labeled P2.7, P2.6, and P2.5 and port 1 pin $\overline{GCS7}$ as input signals to up to four solid-state relays. The three lines to the keypad columns can be connected to three of the unused INT inputs (such as INT1, INT2, and INT3). Recall that the state of these inputs can be sampled as input pins using the interrupt control registers described in Chapter 12.

Software

The software for this project is listed in Example 15–1. Because of the type of EPROM programmer in use, and the loss of the EXE2BIN program with newer versions of DOS, this program is assembled (using MASM 6.11) as a .COM file. Most EPROM programming systems require a binary (which is a .COM file) as input. Because a .COM is used, and because it must start at 100H for proper assembly by the assembly language program, the addresses in the program are biased by a 100H. This makes the reset address location 7F0H + 100H or 8F0H in the software listing. It also causes all addresses referenced in the EPROM to require a −100H after the label for proper addressing. The .COM file contains the instruction at the reset location at address 7F0H as it should for proper programming.

Because the microprocessor resets at a location that is 16 bytes from the top of memory, the first instruction executed is a memory address FFFF0H. This corresponds to EPROM address 7F0H. The EPROM is decoded and resides at memory addresses FF800H–FFFFFH, which causes EPROM address 7F0H to be located at physical address FFFF0H. These "reset" instructions are located at the end of the program in Example 15–1. This is where you should start reading this program.

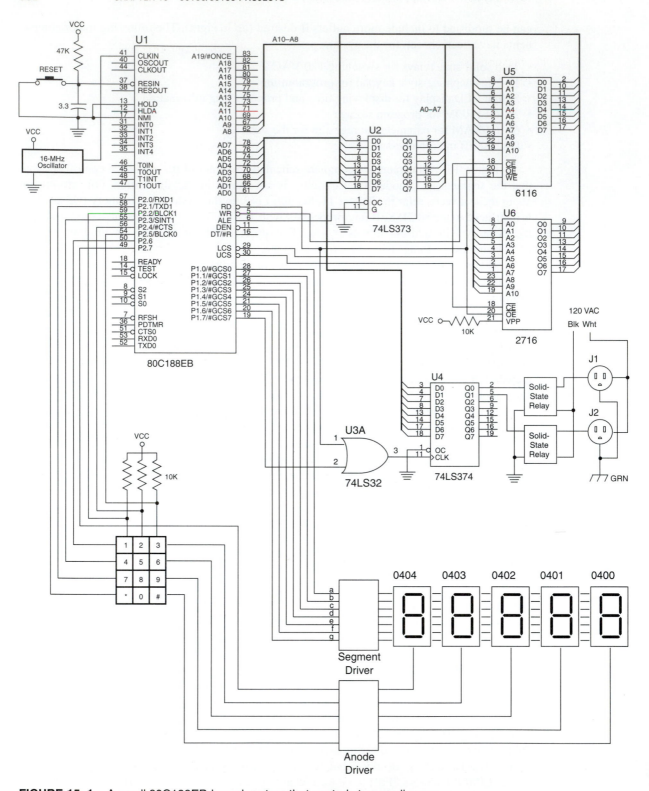

FIGURE 15–1   A small 80C188EB-based system that controls two appliances.

EXAMPLE 15–1

```
.186                                        ;switch to the 80186/80188 instruction set
CODE            SEGMENT 'code'
                ASSUME CS:CODE
;
;Program for the 80188EB microprocessor trainer.
;YOU MUST USE MASM 6.11 to assemble it.
;command line syntax = ML /AT XXXXXXXX.ASM
;
;This first part programs the microprocessor for memory at
;00000H-007FFH (SRAM), FF800H-FFFFFH (EPROM),
;and I/O at location 0000H.
;
                ORG     0100H               ;set origin to start of EPROM
MAIN:
                MOV     DX,0FFA6H           ;address stop for UCS
                MOV     AX,000EH            ;set upper address as FFFFFH
                OUT     DX,AL

                MOV     DX,0FFA0H           ;address start for LCS
                MOV     AX,0000H            ;set lower address at 00000H, no waits
                OUT     DX,AL

                MOV     DX,0FFA2H           ;address stop for LCS
                MOV     AX,008AH            ;set lower address at 00800H
                OUT     DX,AL

                MOV     DX,0FF9CH           ;address start for GCS7
                MOV     AX,0000H            ;set lower I/O at 0000H, no waits
                OUT     DX,AL

                MOV     DX,0FF9EH           ;address stop for GCS7
                MOV     AX,0048H            ;set upper I/O at 003FH
                OUT     DX,AL

                MOV     DX,0FF54H           ;address port 1 control
                MOV     AX,0080H            ;P1.0-P1.6 are outputs
                OUT     DX,AL

                MOV     DX,0FF5CH           ;address port 2 control
                MOV     AX,0H               ;P2.0-P2.7 are all available
                OUT     DX,AL

                MOV     DX,0FF58H           ;address Port 2 direction
                MOV     AX,00E0H            ;set P2.5-P2.7 as IN
                OUT     DX,AL

                MOV     AX,0                ;address segment 0000H
                MOV     DS,AX
                MOV     ES,AX
                MOV     SS,AX
                MOV     SP,0800H            ;set up stack pointer

                MOV     AL,0
                MOV     DS:[407H],AL        ;save port 0000H image
                OUT     00H,AL              ;turn-off outlets

                MOV     WORD PTR DS:[4EH],0F800H    ;save interrupt vector for Timer 2
                MOV     AX,OFFSET TIM2-100H         ;note the bias of -100H (see text)
                MOV     DS:[4CH],AX

                MOV     WORD PTR DS:[408H],07F04H   ;load select and pointer for display

                MOV     DX,0FF40H           ;address timer 2 count
                MOV     AX,0                ;clear counter to 0000H
                OUT     DX,AL
```

```
        MOV     DX,0FF42H           ;address timer 2 compare
        MOV     AX,2000             ;set to 2000 (for 1 ms count)
        OUT     DX,AL

        MOV     DX,0FF46H           ;address timer 2 control
        MOV     AX,0E001H           ;set for interrupt and continuous
        OUT     DX,AL

        MOV     DX,0FF12H           ;address interrupt control for timers
        MOV     AX,0               ;enable timer interrupt
        OUT     DX,AL

        MOV     DX,0FF08H           ;address interrupt mask register
        MOV     AX,00FCH            ;disable all except for timer
        OUT     DX,AL
        STI                         ;enable interrupts
;
;;;;;;;;;;;;;;;;;;;;;;;;;;;;;;;;;;;;;;;;;;;;;;;;;;;;;;;;;;;;;;;;;;;;;;;;;;;;;;;;;;;;
;                       SYSTEM SOFTWARE HERE                                    ;
;;;;;;;;;;;;;;;;;;;;;;;;;;;;;;;;;;;;;;;;;;;;;;;;;;;;;;;;;;;;;;;;;;;;;;;;;;;;;;;;;;;;
;
        MOV     AL,0
        MOV     DS:[40FH],AL        ;turn off clock display
        MOV     AX,0FFFFH           ;disable both outlets
        MOV     DS:[410H],AX        ;P1 start
        MOV     DS:[412H],AX        ;P1 stop
        MOV     DS:[414H],AX        ;P2 start
        MOV     DS:[416H],AX        ;P2 stop
        MOV     AX,4040H            ;set displays to -----
        MOV     DS:[400H],AX
        MOV     DS:[402H],AX
        MOV     DS:[404H],AL
        .WHILE 1                    ;main command loop
        CALL    KEY                 ;test keyboard
        .IF AL == 11                ;test for * key
        CALL    PROG                ;program time, P1, or P2
        .ENDIF
.ENDW
;
;;;;;;;;;;;;;;;;;;;;;;;;;;;;;;;;;;;;;;;;;;;;;;;;;;;;;;;;;;;;;;;;;;;;;;;;;;;;;;;;;;;;
;                       SYSTEM PROCEDURES                                       ;
;;;;;;;;;;;;;;;;;;;;;;;;;;;;;;;;;;;;;;;;;;;;;;;;;;;;;;;;;;;;;;;;;;;;;;;;;;;;;;;;;;;;
;
;Procedure that handles programming.
;
PROG        PROC    NEAR

        MOV     DS:[40FH],0         ;turn off time display
        MOV     AX,0               ;display -
        MOV     DS:[400H],AX
        MOV     DS:[402H],AX
        MOV     DS:[404H},40H
        CALL    KEY                 ;read second key
        .IF AL == 0                 ;*0 program time
                MOV     DS:[404H},58H ;display c
                CALL    GETT        ;get time
                MOV     DS:[40DH],DX ;save time

        .ELSEIF AL == 1             ;*1 program plug 1
                MOV     DS:[404H},6  ;display 1
                CALL    GETT        ;get start time
                PUSH    DX
                MOV     AX,0        ;display 1
                MOV     DS:[400H],AX
                MOV     DS:[402H],AX
                CALL    GETT        ;get stop time
                MOV     DS:[412H],DX ;save stop time
                POP     DX
```

```
                        MOV     DS:[410H],DX       ;save start time
                .ELSEIF AL == 2                    ;*2 program plug 2
                        MOV     DS:[404H},5BH      ;display 2
                        CALL    GETT               ;get start time
                        PUSH    DX
                        MOV     AX,0               ;display 2
                        MOV     DS:[400H],AX
                        MOV     DS:[402H],AX
                        CALL    GETT               ;get stop time
                        MOV     DS:[416H],DX       ;save stop time
                        POP     DX
                        MOV     DS:[414H],DX       ;save start time
                .ENDIF
                MOV     DS:[404H},0                ;blank left display
                MOV     AX,DS:[40DH]
                MOV     BX,400H
                MOV     DS:[40FH],0FFH             ;display time
                CALL    DISP16                     ;force the time to display
                RET

PROG            ENDP
;
;Procedure that reads a time from keypad.
;
GETT            PROC    NEAR
                MOV     DX,0                       ;start time at 0000
        GETT1:
                MOV     AX,DX
                MOV     BX,400H
                CALL    DISP16                     ;display time
                CALL    KEY                        ;read key
                .IF AL > 9
                        .IF AL = 1                 ;if # key set time to FFFFH
                                MOV     DX,0FFFFH
                        .ENDIF
                        RET
                .ENDIF
                SHL     DX,4
                ADD     DL,AL                      ;add in new digit
                JMP     GETT1

GETT            ENDP
;
;Procedure that reads a key from the 12-key keypad.
;
KEY             PROC    NEAR USES AX BX CX DX

        KEY1:                                      ;wait for release
                CALL    SCAN                       ;check keypad
                JNZ     KEY1                       ;key still pressed
                CALL    DELAY                      ;wait for 15 ms
                CALL    SCAN
                JNZ     KEY1
        KEY2:                                      ;wait for key press
                CALL    SCAN
                JZ      KEY2
                CALL    DELAY
                CALL    SCAN
                JZ      KEY2
                PUSH    AX
                MOV     BL,DS:[408H}               ;get row number in BL
                MOV     AL,3                       ;multiply row times 3
                MUL     BL
                MOV     BX,AX                      ;move 3 x row into BX
                DEC     BX                         ;adjust BX
                POP     AX
                AND     AL,34H                     ;strip output bits
                ADD     AL,4                       ;fix AL
```

```
                SHL     AL,2
                .REPEAT                         ;find column
                        SHL     AL,1
                        INC     BX
                .UNTIL !CARRY?                   ;until no carry
                MOV     AL,CS:TABLE1[BX-100H]        ;note bias of -100H (see text)
                RET

KEY             ENDP
;
;Look up table for keypad.
;
TABLE1          DB      11,0,10                 ;*,0,#
                DB      9,8,7
                DB      6,5,4
                DB      3,2,1
;
;Procedure that tests for a keystroke.
;
SCAN            PROC    NEAR
                MOV     CL,5                    ;set count for all 5 outputs of port 2
        SCAN1:
                HLT                             ;wait for an interrupt
                MOV     DX,0FF5AH               ;address port 2
                IN      AL,DX                   ;read port 2 keypad columns
                OR      AL,1FH                  ;set unused bits
                CMP     AL,0FFH                 ;test for no keys
                JNE     SCAN2                   ;if key detected
                DEC     CL                      ;(can't use LOOP here)
                JNZ     SCAN1                   ;repeat 5 times
        SCAN2:
                RET

SCAN            ENDP
;
;Procedure for a 15 ms time delay.
;
DELAY           PROC    NEAR
                MOV     CX,15
                .REPEAT
                        HLT
                .UNTILCXZ
                RET

DELAY           ENDP
;
;Interrupt service procedure for timer 2 interrupt.  (once per millisecond)
;
TIM2            PROC    FAR USES DS AX BX DX

                MOV     AX,0                    ;address segment 0000H
                MOV     DS,AX
                MOV     BX,DS:[408H]            ;get pointer and select code
                ROR     BH,1                    ;rotate select right
                .IF BH == 0DFH                  ;skip P2.5 and P2.4
                        ROR     BH,1
                        ROR     BH,1
                .ENDIF
                .IF BH == 0FBH                  ;skip P2.2
                        ROR     BH,1
                .ENDIF
                DEC     BL
                JNS     TIM2A                   ;if BL is still 0-4
                MOV     BX,07F04H               ;reset BX
TIM2A:
                MOV     DS:[408H],BX            ;save it for next interrupt
                MOV     AL,BH                   ;select digit
                MOV     DX,0FF5EH
```

```
            OUT     DX,AL                   ;send select to port 2
            MOV     BH,4                    ;address display RAM
            MOV     AL,[BX]                 ;get 7-segment code
            MOV     DX,0FF56H
            OUT     DX,AL                   ;send code to port 1

            MOV     BX,40AH                 ;address real-time clock
            MOV     AH,0                    ;load modulus 100
            CALL    UPDATE                  ;update divide by 100 counter (40AH)
            JNZ     TIM2B                   ;if no roll-over
            MOV     AH,10H                  ;load modulus of 10
            CALL    UPDATE                  ;update divide by 10 counter (1/10 sec) (40BH)
            JNZ     TIM2B                   ;if no roll-over
            MOV     AH,60H                  ;load modulus 60
            CALL    UPDATE                  ;update seconds counter (40CH)
            JNZ     TIM2B                   ;if no roll-over
            CALL    UPDATE                  ;update minutes counter (40DH)
            CALL    DISPLAY                 ;display time
            CALL    TIMER                   ;test timers
            JNZ     TIM2B                   ;if no roll-over
            MOV     AH,24H                  ;load modulus 24
            CALL    UPDATE                  ;update hours counter (40EH)
            CALL    DISPLAY                 ;display time
TIM2B:
            MOV     DX,0FF02H               ;address end-of-interrupt register
            MOV     AX,8000H                ;specific end-of-timer interrupt
            OUT     DX,AL

            IRET

TIM2        ENDP
;
;Procedure that tests the timers for plugs 1 and 2.
;
TIMER       PROC    NEAR USES BX AX         ;test appliance timers

            PUSHF
            MOV     AH,1                    ;indicate P1
            MOV     BX,410H
            CALL    TEST                    ;test P1 timer
            MOV     AH,2                    ;indicate P2
            MOV     BX,414H
            CALL    TEST                    ;test P2 timer
            POPF
            RET

TIMER       ENDP
;
;Procedure that tests one timer. AH = timer bit position, BX = address
;
TEST        PROC    NEAR

            CMP     WORD PTR [BX],0FFFFH
            JE      TESTE
            MOV     DX,[BX]                 ;get start time
            .IF WORD PTR DS:[40DH] >= DX
                    MOV     DX,[BX+2]       ;get stop time
                    .IF WORD PTR DS:[40DH] < DX
                            OR      DS:[407H],AH    ;turn on appliance
                    .ELSE
                            NOT     AH
                            AND     DS:[407H],AH    ;turn off appliance
                    .ENDIF
            .ENDIF
            MOV     AL,DS:[407H]
            OUT     00H,AL                  ;change appliance data
TESTE:
            RET
```

```
TEST            ENDP
;
;Procedure that updates the time-of-day clock.
;
UPDATE          PROC    NEAR

                MOV     AL,[BX]                 ;get counter
                INC     BX                      ;address next counter
                ADD     AL,1                    ;increment time
                DAA                             ;make it BCD
                MOV     [BX-1],AL               ;save new counter
                SUB     AL,AH                   ;check modulus
                JNZ     UPDATE                  ;if no roll-over
                MOV     [BX-1],AL               ;clear counter
        UPDATE1:
                RET

UPDATE          ENDP
;
;Procedure that displays the time of day if 40FH <> 0.
;
DISPLAY         PROC NEAR       USES BX         ;display time of day

                PUSHF
                .IF BYTE PTR DS:[40FH] != 0
                        MOV     BX,400H         ;for minutes
                        MOV     AL,DS:[40DH]    ;get minutes
                        CALL    DISP8           ;display minutes
                        MOV     BX,402H         ;address for hours
                        MOV     AL,DS:[40EH]    ;get hours
                        CALL    DISP8           ;display hours
                        MOV     BYTE PTR DS:[404H],40H
                .ENDIF
                POPF
                RET

DISPLAY         ENDP
;
;Procedure that displays a 2-digit number on the displays.
;
DISP8           PROC    NEAR                    ;display a 2-digit number from AL at position BX

                PUSH    AX
                CALL    LOOK                    ;look up 7-segment code
                POP     AX
                ROR     AL,4                    ;position digit
                CALL    LOOK
                RET

DISP8           ENDP
;
;Procedure that displays a 4-digit number on the displays.
;
DISP16          PROC    NEAR

                PUSH    AX
                MOV     AL,AH
                CALL    DISP8
                POP     AX
                CALL    DISP8
                RET

DISP16          ENDP
;
;Look up table procedure for BCD to 7-segment code conversion.
;
LOOK            PROC    NEAR                    ;look up 7-segment code and save it

                PUSH    BX
```

```
              MOV     BX,OFFSET TABLE-100H
              AND     AL,0FH                  ;mask AL
              XLAT    CS:TABLE                ;look up 7-segment code
              POP     BX
              MOV     [BX],AL                 ;display digit
              INC     BX                      ;address next digit
              RET

LOOK          ENDP
;
;BCD to 7-segment lookup table.
;
TABLE         DB      3FH,6,5BH,4FH           ;0, 1, 2, and 3
              DB      66H,6DH,7DH,7           ;4, 5, 6, and 7
              DB      7FH,6FH                 ;8, 9
;
;Software for a system RESET.
;
              ORG     008F0H                  ;get to reset location
RESET:
              MOV     DX,0FFA4H               ;address start for UCS
              MOV     AX,0FF82H               ;start address of FF800H, 2 waits
              OUT     DX,AL
              DB      0EAH                    ;JMP   0F000:F800        (FF800H)
              DW      0F800H,0F000H

CODE          ENDS
              END MAIN
```

The Initialization Operation.   The first thing that occurs at the reset location is that the starting address for the EPROM, which is connected to the $\overline{\text{UCS}}$ pin, is programmed as a 2K memory device. Next, a far JMP (using DB directives to the assembler) is forced, causing the program to continue at the very first location on the EPROM, which is at the start of the program listing. At the start of the program listing (ORG 100H), the ending address of the EPROM is next programmed as FFFFFH. The microprocessor clock is 8 MHz in this example because of the 16-MHz crystal oscillator module. At this frequency, the microprocessor allows approximately 295 ns for memory access. Because standard EPROM has an access time of 450 ns, two wait states (250 ns) are programmed for the $\overline{\text{UCS}}$ pin. This stretches the access time for the EPROM from 295 ns to 545 ns, which is more than adequate for the EPROM.

Because the SRAM (6116) is connected to one of the chip selection pins ($\overline{\text{LCS}}$), it must also be programmed. The SRAM is programmed with a starting address of 00000H and an ending address of 007FFH, or 2K bytes of memory space. Note that no wait states are required for the SRAM because a standard SRAM has an access time of 250 ns.

Once the two memory devices' starting and ending addresses are programmed, the I/O address range for the latch (U4) is programmed for the $\overline{\text{GCS7}}$ pin. As with the SRAM, no wait states are required. The I/O port address selected for the latch is 0000H. This is only one possible address out of the programmed address range (0000H–003FH) for this device.

The 80188 I/O port pins are next programmed so that port 1 is programmed as output pins, except for the $\overline{\text{GCS7}}$ pin, which is programmed as a peripheral pin. Port 2 is programmed so pins P2.0 through P2.4 are programmed as output pins, and P2.5 through P2.7 are programmed as input pins. The output pins from port 2 are used to select the displays and the keypad rows, and the input pins are used to read the keypad columns.

After programming the port 1 and 2 pins on the chip, the contents of all segment registers, except CS, are loaded with 0000H. This places the data type segments at memory locations 00000H–0FFFFH, of which only the first 2K (00000–007FFH) are actually used in the system. The stack pointer is loaded with an 800H, placing the stack area at the very top of the system SRAM. Note that the PUSH, POP, and CALL instructions never use the initial location on the stack. For example, a PUSH BX instruction would place the contents of register BH into location 7FFH and BL into location 7FEH before decrementing the stack point to 7FEH.

Before the system section of the software is started, the function of timer 2 is programmed to cause an interrupt at the rate of 1,000 times per second. The timer 2 interrupt is used in this system to maintain the time-of-day for the system, and to generate time delays for key switch de-bouncing. Timer 2 is programmed by clearing the count register to 0000H, placing a 2000_{10} into the compare register, and starting the timer so it generates an interrupt after every 2,000 clock inputs to timer 2. The clock to timer 2 in this example system is 8 MHz ÷ 4 or 2 MHz. Dividing 2 MHz by 2,000 causes an interrupt one time per millisecond.

The very last thing that occurs before the system software section of the program is entered is that the interrupts are unmasked and enabled. This is accomplished by addressing and programming the internal interrupt controller to enable the timer 2 interrupt. Next the STI instruction is executed to enable the interrupt structure of the system.

The System Software.   The first portion of the system software sets all of the timer start and stop times to FFFFH, which disables all appliance outlets until the start and stop times are programmed. Next the system software displays a ----- on the displays so the user knows that the system has just been powered. Also note that the time-of-day display is disabled by placing a zero into memory location 40FH.

Next, the main part of the system software is created by an infinite .WHILE–.ENDW loop, which is very short, and processes keypad data if typed. The main program loop calls the KEY procedure to determine if the * key has been pressed. The * key is used to enter into the programming mode for the system (CALL PROG). If the * key is not typed, the program returns and continues to execute the CALL KEY instruction. Notice that the main program has but one task: to check for the * (program) key.

System Procedures.   Because this is a substantial system, at least by textbook standards, our "system procedures" section is fairly long. The software was written so it could be followed from the top toward the bottom in as many cases as possible. The first procedure in the system procedure section of Example 15–1 is the PROG procedure, which is called from the main program loop. The purpose of the PROG procedure is to allow the time of day or any of the appliance outlets to be programmed with a start and stop time. The command sequence selected for this system is a *0 for programming the clock, a *1 for programming appliance plug 1, and a *2 for programming appliance plug 2. Notice how the .IF and .ELSEIF statements are used to form a high-level language style case statement that selects these three programming functions.

The GETT procedure is called to set the time of day or to program the start or stop time for an appliance plug. The GETT procedure returns with the time as it is typed on the keypad or with an FFFFH in the DX register. The time is displayed as it is entered using right-hand entry. The FFFFH is a cancel signal for an appliance timer and is obtained by typing a # symbol instead of a * for enter. For example, a *1# would program appliance plug 1 so it never activates. A *12000*2100* would program the start time for appliance plug 1 at 20:00 hours and a stop time of 21:00 hours. After returning from programming, the system displays the time of day.

Most of the other procedures are covered elsewhere in this textbook. For example, the KEY subroutine is covered in Chapter 11, Section 3. The only difference between the KEY procedure in Chapter 11 and the one used in this system is the way that the keys are scanned. Here the interrupt service procedure, which is explained later, changes the rows that are selected on the keyboard. The HLT instruction waits for the interrupt to occur so each row on the keypad is checked for a keystroke. If a return occurs from scan with a not-equal or not-zero condition, then a key on the keypad was detected.

The time delay program, used for de-bouncing the keypad, is also different in this system. Because the HLT instruction halts processing until an interrupt occurs, and because the interrupts occur once per millisecond, the HLT is used to cause a 1-ms time delay. In the DELAY procedure, HLT is executed 15 times to cause a 15-ms time delay for de-bouncing the keypad.

The interrupt service procedure (TIM2) is the basis for all of the background processing that occurs in this system. The TIM2 procedure is responsible for multiplexing the displays, for keeping an accurate time-of-day clock, and for tracking the appliance timers, which control power through the appliance plugs. Timer 2 is programmed to interrupt the microprocessor once per millisecond. Each time the interrupt occurs, it calls the TIM2 procedure, which saves all registers used by the procedure on the stack. After saving the registers, TIM2 addresses the video display memory and switches to the next display position by rotating a selection code and decrementing a pointer. After doing this, TIM2 selects a new display position and keypad row by outputting the selection code to port 2. It then sends out the 7-segment code from one of the display memory locations selected by the pointer to port 1 for display.

Once the display has been handled by TIM2, the real-time clock is updated by accessing a series of memory locations that act as counters. The first location divides the 1-ms interrupt by a factor of 100, the second by a factor of 10, and so forth, until reaching the last counter, which counts hours. The hour counter is accessed and incremented only once per hour, the minute counter is accessed and incremented only once per minute, and so forth.

At each minute or hour change, an additional procedure is called by TIM2 to possibly display the time. The time is displayed by the DISPLAY provided the contents of memory location 40FH is not a zero. If 40FH contains a zero, then no time is displayed so the system can display other data, such as that displayed during programming.

Finally, the last thing that TIM2 accomplishes in the background is a test of the two appliance timers to determine if it is time to enable or disable power to one of the electrical outlets for the appliances. This is handled by a procedure called TIMER, which calls the TEST procedure once for each appliance outlet. If the number stored in the starting counter for an outlet is FFFFH, the corresponding power is disabled for the outlet. If the start counter contains a time, it is then checked along with the ending counter to determine if power should be applied to the outlet for the appliance.

This is by all means not a complete accounting of this software, but enough so that you can carefully go through it to learn how this system functions. You should study each procedure to learn how it functions and how it affects the operation of the system.

15–2   USING THE 80C188XL TO CONTROL TRAFFFIC LIGHTS

Another simple application that applies to a microcontroller such as 80188 is a simple traffic light controller. The unit devised in this section uses some of the ideas learned in the last, such as setting the time, but it also uses a timer to control the duration and direction of traffic lights at a simple intersection. Another difference from Section 15–1 is that no keypad is used for setting the times. Instead of a keypad, a few push-buttons are used to enter times and commands to the controller. Another additional feature is a trip plate so traffic can change the light at certain times of the day. Also chosen for this application is a 2-line by 16-character LCD display.

Hardware

Figure 15–2 illustrates the complete schematic of the traffic light controller, which features the 80C188EB microprocessor at its heart. Notice the inclusion of a 2-line by 16-character LCD display (Optrex DMC-16207). The system requires the following components:

- Six solid-state relays for switching 120 VAC to the traffic lamps
- Two push-buttons for programming
- 2-line by 16-character LCD display for displaying time and programming information
- 2732 EPROM for program storage
- 6116 SRAM for parameter storage and for the stack
- Trip plate with controller circuit
- 80C188EB microprocessor

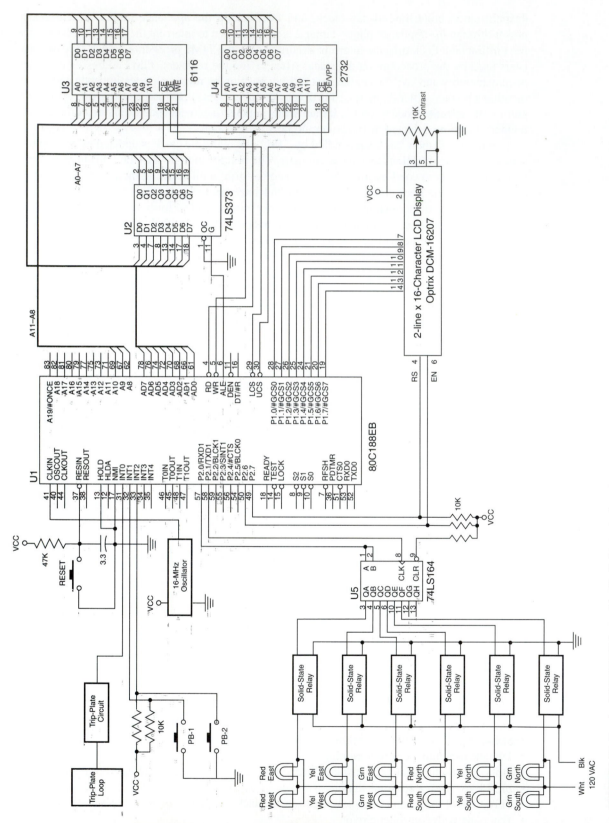

FIGURE 15–2 An 80C188EB-based traffic light control system.

FIGURE 15–3 The intersection used with the traffic light controller illustrated in Figure 15–2.

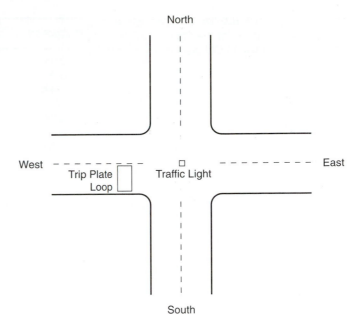

The system assumes that there is an intersection to be controlled that is north-south and east-west, with a trip plate located on the western leg of the intersection. Figure 15–3 illustrates the configuration of the intersection for this system.

The trip plate is actually a loop of wire that is attached to the surface of, or below the surface of, the street. The loop of wire is part of an oscillator tank circuit that detects any metal object placed over it. The metal in a car or truck is sufficient to change the resonant frequency of the tank circuit, which is detected to trip the light. Note that a bicycle or small motorcycle may not contain enough metal to trip the light. Although the trip-plate detector circuit is not shown, it is basically a metal detector that generates a TTL-compatible logic 1 level signal when any large metal object is placed near it. In areas of heavy bicycle traffic, the system could use an ultrasonic or laser-style detector to sense traffic. This system does not contain pedestrian push-buttons that are found at some intersections, but they could easily be included in the system.

The system provides commands and data (ASCII) to the 2-line by 16-character display through port 1 from the 80C188EB microprocessor. Two port 2 pins are used to provide the RS (register select) and EN (enable) control signals from the microprocessor. The display also has a R/$\overline{W}$ connection that has been placed at a logic 0 level in this example, because the display is never read by the software. When the RS input is a logic 0, the display accepts an instruction and when the RS input is a logic 1, the display accepts ASCII data. The EN input clocks the data or instruction into the display. The EN signal is active high and must be pulsed high to enter an instruction or ASCII data into the display. Table 15–1 lists the instructions available to the Optrex DMC-16207 LCD display.

The display is cleared at the start of system operation and programmed (using the 38H command) to enable the display. The cursor is then homed and data can be sent to the display. Note that the display functions so that after each ASCII character is sent, the cursor increments to the next position. The cursor can be turned off with the 0CH command at any time, but when the cursor is moved or new data is sent to the display the cursor reappears.

The signal from the trip-plate and the push-button switches (PB-1 and PB-2) are connected to the INT0, INT1, and INT2 input pins. In this system these pins are not used to cause an interrupt, but are used as additional input pins to the microprocessor. These pins are read by inputting the contents of the interrupt request register, which reflects the logic levels on these pins.

TABLE 15–1   The instructions for the Optrex DMC-1620 LCD display

Instruction	Function
01H	Clear display
02H	Home cursor
06H	Increment cursor
0CH	Display on, cursor off, blink off
38H	Set 8-bit data, 2-line display, 5×7 font
80H	Move to start of line 1
C0H	Move to start of line 2

The solid-state relays, which control the lamps in the traffic light, are connected to six of the pins on port 2 of the microprocessor. Port 2 is used to control the signal lights.

Timer 2, within the microprocessor, is used in this system for providing the time of day. It is also used to time events, such as length of the traffic light sequence, and the effect of a signal from the trip plate.

Software

Example 15–2 illustrates the software for this system. Again, as in Example 15–1, the system software is divided into three main parts: initialization, main program, and procedures. This system is placed on a larger EPROM (4K $\times$ 8) so there is a slight difference in the way that the reset origin appears. The reset address for this system is at location FF0H instead of 7F0H because the EPROM is larger.

EXAMPLE 15–2

```
.186                                      ;switch to the 80186/80188 instruction set
CODE           SEGMENT 'code'
               ASSUME  CS:CODE
;
;Program for the 80188EB microprocessor trainer.
;YOU MUST USE MASM 6.11 to assemble it.
;command line syntax = ML /AT XXXXXXXX.ASM
;
;This first part programs the microprocessor for memory at
;00000H-007FFH (SRAM), FF000H-FFFFFH (EPROM).
;
               ORG     0100H                ;set origin to start of EPROM
MAIN:
               MOV     DX,0FFA6H            ;address stop for UCS
               MOV     AX,000EH             ;set upper address as FFFFFH
               OUT     DX,AL

               MOV     DX,0FFA0H            ;address start for LCS
               MOV     AX,0000H             ;set lower address at 00000H, no waits
               OUT     DX,AL

               MOV     DX,0FFA2H            ;address stop for LCS
               MOV     AX,008AH             ;set lower address at 00800H
               OUT     DX,AL

               MOV     DX,0FF54H            ;address port 1 control
               MOV     AX,0000H             ;P1.0-P1.7 are outputs
               OUT     DX,AL

               MOV     DX,0FF5CH            ;address port 2 control
               MOV     AX,0H                ;P2.0-P2.7 are all available
```

```
        OUT     DX,AL

        MOV     DX,0FF58H               ;address port 2 direction
        MOV     AX,0000H                ;set P2.0-P2.7 as OUT
        OUT     DX,AL

        MOV     AX,0                    ;address segment 0000H
        MOV     DS,AX
        MOV     ES,AX
        MOV     SS,AX
        MOV     SP,0800H                ;set up stack pointer

        MOV     AL,11H                  ;blink pattern
        MOV     DS:[400H],AL            ;save image
        MOV     AL,0                    ;clear port 2
        PLAY
        MOV     BYTE PTR DS:[401H],1 ;enable blink

        MOV     WORD PTR DS:[4EH],0F800H    ;save interrupt vector for timer 2
        MOV     AX,OFFSET TIM2-100H         ;note the bias of -100H (see text)
        MOV     DS:[4CH],AX

        MOV     DX,0FF40H               ;address timer 2 count
        MOV     AX,0                    ;clear counter to 0000H
        OUT     DX,AL

        MOV     DX,0FF42H               ;address timer 2 compare
        MOV     AX,2000                 ;set to 2000 (for 1 ms count)
        OUT     DX,AL

        MOV     DX,0FF46H               ;address timer 2 control
        MOV     AX,0E001H               ;set for interrupt and continuous
        OUT     DX,AL

        MOV     DX,0FF12H               ;address interrupt control for timers
        MOV     AX,0                    ;enable timer interrupt
        OUT     DX,AL

        MOV     DX,0FF08H               ;address interrupt mask register
        MOV     AX,00FCH                ;disable all except for timer
        OUT     DX,AL
        STI                             ;enable interrupts
;
;;;;;;;;;;;;;;;;;;;;;;;;;;;;;;;;;;;;;;;;;;;;;;;;;;;;;;;;;;;;;;;;;;;;;;;;;;;;;;;;;;
;                       SYSTEM SOFTWARE HERE                                    ;
;;;;;;;;;;;;;;;;;;;;;;;;;;;;;;;;;;;;;;;;;;;;;;;;;;;;;;;;;;;;;;;;;;;;;;;;;;;;;;;;;;
PLAY    MACRO                           ;send AL to lights
        PUSH    CX
        PUSH    DX
        MOV     DX,0FF5E                 ;address port 2
        MOV     AH,AL
        MOV     CX,8
        .REPEAT
            ROL     AH,1
            MOV     AL,AH
            AND     AL,1                ;set shift input data
            OUT     DX,AL
            OR      AL,2                ;send clock
            OUT     DX,AL
            AND     AL,1                ;send clear
            OUT     DX,AL
        .UNTILCXZ
        ENDM
        MOV     BYTE PTR DS:[40FH],0 ;disable time display
        PUSH    DS
        MOV     AX,CS
```

```
        MOV     DS,AX
        MOV     SI,OFFSET MES1-100H  ;address MES1
        CALL    STRING               ;display initial message
        POP     DS
        CALL    PROG                 ;program traffic light controller
        .WHILE 1
                .IF BYTE PTR DS:[401H],0   ;if blink is off
                        HLT
                        .IF BYTE PTR DS:[40AH] == 0 && BYTE PTR DS:[40BH] == 0
                                MOV     AX,DS:[402H]   ;get timer
                                ADD     AL,1
                                DAA
                                .IF AL == 60H          ;increment minutes
                                        MOV     AL,1
                                        XCHG    AH,AL
                                        ADD     AL,1
                                        DAA
                                        XCHG    AH,AL
                                .ENDIF
                                MOV     DS:[402H],AX   ;save timer
                                .IF BYTE PTR DS:[404H] == 0
                                        .IF AX == DS:[412H]
                                                MOV     AL,11H          ;EW = RED, NS = YEL
                                                MOV     DS:[400H],AL
                                                PLAY
                                        .ENDIF
                                        .IF AX > DS:[412H]
                                                MOV     WORD PTR DS:[402H],0
                                                MOV     BYTE PTR DS:[400H],0CH
                                                MOV     AL,0CH          ;EW=GRN, NS = RED
                                                PLAY
                                                MOV     BYTE PTR DS:[404H],1

                                        .ENDIF
                                .ENDIF
                                .ELSEIF BYTE PTR DS:[404H] == 1
                                        MOV     DX,0FF0EH       ;test trip plate
                                        IN      AL,DX
                                        SHR     AL,5            ;trip to carry
                                        .IF CARRY?
                                                MOV     AX,DS:[406H]
                                                ADD     AL,1
                                                DAA
                                                .IF AL == 60H
                                                        MOV     AL,0
                                                        XCHG    AH,AL
                                                        ADD     AL,1
                                                        DAA
                                                        XCHG    AH,AL
                                                .ENDIF
                                                MOV     DS:[406H],AX
                                                .IF AX >= DS:[418H]
                                                        .IF AX < DS:[41AH]
                                                                MOV     AX:[402H]
                                                                SUB     AL,1
                                                                DAS
                                                                .IF     AL == 65H
                                                                        MOV     AL,0
                                                                        XCHG    AH,AL
                                                                        SUB     1,AL
                                                                        DAS
                                                                        XCHG    AH,AL
                                                                .ENDIF
                                                        .ENDIF
                                                .ENDIF
                                        .IF AX == DS:[410H]
```

```
                                            MOV     AL,0AH          ;EW = YEL, NS = RED
                                            MOV     DS:[400H],AL
                                            PLAY
                                    .ENDIF
                                    .IF AX > DS:[410H]
                                            MOV     WORD PTR DS:[402H],0
                                            MOV     BYTE PTR DS:[400H],21H
                                            MOV     AL,21H          ;EW=RED, NS = GRN
                                            PLAY
                                            MOV     BYTE PTR DS:[404H],0
                                            MOV     WORD PTR DS:[406H],0 ;clear trip
                                    .ENDIF
                            .ENDIF
                    .ENDIF
            .ENDIF
        .ENDW
;
;;;;;;;;;;;;;;;;;;;;;;;;;;;;;;;;;;;;;;;;;;;;;;;;;;;;;;;;;;;;;;;;;;;;;;;;;;;;;;;;;;;;;;;;
;                             SYSTEM PROCEDURES                                        ;
;;;;;;;;;;;;;;;;;;;;;;;;;;;;;;;;;;;;;;;;;;;;;;;;;;;;;;;;;;;;;;;;;;;;;;;;;;;;;;;;;;;;;;;;
;
;Procedure that reads PB-1 and PB-2.
;
BUTTON          PROC    NEAR

        BUTTON1:                                ;wait for release
                CALL    SCAN
                JNZ     BUTTON1
                CALL    DELAY
                CALL    SCAN
                JNZ     BUTTON1
                MOV     CX,5000                 ;for 5-second time delay
        BUTTON2:                                ;wait for key
                HLT
                DEC     CX
                .IF CX != 0
                        CALL    SCAN
                        JZ      BUTTON2
                        CALL    DELAY
                        CALL    SCAN
                        JZ      BUTTON2
                        NOT     AL
                .ELSE
                        MOV     AL,0
                .ENDIF
                RET                             ;AL = 1 for PNB-1 and 2 for PB-2
                                                ;AL = 0 for time-out of 5 seconds
BUTTON          ENDP
;
;Key scanning procedure.
;
SCAN            PROC    NEAR USES DX

                MOV     DX,0FF0EH               ;address request register
                IN      AL,DX                   ;read keys
                SHR     AL,5
                OR      AL,0FCH
                CMP     AL,0FFH
                RET

SCAN            ENDP


;
;Keyboard time delay.
;
DELAY           PROC    NEAR USES CX
```

```
                MOV     CX,20
                .REPEAT
                        HLT
                .UNTILCXZ
                RET

DELAY           ENDP
;
;Procedure that programs the system.
;
PROG            PROC    NEAR USES DI

                MOV     BYTE PTR DS:[40FH],0 ;disable clock
                MOV     DI,40DH
                MOV     SI,OFFSET MES2-100H
                CALL    SETC                ;set clock
                MOV     DI,410H
                MOV     SI,OFFSET MES3-100H
                CALL    SETC                ;set NS red time
                MOV     DI,412H
                MOV     SI,OFFSET MES4-100H
                CALL    SETC                ;set EW red time
                MOV     DI,414H
                MOV     SI,OFFSET MES5-100H
                CALL    SETC                ;set blink start
                MOV     DI,416H
                MOV     SI,OFFSET MES6-100H
                CALL    SETC                ;set blink stop
                MOV     DI,418H
                MOV     SI,OFFSET MES7-100H
                CALL    SETC                ;set trip minimum time
                MOV     DI,41AH
                MOV     SI,OFFSET MES8-100H
                CALL    SETC                ;set trip maximum time
                CALL    CLEAR
                MOV     BYTE PTR DS:[40FH],0FFH     ;enable clock
                MOV     BX,DS:[40DH]
                CALL    DISPBX              ;display time
                RET

PROG            ENDP
;
;Set a time procedure.
;
SETC            PROC    NEAR

PUSH            DS
                MOV     AX,CS
                MOV     DS,AX
                CALL    STRING
                POP     DS
                CALL    GETT                ;get a time
                .IF AL == 0 && BX != -1
                        MOV     [DI],BX     ;save time
                .ENDIF
                RET

SETC            ENDP


;
;Procedure that reads a time from the PB-1 and PB-2.
;
;If nothing is typed for 5 seconds, a return with AL == 00 occurs.
;PB-1 advances the hours (minutes) time
;PB-2 advances the minutes (seconds) time
;
```

```
GETT            PROC    NEAR USES CL

                MOV     CL,1
                MOV     BX,0
                .WHILE 1
                CALL    BUTTON                  ;read PB-1 or PB-2
                .IF AL == 1                     ;if PB-1
                        MOV     CL,0
                        MOV     AL,BH           ;increment hours (minutes)
                        ADD     AL,1
                        DAA
                        MOV     BH,AL
                        .IF BH == 60H
                        MOV     BH,0
                        .ENDIF
                        CALL    DISPBX          ;display BX
                .ELSEIF AL == 2
                        MOV     CL,0
                        MOV     AL,BL           ;increment minute (seconds)
                        ADD     AL,1
                        DAA
                        MOV     BL,AL
                        .IF BL == 60H
                                MOV     BL,0
                        .ENDIF
                        CALL    DISPBX
                .ENDIF
                .BREAK .IF AL == 0
                .ENDW
                .IF CL == 1
                        MOV     BX,-1
                .ENDIF
                RET

GETT            ENDP
;
;Procedure that displays BX on line 2 as XX:XX.
;
DISPBX          PROC    USES DI SI BX

                MOV     DI,500H                 ;address location 500H
                MOV     AL,2
                STOSB
                MOV     AL,BH
                SHR     AL,4
                ADD     AL,30H                  ;make it ASCII
                STOSB
                MOV     AL,BH
                AND     AL,0FH
                ADD     AL,30H
                STOSB
                MOV     AL,':'
                STOSB
                MOV     AL,BL
                SHR     AL,4
                ADD     AL,30H
                STOSB
                MOV     AL,BL
                AND     AL,0FH
                ADD     AL,30H
                STOSB
                MOV     AX,0
                STOSW
                MOV     SI,500H
                CALL    STRING
                RET
```

```
DISPBX          ENDP
;
;Messages
;
MES1            DB      1,'Power Startup',0
                DB      2,'Please Program',0,0
MES2            DB      1,'Program Clock',0,0
MES3            DB      1,'Set NS Time',0,0
MES4            DB      1,'Set EW Time',0,0
MES5            DB      1,'Set Blink Start',0,0
MES6            DB      1,'Set Blink Stop',0,0
MES7            DB      1,'Set Trip Min',0,0
MES8            DB      1,'Set Trip Max',0,0
;
;Procedure that displays a message on the LCD display.
;
STRING          PROC    NEAR

                CALL    CLEAR                   ;clear screen
                .WHILE 1
                        LODSB
                        .BREAK .IF AL == 0
                        .IF AL == 1
                                CALL    SETLINE         ;select line
                                .WHILE 1
                                        LODSB
                                        .BREAK .IF AL == 0
                                        CALL    CHAR    ;display character
                                .ENDW
                        .ELSE
                                CALL    SETLINE
                                .WHILE 1
                                        LODSB
                                        .BREAK .IF AL == 0
                                        CALL    CHAR
                                .ENDW
                        .ENDIF
                .ENDW
                MOV     AL,0CH                  ;turn off cursor
                MOV     DX,0FF56H               ;address Port 1
                OUT     DX,AL
                MOV     AX,DS:[400H]            ;get image
                AND     AL,7FH                  ;clear RS for command
                MOV     DS:[400H],AL
                CALL    PULSE
                RET

STRING          ENDP


;
;Procedure that clears the display.
;
CLEAR           PROC    NEAR USES DX CX

                MOV     AL,01H                  ;clear display command
                MOV     DX,0FF56H               ;address port 1
                OUT     DX,AL
                MOV     AX,DS:[400H]            ;get image
                AND     AL,7FH                  ;clear RS for command
                MOV     DS:[400H],AL
                CALL    PULSE
                MOV     CX,15
                .REPEAT                         ;15 ms time delay
                        HLT
                .UNTILCXZ
                RET
```

```
CLEAR           ENDP
;
;Procedure moves the cursor to line 1 or line 2.
;
SETLINE         PROC    NEAR USES DX

                .IF AL == 1
                        MOV     AL,80H
                .ELSE
                        MOV     AL,0C0H
                .ENDIF
                MOV     DX,0FF56H               ;address port 1
                OUT     DX,AL
                MOV     AX,DS:[400H]            ;get image
                AND     AL,7FH                  ;clear RS for command
                MOV     DS:[400H],AL
                CALL    PULSE
                RET

SETLINE         ENDP
;
;Procedure that pulses the EN pin on the LCD display.
;
PULSE           PROC    NEAR USES DX

                MOV     AL,DS:[400H]            ;get image
                OR      AL,40H                  ;set EN bit
                MOV     AL,DS:[400H]
                MOV     DX,0FF5EH               ;address port 2
                OUT     DX,AL                   ;EN = 1
                NOP
                NOP
                AND     AL,0BFH                 ;clear EN
                MOV     DS:[400H],AL            ;save image
                OUT     DX,AL                   ;EN = 0
                RET

PULSE           ENDP
;
;Interrupt service procedure for TIMER 2 interrupt.  (once per millisecond)
;
TIM2            PROC    FAR USES DS AX BX DX

                MOV     AX,0                    ;address segment 0000H
                MOV     DS,AX
                MOV     BX,40AH                 ;address real-time clock
                MOV     AH,0                    ;load modulus 100
                CALL    UPDATE                  ;update divide by 100 counter (40AH)
                JNZ     TIM1B                   ;if no roll-over
                MOV     AH,10H                  ;load modulus of 10
                CALL    UPDATE                  ;update divide by 10 counter (1/10 sec) (40BH)
                JNZ     TIM2B                   ;if no roll-over
                MOV     AH,60H                  ;load modulus 60
                CALL    UPDATE                  ;update seconds counter (40CH)
                CALL    BLINK                   ;blink lights
                CALL    TRIP                    ;test trip-plate
                JNZ     TIM1B                   ;if no roll-over
                CALL    UPDATE                  ;update minutes counter (40DH)
                CALL    DISPLAY                 ;display time
                JNZ     TIM1B                   ;if no roll-over
                MOV     AH,24H                  ;load modulus 24
                CALL    UPDATE                  ;update hours counter (40EH)
                CALL    DISPLAY                 ;display time
TIM1B:
                MOV     DX,0FF02H               ;address end of interrupt register
```

```
                MOV     AX,8000H                ;specific end of timer interrupt
                OUT     DX,AL

                IRET

TIM2            ENDP
;
;Procedure that tests the trip plate.
;
TRIP            PROC    NEAR

                PUSHF
                .IF BYTE PTR DS:[401H] == 0                 ;if blink Off
                    .IF BYTE PTR DS:[404H] == 0             ;if east-west is RED
                        MOV     DX,0FF0EH                    ;read trip plate
                        IN      AL,DX
                        SHR     AL,5                         ;trip to carry
                        .IF CARRY?
                            MOV     AX,DS:[406H]    ;get trip counter
                            ADD     AL,1            ;increment it
                            DAA
                            .IF AL = 60H
                                MOV     AL,0
                                XCHG    AH,AL
                                ADD     AL,1
                                DAA
                                XCHG    AH,AL
                            .ENDIF
                            MOV     DS:[406H],AX
                            .IF AX = DS:[418H]              ;if trip time
                                MOV     AX,DS:[410H]
                                MOV     DS:[402H],AX        ;change light
                            .ENDIF
                        .ELSE
                            MOV     WORD PTR DS:[406H],0  ;clear trip counter
                        .ENDIF
                    .ENDIF
                .ENDIF
                POPF
                RET

TRIP            ENDP


;
;Procedure that updates the time-of-day clock.
;
UPDATE          PROC    NEAR

                MOV     AL,[BX]                 ;get counter
                INC     BX                      ;address next counter
                ADD     AL,1                    ;increment time
                DAA                             ;make it BCD
                MOV     [BX-1],AL               ;save new counter
                SUB     AL,AH                   ;check modulus
                JNZ     UPDATE1                 ;if no roll-over
                MOV     [BX-1],AL               ;clear counter
        UPDATE1:
                RET

UPDATE          ENDP
;
;Procedure that blinks the traffic lights one time per second.
;
BLINK           PROC    NEAR
                PUSHF
                .IF BYTE PTR DS:[401H] == 1
```

```
                        MOV     AL,DS:[400H]
                        AND     AL,3FH
                        MOV     DS:[400H],AL
                        .IF AL == 11H
                                MOV     AL,0AH
                        .ENDIF
                        .IF AL == 0AH
                                MOV     AL,11H
                        .ENDIF
                        OR      AL,DS:[400H]
                        MOV     DS:[400H],AL
                        PLAY
                        MOV     AX,DS:[40DH]              ;get time
                        .IF AX >= DS:[416H] && AX < DS:[414H]
                                MOV     BYTE PTR DS:[401H],0           ;blink off
                                MOV     AL,21H       ;EW = RED NS = GRN
                                MOV     DS:[400H],AL
                                PLAY
                                MOV     WORD PTR DS:[402H],0
                                MOV     BYTE PTR DS:[404H],1
                                MOV     WORD PTR DS:[406H],0           ;clear trip timer
                        .ENDIF
                .ELSE
                        MOV     AX,DS:[40DH]             ;get time
                        .IF AX >=DS:[414H] && AX == DS:[416H]
                                MOV     BYTE PTR DS:[401H],1           ;blink on
                                MOV     AL,11H
                                MOV     DS:[400H],AL
                                MOV     DX,0FF5EH
                                OUT     DX,AL
                        .ENDIF

                .ENDIF
                POPF
                RET

BLINK           ENDP


;
;Procedure that displays the time of day if 40FH <> 0.
;
DISPLAY         PROC    NEAR    USES BX        ;display time of day

                PUSHF
                .IF BYTE PTR DS:[40FH] != 0
                        MOV     BX,DS:[40DH] ;get time hh:mm
                        CALL    DISPBX         ;display BX
                .ENDIF
                POPF
                RET

DISPLAY         ENDP
;
;Software for a system RESET.
;
                ORG     010F0H                  ;get to reset location
RESET:
                MOV     DX,0FFA4H               ;address start for UCS
                MOV     AX,0FF02H               ;start address of FF000H, 2 waits
                OUT     DX,AL
                DB      0EAH                    ;JMP   0F000:F000     (FF000H)
                DW      0F000H,0F000H

CODE            ENDS
                END MAIN
```

Initialization Software.   The initialization software for this system is similar to the one illustrated in Section 15–1. The main difference in this system is that all of port 1 is used as an output port, while the system in Section 15–1 used port 1 for output data and to select a latch for enabling the appliance outlets.

After programming the upper and lower chip select pins for the EPROM and SRAM, the initialization section of the software programs both ports 1 and 2 as output ports. The timer, which is programmed next, is used for a real-time clock. On each timer interrupt, the microprocessor advances the time in the real-time clock and checks to see if it is time to blink the traffic lights or if the trip plate is active to change the direction of traffic flow.

System Software.   The system software is responsible for programming the times into the system and also for general traffic light control once the times have been programmed. The first section of the system software addresses and displays message number 1 (MES1) to indicate that the system is powered up and it is time to begin programming. Programming is handled by the PROG procedure, which is called before the traffic control loop is entered.

The PROG procedure contains a series of calls to the SETC procedure, which sets the time for one of the system timers. The SETC procedure also displays a message on the LCD display that identifies the time that is being programmed. Once all times have been programmed, the system software is reentered at the infinite .WHILE loop, which is responsible for controlling the traffic lights.

Notice that the main part of the program really has no function if the traffic lights are in the blink mode. Under non-blink operation, the first part of the program increments the word timer at location 402H, provided that one second of time has passed. Notice how the system clock is checked for zero in both the ÷ 100 counter and the ÷ 10 counter. This condition occurs once per second. After the timer at 402H is incremented, byte 404H is checked for a zero or a one condition. If it is a zero, the system has a red light on the east-west direction, and if it is a one, the system has a red light in the north-south direction.

The software for the red light in the east-west direction is shorted because it has to time the light only for that direction. The trip plate is detected in the interrupt service procedure for timer 2 whenever timer 2 increments the seconds counter. This is explained later in this section. The first event that occurs in this section of software is that the contents of the timer (402H) are tested against the contents of the time for the east-west red light in word counter 412H. If the time is equal, then the software changes to yellow in the north-south direction for one count (one second). On the next pass, the lights are switched so that east-west is now green and north-south is red. Also notice that when the light turns green, the contents of location 404H are changed to a one. This switch indicates the system is now red for north-south.

As we mentioned, the second part of the system software is active when the system has a red light in the north-south direction. Because the trip plate is now active to maintain a green in the east-west direction, it is tested for the maximum count so the light can be switched to red. If the trip plate is not active, the system uses the normal red time for the north-south direction in the counter stored at location 410H.

System Procedures.   The system procedures section contains procedures that read the pushbuttons used for input and also procedures that control what is displayed on the LCD display system. The timer 2 software in this section maintains the real time for the systems as well as the blink and trip-plate detection.

The BUTTON procedure tests for PB-1 and PB-2 during the initial programming phase of operation and is used only within the SETC procedure. A return from the BUTTON procedure occurs if PB-1 or PB-2 is pressed or if a time-out of 5 seconds occurs. The AL register indicates the condition to the calling software by returning a 01H for PB-1, a 02H for PB-2, or a 00H for a time-out of 5 seconds. Note that the BUTTON procedure uses the SCAN and DELAY procedures for de-bouncing the push-buttons.

The DELAY procedure uses the HLT instruction to wait for a system interrupt. Because the system interrupt for timer 2 occurs once per millisecond, the HLT instruction waits for 1 ms. The DELAY procedure uses a counter to wait for 20 HLT instructions to increment, which causes a time delay of between 19 and 20 ms. Note that the delay could be slightly longer than 19 ms if the software is called just before the timer 2 interrupt occurs. The time delay is 20 ms if the DELAY procedure is called just after the timer 2 interrupt has occurred. In either case, 19 or 20 ms is ample time to de-bounce a push-button switch.

The SCAN procedure is very short, but notice how it reads the interrupt request register (FF0EH) to read the states of PB-1 and PB-2. The shift is needed to properly place the INT1 (PB-1) and INT2 (PB-2) bits to the far right of the AL register. The OR instruction forces the other 6 bits of AL to logic 1 levels so the push-buttons can be tested with the CMP AL,0FFH instruction. If no buttons are pressed, the return occurs with a zero or equal condition. Otherwise, if a button is pressed, the return occurs with a not-zero or not-equal condition, which is tested in the BUTTON procedure.

The SETC procedure displays a message and calls the GETT procedure to enter a time for one of the programmable features of this system. Note that upon return from GETT, the time is stored only if AL equals zero and BX is not a –1.

The GETT procedure calls BUTTON until the 5-second time-out occurs. This time-out acts as an enter for the system. If a return from GETT occurs with AL = 01H, the hours counter (BH) increments. If the return occurs with AL = 02H, the minutes part of the counter (BL) is in-cremented. After incrementing either the minutes or hours part, the contents of BX are dis-played with the DISPBX procedure, which places the time on the second line of the display. To set a timer for 01:04, the PB-1 push-button would be pressed once, and the PB-2 push-button would be pressed four times. This would be followed by a 5-second pause to enter this time into the system.

The DISPBX procedure addresses memory location 500H in the system SRAM where the ASCII code for the contents of BX are stored for later display by the STRING procedure. Notice how the contents of BX are converted to ASCII code and how a colon is placed between the hours (BH) and minutes (BL) portion.

The STRING procedure displays the message addressed by DS:SI. The format for each message is a line number (1 or 2) for the LCD display, followed by the test, in ASCII code, for the line. The end of a line is signed by a single 00H, and the end of the message is signaled by a double 00H. Refer to the messages stored in the EPROM just above the STRING procedure.

The CLEAR, SETLINE, and PULSE procedures are used to transfer commands and data to the LCD display. The display functions as follows: (1) data is sent to the data pins via port 1 on the microprocessor, (2) the RS bit is placed at a logic 1 for commands and a logic 0 for ASCII data, and (3) the EN pin is pulsed to write the data from port 1 to the LCD display. The PULSE procedure produces a pulse on the EN pin to the display by setting and then clearing the port 1.6 pin, which is connected to the EN pin. The NOP instructions were added for a short time delay of about 2 μs in this system. This is required by the LCD display.

The TIM2 procedure is accessed each time that timer 2 causes an interrupt. This occurs once per millisecond in this system. The main function of this procedure is to increment the real-time clock in the system. It also calls BLINK to test for blink time, TRIP to test the trip plate, and DISPLAY to display the time of day if it is enabled. The time only displays if the contents of memory location 40FH are not a zero. This feature is used so the system can be programmed. After programming, the LCD display displays the time of day. The BLINK procedure compares the time of day with the start and stop blink times that were programmed through the PROG pro-cedure on powering the system. The TRIP procedure is active only when the east-west direction has a red light; this procedure is responsible only for timing the minimum trip time. If the min-imum trip time is 10 seconds, then 10 seconds must elapse before the traffic light in the east-west direction turns from red to green.

15–3   A DATA CONCENTRATOR

Our next example is part of a data acquisition system, which is designed to remotely monitor a few analog events and then transfer the findings across an RS-232C cable to a remote system for monitoring. This example allows some practice with the serial port located in the 80186/80188 microprocessors for transferring serial data. RS-232C has been chosen, but other serial bus standards could be used to transfer data over a twisted pair of wires for distances of a few miles.

Hardware

Before the software can be written for this system, the hardware must be developed. Our system contains two analog-to-digital converters for sensing analog events. This can be modified to include more converters if desired and can even include multiple-channel converters. The following major components have been chosen for the system:

- Two ADC-0804 analog-to-digital converters
- 2716 EPROM for program storage
- 62256 SRAM (32K × 8) for parameter storage and for the stack
- 80C188EB microprocessor

   In this system a pair of analog-to-digital converters are interfaced to the microprocessor in order to remotely sample two analog events. The task of the system is to collect data and store it in the 32K-byte SRAM. Samples are taken one time per second, so the capacity of the memory is about 16K seconds, which amounts to $4^1/_2$ hours of data storage before it must be sent to the host system through the serial interface. Although the system could be programmable from the host, the software presented here transmits data only when polled by the host. In a sophisticated system, the sampling rate could be programmable and algorithms could be performed on the data locally.

Software

This system uses a pair of ADC0804 converters to acquire analog signals and store them in the 32K-byte SRAM. The clock is provided to the converters by the use of timer 1. Port 1 is programmed to select the pair of converters, and port 2 provides the RTS (request to send) signal for the serial port. The other two pins of port 2 are used to poll the converters to determine if a conversion is complete. Example 15–3 illustrates the software for this system.

EXAMPLE 15–3

```
.186                                    ;switch to the 80186/80188 instruction set
CODE          SEGMENT 'code'
              ASSUME  CS:CODE
;
;Program for the 80188EB microprocessor trainer.
;YOU MUST USE MASM 6.11 to assemble it.
;command line syntax = ML /AT XXXXXXXX.ASM
;
;This first part programs the microprocessor for memory at
;00000H-07FFFH (SRAM), FF800H-FFFFFH (EPROM), and I/O at
;0000H-003FH for converter U6 and 0040H-007FH for converter U6.
;
              ORG     0100H               ;set origin to start of EPROM
MAIN:
              MOV     DX,0FFA6H           ;address stop for UCS
              MOV     AX,000EH            ;set upper address as FFFFFH
              OUT     DX,AL
```

```
        MOV     DX,0FFA0H               ;address start for L̄C̄S̄
        MOV     AX,0000H                ;set lower address at 00000H, no waits
        OUT     DX,AL

        MOV     DX,0FFA2H               ;address stop for L̄C̄S̄
        MOV     AX,080AH                ;set lower address at 08000H
        OUT     DX,AL

        MOV     DX,0FF80H               ;address U6 chip select pin
        MOV     AX,0000H                ;set start I/O address at 0000H, no waits
        OUT     DX,AL

        MOV     DX,0FF82H               ;address stop for U6 chip select
        MOV     AX,0048H                ;at I/O location 003FH
        OUT     DX,AL

        MOV     DX,0FF84H               ;address start for U7 chip select
        MOV     AX,0040H                ;at I/O address 0040H, no waits
        OUT     DX,AL

        MOV     DX,0FF86H               ;address stop for U7 chip select
        MOV     AX,0088H                ;at I/O address 007FH
        OUT     DX,AL

        MOV     DX,0FF54H               ;address port 1 control
        MOV     AX,0003H                ;P1.2-P1.7 are outputs
        OUT     DX,AL

        MOV     DX,0FF5CH               ;address port 2 control
        MOV     AX,0H                   ;P2.0-P2.7 are all available
        OUT     DX,AL

        MOV     DX,0FF58H               ;address port 2 direction
        MOV     AX,00C0H                ;set P2.0-P2.5 as OUT
        OUT     DX,AL                   ;and P2.6 and P2.7 as IN

        MOV     AX,0                    ;address segment 0000H
        MOV     DS,AX
        MOV     ES,AX

        MOV     DX,0FF30H               ;address timer 0 count
        MOV     AX,0                    ;clear count
        OUT     DX,AL

        MOV     DX,0FF32H               ;address timer 0 compare A register
        MOV     AX,2                    ;divide the input clock (2 MHz) by 2
        OUT     DX,AL

        MOV     DX,0FF36H               ;address timer 0 control register
        MOV     AX,0C001H               ;continuous, no interrupt
        OUT     DX,AL

        MOV     DX,0FF40H               ;address timer 2 count
        MOV     AX,0                    ;clear timer 2 counter
        OUT     DX,AL

        MOV     DX,0FF42H               ;address timer 2 compare A register
        MOV     AX,2000                 ;divide down to 1 ms
        OUT     DX,AL

        MOV     DX,0FF46H               ;address timer 2 control register
        MOV     AX,0C001H               ;program for continuous, no interrupt
        OUT     DX,AL

        MOV     DX,0FF38H               ;address timer 1 count register
        MOV     AX,0
        OUT     DX,AL
```

```
        MOV     DX,0FF3AH               ;address timer 1 compare A register
        MOV     AX,1000                 ;divide down to 1 sec
        OUT     DX,AL

        MOV     DX,0FF3EH               ;address timer 1 control register
        MOV     AX,0C009H               ;use timer 2 as input
        OUT     DX,AL

        MOV     DX,0FF60H               ;address serial port 0 Baud count
        MOV     AX,0                    ;clear Baud count
        OUT     DX,AL

        MOV     DX,0FF62H               ;address serial port 0 Baud compare
        MOV     AX,8067H                ;program for 9,600 Baud
        OUT     DX,AL

        MOV     DX,0FF64H               ;address serial port 0 control
        MOV     AX,0021H                ;mode 1, 8-data, no parity, CTS off
        OUT     DX,AL
;
;;;;;;;;;;;;;;;;;;;;;;;;;;;;;;;;;;;;;;;;;;;;;;;;;;;;;;;;;;;;;;;;;;;;;;;;;;;;;;;;;;;
;                            SYSTEM SOFTWARE HERE                                ;
;;;;;;;;;;;;;;;;;;;;;;;;;;;;;;;;;;;;;;;;;;;;;;;;;;;;;;;;;;;;;;;;;;;;;;;;;;;;;;;;;;;
;
        MOV     SI,0000H                ;address start of SRAM queue
        MOV     DI,0000H                ;address start of SRAM queue
        .WHILE 1
                .REPEAT                 ;wait 1 second
                        MOV     DX,0FF66H       ;address serial port status
                        IN      AL,DX
                        TEST    AX,40H          ;test RI bit
                        .IF !ZERO?              ;data waiting
                                MOV     DX,0FF68H
                                IN      AL,DX           ;read data
                                CMP     AL,10H          ;send 100 bytes
                                MOV     CX,100
                                MOV     AX,SI
                                ADD     AX,CX
                                AND     AX,7FFFH
                                .IF AX <= DI
                                REPEAT
                                        .REPEAT
                                                MOV     DX,0FF66H
                                                IN      AL,DX ;test TI bit
                                                TEST    AL,20H
                                        .UNTIL !ZERO?
                                        LODSB
                                        AND     SI.7FFFH
                                        MOV     DX,0FF6AH
                                        OUT     DX,AL
                                .UNTILCXZ
                        .ENDIF
                        MOV     DX,0FF3EH       ;address timer 1 control register
                        IN      AX,DX
                        TEST    AX,20H
                .UNTIL !ZERO?
                MOV     AX,0009H        ;clear MC bit
                OUT     DX,AL
                OUT     00H,AL          ;start U6 converter
                OUT     40H,AL          ;start U7 converter
                MOV     DX,0FF5AH       ;address port 2 pins
                .REPEAT                 ;wait for completion
                        IN      AL,DX
                        OR      AL,0FCH
                        CMP     AL,0FFH
                .UNTIL ZERO?
                IN      AL,00H          ;read U6
```

```
                STOSB
                IN      AL,40H          ;read U7
                STOSB
                AND     SI,7FFFH        ;keep address at 0000H-7FFFH
          .ENDW
;
;Software for a system RESET.
;
                ORG     008F0H                  ;get to reset location
RESET:
                MOV     DX,0FFA4H               ;address start for UCS
                MOV     AX,0FF02H               ;start address of FF000H, 2 waits
                OUT     DX,AL
                DB      0EAH                    ;JMP    F000:F800       (FF800H)
                DW      0F800H,0F000H

CODE            ENDS
                END MAIN
```

Initialization Software.   As with the prior examples in this chapter, the initialization software begins at the end of the programming listing at the RESET location. The starting address for the
EPROM is programmed on the upper chip select pin before the jump to the start of the EPROM.
Next, the addresses for the SRAM and the two analog-to-digital converters are programmed. In
this example, I/O port 0000H is used for converter U6 and port 0040H is used for converter U7.
This example uses timer 0 to provide a 1-MHz clock signal to the analog-to-digital converters by
dividing the timer 0 input clock by 2. Timer 2 is used to generate an internal 1,000-Hz clock
signal for timer 1. Timer 1 divides this by 1,000 to produce a 1-second timer delay that is tested
by software.

　　　The last circuit programmed is the serial port 0 that is used to interface to a remote system.
The mode 1 operation is selected so 8 data bits can be sent and received by serial port 0. Notice
from the schematic in Figure 15–4 that serial port 0 is interfaced to a 9-pin connector through the
HIN232 level converter. This circuit converted the TTL level signals from the microprocessor
into RS-232C level signals for connection to a serial (COM) port on a personal computer system.
You must use a null modem cable for this connection.

System Software.   This system program is different from the others listed in this chapter because it does not use a stack or any RAM except for data storage. This allows all 32K bytes of the
SRAM to store data read from the analog-to-digital converters. The SRAM is used for a 32K-
byte queue memory that is addressed by SI (exit pointer) and DI (entrance pointer). As mentioned in the hardware section, the 32K-byte queue can store about $4^1/_2$ hours' worth of data
from the analog-to-digital converters before data are lost.

　　　One time per second, the converters are started by writing. Next the interrupt request outputs are polled from port 2, where the program waits until both converters have completed the
conversion. Then the contents of both converters are stored in memory at the location addressed
by the DI register or entrance pointer. The address in DI is maintained in the range of 0000H
through 7FFFH by the use of an AND instruction. After the contents of both converters are
stored in memory, the program jumps to near the start of the program section.

　　　At the start of the program second there is a repeat-until loop that waits for timer 1 to count
out one second. Within this loop, serial port 0 is tested to see if a command has been received
from the external host or personal computer. If the number 10H command is received, the program proceeds to check if 100 bytes of data are stored in the queue. If not, the program continues
to wait for the one-second time-out before reading additional data from the converters. The host
system must look for this time-out and use it to indicate that less than 100 bytes of data are available for transfer. If at least 100 bytes are available for transfer, the software sends 100 bytes to
the host system before timing out the one-second sample interval.

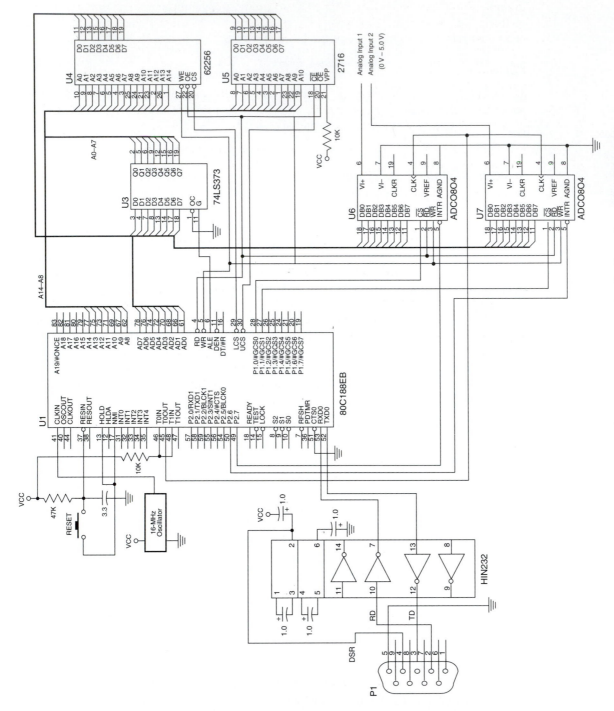

FIGURE 15–4 A simple data acquisition system that connects to a remote site through an RS-232V cable.

15–4 PROCESS CONTROL EXAMPLE

Many systems require a series of processes for control. This section presents a process control example with the techniques used for its design. Our example illustrates a controller that might be found in a dishwasher. Although we will not design the dishwasher itself, we will design the controller for a dishwasher.

Hardware

The dishwasher is a simple mechanical device with only a few working parts that lend themselves to control from the microprocessor. The dishwasher presented here contains the following parts that are controlled by a series of solid-state relays:

- Pump motor to force water out of the drain or through the inside of the dishwasher
- Diverter valve that forces the output of the pump motor out the drain
- Heating element to heat the water and dry the dishes
- Soap dispenser solenoid
- Rinsing agent dispenser solenoid
- Fan motor for air-drying the dishes
- Fill valve solenoid for filling the dishwasher with water

This system has three cycles: (1) normal wash, (2) sterile wash, and (3) extended wash. It also has a push-button to select heated or air-drying. The front panel of the dishwasher has five push-buttons to accomplish these functions. The hardware for the systems consists of an EPROM, five push-button switches, a door interlock switch, the 80C188EB microprocessor, and seven solid-state relays to control the motor, heating element, fan, and various solenoids. Figure 15–5 illustrates the complete system schematic. Note that this system does not have a RAM, so all software must be accomplished without using the stack or memory for data storage.

Once the hardware has been selected, the next task is to develop a timeline diagram for each of the cycles showing each event with reference to time. Figure 15–6 shows the timeline for each cycle and both the air and heated dry options. Notice how clearly each event appears in the timeline. This is an excellent way to develop any system that controls a process.

Each event is illustrated by a dark bar across the timeline chart. During the first minute of any of the charts illustrated, the fill solenoid is activated, which allows water into the dishwasher. Next (second minute) the pump motor is started to pump water around inside the dishwasher. Then during the last minute that the pump motor runs, the drain diverter solenoid is activated to allow the pump to drain the dishwasher. Other parts of each cycle and the events that occur are illustrated in a like manner in the timeline chart.

Software

The software for this system is fairly simple because there are only ports 1 and 2 to contend with. Port 1 contains the seven solid-state relays and port 2 contains the switches for mode selection and the interlock in case the door is opened. Example 15–4 lists the software for controlling the dishwasher.

EXAMPLE 15–4

```
.186                                    ;switch to the 80186/80188 instruction set
CODE        SEGMENT 'code'
            ASSUME  CS:CODE
;
;Program for the 80188EB microprocessor trainer.
```

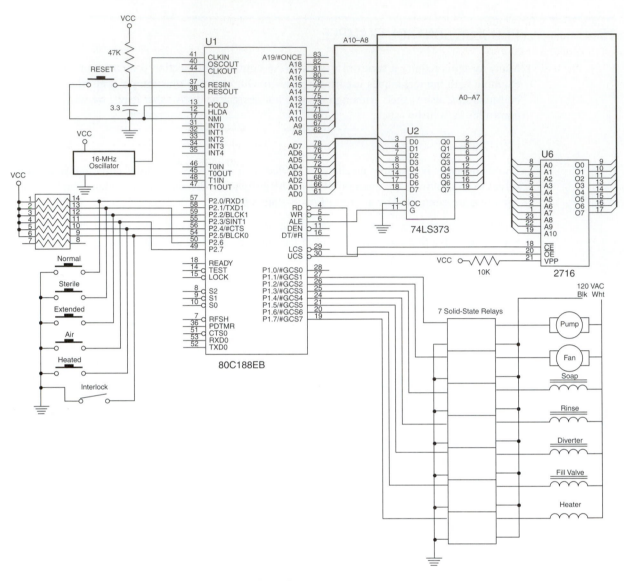

FIGURE 15–5   The circuit for controlling a dishwasher.

```
;YOU MUST USE MASM 6.11 to assemble it.
;command line syntax = ML /AT XXXXXXXX.ASM
;
;This first part programs the microprocessor for memory at
;00000H-07FFFH (SRAM), FF800H-FFFFFH (EPROM), and I/O at
;0000H-003FH for converter U6 and 0040H-007FH for converter U6.
;
                ORG     0100H                   ;set origin to start of EPROM
MAIN:
                MOV     DX,0FFA6H               ;address stop for UCS
                MOV     AX,000EH                ;set upper address as FFFFFH
                OUT     DX,AL

                MOV     DX,0FF54H               ;address port 1 control
                MOV     AX,0000H                ;P1.0-P1.7 are outputs
                OUT     DX,AL

                MOV     DX,0FF5CH               ;address port 2 control
                MOV     AX,0H                   ;P2.0-P2.7 are all available
```

FIGURE 15–6 Timelines for the dishwasher. Note that not all combinations of air and heated drying are shown.

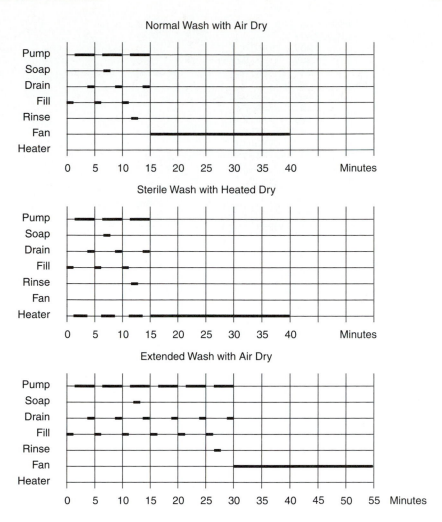

```
        OUT     DX,AL

        MOV     DX,0FF58H           ;address port 2 direction
        MOV     AX,00FFH            ;set P2.0-P2.7 as IN
        OUT     DX,AL

        MOV     AX,0                ;address segment 0000H
        MOV     DS,AX
        MOV     ES,AX
;
;;;;;;;;;;;;;;;;;;;;;;;;;;;;;;;;;;;;;;;;;;;;;;;;;;;;;;;;;;;;;;;;;;;;;;;;;;;;;;;;;;;;;
;                           SYSTEM SOFTWARE HERE                                   ;
;;;;;;;;;;;;;;;;;;;;;;;;;;;;;;;;;;;;;;;;;;;;;;;;;;;;;;;;;;;;;;;;;;;;;;;;;;;;;;;;;;;;;
;
TASK            MACRO   NUMBER      ;;program port 1 with task

        MOV     AH,NUMBER
        MOV     AL,AH
        MOV     DX,0FF56H
        OUT     DX,AL
        ENDM

INTER           MACRO               ;;test interlock macro

        MOV     DX,0FF5AH
        IN      AL,DX               ;;read port 2
```

```
        TEST    AL,20H                  ;;test P2.5
        .IF !ZERO?
                MOV     AL,0
                MOV     DX,0FF56H
                OUT     DX,AL           ;stop all functions
                .REPEAT
                MOV     DX,0FF5AH
                IN      AL,DX
                TEST    AL,20H
                .UNTIL ZERO?
                MOV     AL,AH
                MOV     DX,0FF56H
                OUT     DX,AL           ;restore function
        .ENDIF
        ENDM
SECOND  MACRO   NUMBER                  ;;second macro
        LOCAL   S1
        MOV     BX,NUMBER
    S1:
        MOV     CX,58000
        .REPEAT
                INTER                   ;;test interlock
                AAM                     ;;waste 2.375 µs
                AAM
                AAM
                AAM
                AAM
        .UNTILCXZ
        DEC     BX
        JNZ     S1
        ENDM


MINUTE  MACRO   NUMBER                  ;;minute macro
        LOCAL   M1
        MOV     DX,NUMBER
    M1:
        SECOND 1
        DEC     DX
        JNZ     M1
        ENDM


        MOV     AH,0
        .WHILE 1
                INTER                   ;check interlock
                MOV     DX,0FF5AH
                IN      AL,DX
                TEST    AL,1
                .IF ZERO?               ;do normal cycle
                        TASK    40H     ;fill machine
                        MINUTE 1        ;wait one minute
                        TASK    2       ;pump motor on
                        MINUTE 3        ;wait three minutes
                        TASK    22H     ;pump on and drain on
                        MINUTE 1        ;wait one minute
                        TASK    40H     ;fill machine
                        MINUTE 1
                        TASK    0AH     ;soap and pump on
                        SECOND 1
                        TASK    2       ;pump on
                        MINUTE 3
                        TASK    22H     ;pump on and drain on
                        MINUTE 1
                        TASK    40H     ;fill machine
                        MINUTE 1        ;wait one minute
                        TASK    12H     ;pump & rinse on
                        SEDOND 1
                        TASK    2       ;pump on
                        MINUTE 3        ;wait three minutes
```

```
                TASK    22H     ;pump on and drain on
                MINUTE  1       ;wait one minute
                MOV     DX,0FF5AH
                IN      AL,DX   ;test heat/air
                TEST    AL,80H
                .IF ZERO?
                        TASK    4       ;air on
                .ELSE
                        TASK    80H     ;heat on
                .ENDIF
                MINUTE  25
        .ENDIF
        TEST    AL,40H
        .IF ZERO?               ;do sterile cycle
                TASK    40H     ;fill machine
                MINUTE  1       ;wait one minute
                TASK    82H     ;pump & heat on
                MINUTE  3       ;wait three minutes
                TASK    22H     ;pump & drain on, heat off
                MINUTE  1       ;wait one minute
                TASK    40H     ;fill machine
                MINUTE  1
                TASK    0AH     ;soap and pump on
                SECOND  1
                TASK    82H     ;pump & heat
                MINUTE  3
                TASK    22H     ;pump on and drain on
                MINUTE  1
                TASK    40H     ;fill machine
                MINUTE  1       ;wait one minute
                TASK    12H     ;pump & rinse on
                SEDOND  1
                TASK    82H     ;pump & heat on
                MINUTE  3       ;wait three minutes
                TASK    22H     ;pump on and drain on
                MINUTE  1       ;wait one minute
                MOV     DX,0FF5AH
                IN      AL,DX   ;test heat/air
                TEST    AL,80H
                .IF ZERO?
                        TASK    4       ;air on
                .ELSE
                        TASK    80H     ;heat on
                .ENDIF
                MINUTE  25

        .ENDIF
        TEST    AL,4
        .IF ZERO?               ;do extended cycle
                TASK    40H     ;fill machine
                MINUTE  1       ;wait one minute
                TASK    2       ;pump motor on
                MINUTE  3       ;wait three minutes
                TASK    22H     ;pump on and drain on
                MINUTE  1       ;wait one minute
                TASK    40H     ;fill machine
                MINUTE  1       ;wait one minute
                TASK    2       ;pump motor on
                MINUTE  3       ;wait three minutes
                TASK    22H     ;pump on and drain on
                MINUTE  1       ;wait one minute
                TASK    40H     ;fill machine
                MINUTE  1
                TASK    0AH     ;soap and pump on
                SECOND  1
                TASK    2       ;pump on
                MINUTE  3
                TASK    22H     ;pump on and drain on
```

```
                    MINUTE  1
                    TASK    40H    ;fill machine
                    MINUTE 1       ;wait one minute
                    TASK    2      ;pump motor on
                    MINUTE 3       ;wait three minutes
                    TASK    22H    ;pump on and drain on
                    MINUTE 1       ;wait one minute
                    TASK    40H    ;fill machine
                    MINUTE 1       ;wait one minute
                    TASK    2      ;pump motor on
                    MINUTE 3       ;wait three minutes
                    TASK    22H    ;pump on and drain on
                    MINUTE 1       ;wait one minute
                    TASK    40H    ;fill machine
                    MINUTE 1       ;wait one minute
                    TASK    12H    ;pump & rinse on
                    SEDOND 1
                    TASK    2      ;pump on
                    MINUTE 3       ;wait three minutes
                    TASK    22H    ;pump on and drain on
                    MINUTE 1       ;wait one minute
                    MOV     DX,0FF5AH
                    IN      AL,DX  ;test heat/air
                    TEST    AL,80H
                    .IF ZERO?
                            TASK    4       ;air on
                    .ELSE
                            TASK    80H     ;heat on
                    .ENDIF
                    MINUTE  25

            .ENDIF
        .ENDW
;
;Software for a system RESET
;
        ORG     008F0H                  ;get to reset location
RESET:
        MOV     DX,0FFA4H               ;address start for UCS
        MOV     AX,0FF02H               ;start address of FF000H, 2 waits
        OUT     DX,AL
        DB      0EAH                    ;JMP   F000:F800      (FF800H)
        DW      0F800H,0F000H

CODE    ENDS
        END MAIN
```

Initialization Software.   This is the shortest initialization software in this chapter because only the upper chip select and the two I/O ports need to be programmed. Input/output port 1 is programmed as an output port for controlling the solid-state relays and devices attached to them; I/O port 2 is programmed as an input port to read the switches that select function and act as an interlock.

System Software.   The system software is fairly short because a few macro sequences are used to check the interlock and develop time delays. Another macro sequence is used to select and program a task at port 1 to control the solid-state relays.

The TASK macro sends the pattern associated with it to port 1 and also saves the pattern in the AH register in case the door is opened. When the door is opened, the INTER macro disables the system until the door is closed again.

The INTER macro is used in all time delays so if the door is opened it will be instantly detected and the dishwasher will be shut down. Notice that if the door is opened, a 00H is sent to port 1 to shut off all solid-state relays. The interlock switch is then tested and nothing happens until it is closed again. When the door is closed, the system takes the image for port 1 from the AH register and sends it to port 1 to restore operation at the point of interruption.

The SECOND macro is used to dispense soap and the rinsing agent; it is also used as a basis for the longer minute time delay. The SECOND time delay activates the INTER macro to check the interlock. Note that the interlock is tested many times per second and that the count in CX is used to accurately time a one-second time delay. The NUMBER variable associated with the SECOND macro is used to select delays of more than one second. This feature is used in the MINUTE time delay to obtain a one-minute time delay.

The system software begins at the .WHILE statement where the three function switches to select a normal, sterile, or extended wash are tested. Each wash type is activated by a series of TASK, SECOND, and MINUTE macro calls. These match the timing information developed in Figure 15–6, the timeline chart. Notice how the switches are again tested to determine if the drying cycle is air or heated.

15–5 SUMMARY

1. This chapter provides four design examples. Each was chosen to be as simple as possible to illustrate a programming or interface technique.
2. The first design example illustrates a system that controls appliances. It also illustrates the interface of a keypad and a series of five 7-segment LED displays. The hardware for the system makes maximum use of the ports available on the 80C188EB microprocessor and uses the internal timer to cause a periodic interrupt for keeping the time.
3. The second design example is similar to the first, but instead of using 7-segment LED displays, this example uses a 2-line by 16-character LCD display. In place of a keypad, a few push-button switches function to program this traffic light controller. The trip-plate circuitry is not illustrated, but you can locate a design example on the Internet by searching for "metal detector."
4. The third design example illustrates how a serial port can be used to connect the microprocessor-based system to a personal computer. This system collects analog data through a pair of analog-to-digital converters and stores them in a queue set up in a 32K-byte SRAM. The host system (personal computer) polls the microprocessor system through the serial interface to access and read 100 readings from the queue.
5. The final design example illustrates the design of a process control system. The dishwasher was chosen because it is a simple system that is easily understood. As with the other systems, the complete software and hardware for this device is fully developed in the chapter.

15–6 QUESTIONS AND PROBLEMS

1. Explain how the starting and ending memory addresses are programmed, in Example 15–1, for the $\overline{UCS}$ pin that selects the EPROM.
2. Refer to Example 15–1 and explain how timer 2 is programmed to develop a 1-ms interrupt.
3. Refer to Example 15–1 and explain how and where the interrupt vector for timer 2 is stored in the memory system.
4. Why is the RESET address in every example biased by a 100H?
5. Explain how a RESET origin of 8F0H causes the EPROM in Figure 15–1 to be located at physical address FFFF0H.
6. Detail the operation of the UPDATE procedure in Example 15–1.
7. Why must all addresses accessed as data on the EPROM be biased by 100H?

8. Describe how the DISP8 procedure functions in Example 15–1.

9. Example 15–2 programs the address for the EPROM at which memory locations?

10. What are the starting and ending addresses for the $\overline{\text{LCS}}$ pin programmed in Example 15–2?

11. What is the purpose of the HLT instruction as it is used in Example 15–2?

12. Describe how the KEY procedure in Example 15–2 returns the numbers 00H, 01H, or 02H for the push-buttons.

13. Describe how serial port 0 is programmed for operation in Example 15–3.

14. What are the purposes of the TI and RI bits in the status register of serial port 0?

15. Detail the use of SI and DI as pointers for the queue in Example 15–3.

16. Redraw the hardware schematic of Figure 15–1 using the 80C188XL version of the microprocessor.

17. In Example 15–4, explain how a time delay is obtained with the MINUTE macro.

18. In Example 15–4, explain how the INTER macro detects whether the interlock is open or closed.

19. For the system in Section 15–4, draw a timeline for a new cycle called pot scrubber. The difference between this new cycle and the extended cycle, which it replaces, is that the washing times are extended from 3 minutes each to 5 minutes.

20. Would it be useful to include an SRAM in the dishwasher system so procedures could be used in place of macros? Support your answer.

21. Using the Internet, locate data on LCD display devices and write a short report detailing what is available.

22. Using the Internet, locate sources of solid-state relays and list them.

23. Using the Internet, locate companies that build traffic light control systems and list them.

24. Using the Internet, download a program from Intel called APBuilder. After looking at this program, do you think it would be useful for programming the 80186/80188 family of microprocessors?

CHAPTER 16

The 80386EX Microprocessor

INTRODUCTION

In this chapter we discuss the 80386EX microprocessor, which Intel has dubbed the "embedded personal computer" because of its similarity to the personal computer. The 80386EX can address 64M of memory and contains many of the internal components found on the personal computer main board. In fact, the 80386EX contains just about everything in the personal computer except the video and disk interfaces. The 80386EX has timers, DMA controllers, interrupt controllers, serial ports, and parallel ports. It is also similar to the 80186/80188 family of embedded microprocessors because it, too, includes chip selection logic pins so usually no external decoder for memory or I/O is needed.

CHAPTER OBJECTIVES

Upon completion of this chapter, you will be able to:

1. Use the 80386EX in a system.
2. Write software to program the 80386EX.
3. Interface the 80386EX to memory and I/O components.
4. Create systems that use the 80386EX as the core element.

INTRODUCTION TO THE 80386EX MICROPROCESSOR

This section details the pin-out and function of each internal 80386EX component. You may want to compare this to the same information presented for the 80186/80188 to see the many similarities. As we mentioned in the introduction to this chapter, Intel has dubbed the 80386EX a personal computer on a chip. After reading the preceding 15 chapters, you should know the components found within the personal computer system and how programmable features function. Please recall the major components of the personal computer as you read this chapter.

Architecture

Figure 16–1 illustrates the pin-out diagram of the 80386EX microprocessor. The 80386EX microprocessor is available in a 132-pin FQFP package and in a 144-pin TQFP package. Only the pin-out of the 132-pin version appears in Figure 16–1. The 144-pin version has exactly the same pins, except the pin numbers differ. The 80386EX has a 26-bit address bus, allowing it to address 64M bytes of memory. It also contains a 16-bit data bus, but can be configured with an 8-bit data bus if the $\overline{BS8}$ pin is grounded. The 80386EX also contains three multiple-purpose I/O ports. Recall that the 80186/80188 (certain family members) contain two multiple-purpose I/O ports.

Electrical Characteristics.   Like any digital circuit, the 80386EX has a series of electrical specifications that you must know before it can be interfaced to memory and I/O devices. Because this microprocessor operates at any supply voltage from 2.7 V to 5.5 V, the logic levels are defined differently than for standard TTL parts. The logic 0 level is 0.4 V (maximum) for any power supply voltage. The logic 1 level is Vcc–0.8 V for a Vcc power supply voltage of 4.5 V to 5.5 V,

FIGURE 16–1   The pin-out of the 132-pin package of the 80386EX microprocessor.

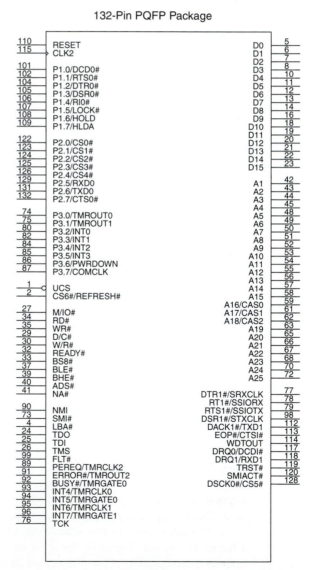

132-Pin PQFP Package

80386EX

and Vcc – 0.6 V for a VCC of 2.7 V to 3.6 V. If a 5.0-V power supply is used, this translates to a logic 0 voltage level of 0.4 V maximum and a logic 1 level of 4.2 V minimum. If the supply is 3.3 V, then the logic 0 voltage level is 0.4 V maximum and the logic 1 level is 2.6 V minimum. Recall that a standard TTL part has a logic 0 level of 0.4 V maximum and a logic 1 level of 2.4 V minimum. This means the 80386EX microprocessor is compatible with all standard TTL logic components at any of the microprocessor supply voltages of 3.3 V and 5.0 V.

The input voltage levels to the 80386EX microprocessor are also dependent on the power supply voltage. If the supply voltage is 5.0 V, the logic 0 input voltage is 1.5 V maximum and the logic 1 input voltage is 3.5 V minimum. These are TTL compatible. If the supply voltage is 3.3 V, the logic 0 input voltage is 1.0 V maximum and the logic 1 input voltage is 2.3 V minimum. All of these are TTL compatible except that the logic 1 input voltage yields a noise immunity of only 0.1 V with a 3.3 V power supply, which could cause noise problems in certain circumstances.

Output currents are much higher than for the 80186/80188 microprocessors. The 80386EX can supply 8 mA of current on all pins except port 3, which can supply 16 mA of current. This assumes that the power supply voltage is 5.0 V. These currents apply to either a logic 0 or a logic 1 output. If a 3.3-V power supply is used, the output currents are reduced to 4 mA from all pins, except for port 3 which is 8 mA. These currents are the same as those produced by a 74LS or 74HC TTL family member. With these standard families, the fan-out from the microprocessor is 10. If different family members are interfaced, the fan-out may be less. In any case, because 74LS, 74ALS, and 74HC families are most prevalent today, the maximum fan-out to most circuits interfaced to the 80386EX is 10.

The amount of power supply current required for the 80386EX microprocessor varies with operating frequency and voltages applied to Vcc. For example, a power supply voltage of 3.3 V with an operating frequency of 16 MHz requires 100 mA of current from the power supply. If the operating frequency is increased to 25 MHz and a 5.0 V power supply is in use, then the power supply current is 250 mA. These are low currents compared to some earlier microprocessors. For example, the 8-bit 8085 microprocessor required 270 mA of current when operated at 3.072 MHz. The 8085 microprocessor is a tiny microprocessor compared to the 80386EX microprocessor.

Internal Architecture. This is a complex microprocessor, although when compared to the 80C186EC version, the block diagram looks similar. Figure 16–2 illustrates the block diagram of the internal architecture of the 80386EX. What does not appear in the block diagram is the internal structure of the 80386EX, which is much more powerful than the 80186/80188 family. Details on the internal architecture of each control unit follows.

The clock generation circuit includes a divide-by-2 counter as do the 80186/80188, a programmable divider for generating a prescaled clock (PSCLK), a divide-by-2 counter for generating Baud-rate clock inputs and reset circuitry. The 80386EX microprocessor does not contain an internal oscillator, as do the 80186/80188. Instead, the CLK2 input provides system clock input. The CLK2 signal is divided by 2 internally to generate the operational clock signal for the microprocessor. If a 32-MHz clock is attached to CLK2, the microprocessor operates at 16 MHz.

The chip selection unit decodes memory and I/O addresses, which select memory and peripheral components interfaced to the chip select pins. The chip selection unit has eight separate chip selection pins that select memory or I/O. This is very similar to the 80186/80188, which have nine chip selection pins.

The interrupt control unit (ICU) contains two 8259A connected in cascade. The 8259As support up to eight external (INT0–INT7) inputs and up to eight internal interrupt request signals. The personal computer system also contains a pair of 8259A interrupt controllers.

The timer unit has the same basic functionality as the 8254 counter/timer described in Chapter 11. As with the 8254, the timer unit of the 80386EX contains three independent 16-bit counters, each capable of handling clock inputs with frequencies of up to 8 MHz. In addition to the timer unit, a watchdog timer is also provided in the 80386EX microprocessor. The watchdog

FIGURE 16–2 The internal block diagram of the 80386EX microprocessor.

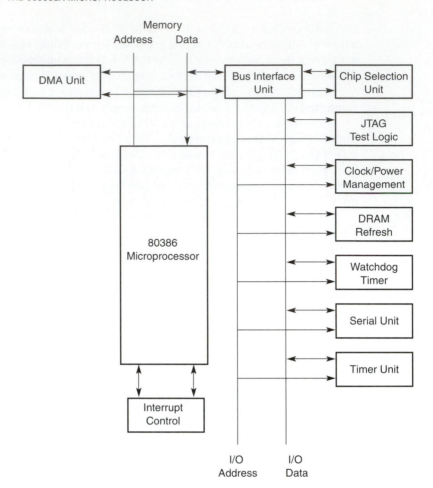

timer is a 32-bit down counter that decrements once per clock. The WDTOUT pin is driven high for 16 CLK2 cycles when the down counter reaches zero. The WDTOUT signal can be used to reset the chip, to request an interrupt, or to indicate to the user that a ready-hang situation has occurred.

The asynchronous serial unit contains a UART. The UART in the 80386EX is equivalent to the 16450, which is similar to the 16550 discussed in Chapter 11. The 80386EX microprocessor contains two full-duplex, asynchronous serial channels, just like the personal computer's COM1 and COM2 ports. Also included for serial data transfer is the synchronous serial unit that provides simultaneous, bi-directional synchronous communications.

The 80386EX microprocessor contains three 8-bit, general-purpose I/O ports. Unlike the 80186/80188, all port pins are bi-directional. Each port pin can be used for I/O provided the peripheral function for the pin is not used.

The DMA controller contains two channels. Although the DMA controller is similar to the 8237, it has enhancements that are explained later in this chapter. Each channel can transfer data between any combination of memory and I/O with any combination (8 or 16 bits) of data path widths.

The refresh control unit simplifies dynamic memory system design with its integrated address and clock counters. Integrating the refresh circuitry into the processor allows an external DRAM controller to use chip-selects, wait state logic, and status lines. The refresh control unit does not contain the address multiplexer, which must be added to interface DRAM. This refresh unit is similar to the refresh control unit within the 80186/80188.

The JTAG test unit is fully compliant with the IEEE 1149.1 standard and thus interfaces with five dedicated pins: $\overline{\text{TRST}}$, TCK, TMS, TDI, and TDO. It contains all of the logic required to generate clock and control signals for the boundary scan chain. (More detail is proved later in this chapter.)

System Timing

In order to interface to the microprocessor, the timing must be known. The timing for this microprocessor differs from the 80186/80188. The 80186/80188 use a basic bus cycle of four clocks; the 80386EX uses a basic bus cycle of two clocks. The basic read and write operation timing is illustrated in Figure 16–3. The timing illustrated assumes that no wait states are inserted and is referenced to the internal clock which is half the CLK2 input frequency.

In Figure 16–3, each bus cycle consists of two clocking periods or T-states. The purpose of T1 is to output an address to the system through the address bus and send the active low $\overline{\text{ADS}}$ signal to the system. Even though the read ($\overline{\text{RD}}$) and write ($\overline{\text{WR}}$) strobes start in T1, they do not cause a data transfer until T2. The purpose of T2 is to transfer data between the microprocessor and its memory and I/O systems. The first two clocks show a write cycle and the next two show a read cycle. The difference between the read and write cycles is the control signals and the point at where data appears on the data bus. During a write cycle data is presented to the data bus at the middle of T1 and throughout T2. The data is actually written to memory at the end of T2. During a read cycle, data appears on the data bus from the memory or I/O and must be valid at the end of T2 for the microprocessor to correctly read the data. Notice how the W/$\overline{\text{R}}$ signal identifies the entire bus cycle as a read or a write cycle. The $\overline{\text{READY}}$ signal must be asserted during T2 to show that no wait states are required.

Memory Access Time. The most important timing specification is the amount of memory access time allowed by the microprocessor. Refer to the timing illustrated in Figure 16–3 and note that the memory address appears at the start of each T1 state. Although the address appears to start at the beginning of T1, there is a delay time associated with the appearance of the address information. The address appears at address connections between 4 and 42 ns after the start of state T1. The exact time delays, for various frequency and voltage operation of the 80386EX, appear in Table 16–1 as the address delay time. Another important time when finding the memory

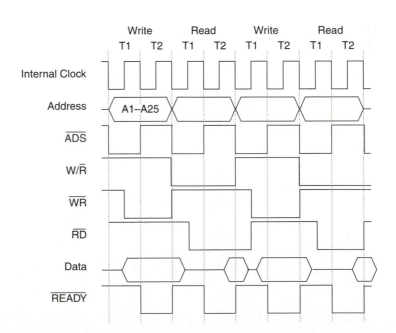

FIGURE 16–3 The timing diagram for read and write operations without any wait states.

TABLE 16–1   Timing for the 80386EX microprocessor

Microprocessor	Address Delay	Data Setup	Data Hold	Access Time
25 MHz, 5.0 V	29 ns	7 ns	8 ns	44 ns
20 MHz, 3.3 V	36 ns	9 ns	6 ns	55 ns
16 MHz, 3.3 V	46 ns	9 ns	6 ns	70 ns

access time is the data sample window located at the end of the T2 timing state. The data sample window begins at a point just before the end of T2 and ends just after the end of T2. The setup time varies between 7 and 9 ns and the hold time varies between 5 and 6 ns. Refer to Table 16–1 for these times versus frequency and voltage operation for the 80386EX.

The actual access time is determined by finding twice the clock period and then subtracting the address delay time and data setup time. For example, the access time for the 16-MHz version of the 80386EX microprocessor is

$$2 \times \frac{1}{16 \text{ MHz}} - 46 - 9 = 70 \text{ ns}$$

For operation at this frequency, only the fastest memory components will require zero wait states. If a 100-ns EPROM is interfaced to the microprocessor, it will not function unless additional time or wait states are inserted into the timing. Access times are adjusted by changing the clock frequency to lower values or by inserting wait states into the timing. For example, if the clock frequency is reduced to 8 MHz, the access time becomes 250 – 55 ns or 195 ns. If the clock frequency is reduced below 16 MHz, use the address delay and setup times listed in Table 16–1 for the 16-MHz version.

Wait States.   Wait states are additional clock periods inserted into the timing to delay the data sample point at the end of T2. These additional wait states (TW) are inserted by controlling the application of the $\overline{\text{READY}}$ signal. Wait states are additional T2 clocking periods that delay the read sample point at the end of the last T2 state in a bus cycle. Figure 16–4 shows the insertion

FIGURE 16–4 The timing diagram for read and write operations with wait states.

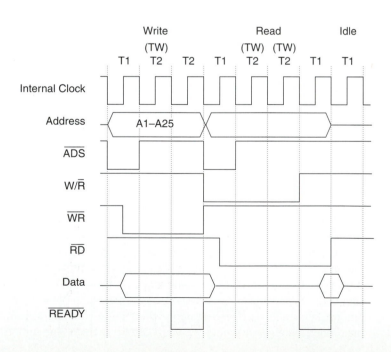

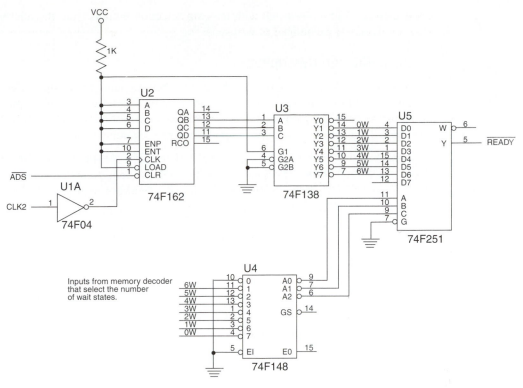

FIGURE 16–5 A wait state generator that is required if external address decoders are in use by the 80386EX microprocessor.

of wait states into the timing diagram. The application of the $\overline{\text{READY}}$ signal is applied in the final T2 period. In almost all instances the internal wait state generator, covered later in this chapter, allows 0–31 wait states to be inserted into the timing. If an external wait state generator is employed, the $\overline{\text{ADS}}$ signal is used to start the external wait state generator.

An external wait state generator can be clocked with the CLK2 signal developed by a crystal oscillator module. The end of a T-state corresponds to the rising edge of the CLK2 input to the microprocessor. Also, a T-state consists of two complete cycles from the CLK2 oscillator. Knowing this, and knowing that $\overline{\text{ADS}}$ signals the start of a bus cycle, a wait state generator can be designed. Figure 16–5 depicts a simple wait state generator that can be externally triggered by memory selection logic to insert wait states into the timing by controlling the $\overline{\text{READY}}$ line.

The counter (U2) is a simple 4-bit binary counter that clears whenever the $\overline{\text{ADS}}$ signal is a logic 0. Because this is a synchronous counter connected to operate from the negative edge of the CLK2 signal, the counter always clears just before the end of T1. During the next T2 (TW) state the counter counts to 1 and 2. The next T2 (TW) state causes it to count to 3 and 4. Because the decoder (U3) is connected to counter outputs QB, QC, and QD, this causes output Y1 of the decoder to become a logic 0 from the middle of T2 through the middle of the next T-state (T1 or T2). The decoder outputs are steered through an 8-to-1 line multiplexer whose input connection from the decoder is selected by the priority encoder (U4). If the priority encoder receives a chip select signal on its $\overline{7}$ input pin, the three outputs become 000. This selects input D0 on the 8-to-1 line multiplexer and the 0W output of the decoder. In this case the $\overline{\text{READY}}$ pin becomes a logic 0 from the middle of T2 to the middle of the following T1 state. Because the $\overline{\text{READY}}$ input is sampled at the end of T2, zero wait states are inserted. The setup and hold times for the $\overline{\text{READY}}$ pin are 19 ns and 4 ns, respectively, which allow this circuit to function without problem. As mentioned, though, it would be very rare to include this circuit because in most cases the memory and I/O are selected with the chip selection outputs from the microprocessor, which internally select the number of wait states.

System Bus Pin Descriptions

The following is a description of the control bus signals. These signals connect to external memory and I/O components in the system and represent only some of the pins on the 80386EX microprocessor. The peripheral pins, which represent most of the pins on the microprocessor, are explained later in this chapter when peripheral programming is discussed.

A1–A25	Address pins select any of the possible 64M bytes of memory or any of the 64K I/O devices connected to the system. Note that the I/O instructions present undefined information on address connections A16–A25.
$\overline{ADS}$	Address strobe becomes a logic 0 during T1 to indicate a valid bus cycle. Note that in the pin-out of Figure 16–1 an active low-level signal is denoted as ADS#.
$\overline{BHE}$, $\overline{BLE}$	Bus enable signals select either the upper (high) or lower (low) memory banks during a data transfer. Note that $\overline{BLE}$ functions as A0 during 8-bit bus operation.
$\overline{BS8}$	Bus size configures the 80386EX with an 8-bit data bus. The state of this pin can be changed during operation. For example, memory could be 16 bits wide and all I/O could be 8 bits wide.
D0–D15	Data bus connections.
$\overline{LBA}$	Local bus access is active to indicate that the microprocessor is providing the ready signal internally.
$\overline{LOCK}$	Generated by prefixing an instruction with the LOCK prefix. Prevents other bus masters from gaining access to the system.
M/$\overline{IO}$	Indicates a memory (1) or I/O (0) operation.
D/$\overline{C}$	Data/control informs the system of a data transfer (1) or control transfer (0) such as interrupt acknowledge.
W/$\overline{R}$	Indicates a write (1) or read (0) cycle.
$\overline{REFRESH}$	If refresh is a logic 0, the system is executing a refresh cycle for the DRAM.
$\overline{NA}$	Requests pipeline access.
$\overline{RD}$	Signals the system that the microprocessor is reading data.
$\overline{READY}$	Ready is an output when the microprocessor is addressing memory or I/O through its chip select pins, or an input that is sampled at the end of T2 to cause wait states.
$\overline{WR}$	Write signals the system that the microprocessor is writing data.

16–2   INTERFACING TO THE 80183EX BUS

This section details interfacing to the 80386EX system bus using standard memory, I/O, and decoding techniques. Usually standard components are interfaced through the peripheral control units in the 80386EX, but there are occasions where standard components still need to be connected to the microprocessor.

The Interface

Because the 80386EX can be structured as a 16-bit or an 8-bit data bus system, the application should dictate which bus structure is selected. Recall that the $\overline{BS8}$ selects 8-bit data bus operation when placed at a logic 0 level. Both the 8- and 16-bit interfaces are provided, but in most cases the peripheral devices would be 8 bits and the memory could be 8 or 16 bits.

The 8-Bit Interface. Figure 16–6 shows the 80386EX connected to a system that contains memory organized for an 8-bit data bus. Notice in this illustration that the $\overline{BLE}$ signal provides the A0 address bit during 8-bit data bus operation. The EPROM is selected using the $\overline{UCS}$ pin from the microprocessor. Because EPROM is relatively slow, wait states must be inserted for its proper operation. Assuming the access time for the EPROM is 250 ns, three or four wait states are required. If three waits are inserted, the access time is 70 ns (see Section 16–1 for the access time for a 16-MHz 80386EX) plus 62.5 ns times 3. This provides a total access time of 257.5 ns, which should be adequate for a 250-ns memory component. This is cutting it close, so if trouble is encountered (which is unlikely) four wait states could be inserted. Note that the 8255 interface has an access time of 250 ns, so it also requires three wait states. SRAM can be purchased with an access time of 100 ns, so it requires one wait state for proper operation. A single wait state allows an access time of 70 ns plus 62.5 ns or 132.5 ns.

The reset circuitry has been modified slightly because the 80386EX microprocessor has an active high reset input. The inverter is added to cause reset to the microprocessor and the 8255 to become a logic 1 at power up or whenever the reset push-button is pressed. A 74LS14 Schmidt trigger circuit is added to allow reliable operation and buffering for the push-button reset circuitry.

Because the 80386EX is a programmable device, we also list the initialization dialog for programming the microprocessor so it functions as illustrated in Figure 16–6. (Refer also to Chapter 10, Section 10–4 for a discussion of the chip selection unit within the 80386EX.)

Example 16–1 lists the software used to program the address of the 27512 (64K × 8 EPROM) so it resides at addresses 3FF0000H–3FFFFFFH, the top 64K-byte section of the memory system. It also lists the software that places U3 (a 32K × 8 SRAM) at locations 0000000H–0007FFFH and U4 at locations 0008000H–000FFFFH. The 8255 is decoded at I/O ports 0040H–0043H for this application. Note that because the number of I/O port address bits decoded by the chip selection unit include only A15–A6, the actual I/O port address range is location 0040H–007FH.

EXAMPLE 16–1

```
0017 BA F43E        MOV    DX,0F43EH       ;address UCS high address
001A B8 03FF        MOV    AX,03FFH        ;start address = 3FF0000H
001D EF             OUT    DX,AX

001E BA F43C        MOV    DX,0F43CH       ;address UCS low address
0021 B8 0104        MOV    AX,0104H        ;4 wait states
0024 EF             OUT    DX,AX           ;8-bit bus

0025 BA F43A        MOV    DX,0F43AH       ;address UCS high mask
0028 B8 001F        MOV    AX,1FH          ;address = 3FF0000H-3FFFFFFH
002B EF             OUT    DX,AX           ;for a 64K device

002C BA F438        MOV    DX,0F438H       ;address UCS low mask
002F B8 0001        MOV    AX,1            ;enable pin
0032 EF             OUT    DX,AX

0033 BA F406        MOV    DX,0F406H       ;address GCS0 high address
0036 B8 0000        MOV    AX,0H           ;start address = 0000000H
0039 EF             OUT    DX,AX           ;for U3

003A BA F404        MOV    DX,0F404H       ;address GCS0 low address
003D B8 0101        MOV    AX,301H         ;1 wait
0040 EF             OUT    DX,AX

0041 BA F402        MOV    DX,0F402H       ;address GCS0 high mask
0044 B8 000F        MOV    AX,0FH          ;address = 0000000H-0007FFFH
0047 EF             OUT    DX,AX           ;for a 32K device

0048 BA F400        MOV    DX,0F400H       ;address GCS0 low mask
004B B8 0001        MOV    AX,1            ;enable pin
```

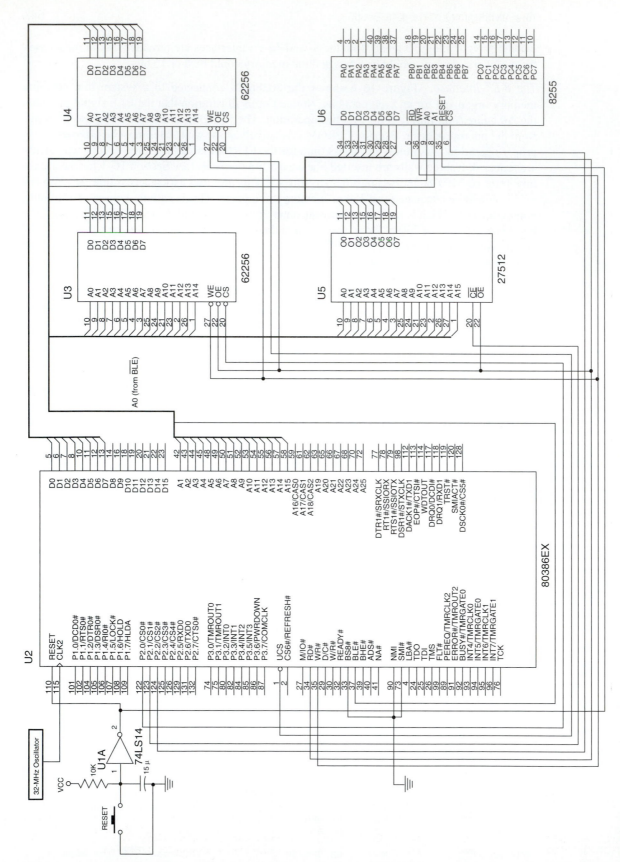

FIGURE 16–6 The 80386EX connected for operation with 8-bit peripherals.

```
004E EF                     OUT   DX,AX

004F BA F40E                MOV   DX,0F40EH    ;address GCS1 high address
0052 B8 0000                MOV   AX,0H        ;start address = 0008000H
0055 EF                     OUT   DX,AX        ;for U4

0056 BA F40C                MOV   DX,0F40CH    ;address GCS1 low address
0059 B8 0901                MOV   AX,901H      ;1 wait
005C EF                     OUT   DX,AX

005D BA F40A                MOV   DX,0F40AH    ;address GCS1 high mask
0060 B8 000F                MOV   AX,0FH       ;address = 0000000H-000FFFFH
0063 EF                     OUT   DX,AX

0064 BA F408                MOV   DX,0F408H    ;address GCS1 low mask
0067 B8 0001                MOV   AX,1         ;enable pin
0069 EF                     OUT   DX,AX

006A BA F416                MOV   DX,0F416H    ;address GCS2 high address
006D B8 0001                MOV   AX,1         ;start port = 0040H
0070 EF                     OUT   DX,AX        ;for U6

0071 BA F414                MOV   DX,0F414H    ;address GCS2 low address
0074 B8 0004                MOV   AX,004H      ;4 waits
0077 EF                     OUT   DX,AX

0078 BA F412                MOV   DX,0F412H    ;address GCS2 high mask
007B B8 0000                MOV   AX,0H        ;address = 0040H-007FH
007E EF                     OUT   DX,AX

007F BA F410                MOV   DX,0F410H    ;address GCS2 low mask
0082 B8 0001                MOV   AX,1         ;enable pin
0085 EF                     OUT   DX,AX
```

The 16-Bit Interface.   Figure 16–7 shows a system that contains 16-bit wide memory and 8-bit wide peripherals. This is an alternate to the connection in Figure 16–6 that would be used in systems that are memory intensive. The main advantage of the 8-bit bus connection is a slight simplification in the interconnection of system components.

Because the memory system is 16 bits wide in Figure 16–7, two EPROMs and two SRAM devices are required, unless a 16-bit wide EPROM is selected. As before, the EPROM and 8255 require four wait states and the SRAM requires one wait state.

The main difference between this system and the one illustrated in Figure 16–6 is that the $\overline{CS}$ pins of the two SRAM devices (U3 and U4) have been tied common to make them into a 16-bit wide memory device connected to the $\overline{GCS0}$ pin from the microprocessor. The two EPROM devices have their $\overline{CE}$ pins tied common to form a 16-bit wide EPROM for the system.

Example 16–2 lists the software required for this system. Compare this program with Example 16–1 to see the differences in programming the chip selection pins. The EPROM begins at memory address 3FE0000H and ends at location 3FFFFFFH. The SRAM appears at locations 0000000H–000FFFFH. As before, the 8255 is decoded at I/O ports 0040H–0043H.

EXAMPLE 16–2

```
0017 BA F43E        MOV   DX,0F43EH    ;address UCS high address
001A B8 03FE        MOV   AX,03FEH     ;start address = 3FE0000H
001D EF             OUT   DX,AX

001E BA F43C        MOV   DX,0F43CH    ;address UCS low address
0021 B8 0104        MOV   AX,0104H     ;4 wait states
0024 EF             OUT   DX,AX        ;8-bit bus

0025 BA F43A        MOV   DX,0F43AH    ;address UCS high mask
0028 B8 003F        MOV   AX,03FH      ;address = 3FE0000H-3FFFFFFH
```

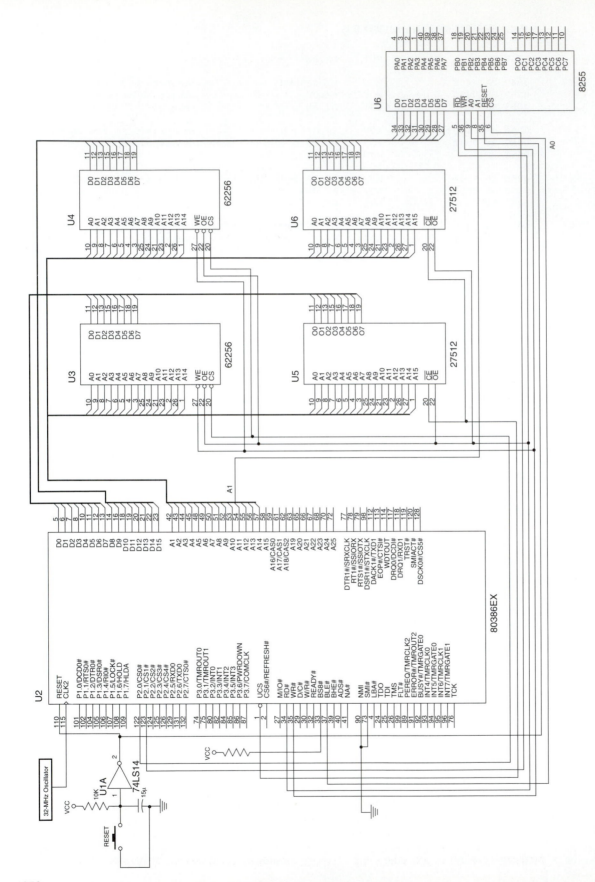

FIGURE 16—7 The 80386EX connected for operation with 8-bit peripherals and 16-bit memory.

```
002B EF                       OUT     DX,AX              ;for a 128K device

002C BA F438                  MOV     DX,0F438H          ;address UCS low mask
002F B8 0001                  MOV     AX,1               ;enable pin
0032 EF                       OUT     DX,AX

0033 BA F406                  MOV     DX,0F406H          ;address GCS0 high address
0036 B8 0000                  MOV     AX,0H              ;start address = 0000000H
0039 EF                       OUT     DX,AX              ;for U3

003A BA F404                  MOV     DX,0F404H          ;address GCS0 low address
003D B8 0101                  MOV     AX,301H            ;1 wait
0040 EF                       OUT     DX,AX

0041 BA F402                  MOV     DX,0F402H          ;address GCS0 high mask
0044 B8 001F                  MOV     AX,1FH             ;address = 0000000H-000FFFFH
0047 EF                       OUT     DX,AX              ;for a 64K device

0048 BA F400                  MOV     DX,0F400H          ;address GCS0 low mask
004B B8 0001                  MOV     AX,1               ;enable pin
004E EF                       OUT     DX,AX

006A BA F416                  MOV     DX,0F416H          ;address GCS1 high address
006D B8 0001                  MOV     AX,1               ;start port = 0040H
0070 EF                       OUT     DX,AX              ;for U6

0071 BA F414                  MOV     DX,0F414H          ;address GCS1 low address
0074 B8 0004                  MOV     AX,004H            ;4 waits
0077 EF                       OUT     DX,AX

0078 BA F412                  MOV     DX,0F412H          ;address GCS1 high mask
007B B8 0000                  MOV     AX,0H              ;address = 0040H-007FH
007E EF                       OUT     DX,AX

007F BA F410                  MOV     DX,0F410H          ;address GCS1 low mask
0082 B8 0001                  MOV     AX,1               ;enable pin
0085 EF                       OUT     DX,AX
```

The Peripheral Control Addresses

Unlike the 80186/80188, the 80386EX contains a more extensive set of peripheral control registers because of its more complex structure. The 80386EX operates in two different I/O address space modes (DOS and expanded) as selected by the address configuration register at I/O port location 0022H. Figure 16–8 shows the structure of the address configuration register.

Table 16–2 gives the I/O port assignments for the peripheral registers used in programming the 80386EX. Note that there are two sets of I/O addresses: The expanded addresses are en-

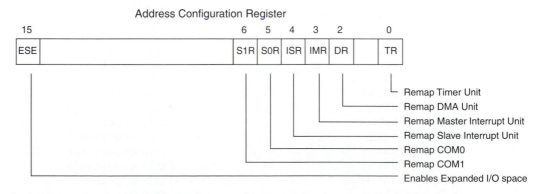

FIGURE 16–8   The structure of the 80386EX addresses' configuration register.

TABLE 16–2   The peripheral addresses for the 80386EX microprocessor

Expanded Address	PC/AT Address	Function	Reset Value
F000H	0000H	DMA0 Target Address0/1	??
F001H	0001H	DMA0 Count0/1	??
F002H	0002H	DMA1 Target Address0/1	??
F003H	0003H	DMA1 Count0/1	??
F008H	0008H	DMA Command1	00H
F009H	0009H	DMA Software Request	00H
F00AH	000AH	DMA Mask	04H
F00BH	000BH	DMA Mode1	00H
F00CH	000CH	DMA Clear Byte Pointer	None
F00DH	000DH	Clear DMA Controller	None
F00EH	000EH	Clear DMA Mask	None
F00FH	000FH	DMA Group Mask	03H
F010H		DMA0 Request0/1	00H
F011H		DMA0 Request2/3	00H
F012H		DMA1 Request0/1	00H
F013H		DMA1 Request2/3	00H
F018H		DMA Bus Size	F0H
F019H		DMA Chaining	00H
F01AH		DMA Command2	08H
F01BH		DMA Mode2	00H
F01CH		DMA Interrupt Enable	00H
F01DH		DMA Overflow Enable	0AH
F01EH		DMA Clear Transfer	None
F020H	0020H	8259A Master	??
F021H	0021H	8259A Master	??
F022H	0022H	Address Configuration	0000H
F040H	0040H	Timer 0	??
F041H	0041H	Timer 1	??
F042H	0042H	Timer 3	??
F043H	0043H	Timer Control	00H
F083H	0083H	DMA1 Target Address2	??
F085H		DMA1 Target Address3	??
F086H		DMA0 Target Address3	??
F087H		DMA0 Target Address2	??
F098H		DMA0 Count2	00H
F099H		DMA1 Count2	00H
F092H	0092H	Port92	00H
F0A0H	00A0H	8259A Slave	??
F0A1H	00A1H	8259A Slave	??
F400H		CS0 Address Low	0000H
F402H		CS0 Address High	0000H
F404H		CS0 Mask Low	0000H
F406H		CS0 Mask High	0000H
F408H		CS1 Address Low	0000H
F40AH		CS1 Address High	0000H
F40CH		CS1 Mask Low	0000H
F40EH		CS1 Mask High	0000H
F410H		CS2 Address Low	0000H
F412H		CS2 Address High	0000H
F414H		CS2 Mask Low	0000H
F416H		CS2 Mask High	0000H

TABLE 16–2   *(continued)*

Expanded Address	PC/AT Address	Function	Reset Value
F418H		CS3 Address Low	0000H
F41AH		CS3 Address High	0000H
F41CH		CS3 Mask Low	0000H
F41EH		CS3 Mask High	0000H
F420H		CS4 Address Low	0000H
F422H		CS4 Address High	0000H
F424H		CS4 Mask Low	0000H
F426H		CS4 Mask High	0000H
F428H		CS5 Address Low	0000H
F42AH		CS5 Address High	0000H
F42CH		CS5 Mask Low	0000H
F42EH		CS5 Mask High	0000H
F430H		CS6 Address Low	0000H
F432H		CS6 Address High	0000H
F434H		CS6 Mask Low	0000H
F436H		CS6 Mask High	0000H
F438H		UCS Address Low	FF6FH
F43AH		UCS Address Low	FFFFH
F43CH		UCS Mask Low	FFFFH
F43EH		UCS Mask High	FFFFH
F480H		Sync Transmit Buffer	0000H
F482H		Sync Receive Buffer	0000H
F484H		Sync Baud	00H
F486H		Sync Control1	C0H
F488H		Sync Control2	00H
F48AH		Sync Count Down	00H
F4A0H		Refresh Base Address	0000H
F4A2H		Refresh Clock Interval	0000H
F4A4H		Refresh Control	0000H
F4A6H		Refresh Address	00FFH
F4C0H		Watchdog Value Low	0000H
F4C2H		Watchdog Value High	FFFFH
F4C4H		Watchdog Counter Low	0000H
F4C6H		Watchdog Counter High	FFFFH
F4C8H		Watchdog Clear	None
F4CAH		Watchdog Status	00H
F4F8H	03F8H	COM1 Data Buffer	FFH
F4F9H	03F9H	COM1 DLH/IE	FFH
F4FAH	03FAH	COM1 Interrupt ID	01H
F4FBH	03FBH	COM1 Line Control	00H
F4FCH	03FCH	COM1 Modem Control	00H
F4FDH	03FDH	COM1 Line Status	60H
F4FEH	03FEH	COM1 Modem Status	?0H
F4FFH	03FFH	COM1 Scratch Pad	??
F800H		Power Control	00H
F804H		Clock Prescale	0000H
F820H		Port 1 Configuration	00H
F822H		Port 2 Configuration	00H
F824H		Port 3 Configuration	00H
F826H		Pin Configuration	00H
F830H		DMA Configuration	00H

(continued on next page)

TABLE 16–2 *(continued)*

Expanded Address	PC/AT Address	Function	Reset Value
F832H		Interrupt Configuration	00H
F834H		Timer Configuration	00H
F836H		Serial Configuration	00H
F860H		Port 1 Pin	??
F862H		Port 1 Latch	FFH
F864H		Port 1 Direction	FFH
F868H		Port 2 Pin	??
F86AH		Port 2 Latch	FFH
F86CH		Port 2 Direction	FFH
F870H		Port 3 Pin	??
F872H		Port 3 Latch	FFH
F874H		Port 3 Direction	FFH
F8F8H	02F8H	COM2 Data Buffer	FFH
F8F9H	02F9H	COM2 DLH/IE	FFH
F8FAH	02FAH	COM2 Interrupt ID	01H
F8FBH	02FBH	COM2 Line Control	00H
F8FCH	02FCH	COM2 Modem Control	00H
F8FDH	02FDH	COM2 Line Status	60H
F8FEH	02FEH	COM2 Modem Status	?0
F8FFH	02FFH	COM2 Scratch Pad	??

abled by the ESE bit in the address configuration register, and the PC/AT addresses are selected by default whenever the microprocessor is reset. In both cases, the location of the address configuration register is at I/O port 0022H. Input/output addresses can be byte, word, or double-word, depending on the function of the register that is programmed in the peripheral address space.

16–3   CONTROLLING THE 80386EX

Just the size of Table 16–2 is overwhelming, but much of its contents have already been discussed in previous chapters. One main difference is the configuration of the 80386EX, which is controlled by the device configuration registers located at I/O ports F820H through F836H.

Configuration of the 80386EX

Before the 80386EX is used for an application, it must be configured. Configuration requires that a number of registers be programmed to select the way that the microprocessor functions in the system. As in the 80186/80188, the port configuration registers select whether a port pin is used for I/O or as a peripheral signal. For example, a logic 0 in bit position 0 of the port 2 configuration register selects I/O operation for P2.0. If a logic 1 is placed in this bit of the configuration register, the P2.0 pin functions as the CS0# pin. Likewise all the pins of ports 1, 2, and 3 configuration registers are programmed in the same manner—a logic 1 selects peripheral and a logic 0 selects I/O operation.

The pin configuration register (F826H) is not as easy to understand, so refer to Figure 16–9 for the bit positions in this register. Notice that this register controls how some of the pins function on the microprocessor. Bit position 7 is not used at this time, but may come into use when the 80486 and Pentium are released in embedded form.

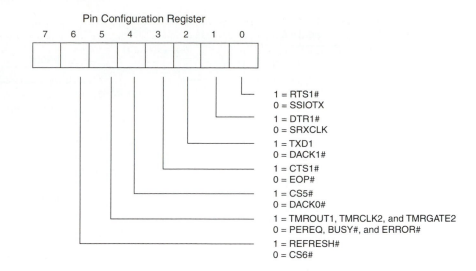

FIGURE 16–9 The pin configuration register at I/O address F826H.

The timer configuration register (F834H) and a diagram illustrating the timer unit appear in Figure 16–10. The switches shown in this illustration are opened or closed by programming bits in the timer configuration register and in the port 3 configuration register. The notation used in the illustration is TMRCFG.4 to indicate timer configuration register bit position 4.

Each timer can be clocked from the output of the internal prescaler or from the external timer clock input pin. Each timer can also present the count at an output pin, and each timer has a gate input that can be connected to an external gate input or to Vcc.

The interrupt unit also contains its own configuration register (F832H) that determines how the interrupt controller functions in the system. Figure 16–11 shows both the control register and the structure of the interrupt control unit. The 80386EX microprocessor contains a pair of 8259A interrupt controllers, as does the personal computer. Most of the bit positions in the interrupt configuration register determine how the inputs to the 8259As function.

If needed, the pair of 8259A interrupt controllers can be expanded by using address connections A16, A17, and A18 to cascade to external 8259As. This feature is controlled by bit 7 in the interrupt configuration register. Recall from Chapter 12 that the cascade lines contain the address of the slave unit during an interrupt acknowledge cycle. Since the master controller is located within the 80386EX, it controls cascading to the slaves, whether internal or external.

Some inputs to the master 8259A and the slave originate internally and some are connected to external pins. Timers 0, 1, and 2 connect to the master at IR0 and to the slave at IR2 and IR3, while the two serial ports have interrupt connections connected to the master at IR3 and IR4. IR3 and IR4 correspond to personal computer interrupts IRQ3 and IRQ4, which are normally connected to the two COM ports in a personal computer system. The synchronous serial unit shares IR1 on the slave with the INT6 interrupt input pin.

The DMA configuration register (F830H) controls the configuration of the two-channel DMA controller. The DMA configuration register does not control the operation of the bus arbiter and refresh control unit. They are presented here because the bus arbiter controls the HOLD and HLDA signals. Figure 16–12 shows the internal structure of the DMA controller and the structure of the DMA control register. As with other configurations, switches are used to show how bits in the control register select various features.

There are two asynchronous communications ports within the 80386EX microprocessor, both controlled by the SIO configuration register (F836H). Figures 16–13 and 16–14 show the structure of each communications port. Figure 16–14 also shows the structure of the control register used to configure both ports.

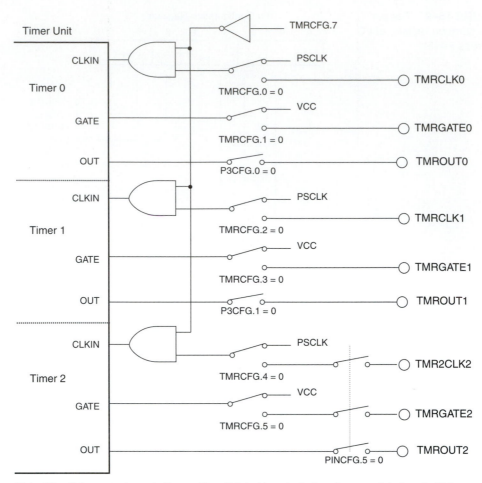

Note: All switches are shown in the position dictated by a logic 0 on the associated control bit.

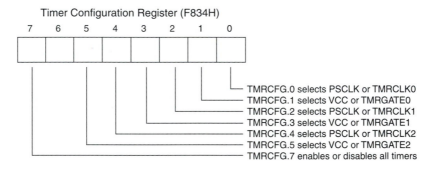

FIGURE 16–10   The internal structure and control register for the timer unit within the 80386EX.

The SIO configuration register also configures the Baud rate clock input to the synchronous serial unit. Figure 16-15 shows the connections to the synchronous serial unit and the effect of programming the SIO configuration register and the pin configuration register.

The final internal configuration register is port 92H. Port 92H and the pin configuration register program the core unit of the microprocessor. Figure 16–16 shows the internal structure of the core unit of the 80386EX microprocessor. Only two bits of port 92H are used to program

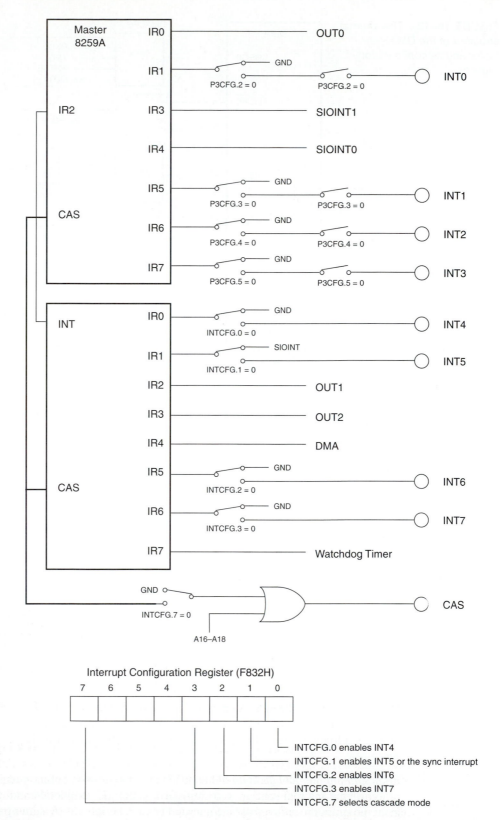

FIGURE 16–11 The structure of the interrupt controller and the interrupt configuration register.

FIGURE 16–12   The internal structure of the DMA controller and its configuration register.

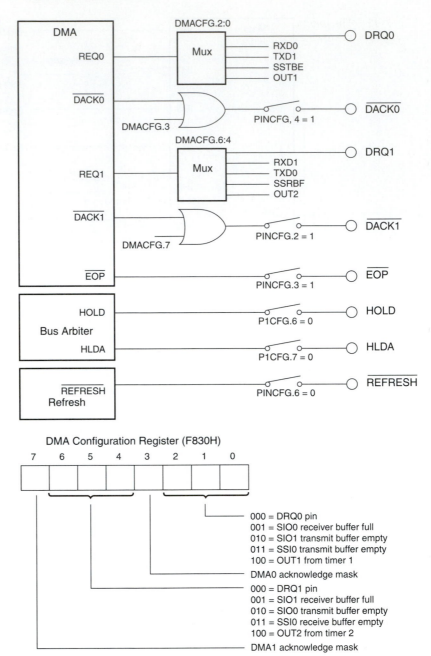

the core unit. Bit position 0 resets the microprocessor when placed at a logic 1 level. The reset will not reset the external peripherals in the system. Bit position 1 of port 92H programs the operation of the A20 address bit. If bit position 1 is a logic 0, A20 is a logic 0. If bit position 1 is a logic 1, the address presented at A20 is the one normally generated by the microprocessor.

Programming the Configuration of the 80386EX.   Obviously, before you program the configuration of the microprocessor, you must first select all peripherals and decide which signals from the microprocessor will be connected to each peripheral. Assuming this has been accomplished, the first step in programming the configuration is to program the pin configuration register.

FIGURE 16–13 The internal structure of serial I/O unit 0.

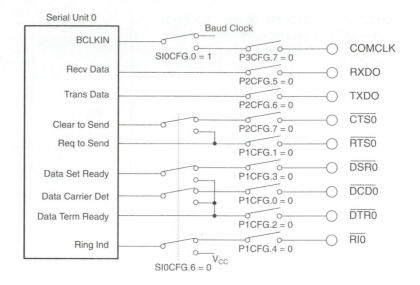

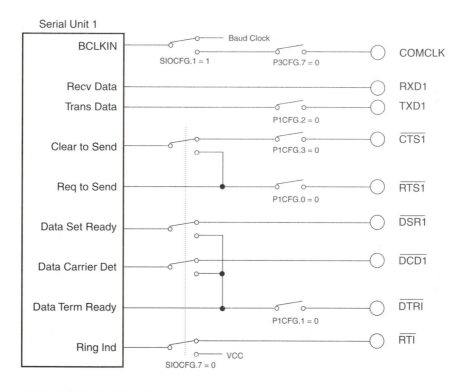

FIGURE 16–14 The internal structure of serial unit 1 and the serial configuration register.

FIGURE 16–15   The serial synchronous unit internal structure.

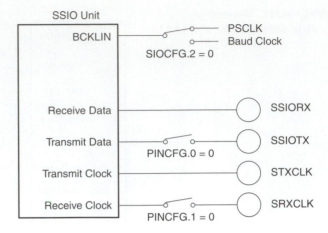

FIGURE 16–16   The core configuration of the 80386EX microprocessor.

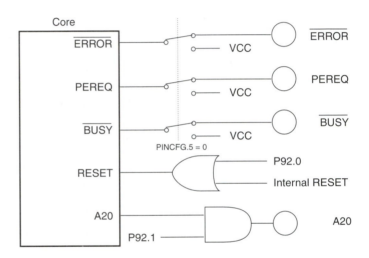

For example, suppose a system will contain a 32K EPROM, a 32K SRAM, a serial port, and a timer to generate a periodic interrupt at the rate of one time per 100 ms. In this example you would start by looking at the timer configuration register and the serial communications configuration register and pin connections for both units. Once you have decided on which pins are used for your system, you would look to the interrupt configuration register to select a timer interrupt and an interrupt (if needed) from the serial unit. You can now program the pin configuration register and the other configuration registers for each unit to set up the system. An example of configuring the system appears in the next section.

The Interrupt Control Unit

The interrupt control unit is comprised of a pair of 8259A interrupt controllers. This device is presented in Chapter 12, Section 12–4, which you should review if necessary. The connections to the 80386EX interrupt unit are presented in Figure 16–11 and should be referred to as you're reading this section.

Programming the interrupt unit requires that the port 3 configuration register, interrupt configuration register, initialization command words (ICWs), and operation command words (OCWs) are programmed. As we mentioned earlier, the first step is to program the configuration registers in the configuration dialog of a system program. Once the port 3 and interrupt configuration registers are programmed, the ICWs are programmed to control how the interrupt unit

FIGURE 16–17 The interrupt unit initialization command words for the 80386EX.

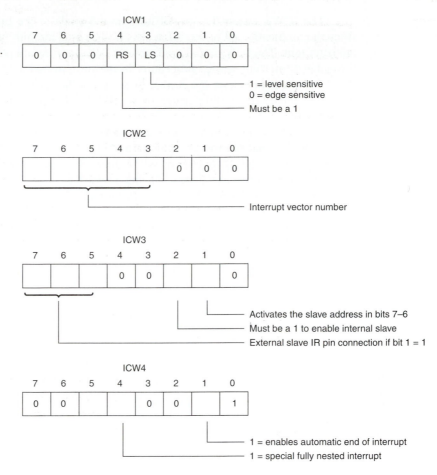

functions and at which interrupt vector the 8259A will function. Not all systems require the use of both the master and slave interrupt controllers. Often only the master is required, which reduces the amount of software needed for initialization. In any case, refer to Figure 16–17 for the contents of the ICW register used for programming both the master and slave interrupt controllers.

If the ICWs are compared to the ones presented in Chapter 12, some differences are noted. The ICW1 contains only one active bit. The LS bit is used to select the type of input, either level or edge sensitive. The RS bit is a logic 1 to select ICW1 and all other bits are logic 0's.

The ICW2 selects the base interrupt vector address. Intel suggests that the base address be placed between interrupt vector numbers 20H and F8H. Notice that the rightmost three bits are logic 0's. This is because the base interrupt vector address points to the starting interrupt vector of a group of eight interrupt vectors. If the interrupt vector is programmed for 40H, then the interrupt vectors generated by the 8259A will be 40H through 47H; IR0 will generate vector 40H, IR1 will generate 41H, and so forth.

The ICW3 register programs information about the master and slave configuration. ICW3 is programmed with the pin number that is connected to the slave for cascading. For example, in the personal computer pin IR2 is connected to the slave 8259A. The address programmed into the ICW3 would be 010 for the personal computer. In the interrupt unit of the 80386EX, the IR2 pin is permanently connected to the slave. For this reason, you must set bit position 1 of ICW3 to indicate this to the master. Note that the contents of the slave interrupt controller, within the microprocessor, must be programmed with ID 010. If external slaves are attached, then the master and each external slave must also be programmed.

The ICW4 selects the special fully nested mode or the fully nested mode. It also enables the automatic end of interrupt, which is most commonly used when interfacing to the interrupt unit. The fully nested mode allows only an interrupt of a higher level to interrupt a lower level interrupt service procedure. In the fully nested mode, the end-of-interrupt command automatically clears the current interrupt. In the special fully nested mode, an equal or higher interrupt input will interrupt a lower level interrupt. In most systems, the fully nested mode is used.

In addition to the initialization command words, there is a set of operational command words (OCWs) that control the interrupt unit and its pair of 8259A interrupt controllers. Figure 16–18 shows the contents of the OCW registers. The purpose of the OCW1 register is to enable the interrupt request inputs. Placing a logic 0 into a bit enables the corresponding interrupt input. The OCW1 is called the interrupt mask register because a logic 1 in a bit masks (disables) an interrupt input.

The OCW2 changes the priority structure of the interrupt controller and issues end-of-interrupt (EOI) commands. OCW3 enables the special mask mode, issues a poll command, and

FIGURE 16–18 The operational command words for the interrupt unit in the 80386EX.

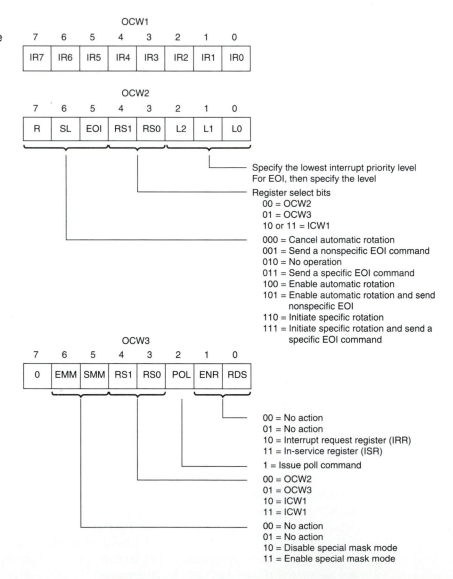

accesses the interrupt in-service register (ISR) and request register (IRR). The poll status byte indicates which interrupts are pending. Note that the OCW3 register is used to issue a poll; the poll status byte is then read to determine the outcome of the poll.

Timer and Watchdog Unit

The timer unit contains three 16-bit counters that each can interrupt the microprocessor and generate an output signal. These timers are comparable to the timers explained in Section 11–6 and are not comparable to the timers in the 80186/80188. Refer to Section 11–6 for complete programming information for all three timers and for the control register. Each timer can be clocked from an external clock signal or from the internal prescaled clock signal. An output pin that is available from each timer can also be used to cause an interrupt. The gate input is used to start and stop the timer with an external signal if connected. The function of the prescaler is discussed later in this section.

The watchdog timer is also part of the internal structure of the microprocessor. The watchdog timer is a 32-bit down counter that decrements for each system clock. When the watchdog reaches zero, it triggers an output from the watchdog timer output pin (WDTOUT) that becomes a logic 1 for 8 system clocks. The watchdog can be used to cause an interrupt that may recover from system hang. The reload value register is used to program the reload value when the counter reaches zero. This allows the watchdog output to be retriggered from 1 to 4G clocks after it times out. The contents of the watchdog can be read through the counter ports. The contents of the watchdog status register are shown in Figure 16–19.

Asynchronous Serial Unit

The asynchronous serial unit is similar to the one found in the EB and EC versions of the 80186 and 80188. The Baud rate clock is programmed by setting a 16-bit divisor for each serial unit. The equation used to program the serial unit Baud rate generator is

$$\text{Baud Rate} = \frac{\text{BCLKIN}}{16 \times \text{divisor}}$$

The BLCKIN signal is obtained from the COMCLK pin or from CLK2 ÷ 4. For example, if the desired Baud rate is 21.4K, the divisor is 23H if the 80386EX is operating at 8 MHz (CLK2 = 16 MHz).

Programming is accomplished as if the serial port were a 14450. The 14450/14550 is covered in Chapter 11, which should be referred to for programming the 80386EX serial unit. In any case, you must program the SIO (serial I/O) configuration register as detailed a few pages back. Ports 1, 2, and 3 configuration registers enable the pin connections to the serial unit (as illustrated in Figures 16–13 and 16–14) and must be programmed.

FIGURE 16–19 The watchdog status register for the 80386EX.

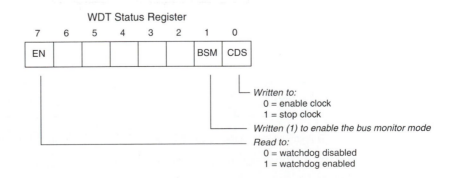

Synchronous Serial Unit

The synchronous serial is not present in the 80186/80188 and not explained elsewhere. Figure 16–20 illustrates the four modes of operation for the synchronous serial unit. Notice that the transmitter and receiver can be placed into master and slave modes. The master mode uses the internal Baud rate generator for clocking the transmitter or receiver. The slave mode uses an external clock input from an external synchronous serial interface as the clock input.

The synchronous serial unit will function at frequencies of up to 6.25 MHz if the microprocessor is operated with a 25-MHz clock. Programming the Baud rate divisor with a count zero generates the 6.25-MHz rate.

$$\text{Baud Rate Output Frequency} = \frac{\text{BCLKIN}}{2 \times \text{divisor} + 2}$$

As with the serial unit, the synchronous serial unit requires that the pin configuration register and SIO/SSIO configuration register are programmed for operation. The transmitter and receiver use 16-bit data formats for transmission and reception. Any protocols are left to the user.

The registers used with the SSIO (synchronous serial I/O unit) are illustrated in Figure 16–21. Also shown is the prescale control register used to program the internal clock prescaler discussed earlier. Note that all bits in the SSI control registers are active high.

DMA Unit

The DMA unit contains a two-channel DMA controller that can be programmed to operate as an 8237 (see Section 13–2) or a more powerful version that is located within the 80386EX. The enhancement provided by the 80386EX DMA controller include chaining, two-cycle DMA transfers, transfers between any combination of memory and I/O, and address registers for both the

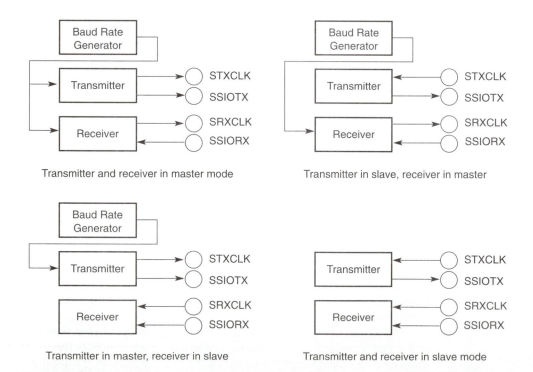

FIGURE 16–20   The master and slave mode configurations for the synchronous serial unit.

FIGURE 16–21 The control register for the synchronous serial unit and the prescaler control register.

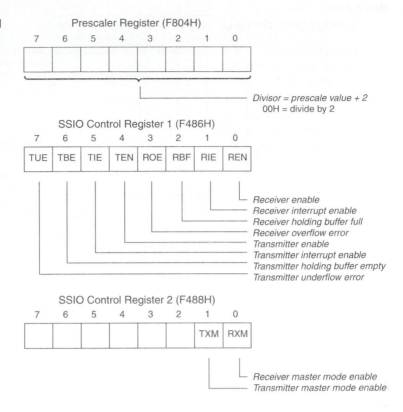

target and requester of a DMA action. Each DMA channel contains a requester address register, a target address register, a count register, and various control, command, and status registers. The width of the address registers is 26 bits and the count register is a 24-bit register. The wider address width allows DMA transfers to any memory location. The count register allows counts to 16M.

Programming the DMA unit is similar to programming the 8237 DMA controller. When used, the DMA channel is programmed with the target and requester addresses, a count, and then the control registers are programmed. The command register is then used to start a DMA action, or an external signal starts the action. Note that the DMA configuration register must also be programmed, as described earlier in this section.

Figure 16–22 illustrates the contents of the two DMA mode registers. The DMA mode 1 register selects the DMA transfer mode, determines whether the target address increments or decrements, selects auto-initialization, chooses the direction of data flow for the target, and selects which channel is being programmed. The DMA mode 2 register performs many of the same tasks for the requester and provides additional features for the target. Both registers are programmed before a DMA transfer is started.

Once the DMA mode registers are programmed for the desired operation of the channel target, request register, and other features, the channel group registers are programmed to format the channel for the DMA transfer. Figure 16–23 illustrates the contents of the channel mask, channel group mask, and DMA bus size register. The bus size register allows DMA transfers to either 8- or 16-bit I/O or memory devices. In many cases I/O transfers are 8 bits and memory-to-memory transfers are 16 bits.

Additional registers are provided to control interrupts and chaining, and to read the status of the DMA controller. The control registers are programmed before a DMA transfer can be started, but the status register is a read-only register. Figure 16–24 illustrates the contents of the DMA chaining register, DMA interrupt enable register, and the DMA status register.

FIGURE 16–22 The contents of the DMA control registers.

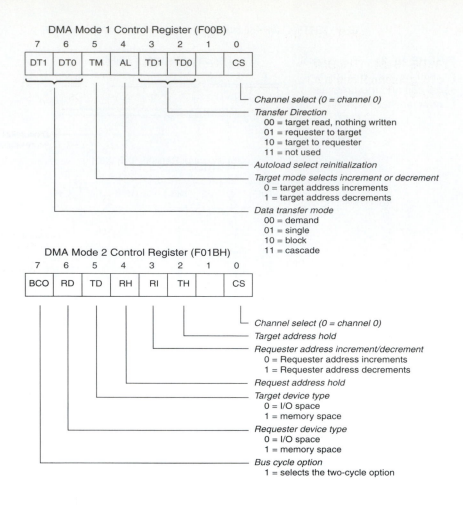

DMA Mode 1 Control Register (F00B)

7	6	5	4	3	2	1	0
DT1	DT0	TM	AL	TD1	TD0		CS

Channel select (0 = channel 0)
Transfer Direction
    00 = target read, nothing written
    01 = requester to target
    10 = target to requester
    11 = not used
Autoload select reinitialization
Target mode selects increment or decrement
    0 = target address increments
    1 = target address decrements
Data transfer mode
    00 = demand
    01 = single
    10 = block
    11 = cascade

DMA Mode 2 Control Register (F01BH)

7	6	5	4	3	2	1	0
BCO	RD	TD	RH	RI	TH		CS

Channel select (0 = channel 0)
Target address hold
Requester address increment/decrement
    0 = Requester address increments
    1 = Requester address decrements
Request address hold
Target device type
    0 = I/O space
    1 = memory space
Requester device type
    0 = I/O space
    1 = memory space
Bus cycle option
    1 = selects the two-cycle option

FIGURE 16–23 The DMA mask and bus size registers.

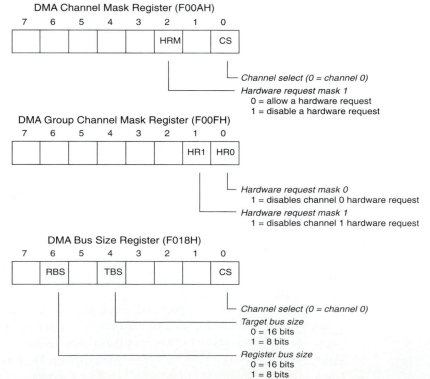

DMA Channel Mask Register (F00AH)

7	6	5	4	3	2	1	0
					HRM		CS

Channel select (0 = channel 0)
Hardware request mask 1
    0 = allow a hardware request
    1 = disable a hardware request

DMA Group Channel Mask Register (F00FH)

7	6	5	4	3	2	1	0
						HR1	HR0

Hardware request mask 0
    1 = disables channel 0 hardware request
Hardware request mask 1
    1 = disables channel 1 hardware request

DMA Bus Size Register (F018H)

7	6	5	4	3	2	1	0
	RBS		TBS				CS

Channel select (0 = channel 0)
Target bus size
    0 = 16 bits
    1 = 8 bits
Register bus size
    0 = 16 bits
    1 = 8 bits

FIGURE 16–24 The DMA chaining, interrupt and status registers.

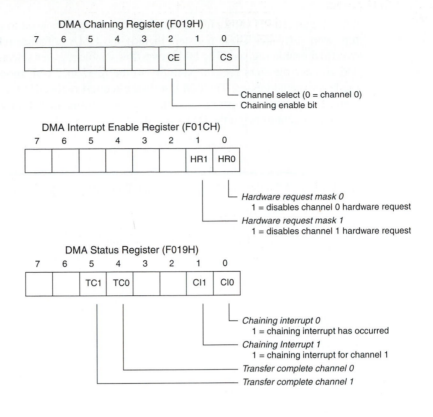

FIGURE 16–25 The software requests and overflow registers.

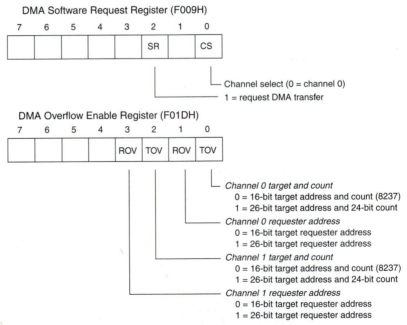

DMA chaining allows a series of DMA commands to be stored in a memory buffer. The commands reprogram the DMA channel each time a transfer is complete, so a chain of events can be set up in the memory buffer. Memory buffer access is controlled by an interrupt that is generated and controlled by the DMA chaining register. The interrupt service procedure tests the DMA status register to determine the cause of the interrupt.

Figure 16–25 shows the various command registers used to control and start DMA transfers with software. These registers include the DMA software request register and the DMA overflow enable register. The DMA overflow enable register selects 16-bit mode for the address and counter registers for compatibility to the 8237 or 24-bit mode for the counter and 26-bit mode for the addresses. The DMA software request register initiates a DMA transfer with software. This is used for operations such as memory-to-memory transfers where there is no external hardware event to begin the DMA transfer.

16–4   AN EXAMPLE PROJECT BASED ON THE 80386EX

In this section we look at an example that develops our skill in preparing a project based on the 80386EX. Because the 80386EX embedded microprocessor is very powerful when compared to the 80186/80188, it is beginning to replace some of the applications formerly piloted by the 80186/80188. The 80386EX is only slightly more expensive, it addresses much more memory, and it is faster in both clock speeds and instruction execution. As the requirement for speed increases, the future will bring the 80486 and Pentium microprocessors in the form of embedded microprocessors.

The Application

A simple application is chosen to reduce the amount of software required for implementation. Although this application is simple, it does illustrate how some of the internal and external features of the 80386EX are used. The application chosen is a printer spooler that connects the printer outputs of two personal computers to one printer. This system could be used in an office to share a printer. The interface accepts data from two computer parallel printer ports, stores the data into a memory queue, and then sends the data to the printer through a parallel printer port. Although the power of the 80386EX may not be needed for such a simple system, the system could be expanded to connect up to 16 or more computers to one printer without a great deal of additional software or hardware. In this application, an 8255 PIA is used as the interface to a pair of computers. This is chosen so that additional computer connections can be added without re-creating the system. The parallel port to the printer is developed on the 80386EX by using port 1, because this will not need to be expanded if additional computer inputs are connected.

The memory requirements are variable. If a large number of computer connections is needed that produce many printed pages of data, a large memory system may be required. In this example system, we use a relatively small amount of SRAM (256K) to act as a queue to store information for the printer. The program is relatively short, and easily fits onto a small 32K-byte EPROM.

Figure 16–26 shows the schematic diagram of the system including all interfaces. The power supply is not shown, but could be a transformer-type supply—like the one that comes with most portable equipment today.

The printer interface uses the Centronics standard, which is described in Chapter 11. There are three DB-25 connectors used connect the printer and the two computers. The printer error signals are connected to the 80386EX, but are not passed to the computers. If a printer error occurs, the 80386EX does not accept data from the computers once the print buffer memory is filled.

This system uses a 256K-byte SRAM for the stack, a few bytes of temporary data to track the print buffer queue, and memory for the print buffer queue. The software listing shows the equates, which illustrate the memory locations used to track the queue. If a larger queue is used, the microprocessor must be operated in the protected mode as described in Chapter 2.

Example 16–3 lists the software for the system in Figure 16–26. The first portion of the software programs and initializes the 80386EX and the 8255 peripheral interface. The 8255 is operated in strobed input mode 1 for each of the two I/O ports (A and B). These ports are connected to the

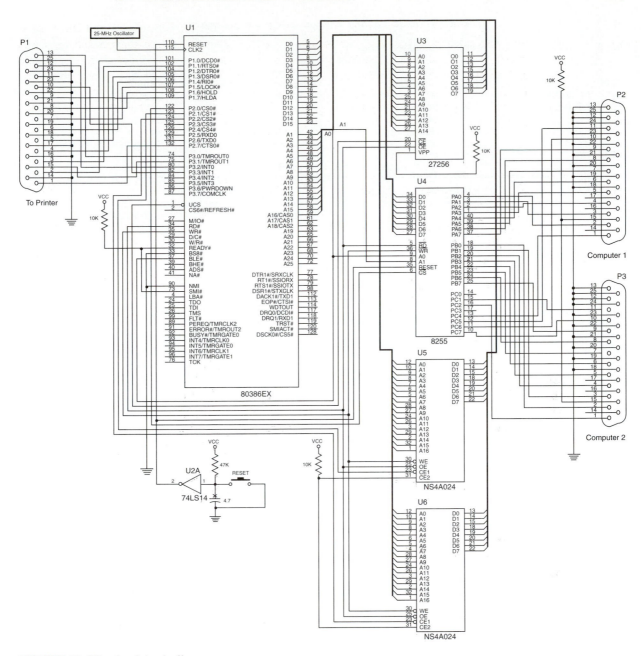

FIGURE 16–26 A printer buffer.

two computers and accept data from these computers for spooling into the printer queue. The 80386EX is initialized so port 1 is programmed as an output port to send data to the printer. Port 2 is programmed to select the memory and 8255 and to perform control functions for the printer. Port 3 is also programmed for controlling the printer. Timer 0 is used internally for time delays. Timer 0 is programmed for continuous counting from 0000H to FFFFH at a rate of 12.5 KHz.

EXAMPLE 16–3

```
;System software for the printer queue of Figure 16-26.
;
;This software must be assembled as a .COM file for a 32K EPROM.
;The .COM file is then in binary format, which allows it to be
```

```
                              ;used as input to any common EPROM programmer.
                              ;
0000                          CODE    SEGMENT 'code'
                                      ASSUME  CS:CODE
                                      ORG     100H              ;must be 100H for .COM file
0100                          MAIN:                             ;physical address = 00F8000H
0100 BA F43A                          MOV     DX,0F43AH         ;address UCS high address
0103 B8 000F                          MOV     AX,000FH
0106 EF                                OUT     DX,AX             ;program address = 00FXXXXH
0107 BA F438                          MOV     DX,0F438H         ;address UCS low address
010A B8 8504                          MOV     AX,8504H          ;4 waits for memory = 00F8000H
010D EF                                OUT     DX,AX
010E BA F43E                          MOV     DX,0F43EH         ;address UCS mask high
0111 B8 000F                          MOV     AX,000FH          ;select 32K EPROM
0114 EF                                OUT     DX,AX
0115 BA F43C                          MOV     DX,0F43CH         ;address UCS mask low
0118 B8 FFFF                          MOV     AX,0FFFFH         ;EPROM is at 00F8000H-00FFFFFH
011B EF                                OUT     DX,AX

011C BA F402                          MOV     DX,0F402H         ;address CS0 address high
011F B8 0000                          MOV     AX,0000H          ;program I/O address = 0000H
0122 8B D0                            MOV     DX,AX
0124 EF                                OUT     DX,AX
0125 BA F400                          MOV     DX,0F400H         ;address CS0 address low
0128 B8 0404                          MOV     AX,0404H          ;4 waits for I/O = 0000H
012B EF                                OUT     DX,AX
012C BA F406                          MOV     DX,0F406H         ;address CS0 mask high
012F B8 0000                          MOV     AX,0000H          ;select 8255 size of 4 bytes
0132 EF                                OUT     DX,AX
0133 BA F404                          MOV     DX,0F404H         ;address CS0 mask low
0136 B8 FFFF                          MOV     AX,0FFFFH         ;8255 at I/O address 0000H-0003H
0139 EF                                OUT     DX,AX

013A BA F40A                          MOV     DX,0F40AH         ;address CS1 address high
013D B8 0000                          MOV     AX,0000H          ;program memory at 0000000H
0140 EF                                OUT     DX,AX
0141 BA F408                          MOV     DX,0F408H         ;address CS1 address low
0144 B8 0500                          MOV     AX,0500H          ;0 waits for memory = 0000000H
0147 EF                                OUT     DX,AX
0148 BA F40E                          MOV     DX,0F40EH         ;address CS1 mask high
014B B8 0001                          MOV     AX,0001H          ;RAM size = 128K
014E EF                                OUT     DX,AX
014F BA F40C                          MOV     DX,0F40CH         ;address CS1 mask low
0152 B8 FFFF                          MOV     AX,0FFFFH         ;SRAM1 at 0000000H-001FFFFH
0155 EF                                OUT     DX,AX

0156 BA F412                          MOV     DX,0F412H         ;address CS2 address high
0159 BA 0002                          MOV     DX,0002H          ;program memory at 0020000H
015C EF                                OUT     DX,AX
015D BA F410                          MOV     DX,0F410H         ;address CS2 address low
0160 B8 0500                          MOV     AX,0500H          ;0 waits for memory = 0020000H
0163 EF                                OUT     DX,AX
0164 BA F416                          MOV     DX,0F416H         ;address CS2 mask high
0167 B8 0001                          MOV     AX,0001H          ;RAM size = 128K
016A EF                                OUT     DX,AX
016B BA F414                          MOV     DX,0F414H         ;address CS2 mask low
016E B8 FFFF                          MOV     AX,0FFFFH         ;SRAM2 at 0020000H-003FFFFH
0171 EF                                OUT     DX,AX

0172 BA F820                          MOV     DX,0F820H         ;port 1 is I/O
0175 B8 0000                          MOV     AX,0
0178 EF                                OUT     DX,AX
0179 BA F864                          MOV     DX,0F864H         ;port 1 = output
017C B8 00FF                          MOV     AX,0FFH
017F EE                                OUT     DX,AL

0180 BA 0F82                          MOV     DX,0F82H          ;port 2 is CS and I/O
0183 B8 0007                          MOV     AX,07H
0186 EE                                OUT     DX,AL
```

```
0187 BA F86C                    MOV     DX,0F86CH        ;port 2 = CS0, CS1, and CS2
018A B0 FF                      MOV     AL,0FFH          ;and output
018C EE                         OUT     DX,AL

018D BA F824                    MOV     DX,0F824H        ;port 3 = I/O and INT1
0190 B8 0008                    MOV     AX,8
0193 EE                         OUT     DX,AL
0194 BA F874                    MOV     DX,0F874H        ;port 3 = input
0197 B8 0000                    MOV     AX,0
019A EE                         OUT     DX,AL

019B BA F834                    MOV     DX,0F834H        ;address timer configuration
019E B8 0000                    MOV     AX,00H           ;clocks are internal
01A1 EE                         OUT     DX,AL            ;clock frequency is 25 MHz/4 or 6.25 MHz
01A2 BA F804                    MOV     DX,0F804H        ;address prescaler
01A5 B8 01F4                    MOV     AX,500           ;divide by 500
01A8 E                          OUT     DX,AL            ;timer 0 input is 12.5 KHz
01A9 BA F043                    MOV     DX,0F043H        ;address timer control
01AC B8 0030                    MOV     AX,30H           ;program timer 0 for continuous count
01AF EE                         OUT     DX,AL
01B0 BA F040                    MOV     DX,0F040H        ;program timer 0 count to FFFFH
01B3 B0 FF                      MOV     AL,0FFH
01B5 EE                         OUT     DX,AL
01B6 EE                         OUT     DX,AL            ;timer 0 counts at 6.25 MHz rate

01B7 B0 B6                      MOV     AL,0B6H          8255 Port A and B strobed input
01B9 E6 03                      OUT     3,AL             ;port C = output
01BB B0 FF                      MOV     AL,0FF           ;set port C for busy
01BD E6 02                      OUT     2,AL

01BF B8 0000                    MOV     AX,0             ;set top of stack to 00000FFH
01C2 8E D8                      MOV     DS,AX
01C4 8E D0                      MOV     SS,AX            ;small stack area
01C6 BC 0100                    MOV     SP,100H

01C9 BA F832                    MOV     DX,0F832H        ;address interrupt configuration
01CC B8 0000                    MOV     AX,0             ;single controller
01CF EE                         OUT     DX,AL
01D0 BA F020                    MOV     DX,0F020H        ;address master interrupt controller
01D3 B8 0001                    MOV     AX,01H           ;write ICW1 (edge triggered)
01D6 EE                         OUT     DX,AL
01D7 BA F021                    MOV     DX,0F021H        ;address master interrupt controller
01DA B8 0020                    MOV     AX,20H           ;select vectors 20H–27H
01DD EE                         OUT     DX,AL
01DE B0 00                      MOV     AL,0
01E0 EE                         OUT     DX,AL            ;no slaves
01E1 B0 03                      MOV     AL,03H           ;select automatic EOI
01E3 EE                         OUT     DX,AL
01E4 BA F021                    MOV     DX,0F021H
01E7 B8 00FD                    MOV     AX,0FDH          ;enable only INT1
01EA EE                         OUT     DX,AL

01EB BB 0084                    MOV     BX,84H           ;address INT1 vector
01EE B8 01D9 R                  MOV     AX,OFFSET INT1–100H
01F1 89 07                      MOV     DS:[BX],AX       ;save interrupt procedure address
01F3 83 C3 02                   ADD     BX,2
01F6 B8 F800                    MOV     AX,0F800H
01F9 89 07                      MOV     DS:[BX],AX
01FB B8 0100                    MOV     AX,100H
01FE A3 0084                    MOV     DS:[0084H],AX    ;input pointer for queue
0201 B8 0000                    MOV     AX,0
0204 A2 0086                    MOV     DS:[0086H],AL
0207 B8 0100                    MOV     AX,100H
020A A3 0088                    MOV     DS:[0088H],AX    ;output pointer for queue
020D B8 0000                    MOV     AX,0
0210 A2 008A                    MOV     DS:[008AH],AL
                                ;
                                ;System software begins here.
                                ;
```

```
0213                          SYSTEM:
0213 E4 03                        IN      AL,3             ;get port C from 8255
0215 A8 22                        TEST    AL,22H           ;check both IBF signals
0217 74 FA                        JZ      SYSTEM           ;wait for data from any computer
0219 A8 02                        TEST    AL,2             ;check for computer
021B 75 17                        JNE     COMP2            ;if computer 2
021D                          COMP1:                       ;else computer 1
021D E8 005A                      CALL    FULL             ;test for full condition
0220 E4 01                        IN      AL,1             ;read data from computer 1
0222 B3 0C                        MOV     BL,12
0224 E8 007F                      CALL    ACK              ;send ACK to computer 1
0227 E8 0021                      CALL    SAVE             ;save data in queue
022A FB                           STI                      ;enable printing
022B B3 20                        MOV     BL,20H           ;set port flag
022D E8 0081                      CALL    TIME
0230 74 E1                        JZ      SYSTEM           ;if port timed out
0232 EB E9                        JMP     COMP1            ;get next byte
0234                          COMP2:                       ;for computer 2
0234 E8 0043                      CALL    FULL             ;test for full condition
0237 E4 02                        IN      AL,2             ;read data from computer 2
0239 B3 0E                        MOV     BL,14
023B E8 0068                      CALL    ACK              ;send ACK to computer 2
023E E8 000A                      CALL    SAVE
0241 FB                           STI
0242 B3 02                        MOV     BL,2
0244 E8 006A                      CALL    TIME
0247 74 CA                        JZ      SYSTEM
0249 EB E9                        JMP     COMP2


                             ;
                             ;Procedure that saves AL into the queue.
                             ;
024B                         SAVE    PROC    NEAR

024B 50                      PUSH    AX
024C A1 0086                     MOV     AX,DS:[0086H]        ;get segment
024F 8E D8                       MOV     DS,AX
0251 8B 1E 0084                  MOV     BX,DS:[0084H]        ;get offset
0255 58                          POP     AX
0256 88 07                       MOV     [BX],AL              ;save data in queue
0258 83 06 0084 01               ADD     WORD PTR DS:[0084H],1 ;increment offset
                             .IF ZERO?
025F 81 06 0086 1000               ADD     WORD PTR DS:[0086H],1000H    ;get next segment
                                   .IF  WORD PTR DS:[0086H]==4000H
026D C7 06 0086 0000                 MOV    WORD PTR DS:[0086H],0
0273 C7 06 0084 0100                 MOV    WORD PTR DS:[0084H],100H
                                   .ENDIF
                             .ENDIF
0279 C3                      RET

027A                         SAVE    ENDP
                             ;
                             ;Check the queue for a full condition
                             ;and get stuck if full.
                             ;
027A                         FULL    PROC    NEAR

027A 8B 1E 0084                  MOV     BX,DS:[0084H]
027E 8B 0E 0086                  MOV     CX,DS:[0086H]
0282 83 C3 01                    ADD     BX,1
                             .IF BX == 0
0289 81 C1 1000                    ADD     CX,1000H
                                 .IF CX == 4000H
0293 BB 0100                         MOV BX,100H
0296 B9 0000                         MOV CX,0
                                 .ENDIF
                             .ENDIF
0299 3B 1E 0088                  CMP     BX,DS:[0088H]
```

```
029D 75 06                      JNE     FULL1           ;not full
029F 3B 0E 008A                 CMP     CX,DS:[008AH]
02A3 74 D5                      JE      FULL            ;if full
02A5                    FULL1:
02A5 C3                         RET

02A6                    FULL    ENDP
                        ;
                        ;Send ACK to computer.
                        ;
02A6                    ACK     PROC    NEAR    USES AX

02A7 8A C3                      MOV     AL,BL
02A9 E6 03                      OUT     3,AL            ;clear bit
02AB FE C0                      INC     AL
02AD E6 03                      OUT     3,AL            ;set bit
                                RET

02B1                    ACK     ENDP
                        ;
                        ;Procedure that times out a computer.
                        ;If no data is received within 1 second,
                        ;a return zero occurs.
                        ;
02B1                    TIME    PROC    NEAR

02B1 BA F040                    MOV     DX,0F040H       ;address timer 0
02B4 EC                         IN      AL,DX           ;get count
02B5 8A C8                      MOV     CL,AL
02B7 EC                         IN      AL,DX
02B8 8A E8                      MOV     CH,AL
02BA 81 C1 30D4                 SUB     CX,12500        ;bias count by 1 second
02BE                    TIME1:
02BE E4 03                      IN      AL,3            ;test port C
02C0 84 C3                      TEST    AL,BL           ;check for buffer full
02C2 75 14                      JNZ     TIME2           ;if data present from computer
02C4 EC                         IN      AL,DX
02C5 8A E0                      MOV     AH,AL
02C7 EC                         IN      AL,DX
02C8 86 C4                      XCHG    AL,AH
02CA 3B C1                      CMP     AX,CX
02CC 75 F0                      JNZ     TIME1           ;if no time out
02CE E8 FFA9                    CALL    FULL
02D1 B0 0F                      MOV     AL,0FH          ;form feed character
02D3 E8 FF75                    CALL    SAVE
02D6 2A C0                      SUB     AL,AL           ;clear zero flag
02D8                    TIME2:
02D8 C3                         RET

02D9                    TIME    ENDP

02D9                    INT1    PROC    FAR

02D9 50                         PUSH    AX
02DA 52                         PUSH    DX
02DB 53                         PUSH    BX
02DC 1E                         PUSH    DS
02DD BA F870                    MOV     DX,0F870H       ;address port 3
02E0                    ERR:
02E0 EC                         IN      AL,DX
02E1 A8 05                      TEST    AL,5
02E3 74 FB                      JZ      ERR             ;on printer error
02E5 A8 02                      TEST    AL,2
02E7 75 F7                      JNZ     ERR             ;if error
02E9 B8 0000                    MOV     AX,0
02EC 8E D8                      MOV     DS,AX
02EE 8B 1E 0088                 MOV     BX,DS:[0088H]   ;get output pointer
02F2 A1 008A                    MOV     AX,DS:[008AH]   ;get segment
```

```
02F5 3B 06 0086              CMP     AX,DS:[0086H]
02F9 75 10                   JNE     INT11           ;if not empty
02FB 3B 1E 0084              CMP     BX,DS:[0084H]
02FF 75 0A                   JNE     INT11           ;if not empty
0301 55                      PUSH    BP
0302 8B EC                   MOV     BP,SP
0304 80 66 0B FD             AND     BYTE PTR [BP+11],0FDH   ;do CLI
0308 5D                      POP     BP
0309 EB 3A                   JMP     INT12           ;exit, but don't clear interrupt
030B               INT11:                            ;if not empty
030B 8E D8                   MOV     DS,AX
030D 8A 07                   MOV     AL,[BX]         ;get data for printer
030F BA F860                 MOV     DX,0F860H       ;address port 1
0312 EE                      OUT     DX,AL           ;send data to printer
0313 BA F868                 MOV     DX,0F868H       ;address port 2
0316 B0 00                   MOV     AL,0            ;send STB to printer
0318 EE                      OUT     DX,AL
0319 B0 FF                   MOV     AL,0FFH
031B EE                      OUT     DX,AL
031C 83 C3 01                ADD     BX,1
031F 8C D8                   MOV     AX,DS
                             .IF     BX == 0
0325 05 1000                     ADD     AX,1000H
                                 .IF  AX==4000H
032D B8 0000                         MOV  AX,0
0330 BB 0100                         MOV  BX,100H
                                 .ENDIF
                             .ENDIF
0333 BA 0000                 MOV     DX,0
0336 8E DA                   MOV     DS,DX
0338 A3 008A                 MOV     DS:[008AH],AX
033B 89 1E 0088              MOV     DS:[0088H],BX
033F BA F020                 MOV     DX,0F020H
0342 B0 20                   MOV     AL,20H          ;signal EOI
0344 EE                      OUT     DX,AL
0345               INT12:
0345 1F                      POP     DS
0346 5B                      POP     BX
0347 5A                      POP     DX
0348 58                      POP     AX
0349 CF                      IRET

034A               INT1    ENDP

034A               RESET:
                             ORG     80F0H           ;biased reset location (7FF0H + 100H)
                                                     ;16 bytes for reset
80F0 BA 0022                 MOV     DX,0022H        ;configure for expanded I/O space
80F3 B8 8000                 MOV     AX,08000H
80F6 EF                      OUT     DX,AX

80F7 EA                      DB      0EAH            ;force far jump to address 00F8000H
80F8 0000 F800               DW      0,0F800H        ;CS = F800H, IP = 0000H

80FC               CODE    ENDS
                           END     MAIN
```

As with the systems illustrated in the last chapter, this example programs the on-chip features before entering the system software. In this system, the EPROM is located at addresses 00F8000H–00FFFFFH (the top of the first 1M of memory). When the 80386EX is reset, it internally initializes the $\overline{UCS}$ pin to select the entire memory system. The 8255 is located at I/O ports 0000H–0003H. The pair of SRAM devices is located at memory addresses 0000000H–0003FFFFH (256K bytes).

Once the addresses for memory and I/O have been programmed, the software initializes the 8255 as a pair of strobed input ports for connection to the computers and the port pins (ports

1,2, and 3) are programmed. The INT1 interrupt input is connected to the printer so whenever the $\overline{ACK}$ signal returns from the printer, an interrupt is requested. Note that the $\overline{ACK}$ signal is sent from the printer after it prints a character and after the printer is first enabled. In this system it is important that the 80386EX be powered up before the printer.

Once the port pins and interrupt input are initialized, the timer 1 is programmed to count from FFFFH down to 0000H. This count sequence occurs at a rate of 12.5 KHz. In the system software the timer is used to generate a one-second time delay so a computer can be timed out after printing.

The system software is very short because of the CALL instruction to procedures. The system starts by checking to see if the buffer full flag, within the 8255, is set. If it is, then data has been sent by one of the two computers for printing.

Once the computer that sent data is detected, the software continues at either COMP1 or COMP2. The COMP1 and COMP2 sections of the software are almost identical. The first event that occurs is a test to determine if the queue is full. If full, the FULL procedure waits until a printer interrupt occurs to remove a byte from the queue, which allows a return from FULL so additional data may be retrieved from the computer for printing.

Following the CALL FULL instruction, the software inputs data from the computer through either port A or port B of the 8255. The next step is to call the ACK software that sends an acknowledge pulse to the computer that sent data to the 80386EX.

The CALL SAVE procedure stores the data from the computer into the queue for the printer. A close look at SAVE shows that it first retrieves the input pointer from the memory. The queue itself is at locations 0000:0100 through 3000:FFFF. This is almost 256K bytes of memory. The contents of the input pointer are incremeted and adjusted by a few conditional .IF statements.

After saving the data into the queue, the program continues and enables interrupts. This must be done for the case where this is the first character to be printed. If the printer has been turned on, this will cause an immediate interrupt to the INT1 procedure.

The very last thing that occurs in the system software section is that the TIME procedure is called. TIME copies the number from timer 0 into a pair of 16-bit registers, where it is biased by subtracting 12,500. Because timer 0 is a down counter, and because it is clocked at a 12.5-KHz rate, it will take one second for the timer to count down to the biased value. As the TIME procedure tests the computer interface for another buffer full flag, it checks the contents of timer 0 to see if one second has elapsed. If not, a return occurs with the flags indicating a not-zero condition. If one second has elapsed, a form-feed is sent to the printer, and a return zero occurs.

The INT1 interrupt service procedure must check for errors from the printer, test the queue for an empty condition, and send data and the $\overline{DS}$ signal to the printer. If an error is detected, the system locks up inside the interrupt service procedure until the error is corrected. If the queue is empty, a return occurs without re-enabling future interrupts, so when data is placed in the queue, the STI instruction in the system software causes an immediate interrupt so printing can resume.

16–5   SUMMARY

1. The 80386EX is similar to the 80186/80188 except that it contains a more powerful micro-processor that can address up to 64M bytes of data.
2. The 80386EX is programmed by first configuring an internal unit and then by programming the function of the unit. Although this seems different from the 80186/80188, it is essentially the same.
3. An 8- or a 16-bit system can be interfaced to the 80386EX, but both sizes of devices can be placed on the same bus system. This allows a 16-bit memory system and an 8-bit I/O system, for example.

4. The 80386EX is packaged in either a 133- or 144-pin device. This is about 25 percent larger that the 64-, 88-, or 100-pin packages for the 80186/80188 family.

5. Memory access time is based on a two-clock cycle for the 80386EX; it is based on a four-clock cycle in the 80186/80188.

6. The CLK2 input pin is provided with a clock that has twice the operating frequency of the microprocessor.

7. As with the 80186/80188, wait states are programmable, but in the 80386EX up to 31 wait states can be selected for each chip select pin.

8. The timer structure of the 80386EX is different from the 80186/80188 in that the tier 2 module in the 80186/80188 acts as a prescaler. In the 80386EX, a separate prescaler is provided so all three timers can be accessed through the pin structure of the microprocessor.

9. Interrupts are expanded on the 80386EX to include a pair of 8259A-compatible interrupt controllers. This allows up to eight external interrupt inputs and eight internal interrupts.

10. The DMA controller for the 80386EX is provided with address registers that are 26 bits wide. This allows a DMA action to any of the addressable memory space.

11. A synchronous serial data unit has been added to the pair of asynchronous data units within the 80386EX. The synchronous unit passes 16 bits of data at a time and can generate the synchronous clock signal for the system.

12. Programming the chip selection pins has changed for the 80386EX. Included are a pair of 16-bit address registers and a pair of 16-bit mask registers for each chip selection pin. The address registers locate the starting memory or I/O address, and the mask register determines the size of the memory or I/O device. The smallest memory device is 2K and the smallest I/O device is two bytes.

16–6   QUESTIONS AND PROBLEMS

1. How many pins are found on the 80386EX microprocessor package?

2. The 80386EX provided up to _____ of current on most of its output pins.

3. What is the purpose of the W/$\overline{R}$ pin?

4. How many memory locations can be addressed by the 80386EX?

5. Describe how the 80386EX can be operated with an 8-bit data bus.

6. How many wait states can be programmed into a chip select pin on the 80386EX?

7. How many chip selection pins are available on the 80386EX?

8. Which 80386EX pin functions as address line A0 for 8-bit bus operation?

9. What is the purpose of the $\overline{BS8}$ pin?

10. What could be stated about the difference in the address and data connections when comparing the 80386EX with the 80186 or 80188?

11. What is the purpose of the configuration registers within the 80386EX?

12. Describe the I/O port structure of the 80386EX.

13. Develop a short sequence of instructions that program the $\overline{CS3}$ pin to select a memory device located at addresses 0010000H–001FFFFH. The device requires three wait states for proper operation.

14. Develop a short sequence of instructions that program the $\overline{CS2}$ pin to select an I/O device located at addresses 1000H–1007H. The device requires five wait states for proper operation.

15. Exactly what does the mask register accomplish when programming a chip selection pin?

16. Program all the pins and registers associated with timer 2 so it uses an internal clock, and generates a square wave at its output that is high for 100 counts and low for 100 counts.

17. Program all the pins and registers associated with the INT3 input so it is an active high (level) triggered input, uses interrupt vector A3H, and uses automatic end-of-interrupt.

18. How is the prescaler programmed and what is its maximum count?
19. Sketch a simple system that uses the INT2 input and timer 1 to measure the width of some unknown active high input signal.
20. Develop software that programs the simple system in Question 19 so it can measure pulses with an accuracy to within 1 μs.

APPENDIX A

The Assembler, Disk Operating System, Basic I/O System, Mouse, and DPMI Memory Manager

This appendix explains how to use the assembler and shows the DOS (disk operating system), BIOS (basic I/O system), and mouse function calls that are used by assembly language to control the personal computer. The function calls control everything from reading and writing disk data, to managing the keyboard and displays, to controlling the mouse. The assembler represented here is the Microsoft ML (version 6.X) and MASM (version 5.10) macro assembler programs. It is fairly important that version 6.X is used instead of the dated version 5.10. Also presented is the DPMI memory manager used when shelling out of Windows.

USING THE ASSEMBLER

The assembler program requires that a symbolic program be written first, using a word processor, text editor, or the Workbench program provided in the assembler package. The editor provided with version 5.10 is M.EXE and it is strictly a full-screen editor. The editor provided with version 6.X is PWB.EXE and is a fully integrated development system that contains extensive help. Refer to the documentation that accompanies your assembler package for details on the operation of the editor program. If at all possible, use version 6.X of the assembler because it contains a detailed help file that guides you through assembly language statements, directives, and even the DOS and BIOS interrupt function calls.

If you are using a word processor to develop your software, make sure that it is initialized to generate a pure ASCII file. The source file that you generate must use the extension .ASM that is required for the assembler to properly identify your source program.

Once your source file is prepared, it must be assembled. If you are using the Workbench provided with version 6.X, this is accomplished by selecting the compile feature with your mouse. If you are using a word processor and DOS command lines with version 5.10, see Example A–1 for the dialog to assemble a file called FROG.ASM.

EXAMPLE A–1

```
A>MASM

Microsoft (R) Macro Assembler Version 5.10
Copyright (C) Microsoft Corp 1981, 1989.  All rights reserved.

Source filename [.ASM]:FROG
Object filename [FROG.OBJ]:FROG
List filename [NUL.LST]:FROG
Cross reference [NUL.CRF]:FROG
```

Once a program is assembled, it must be linked before it can be executed. The linker converts the object file into an executable file (.EXE). Example A–2 shows the dialog required for the linker using an MASM version 5.10 object file. If the ML version 6.X assembler is in use, it automatically assembles and links a program using the COMPILE or BUILD command from Workbench. After compiling with ML, Workbench allows the program to be debugged with a debugging tool called CodeView. CodeView is also available with MASM, but you must type CV at the DOS command line to access it.

EXAMPLE A–2

```
A:\>LINK

Microsoft (R) Overlay Linker Version 3.64
Copyright (C) Microsoft Corp 1983-1988.  All rights reserved.

Object modules [.OBJ]:FROG
Run file [FROG.EXE]:FROG
List file [NUL.MAP]:FROG
Libraries [.LIB]:SUBR
```

If MASM version 6.X is in use, the command-line syntax differs from version 5.10. Example A–3 shows the command-line syntax for ML, the assembler, and linker for MASM version 6.X.

EXAMPLE A–3

```
C:\>ML /FlTEST.LST TEST.ASM

Microsoft (R) Macro Assembler Version 6.11
Copyright (C) Microsoft Corp 1981-1993.  All rights reserved.

    Assembling: TEST.ASM

Microsoft (R) Segmented-Executable Linker  Version 5.13
Copyright (C) Microsoft Corp 1984-1993.  All rights reserved.

Object Modules [.OBJ]: TEST.obj/t
Run File [TEST.com]: "TEST.com"
List File [NUL.MAP]: NUL
Libraries [.LIB]:
Definitions File [NUL.DEF]: ;
```

Version 6.X of the Microsoft MASM program contains the Programmer's Workbench program. Programmer's Workbench allows an assembly language program to be developed with its full screen editor and toolbar. Figure A–1 illustrates the display found with Programmer's Workbench. To access this program, type PWB at the DOS prompt. The make option allows a program to be automatically assembled and linked making these tasks simple in comparison to version 5.10 of the assembler.

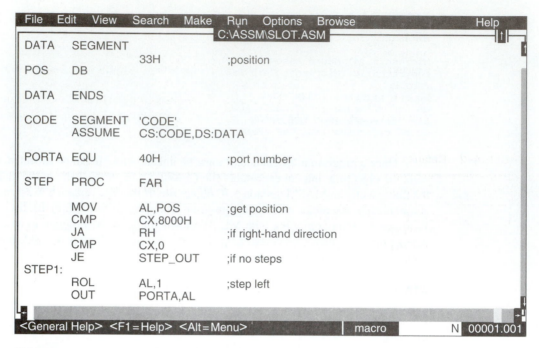

FIGURE A–1   The edit screen from Programmer's Workbench used to develop assembly language programs.

ASSEMBLER MEMORY MODELS

Memory models and the .MODEL statement are introduced in Chapter 4 and used extensively throughout the text. Here we completely define the memory models available for software development. Each model defines the way that a program is stored in the memory system. Table A–1 lists the different models available with MASM and ML.

TABLE A–1   Memory models for the assembler

Model Type	Description
Tiny	All data and code must fit into one segment. Tiny programs are written in .COM format, which means that the program must be originated at location 100H.
Small	This model contains two segments: one data segment of 64K bytes and one code segment of 64K bytes.
Medium	This model contains one data segment of 64K bytes and any number of code segments for large programs.
Compact	One code segment contains the program, and any number of data segments contain the data.
Large	The large model allows any number of code and data segments.
Huge	This model is the same as large, but the data segments may contain more than 64K bytes each.
Flat	Only available to MASM 6.X. The flat model uses one segment of 512K bytes to store all data and code. Note that this model is mainly used with Windows NT.

Note that the tiny model is used to create a .COM file instead of an execute file. The .COM file is different because all data and code fit into one code segment. A .COM file must use an origin of offset address 0100H as the start of the program. A .COM file loads from the disk and executes faster than the normal execute (.EXE) file. For most applications we normally use the execute file (.EXE) and the small memory model.

When models are used to create a program, certain defaults apply as listed in Table A–2. The directive in this table is used to start a particular type of segment for the models listed in

TABLE A–2 Defaults for the .MODEL directive

Model	Directives	Name	Align	Combine	Class	Group
Tiny	.CODE	_TEXT	word	PUBLIC	'CODE'	DGROUP
	.FARDATA	FAR_DATA	para	private	'FAR_DATA'	
	.FARDATA?	FAR_BSS	para	private	'FAR_BSS'	
	.DATA	_DATA	word	PUBLIC	'DATA'	DGROUP
	.CONST	CONST	word	PUBLIC	'CONST'	DGROUP
	.DATA?	_BSS	word	PUBLIC	'BSS'	DGROUP
Small	.CODE	_TEXT	word	PUBLIC	'CODE'	
	.FARDATA	FAR_DATA	para	private	'FAR_DATA'	
	.FARDATA?	FAR_BSS	para	private	'FAR_BSS'	
	.DATA	_DATA	word	PUBLIC	'DATA'	DGROUP
	.CONST	CONST	word	PUBLIC	'CONST'	DGROUP
	.DATA?	_BSS	word	PUBLIC	'BSS'	DGROUP
	.STACK	STACK	para	STACK	'STACK'	DGROUP
Medium	CODE	name_TEXT	word	PUBLIC	'CODE'	
	.FARDATA	FAR_DATA	para	private	'FAR_DATA'	
	.FARDATA?	FAR_BSS	para	private	'FAR_BSS'	
	.DATA	_DATA	word	PUBLIC	'DATA'	DGROUP
	.CONST	CONST	word	PUBLIC	'CONST'	DGROUP
	.DATA?	_BSS	word	PUBLIC	'BSS'	DGROUP
	.STACK	STACK	para	STACK	'STACK'	DGROUP
Compact	CODE	_TEXT	word	PUBLIC	'CODE'	
	.FARDATA	FAR_DATA	para	private	'FAR_DATA'	
	.FARDATA?	FAR_BSS	para	private	'FAR_BSS'	
	.DATA	_DATA	word	PUBLIC	'DATA'	DGROUP
	.CONST	CONST	word	PUBLIC	'CONST'	DGROUP
	.DATA?	_BSS	word	PUBLIC	'BSS'	DGROUP
	.STACK	STACK	para	STACK	'STACK'	DGROUP
Large or Huge	CODE	name_TEXT	word	PUBLIC	'CODE'	
	.FARDATA	FAR_DATA	para	private	'FAR_DATA'	
	.FARDATA?	FAR_BSS	para	private	'FAR_BSS'	
	.DATA	_DATA	word	PUBLIC	'DATA'	DGROUP
	.CONST	CONST	word	PUBLIC	'CONST'	DGROUP
	.DATA?	_BSS	word	PUBLIC	'BSS'	DGROUP
	.STACK	STACK	para	STACK	'STACK'	DGROUP
Flat	CODE	_TEXT	dword	PUBLIC	'CODE'	
	.FARDATA	_DATA	dword	PUBLIC	'DATA'	
	.FARDATA?	_BSS	dword	PUBLIC	'FBSS'	
	.DATA	_DATA	dword	PUBLIC	'DATA'	DGROUP
	.CONST	CONST	dword	PUBLIC	'CONST'	DGROUP
	.DATA?	_BSS	dword	PUBLIC	'BSS'	DGROUP
	.STACK	STACK	dword	STACK	'STACK'	DGROUP

the table. If the .CODE directive is placed in a program, it indicates the beginning of the code segment. Likewise, .DATA indicates the start of a data segment. The name column indicates the name of the segment. Align indicates whether the segment is aligned on a word, doubleword, or a 16-byte paragraph. Combine indicates the type of segment created. The class indicates the class of the segment, such as 'CODE' or 'DATA'. The group indicates the group type of the segment.

The directive from Table A–2 selects the type of information in a program. For example, .CODE is placed before the code. The name column is used if full segment descriptions are mixed with the programming models for reference. The alignment specifies how the data in the segment is aligned. A para (paragraph) alignment starts a segment at the next paragraph, that is, the next hexadecimal address ending in a 0H. The combine column indicates how various segments are combined and labeled (PUBLIC or private). The class is the actual segment name and the group is the grouping of segments.

Example A–4 uses the small model, which is used for programs that contain one DATA and one CODE segment. This applies to many programs that are developed. Notice that not only is the program listed, but so is all the information generated by the assembler. Here the .DATA and .CODE directives indicate the start of each segment. Also notice how the DS register is loaded in this program. As presented throughout the text, the .STARTUP directive can be used to load the data segment register, set up the stack, and define the starting address of a program. In this example, an alternate method is illustrated for loading the data segment register and defining the starting address of the program (END BEGIN).

EXAMPLE A–4

```
Microsoft (R) Macro Assembler Version 6.11

                            .MODEL SMALL
                            .STACK 100H
0000                        .DATA

0000 0A            FROG     DB      10
0001 0064 [        DATA1    DB      100 DUP (2)
         02
          ]

0000                        .CODE

0000 B8 ---- R     BEGIN: MOV    AX,DGROUP          ;set up DS
0003 8E D8                MOV    DS,AX
                           .
                           .
                           .
                         END    BEGIN

Segments and Groups:

        N a m e                  Size    Length  Align  Combine   Class

DGROUP . . . . . . . . . . . . GROUP
_DATA  . . . . . . . . . . . . 16 Bit   0065    Word   Public    'DATA'
STACK  . . . . . . . . . . . . 16 Bit   0100    Para   Stack     'STACK'
_TEXT  . . . . . . . . . . . . 16 Bit   0005    Word   Public    'CODE'


Symbols:

        N a m e                  Type    Value   Attr

@CodeSize . . . . . . . . . . . Number  0000h
@DataSize . . . . . . . . . . . Number  0000h
```

```
@Interface . . . . . . . . . . . Number 0000h
@Model . . . . . . . . . . . . . Number 0002h
@code  . . . . . . . . . . . . . Text              _TEXT
@data  . . . . . . . . . . . . . Text              DGROUP
@fardata?  . . . . . . . . . . . Text              FAR_BSS
@fardata . . . . . . . . . . . . Text              FAR_DATA
@stack . . . . . . . . . . . . . Text              DGROUP
BEGIN  . . . . . . . . . . . . . L Near 0000       _TEXT
DATA1  . . . . . . . . . . . . . Byte   0001       _DATA
FROG . . . . . . . . . . . . . . Byte   0000       _DATA


          0 Warnings
          0 Errors
```

Example A–5 uses the large model. Notice how it differs from the small model program of Example A–4. Models can be very useful in developing software, but often we use full segment descriptions in our examples.

EXAMPLE A–5

```
Microsoft (R) Macro Assembler Version 6.11


                               .MODEL LARGE
                               .STACK 1000H
0000                           .FARDATA?

0000 00              FROG    DB      ?
0001 0064 [          DATA1   DW      100 DUP (?)
       0000
           ]

0000                           .CONST

0000 54 68 69 73 20 69  MES1    DB      'This is a character string'
     73 20 61 20 63 68
     61 72 61 63 74 65
     72 20 73 74 72 69
     6E 67
001A 53 6F 20 69 73 20  MES2    DB      'So is this!'
     74 68 69 73 21

0000                           .DATA

0000 000C             DATA2   DW      12
0002 00C8 [           DATA3   DB      200 DUP (1)
       01
         ]

0000                           .CODE

0000                 FUNC    PROC    FAR
                             .
                             .
                             .
0000 CB                      RET

0001                 FUNC    ENDP

                     END     FUNC

Segments and Groups:
```

```
             N a m e              Size      Length  Align Combine  Class

DGROUP . . . . . . . . . . . . GROUP
_DATA  . . . . . . . . . . . . 16 Bit    00CA    Word  Public   'DATA'
STACK  . . . . . . . . . . . . 16 Bit    1000    Para  Stack    'STACK'
CONST  . . . . . . . . . . . . 16 Bit    0025    Word  Public   'CONST'  ReadOnly
EXA_TEXT . . . . . . . . . . . 16 Bit    0001    Word  Public   'CODE'
FAR_BSS  . . . . . . . . . . . 16 Bit    00C9    Para  Private  'FAR_BSS'
_TEXT  . . . . . . . . . . . . 16 Bit    0000    Word  Public   'CODE'


Procedures, parameters, and locals:

             N a m e              Type    Value   Attr

FUNC . . . . . . . . . . . . . P Far    0000     EXA_TEXT Length= 0001 Public

Symbols:

             N a m e            Type    Value   Attr

@CodeSize  . . . . . . . . . . Number 0001h
@DataSize  . . . . . . . . . . Number 0001h
@Interface . . . . . . . . . . Number 0000h
@Model . . . . . . . . . . . . Number 0005h
@code  . . . . . . . . . . . . Text            EXA_TEXT
@data  . . . . . . . . . . . . Text            DGROUP
@fardata?  . . . . . . . . . . Text            FAR_BSS
@fardata . . . . . . . . . . . Text            FAR_DATA
@stack . . . . . . . . . . . . Text            DGROUP
DATA1  . . . . . . . . . . . . Word    0001    FAR_BSS
DATA2  . . . . . . . . . . . . Word    0000    _DATA
DATA3  . . . . . . . . . . . . Byte    0002    _DATA
FROG . . . . . . . . . . . . . Byte    0000    FAR_BSS
MES1 . . . . . . . . . . . . . Byte    0000    CONST
MES2 . . . . . . . . . . . . . Byte    001A    CONST


        0 Warnings
        0 Errors
```

DOS FUNCTION CALLS

To use DOS function calls, always place the function number into register AH and load all other pertinent information into the registers listed as entry data in Table A–3. Once this is accomplished, follow with an INT 21H to execute the DOS function. Example A–6 shows how to display an ASCII A on the CRT screen at the current cursor position with a DOS function call. Table A–3 is a complete listing of the DOS function calls. Note that some function calls require a segment and offset address indicated as DS:DI, for example. This means the data segment is the segment address and DI is the offset address. All of the function calls use INT 21H, and AH contains the function call number. Functions marked with an @ should not be used unless DOS version 2.XX is in use. Note that not all function numbers are implemented. As a rule, DOS function calls save all registers not used as exit data, but in certain cases some registers may change. In order to prevent problems, it is advisable to save registers where problems occur.

EXAMPLE A–6

```
0000 B4 06              MOV    AH,6        ;load function 06H
0002 B2 41              MOV    DL,'A'      ;select letter 'A'
0004 CD 21              INT    21H         ;call DOS function
```

TABLE A–3   DOS function calls (pp. 685–707)

00H	TERMINATE A PROGRAM
Entry	AH = 00H CS = program segment prefix address
Exit	DOS is entered

01H	READ THE KEYBOARD
Entry	AH = 01H
Exit	AL = ASCII character
Notes	If AL = 00H, the function call must be invoked again to read an extended ASCII character. Refer to Chapter 9, Table 9–1, for a listing of the extended ASCII keyboard codes. This function call automatically echoes whatever is typed to the video screen.

02H	WRITE TO STANDARD OUTPUT DEVICE
Entry	AH = 02H DL = ASCII character to be displayed
Notes	This function call normally displays data on the video display.

03H	READ CHARACTER FROM COM1
Entry	AH = 03H
Exit	AL = ASCII character read from the communications port
Notes	This function call reads data from the serial communications port.

04H	WRITE TO COM1
Entry	AH = 04H DL = character to be sent out of COM1
Notes	This function transmits data through the serial communications port. The COM port assignment can be changed to use other COM ports with functions 03H and 04H by using the DOS MODE command to reassign COM1 to another COM port.

05H	WRITE TO LPT1
Entry	AH = 05H DL = ASCII character to be printed
Notes	Prints DL on the line printer attached to LPT1. Note that the line printer ports can be changed with the DOS MODE command.

06H	DIRECT CONSOLE READ/WRITE
Entry	AH = 06H DL = 0FFH or DL = ASCII character
Exit	AL = ASCII character
Notes	If DL = 0FFH on entry, then this function reads the console. If DL = ASCII character, then this function displays the ASCII character on the console (CON) video screen. If a character is read from the console keyboard, the zero flag (ZF) indicates whether a character was typed. A zero condition indicates no key was typed, and a not-zero condition indicates that AL contains the ASCII code of the key or a 00H. If AL = 00H, the function must be invoked again to read an extended ASCII character from the keyboard. Note that the key does not echo to the video screen.

07H	DIRECT CONSOLE INPUT WITHOUT ECHO
Entry	AH = 07H
Exit	AL = ASCII character
Notes	This functions exactly as function number 06H with DL = 0FFH, but it will not return from the function until the key is typed.

08H	READ STANDARD INPUT WITHOUT ECHO
Entry	AH = 08H
Exit	AL = ASCII character
Notes	Performs as function 07H, except that it reads the standard input device. The standard input device can be assigned as either the keyboard or the COM port. This function also responds to a control-break, whereas function 06H and 07H do not. A control-break causes INT 23H to execute. By default, this functions the same as function 07H.

09H	DISPLAY A CHARACTER STRING
Entry	AH = 09H DS:DX = address of the character string
Notes	The character string must end with an ASCII $ (24H). The character string can be of any length and may contain control characters such as carriage return (0DH) and line feed (0AH).

0AH	BUFFERED KEYBOARD INPUT
Entry	AH = 0AH DS:DX = address of keyboard input buffer
Notes	The first byte of the buffer contains the size of the buffer (up to 255). The second byte is filled with the number of characters typed upon return. The third byte through the end of the buffer contains the character string typed, followed by a carriage return (0DH). This function continues to read the keyboard (displaying data as typed) until either the specified number of characters or the enter key is typed.

0BH	TEST STATUS OF THE STANDARD INPUT DEVICE
Entry	AH = 0BH
Exit	AL = status of the input device
Notes	This function tests the standard input device to determine if data is available. If AL = 00, no data is available. If AL = 0FFH, then data is available that must be input using function number 08H.

0CH	CLEAR KEYBOARD BUFFER AND INVOKE KEYBOARD FUNCTION
Entry	AH = 0CH AL = 01H, 06H, 07H, or 0AH
Exit	See exit for functions 01H, 06H, 07H, or 0AH
Notes	The keyboard buffer holds keystrokes while programs execute other tasks. This function empties or clears the buffer and then invokes the keyboard function located in register AL.

0DH	FLUSH DISK BUFFERS
Entry	AH = 0DH
Notes	Erases all filenames stored in disk buffers. This function does not close the files specified by the disk buffers, so care must be exercised in its usage.

0EH	SELECT DEFAULT DISK DRIVE
Entry	AH = 0EH DL = desired default disk drive number
Exit	AL = the total number of drives present in the system
Notes	Drive A = 00H, drive B = 01H, drive C = 02H, and so forth.

0FH	@OPEN FILE WITH FCB
Entry	AH = 0FH DS:DX = address of the unopened file control block (FCB)
Exit	AL = 00H if file found AL = 0FFH if file not found
Notes	The file control block (FCB) is used only with early DOS software and should never be used with new programs. File control blocks do not allow path names as do the newer file access function codes presented later. Figure A–2 (p. 707) illustrates the structure of the FCB. To open a file, the file must either be present on the disk or be created with function call 16H.

10H	@CLOSE FILE WITH FCB
Entry	AH = 10H DS:DX = address of the opened file control block (FCB)
Exit	AL = 00H if file closed AL = 0FFH if error found
Notes	Errors that occur usually indicate either that the disk is full or the media is bad.

11H	@SEARCH FOR FIRST MATCH (FCB)
Entry	AH = 11H DS:DX = address of the file control block to be searched
Exit	AL = 00H if file found AL = 0FFH if file not found
Notes	Wild card characters (? or *) may be used to search for a filename. The ? wild card character matches any character and the * matches any name or extension.

12H	@SEARCH FOR NEXT MATCH (FCB)
Entry	AH = 12H DS:DX = address of the file control block to be searched
Exit	AL = 00H if file found AL = 0FFH if file not found
Notes	This function is used after function 11H finds the first matching filename.
13H	**@DELETE FILE USING FCB**
Entry	AH = 13H DS:DX = address of the file control block to be deleted
Exit	AL = 00H if file deleted AL = 0FFH if error occurred
Notes	Errors that most often occur are defective media errors.
14H	**@SEQUENTIAL READ (FCB)**
Entry	AH = 14H DS:DX = address of the file control block to be read
Exit	AL = 00H if read successful AL = 01H if end of file reached AL = 02H if DTA had a segment wrap AL = 03H if less than 128 bytes were read
15H	**@SEQUENTIAL WRITE (FCB)**
Entry	AH = 15H DS:DX = address of the file control block to be written
Exit	AL = 00H if write successful AL = 01H if disk is full AL = 02H if DTA had a segment wrap
16H	**@CREATE A FILE (FCB)**
Entry	AH = 16H DS:DX = address of an unopened file control block
Exit	AL = 00H if file created AL = 01H if disk is full

17H	@RENAME A FILE (FCB)
Entry	AH = 17H DS:DX = address of a modified file control block
Exit	AL = 00H if file renamed AL = 01H if error occurred
Notes	Refer to Figure A–3 (p. 707) for the modified FCB used to rename a file.

19H	RETURN CURRENT DRIVE
Entry	AH = 19H
Exit	AL = current drive
Notes	AL = 00H for drive A, 01H for drive B, and so forth.

1AH	SET DISK TRANSFER AREA
Entry	AH = 1AH DS:DX = address of new DTA
Notes	The disk transfer area is normally located within the program segment prefix at offset address 80H. The DTA is used by DOS for all disk data transfers using file control blocks.

1BH	GET DEFAULT DRIVE FILE ALLOCATION TABLE (FAT)
Entry	AH = 1BH
Exit	AL = number of sectors per cluster DS:BX = address of the media-descriptor CX = size of a sector in bytes DX = number of clusters on drive
Notes	Refer to Figure A–4 (p. 708) for the format of the media-descriptor byte. The DS register is changed by this function, so make sure to save it before using this function.

1CH	GET ANY DRIVE FILE ALLOCATION TABLE (FAT)
Entry	AH = 1CH DL = disk drive number
Exit	AL = number of sectors per cluster DS:BX = address of the media-descriptor CX = size of a sector in bytes DX = number of clusters on drive

21H	@RANDOM READ USING FCB
Entry	AH = 21H DS:DX = address of opened FCB
Exit	AL = 00H if read successful AL = 01H if end of file reached AL = 02H if the segment wrapped AL = 03H if less than 128 bytes read
22H	@RANDOM WRITE USING FCB
Entry	AH = 22H DS:DX = address of opened FCB
Exit	AL = 00H if write successful AL = 01H if disk full AL = 02H if the segment wrapped
23H	@RETURN NUMBER OF RECORDS (FCB)
Entry	AH = 23H DS:DX = address of FCB
Exit	AL = 00H number of records AL = 0FFH if file not found
24H	@SET RELATIVE RECORD SIZE (FCB)
Entry	AH = 24H DS:DX = address of FCB
Notes	Sets the record field to the value contained in the FCB.
25H	SET INTERRUPT VECTOR
Entry	AH = 25H AL = interrupt vector number DS:DX = address of new interrupt procedure
Notes	Before changing the interrupt vector, it is suggested that the current interrupt vector first be saved using DOS function 35H. This allows a back-link so the original vector can later be restored.
26H	CREATE NEW PROGRAM SEGMENT PREFIX
Entry	AH = 26H DX = segment address of new PSP
Notes	Figure A–5 (p. 708) illustrates the structure of the program segment prefix.

27H	**@RANDOM FILE BLOCK READ (FCB)**
Entry	AH = 27H CX = the number of records DS:DX = address of opened FCB
Exit	AL = 00H if read successful AL = 01H if end of file reached AL = 02H if the segment wrapped AL = 03H if less than 128 bytes read CX = the number of records read
28H	**@RANDOM FILE BLOCK WRITE (FCB)**
Entry	AH = 28H CX = the number of records DS:DX = address of opened FCB
Exit	AL = 00H if write successful AL = 01H if disk full AL = 02H if the segment wrapped CX = the number of records written
29H	**@PARSE COMMAND LINE (FCB)**
Entry	AH = 29H AL = parse mask DS:SI = address of FCB DS:DI = address of command line
Exit	AL = 00H if no filename characters found AL = 01H if filename characters found AL = 0FFH if drive specifier incorrect DS:SI = address of character after name DS:DI = address first byte of FCB
2AH	READ SYSTEM DATE
Entry	AH = 2AH
Exit	AL = day of the week CX = the year (1980–2099) DH = the month DL = day of the month
Notes	The day of the week is encoded as Sunday = 00H through Saturday = 06H. The year is a binary number equal to 1980 through 2099.

2BH	SET SYSTEM DATE
Entry	AH = 2BH CX = the year (1980–2099) DH = the month DL = day of the month
2CH	**READ SYSTEM TIME**
Entry	AH = 2CH
Exit	CH = hours (0–23) CL = minutes DH = seconds DL = hundredths of seconds
Notes	All times are returned in binary form, and hundredths of seconds may not be available.
2DH	**SET SYSTEM TIME**
Entry	AH = 2DH CH = hours CL = minutes DH = seconds DL = hundredths of seconds
2EH	**DISK VERIFY WRITE**
Entry	AH = 2EH AL = 00H to disable verify on write AL = 01H to enable verify on write
Notes	By default, disk verify is disabled.
2FH	**READ DISK TRANSFER AREA ADDRESS**
Entry	AH = 2FH
Exit	ES:BX = contains DTA address
30H	**READ DOS VERSION NUMBER**
Entry	AH = 30H
Exit	AH = fractional version number AL = whole number version number
Notes	For example, DOS version number 3.2 is returned as a 3 in AL and a 14H in AH.

31H	TERMINATE AND STAY RESIDENT (TSR)
Entry	AH = 31H AL = the DOS return code DX = number of paragraphs to reserve for program
Notes	A paragraph is 16 bytes and the DOS return code is read at the batch file level with ERRORCODE.

33H	TEST CONTROL-BREAK
Entry	AH = 33H AL = 00H to request current control-break AL = 01H to change control-break DL = 00H to disable control-break DL = 01H to enable control-break
Exit	DL = current control-break state

34H	GET ADDRESS OF InDOS FLAG
Entry	AH = 34H
Exit	ES:BX = address of InDOS flag
Notes	The InDOS flag is available in DOS versions 3.2 or newer and indicates DOS activity. If InDOS = 00H, DOS is inactive or 0FFH if DOS is active and pursuing another operation.

35H	READ INTERRUPT VECTOR
Entry	AH = 35H AL = interrupt vector number
Exit	ES:BX = address stored at vector
Notes	This DOS function is used with function 25H to install/remove interrupt handlers.

36H	DETERMINE FREE DISK SPACE
Entry	AH = 36H DL = drive number
Exit	AX = FFFFH if drive invalid AX = number of sectors per cluster BX = number of free clusters CX = bytes per sector DX = number of clusters on drive
Notes	The default disk drive is DL = 00H, drive A = 01H, drive B = 02H, and so forth.

38H	RETURN COUNTRY CODE
Entry	AH = 38H AL = 00H for current country code BX = 16-bit country code DS:DX = data buffer address
Exit	AX = error code if carry set BX = counter code DS:DX = data buffer address
39H	CREATE SUBDIRECTORY
Entry	AH = 39H DS:DX = address of ASCII-Z string subdirectory name
Exit	AX = error code if carry set
Notes	The ASCII-Z string is the name of the subdirectory in ASCII code ended with a 00H instead of a carriage return/line feed.
3AH	ERASE SUBDIRECTORY
Entry	AH = 3AH DS:DX = address of ASCII-Z string subdirectory name
Exit	AX = error code if carry set
3BH	CHANGE SUBDIRECTORY
Entry	AH = 3BH DS:DX = address of new ASCII-Z string subdirectory name
Exit	AX = error code if carry set
3CH	CREATE A NEW FILE
Entry	AH = 3CH CX = attribute word DS:DX = address of ASCII-Z string filename
Exit	AX = error code if carry set AX = file handle if carry cleared
Notes	The attribute word can contain any of the following (added together): 01H read-only access, 02H = hidden file or directory, 04H = system file, 08H = volume label, 10H = subdirectory, and 20H = archive bit. In most cases, a file is created with 0000H.

3DH	OPEN A FILE
Entry	AH = 3DH AL = access code DS:DX = address of ASCII-Z string filename
Exit	AX = error code if carry set AX = file handle if carry cleared
Notes	The access code in AL = 00H for a read-only access, AL = 01H for a write-only access, and AL = 02H for a read/write access. For shared files in a network environment, bit 4 of AL = 1 will deny read/write access, bit 5 of AL = 1 will deny a write access, bits 4 and 5 of AL = 1 will deny read access, bit 6 of AL = 1 denies none, bit 7 of AL = 0 causes the file to be inherited by child; if bit 7 of AL = 1, file is restricted to current process.
3EH	CLOSE A FILE
Entry	AH = 3EH BX = file handle
Exit	AX = error code if carry set
3FH	READ A FILE
Entry	AH = 3FH BX = file handle CX = number of bytes to be read DS:DX = address of file buffer to hold data read
Exit	AX = error code if carry set AX = number of bytes read if carry cleared
40H	WRITE A FILE
Entry	AH = 40H BX = file handle CX = number of bytes to write DS:DX = address of file buffer that holds write data
Exit	AX = error code if carry set AX = number of bytes written if carry cleared
41H	DELETE A FILE
Entry	AH = 41H DS:DX = address of ASCII-Z string filename
Exit	AX = error code if carry set

42H	MOVE FILE POINTER
Entry	AH = 42H AL = move technique BX = file handle CX:DX = number of bytes pointer moved
Exit	AX = error code if carry set AX:DX = bytes pointer moved
Notes	The move technique causes the pointer to move from the start of the file if AL = 00H, from the current location if AL = 01H, and from the end of the file if AL = 02H. The count is stored so DX contains the least-significant 16 bits and either CX or AX contains the most-significant 16 bits.
43H	READ/WRITE FILE ATTRIBUTES
Entry	AH = 43H AL = 00H to read attributes AL = 01H to write attributes CX = attribute word (see function 3CH) DS:DX = address of ASCII-Z string filename
Exit	AX = error code if carry set CX = attribute word of carry cleared
44H	I/O DEVICE CONTROL (IOTCL)
Entry	AH = 44H AL = subfunction code (see notes)
Exit	AX = error code (see function 59H) if carry set

Notes	The subfunction codes found in AL are as follows:
	
	00H = read device status
	    Entry: BX = file handle
	    Exit: DX = status
	01H = write device status
	    Entry: BX = file handle, DH = 0, DL = device information
	    Exit: AX = error code if carry set
	02H = read control data from character device
	    Entry: BX = file handle, CX = number of bytes, DS:DX = I/O
	              buffer address
	    Exit: AX = number of bytes read
	03H = write control data to character device
	    Entry: BX = file handle, CX = number of bytes, DS:DX = I/O
	              buffer address
	    Exit: AX = number of bytes written

04H = read control data from block device
    Entry: BL = drive number (0 = default, 1 = A, 2 = B, etc),
              CX = number of bytes, DS:DX = I/O buffer address
    Exit: AX = number of bytes read
05H = write control data to block device
    Entry: BL = drive number, CX = number of bytes,
              DS:DX = I/O buffer address
    Exit: AX = number of bytes written
06H = check input status
    Entry: BX = file handle
    Exit: AL = 00H ready or FFH not ready
07H = check output status
    Entry: BX = file handle
    Exit: AL = 00H ready or FFH not ready
08H = removable media?
    Entry: BL = drive number
    Exit: AL = 00H removable, 01H fixed
09H = network block device?
    Entry: BL = drive number
    Exit: bit 12 of DX set for network block device
0AH = local or network character device?
    Entry: BX = file handle
    Exit: bit 15 of DX set for network character device
0BH = change entry count (must have SHARE.EXE loaded)
    Entry: CX = delay loop count, DX = retry count
    Exit: AX = error code if carry set
0CH = generic I/O control for character devices
    Entry: BX = file handle, CH = category, CL = function
                 Categories: 00H = unknown, 01H = COM port, 02H =
                 CON, 05H = LPT ports
    Function:
                 CL = 45H; set iteration count
                 CL = 4AH; select code page
                 CL = 4CH; start code page preparation
                 CL = 4DH; end code page preparation
                 CL = 5FH; set display information
                 CL = 65H; get iteration count
                 CL = 6AH; query selected code page
                 CL = 6BH; query preparation list
                 CL = 7FH; get display information
0DH = generic I/O control for block devices
    Entry: BL = drive number, CH = category, CL = function,
              DS:DX = address of parameter block
    Category: 08H = disk drive
    Function:
                 CL = 40H; set device parameters
                 CL = 41H; write track
                 CL = 42H; format and verify track
                 CL = 46H, set media ID code
                 CL = 47H; set access flag
                 CL = 60H; get device parameters
                 CL = 61H; read track
                 CL = 62H; verify track
                 CL = 66H; get media ID code
                 CL = 67H; get access code

	0EH = return logical device map
	    Entry: BL = drive number
	    Exit: AL = number of last device
	0FH = change logical device map
	    Entry: BL = drive number
	    Exit: AL = number of last device

45H	DUPLICATE FILE HANDLE
Entry	AH = 45H BX = current file handle
Exit	AX = error code if carry set AX = duplicate file handle
46H	FORCE DUPLICATE FILE HANDLE
Entry	AH = 46H BX = current file handle CX = new file handle
Exit	AX = error code if carry set
Notes	This function works like function 45H except that function 45H allows DOS to select the new handle, while this function allows the user to select the new handle.
47H	READ CURRENT DIRECTORY
Entry	AH = 47H DL = drive number DS:SI = address of a 64-byte buffer for directory name
Exit	DS:SI addresses current directory name if carry cleared
Notes	Drive A = 00, drive B = 01, and so forth
48H	ALLOCATE MEMORY BLOCK
Entry	AH = 48H BX = number of paragraphs to allocate CX = new file handle
Exit	BX = largest block available if carry cleared
49H	RELEASE ALLOCATED MEMORY BLOCK
Entry	AH = 49H ES = segment address of block to be released CX = new file handle
Exit	Carry indicates an error if set

4AH	MODIFY ALLOCATED MEMORY BLOCK
Entry	AH = 4AH BX = new block size in paragraphs ES = segment address of block to be modified
Exit	BX = largest block available if carry cleared

4BH	LOAD OR EXECUTE A PROGRAM
Entry	AH = 4BH AL = function code ES:BX = address of parameter block DS:DX = address ASCII-Z string command
Exit	Carry indicates an error if set
Notes	The function codes are AL = 00H to load and execute a program, AL = 01H to load a program but not execute it, AL = 03H to load a program overlay, and AL = 05H to enter the EXEC state. Figure A–6 (p. 709) shows the parameter block used with this function.

4CH	TERMINATE A PROCESS
Entry	AH = 4CH AL = error code
Exit	Returns control to DOS
Notes	This function returns control to DOS with the error code saved so it can be obtained using DOS ERROR LEVEL batch processing system. We normally use this function with an error code of 00H to return to DOS.

4DH	READ RETURN CODE
Entry	AH = 4DH
Exit	AX = return error code
Notes	This function is used to obtain the return status code created by executing a program with DOS function 4BH. The return codes are AX = 0000H for a normal-no error-termination, AX = 0001H for a control-break termination, AX = 0002H for a critical device error, and AX = 0003H for a termination by an INT 31H.

4EH	FIND FIRST MATCHING FILE
Entry	AH = 4EH CX = file attributes DS:DX = address ASCII-Z string filename
Exit	Carry is set for file not found
Notes	This function searches the current or named directory for the first matching file. Upon exit, the DTA contains the file information. See Figure A–7 (p. 709) for the disk transfer area (DTA).
4FH	FIND NEXT MATCHING FILE
Entry	AH = 4FH
Exit	Carry is set for file not found
Notes	This function is used after the first file is found with function 4EH.
50H	SET PROGRAM SEGMENT PREFIX (PSP) ADDRESS
Entry	AH = 50H BX = offset address of the new PSP
Notes	Extreme care must be used with this function because no error recovery is possible.
51H	GET PSP ADDRESS
Entry	AH = 51H
Exit	BX = current PSP segment address
54H	READ DISK VERIFY STATUS
Entry	AH = 54H
Exit	AL = 00H if verify off AL = 01H if verify on
56H	RENAME FILE
Entry	AH = 56H ES:DI = address of ASCII-Z string containing new filename DS:DX = address of ASCII-Z string containing file to be renamed
Exit	Carry is set for error condition

57H	READ FILE'S DATE AND TIME STAMP
Entry	AH = 57H AL = function code BX = file handle CX = new time DX = new date
Exit	Carry is set for error condition CX = time if carry cleared DX = date if carry cleared
Notes	AL = 00H to read date and time or 01H to write date and time.

59H	GET EXTENDED ERROR INFORMATION
Entry	AH = 59H BX = 0000H for DOS version 3.X
Exit	AX = extended error code BH = error class BL = recommended action CH = locus
Notes	Following are the extended error codes found in AX: 0001H = invalid function number 0002H = file not found 0003H = path not found 0004H = no file handles available 0005H = access denied 0006H = file handle invalid 0007H = memory control block failure 0008H = insufficient memory 0009H = memory block address invalid 000AH = environment failure 000BH = format invalid 000CH = access code invalid 000DH = data invalid 000EH = unknown unit 000FH = disk drive invalid 0010H = attempted to remove current directory 0011H = not same device 0012H = no more files 0013H = disk write-protected 0014H = unknown unit 0015H = drive not ready 0016H = unknown command 0017H = data error (CRC check error) 0018H = bad request structure length 0019H = seek error 001AH = unknown media type 001BH = sector not found

001CH = printer out of paper
001DH = write fault
001EH = read fault
001FH = general failure
0020H = sharing violation
0021H = lock violation
0022H = disk change invalid
0023H = FCB unavailable
0024H = sharing buffer exceeded
0025H = code page mismatch
0026H = handle end of file operation not completed
0027H = disk full
0028H–0031H reserved
0032H = unsupported network request
0033H = remote machine not listed
0034H = duplicate name on network
0035H = network name not found
0036H = network busy
0037H = device no longer exists on network
0038H = netBIOS command limit exceeded
0039H = error in network adapter hardware
003AH = incorrect response from network
003BH = unexpected network error
003CH = remote adapter is incompatible
003DH = print queue is full
003EH = not enough room for print file
003FH = print file was deleted
0040H = network name deleted
0041H = network access denied
0042H = incorrect network device type
0043H = network name not found
0044H = network name exceeded limit
0045H = netBIOS session limit exceeded
0046H = temporary pause
0047H = network request not accepted
0048H = print or disk redirection pause
0049H–004FH reserved
0050H = file already exists
0051H = duplicate FCB
0052H = cannot make directory
0053H = failure in INT 24H (critical error)
0054H = too many redirections
0055H = duplicate redirection
0056H = invalid password
0057H = invalid parameter
0058H = network write failure
0059H = function not supported by network
005AH = required system component not installed
0065H = device not selected

Following are the error class codes found in BH:

01H = no resources available
02H = temporary error
03H = authorization error

	04H = internal software error 05H = hardware error 06H = system failure 07H = application software error 08H = item not found 09H = invalid format 0AH = item blocked 0BH = media error 0CH = item already exists 0DH = unknown error Following is the recommended action found in BL: 01H = retry operation 02H = delay and retry operation 03H = user retry 04H = abort processing 05H = immediate exit 06H = ignore error 07H = retry with user intervention Following is a list of loci in CH: 01H = unknown source 02H = block device error 03H = network area 04H = serial device error 05H = memory error
5AH	CREATE UNIQUE FILE NAME
Entry	AH = 5AH CX = attribute code DS:DX = address of the ASCII-Z string directory path
Exit	Carry is set for error condition AX = file handle if carry cleared DS:DX = address of the appended directory name
Notes	The ASCII-Z file directory path must end with a backslash (\). On exit, the directory name is appended with a unique filename.
5BH	CREATE A DOS FILE
Entry	AH = 5BH CX = attribute code DS:DX = address of the ASCII-Z string contain the filename
Exit	Carry is set for error condition AX = file handle if carry cleared
Notes	The function works only in DOS version 3.X or higher. It is almost identical to function 3CH, except that function 3CH erases the file if it already exists, while function 5BH reports that the file exists without erasing it.

5CH	LOCK/UNLOCK FILE CONTENTS
Entry	AH = 5CH BX = file handle CX:DX = offset address of locked/unlocked area SI:DI = number of bytes to lock or unlock beginning at offset
Exit	Carry is set for error condition

5DH	SET EXTENDED ERROR INFORMATION
Entry	AH = 5DH AL = 0AH DS:DX = address of the extended error data structure
Notes	This function is used by DOS version 3.1 or higher to store extended error information.

5EH	NETWORK/PRINTER
Entry	AH = 5EH AL = 00H (get network name) DS:DX = address of the ASCII-Z string containing network name
Exit	Carry is set for error condition CL = netBIOS number if carry cleared
Entry	AH = 5EH AL = 02H (define network printer) BX = redirection list CX = length of setup string DS:DX = address of printer setup buffer
Exit	Carry is set for error condition
Entry	AH = 5EH AL = 03H (read network printer setup string) BX = redirection list DS:DX = address of printer setup buffer
Exit	Carry is set for error condition CX = length of setup string if carry cleared ES:DI = address of printer setup buffer

62H	GET PSP ADDRESS
Entry	AH = 62H
Exit	BX = segment address of the current program
Notes	The function works only in DOS version 3.0 or higher.

65H	GET EXTENDED COUNTRY INFORMATION
Entry	AH = 65H AL = function code ES:DI = address of buffer to receive information
Exit	Carry is set for error condition CX = length of country information
Notes	The function works only in DOS version 3.3 or higher.

66H	GET/SET CODE PAGE
Entry	AH = 66H AL = function code BX = code page number
Exit	Carry is set for error condition BX = active code page number DX = default code page number
Notes	A function code in AL of 01H gets the code page number, and a code of 02H sets the code page number.

67H	SET HANDLE COUNT
Entry	AH = 67H BX = number of handles desired
Exit	Carry is set for error condition
Notes	This function is available for DOS version 3.3 or higher.

68H	COMMIT FILE
Entry	AH = 68H BX = handle number
Exit	Carry is set for error condition Else the date and time stamp is written to directory.
Notes	This function is available for DOS version 3.3 or higher.

6CH	EXTENDED OPEN FILE
Entry	AH = 6CH AL = 00H BX = open mode CX = attributes DX = open flag DS:SI = address of ASCII-Z string filename
Exit	AX = error code if carry is set AX = handle if carry is cleared CX = 0001H file existed and was opened CX = 0002H file did not exist and was created
Notes	This function is available for DOS version 4.0 or higher.

FIGURE A–2 Contents of the file control block (FCB).

Offset	Contents
00H	Drive
01H	8-character filename
09H	3-character file extension
0CH	Current block number
0EH	Record size
10H	File size
14H	Creation date
16H	Reserved space
20H	Current record number
21H	Relative record number

FIGURE A–3 Contents of the modified file control block (FCB).

Offset	Contents
00H	Drive
01H	8-character filename
09H	3-character extension
0CH	Current block number
0EH	Record size
10H	File size
14H	Creation date
16H	Second filename

FIGURE A–4 Contents of the media-descriptor byte.

7	6	5	4	3	2	1	0
?	?	?	?	?	?	?	?

Bit 0 = 0 if not two-sided
     = 1 if two-sided

Bit 1 = 0 if not eight sectors per track
     = 1 if eight sectors per track

Bit 2 = 0 if nonremovable
     = 1 if removable

FIGURE A–5 Contents of the program segment prefix (PSP).

Offset	Contents
00H	INT 20H
02H	Top of memory
04H	Reserved
05H	Opcode
06H	Number of bytes in segment
0AH	Terminate address (offset)
0CH	Terminate address (segment)
0EH	Control-break address (offset)
10H	Control-break address (segment)
12H	Critical error address (offset)
14H	Critical error address (segment)
16H	Reserved
2CH	Environment address (segment)
2EH	Reserved
50H	DOS call
52H	Reserved
5CH	File control block 1
6CH	File control block 2
80H	Command line length
81H	Command line

FIGURE A–6 The parameter blocks used with function 4BH (EXEC). (a) For function code 00H. (b) For function code 03H.

(a)

Offset	Contents
00H	Environment address (segment)
02H	Command line address (offset)
04H	Command line address (segment)
06H	File control block 1 address (offset)
08H	File control block 1 address (segment)
0AH	File control block 2 address (segment)
0CH	File control block 2 address (offset)

(b)

Offset	Contents
00H	Overlay destination segment address
02H	Relocation factor

FIGURE A–7 Data transfer area (DTA) used to find a file.

Offset	Contents
15H	Attributes
16H	Creation time
18H	Creation date
1AH	Low word file size
1CH	High word file size
1EH	Search filename

BIOS FUNCTION CALLS

In addition to DOS function call INT 21H, some other BIOS function calls are useful in controlling the I/O environment of the computer. Unlike INT 21H, which exists in the DOS program, the BIOS function calls are stored in the system and video BIOS ROMs. These BIOS ROM functions directly control the I/O devices with or without DOS loaded into a system.

INT 10H

The INT 10H BIOS interrupt is often called the video services interrupt because it directly controls the video display in a system. The INT 10H instruction uses register AH to select the video service provided by this interrupt. The video BIOS ROM is located on the video board and varies from one video card to another.

Video Mode Selection.   The mode of operation for the video display is accomplished by placing a 00H into AH followed by one of many mode numbers in AL. Table A–4 lists the modes of operation found in video display systems using standard video modes. The VGA can use any mode listed, while the other displays are more restrictive in use. Additional higher resolution modes are explained later in this section.

TABLE A–4   Video display modes

Mode	Type	Columns	Rows	Resolution	Standard	Colors
00H	Text	40	25	320×200	CGA	2
00H	Text	40	25	320×250	EGA	2
00H	Text	40	25	360×400	VGA	2
01H	Text	40	25	320×200	CGA	16
01H	Text	40	25	320×350	EGA	16
01H	Text	40	25	360×640	VGA	16
02H	Text	80	25	640×200	CGA	2
02H	Text	80	25	640×350	EGA	2
02H	Text	80	25	720×400	VGA	2
03H	Text	80	25	640×200	CGA	16
03H	Text	80	25	640×350	EGA	16
03H	Text	80	25	720×400	VGA	16
04H	Graphics	80	25	320×200	CGA	4
05H	Graphics	80	25	320×350	CGA	2
06H	Graphics	80	25	640×200	CGA	2
07H	Text	80	25	720×350	EGA	4
07H	Text	80	25	720×400	VGA	4
0DH	Graphics	80	25	320×200	CGA	16
0EH	Graphics	80	25	640×200	CGA	16
0FH	Graphics	80	25	640×350	EGA	4
10H	Graphics	80	25	640×350	EGA	16
11H	Graphics	80	30	640×480	VGA	2
12H	Graphics	80	30	640×480	VGA	16
13H	Graphics	40	25	320×200	VGA	256

Example A–7 lists a short sequence of instructions that place the video display into mode 03H operation. This mode, available on CGA, EGA, and VGA displays, allows the display to draw test data with 16 colors at various resolutions depending upon the display adapter.

EXAMPLE A–7

```
0000 B4 00              MOV    AH,0           ;select mode
0002 B0 03              MOV    AL,3           ;mode is 03H
0004 CD 10              INT    10H
```

Cursor Control and Other Standard Features.   Table A–5 shows the function codes (placed in AH) used to control the cursor on the video display. These cursor control functions will work on any video display from the CGA display to the latest super VGA display. It also lists the functions used to display data and change to a different character set.

TABLE A–5   Video BIOS (INT 10H) functions (pp. 710–713)

00H	SELECT VIDEO MODE
Entry	AH = 00H AL = mode number
Exit	Mode changed and screen cleared

01H	SELECT CURSOR TYPE
Entry	AH = 01H CH = starting line number CL = ending line number
Exit	Cursor size changed

02H	SELECT CURSOR POSITION
Entry	AH = 02H BH = page number (usually 0) DH = row number (beginning with 0) DL = column number (beginning with 0)
Exit	Changes cursor to new position

03H	READ CURSOR POSITION
Entry	AH = 03H BH = page number
Exit	CH = starting line (cursor size) CL = ending line (cursor size) DH = current row DL = current column

04H	READ LIGHT PEN
Entry	AH = 04H (not supported in VGA)
Exit	AH = 0, light pen triggered BX = pixel column CX = pixel row DH = character row DL = character column

05H	SELECT DISPLAY PAGE
Entry	AH = 05H AL = page number
Exit	Page number selected. Following are the valid page numbers: Mode 0 and 1 support pages 0–7 Mode 2 and 3 support pages 0–7 Mode 4, 5, and 6 support page 0 Mode 7 and D support pages 0–7 Mode E supports pages 0–3 Mode F and 10 support pages 0–1 Mode 11, 12, and 13 support page 0
Notes	Most modern displays use page 0 for most operations.

06H	SCROLL PAGE UP
Entry	AH = 06H AL = number of lines to scroll (0 clears window) BH = character attribute for new lines CH = top row of scroll window CL = left column of scroll window DH = bottom row of scroll window DL = right column of scroll window
Exit	Scrolls window from the bottom toward the top of the screen. Blank lines fill the bottom using the character attribute in BH.
07H	SCROLL PAGE DOWN
Entry	AH = 07H AL = number of lines to scroll (0 clears window) BH = character attribute for new lines CH = top row of scroll window CL = left column of scroll window DH = bottom row of scroll window DL = right column of scroll window
Exit	Scrolls window from the top toward the bottom of the screen. Blank lines fill from the top using the character attribute in BH.
08H	READ ATTRIBUTE/CHARACTER AT CURRENT CURSOR POSITION
Entry	AH = 08H BH = page number
Exit	AL = ASCII character code AH = character attribute
Notes	This function does not advance the cursor.
09H	WRITE ATTRIBUTE/CHARACTER AT CURRENT CURSOR POSITION
Entry	AH = 09H AL = ASCII character code BH = page number BL = character attribute CX = number of characters to write
Notes	This function does not advance the cursor.

0AH	WRITE CHARACTER AT CURRENT CURSOR POSITION
Entry	AH = 0AH AL = ASCII character code BH = page number CX = number of characters to write
Note	This function does not advance the cursor.

0FH	READ VIDEO MODE
Entry	AH = 0FH
Exit	AL = current video mode AH = number of character columns BH = page number

10H	SET VGA PALETTE REGISTER
Entry	AH = 10H AL = 10H BX = color number (0–255) CH = green (0–63) CL = blue (0–63) DH = red (0–63)
Exit	Palette register color is changed. Note: The first 16 colors (0–15) are used in the 16-color VGA text mode and other modes.

10H	READ VGA PALETTE REGISTER
Entry	AH = 10H AL = 15H BX = color number (0–255)
Exit	CH = green CL = blue DH = red

11H	GET ROM CHARACTER SET
Entry	AH = 11H AL = 30H BH = 2 = ROM 8 × 14 character set BH = 3 = ROM 8 × 8 character set BH = 4 = ROM 8 × 8 extended character set BH = 5 = ROM 9 × 14 character set BH = 6 = ROM 8 × 16 character set BH = 7 = ROM 9 × 16 character set
Exit	CX = bytes per character DL = rows per character ES:BP = address of character set

If an SVGA (super VGA), EVGA (extended VGA), or XVGA (also extended VGA) adapter is available, the super VGA mode is set by using INT 10H function call AX = 4F02H with BX equal to the VGA mode for these advanced display adapters. This conforms to the VESA standard for VGA adapters. Table A–6 shows the modes selected by register BX for this INT 10H function call. Most video cards are equipped with a driver called VVESA.COM or VVESA.SYS that conforms the card to the VESA standard functions.

INT 11H

This function is used to determine the type of equipment installed in the system. To use this call, the AX register is loaded with an FFFFH and then the INT 11H instruction is executed. In return, an INT 11H provides information as shown in Figure A–8.

INT 12H

The memory size is returned by the INT 12H instruction. After executing the INT 12H instruction, the AX register contains the number of 1K-byte blocks of memory (conventional memory in the first 1M byte of address space) installed in the computer.

INT 13H

This call controls the diskettes ($5^1/_4''$ or $3^1/_2''$) and also fixed or hard disk drives attached to the system. Table A–7 lists the functions available to this interrupt via register AH. The direct control of a floppy or hard disk can lead to problems. Therefore we only provide a listing of the functions without detail on their usage. Before using these functions, refer to the BIOS literature available from the company that produced your version of the BIOS ROM. Never use these functions for normal disk operations.

TABLE A–6   Extended VGA functions

BX	Extended Mode
100H	640×400 with 256 colors
101H	640×480 with 256 colors
102H	800×600 with 16 colors
103H	800×600 with 256 colors
104H	$1,024 \times 768$ with 16 colors
105H	$1,024 \times 768$ with 256 colors
106H	$1,280 \times 1,024$ with 16 colors
107H	$1,280 \times 1,024$ with 256 colors
108H	80×60 in text mode
109H	132×25 in text mode
10AH	132×43 in text mode
10BH	132×50 in text mode
10CH	132×60 in text mode

FIGURE A–8   The contents of AX as it indicates the equipment attached to the computer.

P1, P0 = number of parallel ports
G = 1 if game I/O attached
S2, S1, S0 = number of serial ports
D2, D1 = number of disk drives

TABLE A–7 Disk I/O functions via INT 13H

AH	Function
00H	Reset the system disk
01H	Read disk status to AL
02H	Read sector
03H	Write sector
04H	Verify sector
05H	Format track
06H	Format bad track
07H	Format drive
08H	Get drive parameters
09H	Initialize fixed disk characteristics
0AH	Read long sector
0BH	Write long sector
0CH	Seek
0DH	Reset fixed disk system
0EH	Read sector buffer
0FH	Write sector buffer
10H	Get drive status
11H	Recalibrate drive
12H	Controller RAM diagnostics
13H	Controller drive diagnostics
14H	Controller internal diagnostics
15H	Get disk type
16H	Get disk changed status
17H	Set disk type
18H	Set media type
19H	Park heads
1AH	Format ESDI drive

INT 14H

Interrupt 14H controls the serial COM (communications) ports attached to the computer. The computer system contains two COM ports, COM1 and COM2, unless you have a newer AT-style machine where the number of communications ports are extended to COM3 and COM4. Communications ports are normally controlled with software packages that allow data transfer through a modem and the telephone lines. The INT 14H instruction controls these ports as listed in Table A–8.

TABLE A–8 COM port interrupt INT 14H

AH	Function
00H	Initialize communications port
01H	Send character
02H	Receive character
03H	Get COM port status
04H	Extended initialize communications port
05H	Extended communications port control

INT 15H

The INT 15H instruction controls many of the various I/O devices interfaced to the computer. It also allows access to protected-mode operation and the extended memory system on an 80286–Pentium Pro, but it is not recommended. Table A–9 lists the functions supported by INT 15H.

INT 16H

The INT 16H instruction is used as a keyboard interrupt. This interrupt is accessed by DOS interrupt INT 21H, but can also be accessed directly. Table A–10 lists the functions performed by INT 16H.

TABLE A–9   The I/O subsystem interrupt INT 15H

AH	Function
00H	Cassette motor on
01H	Cassette motor off
02H	Read cassette
03H	Write cassette
0FH	Format ESDI periodic interrupt
21H	Keyboard intercept
80H	Device open
81H	Device closed
82H	Process termination
83H	Event wait
84H	Read joystick
85H	System request key
86H	Delay
87H	Move extended block of memory
88H	Get extended memory size
89H	Enter protected mode
90H	Device wait
91H	Device power on self-test (POST)
C0H	Get system environment
C1H	Get address of extended BIOS data area
C2H	Mouse pointer
C3H	Set watchdog timer
C4H	Programmable option select

TABLE A–10   Keyboard interrupt INT 16H

AH	Function
00H	Read keyboard character
01H	Get keyboard status
02H	Get keyboard flags
03H	Set repeat rate
04H	Set keyboard click
05H	Push character and scan code

INT 17H

The INT 17H instruction accesses the parallel printer port usually labeled LPT1 in most systems. Table A–11 lists the three functions available for the INT 17H instruction.

DOS Low-Memory Assignments

Table A–12 shows the low memory assignments (00000H–005FFH) for the DOS-based microprocessor system. This area of memory contains the interrupt vectors, BIOS data area, and the DOS/BIOS data.

TABLE A–11 Parallel printer interrupt INT 17H

AH	Function
00H	Print character
01H	Initialize printer
02H	Get printer status

TABLE A–12 DOS low-memory assignments

Location	Purpose
00000H–002FFH	System interrupt vectors
00300H–003FFH	System interrupt vectors, power on, and bootstrap area
00400H–00407H	COM1–COM4 I/O port base addresses
00408H–0040FH	LPT1–LPT4 I/O port base addresses
00410H–00411H	Equipment flag word, returned in AX by INT 11H (refer to Figure A–8)
00412H	Reserved
00413H–00414H	Memory size in K byte (0–640K)
00415H–00416H	Reserved
00417H	Keyboard control byte

Bit	Purpose
7	Insert locked
6	Caps locked
5	Numbers locked
4	Scroll locked
3	Alternate key pressed
2	Control key pressed
1	Left shift key pressed
0	Right shift key pressed

Location	Purpose
00418H	Keyboard control byte

Bit	Purpose
7	Insert locked
6	Caps locked
5	Numbers locked
4	Scroll locked
3	Pause key pressed
2	System request key pressed
1	Left alternate key pressed
0	Right control key pressed

(continued on the next page)

TABLE A–12   *(continued)*

Location	Purpose
00419H	Alternate keyboard entry
0041AH–0041BH	Keyboard buffer header pointer
0041CH–0041DH	Keyboard buffer tail pointer
0041EH–0043DH	32-byte keyboard buffer area
0043EH–00448H	Disk drive control area
00449H–00466H	Video control area
00467H–0046BH	Reserved
0046CH–0046FH	Timer counter
00470H	Timer overflow
00471H	Break key state
00472H–00473H	Reset flag
00474H–00477H	Hard disk drive data area
00478H–0047BH	LPT1–LPT4 time-out area
0047CH–0047FH	COM1–COM4 time-out area
00480H–00481H	Keyboard buffer start offset pointer
00482H–00483H	Keyboard buffer end offset pointer
00484H–0048AH	Video control data area
0048BH–00495H	Hard disk control area
00496H	Keyboard mode, state, and type flags
00497H	Keyboard LED flags
00498H–00499H	Offset address of user wait complete flag
0049AH–0049BH	Segment address of user wait complete flag
0049CH–0049FH	User wait count
004A0H	Wait active flag
004A1H–004A7H	Reserved
004A8H–004ABH	Pointer to video parameters
004ACH–004EFH	Reserved
004F0H–004FFH	Applications program communications area
00500H	Print screen status
00501H–00503H	Reserved
00504H	Single-drive mode status
00505H–0050FH	Reserved
00510H–00521H	Used by ROM BASIC
00522H–0052FH	Used by DOS for disk initialization
00530H–00533H	Used by the MODE command
00534H–005FFH	Reserved

MOUSE FUNCTIONS

The mouse is controlled and adjusted with INT 33H function call instructions. These functions provide complete control over the mouse and information provided by the mouse driver program. Table A–13 lists the mouse INT 33H functions by number and details the parameters required and any note needed to use them. Refer to Chapter 8 for a discussion of the mouse driver and some example programs that access the mouse functions through INT 33H.

TABLE A–13   The mouse (INT 33H) functions (pp. 719–730)

00H	RESET MOUSE
Entry	AL = 00H
Exit	BX = number of mouse buttons Both software and hardware are reset to their default values.
Notes	The default values are listed below: CRT Page = 0 Cursor = off Current cursor position = center of screen Minimum horizontal position = 0 Minimum vertical position = 0 Maximum horizontal position = maximum for display mode Maximum vertical position = maximum for display mode Horizontal mickey-to-pixel ratio = 1 to 1 Vertical mickey-to-pixel ratio = 2 to 1 Double-speed threshold = 64 per second Graphics cursor = arrow Text cursor = reverse block Light-pen emulation = on Interrupt call mask = 0
01H	**SHOW MOUSE CURSOR**
Entry	AL = 01H
Exit	Displays the mouse cursor
02H	**HIDE MOUSE CURSOR**
Entry	AL = 02H
Exit	Hides the mouse cursor
Notes	When displaying data on the screen, it is important to hide the mouse cursor. If you don't, problems with the display will occur and the computer may reset and reboot.
03H	**READ MOUSE STATUS**
Entry	AL = 03H
Exit	BX = button status CX = horizontal cursor position DX = vertical cursor position
Notes	The right bit (bit 0) of BX contains the status of the left mouse button and bit 1 contains the status of the right mouse button. A logic 1 indicates the button is active.

04H	SET MOUSE CURSOR POSITION
Entry	AL = 04H CX = horizontal position DX = vertical position

05H	GET BUTTON PRESS INFORMATION
Entry	AL = 05H BX = desired button (0 for left and 1 for right)
Exit	AX = button status BX = number of presses CX = horizontal position of last press DX = vertical position of last press

06H	GET BUTTON RELEASE INFORMATION
Entry	AL = 06H BX = desired button
Exit	AX = button status BX = number of releases CX = horizontal position of last release DX = vertical position of last release

07H	SET HORIZONTAL BOUNDARY
Entry	AL = 07H CX = minimum horizontal position DX = maximum horizontal position
Exit	Horizontal boundary is changed.

08H	SET VERTICAL BOUNDARY
Entry	AL = 08H CX = minimum vertical position DX = maximum vertical position
Exit	Vertical boundary is changed.

09H	SET GRAPHICS CURSOR
Entry	AL = 09H BX = horizontal center CX = vertical center ES:DX = address of 16 x 16 bit map of cursor
Exit	New graphics cursor installed.
Notes	The center is where the mouse pointer position is set. For example, a center of 0,0 is the upper left corner and 15,15 is the lower right corner. The pixel bit-map mask is stored at the address passed through ES:DX and is a 16×16 array. Following the pixel bit mask is the cursor mask, also 16×16. The contents of the bit-map mask are ANDed with a 16×16 portion of the video display, after which the contents of the cursor mask are Exclusive-ORed with a 16×16 portion of the video display to produce a cursor.
0AH	SET TEXT CURSOR
Entry	AL = 0AH BX = cursor type (0 = software, 1 = hardware) CX = pixel bit mask or beginning scan line DX = cursor mask or ending scan line
Exit	Changes the text cursor.
0BH	READ MOTION COUNTERS
Entry	AL = 0BH
Exit	CX = horizontal distance DX = vertical distance
Notes	Returns the distance traveled by the mouse since the last call to this function.

0CH	SET INTERRUPT SUBROUTINE
Entry	AL = 0CH CX = interrupt mask ES:DX = address of interrupt service procedure
Exit	New interrupt handler is installed.
Notes	The interrupt mask defines the actions that request the installed interrupt handler. Following is a list of the actions that cause the interrupt when placed in the interrupt mask. Note that these actions can appear singly or in combination. For example, an interrupt mask of 8 plus 2 or 10 (0AH) causes an interrupt for either the left or right mouse buttons. 1 = any change in cursor position 2 = left mouse button pressed 4 = left mouse button released 8 = right mouse button pressed 16 = right mouse button released After the installation, interrupt is called by one of the selected changes, the interrupt returns the following registers, which contain information about the mouse: AX = interrupt mask BX = button status CX = horizontal position DX = vertical position SI = horizontal change DI = vertical change
0DH	ENABLE LIGHT-PEN EMULATION
Entry	AL = 0DH
Exit	Light-pen emulation is enabled.
Notes	Used whenever the mouse must replace the action of a light pen.
0EH	DISABLE LIGHT-PEN EMULATION
Entry	AL = 0EH
Exit	Light-pen emulation is disabled.
Notes	Used to disable light-pen emulation.

0FH	SET MICKEY-TO-PIXEL RATIO
Entry	AL = 0FH CX = horizontal ratio DX = vertical ratio
Exit	Mickey-to-pixel ratio changed.
Notes	This function allows the screen-tracking speed of the mouse cursor to be changed. The default value is 1. If it is changed to 2, the normal speed will be reduced by half.
10H	BLANK MOUSE CURSOR
Entry	AL = 10H CX = left corner DX = upper position SI = right corner DI = bottom position
Exit	Blanks the cursor in the window specified.
Notes	The mouse pointer is blanked in the area of the screen selected by registers CX, DX, SI, and DI.
13H	SET DOUBLE-SPEED
Entry	AL = 13H DX = threshold value
Exit	Changes the threshold for double-speed pointer movement.
Notes	The default threshold value for mouse pointer acceleration is 64. This default threshold is changed by this function to any other value.
14H	SWAP INTERRUPTS
Entry	AL = 14H CX = interrupt mask ES:DX = interrupt service procedure address
Exit	CX = old interrupt mask ES:DX = old interrupt service procedure address
Notes	As with mouse INT 33H function 0CH, function 14H installs a new interrupt handler. The difference is that function 14H replaces the handler already installed with function 0CH.

15H	GET MOUSE STATE SIZE
Entry	AL = 15H
Exit	BX = size of buffer required to store mouse state
Notes	Indicates the amount of memory required to store the state of the mouse with mouse function INT 33H number 16H.

16H	SAVE MOUSE STATE
Entry	AL = 16H ES:DX = address where state of mouse is stored
Exit	Saves the current state of the mouse.

17H	RELOAD MOUSE STATE
Entry	AL = 17H ES:DX = address where state of mouse is stored
Exit	Reloads the saved state of the mouse.

18H	SET ALTERNATE INTERRUPT SUBROUTINE
Entry	AL = 18H CX = alternate interrupt mask ES:DX = address of alternate interrupt service procedure Carry = 0
Exit	Alternate interrupt handler is installed.
Notes	The alternate interrupt handler is accessed after the primary handler. The alternate interrupt mask defines the actions that request the alternate interrupt handler. Following is a list of actions that cause an alternate interrupt when placed in the alternate interrupt mask: 1 = any change in cursor position 2 = left mouse button pressed 4 = left mouse button released 8 = right mouse button pressed 16 = right mouse button released 32 = shift key with mouse button 64 = control key with mouse button 128 = alternate key with mouse button After the installed interrupt is called by one of the selected changes, the following registers contain information about the mouse: Carry = 1 AX = interrupt mask (see function 0CH) BX = button status CX = horizontal position DX = vertical position SI = horizontal change DI = vertical change

19H	GET ALTERNATE INTERRUPT ADDRESS
Entry	AL = 19H CX = alternate interrupt mask
Exit	AX = −1 if unsuccessful ES:DX = address of interrupt service procedure
Notes	Used to read the address of the alternate interrupt handler.

1AH	SET MOUSE SENSITIVITY
Entry	AL = 1AH BX = horizontal sensitivity CX = vertical sensitivity DX = double-speed threshold
Exit	Sensitivity is changed.
Notes	The default values for vertical and horizontal sensitivity are both 50. The default value for the double-speed threshold is 64. The values for sensitivity range between 1 and 100, which represent ratios of between 33 and 350 percent.

1BH	GET MOUSE SENSITIVITY
Entry	AL = 1BH
Exit	BX = horizontal sensitivity CX = vertical sensitivity DX = double-speed threshold

1CH	SET INTERRUPT RATE
Entry	AL = 1CH BX = number of interrupts per second
Exit	Changes the interrupt rate for the InPort mouse only.
Notes	The default value is 30 interrupts per second, but this can be set to any value listed: 0 = interrupt off 1 = 30 interrupts per second 2 = 50 interrupts per second 3 = 100 interrupts per second 4 = 200 interrupts per second

1DH	SET CRT PAGE
Entry	AL = 1DH BX = CRT page number
Exit	Changes to a new CRT page for mouse cursor support.

1EH	GET CRT PAGE
Entry	AL = 1EH
Exit	BX = CRT page

1FH	DISABLE MOUSE DRIVER
Entry	AL = 1FH
Exit	AX = –1 if unsuccessful ES:BX = address of mouse driver

20H	ENABLE MOUSE DRIVER
Entry	AL = 20H
Exit	The mouse driver is enabled.

21H	SOFTWARE RESET
Entry	AL = 21H
Exit	AX = 0FFFFH if successful BX = 2 if successful

22H	SET LANGUAGE
Entry	AL = 22H BX = language number
Exit	Language is changed only with the International version.
Notes	The languages selected by BX are as follows: 0 = English 1 = French 2 = Dutch 3 = German 4 = Swedish 5 = Finnish 6 = Spanish 7 = Portuguese 8 = Italian

23H	GET LANGUAGE
Entry	AL = 23H
Exit	BX = language number

24H	GET DRIVER VERSION
Entry	AL = 24H
Exit	BH = major version number BL = minor version number CH = mouse type CL = interrupt request number

25H	GET DRIVER INFORMATION
Entry	AL = 25H
Exit	AX contains the following driver information in the bits indicated: Bits 13,12 00 = software text cursor active 01 = hardware text cursor active 1X = graphics cursor active Bit 14 0 = non-integrated mouse display driver 1 = integrated mouse display driver Bit 15 0 = driver is a .SYS file 1 = driver is a .COM file

26H	GET MAXIMUM VIRTUAL COORDINATES
Entry	AL = 26H
Exit	BX = mouse driver status CX = maximum horizontal coordinate DX = maximum vertical coordinate

27H	GET CURSOR MASKS AND COUNTS
Entry	AL = 27H
Exit	AX = screen mask or beginning scanning line BX = cursor mask or ending scan line CX = horizontal mickey count DX = vertical mickey count

28H	SET VIDEO MODE
Entry	AL = 28H CX = video mode DX = font size
Exit	CX = 0 if successful

29H	GET SUPPORTED VIDEO MODES
Entry	AL = 29H CX = search flag
Exit	BX:DX = address of ASCII description of video mode CX = mode number or 0 if unsuccessful
Notes	The search flag is 0 to search for the first video mode, and non-zero to search for the next video mode.

2AH	GET CURSOR HOT SPOT
Entry	AL = 2AH
Exit	AX = display flag BX = horizontal hot spot CX = vertical hot spot DX = mouse type
Notes	The hot spot is the position on the cursor that is returned when a mouse button is clicked. AX (display flag) indicates whether the cursor is active (0) or inactive (1).

2BH	SET ACCELERATION CURVE
Entry	AL = 2BH BX = curve number ES:SI = address of the acceleration curve
Exit	AX = 0 if successful
Notes	Following are the contents of the acceleration curve table that contain curves 1 through 4: Offset Meaning 00H = curve 1 counts 01H = curve 2 counts 02H = curve 3 counts 03H = curve 4 counts 04H = curve 1 mouse count and threshold 24H = curve 2 mouse count and threshold 44H = curve 3 mouse count and threshold 64H = curve 4 mouse count and threshold 84H = curve 1 scale factor array A4H = curve 2 scale factor array C4H = curve 3 scale factor array E4H = curve 4 scale factor array 104H = curve 1 name 114H = curve 2 name 124H = curve 3 name 134H = curve 4 name

2CH	GET ACCELERATION CURVE
Entry	AL = 2CH BX = current curve ES:SI = address of current acceleration curves
Exit	AX = 0 if successful

2DH	GET ACTIVE ACCELERATION CURVE
Entry	AL = 2DH BX = curve number or −1 for current
Exit	AX = 0 if successful BX = curve number ES:SI = address of acceleration curves

2FH	MOUSE HARDWARE RESET
Entry	AL = 2FH
Exit	AX = 0 if unsuccessful

30H	SET/GET BALLPOINT INFORMATION
Entry	AL = 30H BX = rotation angle (+32K) CX = command (0 for read and 1 for write)
Exit	AX = FFFFH if unsuccessful or button state BX = rotation angle CX = button masks

31H	GET VIRTUAL COORDINATES
Entry	AL = 31H
Exit	AX = minimum horizontal BX = minimum vertical CX = maximum horizontal DX = maximum vertical

32H	GET ACTIVE ADVANCED FUNCTIONS
Entry	AL = 32H
Exit	AX = function flag
Notes	The function flags indicate which INT 33H advanced functions are available. The leftmost bit indicates function 25H through the rightmost bit that indicates function 34H.

33H	GET SWITCH SETTING
Entry	AL = 33H CX = buffer length ES:DI = address of buffer
Exit	AX = 0 CX = byte in buffer ES:DI = address of buffer

34H	GET MOUSE.INI
Entry	AL = 34H
Exit	AX = 0 ES:DX = buffer address

DPMI CONTROL FUNCTIONS

The DPMI (DOS protected-mode interface) provides DOS and Windows applications with access to protected mode and also access to real and extended memory. Four functions are accessed through the DOS multiplex interrupt (INT 2FH), and the remaining functions are accessed via INT 31H. The functions accessed by the INT 2FH interrupt release a time slice (AX = 1680H), get the CPU mode (AX = 1686H), get the mode switch entry point (AX = 1687H), and get the API entry point (AX = 168AH). The INT 2FH functions are listed in Table A–14; the INT 31H functions appear in Table A–15.

TABLE A–14   The INT 2FH, DPMI functions (pp. 731–732)

80H	RELEASE TIME SLICE
Entry	AX = 1680H
Exit	AX = 0000H if successful AX <> 1680H if unsuccessful
Notes	A time slice is a portion of time allowed to an application. If the application is idle, it should release the remaining portion of its time slice to other applications using this function.
86H	GET CPU MODE
Entry	AX = 1686H
Exit	AX = 0000H if CPU in protected mode AX <> 0000H if CPU in real or virtual mode
87H	GET MODE ENTRY POINT
Entry	AX = 1687H
Exit	AX = 0000H if successful AX <> 0000H if unsuccessful BX = support flag CL = processor type DX = DPMI version number SI = private paragraph count ES:DI = protected mode entry point
Notes	Used to locate the entry point required to switch the CPU into the protected mode using DPMI. BX is a logic 1 if 32-bit support is provided by DPMI. CL indicates the microprocessor (2, 3, 4, or 5 for the 80286, 80386, 80486, or Pentium/Pentium Pro). DX = the DPMI version number returned binary value in DH (major) and DL (minor). SI = the number of paragraphs required by DPMI for proper operation. ES:DI = the address used to switch to protected mode.

8AH	GET API ENTRY POINT
Entry	AX = 168AH DS:DI = address of vendor name (ASCII-Z)
Exit	AX = 0000H if successful AX = 168AH if unsuccessful ES:DI = API extensions entry point
Notes	Used to reference a specific vendor's extensions to the DPMI interface.

Table A–15   The INT 31H, DPMI functions (pp. 732–751)

0000H	ALLOCATE LDT DESCRIPTOR
Entry	AX = 0000H CX = number of descriptors needed
Exit	AX = base selector, carry = 0 AX = error code, carry = 1
Notes	Allocates one or more local descriptors, with the base or starting descriptor number returned in AX.

0001H	RELEASE LDT DESCRIPTOR
Entry	AX = 0001H BX = selector
Exit	Carry = 0 if successful Carry = 1 if unsuccessful, AX = error code
Notes	Releases one descriptor selected by the selector placed in BX on the call to the INT 31H function.

0002H	MAP REAL SEGMENT TO DESCRIPTOR
Entry	AX = 0002H BX = segment address
Exit	Carry = 0 if successful, AX = selector Carry = 1 if unsuccessful, AX = error code
Notes	Maps a real-mode segment to a protected-mode descriptor. This cannot be released and should be used only to map the start of the descriptor table or other global protected-mode segments.

0003H	GET SEGMENT INCREMENT
Entry	AX = 0003H
Exit	Carry = 0 if successful, AX = increment
Notes	Used with function AX = 0000H to determine the increment of the selector returned as the base selector.

0006H	GET SEGMENT BASE ADDRESS
Entry	AX = 0006H BX = selector
Exit	Carry = 0 if successful Carry = 1 if unsuccessful, AX = error code CX:DX = base address of segment
Notes	Returns the segment address as selected by BX in CX:DX, where CX = the high-order word and DX = the low-order word expressed as a 32-bit linear address.

0007H	SET SEGMENT BASE ADDRESS
Entry	AX = 0007H BX = selector CX:DX = linear base address
Exit	Carry = 0 if successful Carry = 1 if unsuccessful, AX = error code
Notes	Sets the base address to the 32-bit linear address found in CX:DX, where CX = high-order word and DX = low-order word.

0008H	SET SEGMENT LIMIT
Entry	AX = 0008H BX = selector CX:DX = segment limit in bytes
Exit	Carry = 0 if successful Carry = 1 if unsuccessful, AX = error code

0009H	SET DESCRIPTOR ACCESS RIGHTS
Entry	AX = 0009H BX = selector CX = access rights
Exit	Carry = 0 if successful Carry = 1 if unsuccessful, AX = error code
Notes	The access rights determine how a segment is accessed in the protected mode. The definition of each bit of CX follows: *Bit* *Rights* 0 Descriptor access (1) 1 Read/write (1) or read-only (0) 2 Expand segment up (0) or expand segment down (1) 3 Code segment (1) or data segment (0) 4 Must be 1 5, 6 Descriptor desired privilege level 7 Present (1) or not present (0) 8–13 Must be 000000 14 32-bit instruction mode (1) or 16-bit mode (0) 15 Granularity bit is 0 for 1X multiplier and 1 for 4K multiplier for limit field Bits 15-8 must be 0000 0000 for the 80286.
000AH	CREATE ALIAS DESCRIPTOR
Entry	AX = 000AH BX = selector
Exit	Carry = 0 if successful, AX = alias selector Carry = 1 if unsuccessful, AX = error code
Notes	Creates a new "carbon copy" as an alias local descriptor.
000BH	GET DESCRIPTOR
Entry	AX = 000BH BX = selector ES:DI = address of buffer
Exit	Carry = 0 if successful Carry = 1 if unsuccessful, AX = error code
Notes	Copies the 8-byte descriptor into the buffer addressed by ES:DI.

000CH	SET DESCRIPTOR
Entry	AX = 000CH BX = selector ES:DI = address of buffer
Exit	Carry = 0 if successful Carry = 1 if unsuccessful, AX = error code
Notes	Copies the 8-byte descriptor from the buffer to the descriptor table.

000DH	ALLOCATE SPECIFIC LDT DESCRIPTOR
Entry	AX = 000DH BX = selector
Exit	Carry = 0 if successful Carry = 1 if unsuccessful, AX = error code
Notes	Creates a descriptor based in the selector of your choice (4–7CH) with as many as 16 assigned by this function.

000EH	GET MULTIPLE DESCRIPTORS
Entry	AX = 000EH BX = number to get ES:DI = buffer address
Exit	Carry = 0 if successful, CX = number copied to buffer Carry = 1 if unsuccessful, AX = error code
Notes	Copies multiple descriptors to a buffer addressed by ES:DI.

000FH	SET MULTIPLE DESCRIPTORS
Entry	AX = 000FH BX = number to set ES:DI = buffer address
Exit	Carry = 0 if successful, number copied from buffer Carry = 1 if unsuccessful, AX = error code
Notes	Creates descriptors from the buffer at the location addressed by ES:DI. Each entry is 10 bytes in length, with the first 2 bytes containing the selector number, followed by 8 bytes for descriptor data.

0100H	ALLOCATE DOS MEMORY BLOCK
Entry	AX = 0100H DX = paragraphs to allocate
Exit	Carry = 0 if successful 　　AX = real mode segment 　　DX = selector of descriptor for memory block Carry = 1 if unsuccessful 　　AX = error code 　　BX = size of available block
Notes	Allocates a local descriptor for the DOS memory block. Releases with function 0101H only.
0101H	RELEASE DOS MEMORY BLOCK
Entry	AX = 0101H DX = selector
Exit	Carry = 0 if successful Carry = 1 if unsuccessful, AX = error code
0200H	GET REAL-MODE INTERRUPT
Entry	AX = 0200H BX = interrupt number
Exit	CX = vector segment DX = vector offset
Notes	Releases one descriptor selected by the selector placed in BX on the call to the INT 31H function.
0201H	SET REAL-MODE INTERRUPT VECTOR
Entry	AX = 0201H BX = interrupt number CX = vector segment DX = vector offset
0202H	GET EXCEPTION HANDLER VECTOR
Entry	AX = 0202H BX = exception number 0–1FH
Exit	Carry = 0 if successful 　　CX = handler selector 　　DX = handler offset Carry = 1 if unsuccessful 　　AX = error code

0203H	SET EXCEPTION HANDLER VECTOR
Entry	AX = 0203H BX = exception number 0–1FH CX = handler selector DX = handler offset
Exit	Carry = 0 if successful Carry = 1 if unsuccessful, AX = error code
0204H	GET PROTECTED-MODE INTERRUPT VECTOR
Entry	AX = 0204H BX = interrupt number

Exit	Carry = 0 if successful
	    CX = handler selector
	    DX = handler offset
	Carry = 1 if unsuccessful
	    AX = error code

0205H	SET PROTECTED-MODE INTERRUPT VECTOR
Entry	AX = 0205H BX = interrupt number CX = handler selector DX = handler offset
Exit	Carry = 0 if successful Carry = 1 if unsuccessful, AX = error code
0210H	GET EXTENDED REAL EXCEPTION HANDLER VECTOR
Entry	AX = 0210H BX = exception number 0–1FH

Exit	Carry = 0 if successful
	    CX = handler selector
	    DX = handler offset
	Carry = 1 if unsuccessful
	    AX = error code

0211H	GET EXTENDED PROTECTED EXCEPTION HANDLER VECTOR
Entry	AX = 0211H BX = exception number 0–1FH

Exit	Carry = 0 if successful
	    CX = handler selector
	    DX = handler offset
	Carry = 1 if unsuccessful
	    AX = error code

0212H	SET EXTENDED PROTECTED EXCEPTION HANDLER VECTOR
Entry	AX = 0212H BX = exception number 0–1FH CX = handler selector DX = handler offset
Exit	Carry = 0 if successful Carry = 1 if unsuccessful, AX = error code

0213H	SET EXTENDED REAL EXCEPTION HANDLER VECTOR
Entry	AX = 0213H BX = exception number 0–1FH CX = handler selector DX = handler offset
Exit	Carry = 0 if successful Carry = 1 if unsuccessful, AX = error code

0300H	EMULATE REAL-MODE INTERRUPT
Entry	AX = 0300H BX = interrupt number CX = word copy count ES:DI = buffer address
Exit	Carry = 0 if successful, ES:DI = buffer address Carry = 1 if unsuccessful, AX = error code
Notes	Copies the number of words (CX) from the protected-mode stack to the real-mode stack. The ES:DI register combination addresses a memory buffer that specifies the contents of the register when the switch to the real-mode interrupt occurs. The register set is stored as:

Offset	Size	Register
00H	doubleword	EDI
04H	doubleword	ESI
08H	doubleword	EBP
0CH	doubleword	must be 00000000H
10H	doubleword	EBX
14H	doubleword	EDX
18H	doubleword	ECX
1CH	doubleword	EAX
20H	word	flags
22H	word	ES
24H	word	DS
26H	word	FS
28H	word	GS
2AH	word	IP
2CH	word	CS
2EH	word	SP
30H	word	SS

0301H	CALL FOR REAL-MODE PROCEDURE
Entry	AX = 0301H BH = 00H CX = word copy count ES:DI = buffer address
Exit	Carry = 0 if successful, ES:DI = buffer address Carry = 1 if unsuccessful, AX = error code
Notes	Calls a real-mode procedure from protected mode. The buffer contains the register set as defined for function 0300H, which is loaded on the switch to real mode.

0302H	CALL REAL-MODE INTERRUPT PROCEDURE
Entry	AX = 0302H BH = 00H CX = word copy count ES:DI = buffer address
Exit	Carry = 0 if successful, ES:DI = buffer address Carry = 1 if unsuccessful, AX = error code
Notes	Calls the real-mode interrupt procedure from protected mode. The buffer contains the register set as defined for function 0300H, which is loaded on the switch to real mode.

0303H	ALLOCATE REAL-MODE CALLBACK ADDRESS
Entry	AX = 0303H DS:SI = protected-mode procedure address ES:DI = buffer address
Exit	Carry = 0 if successful, CX:DX = callback address Carry = 1 if unsuccessful, AX = error code
Notes	Calls a protected-mode procedure from the real mode. The real-mode address is found in CX (segment) and DX (offset).

0304H	RELEASE REAL-MODE CALLBACK ADDRESS
Entry	AX = 0304H CX:DX = callback address
Exit	Carry = 0 if successful Carry = 1 if unsuccessful, AX = error code
Notes	Releases the callback address allocated by function 0303H.

0305H	GET STATE SAVE AND RESTORE ADDRESS
Entry	AX = 0305H

Exit	Carry = 0 if successful
	    AX = buffer size
	    BX:CX = real-mode procedure address
	    SI:DI = protected-mode procedure address
	Carry = 1 if unsuccessful, AX = error code

Notes	The procedure addresses returned by this function save the real- or protected-mode registers in memory before switching modes. If in the real mode, BX:CX contain the far address or a procedure that saves the protected-mode registers. Likewise, if operating in the protected-mode, save the real-mode register with a call to the protected mode address in SI:DI. The value of AL dictates how the procedure called by the procedure address functions. If AL = 0, the registers are saved. If AL = 1, the registers are restored.

0306H	GET RAW CPU MODE SWITCH ADDRESS
Entry	AX = 0306H

Exit	Carry = 0 if successful
	    BX:CX = switch to protected-mode address
	    SI:DI = switch to real-mode address
	Carry = 1 if unsuccessful, AX = error code

Notes	To switch to protected mode from real mode, use a far JMP to the address found in BX:CX. To switch to real mode from protected mode, use a JMP to the address found in SI:DI.
	
	The register must be pre-loaded as follows before jumping to the switch address:
	
	    AX = new DS
	    BX = new SP
	    CX = new ES
	    DX = new SS
	    SI = new CS
	    DI = new IP

0400H	GET DPMI VERSION
Entry	AX = 0400H
Exit	Carry = 0 if successful AX = DPMI version number BX = implementation flag CL = processor type DH = master interrupt controller base address DL = slave interrupt controller base address Carry = 1 if unsuccessful, AX = error code
Notes	The interrupt controller vector numbers are usually returned as 08H for the master and 70H for the slave.

0401H	GET DPMI CAPABILITIES
Entry	AX = 0401H ES:DI = buffer address
Exit	Carry = 0 if successful, AX = capabilities Carry = 1 if unsuccessful, AX = error code
Notes	This function is supported under version 1.0 of DPMI and uses AX to return the following capabilities: *Bit* *Purpose* 6 write-protect host (DPMI) 5 write-protect client (DOS/Windows) 4 demand zero fill 3 conventional memory mapping 2 device mapping 1 exceptions can be restarted 0 page dirty

0500H	GET FREE MEMORY
Entry	AX = 0500H ES:DI = buffer address
Exit	Carry = 0 if successful Carry = 1 if unsuccessful
Notes	Information about the system memory is returned by this function in the buffer addressed by the buffer address. The contents of the buffer are: *Offset* *Size* *Purpose* 00H doubleword largest free block in bytes 04H doubleword maximum unlocked pages 08H doubleword maximum locked pages 0CH doubleword linear address in pages 10H doubleword total unlocked pages 14H doubleword total free pages 18H doubleword total physical pages 1CH doubleword free linear pages 20H doubleword size of paging file in pages 24H 12 bytes reserved

0501H	ALLOCATE MEMORY BLOCK
Entry	AX = 0501H BX:CX = memory block size in bytes

Exit	Carry = 0 if successful
	    BX:CX = linear base address of block
	    SI:DI = memory block handle
	Carry = 1 if unsuccessful, AX = error code

Notes	Allocates a block of memory whose size in bytes is in BX:CX on the call to this function. On the return, BX:CX = the starting address of the memory block and SI:DI contain the block handle.

0502H	RELEASE MEMORY BLOCK
Entry	AX = 0502H SI:DI = memory block handle
Exit	Carry = 0 if successful Carry = 1 if unsuccessful, AX = error code

0503H	RESIZE MEMORY BLOCK
Entry	AX = 0503H BX:CX = new memory block size in bytes SI:DI = memory block handle

Exit	Carry = 0 if successful
	    BX:CX = linear base address of block
	    SI:DI = memory block handle
	Carry = 1 if unsuccessful, AX = error code

0504H	ALLOCATE LINEAR MEMORY BLOCK
Entry	AX = 0504H EBX = desired linear base address ECX = block size in bytes EDX = action code

Exit	Carry = 0 if successful
	    EBX = linear base address of block
	    ESI = memory block handle
	Carry = 1 if unsuccessful, AX = error code

Notes	Used with a 32-bit DPMI host to allocate a linear memory block. The action code is 0 to create an uncommitted block and 1 to create a committed block.

0505H	RESIZE LINEAR MEMORY BLOCK
Entry	AX = 0505H ES:EBX = buffer address ECX = block size in bytes EDX = action code (see function 0504H) ESI = memory block handle

Exit	Carry = 0 if successful
	    BX:CX = linear base address of block
	    SI:DI = memory block handle
	Carry = 1 if unsuccessful, AX = error code

0506H	GET PAGE ATTRIBUTES
Entry	AX = 0506H EBX = page offset within memory block ECX = page count ES:EDX = buffer address ESI = memory block handle
Exit	Carry = 0 if successful Carry = 1 if unsuccessful, AX = error code
Notes	The memory buffer contains a word of attribute information for each page in the memory block. The attribute word stored in the buffer indicates the following information:

Bit	Function
0	0 = uncommitted and 1 = committed
1	1 = mapped
3	0 = read-only and 1 = read/write
4	0 = dirty bit invalid and 1 = dirty bit valid
5	0 = page unaccessed and 1 = page accessed
6	0 = page unmodified and 1 = page modified

0507H	SET PAGE ATTRIBUTES
Entry	AX = 0507H EBX = page offset within memory block ECX = page count ES:EDX = buffer address ESI = memory block handle
Exit	Carry = 0 if successful Carry = 1 if unsuccessful, AX = error code
Notes	As with function 0506H, the memory buffer contains a word for each page that defines the attribute for the page. Following is the meaning of the bits in the page attribute word: Bit Function 0 0 = make uncommitted and 1 = make committed 1 1 = modify attributes, but not page type 3 0 = make read-only and 1 = make read/write 4 1 = modify dirty bit 5 1 = mark page accessed 6 1 = mark page modified
0508H	MAP DEVICE IN MEMORY BLOCK
Entry	AX = 0508H EBX = page offset within memory block ECX = page count EDX = device address ESI = memory block handle
Exit	Carry = 0 if successful Carry = 1 if unsuccessful, AX = error code
Notes	Assigns the physical address of a device (EDX) to a linear address in a memory block.
0509H	MAP CONVENTIONAL MEMORY
Entry	AX = 0509H EBX = page offset within memory block ECX = page count EDX = linear address of conventional memory ESI = memory block handle
Exit	Carry = 0 if successful Carry = 1 if unsuccessful, AX = error code
Notes	Allocates a linear (real) address to a memory block.

050AH	GET MEMORY BLOCK SIZE
Entry	AX = 050AH SI:DI = memory block handle
Exit	Carry = 0 if successful SI:DI = memory block size Carry = 1 if unsuccessful, AX = error code

050BH	GET MEMORY INFORMATION
Entry	AX = 050BH DI:SI = buffer address
Exit	Carry = 0 if successful Carry = 1 if unsuccessful, AX = error code
Notes	This function fills a 128-byte buffer addressed by DI:SI with information about the memory. Following is the contents of the buffer: Offset Size Function 00H doubleword allocated physical memory 04H doubleword allocated virtual memory (host) 08H doubleword available virtual memory (host) 0CH doubleword allocated virtual memory (machine) 10H doubleword available virtual memory (machine) 14H doubleword allocated virtual memory (client) 18H doubleword available virtual memory (client) 1CH doubleword locked memory (client) 20H doubleword maximum locked memory (client) 24H doubleword maximum linear address (client) 28H doubleword maximum free memory block size 2CH doubleword minimum allocation unit 30H doubleword allocation alignment unit 34H 76 bytes reserved

0600H	LOCK LINEAR REGION
Entry	AX = 0600H BX:CX = linear address of memory to lock SI:DI = number of bytes to lock
Exit	Carry = 0 if successful Carry = 1 if unsuccessful, AX = error code

0601H	UNLOCK LINEAR REGION
Entry	AX = 0601H BX:CX = linear address of memory to unlock SI:DI = number of bytes to unlock
Exit	Carry = 0 if successful Carry = 1 if unsuccessful, AX = error code

0602H	MARK REAL-MODE REGION PAGABLE
Entry	AX = 0602H BX:CX = linear address of memory to mark SI:DI = number of bytes in region
Exit	Carry = 0 if successful Carry = 1 if unsuccessful, AX = error code
0603H	RELOCK REAL-MODE REGION
Entry	AX = 0603H BX:CX = linear address to relock SI:DI = number of bytes to relock
Exit	Carry = 0 if successful Carry = 1 if unsuccessful, AX = error code
0604H	GET PAGE SIZE
Entry	AX = 0604H
Exit	Carry = 0 if successful BX:CX = page size in bytes Carry = 1 if unsuccessful, AX = error code
0702H	MARK PAGE AS DEMAND PAGING CANDIDATE
Entry	AX = 0702H BX:CX = linear address of memory region SI:DI = number of bytes in region
Exit	Carry = 0 if successful Carry = 1 if unsuccessful, AX = error code
0703H	DISCARD PAGE CONTENTS
Entry	AX = 0703H BX:CX = linear address of memory region SI:DI = number of bytes in region
Exit	Carry = 0 if successful Carry = 1 if unsuccessful, AX = error code
Notes	This releases the memory for other uses by DPMI. A discarded page still contains data, but it is undefined.

0800H	PHYSICAL ADDRESS MAPPING
Entry	AX = 0800H BX:CX = base physical address
Exit	Carry = 0 if successful BX:CX = base linear address Carry = 1 if unsuccessful, AX = error code
Notes	Converts a physical address to a linear address.
0801H	RELEASE PHYSICAL ADDRESS MAPPING
Entry	AX = 0801H BX:CX = linear address
Exit	Carry = 0 if successful Carry = 1 if unsuccessful, AX = error code
0900H	GET AND DISABLE VIRTUAL INTERRUPT STATE
Entry	AX = 0900H
Exit	Carry = 0 if successful Carry = 1 if unsuccessful, AX = error code
0901H	GET AND ENABLE VIRTUAL INTERRUPT STATE
Entry	AX = 0901H
Exit	Carry = 0 if successful Carry = 1 if unsuccessful, AX = error code
0902H	GET VIRTUAL INTERRUPT STATE
Entry	AX = 0902H
Exit	Carry = 0 if successful, AL = 0 if disabled Carry = 1 if unsuccessful, AX = error code
0A00H	GET API ENTRY POINT
Entry	AX = 0A00H DS:SI = vendor ASCII offset
Exit	Carry = 0 if successful, AL = 0 if disabled ES:DI = entry point address Carry = 1 if unsuccessful, AX = error code

0B00H	SET DEBUG WATCH-POINT
Entry	AX = 0B00H BX:CX = linear watch-point address DH = watch-point type DL = watch-point size
Exit	Carry = 0 if successful BX = watch-point handle Carry = 1 if unsuccessful, AX = error code
Notes	The BX:CX register provides the watch-point address to the function. The DH register provides the type of watch-point (0 = instruction executed at watch-point address, 1 = memory write to watch-point address, and 2 = a read or write to watch-point address). The DL register holds the size in bytes of the watch-point address for types 1 and 2. When the watch-point is triggered, function AX = 0B02H is used to test for the trigger.
0B01H	CLEAR DEBUG WATCH-POINT
Entry	AX = 0B01H BX = watch-point handle
Exit	Carry = 0 if successful Carry = 1 if unsuccessful, AX = error code
Notes	Erase a watch-point assigned with function 0B00H.
0B02H	GET STATE OF DEBUG WATCH-POINT
Entry	AX = 0B02H BX = watch-point handle
Exit	Carry = 0 if successful AX = watch-point status Carry = 1 if unsuccessful, AX = error code
Notes	Used to detect a watch-point trigger. A status of 0 indicates that a watch-point has not been detected, and 1 indicates that it has.
0B03H	RESET DEBUG WATCH-POINT
Entry	AX = 0B03H BX = watch-point handle
Exit	Carry = 0 if successful Carry = 1 if unsuccessful, AX = error code
Notes	Clears the watch-point status, but does not erase the watch-point.

0C00H	INSTALL RESIDENT SERVICE PROVIDER CALLBACK
Entry	AX = 0C00H ES:DI = buffer address
Exit	Carry = 0 if successful Carry = 1 if unsuccessful, AX = error code
Notes	Used for a protected-mode TSR to notify the DPMI to call the TSR whenever another machine is in the same virtual memory. The buffer contains the following information for DPMI: *Offset* *Size* *Function* 00H quadword 16-bit data segment descriptor 08H quadword 16-bit code segment descriptor 10H word 16-bit callback procedure offset 12H word reserved 14H quadword 32-bit data segment descriptor 1CH quadword 32-bit code segment descriptor 24H doubleword 32-bit callback procedure offset

0C01H	TERMINATE-AND-STAY RESIDENT
Entry	AX = 0C01H BL = exit code DX = number of paragraphs to reserve
Exit	None

0D00H	ALLOCATE SHARED MEMORY
Entry	AX = 0D00H ES:DI = buffer address
Exit	Carry = 0 if successful Carry = 1 if unsuccessful, AX = error code
Notes	The buffer contains the following information for DPMI used to allocate shared memory: *Offset* *Size* *Function* 00H doubleword desired block size in bytes 04H doubleword actual memory block size in bytes 08H doubleword memory block handle 0CH doubleword linear address of memory block 10H doubleword offset of ASCII-Z string name 14H word selector of ASCII-Z string name 16H word reserved 18H doubleword must be set to 00000000H

0D01H	RELEASE SHARED MEMORY
Entry	AX = 0D01H SI:DI = memory block handle
Exit	Carry = 0 if successful Carry = 1 if unsuccessful, AX = error code

0D02H	SERIALIZE ON SHARED MEMORY
Entry	AX = 0D02H DX = option flag SI:DI = memory block handle
Exit	Carry = 0 if successful Carry = 1 if unsuccessful, AX = error code

0D03H	RELEASE SERIALIZATION ON SHARED MEMORY
Entry	AX = 0D03H DX = option flag SI:DI = memory block handle
Exit	Carry = 0 if successful Carry = 1 if unsuccessful, AX = error code
Notes	The DX register contains the following option codes: *Bit* *Function* 0 0 = suspend until serialization and 1 = return error code 1 0 = exclusive serialization and 1 = shared serialization

0E00H	GET COPROCESSOR STATUS
Entry	AX = 0E00H
Exit	Carry = 0 if successful AX = status code Carry = 1 if unsuccessful, AX = error code
Notes	The following is the coprocessor status: *Bit* *Function* 0 1 = coprocessor enabled 1 1 = emulation of coprocessor enabled for client 2 1 = coprocessor present 3 1 = emulation of coprocessor enabled for host 4–7 0000 = no coprocessor 0001 = 80287 present 0010 = 80387 present 0011 = 80487/Pentium present

0E01H	SET COPROCESSOR EMULATION
Entry	AX = 0E01H BX = action code
Exit	Carry = 0 if successful Carry = 1 if unsuccessful, AX = error code
Notes	The BX register contains the following: *Bit Function* 0 1 = enable coprocessor for client 1 1 = client will supply emulation

APPENDIX B

Instruction Set Summary

This appendix contains a complete alphabetical listing of the entire 80186, 80188, and 80386EX instruction sets. The coprocessor instructions are listed in Chapter 14 and are not repeated here.

Each instruction entry lists the mnemonic opcode plus a brief description of its purpose. Also listed is the binary machine language coding of each instruction, plus any other data needed to form the instruction, such as the displacement of immediate data. To the right of each binary machine language version of the instruction are the flag bits and any change that might occur for the instruction. The flags are described in the following manner: a blank indicates no effect or change, a ? indicates a change with an unpredictable outcome, a * indicates a change with a predictable outcome, a 1 indicates the flag is set, and a 0 indicates the flag is cleared. If the flag bits ODITSZAPC are not illustrated with an instruction, the instruction does not modify any of these flags.

Before the instruction listing begins, look at the information in Table B–1 about the binary bit settings in binary machine language versions of the instructions. The table lists the modifier bits, coded as oo in the instruction listing.

Table B–2 lists the memory-addressing modes available using a register field coding of mmm. This table applies to all versions of the microprocessor as long as the operating mode is 16 bits.

Table B–3 lists the register selections provided by the rrr field in an instruction. This table includes the register selections for 8-, 16-, and 32-bit registers.

TABLE B–1 The modifier bits, coded as oo in the instruction listing

oo	Function
00	If mmm = 110, then a displacement follows the opcode; otherwise, no displacement is used.
01	An 8-bit signed displacement follows the opcode.
10	A 16-bit signed displacement follows the opcode (unless it is a 32-bit displacement).
11	mmm specifies a register instead of an addresing mode.

TABLE B–2   The 16-bit register/memory (mmm) field description

mmm	Function
000	DS:[BX+SI]
001	DS:[BX+DI]
010	SS:[BP+SI]
011	SS:[BP+DI]
100	DS:[SI]
101	DS:[DI]
110	SS:[BP]
111	DS:[BX]

TABLE B–3   The register field (rrr) assignment

rrr	W = 0	W = 1 (16-bit)	W = 1 (32-bit)
000	AL	AX	EAX
001	CL	CX	ECX
010	DL	DX	EDX
011	BL	BX	EBX
100	AH	SP	ESP
101	CH	BP	EBP
110	DH	SI	ESI
111	BH	DI	EDI

Table B–4 lists the segment register bit assignment (rrr) found with the MOV, PUSH, and POP instructions.

When the 80386–Pentium Pro are used, some of the definitions provided in Tables B–1 through B–3 change. Refer to Tables B–5 and B–6 for these changes as they apply to the 80386–Pentium Pro microprocessors.

TABLE B–4   Register field assignments (rrr) for the segment registers

rrr	Segment Register
000	ES
001	CS
010	SS
011	DS
100	FS
101	GS

TABLE B–5   Index register specified with rrr for the advanced addressing mode found in the 80386–Pentium Pro microprocessors

rrr	Index Register
000	DS:[EAX]
001	DS:[ECX]
010	DS:[EDX]
011	DS:[EBX]
100	No index (see Table B–6)
101	SS:[EBP]
110	DS:[ESI]
111	DS:[EDI]

TABLE B–6   Possible combinations of oo, mmm, and rrr for the 80386–Pentium Pro microprocessors using 32-bit addressing

oo	mmm	rrr (base in scaled index byte)	Addressing Mode
00	000	—	DS:[EAX]
00	001	—	DS:[ECX]
00	010	—	DS:[EDX]
00	011	—	DS:[EBX]
00	100	000	DS:[EAX+scaled index]
00	100	001	DS:[ECX+scaled index]
00	100	010	DS:[EDX+scaled index]
00	100	011	DS:[EBX+scaled index]
00	100	100	SS:[ESP+scaled index]
00	100	101	DS:[disp32+scaled index]
00	100	110	DS:[ESI+scaled index]
00	100	111	DS:[EDI+scaled index]
00	101	—	DS:disp32
00	110	—	DS:[ESI]
00	111	—	DS:[EDI]
01	000	—	DS:[EAX+disp8]
01	001	—	DS:[ECX+disp8]
01	010	—	DS:[EDX+disp8]
01	011	—	DS:[EBX+disp8]
01	100	000	DS:[EAX+scaled index+disp8]
01	100	001	DS:[ECX+scaled index+disp8]
01	100	010	DS:[EDX+scaled index+disp8]
01	100	011	DS:[EBX+scaled index+disp8]
01	100	100	SS:[ESP+scaled index+disp8]
01	100	101	SS:[EBP+scaled index+disp8]
01	100	110	DS:[ESI+scaled index+disp8]
01	100	111	DS:[EDI+scaled index+disp8]
01	101	—	SS:[EBP+disp8]
01	110	—	DS:[ESI+disp8]
01	111	—	DS:[EDI+disp8]
10	000	—	DS:[EAX+disp32]
10	001	—	DS:[ECX+disp32]
10	010	—	DS:[EDX+disp32]
10	011	—	DS:[EBX+disp32]
10	100	000	DS:[EAX+scaled index+disp32]
10	100	001	DS:[ECX+scaled index+disp32]
10	100	010	DS:[EDX+scaled index+disp32]
10	100	011	DS:[EBX+scaled index+disp32]
10	100	100	SS:[ESP+scaled index+disp32]
10	100	101	SS:[EBP+scaled index+disp32]
10	100	110	DS:[ESI+scaled index+disp32]
10	100	111	DS:[EDI+scaled index+disp32]
10	101	—	SS:[EBP+disp32]
10	110	—	DS:[ESI+disp32]
10	111	—	DS:[EDI+disp32]

Note: disp8 = 8-bit displacement and disp32 = 32-bit displacement.

In order to use the scaled index addressing modes listed in Table B–6, codes oo and mmm must occur in the second byte of the opcode. The scaled index byte is usually the third byte and contains three fields. The leftmost two bits determine the scaling factor (00 = X1, 01 = X2, 10 = X4, or 11 = X8). The next three bits toward the right contain the scaled index register number (this is obtained from Table B–5). The rightmost three bits are from the rrr field listed in Table B–6. For example, the MOV AL,[EBX + 2*ECX] instruction has a scaled index byte of 01 001 011 where 01 = X2, 001 = ECX, and 011 = EBX.

Some instructions are prefixed to change the default segment or to override the instruction mode. Table B–7 lists the segment and instruction mode override prefixes that append the beginning of an instruction if they are used to form the instruction. For example, the MOV AL,ES:[BX] instruction uses the extra segment because of the override prefix ES:.

In the 80186 and 80188 microprocessors the effective address calculation required additional clocks that are added to the times in the instruction set summary. These additional times are listed in Table B–8. No such times are added to the 80386EX.

TABLE B–7 Override prefixes

Prefix Byte	Purpose
26H	ES: segment override prefix
2EH	CS: segment override prefix
36H	SS: segment override prefix
3EH	DS: segment override prefix
64H	FS: segment override prefix
65H	GS: segment override prefix
66H	Operand size instruction mode override
67H	Register size instruction mode override

TABLE B–8 Effective address calculations for the 80186 and 80188 microprocessors

Type	Clocks	Example Instruction
Base or index	5	MOV CL,[DI]
Displacement	3	MOV AL,DATA1
Base plus index	7	MOV AL,[BP+SI]
Displacement plus base or index	9	MOV DH,[DI+20H]
Base plus index plus displacement	11	MOV CL,[BX+DI+2]
Segment override	ea + 2	MOV AL,ED:[DI]

INSTRUCTION SET SUMMARY   (pp. 787–870)

AAA	ASCII adjust AL after addition										

00110111		O ?	D	I	T		S ?	Z ?	A *	P ?	C *
Example		Microprocessor					Clocks				
AAA		80186					8				
		80188					8				
		80386EX					4				

AAD	ASCII adjust AX before division										

11010101 00001010		O ?	D	I	T		S *	Z *	A ?	P *	C ?
Example		Microprocessor					Clocks				
AAD		80186					15				
		80188					15				
		80386EX					19				

AAM	ASCII adjust AX after multiplication										

11010100 00001010		O ?	D	I	T		S *	Z *	A ?	P *	C ?
Example		Microprocessor					Clocks				
AAM		80186					19				
		80188					19				
		80386EX					17				

AAS	ASCII adjust AL after subtraction										

00111111		O ?	D	I	T		S ?	Z ?	A *	P ?	C *
Example		Microprocessor					Clocks				
AAS		80186					7				
		80188					7				
		80386EX					4				

ADC   Addition with carry

000100dw oorrrmmm disp

		O D I T S Z A P C
		* * * * * *
Format	Examples	Microprocessor	Clocks

Format	Examples	Microprocessor	Clocks
ADC reg,reg	ADC AX,BX ADC AL,BL ADC EAX,EBX	80186	3
		80188	3
		80386EX	3
ADC mem,reg	ADC DATAY,AL ADC LIST,SI ADC DATA2[DI],CL	80186	10 + ea
		80188	11 + ea
		80386EX	7
ADC reg,mem	ADC BL,DATA1 ADC SI,LIST1 ADC CL,DATA2[SI]	80186	10 + ea
		80188	11 + ea
		80386EX	6

100000sw oo010mmm disp data

Format	Examples	Microprocessor	Clocks
ADC reg,imm	ADC CX,3 ADC DI,1AH ADC DL,34H	80186	4
		80188	4
		80386EX	2
ADC mem,imm	ADC DATA4,33 ADC LIST,'A' ADC DATA3[DI],2	80186	16 + ea
		80188	17 + ea
		80386EX	7
ADC acc,imm	ADC AX,3 ADC AL,1AH ADC AH,34H	80186	4
		80188	4
		80386EX	2

ADD   Addition

000000dw oorrrmmm disp

		O D I T S Z A P C
		* * * * * *
Format	Examples	Microprocessor	Clocks

Format	Examples	Microprocessor	Clocks

ADD reg,reg	ADD AX,BX
	ADD AL,BL
	ADD EAX,EBX
		80186	3

		80188	3
		80386EX	2

ADD mem,reg	ADD DATAY,AL
	ADD LIST,SI
	ADD DATA6[DI],CL
		80186	10 + ea

		80188	11 + ea
		80386EX	7

ADD reg,mem	ADD BL,DATA2
	ADD SI,LIST3
	ADD CL,DATA2[DI]
		80186	10 + ea

		80188	11 + ea
		80386EX	6

100000sw oo000mmm disp data

Format	Examples	Microprocessor	Clocks

ADD reg,imm	ADD CX,3
	ADD DI,1AH
	ADD DL,34H
		80186	4

		80188	4
		80386EX	2

ADD mem,imm	ADD DATA4,33
	ADD LIST,'A'
	ADD DATA3[DI],2
		80186	10 + ea

		80188	11 + ea
		80386EX	7

ADD acc,imm	ADD AX,3
	ADD AL,1AH
	ADD AH,34H
		80186	3

		80188	3
		80386EX	2

AND   Logical AND

001000dw oorrrmmm disp

		O D I T S Z A P C 0 * * ? * 0	
Format	Examples	Microprocessor	Clocks
AND reg,reg	AND CX,BX AND DL,BL AND ECX,EBX	80186	3
		80188	3
		80386EX	3
AND mem,reg	AND BIT,AL AND LIST,DI AND DATAZ[BX],CL	80186	10 + ea
		80188	11 + ea
		80386EX	7
AND reg,mem	AND BL,DATAW AND SI,LIST AND CL,DATAQ[SI]	80186	10 + ea
		80188	11 + ea
		80386EX	6

100000sw oo100mmm disp data

Format	Examples	Microprocessor	Clocks
AND reg,imm	AND BP,1 AND DI,10H AND DL,34H	80186	4
		80188	4
		80386EX	2
AND mem,imm	AND DATA4,33 AND LIST,'A' AND DATA3[DI],2	80186	16 + ea
		80188	17 + ea
		80386EX	7

AND acc,imm	AND AX,3 AND AL,1AH AND AH,34H	80186	4
		80188	4
		80386EX	2

ARPL   Adjust requested privilege level

01100011 oorrrmmm disp

Format	Examples	Microprocessor	Clocks
		O D I T S Z A P C *	
ARPL reg,reg	ARPL AX,BX ARPL BX,SI	80186	—
		80188	—
		80386EX	20
ARPL mem,reg	ARPL DATAY,AX ARPL LIST,DI ARPL DATA3[DI],CX	80186	—
		80188	—
		80386EX	21

BOUND   Check array against boundary

01100010 oorrrmmm disp

Format	Examples	Microprocessor	Clocks
BOUND reg,mem	BOUND AX,BETS BOUND BP,LISTG BOUND CX,DATAX	80186	33
		80188	35
		80386EX	10

BSF   Bit scan forward

00001111 10111100 oorrrmmm disp

Format	Examples	Microprocessor	Clocks
		O D I T S Z A P C ? ? * ? ? ?	
BSF reg,reg	BSF AX,BX BSF BX,SI BSF EAX,EDX	80186	—
		80188	—
		80386EX	10 + 3n
BSF reg,mem	BSF AX,DATAY BSF SI,LIST BSF CX,DATA3[DI]	80186	—
		80188	—
		80386EX	10 + 3n

BSR — Bit scan reverse

00001111 10111101 oorrrmmm disp		O D I T S Z A P C	
		? ? * ? ? ?	
Format	Examples	Microprocessor	Clocks
BSR reg,reg	BSR AX,BX BSR BX,SI BSR EAX,EDX	80186	—
		80188	—
		80386EX	10 + 3n
BSR reg,mem	BSR AX,DATAY BSR SI,LIST BSR CX,DATA3[DI]	80186	—
		80188	—
		80386EX	10 + 3n

BT — Bit test

00001111 10111010 oo100mmm disp data		O D I T S Z A P C	
		*	
Format	Examples	Microprocessor	Clocks
BT reg,imm8	BT AX,2 BT CX,4 BT BP,10H	80186	—
		80188	—
		80386EX	3
BT mem,imm8	BT DATA1,2 BT LIST,2 BT DATA2[DI],3	80186	—
		80188	—
		80386EX	6

00001111 10100011 disp			
Format	Examples	Microprocessor	Clocks
BT reg,reg	BT AX,CX BT CX,DX BT BP,AX	80186	—
		80188	—
		80386EX	3
BT mem,reg	BT DATA4,AX BT LIST,BX BT DATA3[DI],CX	80186	—
		80188	—
		80386EX	12

BTC — Bit test and complement

00001111 10111010 oo111mmm disp data		O D I T S Z A P C	
		*	
Format	Examples	Microprocessor	Clocks
BTC reg,imm8	BTC AX,2 BTC CX,4 BTC BP,10H	80186	—
		80188	—
		80386EX	6

BTC mem,imm8	BTC DATA1,2 BTC LIST,2 BTC DATA2[DI],3	80186	—
		80199	—
		80386EX	7 or 8

00001111 10111011 disp Format	Examples	Microprocessor	Clocks
BTC reg,reg	BTC AX,CX BTC CX,DX BTC BP,AX	80186	—
		80188	—
		80386EX	6
BTC mem,reg	BTC DATA4,AX BTC LIST,BX BTC DATA3[DI],CX	80186	—
		80188	—
		80386EX	13

BTR   Bit test and reset

| 00001111 10111010 oo110mmm disp data | | O D I T   S Z A P C | |
| | | | * |
Format	Examples	Microprocessor	Clocks
BTR reg,imm8	BTR AX,2 BTR CX,4 BTR BP,10H	80186	—
		80188	—
		80386EX	6
BTR mem,imm8	BTR DATA1,2 BTR LIST,2 BTR DATA2[DI],3	80186	—
		80188	—
		80386EX	8

00001111 10110011 disp Format	Examples	Microprocessor	Clocks
BTR reg,reg	BTR AX,CX BTR CX,DX BTR BP,AX	80186	—
		80188	—
		80386EX	6
BTR mem,reg	BTR DATA4,AX BTR LIST,BX BTR DATA3[DI],CX	80186	—
		80188	—
		80386EX	13

BTS   Bit test and set

| 00001111 10111010 oo101mmm disp data | | O D I T   S Z A P C | |
| | | | * |
Format	Examples	Microprocessor	Clocks
BTS reg,imm8	BTS AX,2 BTS CX,4 BTS BP,10H	80186	—
		80188	—
		80386EX	6

BTS mem,imm8	BTS DATA1,2 BTS LIST,2 BTS DATA2[DI],3	80186	—
		80188	—
		80386EX	8

00001111 10101011 disp

Format	Examples	Microprocessor	Clocks
BTS reg,reg	BTS AX,CX BTS CX,DX BTS BP,AX	80186	—
		80188	—
		80386EX	6
BTS mem,reg	BTS DATA4,AX BTS LIST,BX BTS DATA3[DI],CX	80186	—
		80188	—
		80386EX	13

CALL   Call procedure (subroutine)

11101000 disp

Format	Examples	Microprocessor	Clocks
CALL label (near)	CALL FOR_FUN CALL HOME CALL ET	80186	15
		80188	15
		80386EX	3

10011010 disp

Format	Examples	Microprocessor	Clocks
CALL label (far)	CALL FAR PTR DATES CALL WHAT CALL WHERE	80186	23
		80188	25
		80386EX	17

11111111 oo010mmm

Format	Examples	Microprocessor	Clocks
CALL reg (near)	CALL AX CALL BX CALL CX	80186	13
		80188	13
		80386EX	7
CALL mem (near)	CALL ADDRESS CALL NEAR PTR [DI] CALL DATA1	80186	19 + ea
		80188	19 + ea
		80386EX	10

11111111 oo011mmm

Format	Examples	Microprocessor	Clocks
CALL mem (far)	CALL FAR_LIST[SI] CALL FROM_HERE CALL TO_THERE	80186	38
		80188	38
		80386EX	7

CBW   Convert byte to word (AL $\Rightarrow$ AX)

10011000 Example			Microprocessor	Clocks
CBW			80186	2
			80188	2
			80386EX	3

CDQ   Convert doubleword to quadword
(EAX $\Rightarrow$ EDX:EAX)

11010100 00001010 Example			Microprocessor	Clocks
CDQ			80186	—
			80188	—
			80386EX	2

CLC   Clear carry flag

11111000			O D I T S Z A P C 0	
Example			Microprocessor	Clocks
CLC			80186	2
			80188	2
			80386EX	2

CLD   Clear direction flag

11111100			O D I T S Z A P C 0	
Example			Microprocessor	Clocks
CLD			80186	2
			80188	2
			80386EX	2

CLI   Clear interrupt flag

11111010			O D I T S Z A P C 0	
Example			Microprocessor	Clocks
CLI			80186	2
			80188	2
			80386EX	3

CLTS	Clear task switched flag (CR0)		

00001111 00000110			
Example		Microprocessor	Clocks
CLTS		80186	—
		80188	—
		80386EX	5

CMC	Complement carry flag		

10011000		O D I T S Z A P C *	
Example		Microprocessor	Clocks
CMC		80186	2
		80188	2
		80386EX	2

CMP	Compare		

001110dw oorrmmm disp		O D I T S Z A P C * * * * * *	
Format	Examples	Microprocessor	Clocks

CMP reg,reg	CMP AX,BX
	 CMP AL,BL
	 CMP EAX,EBX
		80186	3

		80188	3
		80386EX	2

CMP mem,reg	CMP DATAY,AL
	 CMP LIST,SI
	 CMP DATA6[DI],CL
		80186	10 + ea

		80188	10 + ea
		80386EX	5

CMP reg,mem	CMP BL,DATA2
	 CMP SI,LIST3
	 CMP CL,DATA2[DI]
		80186	10 + ea

		80188	10 + ea
		80386EX	6

100000sw oo111mmm disp data			
Format	Examples	Microprocessor	Clocks

CMP reg,imm	CMP CX,3
	 CMP DI,1AH
	 CMP DL,34H
		80186	3

		80188	3
		80386EX	2

CMP mem,imm	CMP DATAS,3
	 CMP BYTE PTR[EDI],1AH
	 CMP DADDY,34H
		80186	10 + ea

		80188	10 + ea
		80386EX	5

0001111w data Format	Examples	Microprocessor	Clocks

CMP acc,imm	CMP AX,3
	CMP AL,1AH
	CMP AH,34H
		80186	3

		80188	4
		80386EX	2

CMPS   Compare strings

1010011w Format	Examples	O D I T S Z A P C * * * * * * Microprocessor	Clocks
CMPSB CMPSW CMPSD	CMPSB CMPSW CMPSD	80186	22
		80188	22
		80386EX	10

CWD   Convert word to doubleword (AX ⇒ DX:AX)

10011000 Example		Microprocessor	Clocks
CWD		80186	4
		80188	4
		80386EX	2

CWDE   Convert word to extended doubleword
(AX ⇒ EAX)

10011000 Example		Microprocessor	Clocks
CWDE		80186	—
		80188	—
		80386EX	3

DAA   Decimal adjust AL after addition

00100111 Example		O D I T S Z A P C ? * * * * * Microprocessor	Clocks
DAA		80186	4
		80188	4
		80386EX	4

DAS	Decimal adjust AL after subtraction			

00101111		O D I T S Z A P C	
		? * * * * *	
Example		Microprocessor	Clocks
DAS		80186	4
		80188	4
		80386EX	4

DEC	Decrement			

1111111w oo001mmm disp		O D I T S Z A P C	
		* * * * *	
Format	Examples	Microprocessor	Clocks
DEC reg8	DEC BL	80186	3
	DEC BH	80188	3
	DEC CL	80386EX	2
DEC mem	DEC DATAY	80186	15 + ea
	DEC LIST	80188	16 + ea
	DEC DATA6[DI]	80386EX	6

01001rrr			
Format	Examples	Microprocessor	Clocks
DEC reg16	DEC CX	80186	3
DEC reg32	DEC DI	80188	3
	DEC EDX	80386EX	2

DIV	Divide			

1111011w oo110mmm disp		O D I T S Z A P C	
		? ? ? ? ? ?	
Format	Examples	Microprocessor	Clocks
DIV reg	DIV BL	80186	29–38
	DIV BH	80188	29–38
	DIV ECX	80386EX	38
DIV mem	DIV DATAY	80186	35–44
	DIV LIST	80188	35–44
	DIV DATA6[DI]	80386EX	41

ENTER   Create a stack frame

11001000 data

Format	Examples	Microprocessor	Clocks
ENTER imm,0	ENTER 4,0 ENTER 8,0 ENTER 100,0	80186	15
		80188	15
		80386EX	10
ENTER imm,1	ENTER 4,1 ENTER 10,1	80186	25
		80188	25
		80386EX	12
ENTER imm,imm	ENTER 3,6 ENTER 100,3	80186	22 + 16n
		80188	22 + 16n
		80386EX	15

ESC   Escape (obsolete–see coprocessor)

HLT   Halt

11110100

Example		Microprocessor	Clocks
HLT		80186	2
		80188	2
		80386EX	5

IDIV   Integer (signed) division

1111011w oo111mmm disp

		O D I T S Z A P C ? ? ? ? ? ?	
Format	Examples	Microprocessor	Clocks
IDIV reg	IDIV BL IDIV BH IDIV ECX	80186	29–38
		80188	29–38
		80386EX	43
IDIV mem	IDIV DATAY IDIV LIST IDIV DATA6[DI]	80186	35–44
		80188	35–44
		80386EX	46

IMUL	Integer (signed) multiplication

1111011w oo101mmm disp

		O D I T	S Z A P C
		*	? ? ? ? *
Format	Examples	Microprocessor	Clocks
IMUL reg	IMUL BL	80186	25–37
	IMUL CX	80188	25–37
	IMUL ECX	80386EX	38
IMUL mem	IMUL DATAY	80186	32–43
	IMUL LIST	80188	32–43
	IMUL DATA6[DI]	80386EX	41

011010s1 oorrmmm disp data

Format	Examples	Microprocessor	Clocks
IMUL reg,imm	IMUL CX,16	80186	22–25
	IMUL DI,100	80188	22–25
	IMUL EDX,20	80386EX	38
IMUL reg,reg,imm	IMUL DX,AX,2	80186	29–32
	IMUL CX,DX,3	80188	29–32
	IMUL BX,AX,33	80386EX	38
IMUL reg,mem,imm	IMUL CX,DATAY,99	80186	29–32
		8088	29–32
		80386EX	38

00001111 10101111 oorrmmm disp

Format	Examples	Microprocessor	Clocks
IMUL reg,reg	IMUL CX,DX	80186	—
	IMUL DI,BX	80188	—
	IMUL EDX,EBX	80386EX	38
IMUL reg,mem	IMUL DX,DATAY	80186	—
	IMUL CX,LIST	80188	—
	IMUL ECX,DATA6[DI]	80386EX	41

IN	Input data from port

1110010w port#

Format	Examples	Microprocessor	Clocks
IN acc,pt	IN AL,12H	80186	10
	IN AX,12H	80188	10
	IN AL,0FFH	80386EX	12

1110110w

Format	Examples	Microprocessor	Clocks
IN acc, DX	IN AL,DX IN AX,DX IN EAX,DX	80186	8
		80188	8
		80386EX	13

INC   Increment

1111111w oo000mmm disp

		O D I T	S Z A P C
		*	* * * *

Format	Examples	Microprocessor	Clocks
INC reg8	INC BL INC BH INC AL	80186	3
		80188	3
		80386EX	2
INC mem	INC DATA3 INC LIST INC COUNT	80186	15 + ea
		80188	16 + ea
		80386EX	6
INC reg16 INC reg32	INC CX INC DX INC EBP	80186	3
		80188	3
		80386EX	2

INS   Input string from port

0110110w

Format	Examples	Microprocessor	Clocks
INSB INSW INSD	INSB INSW INSD	80186	14
		80188	14
		80386EX	15

INT   Interrupt

11001101 type

Format	Examples	Microprocessor	Clocks
INT type	INT12H INT15H INT 21H	80186	47
		80188	47
		80386EX	37

INT 3   Interrupt 3

11001100 Example		Microprocessor	Clocks
INT 3		80186	45
		80188	72
		80386EX	33

INTO   Interrupt on overflow

11001110 Example		Microprocessor	Clocks
INTO		80186	48
		80188	48
		80386EX	35

IRET/IRETD   Return from interrupt

11001101 data		O D I T S Z A P C * * * * * * * * *	
Format	Examples	Microprocessor	Clocks
IRET IRETD	IRET IRETD IRET 100	80186	28
		80188	28
		80386EX	22

Jcondition   Conditional jump

0111cccc disp Format	Examples	Microprocessor	Clocks
Jcnd label (8-bit disp)	JA ABOVE JB BELOW JG GREATER	80186	13/4
		80188	13/4
		80386EX	7/3

00001111 1000cccc disp Format	Examples	Microprocessor	Clocks
Jcnd label (16-bit disp)	JNE NOT_MORE JLE LESS_OR_SO	80186	—
		80188	—
		80386EX	7/3

Condition Codes	Mnemonic	Flag	Description
0000	JO	O = 1	Jump if overflow
0001	JNO	O = 0	Jump if no overflow
0010	JB/NAE	C = 1	Jump if below
0011	JAE/JNB	C = 0	Jump if above or equal
0100	JE/JZ	Z = 1	Jump if equal/zero
0101	JNE/JNZ	Z = 0	Jump if not equal/zero
0110	JBE/JNA	C = 1 + Z = 1	Jump if below or equal
0111	JA/JNBE	C = 0 • Z = 0	Jump if above
1000	JS	S = 1	Jump if sign
1001	JNS	S = 0	Jump if no sign
1010	JP/JPE	P = 1	Jump if parity
1011	JNP/JPO	P = 0	Jump if no parity
1100	JL/JNGE	S • O	Jump if less than
1101	JGE/JNL	S = 0	Jump if greater than or equal
1110	JLE/JNG	Z = 1 + S • O	Jump if less than or equal
1111	JG/JNLE	Z = 0 + S = O	Jump if greater than

JCXZ/JECXZ Jump if CX (ECX) equals zero

11100011

Format	Examples	Microprocessor	Clocks
JCXZ label JECXZ label	JCXZ ABOVE JCXZ BELOW JECXZ GREATER	80186	16/6
		80188	16/6
		80386EX	9/5

JMP   Jump

11101011 disp

Format	Examples	Microprocessor	Clocks
JMP label (short)	JMP SHORT UP JMP SHORT DOWN JMP SHORT OVER	80186	14
		80188	14
		80386EX	7

11101001 disp

Format	Examples	Microprocessor	Clocks
JMP label (near)	JMP VERS JMP FROG JMP UNDER	80186	14
		80188	14
		80386EX	7

11101010 disp

Format	Examples	Microprocessor	Clocks
JMP label (far)	JMP NOT_MORE JMP UNDER JMP AGAIN	80186	14
		80188	14
		80386EX	12

11111111 oo100mmm			
Format	Examples	Microprocessor	Clocks
JMP reg (near)	JMP AX JMP EAX JMP CX	80186	26
		80188	26
		80386EX	7
JMP mem (near)	JMP VERS JMP FROG JMP CS:UNDER	80186	26 + ea
		80188	26 + ea
		80386EX	10

11111111 oo101mmm			
Format	Examples	Microprocessor	Clocks
JMP mem (far)	JMP WAY_OFF JMP TABLE JMP UP	80186	17 + ea
		80188	17 + ea
		80386EX	12

LAHF   Load AH from flags

10011111		
Example	Microprocessor	Clocks
LAHF	80186	2
	80188	2
	80386EX	2

LAR   Load access rights byte

00001111 00000010 oorrmmm disp		O D I T S Z A P C
		*
Format	Examples	Microprocessor	Clocks
LAR reg,reg	LAR AX,BX LAR CX,DX LAR ECX,EDX	80186	—
		80188	—
		80386EX	15
LAR reg,mem	LAR CX,DATA1 LAR AX,LIST3 LAR ECX,TOAD	80186	—
		80188	—
		80386EX	16

LDS   Load far pointer to DS and register

11000101 oorrmmm			
Format	Examples	Microprocessor	Clocks
LDS reg,mem	LDS DI,DATA3 LDS SI,LIST2 LDS BX,ARRAY_PTR	80186	18 + ea
		80188	19 + ea
		80386EX	7

LEA   Load effective address

10001101 oorrrmmm disp

Format	Examples	Microprocessor	Clocks
LEA reg,mem	LEA DI,DATA3	80186	6 + ea
	LEA SI,LIST2	80188	6 + ea
	LEA BX,ARRAY_PTR	80386EX	2

LEAVE   Leave high-level procedure

11001001

Example		Microprocessor	Clocks
LEAVE		80186	8
		80188	8
		80386EX	4

LES   Load far pointer to ES and register

11000100 oorrrmmm

Format	Examples	Microprocessor	Clocks
LES reg,mem	LES DI,DATA3	80186	18 + ea
	LES SI,LIST2	80188	18 + ea
	LES BX,ARRAY_PTR	80386EX	7

LFS   Load far pointer to FS and register

00001111 10110100 oorrrmmm disp

Format	Examples	Microprocessor	Clocks
LFS reg,mem	LFS DI,DATA3	80186	—
	LFS SI,LIST2	80188	—
	LFS BX,ARRAY_PTR	80386EX	7

LGDT   Load global descriptor table

00001111 00000001 oo010mmm disp

Format	Examples	Microprocessor	Clocks
LGDT mem64	LGDT DESCRIP	80186	—
	LGDT TABLED	80188	—
		80386EX	11

LGS — Load far pointer to GS and register

00001111 10110101 oorrrmmm disp

Format	Examples	Microprocessor	Clocks
LGS reg,mem	LGS DI,DATA3 LGS SI,LIST2 LGS BX,ARRAY_PTR	80186	—
		80188	—
		80386EX	7

LIDT — Load interrupt descriptor table

00001111 00000001 oo011mmm disp

Format	Examples	Microprocessor	Clocks
LIDT mem64	LIDT DATA3 LIDT LIST2	80186	—
		80188	—
		80386EX	11

LLDT — Load local descriptor table

00001111 00000000 oo010mmm disp

Format	Examples	Microprocessor	Clocks
LLDT reg	LLDT BX LLDT DX LLDT CX	80186	—
		80188	—
		80386EX	20
LLDT mem	LLDT DATA1 LLDT LIST3 LLDT TOAD	80186	—
		80188	—
		80386EX	24

LOCK — Lock the bus

11110000

Format	Examples	Microprocessor	Clocks
LOCK:inst	LOCK:XCHG AX,BX LOCK:ADD AL,3	80186	2
		80188	2
		80386EX	0

LODS — Load string operand

1010110w

Format	Examples	Microprocessor	Clocks
LODSB LODSW LODSD	LODSB LODSW LODSD	80186	12
		80188	13
		80386EX	5

LOOP/LOOPD   Loop until CX = 0 or ECX = 0

11100010 disp

Format	Examples	Microprocessor	Clocks
LOOP label LOOPD label	LOOP NEXT LOOP BACK LOOPD LOOPS	80186	16/6
		80188	16/6
		80386EX	11

LOOPE/LOOPED   Loop while equal

11100001 disp

Format	Examples	Microprocessor	Clocks
LOOPE label LOOPED label LOOPZ label	LOOPE AGAIN LOOPED UNTIL LOOPZ ZORRO	80186	16/5
		80188	16/5
		80386EX	11

LOOPNE/LOOPNED   Loop while not equal

11100000 disp

Format	Examples	Microprocessor	Clocks
LOOPNE label LOOPNED label LOOPNZ label	LOOPNE FORWARD LOOPNED UPS LOOPNZ TRY_AGAIN	80186	16/5
		80188	16/5
		80386EX	11

LSL   Load segment limit

00001111 00000011 oorrrmmm disp

		O D I T	S Z A P C *

Format	Examples	Microprocessor	Clocks
LSL reg,reg	LSL AX,BX LSL CX,BX LSL EDX,EAX	80186	—
		80188	—
		80386EX	25
LSL reg,mem	LSL AX,LIMIT LSL EAX,NUM	80186	—
		80188	—
		80386EX	26

LSS — Load far pointer to SS and register

00001111 10110010 oorrrmmm disp

Format	Examples	Microprocessor	Clocks
LSS reg,mem	LSS DI,DATA1 LSS SP,STACK_TOP LSS CX,ARRAY	80186	—
		80188	—
		80386EX	7

LTR — Load task register

00001111 00000000 oo001mmm disp

Format	Examples	Microprocessor	Clocks
LTR reg	LTR AX LTR CX LTR DX	80186	—
		80188	—
		80386EX	23
LTR mem16	LTR TASK LTR NUM	80186	—
		80188	—
		80386EX	27

MOV — Move data

100010dw oorrrmmm disp

Format	Examples	Microprocessor	Clocks

MOV reg,reg	MOV CL,CH
	MOV BH,CL
	MOV CX,DX
		80186	2

		80188	2
		80386EX	2
MOV mem,reg	MOV DATA7,DL MOV NUMB,CX	80186	9 + ea
		80188	10 + ea
		80386EX	2

MOV reg,mem	MOV DL,DATA8
	MOV DX,NUMB
	MOV EBX,TEMP+3
		80186	12 + ea

		80188	13 + ea
		80386EX	4

1100011w oo000mmm disp data

Format	Examples	Microprocessor	Clocks

MOV mem,imm	MOV DATAF,23H
	MOV LIST,12H
	MOV BYTE PTR [DI],2
		80186	13 + ea

		80188	13 + ea
		80386EX	2

1011wrrr data

Format	Examples	Microprocessor	Clocks

MOV reg,imm	MOV BX,22H
	MOV CX,12H
	MOV CL,2
		80186	12

		80188	12
		80386EX	2

101000dw disp

Format	Examples	Microprocessor	Clocks

MOV mem,acc	MOV DATAF,AL
	MOV LIST,AX
	MOV NUMB,EAX
		80186	8

		80188	8
		80386EX	2

MOV acc,mem	MOV AL,DATAE
	MOV AX,LIST
	MOV EAX,LUTE
		80186	9

		80188	9
		80386EX	4

100011d0 oosssmmm disp

Format	Examples	Microprocessor	Clocks

MOV seg,reg	MOV SS,AX
	MOV DS,DX
	MOV ES,CX
		80186	2

		80188	2
		80386EX	2

MOV seg,mem	MOV SS,STACK_TOP
	MOV DS,DATAS
	MOV ES,TEMP1
		80186	11 + ea

		80188	11 + ea
		80386EX	2

MOV reg,seg	MOV BX,DS
	MOV CX,FS
	MOV CX,ES
		80186	2

		80188	2
		80386EX	2

MOV mem,seg	MOV DATA2,CS
	MOV TEMP,DS
	MOV NUMB1,SS
		80186	12 + ea

		80188	13 + ea
		80386EX	2

00001111 001000d0 11rrrmmm

Format	Examples	Microprocessor	Clocks

MOV reg,cr	MOV EBX,CR0
	MOV ECX,CR2
	MOV EBX,CR3
		80186	—

		80188	—
		80386EX	6

MOV cr,reg	MOV CR0,EAX
	MOV CR1,EBX
	MOV CR3,EDX
		80186	—

		80188	—
		80386EX	10

00001111 001000d1 11rrrmmm			
Format	Examples	Microprocessor	Clocks

MOV reg,dr	MOV EBX,DR6
	MOV ECX,DR7
	MOV EBX,DR1
		80186	—

		80188	—
		80386EX	22

MOV dr,reg	MOV DR0,EAX
	MOV DR1,EBX
	MOV DR3,EDX
		80186	—

		80188	—
		80386EX	22

00001111 001001d0 11rrrmmm			
Format	Examples	Microprocessor	Clocks
MOV reg,tr	MOV EBX,TR6 MOV ECX,TR7	80186	—
		80188	—
		80386EX	12
MOV tr,reg	MOV TR6,EAX MOV TR7,EBX	80186	—
		80188	—
		80386EX	12

MOVS   Move string data

1010010w			
Format	Examples	Microprocessor	Clocks
MOVSB MOVSW MOVSD	MOVSB MOVSW MOVSD	80186	14
		80188	14
		80386EX	7

MOVSX   Move with sign extend

00001111 1011111w oorrrmmm disp			
Format	Examples	Microprocessor	Clocks
MOVSX reg,reg	MOVSX BX,AL MOVSX EAX,DX	80186	—
		80188	—
		80386EX	3
MOVSX reg,mem	MOVSX AX,DATA34 MOVSX EAX,NUMB	8086	—
		8088	—
		80286	6

MOVZX	Move with zero extend

00001111 1011011w oorrrmmm disp

Format	Examples	Microprocessor	Clocks
MOVZX reg,reg	MOVZX BX,AL MOVZX EAX,DX	80186	—
		80188	—
		80386EX	3
MOVZX reg,mem	MOVZX AX,DATA34 MOVZX EAX,NUMB	80186	—
		80188	—
		80386EX	6

MUL	Multiply

1111011w oo100mmm disp

		O	D	I	T		S	Z	A	P	C
		*						?	?	?	*

Format	Examples	Microprocessor	Clocks
MUL reg	MUL BL MUL CX MUL EDX	80186	26–37
		80188	26–37
		80386EX	38
MUL mem	MUL DATA9 MUL WORD PTR [ESI]	80186	32–43
		80188	32–43
		80386EX	41

NEG	Negate

1111011w oo011mmm disp

		O	D	I	T		S	Z	A	P	C
		*					*	*	*	*	*

Format	Examples	Microprocessor	Clocks
NEG reg	NEG BL NEG CX NEG EDI	80186	3
		80188	3
		80386EX	2
NEG mem	NEG DATA9 NEG WORD PTR [ESI]	80186	16 + ea
		80188	24 + ea
		80386EX	6

NOP	No operation		

10010000 Example		Microprocessor	Clocks
NOP		80186	3
		80188	3
		80386EX	3

NOT	One's complement		

1111011w oo010mmm disp Format	Examples	Microprocessor	Clocks
NOT reg	NOT BL NOT CX NOT EDI	80186	3
		80188	3
		80386EX	2
NOT mem	NOT DATA9 NOT WORD PTR [ESI]	80186	16 + ea
		80188	24 + ea
		80386EX	6

OR	Inclusive-OR		

000010dw oorrrmmm disp		O D I T S Z A P C 0 * * ? * 0	
Format	Examples	Microprocessor	Clocks
OR reg,reg	OR AX,BX OR AL,BL OR EAX,EBX	80186	3
		80188	3
		80386EX	2
OR mem,reg	OR DATAY,AL OR LIST,SI OR DATA2[DI],CL	80186	10 + ea
		80188	11 + ea
		80386EX	7
OR reg,mem	OR BL,DATA1 OR SI,LIST1 OR CL,DATA2[SI]	80186	10 + ea
		80188	10 + ea
		80386EX	6

100000sw oo001mmm disp data Format	Examples	Microprocessor	Clocks
OR reg,imm	OR CX,3 OR DI,1AH OR DL,34H	80186	4
		80188	4
		80386EX	2

OR mem,imm	OR DATAS,3 OR BYTE PTR[EDI],1AH OR DADDY,34H	80186	10 + ea
		80188	10 + ea
		80386EX	7

0000110w data Format	Examples	Microprocessor	Clocks
OR acc,imm	OR AX,3 OR AL,1AH OR AH,34H	80186	3
		80188	4
		80386EX	2

OUT   Output data to port

1110011w port# Format	Examples	Microprocessor	Clocks
OUT pt,acc	OUT 12H,AL OUT 12H,AX OUT 0FFH,AL	80186	9
		80188	9
		80386EX	10

1110111w Format	Examples	Microprocessor	Clocks
OUT DX,acc	OUT DX,AL OUT DX,AX OUT DX,EAX	80186	7
		80188	7
		80386EX	11

OUTS   Output string to port

0110111w Format	Examples	Microprocessor	Clocks
OUTSB OUTSW OUTSD	OUTSB OUTSW OUTSD	80186	14
		80188	14
		80386EX	14

POP   Pop data from stack

01011rrr Format	Examples	Microprocessor	Clocks
POP reg	POP CX POP AX POP EDI	80186	10
		80188	10
		80386EX	4

| 10001111 oo000mmm disp | | | |
Format	Examples	Microprocessor	Clocks
POP mem	POP DATA1 POP LISTS POP NUMBS	80186	20 + ea
		80188	20 + ea
		80386EX	5

| 00sss111 | | | |
Format	Examples	Microprocessor	Clocks
POP seg	POP DS POP ES POP SS	80286	8
		80288	8
		80386EX	7

| 00001111 10sss001 | | | |
Format	Examples	Microprocessor	Clocks
POP seg	POP FS POP GS	80186	—
		80188	—
		80386EX	7

POPA/POPAD   Pop all registers from stack

| 01100001 | | |
Example	Microprocessor	Clocks
POPA POPAD	8086	51
	8088	51
	80286	24

POPF/POPFD   Pop flags from stack

| 10010000 | O D I T   S Z A P C
* * * *   * * * * * | |
Example	Microprocessor	Clocks
POPF POPFD	80186	10
	80188	10
	80386EX	5

PUSH   Push data onto stack

| 01010rrr | | | |
Format	Examples	Microprocessor	Clocks
PUSH reg	PUSH CX PUSH AX PUSH EDI	80186	10
		80188	10
		80386EX	2

11111111 oo110mmm disp			
Format	Examples	Microprocessor	Clocks
PUSH mem	PUSH DATA1 PUSH LISTS PUSH NUMBS	80186	16 + ea
		80188	16 + ea
		80386EX	5

00ss110			
Format	Examples	Microprocessor	Clocks
PUSH seg	PUSH ES PUSH CS PUSH DS	80186	9
		80188	9
		80386EX	2

00001111 10sss000			
Format	Examples	Microprocessor	Clocks
PUSH seg	PUSH FS PUSH GS	80186	—
		80188	—
		80386EX	2

011010s0 data			
Format	Examples	Microprocessor	Clocks
PUSH imm	PUSH 2000H PUSH 53220 PUSHW 10H	80186	10
		80188	10
		80386EX	2

PUSHA/PUSHAD   Push all registers onto stack

01100000		
Example	Microprocessor	Clocks
PUSHA PUSHAD	80186	36
	80188	36
	80386EX	18

PUSHF/PUSHFD   Push flags onto stack

10011100		
Example	Microprocessor	Clocks
PUSHF PUSHFD	80186	10
	80188	10
	80386EX	4

RCL/RCR/ROL/ROR Rotate

1101000w ooTTTmmm disp

O	D	I	T	S	Z	A	P	C
*								*

TTT = 000 = ROL, TTT = 001 = ROR, TTT = 010 = RCL, and TTT = 011 = RCR

Format	Examples	Microprocessor	Clocks
ROL reg,1 ROR reg,1	ROL CL,1 ROL DX,1 ROR CH,1	80186	2
		80188	2
		80386EX	3
RCL reg,1 RCR reg,1	RCL CL,1 RCL SI,1 RCR AH,1	80186	2
		80188	2
		80386EX	9
ROL mem,1 ROR mem,1	ROL DATAY,1 ROL LIST,1 ROR DATA2[DI],1	80186	15 + ea
		80188	15 + ea
		80386EX	7
RCL mem,1 RCR mem,1	RCL DATA1,1 RCL LIST,1 RCR DATA2[SI],1	80186	15 + ea
		80188	15 + ea
		80386EX	10

1101001w ooTTTmmm disp

Format	Examples	Microprocessor	Clocks
ROL reg,CL ROR reg,CL	ROL CH,CL ROL DX,CL ROR AL,CL	80186	5 + n
		80188	5 + n
		80386EX	3
RCL reg,CL RCR reg,CL	RCL CH,CL RCL SI,CL RCR AH,CL	80186	5 + n
		80188	5 + n
		80386EX	9
ROL mem,CL ROR mem,CL	ROL DATAY,CL ROL LIST,CL ROR DATA2[DI],CL	80186	17 + n
		80188	17 + n
		80386EX	7
RCL mem,CL RCR mem,CL	RCL DATA1,CL RCL LIST,CL RCR DATA2[SI],CL	80186	17 + n
		80188	17 + n
		80386EX	10

1100000w ooTTTmmm disp data

Format	Examples	Microprocessor	Clocks
ROL reg,imm ROR reg,imm	ROL CH,4 ROL DX,5 ROR AL,2	80186	5 + n
		80188	5 + n
		80386EX	3
RCL reg,imm RCR reg,imm	RCL CL,2 RCL SI,12 RCR AH,5	80186	5 + n
		80188	5 + n
		80386EX	9
ROL mem,imm ROR mem,imm	ROL DATAY,4 ROL LIST,3 ROR DATA2[DI],7	80186	17 + n
		80188	17 + n
		80386EX	7
RCL mem,imm RCR mem,imm	RCL DATA1,5 RCL LIST,3 RCR DATA2[SI],9	80186	17 + n
		80188	17 + n
		80386EX	10

REP   Repeat prefix

11110011 1010010w

Format	Examples	Microprocessor	Clocks
REP MOVS	REP MOVSB REP MOVSW REP MOVSD	80186	8 + 8n
		80188	9 + 8n
		80386EX	8 + 4n

11110011 1010101w

Format	Examples	Microprocessor	Clocks
REP STOS	REP STOSB REP STOSW REP STOSD	80186	6 + 9n
		80188	6 + 9n
		80386EX	5 + 5n

11110011 0110110w

Format	Examples	Microprocessor	Clocks
REP INS	REP INSB REP INSW REP INSD	80186	8 + 8n
		80188	8 + 8n
		80386EX	12 + 5n

11110011 0110111w

Format	Examples	Microprocessor	Clocks
REP OUTS	REP OUTSB REP OUTSW REP OUTSD	80186	8 + 8n
		80188	8 + 8n
		80386EX	12 + 5n

REPE/REPNE Repeat conditional

11110011 1010011w

Format	Examples	Microprocessor	Clocks
REPE CMPS	REPE CMPSB REPE CMPSW REPE CMPSD	80186	5 + 22n
		80188	5 + 22n
		80386EX	5 + 9n

11110011 1010111w

Format	Examples	Microprocessor	Clocks
REPE SCAS	REPE SCASB REPE SCASW REPE SCASD	80186	5 + 15n
		80188	5 + 15n
		80386EX	5 + 8n

11110010 1010011w

Format	Examples	Microprocessor	Clocks
REPNE CMPS	REPNE CMPSB REPNE CMPSW REPNE CMPSD	80186	5 + 22n
		80188	5 + 22n
		80386EX	5 + 9n

11110010 101011w

Format	Examples	Microprocessor	Clocks
REPNE SCAS	REPNE SCASB REPNE SCASW REPNE SCASD	80186	5 + 15n
		80188	5 + 15n
		80386EX	5 + 8n

RET Return from procedure

11000011

Example		Microprocessor	Clocks
RET (near)		80186	16
		80188	16
		80386EX	10

11000010 data

Format	Examples	Microprocessor	Clocks
RET imm (near)	RET 4 RET 100H	80186	18
		80188	18
		80386EX	10

11001011 Example		Microprocessor	Clocks
RET (far)		80186	22
		80188	22
		80386EX	18

11001010 data Format	Examples	Microprocessor	Clocks
RET imm (far)	RET 4 RET 100H	80186	25
		80188	25
		80386EX	10

SAHF   Store AH into flags

10011110	O D I T	S Z A P C
		* * * * *

Example		Microprocessor	Clocks
SAHF		80186	3
		80188	3
		80386EX	3

SAL/SAR/SHL/SHR   Shift

1101000w ooTTTmmm disp	O D I T	S Z A P C
	*	* * ? * *

TTT = 100 = SHL/SAL , TTT = 101 = SHR, and TTT = 111 = SAR

Format	Examples	Microprocessor	Clocks
SAL reg,1 SHL reg,1 SHR reg,1	SAL CL,1 SHL DX,1 SAR CH,1	80186	2
		80188	2
		80386EX	3
SHL mem,1 SHL mem,1 SHR mem,1	SAL DATA1,1 SHL BYTE PTR [DI],1 SAR NUMB,1	80188	15 + ea
		80186	15 + ea
		80386EX	7

1101001w ooTTTmmm disp Format	Examples	Microprocessor	Clocks
SAL reg,CL SHL reg,CL SAR reg,CL	SAL CH,CL SHL DX,CL SAR AL,CL	80186	5 + n
		80188	5 + n
		80386EX	3
SAL mem,CL SHL mem,CL SAR mem,CL	SAL DATAU,CL SHL BYTE PTR [ESI],CL SAR NUMB,CL	80186	17 + n
		80188	17 + n
		80386EX	7

1100000w ooTTTmmm disp data

Format	Examples	Microprocessor	Clocks
SAL reg,imm SHL reg,imm SAR reg,imm	SAL CH,4 SHL DX,10 SAR AL,2	80186	5 + n
		80188	5 + n
		80386EX	3
SAL mem,imm SHL mem,imm SAR mem,imm	SAL DATAU,3 SHL BYTE PTR [ESI],15 SAR NUMB,3	8086	17 + n
		8088	17 + n
		80386EX	7

SBB   Subtract with borrow

000110dw oorrrmmm disp

	O D I T	S Z A P C
	*	* * * * *

Format	Examples	Microprocessor	Clocks
SBB reg,reg	SBB CL,DL SBB AX,DX SBB CH,CL	80186	3
		80188	3
		80386EX	2
SBB mem,reg	SBB DATAJ,CL SBB BYTES,CX SBB NUMBS,ECX	80186	10 + ea
		80188	10 + ea
		80386EX	6
SBB reg,mem	SBB CL,DATAL SBB CX,BYTES SBB ECX,NUMBS	80186	10 + ea
		80188	11 + ea
		80386EX	7

100000sw oo011mmm disp data

Format	Examples	Microprocessor	Clocks
SBB reg,imm	SBB CX,3 SBB DI,1AH SBB DL,34H	80186	4
		80188	4
		80386EX	2
SBB mem,imm	SBB DATAS,3 SBB BYTE PTR[EDI],1AH SBB DADDY,34H	80186	16 + ea
		80188	16 + ea
		80386EX	7

0001110w data

Format	Examples	Microprocessor	Clocks
SBB acc,imm	SBB AX,3 SBB AL,1AH SBB AH,34H	80186	3
		80188	4
		80386EX	2

SCAS — Scan string

1010111w			O D I T S Z A P C	
			* * * * * *	
Format	Examples		Microprocessor	Clocks
SCASB	SCASB		80186	15
SCASW	SCASW		80188	15
SCASD	SCASD		80386EX	7

SETcondition — Conditional set

00001111 1001cccc oo000mmm				
Format	Examples		Microprocessor	Clocks
SETcnd reg8	SETA BL		80186	—
	SETB CH		80188	—
	SETG DL		80386EX	4
SETcnd mem8	SETE DATAK		80186	—
	SETAE LESS_OR_SO		80188	—
			80386EX	5

Condition Codes	Mnemonic	Flag	Description
0000	SETO	O = 1	Set if overflow
0001	SETNO	O = 0	Set if no overflow
0010	SETB/SETAE	C = 1	Set if below
0011	SETAE/SETNB	C = 0	Set if above or equal
0100	SETE/SETZ	Z = 1	Set if equal/zero
0101	SETNE/SETNZ	Z = 0	Set if not equal/zero
0110	SETBE/SETNA	C = 1 + Z = 1	Set if below or equal
0111	SETA/SETNBE	C = 0 • Z = 0	Set if above
1000	SETS	S = 1	Set if sign
1001	SETNS	S = 0	Set if no sign
1010	SETP/SETPE	P = 1	Set if parity
1011	SETNP/SETPO	P = 0	Set if no parity
1100	SETL/SETNGE	S • O	Set if less than
1101	SETGE/SETNL	S = 0	Set if greater than or equal
1110	SETLE/SETNG	Z = 1 + S • O	Set if less than or equal
1111	SETG/SETNLE	Z = 0 + S = O	Set if greater than

SGDT/SIDT/SLDT — Store descriptor table registers

00001111 00000001 oo000mmm disp				
Format	Examples		Microprocessor	Clocks
SGDT mem	SGDT MEMORY		80186	—
	SGDT GLOBAL		80188	—
			80386EX	9

00001111 00000001 oo001mmm disp			
Format	Examples	Microprocessor	Clocks
SIDT mem	SIDT DATAS	80186	—
	SIDT INTERRUPT	80188	—
		80386EX	9

00001111 00000000 oo000mmm disp			
Format	Examples	Microprocessor	Clocks
SLDT reg	SLDT CX	80186	—
	SLDT DX	80188	—
		80386EX	2
SLDT mem	SLDT NUMBS	80186	—
	SLDT LOCALS	80188	—
		80386EX	2

SHLD/SHRD Double precision shift

00001111 10100100 oorrrmmm disp data		O D I T S Z A P C ? * * ? * *	
Format	Examples	Microprocessor	Clocks
SHLD	SHLD AX,CX,10	80186	—
reg,reg,imm	SHLD DX,BX,8	80188	—
	SHLD CX,DX,2	80386EX	3
SHLD	SHLD DATAQ,CX,8	80186	—
mem,reg,imm		80188	—
		80386EX	7

00001111 10101100 oorrrmmm disp data			
Format	Examples	Microprocessor	Clocks
SHRD	SHRD CX,DX,2	80186	—
reg,reg,imm		80188	—
		80386EX	3
SHRD	SHRD DATAZ,DX,4	80186	—
mem,reg,imm		80188	—
		80386EX	7

00001111 10100101 oorrrmmm disp			
Format	Examples	Microprocessor	Clocks
SHLD	SHLD BX,DX,CL	8086	—
reg,reg,CL		8088	—
		80286	3

SHLD mem,reg,CL	SHLD DATAZ,DX,CL	80186	—
		80188	—
		80386EX	7

00001111 10101101 oorrrmmm disp Format	Examples	Microprocessor	Clocks
SHRD reg,reg,CL	SHRD AX,DX,CL	80186	—
		80188	—
		80386EX	3
SHRD mem,reg,CL	SHRD DATAZ,DX,CL	80186	—
		80188	—
		80386EX	7

STC   Set carry flag

11111001		O D I T S Z A P C 1
Example		Microprocessor Clocks

STC	80186	2
	80188	2
	80386EX	2

STD   Set direction flag

11111101		O D I T S Z A P C 1
Example		Microprocessor Clocks

STD	80186	2
	80188	2
	80386EX	2

STI   Set interrupt flag

11111011		O D I T S Z A P C 1
Example		Microprocessor Clocks

STI	8086	2
	8088	2
	80286	3

STOS — Store string data

1010101w

Format	Examples	Microprocessor	Clocks
STOSB	STOSB	80186	10
STOSW	STOSW	80188	10
STOSD	STOSD	80386EX	40

STR — Store task register

00001111 00000000 oo001mmm disp

Format	Examples	Microprocessor	Clocks
STR reg	STR AX	81086	—
	STR DX	80188	—
	STR BP	80386EX	2
STR mem	STR DATA3	80186	—
		80188	—
		80386EX	2

SUB — Subtract

000101dw oorrrmmm disp

	O	D	I	T		S	Z	A	P	C
	*					*	*	*	*	*

Format	Examples	Microprocessor	Clocks
SUB reg,reg	SUB CL,DL	80186	3
	SUB AX,DX	80188	3
	SUB CH,CL	80386EX	2
SUB mem,reg	SUB DATAJ,CL	80186	10 + ea
	SUB BYTES,CX	80188	11 + ea
	SUB NUMBS,ECX	80386EX	6
SUB reg,mem	SUB CL,DATAL	80186	10 + ea
	SUB CX,BYTES	80188	11 + ea
	SUB ECX,NUMBS	80386EX	7

100000sw oo101mmm disp data

Format	Examples	Microprocessor	Clocks
SUB reg,imm	SUB CX,3	80186	4
	SUB DI,1AH	80188	4
	SUB DL,34H	80386EX	2

SUB mem,imm	SUB DATAS,3
	SUB BYTE PTR[EDI],1AH
	SUB DADDY,34H
		80186	15 + ea

		80188	16 + ea
		80386EX	7

0010110w data

Format	Examples	Microprocessor	Clocks

SUB acc,imm	SUB AL,3
	SUB AX,1AH
	SUB EAX,34H
		80186	3

		80188	4
		80386EX	2

TEST   Test operands (logical compare)

1000001w oorrrmmm disp

		O D I T	S Z A P C
		0	* * ? * 0
Format	Examples	Microprocessor	Clocks
TEST reg,reg	TEST CL,DL TEST BX,DX TEST DH,CL	80186	3
		80188	3
		80386EX	2
TEST mem,reg reg,mem	TEST DATAJ,CL TEST BYTES,CX TEST NUMBS,ECX	80186	10 + ea
		80188	10 + ea
		80386EX	5

1111011sw oo000mmm disp data

Format	Examples	Microprocessor	Clocks
TEST reg,imm	TEST BX,3 TEST DI,1AH TEST DH,44H	80186	4
		80188	4
		80386EX	2
TEST mem,imm	TEST DATAS,3 TEST BYTE PTR[EDI],1AH TEST DADDY,34H	80186	10 + ea
		80188	10 + ea
		80386EX	5

1010100w data

Format	Examples	Microprocessor	Clocks
TEST acc,imm	TEST AL,3 TEST AX,1AH TEST EAX,34H	80186	3
		80188	4
		80386EX	2

VERR/VERW Verify read/write

00001111 00000000 oo100mmm disp		O D I T S Z A P C *	
Format	Examples	Microprocessor	Clocks
VERR reg	VERR CX VERR DX VERR DI	80186	—
		80188	—
		80386EX	10
VERR mem	VERR DATAJ VERR TESTB	80186	—
		80188	—
		80386EX	11

00001111 00000000 oo101mmm disp			
Format	Examples	Microprocessor	Clocks
VERW reg	VERW CX VERW DX VERW DI	80186	—
		80188	—
		80386EX	15
VERW mem	VERW DATAJ VERW TESTB	8086	—
		8088	—
		80386EX	16

WAIT   Wait for coprocessor

10011011		
Example	Microprocessor	Clocks
WAIT FWAIT	80186	6
	80188	6
	80386EX	6

XCHG   Exchange

1000011w oorrrmmm			
Format	Examples	Microprocessor	Clocks
XCHG reg,reg	XCHG CL,DL XCHG BX,DX XCHG DH,CL	80186	4
		80188	4
		80386EX	3
XCHG mem,reg reg,mem	XCHG DATAJ,CL XCHG BYTES,CX XCHG NUMBS,ECX	80186	17 + ea
		80188	17 + ea
		80386EX	5

10010reg Format	Examples	Microprocessor	Clocks
XCHG acc,reg reg,acc	XCHG BX,AX XCHG AX,DI XCHG DH,AL	80186	3
		80188	3
		80386EX	3

XLAT   Translate

11010111 Example	Microprocessor	Clocks
XLAT	80186	11
	80188	11
	80386EX	3

XOR   Exclusive-OR

000110dw oorrrmmm disp

	O	D	I	T		S	Z	A	P	C
	0					*	*	?	*	0

Format	Examples	Microprocessor	Clocks
XOR reg,reg	XOR CL,DL XOR AX,DX XOR CH,CL	80186	3
		80188	3
		80386EX	2
XOR mem,reg	XOR DATAJ,CL XOR BYTES,CX XOR NUMBS,ECX	80186	10 + ea
		80188	11 + ea
		80386EX	6
XOR reg,mem	XOR CL,DATAL XOR CX,BYTES XOR ECX,NUMBS	80186	10 + ea
		80188	11 + ea
		80386EX	7

100000sw oo110mmm disp data

Format	Examples	Microprocessor	Clocks
XOR reg,imm	XOR CX,3 XOR DI,1AH XOR DL,34H	80186	4
		80188	4
		80386EX	2
XOR mem,imm	XOR DATAS,3 XOR BYTE PTR[EDI],1AH XOR DADDY,34H	80186	10 + ea
		80188	10 + ea
		80386EX	7

0010101w data			
Format	Examples	Microprocessor	Clocks
XOR acc,imm	XOR AL,3 XOR AX,1AH XOR EAX,34H	80186	3
		80188	4
		80386EX	2

APPENDIX C

Flag-Bit Changes

This appendix shows only the instructions that actually change the flag bits. Any instruction not listed here does not affect any of the flag bits.

Instruction	O	D	I	T	S	Z	A	P	C
AAA	?				?	?	*	?	*
AAD	?				*	*	?	*	?
AAM	?				*	*	?	*	?
AAS	?				?	?	*	?	*
ADC	*				*	*	*	*	*
ADD	*				*	*	*	*	*
AND	0				*	*	?	*	0
ARPL						*			
BSF						*			
BSR						*			
BT									*
BTC									*
BTR									*
BTS									*
CLC									0
CLD		0							
CLI			0						
CMC									*
CMP	*				*	*	*	*	*
CMPS	*				*	*	*	*	*
CMPXCHG	*				*	*	*	*	*
CMPXCHG8B						*			
DAA	?				*	*	*	*	*
DAS	?				*	*	*	*	*
DEC	*				*	*	*	*	
DIV	?				?	?	?	?	?
IDIV	?				?	?	?	?	?
IMUL	*				?	?	?	?	*
INC	*				*	*	*	*	

(Continued on next page)

Instruction	O	D	I	T	S	Z	A	P	C
IRET	*	*	*	*	*	*	*	*	*
LAR						*			
LSL						*			
MUL	*				?	?	?	?	*
NEG	*				*	*	*	*	*
OR	0				*	*	?	*	0
POPF	*	*	*	*	*	*	*	*	*
RCL/RCR	*								*
REPE/REPNE						*			
ROL/ROR	*								*
SAHF					*	*	*	*	*
SAL/SAR	*				*	*	?	*	*
SHL/SHR	*				*	*	?	*	*
SBB	*				*	*	*	*	*
SCAS	*				*	*	*	*	*
SHLD/SHRD	?				*	*	?	*	*
STC									1
STD		1							
STI			1						
SUB	*				*	*	*	*	*
TEST	0				*	*	?	*	0
VERR/VERW						*			
XADD	*				*	*	*	*	*
XOR	0				*	*	?	*	0

INDEX